# Engineering Problems for Undergraduate Students

Xian Wen Ng

# Engineering Problems for Undergraduate Students

## Over 250 Worked Examples with Step-by-Step Guidance

Springer

Xian Wen Ng
Singapore, Singapore

ISBN 978-3-030-13855-4 ISBN 978-3-030-13856-1 (eBook)
https://doi.org/10.1007/978-3-030-13856-1

Library of Congress Control Number: 2019935805



This Springer imprint is published by the registered company Springer Nature Switzerland AG
The registered company address is: Gewerbestrasse 11, 6330 Cham, Switzerland

# Preface

***Engineering Problems for Undergraduate Students*** contains over 250 example problems covering key topics in engineering courses. Step-by-step solutions are presented with clear and detailed explanations. This book will support a thorough understanding of fundamental concepts in engineering for tertiary-level students. The problems in this book are quality examples which were carefully selected to demonstrate the application of abstract concepts in solving practical engineering problems, with comprehensive guidance provided in the explanations that follow each step of the solutions.

Topics included in this book are fundamental in the engineering discipline. Hence, they are versatile in their overarching application across various engineering sub-specializations. These topics include thermodynamics, fluid mechanics, separation processes (e.g., flash distillation), reactor design and kinetics (including bioreactor concepts), and engineering mathematics (e.g., Laplace transform, differentiation and integration, Fourier series, statistics).

There is also a section included which summarizes key mathematical formula and other useful data commonly referred to when solving engineering problems. This book will support step-by-step learning for students taking first or second-year undergraduate courses in engineering.

Singapore

Xian Wen Ng

# Acknowledgments

My heartfelt gratitude goes to the team at Springer for their unrelenting support and professionalism throughout the publication process. Special thanks to Michael Luby, Nicole Lowary, and Brian Halm for their kind effort and contributions toward making this publication possible. I am also deeply appreciative of the reviewers for my manuscript who had provided excellent feedback and numerous enlightening suggestions to help improve the book's contents.

Finally, I wish to thank my loved ones who have, as always, offered only patience and understanding throughout the process of making this book a reality.

# Contents

# About the Author

**Xian Wen Ng** graduated with First-Class Honors from the University of Cambridge, UK, with a Master's Degree in Chemical Engineering and Bachelor of Arts in 2011 and was subsequently conferred a Master of Arts in 2014. She was ranked second in her graduating class and was the recipient of a series of college scholarships, including the Samuel Taylor Scholarship, Thomas Ireland Scholarship, and British Petroleum Prize in Chemical Engineering, for top performance in consecutive years of academic examinations. Ng was also one of two students from Cambridge University selected for the Cambridge-Massachusetts Institute of Technology (MIT) exchange program in Chemical Engineering, which she completed with Honors with a cumulative GPA of 4.8 (5.0). During her time at MIT, she was also a part-time tutor for junior classes in engineering and pursued other disciplines including economics, real estate development, and finance at MIT as well as the John F. Kennedy School of Government at Harvard University. Upon graduation, Ng was elected by her College Fellowship to the Title of Scholar, as a mark of her academic distinction.

Since graduation, Ng has been keenly involved in teaching across various academic levels, doing so both in schools and with smaller groups as a private tutor. Ng's topics of specialization range from secondary-level Mathematics, Physics, and Chemistry to tertiary-level Mathematics and Engineering subjects.

# Mathematics

## Useful Mathematical Formula

**Before we begin to tackle mathematics, we should familiarize ourselves with mathematical formulae or identities that help us observe patterns in problems and hence deduce more efficient approaches to solutions. I have listed below a collection of useful identities and formulae that are worth remembering.**

### *Complex Numbers*

The complex number $z$ can be expressed in the following forms, where $i^2 = -1$.

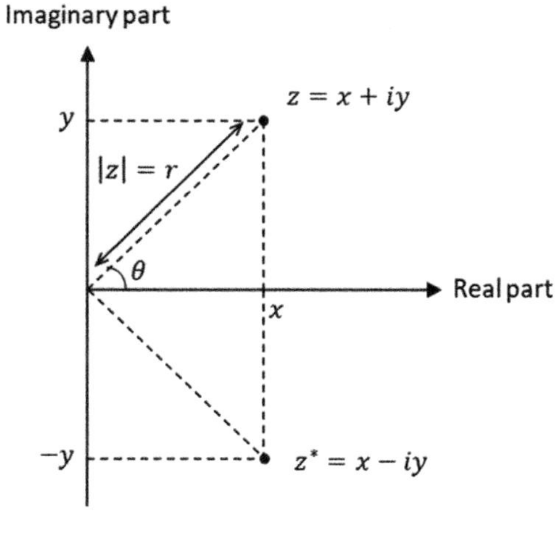




X. W. Ng, *Engineering Problems for Undergraduate Students*,
https://doi.org/10.1007/978-3-030-13856-1_1

In Cartesian form, where $z^*$ is the complex conjugate of $z$.

$$z = x + iy;\ z^* = x - iy$$
$$|z| = x^2 + y^2 = zz^*$$

In polar form, where $\theta$ is the argument of $z$.

$$z = re^{i(\theta+2n\pi)}$$
$$|z| = r$$
$$x = r\cos\theta;\ y = r\sin\theta$$

In trigonometric form

$$z = r(\cos\theta + i\sin\theta)$$
$$\cos\theta = \frac{1}{2}\left(e^{i\theta} + e^{-i\theta}\right)$$
$$\sin\theta = \frac{1}{2i}\left(e^{i\theta} - e^{-i\theta}\right)$$

**De Moivre's Theorem**

$$(\cos\theta + i\sin\theta)^n = e^{in\theta} = \cos n\theta + i\sin n\theta$$

## *Hyperbolic Trigonometric Functions*

$$\cosh z = \cos iz$$
$$i\sinh z = \sin iz$$
$$i\tanh z = \tan iz$$

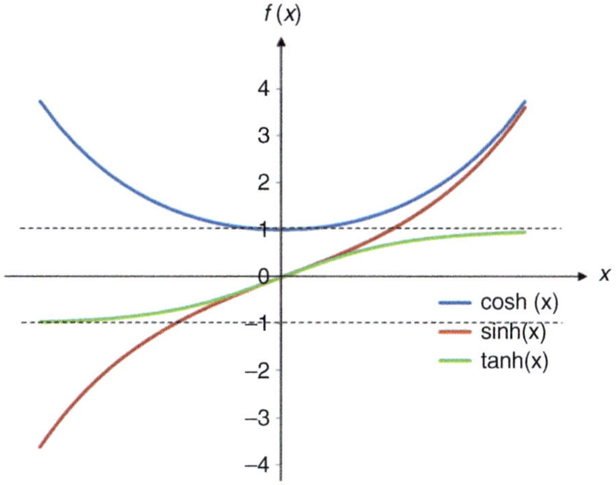


## *Trigonometric Formulae and Identities*

### Double Angle Formula

$$\sin 2A = 2 \sin A \cos A$$

$$\cos 2A = \cos^2 A - \sin^2 A = 2\cos^2 A - 1 = 1 - 2\sin^2 A$$

$$\tan 2A = \frac{2 \tan A}{1 - \tan^2 A}$$

### Negative Angle Formula

It is possible to deduce these results by relating to their graphs. As the cosine curve is an even function symmetric about the y axis, the negative angle formula below makes sense. As for the sine and tangent curves, they are odd functions which lead to the results shown below.

$$\sin(-A) = -\sin A$$

$$\cos(-A) = \cos A$$

$$\tan(-A) = -\tan A$$

**Addition Formula**

$$\sin(A \pm B) = \sin A \cos B \pm \cos A \sin B$$

$$\cos(A \pm B) = \cos A \cos B \mp \sin A \sin B$$

$$\tan(A \pm B) = \frac{\tan A \pm \tan B}{1 \mp \tan A \tan B}$$

**Trigonometric Identities**

$$1 + \tan^2 A = \sec^2 A$$

$$\cos^2 A + \sin^2 A = 1$$

$$1 + \cot^2 A = \operatorname{cosec}^2 A$$

## *Graphical Transformations and Common Functions*

Here are some quick tips about graphical transformations when slight changes are made to a function.

- $f(x) \to f(x + a)$: means the graph is shifted by $a$ units to the left, i.e., negative $x$ direction.
- $f(x) \to f(x) + a$: means the graph is shifted up by $a$ units, i.e., positive $y$ (or) $f(x)$ direction.
- $f(x) \to f(ax)$: means the graph is "compressed"/"expanded" horizontally. It is "compressed" if $a > 1$, and "expanded" if $a < 1$. For the same $y$ value, $x$ values are multiplied by $1/a$. So for example, if the original graph was $y = x$, and we want to transform it to $y = 2x$, we will "compress" the graph of $y = x$ horizontally to obtain a steeper graph of $y = 2x$. For the same $y$ value for $y = x$, the $x$ values of the new graph $y = 2x$ will be halved (i.e., multiplied by ½).
- $f(x) \to af(x)$: means the graph is "compressed"/"stretched" vertically. It is "compressed" if $a < 1$, and "stretched" if $a > 1$. For the same $x$ value, $y$ values are multiplied by $a$. So for example, if the original graph was $y = x^2 + k$, and we want to transform it to $y = 2(x^2 + k)$, we will "stretch" the graph of $y = x^2 + k$ to obtain the graph of $y = 2(x^2 + k)$. For the same $x$ value of $= x^2 + k$, the $y$ values of the new graph $y = 2(x^2 + k)$ will be doubled (i.e., multiplied by 2).
- $f(x) \to -f(x)$: means a reflection about the $x$ axis.
- $f(x) \to f(-x)$: means a reflection about the $y$ axis.
- $f(x) \to f^{-1}(x)$: means a reflection about the line $y = x$.

Note that the inverse of a function $f(x)$, i.e., $f^{-1}(x)$ can be expressed graphically as the reflection of $f(x)$ about the line $y = x$. In the example below, we obtain the graph of $\cosh^{-1}x$ from $\cosh x$.

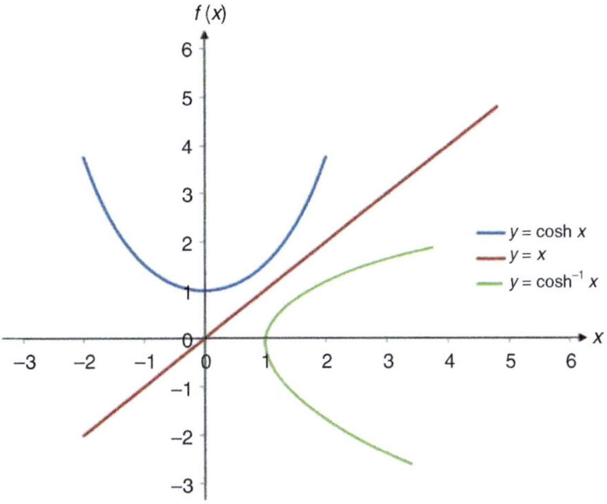


Another useful series of graphs to note is the family of $y = x^n$ and their inverse functions of $y = x^{\frac{1}{n}}$ as shown below. Note that in $y = x^{\frac{1}{n}}$, when the index is an even number, e.g., $y = x^{\frac{1}{2}}, y = x^{\frac{1}{4}}$, the function is only defined for $x \geq 0$. Therefore care has to be taken when deriving inverse functions via a reflection about the line $y = x$, and only the valid region should be considered.

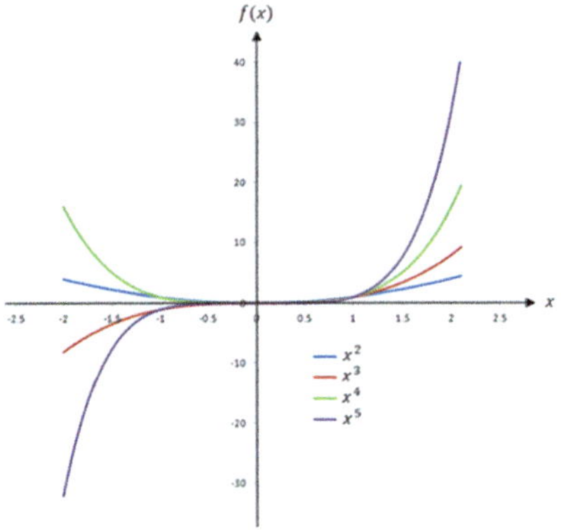

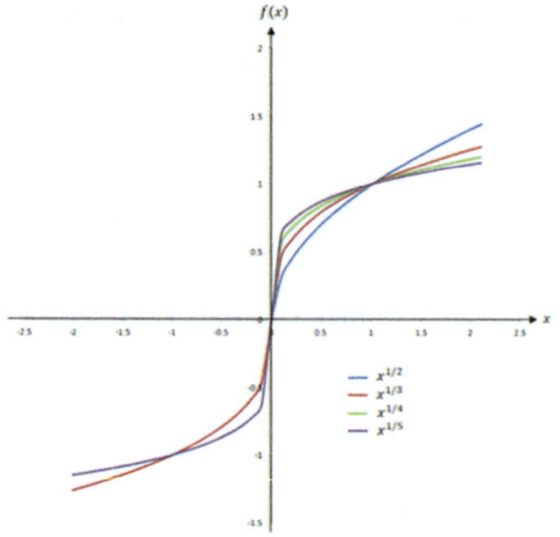


The exponential graph is also often encountered in physical problems, such as exponential behaviors in bacteria growth patterns and residence time distributions of reactors. It is useful to note the shape of the graphs of $f(x) = e^x$ and its inverse function, $f(x) = \ln x$ (which is a reflection of $e^x$ about the line $y = x$).

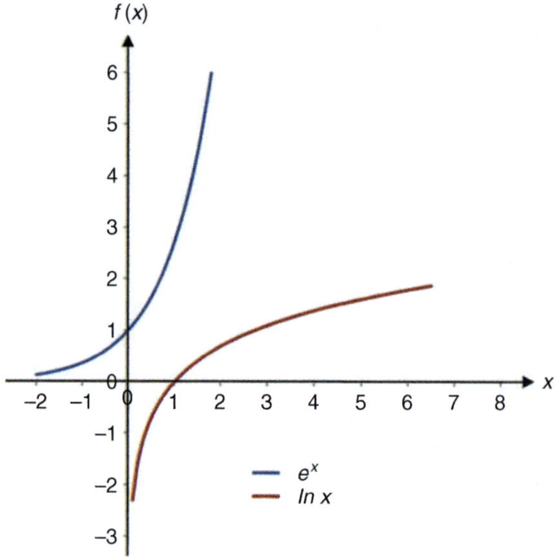


## *Power Series*

### Taylor's Series

When we have a known reference point $x = a$, about which a small deviation of $h$ occurs, we can use the Taylor series to approximate the value of $f(a + h)$ when $h$ is small, by ignoring higher order terms. Note that the Maclaurin's series is simply a special case of the Taylor's series whereby the reference point is $x = 0$:

$$f(a+h) = f(a) + hf'(a) + \frac{h^2}{2!}f''(a) + \ldots + \frac{h^n}{n!}f^n(a)$$

$$f(a+h) \cong f(a) + hf'(a) \text{ for small } h$$

### Other Series

Some other useful power series to note are as follows:

- $e^x = 1 + x + \frac{x^2}{2!} + \frac{x^3}{3!} + \ldots + \frac{x^n}{n!}$
- $e^{-x} = 1 - x + \frac{x^2}{2!} - \frac{x^3}{3!} + \ldots + \frac{x^n}{n!}$
- $\cos x = 1 - \frac{x^2}{2!} + \frac{x^4}{4!} - \frac{x^6}{6!} + \ldots$
- $\cosh x = 1 + \frac{x^2}{2!} + \frac{x^4}{4!} + \frac{x^6}{6!} + \ldots$
- $\sin x = x - \frac{x^3}{3!} + \frac{x^5}{5!} + \ldots$
- $\sinh x = x + \frac{x^3}{3!} + \frac{x^5}{5!} + \ldots$
- $\ln(1+x) = x - \frac{x^2}{2} + \frac{x^3}{3} - \frac{x^4}{4}$, for $-1 < x \leq 1$
- $\ln(1-x) = -x - \frac{x^2}{2} - \frac{x^3}{3} - \frac{x^4}{4}$, for $-1 \leq x < 1$
- $(1+x)^{-1} = 1 - x + x^2 - x^3 + \ldots$ [this can be derived by differentiating the power series for $\ln(1+x)$]
- $(1-x)^{-1} = 1 + x + x^2 + x^3 + \ldots$[this can be derived by differentiating the power series for $-\ln(1-x)$]
- $(1-x)^{-2} = 1 + 2x + 3x^2 + 4x^3 + \ldots$[this can be derived by differentiating the power series for $(1-x)^{-1}$]

## *Fourier Series*

$$f(t) = \frac{a_0}{2} + \sum a_n \cos\frac{n\pi t}{L} + b_n \sin\frac{n\pi t}{L}$$

There are certain shortcuts that are useful when considering specific type of functions.

- For even functions: $f(t) = \frac{a_0}{2} + \sum a_n \cos\frac{n\pi t}{L}$
- For odd functions: $f(t) = \frac{a_0}{2} + \sum b_n \sin\frac{n\pi t}{L}$
- A product of two even functions or two odd functions is even. For even functions, we will be able to convert the limits of the integral without changing the final result as shown below:

- If $f(t)$ is odd, then $b_n = \frac{1}{L}\int_{-L}^{L} f(t)\sin\frac{n\pi t}{L}dt = \frac{2}{L}\int_{0}^{L} f(t)\sin\frac{n\pi t}{L}dt$
- If $f(t)$ is even, then $a_n = \frac{1}{L}\int_{-L}^{L} f(t)\cos\frac{n\pi t}{L}dt = \frac{2}{L}\int_{0}^{L} f(t)\cos\frac{n\pi t}{L}dt$

- A product of an odd and even function is an odd function.
- The value of $\cos n\pi$ can also be expressed as $(-1)^n$, for a positive integer $n \geq 1$.

Note that the difference between Taylor's series and Fourier series is that the Taylor's series approximates values around a reference point, and for small deviations from this reference point, only initial terms to be considered.

For the Fourier series, all terms in the series need to be accounted for as the approximation comprises an integral over the entire interval (based on periodicity of function) considered. Fourier series can be used for functions that are not periodic, by "making them periodic".

## *Differentiation Techniques*

One of the most commonly used techniques for differentiation is the product rule, whereby for two functions $u = u(x)$ and $v = v(x)$, we have

$$\frac{d(uv)}{dx} = u\frac{dv}{dx} + v\frac{du}{dx}$$

We may also use the quotient rule when the form of the expression is already in a quotient form.

$$\frac{d\left(\frac{u}{v}\right)}{dx} = \frac{v\frac{du}{dx} - u\frac{dv}{dx}}{v^2}$$

## *Integration Techniques*

A useful technique in integration is the method of integration by parts which is as follows

$$\int_a^b u dv = [uv]_a^b - v\int_a^b du$$

## *Useful Integrals*

Below are some useful integral values to note as shortcuts in simplifying expressions, due to their exact values.

$$\int_{-\infty}^{\infty} e^{-x^2}\,dx = \sqrt{\pi}$$

$$\int_{-\infty}^{\infty} xe^{-x^2}\,dx = 0, \quad \text{(odd function)}$$

$$\int_{-\infty}^{\infty} x^2 e^{-x^2}\,dx = \frac{\sqrt{\pi}}{2}$$

## Partial Fractions

**It is important to be familiar with converting fractions into partial fractions as it is a convenient form to convert into Laplace transforms or to perform integration or differentiation.**

**Problem 1**

**Determine the partial fractions for the following expressions:**

(a) $\dfrac{x^2+3x}{(x-4)(x^2+x+1)}$

(b) $\dfrac{11-x-x^2}{(x+2)(x-1)^2}$

(c) $\dfrac{7}{-2x(4-x)}$

(d) $\dfrac{x^3+2x^2-x+1}{(x-1)(x+2)}$

(e) $\dfrac{2x^2-7x-1}{(x+1)(x-2)}$

**Solution 1**

***Worked Solution***

(a) In order to express a fraction as a sum of partial fractions, we will first have to factorize the denominator of this fraction. We note that in this example, the fraction's denominator is already in the most factorized form. Therefore we can proceed to express it as a sum of two fractions, with each factor in the denominator becoming the single denominator to each partial fraction as shown below.

$$\frac{x^2+3x}{(x-4)(x^2+x+1)}=\frac{A}{x-4}+\frac{Bx+C}{x^2+x+1}$$

One key point to note is that if the denominator of the partial fraction is first order (or linear) in $x$, then the numerator is one order lower (i.e., a constant term). Therefore we have inserted the constant $A$ in the first partial fraction with a linear denominator "$x-4$". Separately, the second partial fraction has a second-order (or quadratic) denominator "$x^2+x+1$", therefore, we have inserted a numerator which is a linear expression.

In order to solve for the constant $A$, we can apply the cover up rule. This rule is such that we let $x-4=0$ which gives us $x=4$. We cover up $(x-4)$ in the original fraction, and substitute $x=4$ into the remaining fraction. The resulting value will be the value of $A$. This is shown below

$$A=\frac{4^2+3(4)}{(4^2+4+1)}=\frac{4}{3}$$

To solve for $B$ and $C$, we can observe by comparison between the left and right sides of the equation.

$$\frac{x^2+3x}{(x-4)(x^2+x+1)}=\frac{\frac{4}{3}}{x-4}+\frac{Bx+C}{x^2+x+1}$$
$$=\frac{\frac{4}{3}(x^2+x+1)+(Bx+C)(x-4)}{(x-4)(x^2+x+1)}$$

The coefficient of $x^2$ in the numerator on the left-hand side is 1; therefore, $1=\frac{4}{3}+B\rightarrow B=-\frac{1}{3}$.

The constant term in the numerator on the left-hand side is 0; therefore, $0=\frac{4}{3}-4C\rightarrow C=\frac{1}{3}$. [Note that we can also choose to compare the coefficient of the $x$ term but it may require more steps.]

Finally, we have the following partial fractions:

$$\frac{x^2+3x}{(x-4)(x^2+x+1)}=\frac{\frac{4}{3}}{x-4}+\frac{-\frac{1}{3}x+\frac{1}{3}}{x^2+x+1}=\frac{4}{3(x-4)}+\frac{1-x}{3(x^2+x+1)}$$

(b) This example is slightly different as one of its factors is squared. There is a slightly different treatment which will be shown here. In this case, we will need to note that we have two fractions to express for the squared term, one is $\frac{B}{x-1}$ and the other is $\frac{C}{(x-1)^2}$. Also note that the numerators for both fractions are constants, i.e., $B$ and $C$.

$$\frac{11-x-x^2}{(x+2)(x-1)^2}=\frac{A}{x+2}+\frac{B}{x-1}+\frac{C}{(x-1)^2}$$

In this case, the cover up rule will help us find the constants $A$ and $C$.

$$A=\frac{11-(-2)-(-2)^2}{(-2-1)^2}=1$$

$$C=\frac{11-1-1^2}{(1+2)}=3$$

The value of $B$ can be found by comparison of left and right sides of the equation.

$$\frac{11-x-x^2}{(x+2)(x-1)^2}=\frac{1}{x+2}+\frac{B}{x-1}+\frac{3}{(x-1)^2}$$
$$=\frac{(x-1)^2+B(x+2)(x-1)+3(x+2)}{(x+2)(x-1)^2}$$

The coefficient of $x^2$ in the numerator of the left-hand side is $-1$; therefore, $-1=1+B\rightarrow B=-2$.

Finally, we have the following partial fractions:

$$\frac{11-x-x^2}{(x+2)(x-1)^2}=\frac{1}{x+2}-\frac{2}{x-1}+\frac{3}{(x-1)^2}$$

(c) In this example, all of the numerators of the partial fractions can be found using the cover up rule, since they are simple linear factors.

$$\frac{7}{-2x(4-x)}=\frac{A}{-2x}+\frac{B}{4-x}$$

Using cover up rule, we have

$$A=\frac{7}{4-0}=\frac{7}{4}$$

$$B=\frac{7}{-2(4)}=-\frac{7}{8}$$

Finally, we have the following partial fractions:

$$\frac{7}{-2x(4-x)} = \frac{7}{-8x} - \frac{7}{8(4-x)}$$

(d) In this example, we have an improper fraction, where the numerator is of order 3 (or cubic) which is greater than that of the denominator (order 2, or quadratic). The slight difference in approach to this problem is that we need to include a linear function of $x$, i.e., $(x + k)$ to the sum of partial fractions, as this linear function will account for the difference in the orders.

$$\frac{x^3+2x^2-x+1}{(x-1)(x+2)} = (x+k) + \frac{A}{x-1} + \frac{B}{x+2}$$

Using cover up rule, we can find the values of $A$ and $B$,

$$A = \frac{1^3+2(1)^2-1+1}{(1+2)} = 1$$

$$B = \frac{(-2)^3+2(-2)^2-(-2)+1}{(-2-1)} = -1$$

To find $k$, we can compare the left- and right-hand side of the equation,

$$\begin{aligned}\frac{x^3+2x^2-x+1}{(x-1)(x+2)} &= (x+k) + \frac{1}{x-1} - \frac{1}{x+2}\\ &= \frac{(x+k)(x-1)(x+2)+(x+2)-(x-1)}{(x-1)(x+2)}\end{aligned}$$

The constant term in the numerator on the left-hand side is 1; therefore, $1 = -2k + 2 + 1 \rightarrow k = 1$.

Finally, we have the following partial fractions:

$$\frac{x^3+2x^2-x+1}{(x-1)(x+2)} = (x+1) + \frac{1}{x-1} - \frac{1}{x+2}$$

(e) This example is also an improper fraction (proper fraction is when the order of numerator is less than that of denominator). The difference here as compared to part d is that in this case, the term to add is a constant term (order zero).

$$\frac{2x^2-7x-1}{(x+1)(x-2)} = K + \frac{A}{x+1} + \frac{B}{x-2}$$

Using cover up rule, we can find the values of $A$ and $B$,

$$A = \frac{2(-1)^2 - 7(-1) - 1}{(-1-2)} = -\frac{8}{3}$$

$$B = \frac{2(2)^2 - 7(2) - 1}{(2+1)} = -\frac{7}{3}$$

We can find $K$ by comparison,

$$\frac{2x^2 - 7x - 1}{(x+1)(x-2)} = K - \frac{\frac{8}{3}}{x+1} - \frac{\frac{7}{3}}{x-2} = \frac{K(x+1)(x-2) - \frac{8}{3}(x-2) - \frac{7}{3}(x+1)}{(x+1)(x-2)}$$

The coefficient of the $x^2$ term in the numerator on the left-hand side is 2; therefore, $2 = K$.

Finally, we have the following partial fractions:

$$\frac{2x^2 - 7x - 1}{(x+1)(x-2)} = 2 - \frac{8}{3(x+1)} - \frac{7}{3(x-2)}$$

## Differentiation and Integration

**Mathematical techniques are used to solve engineering problems. It is common that we will encounter differentiation and integration problems as engineering often involves breaking up a complex problem into small elements for analysis (differential elements) and integration helps scale the solution for the differential element back to fit the entire system.**

### Problem 2

**Solve the following differential equations by using appropriate substitutions:**

(a) $\frac{dy}{dx} - y = xy^5$

(b) $\frac{1}{y^2}\frac{dy}{dx} + \frac{1}{y} = \cos x - \sin x$

(c) $\frac{dy}{dx} + \frac{(2-3x^2)y}{x^3} = 2$

### Solution 2

***Worked Solution***

In these problems, we show how we can apply useful substitutions to simplify higher order differential equations into equations of order 1 with respect to the variables.

The key observation here is that the term "$a(y)$" is of one order (with respect to variable $y$) higher than the term "$b(y)$" in the following form. Then the substitution can be made for the lower order term $b(y)$. This is demonstrated in the two examples below.

$$a(y)\frac{dy}{dx} + b(y) = c(x)$$

(a) We have the following differential equation.

$$\frac{1}{y^5}\frac{dy}{dx} - \frac{1}{y^4} = x$$

One useful substitution we can make is as follows:

$$v = \frac{1}{y^4} \tag{1}$$

We can then apply the Chain Rule:

$$\frac{dy}{dx} = \frac{dy}{dv}\left(\frac{dv}{dx}\right) = \frac{1}{\frac{dv}{dy}}\left(\frac{dv}{dx}\right)$$

$$\frac{dv}{dy} = -\frac{4}{y^5}$$

$$\frac{dy}{dx} = -\frac{y^5}{4}\left(\frac{dv}{dx}\right) \tag{2}$$

We can now substitute our results in (1) and (2) back into the differential equation,

$$\frac{1}{y^5}\left[-\frac{y^5}{4}\left(\frac{dv}{dx}\right)\right] - v = x$$

$$\frac{dv}{dx} + 4v = -4x$$

One trick to solving ODEs of the form: $\frac{dv}{dx} + Av = f(x)$, where $A$ is a constant, is to let $e^{\int A dx}$. Then the equation can be simplified to the form: $\frac{d}{dx}\left(ye^{\int Adx}\right) = e^{\int Adx}f(x)$ which is easier to solve. Let us demonstrate this in this example. Now, we create the term $e^{\int 4dx} = e^{4x}$. And we can solve the right-hand side of the integral using integration by parts.

$$\frac{d}{dx}\left(ve^{4x}\right) = (-4x)e^{4x}$$

$$ve^{4x} = -\int 4xe^{4x}dx$$

Integration by parts is defined as $\int_a^b udv = [uv]_a^b - \int_a^b vdu$; therefore, we let $u = x$, and $dv = 4e^{4x}dx$

$$ve^{4x} = -\left(xe^{4x} - \int e^{4x}dx\right) = -xe^{4x} + \frac{e^{4x}}{4} + c_1$$

Now we can substitute $v$ with its original expression. $c_1$ can be evaluated if we apply boundary conditions, i.e., for a specific value of $x$, its corresponding $y$ value is known.

$$\frac{1}{y^4}\left(e^{4x}\right) = -xe^{4x} + \frac{e^{4x}}{4} + c_1$$

$$y^4 = \frac{1}{-x + \frac{1}{4} + c_1e^{-4x}}$$

(b)

$$\frac{1}{y^2}\frac{dy}{dx} + \frac{1}{y} = \cos x - \sin x$$

One useful substitution we can make is

$$v = \frac{1}{y} \quad (1)$$

Then we can similarly apply the Chain Rule:

$$\frac{dy}{dx} = \frac{dy}{dv}\left(\frac{dv}{dx}\right) = \frac{1}{\frac{dv}{dy}}\left(\frac{dv}{dx}\right)$$

$$\frac{dv}{dy} = -\frac{1}{y^2}$$

$$\frac{dy}{dx} = -y^2\left(\frac{dv}{dx}\right) \quad (2)$$

Substituting our results in (1) and (2) back into the differential equation,

$$\frac{1}{y^2}\left[-y^2\left(\frac{dv}{dx}\right)\right] + v = \cos x - \sin x$$

$$\frac{dv}{dx} + (-1)v = \sin x - \cos x$$

Now, we create the term $e^{\int(-1)dx} = e^{-x}$. And we can solve the right-hand side of the integral using integration by parts.

$$\frac{d}{dx}(ve^{-x}) = (\sin x - \cos x)e^{-x}$$

$$ve^{-x} = \int(\sin x - \cos x)e^{-x}dx$$

Using integration by parts where $u = e^{-x}$, and $dv = (\sin x - \cos x)dx$

$$ve^{-x} = e^{-x}(-\cos x - \sin x) - \int(-\cos x - \sin x)(-e^{-x})dx$$

$$= e^{-x}(-\cos x - \sin x) - \int(\cos x + \sin x)(e^{-x})dx$$

Using integration by parts again, where $u = e^{-x}$, and $dv = (\cos x + \sin x)dx$

$$ve^{-x} = e^{-x}(-\cos x - \sin x) - \left[e^{-x}(\sin x - \cos x) - \int(\sin x - \cos x)(-e^{-x})dx\right]$$

$$\int(\sin x - \cos x)e^{-x}dx = e^{-x}(-\cos x - \sin x) - e^{-x}(\sin x - \cos x) + \int(\sin x - \cos x)(-e^{-x})dx$$

$$2\int(\sin x - \cos x)e^{-x}dx = e^{-x}(-\cos x - \sin x) - e^{-x}(\sin x - \cos x)$$

$$ve^{-x} = \frac{e^{-x}(-\cos x - \sin x) - e^{-x}(\sin x - \cos x)}{2} + c_1$$

$$v = \frac{1}{y} = \frac{1}{2}(-\cos x - \sin x) - \frac{1}{2}(\sin x - \cos x) + c_1e^x = -\sin x + c_1e^x$$

$$y = \frac{1}{-\sin x + c_1e^x}$$

(c)
$$\frac{dy}{dx} + \frac{(2-3x^2)}{x^3} y = 2$$

Now, we create the term $e^{\int \frac{(2-3x^2)}{x^3} dx}$. This term requires some simplification as shown.

$$e^{\int \frac{(2-3x^2)}{x^3} dx} = e^{\int \frac{2}{x^3} dx - \int \frac{3x^2}{x^3} dx} = e^{-x^{-2} - \ln x^3} = \frac{e^{-x^{-2}}}{x^3} = \frac{1}{x^3 e^{x^{-2}}}$$

Substituting back into the ODE, we have

$$\frac{d}{dx}\left(y \frac{1}{x^3 e^{x^{-2}}}\right) = 2\left(\frac{1}{x^3 e^{x^{-2}}}\right)$$

$$y \frac{1}{x^3 e^{x^{-2}}} = \int \frac{2}{x^3 e^{x^{-2}}} dx + c_1$$

To help solve the integral on the right-hand side of the above equation, we can use another substitution, $u = x^{-2}$, then $x^3 = u^{-3/2}$ and $\frac{du}{dx} = -2x^{-3}$ *or* $dx = -\frac{1}{2}x^3 du = -\frac{1}{2}u^{-3/2} du.$

$$\int \frac{2}{x^3 e^{x^{-2}}} dx + c_1 = \int \frac{2}{u^{-3/2} e^u}\left(-\frac{1}{2}u^{-3/2} du\right) + c_1$$

$$= \int -e^{-u} du + c_1 = e^{-u} + c_1 = e^{-x^{-2}} + c_1$$

Finally, we have the solution

$$y = x^3 e^{x^{-2}}\left(e^{-x^{-2}} + c_1\right) = x^3 + c_1 x^3 e^{x^{-2}}$$

## Problem 3

**Solve the following differential equations by using appropriate substitutions:**

(a) $(y - x)\frac{dy}{dx} + (2x + 3y) = 0$
(b) $(x + y - 2)^2 \frac{dy}{dx} + (x + y - 2)^2 + 3x^4 = 0$

## Solution 3

***Worked Solution***

In these examples, we can make appropriate substitutions that help us simplify the equation into a form that is variable separable.

(a) We are given the following,

$$(y-x)\frac{dy}{dx}+(2x+3y)=0$$
$$\frac{dy}{dx}=\frac{-(2x+3y)}{y-x}$$

One useful substitution we can make is

$$y=vx$$
$$\frac{dy}{dx}=v+x\frac{dv}{dx}$$

Equating both expressions of $\frac{dy}{dx}$, we have

$$\frac{-(2x+3y)}{y-x}=v+x\frac{dv}{dx}$$
$$\frac{-(2x+3vx)}{vx-x}=v+x\frac{dv}{dx}$$
$$-x\frac{dv}{dx}=v+\frac{2+3v}{v-1}$$
$$\frac{v-1}{v^2+2v+2}dv=-\frac{1}{x}dx$$
$$\int\frac{\frac{1}{2}(2v+2)-2}{v^2+2v+2}dv=-\ln x+c_1$$
$$\frac{1}{2}\ln\left(v^2+2v+2\right)-\int\frac{2}{v^2+2v+2}dv=-\ln x+c_1$$

To evaluate the integral $\int\frac{2}{v^2+2v+2}dv$, we can apply the following technique where we make use of the trigonometric identity $1+\tan^2\theta=\sec^2\theta$.

$$\int\frac{2}{v^2+2v+2}dv=2\int\frac{1}{(v+1)^2+1}dv$$

Let $\tan\theta=v+1$, and this means $dv=\sec^2\theta\, d\theta$, so we have the following. [Note that other substitutions that show similar patterns in terms of their trigonometric identities also work, for example, we can also use the substitution, $\sinh\theta=v+1$, since it also helps us simplify the expression with the identity $1+\sinh^2\theta=\cosh^2\theta$ and the fact that $\frac{d}{d\theta}\sinh\theta=\cosh\theta$.]

$$2\int\frac{1}{(v+1)^2+1}dv=2\int\frac{1}{\tan^2\theta+1}\left(\sec^2\theta\right)d\theta=2\int 1d\theta=2\theta=2\tan^{-1}(v+1)$$

Substituting this back into our earlier equation, and replacing $v$ with our original variables, we obtain the solution for $y$.

$$\frac{1}{2}\ln\left(v^2+2v+2\right)-2\tan^{-1}(v+1)=-\ln x+c_1$$

$$\frac{1}{2}\ln\left(\left(\frac{y}{x}\right)^2+2\left(\frac{y}{x}\right)+2\right)-2\tan^{-1}\left(\left(\frac{y}{x}\right)+1\right)=-\ln x+c_1$$

(b)
$$(x+y-2)^2\frac{dy}{dx}+(x+y-2)^2+3x^4=0$$

We can substitute $v = x + y - 2$ to simplify the equation, then $\frac{dv}{dx}=1+\frac{dy}{dx}$

$$v^2\left(\frac{dv}{dx}-1\right)+v^2+3x^4=0$$

$$\int v^2dv=\int -3x^4dx$$

$$\frac{1}{3}v^3=-\frac{3}{5}x^5+c_1$$

$$\frac{1}{3}(x+y-2)^3=-\frac{3}{5}x^5+c_1$$

## Problem 4

**Second-order differential equations may take on several forms. Depending on which form we have, we arrive at solutions that can be either exponential or oscillatory or a combination of both. Let us explore some examples here. Solve the ODEs below and comment on their physical significance.**

(a) $\frac{d^2y}{dx^2}-6\frac{dy}{dx}+8y=0$, **where y(0)=0, y′(0)=1**
(b) $\frac{d^2y}{dx^2}+4y=0$, **where y(0)=0, y′(0)=1**
(c) $\frac{d^2y}{dx^2}+2\frac{dy}{dx}+y=0$, **where y(0)=0, y′(0)=1**
(d) $\frac{d^2y}{dx^2}+4\frac{dy}{dx}+5y=0$, **where y(0)=0, y′(0)=1**

## Solution 4

***Worked Solution***

(a) We can first find the characteristic roots of the equation denoted by $\lambda$. In this example, we have two distinct ($\lambda_1 \neq \lambda_2$) real roots. This scenario yields an exponential solution.

$$\frac{d^2y}{dx^2} - 6\frac{dy}{dx} + 8y = 0$$

$$\lambda^2 - 6\lambda + 8 = 0$$

$$(\lambda - 4)(\lambda - 2) = 0 \rightarrow \lambda = 2 \text{ or } 4$$

Note that in this case since $\lambda > 0$, the solution exponentially "explodes" in value as $x$ increases. Not all systems are physically meaningful when this happens, hence we need to conscious of its significance. In order for stable solutions (i.e., exponential decay), we require $\lambda < 0$. The solution to this problem is $y = Ae^{2x} + Be^{4x}$. We can derive this by guessing a trial solution of exponential form. Let us substitute $y = e^{\lambda x}$ into the ODE.

$$\frac{d^2y}{dx^2} - 6\frac{dy}{dx} + 8y = \lambda^2 e^{\lambda x} - 6\lambda e^{\lambda x} + 8e^{\lambda x} = 0$$

$$e^{\lambda x}\left(\lambda^2 - 6\lambda + 8\right) = 0 \rightarrow \lambda = 2 \text{ or } 4 \text{ (as above)}$$

$$y = Ae^{2x} + Be^{4x}$$

The constants $A$ and $B$ can be evaluated when we apply boundary conditions, e.g., known values of $y$ and $\frac{dy}{dx}$ at known values of $x$ (usually at spatial boundaries for spatial (or positional) coordinate $x$)

$$y(0) = 0$$

$$0 = A + B \tag{1}$$

$$y'(0) = 1$$

$$1 = 2A + 4B \tag{2}$$

Solving the simultaneous Eqs. (1) and (2), we have the solution as follows.

$$1 = -2B + 4B \rightarrow B = \frac{1}{2}, A = -\frac{1}{2}$$

$$y = -\frac{1}{2}e^{2x} + \frac{1}{2}e^{4x}$$

(b) We can first find the characteristic roots of the equation denoted by $\lambda$, where $y = e^{\lambda x}$. In this example, we have distinct ($\lambda_1 \neq \lambda_2$) imaginary roots.

$$\frac{d^2y}{dx^2} + 4y = 0$$

$$e^{\lambda x}\left(\lambda^2 + 4\right) = 0$$

$$\left(\lambda - (2i)^2\right) = 0$$

$$(\lambda + 2i)(\lambda - 2i) = 0 \rightarrow \lambda = \pm 2i$$

This differential equation yields a purely oscillatory solution. Note that this type of solution does not "decay" as it does not have any exponential component with negative power that reduces the solution at large values of $x$. Not all systems are physically meaningful when this happens, hence we need to conscious of its significance:

$$y = Ae^{2ix} + Be^{-2ix}$$

Note that in general for complex numbers, we can express $e^{ni\theta} = \cos n\theta + i \sin n\theta = (\cos \theta + i \sin \theta)^n$, where $\theta$ is the argument, and is commonly defined $\theta = \omega t$ in terms of angular frequency $\omega$ and independent variable of time $t$ in physical problems. Analogously, we can obtain the following sinusoidal form of solution for our problem.

$$\begin{aligned} y &= A(\cos 2x + i \sin 2x) + B(\cos(-2x) + i \sin(-2x)) \\ &= (A + B)\cos 2x + i(A - B)\sin 2x \\ &= (A + B)\left[\frac{e^{2ix} + e^{-2ix}}{2}\right] + i(A - B)\left[\frac{e^{2ix} - e^{-2ix}}{2i}\right] \end{aligned}$$

We know that $\frac{A+B}{2}$ is just another constant, which we can denote as $C$. Similarly for $\frac{A-B}{2} = D$. We also know that $e^{2ix} + e^{-2ix} = \cos 2x$ and $e^{2ix} - e^{-2ix} = \sin 2x$. Therefore,

$$y = C\cos 2x + D\sin 2x$$

As a shortcut, when we see the form of solution, $y = Ae^{2ix} + Be^{-2ix}$, we can express directly as a sum of cosine and sine functions in $y = C\cos 2x + D\sin 2x$. To solve for the constants, we can apply boundary conditions.

$$C = 0$$

$$2D = 1 \rightarrow D = \frac{1}{2}$$

Therefore we have the solution,

$$y = \frac{1}{2}\sin 2x$$

(c) In this problem, we have two identical real roots ($\lambda_1 = \lambda_2$).

$$\frac{d^2y}{dx^2} + 2\frac{dy}{dx} + y = 0$$

$$e^{\lambda x}\left(\lambda^2 + 2\lambda + 1\right) = 0$$

$$(\lambda + 1)^2 = 0 \rightarrow \lambda = -1$$

This scenario also yields an exponential solution, of the form below. Note that in this case we have a stable solution (i.e., exponential decay at large values of $x$) since $\lambda < 0$:

$$y = Ae^{-x} + Bxe^{-x}$$

We can apply boundary conditions to find $A$ and $B$,

$$0 = A$$

$$-A + B = 1 \rightarrow B = 1$$

$$y = xe^{-x}$$

(d) We can first find the characteristic roots of the equation denoted by $\lambda$. In this example, we have distinct ($\lambda_1 \neq \lambda_2$) roots that have both real and imaginary parts. Therefore the solution will have both exponential and oscillatory components.

$$\frac{d^2y}{dx^2} + 4\frac{dy}{dx} + 5y = 0$$

$$e^{\lambda x}\left(\lambda^2 + 4\lambda + 5\right) = 0$$

$$\lambda = \frac{-4 \pm \sqrt{4^2 - 4(1)(5)}}{2(1)} = -2 \pm \frac{\sqrt{4i^2}}{2} = -2 \pm i$$

Applying our derivation from part b, we can deduce a solution of this form

$$y = e^{-2x}(A\cos x + B\sin x)$$

The part of the solution "$e^{-2x}$" damps (reduces amplitude as $x$ increases) the oscillations, via an exponential decay. The other part of the solution "$A$ cos $x$ + $B$ sin $x$" describes the purely oscillatory (sinusoidal) part of the system.

We can apply boundary conditions to find $A$ and $B$,

$$A = 0$$

$$B = 1$$

Therefore

$$y = e^{-2x}(\sin x)$$

## Problem 5

**Solve the following inhomogenous ODEs, using the method of undetermined coefficients.**

(a) $\frac{d^2y}{dx^2} - 4\frac{dy}{dx} + 4y = e^{3x} + e^{2x}$, **where y(0)=0, y′(0)=1**
(b) $\frac{d^2y}{dx^2} + 6\frac{dy}{dx} + 5y = e^{-2x}$, **where y(0)=0, y′(0)=1**
(c) $\frac{d^2y}{dx^2} + 4y = 16$, **where y(0)=0, y′(0)=1**

## Solution 5

***Worked Solution***

In the earlier problem, we encountered second-order differential equations that were homogeneous. They could be expressed in the form $A\frac{d^2y}{dx^2} + B\frac{dy}{dx} + Cy = 0$, where $A$, $B$, and $C$ are constant coefficients, and the right-hand side of the equation is zero. In this problem, we will demonstrate how we can solve inhomogeneous ODEs, whereby the right-hand side is non-zero and is a function of $x$, as shown in the form $A\frac{d^2y}{dx^2} + B\frac{dy}{dx} + Cy = f(x)$.

The general approach to solving such ODEs is to first solve it as if it was a homogeneous equation, i.e., let $f(x) = 0$. We will obtain solutions to this homogeneous equation, called the complementary solution, which we can denote as $y_c(x)$. We will then need to add on to this solution, an additional solution called the particular integral, $y_p(x)$ which will be obtained by analyzing $f(x)$. The particular integral is equivalent to the ***term itself (with a constant coefficient added to it)***, plus ***its derivative***, and the ***derivative of its derivative*** (repeat this until all forms of expressions are represented). Below are two examples of how this process is applied. After considering all relevant terms, we can then omit terms that are already represented in the $y_c(x)$ part of the solution:

- $\cos^2 x \to \boldsymbol{A\cos^2 x + B\sin x\cos x + C\sin^2 x}$
- $(1 + x)e^x \to \boldsymbol{Ae^x + Bxe^x}$

Finally the complete solution will be $y(x) = y_c(x) + y_p(x)$. Any boundary conditions should be applied to the complete solution.

(a) Let us now consider our ODE for part a,

$$\frac{d^2y}{dx^2} - 4\frac{dy}{dx} + 4y = e^{3x} + e^{2x}$$

To solve for $y_c(x)$, we will first consider the right-hand side =0.

$$\frac{d^2y}{dx^2} - 4\frac{dy}{dx} + 4y = 0$$

$$e^{\lambda x}\left(\lambda^2 - 4\lambda + 4\right) = 0$$

$$(\lambda - 2)^2 = 0 \rightarrow \lambda = 2$$

$$y_c(x) = Ae^{2x} + Bxe^{2x}$$

To find out the particular integral, $y_p(x)$, we will look at the right-hand side of the equation. We note that the term $e^{2x}$ is already represented in $y_c(x)$; therefore, we can ignore it in $y_p(x)$.

$$e^{3x} + e^{2x} \rightarrow Ce^{3x} + De^{2x} \rightarrow Ce^{3x}$$

$$y_p(x) = Ce^{3x}$$

However, we have to be cautious when dealing with repeated roots such as in this problem. As there are two roots (although identical), we will have two constants of integration. And we will need one more term to add back to $y_p(x)$. In such cases, a good guess is the term of the form $Dx^2e^{2x}$. So it is now correct to express our particular integral as

$$y_p(x) = Ce^{3x} + Dx^2e^{2x}$$

$$\frac{dy_p}{dx} = 3Ce^{3x} + 2Dx^2e^{2x} + 2Dxe^{2x}$$

$$\begin{aligned}\frac{d^2y_p}{dx^2} &= 9Ce^{3x} + 4Dx^2e^{2x} + 4Dxe^{2x} + 4Dxe^{2x} + 2De^{2x} \\ &= 9Ce^{3x} + 4Dx^2e^{2x} + 8Dxe^{2x} + 2De^{2x}\end{aligned}$$

Substituting these terms into the ODE, we have

$$\frac{d^2y_p}{dx^2} - 4\frac{dy_p}{dx} + 4y_p = e^{3x} + e^{2x}$$

$$9Ce^{3x} + 4Dx^2e^{2x} + 8Dxe^{2x} + 2De^{2x} - 12Ce^{3x} - 8Dx^2e^{2x} - 8Dxe^{2x} + 4Ce^{3x} + 4Dx^2e^{2x} = e^{3x} + e^{2x}$$

By comparing coefficients, we can determine $C$ and $D$

$$Ce^{3x} + 2De^{2x} = e^{3x} + e^{2x}$$

$$C = 1, D = \frac{1}{2}$$

$$y_p(x) = e^{3x} + \frac{x^2}{2}e^{2x}$$

Putting our results together, the full general solution is given by

$$y(x) = y_c(x) + y_p(x) = Ae^{2x} + Bxe^{2x} + e^{3x} + \frac{x^2}{2}e^{2x}$$

We can now apply boundary conditions to find $A$ and $B$,

$$0 = A + 1 \rightarrow A = -1$$

$$y'(x) = 2Ae^{2x} + 2Bxe^{2x} + Be^{2x} + 3e^{3x} + x^2e^{2x} + xe^{2x}$$

$$1 = 2A + B + 3 = -2 + B + 3 \rightarrow B = 0$$

Finally, we obtain our solution as follows:

$$y = -e^{2x} + e^{3x} + \frac{x^2}{2}e^{2x}$$

(b)

$$\frac{d^2y}{dx^2} + 6\frac{dy}{dx} + 5y = e^{-2x}$$

To solve for $y_c(x)$, we will first consider the right-hand side =0.

$$e^{\lambda x}\left(\lambda^2 + 6\lambda + 5\right) = 0$$

$$(\lambda + 5)(\lambda + 1) = 0 \rightarrow \lambda = -1 \text{ or } -5$$

$$y_c(x) = Ae^{-x} + Be^{-5x}$$

To find out the particular integral, $y_p(x)$, we will look at the right-hand side of the equation.

$$e^{-2x} \rightarrow Ce^{-2x}$$

$$y_p(x) = Ce^{-2x}$$

$$\frac{dy_p}{dx} = -2Ce^{-2x}; \ \frac{d^2y_p}{dx^2} = 4Ce^{-2x}$$

Substituting these terms into the ODE, we have

$$\frac{d^2y_p}{dx^2} + 6\frac{dy_p}{dx} + 5y_p = e^{-2x}$$

$$4Ce^{-2x} + 6(-2Ce^{-2x}) + 5Ce^{-2x} = e^{-2x}$$

By comparing coefficients, we can determine $C$

$$-3Ce^{-2x} = e^{-2x}$$

$$C = -\frac{1}{3}$$

$$y_p(x) = -\frac{1}{3}e^{-2x}$$

Putting our results together, the full general solution is given by

$$y(x) = y_c(x) + y_p(x) = Ae^{-x} + Be^{-5x} - \frac{1}{3}e^{-2x}$$

We can now apply boundary conditions to find $A$ and $B$,

$$0 = A + B - \frac{1}{3}$$

$$y'(x) = -Ae^{-x} - 5Be^{-5x} + \frac{2}{3}e^{-2x}$$

$$y'(0) = 1 = -A - 5B + \frac{2}{3}$$

$$A + 5B = -\frac{1}{3}$$

$$B = -\frac{1}{6}, A = \frac{1}{2}$$

Finally, we obtain our solution as follows:

$$y = \frac{1}{2}e^{-x} - \frac{1}{6}e^{-5x} - \frac{1}{3}e^{-2x}$$

(c)

$$\frac{d^2y}{dx^2} + 4y = 16$$

To solve for $y_c(x)$, we will first consider the right-hand side =0.

$$e^{\lambda x}\left(\lambda^2 + 4\right) = 0$$

$$(\lambda \pm 2i) = 0 \rightarrow \lambda = \pm 2i$$

$$y_c(x) = Ae^{2ix} + Be^{-2ix}$$

To find out the particular integral, $y_p(x)$, we will look at the right-hand side of the equation.

$$16 \rightarrow C$$

$$y_p(x) = C$$

$$\frac{dy_p}{dx} = \frac{d^2y_p}{dx^2} = 0$$

Substituting these terms into the ODE, we have

$$\frac{d^2y_p}{dx^2} + 4y_p = 16$$

$$4C = 16 \rightarrow C = 4$$

$$y_p(x) = 4$$

Putting our results together, the full general solution is given by

$$y(x) = y_c(x) + y_p(x) = Ae^{2ix} + Be^{-2ix} + 4$$

We can now apply boundary conditions to find $A$ and $B$,

$$0 = A + B + 4$$

$$y'(x) = 2iAe^{2ix} - 2iBe^{-2ix}$$

$$y'(0) = 1 = 2iA - 2iB$$

$$B = -2 + \frac{i}{4}, A = -2 - \frac{i}{4}$$

Finally, we obtain our solution as follows:

$$y = \left(-2 - \frac{i}{4}\right)e^{2ix} + \left(-2 + \frac{i}{4}\right)e^{-2ix} + 4$$

## Problem 6

**Demonstrate the difference between an exact and inexact differential, and show that for an exact differential of the form $df = Pdx + Qdy$, the following is true.**

$$\left.\frac{\partial P}{\partial y}\right|_x = \left.\frac{\partial Q}{\partial x}\right|_y$$

**Show also that for an inexact differential of the form $df = (\mu P)dx + (\mu Q)dy$, whereby $\mu$ is the integrating factor. Using this result, derive the expression as shown.**

$$\frac{d\mu}{\mu} = \frac{1}{Q}\left[\left.\frac{\partial P}{\partial y}\right|_x - \left.\frac{\partial Q}{\partial x}\right|_y\right]dx$$

## Solution 6

***Worked Solution***

In an exact differential, we note that

$$df = Pdx + Qdy$$

$$P = \left.\frac{\partial f}{\partial x}\right|_y \rightarrow f = \int Pdx \text{ (constant } y\text{)}$$

$$Q = \left.\frac{\partial f}{\partial y}\right|_x \rightarrow f = \int Qdy \text{ (constant } x\text{)}$$

Since $\frac{\partial^2 f}{\partial x \partial y} = \frac{\partial^2 f}{\partial y \partial x}$, therefore $\left.\frac{\partial P}{\partial y}\right|_x = \left.\frac{\partial Q}{\partial x}\right|_y$ (1) for exact differential

In an inexact differential of the form below, $\left.\frac{\partial P}{\partial y}\right|_x \neq \left.\frac{\partial Q}{\partial x}\right|_y$ and instead, we have Eq. (2), whereby $\mu = \mu(x)$,

$$df = P'dx + Q'dy = (\mu P)dx + (\mu Q)dy$$

$$\left.\frac{\partial(\mu P)}{\partial y}\right|_x = \left.\frac{\partial(\mu Q)}{\partial x}\right|_y \qquad (2)$$

Since $\mu = \mu(x)$, Eq. (2) becomes,

$$\left.\frac{\partial \mu}{\partial y}\right|_x = 0 \ ; \ \left.\frac{\partial \mu}{\partial x}\right|_y = \frac{d\mu}{dx}$$

$$\left.\frac{\partial(\mu P)}{\partial y}\right|_x = \mu\left.\frac{\partial P}{\partial y}\right|_x + P\left.\frac{\partial \mu}{\partial y}\right|_x = \left.\frac{\partial(\mu Q)}{\partial x}\right|_y = \mu\left.\frac{\partial Q}{\partial x}\right|_y + Q\left.\frac{\partial \mu}{\partial x}\right|_y$$

$$\mu\left.\frac{\partial P}{\partial y}\right|_x + 0 = \mu\left.\frac{\partial Q}{\partial x}\right|_y + Q\left.\frac{\partial \mu}{\partial x}\right|_y$$

$$Q\left.\frac{\partial \mu}{\partial x}\right|_y = \mu\left[\left.\frac{\partial P}{\partial y}\right|_x - \left.\frac{\partial Q}{\partial x}\right|_y\right]$$

$$\frac{d\mu}{\mu} = \frac{1}{Q}\left[\left.\frac{\partial P}{\partial y}\right|_x - \left.\frac{\partial Q}{\partial x}\right|_y\right]dx$$

**Problem 7**

**Consider the differential equation $df = ydx + 2x^2dy$.**

(a) **Comment on whether the differential equation is exact or inexact and derive the following expressions where $A$ is a constant.**

$$\left.\frac{\partial f}{\partial x}\right|_y = \frac{Ae^{-\frac{1}{2x}}}{x^2}y$$

$$\left.\frac{\partial f}{\partial y}\right|_x = 2Ae^{-\frac{1}{2x}}$$

(b) **Show that for $df = 0$, the following is true.**

$$\frac{dx}{x^2} = -\frac{2dy}{y}$$

**Solution 7**

***Worked Solution***

(a) This is an inexact differential as shown below.

$$Pdx + Qdy \rightarrow P = y,\ Q = 2x^2$$

$$\left.\frac{\partial P}{\partial y}\right|_x = 1\ ;\ \left.\frac{\partial Q}{\partial x}\right|_y = 4x$$

$$\left.\frac{\partial P}{\partial y}\right|_x \neq \left.\frac{\partial Q}{\partial x}\right|_y \text{ (inexact)}$$

Therefore, we assume $\mu = \mu(x)$, whereby

$$df = P'dx + Q'dy = (\mu P)dx + (\mu Q)dy = (\mu y)dx + \left(2\mu x^2\right)dy$$

$$\left.\frac{\partial P'}{\partial y}\right|_x = y\left.\frac{\partial \mu}{\partial y}\right|_x + \mu = 0 + \mu = \mu$$

$$\left.\frac{\partial Q'}{\partial x}\right|_y = 2x^2\left.\frac{\partial \mu}{\partial x}\right|_y + 4x\mu$$

$$\left.\frac{\partial P'}{\partial y}\right|_x = \left.\frac{\partial Q'}{\partial x}\right|_y$$

$$\mu = 2x^2\frac{d\mu}{dx} + 4x\mu$$

$$\frac{d\mu}{dx}2x^2 = \mu(1 - 4x)$$

$$\int \frac{1}{\mu}d\mu = \int \frac{1 - 4x}{2x^2}dx$$

$$\ln \mu = -\frac{1}{2x} - 2\ln x + c_1$$

$$\mu = \frac{c_2 e^{-\frac{1}{2x}}}{x^2}$$

Substituting this result back into the differential equation, and let $c_2 = A$,

$$df = (\mu y)dx + \left(2\mu x^2\right)dy$$

$$df = \left(\frac{c_2 e^{-\frac{1}{2x}}}{x^2}y\right)dx + \left(2\frac{c_2 e^{-\frac{1}{2x}}}{x^2}x^2\right)dy = \left(\frac{c_2 e^{-\frac{1}{2x}}}{x^2}y\right)dx + \left(2c_2 e^{-\frac{1}{2x}}\right)dy$$

$$\left.\frac{\partial f}{\partial x}\right|_y = \frac{Ae^{-\frac{1}{2x}}}{x^2}y$$

$$\left.\frac{\partial f}{\partial y}\right|_x = 2Ae^{-\frac{1}{2x}}$$

(b) For $df = 0$,

$$\left(\frac{y}{x^2}\right)dx = -2dy$$

$$\frac{dx}{x^2} = -\frac{2dy}{y}$$

## Problem 8

**Consider the differential equation $df = (\cos 2x - \sinh x)dx + (\sinh x + \cos 2x)dy$.**

(a) **Comment on whether the differential equation is exact or inexact and derive the following expressions where $A$ is a constant.**

$$\left.\frac{\partial f}{\partial x}\right|_y = A(\sinh x + \cos 2x)^{-1}(\cos 2x - \sinh x)$$

$$\left.\frac{\partial f}{\partial y}\right|_x = A(\sinh x + \cos 2x)^{-1}(\sinh x + \cos 2x)$$

(b) **Show that for $df = 0$, the following is true.**

$$\left(\frac{\cos 2x - \sinh x}{\sinh x + \cos 2x}\right)dx + dy = 0$$

## Solution 8

***Worked Solution***

(a) This is an inexact differential as shown below.

$$Pdx + Qdy \rightarrow P = \cos 2x - \sinh x, Q = \sinh x + \cos 2x$$

$$\left.\frac{\partial P}{\partial y}\right|_x = 0 \ ; \ \left.\frac{\partial Q}{\partial x}\right|_y = \cosh x - 2\sin 2x$$

$$\left.\frac{\partial P}{\partial y}\right|_x \neq \left.\frac{\partial Q}{\partial x}\right|_y \text{ (inexact)}$$

Therefore, we assume $\mu = \mu(x)$, whereby

$$df = P'dx + Q'dy = (\mu P)dx + (\mu Q)dy = \mu(\cos 2x - \sinh x)dx + \mu(\sinh x + \cos 2x)dy$$

$$\left.\frac{\partial P'}{\partial y}\right|_x = (\cos 2x - \sinh x)\left.\frac{\partial \mu}{\partial y}\right|_x + \mu\left.\frac{\partial(\cos 2x - \sinh x)}{\partial y}\right|_x = 0 + 0 = 0$$

$$\left.\frac{\partial Q'}{\partial x}\right|_y = (\sinh x + \cos 2x)\left.\frac{\partial \mu}{\partial x}\right|_y + \mu(\cosh x - 2\sin 2x)$$

$$\left.\frac{\partial P'}{\partial y}\right|_x = \left.\frac{\partial Q'}{\partial x}\right|_y$$

$$0 = (\sinh x + \cos 2x)\left.\frac{\partial \mu}{\partial x}\right|_y + \mu(\cosh x - 2\sin 2x)$$

$$\frac{d\mu}{dx} = -\frac{\mu(\cosh x - 2\sin 2x)}{\sinh x + \cos 2x}$$

$$\int \frac{1}{\mu} d\mu = \int -\frac{\cosh x - 2\sin 2x}{\sinh x + \cos 2x} dx$$

$$\ln \mu = -\ln(\sinh x + \cos 2x) + c_1$$

$$\mu = c_2(\sinh x + \cos 2x)^{-1}$$

Substituting this result back into the differential equation, and let $c_2 = A$,

$$df = \mu(\cos 2x - \sinh x)dx + \mu(\sinh x + \cos 2x)dy$$

$$df = A(\sinh x + \cos 2x)^{-1}(\cos 2x - \sinh x)dx$$
$$+ A(\sinh x + \cos 2x)^{-1}(\sinh x + \cos 2x)dy$$

$$\left.\frac{\partial f}{\partial x}\right|_y = A(\sinh x + \cos 2x)^{-1}(\cos 2x - \sinh x)$$

$$\left.\frac{\partial f}{\partial y}\right|_x = A(\sinh x + \cos 2x)^{-1}(\sinh x + \cos 2x)$$

(b) For $df = 0$,

$$A(\sinh x + \cos 2x)^{-1}(\cos 2x - \sinh x)dx + A(\sinh x + \cos 2x)^{-1}(\sinh x + \cos 2x)dy = 0$$

$$\left(\frac{\cos 2x - \sinh x}{\sinh x + \cos 2x}\right)dx + dy = 0$$

**Problem 9**

**Solve the following differential equations**

**a)** $df = -(\cosh x \cos y + \cosh y \cos x)dx - (\sinh y \sin x - \sinh x \sin y)dy$
**b)** $df = (x + 2y)dx + (2x - y)dy$
**c)** $df = (13x + 6y - 10)dx + (6x + 2y - 5)dy$

**Solution 9** ***Worked Solution***

(a)

$$df = Pdx + Qdy$$

$$P = -(\cosh x \cos y + \cosh y \cos x) = -\cosh x \cos y - \cosh y \cos x$$

$$Q = -(\sinh y \sin x - \sinh x \sin y) = \sinh x \sin y - \sinh y \sin x$$

$$\left.\frac{\partial P}{\partial y}\right|_x = \cosh x \sin y - \cos x \sinh y$$

$$\left.\frac{\partial Q}{\partial x}\right|_y = \cosh x \sin y - \sinh y \cos x$$

$$\left.\frac{\partial P}{\partial y}\right|_x = \left.\frac{\partial Q}{\partial x}\right|_y \quad \text{(exact)}$$

$$\frac{df}{dx} = -\cosh x \cos y - \cosh y \cos x$$

$$f|_y = -\sinh x \cos y - \cosh y \sin x + g(y)$$

$$\frac{df}{dy} = \sinh x \sin y - \sinh y \sin x$$

$$f|_x = -\sinh x \cos y - \cosh y \sin x + h(x)$$

$$f = -\sinh x \cos y - \cosh y \sin x + \text{constant}$$

(b)

$$df = (x + 2y)dx + (2x - y)dy$$

$$P = x + 2y$$

$$Q = 2x - y$$

$$\left.\frac{\partial P}{\partial y}\right|_x = 2$$

$$\left.\frac{\partial Q}{\partial x}\right|_y = 2$$

$$\left.\frac{\partial P}{\partial y}\right|_x = \left.\frac{\partial Q}{\partial x}\right|_y \quad \text{(exact)}$$

$$\frac{df}{dx} = x + 2y$$

$$f|_y = \frac{x^2}{2} + 2yx + g(y)$$

$$\frac{df}{dy} = 2x - y$$

$$f|_x = -\frac{y^2}{2} + 2xy + h(x)$$

$$f = \frac{x^2}{2} + 2yx - \frac{y^2}{2} + \text{constant}$$

(c)

$$df = (13x + 6y - 10)dx + (6x + 2y - 5)dy$$

$$P = 13x + 6y - 10$$

$$Q = 6x + 2y - 5$$

$$\left.\frac{\partial P}{\partial y}\right|_x = 6$$

$$\left.\frac{\partial Q}{\partial x}\right|_y = 6$$

$$\left.\frac{\partial P}{\partial y}\right|_x = \left.\frac{\partial Q}{\partial x}\right|_y \text{ (exact)}$$

$$\frac{df}{dx} = 13x + 6y - 10$$

$$f|_y = \frac{13x^2}{2} + 6yx - 10x + g(y)$$

$$\frac{df}{dy} = 6x + 2y - 5$$

$$f|_x = 6xy + y^2 - 5y + h(x)$$

$$f = \frac{13x^2}{2} + 6yx - 10x + y^2 - 5y + \text{constant}$$

**Problem 10**

**Consider the change of coordinate system for a function, from $(x, y)$ to $(u, v)$. The coordinate axes of $(u, v)$ is at an angle $\theta$ anticlockwise from $(x, y)$. Comment on whether the following equation is valid.**

$$\frac{\partial^2 f}{\partial x^2} + \frac{\partial^2 f}{\partial y^2} = \frac{\partial^2 f}{\partial u^2} + \frac{\partial^2 f}{\partial v^2}$$

## Solution 10

***Worked Solution***

$$f(x,y) = f(u,v)$$

Regardless of coordinate system, *df* is the same

$$df = df$$

$$dx\frac{df}{dx}\bigg|_y + dy\frac{df}{dy}\bigg|_x = du\frac{df}{du}\bigg|_v + dv\frac{df}{dv}\bigg|_u$$

We can express $u$ and $v$ in terms of $x$ and $y$

$$u = x\cos\theta + y\sin\theta$$

$$v = -x\sin\theta + y\cos\theta$$

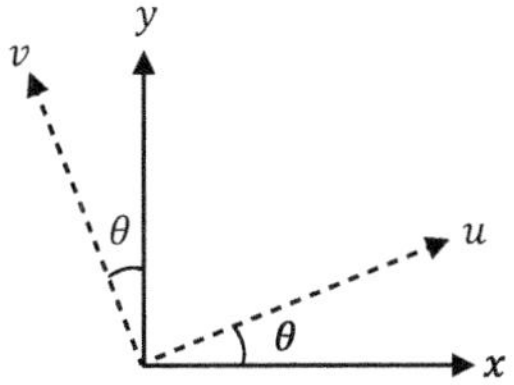


Now we can try to find an expression for $\frac{d^2f}{dx^2}$

$$\frac{du}{dx}\bigg|_y = \cos\theta \; ; \; \frac{dv}{dx}\bigg|_y = -\sin\theta$$

$$\frac{df}{dx}\bigg|_y = \frac{du}{dx}\bigg|_y\frac{df}{du}\bigg|_v + \frac{dv}{dx}\bigg|_y\frac{df}{dv}\bigg|_u - \frac{dy}{dx}\bigg|_y\frac{df}{dy}\bigg|_x = \frac{du}{dx}\bigg|_y\frac{df}{du}\bigg|_v + \frac{dv}{dx}\bigg|_y\frac{df}{dv}\bigg|_u - 0$$

$$\frac{df}{dx}\bigg|_y = \frac{du}{dx}\bigg|_y\frac{df}{du}\bigg|_v + \frac{dv}{dx}\bigg|_y\frac{df}{dv}\bigg|_u = \cos\theta\frac{df}{du}\bigg|_v - \sin\theta\frac{df}{dv}\bigg|_u$$

$$\left[\frac{d}{dx}\bigg|_y\right]f = \left[\cos\theta\frac{d}{du}\bigg|_v - \sin\theta\frac{d}{dv}\bigg|_u\right]f$$

$$\left[\frac{d^2}{dx^2}\bigg|_y\right]f = \left[\cos\theta\frac{d}{du}\bigg|_v - \sin\theta\frac{d}{dv}\bigg|_u\right]\left[\cos\theta\frac{d}{du}\bigg|_v - \sin\theta\frac{d}{dv}\bigg|_u\right]f$$

$$\frac{d^2f}{dx^2} = \cos^2\theta\frac{d^2f}{du^2} - 2\cos\theta\sin\theta\frac{d^2f}{dudv} + \sin^2\theta\frac{d^2f}{dv^2}$$

Similarly, we find an expression for $\frac{d^2f}{dy^2}$

$$\left.\frac{du}{dy}\right|_x = \sin\theta \;;\; \left.\frac{dv}{dy}\right|_x = \cos\theta$$

$$\left.\frac{df}{dy}\right|_x = \left.\frac{du}{dy}\right|_x \left.\frac{df}{du}\right|_v + \left.\frac{dv}{dy}\right|_x \left.\frac{df}{dv}\right|_u - \left.\frac{dx}{dy}\right|_x \left.\frac{df}{dx}\right|_y = \left.\frac{du}{dy}\right|_x \left.\frac{df}{du}\right|_v + \left.\frac{dv}{dy}\right|_x \left.\frac{df}{dv}\right|_u - 0$$

$$\left.\frac{df}{dy}\right|_x = \left.\frac{du}{dy}\right|_x \left.\frac{df}{du}\right|_v + \left.\frac{dv}{dy}\right|_x \left.\frac{df}{dv}\right|_u = \sin\theta \left.\frac{df}{du}\right|_v + \cos\theta \left.\frac{df}{dv}\right|_u$$

$$\left[\left.\frac{d}{dy}\right|_x\right] f = \left[\sin\theta \left.\frac{d}{du}\right|_v + \cos\theta \left.\frac{d}{dv}\right|_u\right] f$$

$$\left[\left.\frac{d^2}{dy^2}\right|_x\right] f = \left[\sin\theta \left.\frac{d}{du}\right|_v + \cos\theta \left.\frac{d}{dv}\right|_u\right]\left[\sin\theta \left.\frac{d}{du}\right|_v + \cos\theta \left.\frac{d}{dv}\right|_u\right] f$$

$$\frac{d^2f}{dy^2} = \sin^2\theta \frac{d^2f}{du^2} + 2\cos\theta\sin\theta \frac{d^2f}{dudv} + \cos^2\theta \frac{d^2f}{dv^2}$$

Combining our results, we have

$$\frac{d^2f}{dx^2} + \frac{d^2f}{dy^2} = \cos^2\theta \frac{d^2f}{du^2} - 2\cos\theta\sin\theta \frac{d^2f}{dudv} + \sin^2\theta \frac{d^2f}{dv^2} + \sin^2\theta \frac{d^2f}{du^2}$$
$$+ 2\cos\theta\sin\theta \frac{d^2f}{dudv} + \cos^2\theta \frac{d^2f}{dv^2}$$

$$\frac{d^2f}{dx^2} + \frac{d^2f}{dy^2} = \left(\cos^2\theta + \sin^2\theta\right)\frac{d^2f}{du^2} + \left(\sin^2\theta + \cos^2\theta\right)\frac{d^2f}{dv^2} = \frac{d^2f}{du^2} + \frac{d^2f}{dv^2}$$

## Problem 11

**In thermodynamics, we may encounter the cyclic rule whereby for variables *T*, *V*, and *P* representing temperature, volume, and pressure, the following is true**

$$\left(\frac{dP}{dT}\right)_V \left(\frac{dT}{dV}\right)_P \left(\frac{dV}{dP}\right)_T = -1$$

**Starting from this equation, show that the above is satisfied.**

$$PTV + P^3 + T^4 + V^5 = 0$$

## Solution 11

### *Worked Solution*

We can first differentiate the given equation

$$PTV + P^3 + T^4 + V^5 = 0$$

$$TVdP + PVdT + PTdV + 3P^2dP + 4T^3dT + 5V^4dV = 0$$

$$dP(TV + 3P^2) + dT(PV + 4T^3) + dV(PT + 5V^4) = 0$$

Let $F = TV + 3P^2$, $G = PV + 4T^3$, and $H = PT + 5V^4$,

$$FdP + GdT + HdV = 0$$

$$\left(\frac{dP}{dT}\right)_V = -\frac{G}{F} = -\frac{PV + 4T^3}{TV + 3P^2}$$

$$\left(\frac{dT}{dV}\right)_P = -\frac{H}{G} = -\frac{PT + 5V^4}{PV + 4T^3}$$

$$\left(\frac{dV}{dP}\right)_T = -\frac{F}{H} = -\frac{TV + 3P^2}{PT + 5V^4}$$

$$\left(\frac{dP}{dT}\right)_V\left(\frac{dT}{dV}\right)_P\left(\frac{dV}{dP}\right)_T = \left(-\frac{G}{F}\right)\left(-\frac{H}{G}\right)\left(-\frac{F}{H}\right) = -1$$

## Problem 12

**Can the following differential equation be solved using the method of Separation of Variables? Show that the solution to this equation takes the form $y = \pm\sqrt{c_1 - 4x^{20}} - x - 1$, where $c_1$ is an integration constant.**

$$(x + y + 1)\frac{dy}{dx} + \left(x + y + 1 + 40x^{19}\right) = 0$$

## Solution 12

### *Worked Solution*

Note that in order to use the method of Separation of Variables (VS), we need the equation to be linear, homogeneous with linear homogeneous boundary conditions. In the general form, the VS method requires a function of the form:

$$\frac{dy}{dx} = f(x,y)$$

$$f(x,y) = g(x)h(y)$$

When the function's variables cannot be separated, we may try manipulating into a form that is separable, such as in this example. We may first observe that the equation in this problem cannot be separated into clear-cut variables $x$ and $y$. Let $z = x + y + 1$

$$\frac{dz}{dx} = 1 + \frac{dy}{dx}$$

$$z\left(\frac{dz}{dx} - 1\right) + \left(z + 40x^{19}\right) = 0$$

$$\int zdz = \int -40x^{19}dx$$

$$\frac{z^2}{2} = -2x^{20} + c_1$$

$$z = \pm\sqrt{c_1 - 4x^{20}}$$

$$z = \pm\sqrt{c_1 - 4x^{20}} = x + y + 1$$

$$y = \pm\sqrt{c_1 - 4x^{20}} - x - 1$$

**Problem 13**

**In heat transfer problems, we may encounter the heat equation, written in the form as follows, where $u$ represents temperature and it is a function of position and time, and $\alpha$ denotes thermal diffusivity:**

$$\frac{\partial u}{\partial t} = \alpha \nabla^2 u$$

**Explain the physical meaning of the following conditions and solutions to the above equation.**

$$\frac{\partial u}{\partial x}(1,0,0,t) = 1$$

$$u(0,0,0,t) = 0$$

$$u(x,y,z,0) = \sin 2\pi x$$

## Solution 13

### *Worked Solution*

We can observe that this is a one-dimensional heat transfer (in $x$ direction using Cartesian coordinates) system. One possible scenario is heat conduction in a solid rod of length $0 \leq x \leq 1$. At the position $x = 0$, $u(0, 0, 0, t) = 0$. This means that there is a constant temperature $u = 0$, at $x = 0$ at all times. This also implies that since temperature is constant, the temperature gradient is zero.

$$\frac{\partial u}{\partial x} = 0, \text{ at } x = 0 \tag{1}$$

$$u = 0, \text{ at } x = 0 \tag{2}$$

At the other end of the rod, i.e., at $x = 1$, there is a fixed heat flux entering the rod, $\frac{\partial u}{\partial x} = 1$ at all times. [Note that heat flux is defined as $q'' = -k\nabla T$, where $k$ is thermal conductivity.]

$$\frac{\partial u}{\partial x} = 1, \text{ at } x = 1 \tag{3}$$

As for the third condition, it represents the initial condition whereby at $t = 0$, the temperature variation along the length of the rod can be represented by a sine function of the form, $\sin 2\pi x$.

At steady state as $t \to \infty$,

$$\frac{\partial u}{\partial t} = 0 = \frac{\partial^2 u}{\partial x^2}$$

$$\frac{du}{dx} = Ax + B \to B = 0, \text{ applying (1)}$$

$$u = \frac{Ax^2}{2} + C \to C = 0, \text{ applying (2)}$$

$$A = 1, \text{ applying (3)}$$

$$u(x, \infty) = \frac{x^2}{2}$$

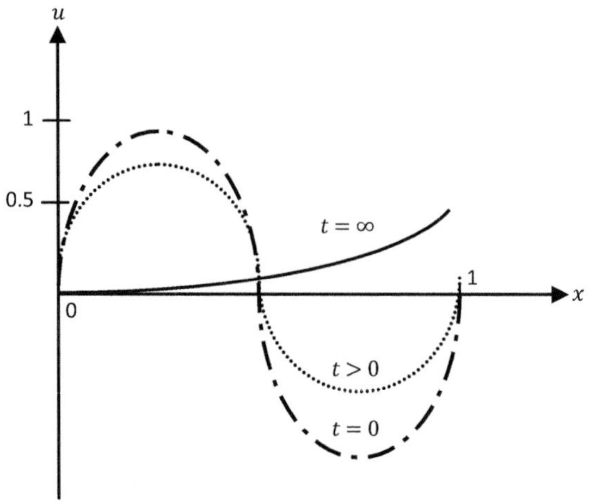


## Problem 14

**Consider the following partial differential equation for $0 \le x \le 1$ and $0 \le t \le \infty$, where $\alpha$ is a constant.**

$$\frac{\partial u}{\partial t} = \alpha \frac{\partial^2 u}{\partial x^2}$$

(a) **Given that the boundary conditions are $u(0,t) = u(1,t) = 0$ and $u(x,0) = 1$. Show that the solution to the equation can be expressed as follows, where $k$ is an integer.**

$$u(x,t) = \sum_{k=1}^{\infty} \left[\frac{4}{(2k-1)\pi}\right] e^{-(2k-1)^2 \pi^2 \alpha t} \sin\left((2k-1)\pi x\right)$$

(b) **What would the solution be like if the initial condition was given as follows.**

$$u(x,0) = \sin 2\pi x + \frac{1}{3}\sin 4\pi x + \frac{1}{5}\sin 6\pi x + \frac{1}{7}\sin 8\pi x$$

## Solution 14

*Worked Solution*

(a) We note that the variables of position $x$ and time $t$ are independent, and the equation is linear and homogeneous; therefore, we can use the method of separation of variables.

$$u(x,t) = X(x)T(t)$$

$$XT' = \alpha X''T$$

$$\frac{T'}{\alpha T} = \frac{X''}{X} = -\beta, \text{ where } \beta \text{ is a constant} > 0$$

$$T' + \beta\alpha T = 0 \tag{1}$$

$$T = A\mathrm{e}^{-\beta\alpha t}, \text{ where } A \text{ is a constant}$$

$$X'' + \beta X = 0$$

$$X = B\sin\sqrt{\beta}x + C\cos\sqrt{\beta}x$$

We know that $u(0, t) = 0$; therefore $C = 0$

$$X = B\sin\sqrt{\beta}x$$

And we know that $u(1, t) = 0$,

$$0 = B\sin\sqrt{\beta}$$

$$B = 0 \text{ or } \sin\sqrt{\beta} = 0$$

$$\sqrt{\beta_m} = m\pi, \text{ where } m \text{ is a positive integer}$$

$$u(x,t) = \sum_{m=1}^{\infty} A_m \mathrm{e}^{-m^2\pi^2\alpha t}\sin(m\pi x)$$

We know that $u(x, 0) = 1$. So at a specific time $t = 0$, we have a function of $x$, call it $f(x)$

$$u(x,0) = f(x) = 1 = \sum_{m=1}^{\infty} A_m \sin(m\pi x)$$

We can find out the coefficient $A_m$ using the principle of orthogonality, which has the general form below for the coefficient of the sine term.

$$A_m = \frac{1}{L}\int_{-L}^{L} f(x)\sin\left(\frac{m\pi x}{L}\right)dx$$

We observe that our function $u(x, 0)$ or $f(x)$ is odd because there are no constant terms or cosine terms, the product of this odd function $f(x)$ with another odd function, $\sin\left(\frac{m\pi x}{L}\right)$ is an even function. Therefore, the integral limits can be expressed as:

$$A_m = \frac{2}{L}\int_0^L f(x)\sin\left(\frac{m\pi x}{L}\right)dx$$

We know that $L = 1$ based on the form of our expression, and at $t = 0$, $f(x) = 1$.

$$A_m = 2\int_0^1 f(x)\sin m\pi x dx = 2\int_0^1 \sin m\pi x dx$$

$$= 2\left[\frac{-\cos m\pi x}{m\pi}\right]_0^1 = 2\left(\frac{1-\cos m\pi}{m\pi}\right)$$

So it follows that when $m = 1, A_1 = \frac{4}{\pi}$, when $m = 2, A_2 = 0, A_3 = \frac{4}{3\pi}, A_4 = 0$, etc.

$$A_m = \begin{cases} \dfrac{4}{m\pi}, & \text{when } m \text{ is odd} \\ 0, & \text{when } m \text{ is even} \end{cases}$$

$$u(x,t) = \sum_{k=1}^{\infty}\left[\frac{4}{(2k-1)\pi}\right]\mathrm{e}^{-(2k-1)^2\pi^2\alpha t}\sin((2k-1)\pi x)$$

(b) If the initial conditions were changed to another form as shown below, then we can deduce that:

$$\sin 2\pi x + \frac{1}{3}\sin 4\pi x + \frac{1}{5}\sin 6\pi x + \frac{1}{7}\sin 8\pi x$$

$$\sum_{m=1}^{\infty} A_m \sin(m\pi x) = \sin 2\pi x + \frac{1}{3}\sin 4\pi x + \frac{1}{5}\sin 6\pi x + \frac{1}{7}\sin 8\pi x$$

By inspection, we can see that the surviving $m$ values are $m = 2$, 4, 6, and 8. And we can observe that $A_2 = 1$, $A_4 = \frac{1}{3}$, $A_6 = \frac{1}{5}$, and $A_8 = \frac{1}{7}$. Therefore the solution would be as follows:

$$u(x,t) = A_2\mathrm{e}^{-2^2\pi^2\alpha t}\sin(2\pi x) + A_4\mathrm{e}^{-4^2\pi^2\alpha t}\sin(4\pi x) + A_6\mathrm{e}^{-6^2\pi^2\alpha t}\sin(6\pi x) + A_8\mathrm{e}^{-8^2\pi^2\alpha t}\sin(8\pi x)$$

$$= \mathrm{e}^{-2^2\pi^2\alpha t}\sin(2\pi x) + \frac{1}{3}\mathrm{e}^{-4^2\pi^2\alpha t}\sin(4\pi x) + \frac{1}{5}\mathrm{e}^{-6^2\pi^2\alpha t}\sin(6\pi x) + \frac{1}{7}\mathrm{e}^{-8^2\pi^2\alpha t}\sin(8\pi x)$$

## Problem 15

**Classify the following partial differential equation and comment on a possible method to solve the equation. Proceed to solve using the boundary conditions and initial conditions provided below.**

$$\frac{\partial u}{\partial t} = \frac{\partial^2 u}{\partial x^2}$$

**Initial condition: When $t = 0$, $u = 0$ for all $x$. Boundary condition: When $x = 0$, $u = 2 \sin t$ for all $t$.**

**You may use the following result as necessary, where $s$ is the Laplace variable, and $\mathcal{L}^{-1}$ refers to inverse Laplace transform:**

$$\mathcal{L}^{-1}\left(e^{-\sqrt{s}x}\right) = \frac{1}{2\sqrt{\pi}}\frac{x}{t^{\frac{3}{2}}}e^{\left(-\frac{x^2}{4t}\right)}$$

## Solution 15

### *Worked Solution*

The equation is linear and homogeneous with constant coefficient; therefore, we can use Laplace transform or separation of variables to solve. In this solution, we will show the method of Laplace transform since initial condition $u(t = 0, x) = 0$, and this simplifies the Laplace transform expression.

From the data booklet, we can find the Laplace transform for derivatives.

$$\frac{du}{dt} \rightarrow s\bar{u} - u(0)$$

Therefore, we have the Laplace transform of the differential equation as follows:

$$s\bar{u} - u(0) = \frac{d^2\bar{u}}{ds^2}$$

$$s\bar{u} = \frac{d^2\bar{u}}{ds^2}$$

$$\alpha^2 - s = 0 \rightarrow \alpha = \pm\sqrt{s}$$

$$\bar{u} = c_1 e^{\sqrt{s}x} + c_2 e^{-\sqrt{s}x}$$

$c_1 = 0$ so that the function decays with increasing $x$, instead of inflating exponentially as $x \rightarrow \infty$ (not physically meaningful).

$$\bar{u} = c_2 e^{-\sqrt{s}x}$$

Given the boundary condition at $x = 0$, $u = 2\sin t$, we can find the Laplace transform of this boundary condition which is $\bar{u}(x = 0) = \frac{2}{s^2+1} = c_2$.

Therefore we have the solution in Laplace variable $s$ as follows,

$$\bar{u} = \frac{2}{s^2 + 1} e^{-\sqrt{s}x}$$

We know that the convolution theorem states that if $\bar{u} = \bar{F}\bar{G}$, then $u(t) = \int_0^t f(t - \tau)g(\tau)d\tau$

In our case,

$$\bar{F} = \frac{2}{s^2 + 1} \xrightarrow{\mathcal{L}^{-1}} f(t) = \frac{2}{\sin t}$$

$$\bar{G} = e^{-\sqrt{s}x} \xrightarrow{\mathcal{L}^{-1}} g(t) = \frac{1}{2\sqrt{\pi}} \frac{x}{t^{\frac{3}{2}}} e^{\left(-\frac{x^2}{4t}\right)}$$

Therefore, the solution in its original variables is as shown below.

$$u(x, t) = \int_0^t \frac{2}{\sin(t - \tau)} \frac{1}{2\sqrt{\pi}} \frac{x}{\tau^{\frac{3}{2}}} e^{\left(-\frac{x^2}{4\tau}\right)} d\tau$$

## Problem 16

**Consider the following differential equation and solve it using the method of characteristics.**

$$\frac{\partial \boldsymbol{u}}{\partial \boldsymbol{t}} = -\frac{\partial \boldsymbol{u}}{\partial \boldsymbol{x}}$$

**The initial condition is such that at $\boldsymbol{t = 0}$, $\boldsymbol{u(x, 0) = \sin x}$.**

## Solution 16

### *Worked Solution*

Before we solve this problem, let us revisit some concepts about the Method of Characteristics, which is a method that provides solutions that move at a velocity $V$. Consider the general form of the equation:

$$\frac{\partial u}{\partial t} = -V\frac{\partial u}{\partial x}$$

We can change our coordinate system from $x$ and $t$ to new coordinates $s$ and $\tau$ such that $s$ varies along the characteristics (shown in purple below) and $\tau$ varies as we go from one characteristic to another.

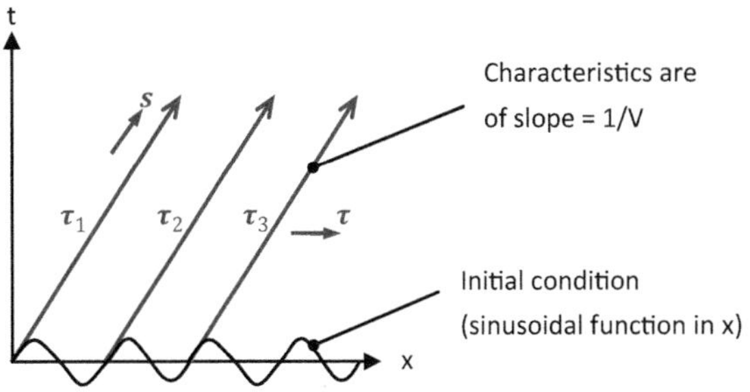


We can express the variables $x$ and $t$ in the form of the new variable, $s$. $\frac{du}{ds}$ represents how u changes along the characteristics in the $s$ direction.

$$\frac{du}{ds} = \frac{dx}{ds}\frac{\partial u}{\partial x} + \frac{dt}{ds}\frac{\partial u}{\partial t}$$

Let $a(x, t) = \frac{dx}{ds}$ and $b(x, t) = \frac{dt}{ds}$, then we have

$$\frac{du}{ds} = a(x, t)\frac{\partial u}{\partial x} + b(x, t)\frac{\partial u}{\partial t}$$

And the slope (as shown in the plot above) of the characteristics along which the solution propagates is given by 1/*V*.

$$\frac{1}{V} = \frac{dt}{dx} = \frac{b}{a}$$

If we are told that the equation is $\frac{\partial u}{\partial t} = -\frac{\partial u}{\partial x}$ as shown in the problem statement, then we have the following:

$$\frac{\partial u}{\partial x} + \frac{\partial u}{\partial t} = \frac{du}{ds} = 0$$

$$a(x, t) = \frac{dx}{ds} = 1 \rightarrow x(s) = s + c_1$$

When $t = 0$, $s = 0$ and $x(0) = c_1 = \tau$

$$b(x, t) = \frac{dt}{ds} = 1 \rightarrow t(s) = s + c_2$$

When $t = 0$, $s = 0$ and $t(0) = c_2 = 0$

Therefore, we have $x = s + \tau$ and $t = s$, and it follows that $x = t + \tau$. This equation describes the lines of characteristics with slope of 1.

We know that when $t = 0$, $u(x, 0) = sinx$. Therefore $u(\tau, 0) = \sin \tau$.

And the solution is of the following form whereby the solution moves to the right with no change in amplitude relative to the initial condition wave.

$$u(x,t) = \sin(x-t)$$

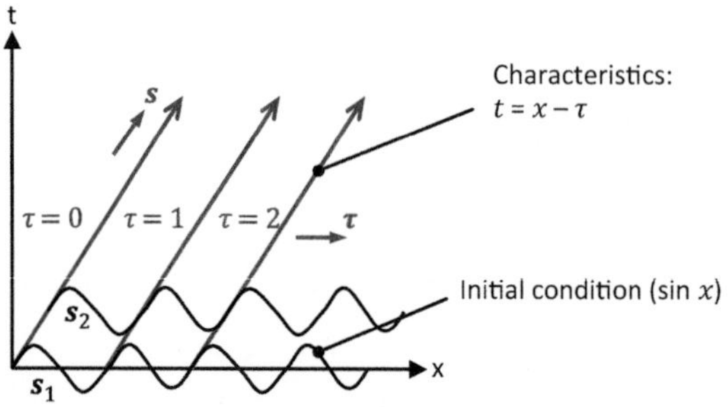


## Problem 17

**Consider the following partial differential equation whereby $0 < t < \infty$, and $-\infty < x < \infty$, and using the method of characteristics, show that the slope of its characteristics is 1. Determine the solution given that the initial condition is that when $t = 0$, $u(x,0) = \sin x$**

$$\frac{\partial u}{\partial x} + \frac{\partial u}{\partial t} + 2u = 0$$

## Solution 17

***Worked Solution***

In terms of the method of characteristics, this time we may observe that our differential equation takes on the general form below where $a$, $b$, and $c$ are functions of $x$ and $t$.

$$a\frac{\partial u}{\partial x} + b\frac{\partial u}{\partial t} + cu = 0$$

Upon transformation into the $s$ and $\tau$ coordinates, we have the following where $\frac{du}{ds} = a\frac{\partial u}{\partial x} + b\frac{\partial u}{\partial t} = \frac{dx}{ds}\frac{\partial u}{\partial x} + \frac{dt}{ds}\frac{\partial u}{\partial t}$.

$$\frac{du}{ds} + c(s,\tau)u = 0$$

So we return to the equation in our problem.

$$\frac{\partial u}{\partial x}+\frac{\partial u}{\partial t}+2u=0$$

$$a=\frac{dx}{ds}=1\rightarrow x=s+c_1$$

When $t=0$, $s=0$ and $x(0)=c_1=\tau$; therefore

$$x=s+\tau$$

$$b=\frac{dt}{ds}=1\rightarrow t=s+c_2$$

When $t=0$, $s=0$ and $c_2=0$; therefore

$$s=t \tag{1}$$

$$\tau=x-t \tag{2}$$

The result in (2) tells us that the equation of the characteristics is $t=x-\tau$ and the slope $\frac{dt}{dx}$ is 1.

In our original equation, we can then convert our PDE into an ODE, by expressing the partial derivatives into a total derivative in terms of variable $s$. In physical terms, the term $\frac{\partial u}{\partial x}$ in our equation refers to a convective part that contributes to the "movement" of the solution at velocity $V$. The part $\frac{\partial u}{\partial t}=-2u$ refers to an exponentially decaying part which diminishes the solution as $t\rightarrow\infty$.

$$\frac{\partial u}{\partial x}+\frac{\partial u}{\partial t}+2u=0$$

$$\frac{du}{ds}+2u=0$$

The initial condition is $u(x,0)=\ \sin x$. Since $s=t$ from our result in (1), therefore

$$u(s=0)=\ \sin x$$

From our result in (2), since $\tau=x-t$, therefore at $t=0$, $\tau=x$,

$$u(s=0)=\ \sin\tau$$

For the ODE of this form, the solution has the form of an exponential decay where $A$ is an integration constant.

$$\frac{du}{ds} + 2u = 0$$

$$u = A\exp(-2t) \rightarrow A = \sin\tau$$

Putting our results together, we have

$$u = \sin\tau\exp(-2t) = \sin(x - t)\exp(-2t)$$

In this case, our solution has a decaying amplitude (due to exponential decay part of equation) as it travels along the characteristics (due to the convective term of the equation) at velocity $V = \frac{1}{\text{slope}} = 1$.

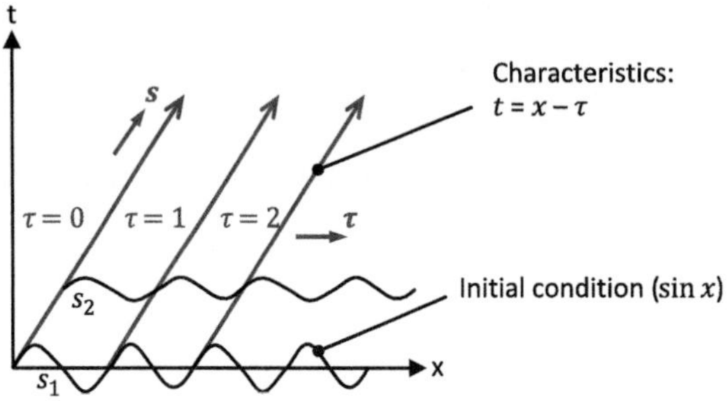


## Problem 18

**Consider the following partial differential equation which has a damping term *tu*. This is a non-linear equation whereby there is a velocity component represented by the term $x\frac{\partial u}{\partial x}$. Given that the initial condition is such that when $t = 0$, $u(x, 0) = F(x)$, determine the solution of this equation and sketch it in a plot of *t* against *x*.**

$$x\frac{\partial u}{\partial x} + \frac{\partial u}{\partial t} + tu = 0$$

## Solution 18

***Worked Solution***

$$x\frac{\partial u}{\partial x} + \frac{\partial u}{\partial t} + tu = 0$$

$$a = \frac{dx}{ds} = x$$

$$b = \frac{dt}{ds} = 1$$

The slope of the characteristics will be $\frac{dt}{dx} = \frac{1}{x}$, and we can expect the characteristics to take on the general form as shown below.

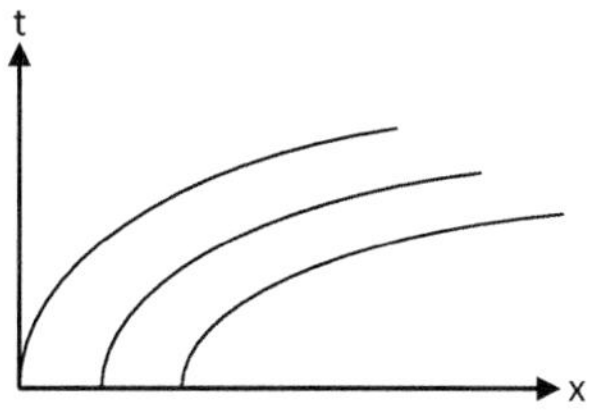


$$\frac{dx}{ds} = x$$

$$x = c_1 \exp(s)$$

$$\frac{dt}{ds} = 1$$

$$t = s + c_2$$

When $t = 0$, $s = 0$ and $x(0) = \tau$; therefore

$$x = c_1 \exp(s)$$

$$\tau = c_1 \exp(0) = c_1$$

$$x = \tau \exp(s)$$

When $t = 0$, $s = 0$ and $c_2 = 0$; therefore

$$s = t$$

The partial differential equation becomes

$$\frac{du}{ds} + tu = 0$$

$$\frac{du}{ds} + su = 0$$

$$\int \frac{du}{u} = \int -sds \rightarrow \ln u = -\frac{s^2}{2} + \text{constant}$$

$$u = A\exp\left(-\frac{s^2}{2}\right)$$

When $t = 0$, $u(x, 0) = F(x)$, and

$$s = t = 0$$

$$x = \tau\exp(s) = \tau$$

$$u(s = 0) = F(\tau)$$

Therefore we have the solutions as shown below in the $s$, $\tau$ coordinates and $x$, $t$ coordinates.

$$u = A\exp\left(-\frac{s^2}{2}\right) = F(\tau)\exp\left(-\frac{s^2}{2}\right)$$

$$u = F(xe^{-t})\exp\left(-\frac{t^2}{2}\right)$$

We may observe that the initial condition is transported along at a non-uniform velocity (slope of characteristics not constant, i.e., not straight lines), and there is a non-linear damping term $\exp\left(-\frac{t^2}{2}\right)$ that reduces amplitude of the solution as it travels. Consider the initial points $A$, $B$, and $C$. They will travel to $A'$, $B'$, and $C'$ from $s = 0$ to $s = 1$, along the curved lines of characteristics.

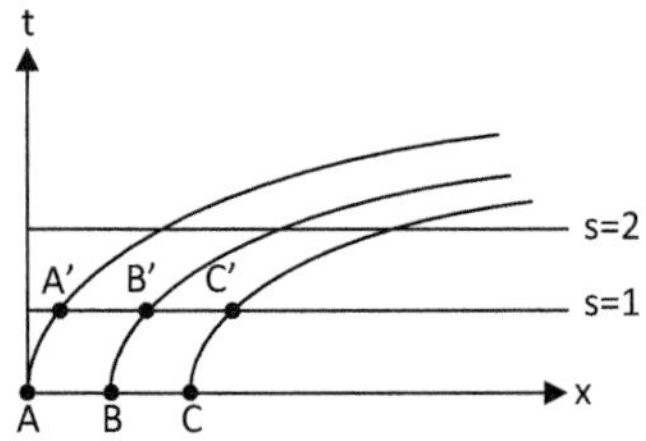


**Problem 19**

**Consider the partial differential equation as follows whereby $1 < t < \infty$, and $-\infty < x < \infty$, and using the method of characteristics, determine the solution in terms of $x$ and $t$. The initial condition is that when $t = 1$, $u(x, 1) = 2 \sin x$:**

$$t\frac{\partial u}{\partial t} + x\frac{\partial u}{\partial x} + 2u = 0$$

## Solution 19

***Worked Solution***

In the same way, we define the expressions for $a$ and $b$, in the form

$$b\frac{\partial u}{\partial t}+a\frac{\partial u}{\partial x}+2u=\frac{du}{ds}+2u=0$$

$$a=\frac{dx}{ds}=x$$

$$x=c_1e^s$$

$$b=\frac{dt}{ds}=t$$

$$t=c_2e^s$$

We can apply initial conditions to find $c_1$ and $c_2$. When $t=1$, $s=0$, $x=\tau$.

$$c_1=\tau$$

$$c_2=1$$

So we obtain the following two expressions that convert the original coordinates in $x$ and $t$ to $s$ and $\tau$.

$$x=\tau e^s$$

$$t=e^s$$

From here, we can obtain the equation for the characteristics to be

$$t=\frac{x}{\tau}$$

We can also proceed to find the solution to the differential equation in the new coordinates $s$ and $\tau$.

$$u(x,1)=2\sin x=2\sin(t\tau)=2\sin\tau$$

$$\frac{du}{ds}=-2u\rightarrow\int\frac{1}{u}du=\int-2ds$$

$$u=A\exp(-2s)$$

When $s=0$, $t=1$, therefore $x=\tau$ and $u(0)=A=2\sin\tau$. So we have the solution of the form in the $s$ and $\tau$ coordinates.

$$u = 2\sin\tau\,\exp(-2s)$$

$$u = 2\sin\left(\frac{x}{t}\right)\exp(-2\ln t) = \frac{2}{t^2}\sin\left(\frac{x}{t}\right)$$

This solution is a moving sine wave along the characteristics, with a decaying frequency$\sim \frac{1}{t}$ and amplitude$\sim \frac{1}{t^2}$ as time increases.

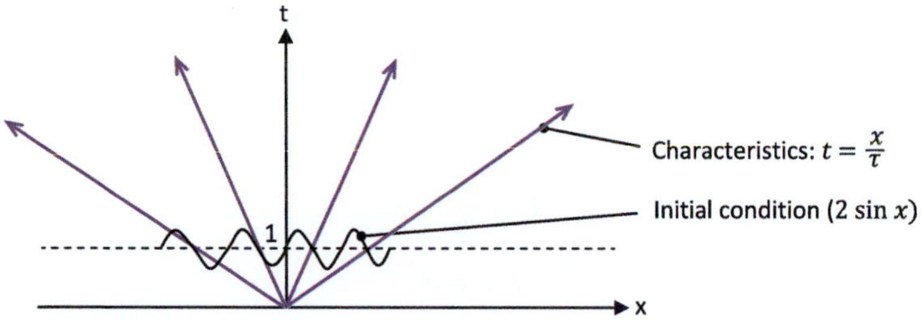


**Problem 20**

**Consider the following equations, and use the Laplace transform method to find the solutions.**

(a) $2\tau\frac{dy}{dt} + \frac{1}{2}y = 1$, $y = 0$ **when** $t = 0$

(b) $3\frac{d^2y}{dt^2} + 2\frac{dy}{dt} + y = 7$, $\frac{dy}{dt} = y = 0$ **when** $t = 0$

**Solution 20**

***Worked Solution***

(a) Referring to the data booklet, we will note that the Laplace transforms as follows for a first-order derivative

$$\frac{dy}{dt} \rightarrow s\bar{y} - y(0)$$

Since we are told that when $t = 0$, $y = 0$, $y(0) = 0$. We can perform the following Laplace transforms for each term.

$$2\tau\frac{dy}{dt} \rightarrow 2\tau s\bar{y}$$

$$\frac{1}{2}y \rightarrow \frac{1}{2}\bar{y}$$

$$1 \rightarrow \frac{1}{s}$$

Putting the results together, we have

$$2\tau s\bar{y} + \frac{1}{2}\bar{y} = \frac{1}{s}$$

$$\bar{y} = \frac{\frac{1}{s}}{2\tau s + \frac{1}{2}} = \frac{1}{2\tau s^2 + \frac{s}{2}} = \frac{1/2\tau}{s\left(s + \frac{1}{4\tau}\right)}$$

We can express this equation in terms of partial fractions

$$\bar{y} = \frac{A}{s} + \frac{B}{\left(s + \frac{1}{4\tau}\right)}$$

There is a quick method called the "cover up rule" that we can use in finding out the values of A and B. The first step is to factorize the denominators such that we obtain the form as shown in the expression above. Then to find $A$, we let $s = 0$, and cover up the "s" term in the denominator to find the value of $A$ as shown below.

$$\frac{1/2\tau}{s\left(s + \frac{1}{4\tau}\right)} \to A = \frac{1/2\tau}{\left(0 + \frac{1}{4\tau}\right)} = 2$$

Similarly, we can find $B$ by letting $s = -\frac{1}{4\tau}$, and cover up the "$\left(s + \frac{1}{4\tau}\right)$" term in the denominator to find the value of $B$, as shown below.

$$\frac{1/2\tau}{s\left(s + \frac{1}{4\tau}\right)} \to B = \frac{1/2\tau}{-\frac{1}{4\tau}} = -2$$

Therefore, we have obtained the partial fractions

$$\bar{y} = \frac{2}{s} - \frac{2}{s + \frac{1}{4\tau}} = 2\left[\frac{1}{s} - \frac{1}{s + \frac{1}{4\tau}}\right]$$

We can now convert the Laplace form back into the original coordinates of $y$ and $t$. And the solution is therefore

$$y = 2\left[1 - e^{-t/4\tau}\right]$$

(b) We have the following differential equation

$$3\frac{d^2y}{dt^2} + 4\frac{dy}{dt} + y = 7, \quad \frac{dy}{dt} = y = 0 \text{ when } t = 0$$

Referring to the data booklet, we will note that the Laplace transforms as follows for derivatives

$$\frac{d^2y}{dt^2} \to s^2\bar{y} - sy(0) - \left[\frac{dy}{dt}\right]_{t=0}$$

$$\frac{dy}{dt} \to s\bar{y} - y(0)$$

Since we are told that when $t = 0$, $\frac{dy}{dt} = y = 0$, $y(0) = 0$ and $\left[\frac{dy}{dt}\right]_{t=0} = 0$. We can perform the following Laplace transforms for each term.

$$3\frac{d^2y}{dt^2} \to 3s^2\bar{y}$$

$$4\frac{dy}{dt} \to 4s\bar{y}$$

$$7 \to \frac{7}{s}$$

Putting the results together, we have

$$\bar{y}\left(3s^2 + 4s + 1\right) = \frac{7}{s}$$

$$\bar{y} = 7\left[\frac{1}{s(3s+1)(s+1)}\right] = 7\left[\frac{A}{s} + \frac{B}{3s+1} + \frac{C}{s+1}\right]$$

We can use cover up rule to determine the numerators of the partial fractions.

$$A = 1$$

$$B = \frac{1}{\left(-\frac{1}{3}\right)\left(\frac{2}{3}\right)} = -\frac{9}{2}$$

$$C = \frac{1}{(-1)(-2)} = \frac{1}{2}$$

Putting the results together, we have

$$\bar{y} = 7\left[\frac{1}{s} - \frac{9/2}{3s+1} + \frac{1/2}{s+1}\right] = 7\left[\frac{1}{s} - \frac{3/2}{s+1/3} + \frac{1/2}{s+1}\right]$$

We can now convert the Laplace form back into the original coordinates of $y$ and $t$. And the solution is therefore

$$y = 7\left[1 - \frac{3}{2}e^{-t/3} + \frac{1}{2}e^{-t}\right]$$

## Laplace Transform and Transfer Functions

**The Laplace transform takes a function of a real variable t (e.g., time) and converts it into a function of a complex variable s (e.g., frequency). While the Fourier transform of a function is a complex function of a real variable (e.g., frequency), the Laplace transform of a function is a complex function of a complex variable. Laplace transforms are usually restricted to functions of t with t $\geq$ 0. Given a defined input or output for a system, the Laplace transform can help translate from the time domain to the frequency domain and in doing so, mathematically simplifying differential equations into algebraic equations.**

### Problem 21

**Find the inverse Laplace transforms for the following expressions:**

(a) $\dfrac{5}{s^2 + 5s + 4}$

(b) $\dfrac{1}{(s + 4)(s + 2)^2}$

(c) $\dfrac{2}{s^2 + 2s + 4}$

(d) $\dfrac{s^2 + 4s}{(s + 2)^2(s + 3)}$

### Solution 21

***Worked Solution***

A useful method to find inverse Laplace transforms is to simplify our expressions by factorizing the denominator and expressing in terms of a sum of partial fractions. This allows us to find the inverse Laplace expressions for each fraction from the data booklet.

(a)

$$\frac{5}{s^2 + 5s + 4} = \frac{5}{(s + 4)(s + 1)}$$

$$= \frac{A}{s + 4} + \frac{B}{s + 1}$$

We can find the numerators using the "cover up rule," as shown below:

$$A = \frac{5}{(-4 + 1)} = -\frac{5}{3}$$

$$B = \frac{5}{(-1+4)} = \frac{5}{3}$$

$$\frac{5}{s^2+5s+4} = \frac{5}{3}\left[\frac{1}{s+1} - \frac{1}{s+4}\right]$$

From the data booklet, we can find the following

$$\mathcal{L}[e^{-at}] = \frac{1}{s+a}$$

Therefore the inverse Laplace transform is

$$\mathcal{L}^{-1}\left\{\frac{5}{3}\left[\frac{1}{s+1} - \frac{1}{s+4}\right]\right\} = \frac{5}{3}\left[e^{-t} - e^{-4t}\right]$$

(b) For this expression, we can similarly express in partial fractions,

$$\frac{1}{(s+4)(s+2)^2} = \frac{A}{s+4} + \frac{B}{s+2} + \frac{C}{(s+2)^2}$$

We can find the numerators $A$ and $C$ using the "cover up rule," as shown below:

$$A = \frac{1}{(-4+2)^2} = \frac{1}{4}$$

$$C = \frac{1}{(-2+4)} = \frac{1}{2}$$

$$\frac{1}{(s+4)(s+2)^2} = \frac{1/4}{s+4} + \frac{B}{s+2} + \frac{1/2}{(s+2)^2} \rightarrow B = -\frac{1}{4}$$

$$\frac{1}{(s+4)(s+2)^2} = \frac{1}{4}\left[\frac{1}{s+4} - \frac{1}{s+2} + \frac{2}{(s+2)^2}\right]$$

From the data booklet, we can find the following

$$\mathcal{L}[e^{-at}] = \frac{1}{s+a}$$

Therefore the inverse Laplace transform is

$$\mathcal{L}^{-1}\left\{\frac{1}{4}\left[\frac{1}{s+4} - \frac{1}{s+2} + \frac{2}{(s+2)^2}\right]\right\} = \frac{1}{4}\left[e^{-4t} - e^{-2t}\right] + \mathcal{L}^{-1}\left\{\frac{1}{2}\left[\frac{1}{(s+2)^2}\right]\right\}$$

From the data booklet, we can find the following

$$\mathcal{L}[t^n] = \frac{n!}{s^{n+1}}$$

$$\mathcal{L}\left[\frac{t^n}{n!}\right] = \frac{1}{s^{n+1}} \text{ is true}$$

$$\mathcal{L}\left[\frac{t^{n-1}}{(n-1)!}\right] = \frac{1}{s^n} \text{ is also true}$$

We also note from the data booklet that

$$\mathcal{L}[e^{-at}y(t)] = \bar{y}(s+a)$$

Therefore, if $y(t) = \frac{t^{n-1}}{(n-1)!}$, then we have the following Laplace transform.

$$\mathcal{L}\left[e^{-at}\frac{t^{n-1}}{(n-1)!}\right] = \frac{1}{(s+a)^n}$$

Applying this result to our earlier expression, we have

$$\frac{1}{4}\left[e^{-4t} - e^{-2t}\right] + \mathcal{L}^{-1}\left\{\frac{1}{2}\left[\frac{1}{(s+2)^2}\right]\right\} = \frac{1}{4}\left[e^{-4t} - e^{-2t}\right] + \left\{\frac{1}{2}\left[e^{-2t}\frac{t^{2-1}}{(2-1)!}\right]\right\}$$

$$= \frac{1}{4}\left[e^{-4t} - e^{-2t} + 2te^{-2t}\right]$$

(c) In this example, we find that the denominator cannot be "fully" factorized; therefore, it can be expressed as a sum as shown.

$$\frac{2}{s^2+2s+4} = \frac{2}{(s+1)^2+3}$$

From the data booklet, we can find the following results, which we can combine to help us solve our problem

$$\mathcal{L}[\sin \omega t] = \frac{\omega}{s^2+\omega^2}$$

$$\mathcal{L}[e^{-at}y(t)] = \bar{y}(s+a) \rightarrow \mathcal{L}[e^{-t}y(t)] = \bar{y}(s+1)$$

We can express our earlier expression in the following form, such that $\omega = \sqrt{3}$

$$\mathcal{L}^{-1}\left[\left(\frac{2}{\sqrt{3}}\right)\frac{(\sqrt{3})}{(s+1)^2+(\sqrt{3})^2}\right] = \left(\frac{2}{\sqrt{3}}\right)e^{-t}\sin\sqrt{3}t$$

(d) For this expression, we have variables in the numerator as well as denominator. We can still express this fraction in terms of partial fractions as shown below

$$\frac{s^2+4s}{(s+2)^2(s+3)} = \frac{A}{s+3} + \frac{B}{s+2} + \frac{C}{(s+2)^2}$$

We can find the numerators $A$ and $C$ using the "cover up rule," as shown below:

$$A = \frac{(-3)^2+4(-3)}{(-3+2)^2} = -3$$

$$C = \frac{(-2)^2+4(-2)}{(-2+3)} = -4$$

$$\frac{s^2+4s}{(s+2)^2(s+3)} = \frac{-3}{s+3} + \frac{B}{s+2} + \frac{-4}{(s+2)^2} \to B = 4$$

$$\frac{s^2+4s}{(s+2)^2(s+3)} = \frac{4}{s+2} - \frac{3}{s+3} - \frac{4}{(s+2)^2}$$

We note from the data booklet the following result which will help us transform the third term $\frac{4}{(s+2)^2}$,

$$\mathcal{L}[t^n] = \frac{n!}{s^{n+1}} \to \mathcal{L}[t] = \frac{1}{s^2}$$

Therefore, we have the inverse Laplace transform as follows.

$$\mathcal{L}^{-1}\left[\frac{4}{s+2} - \frac{3}{s+3} - \frac{4}{(s+2)^2}\right] = 4\mathrm{e}^{-2t} - 3\mathrm{e}^{-3t} - 4t\mathrm{e}^{-2t}$$

## Problem 22

**Determine the Laplace transform for $\mathrm{e}^{-\gamma t}$ where $\gamma$ is positive. Derive the inverse Laplace transform expressions for $\frac{1}{(s+\gamma)^n}$, where $n$ is a positive integer. Show how your result tallies with the following Laplace transform expressions obtained from the data booklet.**

$$\mathcal{L}[t^n] = \frac{n!}{s^{n+1}} \quad \textbf{and} \quad \mathcal{L}[\mathrm{e}^{-at}y(t)] = \bar{y}(s+a)$$

## Solution 22

### *Worked Solution*

The definition of the Laplace transform is as follows:

$$\bar{y} = \mathcal{L}[y(t)] = \int_0^\infty e^{-st} y(t) dt$$

Therefore when we take the Laplace transform of $e^{-\gamma t}$, we can express in terms of the integral

$$\mathcal{L}[e^{-\gamma t}] = \int_0^\infty e^{-st} e^{-\gamma t} dt = \int_0^\infty e^{-(s+\gamma)t} dt$$

$$= \left[\frac{e^{-(s+\gamma)t}}{-(s+\gamma)}\right]_0^\infty = \frac{1}{s+\gamma}$$

We observe that when we add $t$, $t^2$, $t^3 \cdots t^n$ to the exponential term in the integral, we get a pattern. Using integration by parts to integrate the following, we have:

$$\int_0^\infty t e^{-(s+\gamma)t} dt = \left[\frac{t e^{-(s+\gamma)t}}{-(s+\gamma)}\right]_0^\infty - \int_0^\infty \frac{e^{-(s+\gamma)t}}{-(s+\gamma)} dt$$

$$= 0 + \int_0^\infty \frac{e^{-(s+\gamma)t}}{s+\gamma} dt = \left[-\frac{e^{-(s+\gamma)t}}{(s+\gamma)^2}\right]_0^\infty$$

$$\int_0^\infty t e^{-(s+\gamma)t} dt = \frac{1}{(s+\gamma)^2} \quad (1)$$

If we add the term $t^2$ to the product, we get,

$$\int_0^\infty t^2 e^{-(s+\gamma)t} dt = \left[\frac{t^2 e^{-(s+\gamma)t}}{-(s+\gamma)}\right]_0^\infty - \int_0^\infty \frac{(2t) e^{-(s+\gamma)t}}{-(s+\gamma)} dt$$

$$= 0 + \int_0^\infty \frac{(2t) e^{-(s+\gamma)t}}{s+\gamma} dt = \frac{2}{s+\gamma} \int_0^\infty t e^{-(s+\gamma)t} dt$$

We can plug in the result (1) earlier to get

$$\int_0^\infty t^2 e^{-(s+\gamma)t} dt = \frac{2}{s+\gamma} \left[\frac{1}{(s+\gamma)^2}\right]$$

$$\int_0^\infty t^2 e^{-(s+\gamma)t} dt = \frac{2}{(s+\gamma)^3} \quad (2)$$

If we add the term $t^3$ to the product, we get,

$$\int_0^\infty t^3 e^{-(s+\gamma)t} dt = \left[\frac{t^3 e^{-(s+\gamma)t}}{-(s+\gamma)}\right]_0^\infty - \int_0^\infty \frac{(3t^2)e^{-(s+\gamma)t}}{-(s+\gamma)} dt$$

$$= 0 + \int_0^\infty \frac{(3t^2)e^{-(s+\gamma)t}}{s+\gamma} dt = \frac{3}{s+\gamma}\int_0^\infty t^2 e^{-(s+\gamma)t} dt$$

We can plug in the result (2) earlier to get

$$\int_0^\infty t^3 e^{-(s+\gamma)t} dt = \frac{3}{s+\gamma}\left[\frac{2}{(s+\gamma)^3}\right]$$

$$\int_0^\infty t^3 e^{-(s+\gamma)t} dt = \frac{3!}{(s+\gamma)^4} \quad (3)$$

Therefore, we can see that the following general form of the integral is true:

$$\int_0^\infty t^{n-1} e^{-(s+\gamma)t} dt = \frac{(n-1)!}{(s+\gamma)^n}$$

$$\frac{1}{(n-1)!}\int_0^\infty t^{n-1} e^{-\gamma t} e^{-st} dt = \frac{1}{(s+\gamma)^n}$$

$$\mathcal{L}\left[\frac{1}{(n-1)!} t^{n-1} e^{-\gamma t}\right] = \frac{1}{(s+\gamma)^n}$$

Therefore we note that the inverse Laplace transform of $\frac{1}{(s+\gamma)^n}$ is $\frac{1}{(n-1)!} t^{n-1} e^{-\gamma t}$. Now, we can verify our result with the Laplace expressions obtained from the data booklet. We can make some simple manipulations to obtain the form of the expression that we want (i.e., $\frac{1}{(s+\gamma)^n}$).

We note from the data booklet that $\mathcal{L}[t^n] = \frac{n!}{s^{n+1}}$; therefore

$$\mathcal{L}\left[\frac{t^n}{n!}\right] = \frac{1}{s^{n+1}} \text{ is true}$$

$$\mathcal{L}\left[\frac{t^{n-1}}{(n-1)!}\right] = \frac{1}{s^n} \text{ is also true}$$

We also note from the data booklet that $\mathcal{L}[e^{-at}y(t)] = \bar{y}(s+a)$; therefore, if $y(t) = \frac{t^{n-1}}{(n-1)!}$ from above, then we have the following where $a = \gamma$ in our expression, which tallies with our earlier result.

$$\mathcal{L}\left[e^{-\gamma t}\frac{t^{n-1}}{(n-1)!}\right] = \frac{1}{(s+\gamma)^n}$$

## Problem 23

**An industrial process can be modeled using the following ordinary differential equation.**

$$\frac{d^2y}{dt^2} + 5\frac{dy}{dt} + 4y = x$$

(a) **If we charge the system with an input that follows the sinusoidal function $x = \sin \omega t$, what would be the output function for large values of time $t$ (assuming that it does not decay at large times)? Also you may assume that at $t = 0$, $\frac{dy}{dt} = y = 0$.**

(b) **Show that the transfer function $\bar{G}$ that converts input to output has the following magnitude and argument, where the Laplace variable $s = i\omega$:**

$$\left|\bar{G}(i\omega)\right| = \frac{1}{\sqrt{(\omega^2 + 16)(\omega^2 + 1)}}$$

$$\arg\left(\bar{G}\right) = -\tan^{-1}\left(\frac{5\omega}{4 - \omega^2}\right)$$

## Solution 23

***Worked Solution***

(a) Equations like the one here are typical of input and output systems, whereby the input is the $x$ term which is a function of $t$ on the right-hand side of the equation here, and the response output is $y(t)$.

When we have an inhomogeneous equation as in this question, we need to first solve the homogeneous case, which is when:

$$\frac{d^2y}{dt^2} + 5\frac{dy}{dt} + 4y = 0$$

This is a linear second-order ODE with constant coefficients. We can assume the general solution as follows, where $y_1$ and $y_2$ are solutions. Initial conditions are satisfied by the values of $c_1$ and $c_2$.

$$y = c_1y_1 + c_2y_2$$

The characteristic equation of the ODE is as follows,

$$r^2 + 5r + 4 = 0$$

$$(r + 4)(r + 1) = 0$$

$$r = -4, -1$$

Therefore we have two real roots, $r_1 = -4$ and $r_2 = -1$, and the general solution is

$$y = c_1 e^{r_1 t} + c_2 e^{r_2 t} = c_1 e^{-4t} + c_2 e^{-t}$$

In this solution, we can see that the solutions decay as time tends to large values of *t*.

Instead we try solution of the form $y = c_1 \cos \omega t + c_2 \sin \omega t$ which does not decay as they are pure oscillations. Then it follows that their derivatives are as shown:

$$\frac{dy}{dt} = -c_1 \omega \sin \omega t + c_2 \omega \cos \omega t$$

$$\frac{d^2y}{dt^2} = -c_1 \omega^2 \cos \omega t - c_2 \omega^2 \sin \omega t$$

Putting the expressions back into the ODE, we have

$$\frac{d^2y}{dt^2} + 5\frac{dy}{dt} + 4y = 0$$

$$-c_1\omega^2 \cos \omega t - c_2\omega^2 \sin \omega t + 5(-c_1\omega \sin \omega t + c_2\omega \cos \omega t) + 4(c_1 \cos \omega t + c_2 \sin \omega t) = 0$$

Now rearranging the left-hand side, and inserting the input function of sin$\omega t$, we have:

$$\left(-c_1\omega^2 + 5c_2\omega + 4c_1\right)\cos \omega t + \left(-c_2\omega^2 - 5c_1\omega + 4c_2\right)\sin \omega t = \sin \omega t$$

$$-c_1\omega^2 + 5c_2\omega + 4c_1 = 0$$

$$c_2 = \frac{-\omega^2 + 4}{(\omega^2 + 16)(\omega^2 + 1)}$$

$$-c_2\omega^2 - 5c_1\omega + 4c_2 = 1$$

$$c_1 = \frac{-5\omega}{(\omega^2 + 16)(\omega^2 + 1)}$$

Substituting the constants back into the equation for the general solution, we have the following solution for *y* which provides the output values at large times.

$$y = \left[\frac{-5\omega}{(\omega^2 + 16)(\omega^2 + 1)}\right]\cos \omega t + \left[\frac{4 - \omega^2}{(\omega^2 + 16)(\omega^2 + 1)}\right]\sin \omega t$$

$$y = \frac{-5\omega \cos \omega t + (4 - \omega^2) \sin \omega t}{(\omega^2 + 16)(\omega^2 + 1)}$$

(b) We can express the solution as a single trigonometric function, using the addition formula.

$$y = R \sin(\omega t + \theta) = R \sin \omega t \cos \theta + R \cos \omega t \sin \theta$$

$$R \cos \theta = \frac{4 - \omega^2}{(\omega^2 + 16)(\omega^2 + 1)}$$

$$R \sin \theta = \frac{-5\omega}{(\omega^2 + 16)(\omega^2 + 1)}$$

$$R^2 = (R \cos \theta)^2 + (R \sin \theta)^2 = \frac{25\omega^2 + \omega^4 - 8\omega^2 + 16}{(\omega^2 + 16)^2 (\omega^2 + 1)^2} = \frac{1}{(\omega^2 + 16)(\omega^2 + 1)}$$

$$R = \frac{1}{\sqrt{(\omega^2 + 16)(\omega^2 + 1)}}$$

$$\tan \theta = \frac{-5\omega}{4 - \omega^2}$$

The transfer function in Laplace variable $s$ that transforms the input $\bar{x}$ to output response $\bar{y}$ can be expressed as:

$$\bar{G}(s) = \frac{\bar{y}}{\bar{x}}$$

$\bar{y}$ is the Laplace transform of the left-hand side of the ODE, and $\bar{x}$ is the Laplace transform of the right-hand side of the equation.

$$\frac{d^2y}{dt^2} + 5\frac{dy}{dt} + 4y = x \rightarrow s^2\bar{y} + 5s\bar{y} + 4\bar{y} = \bar{x}$$

$$\bar{G}(s) = \frac{1}{s^2 + 5s + 4}$$

We substitute the Laplace variable $s$ with $i\omega$, and we obtain

$$\bar{G}(i\omega) = \frac{1}{(4 - \omega^2) + 5\omega i}$$

$$|\bar{G}(i\omega)| = \frac{1}{\sqrt{16 - 8\omega^2 + \omega^4 + 25\omega^2}} = \frac{1}{\sqrt{(\omega^2 + 16)(\omega^2 + 1)}}$$

$$\arg(\bar{G}) = -\tan^{-1}\left(\frac{5\omega}{4 - \omega^2}\right)$$

## Problem 24

**Consider the system below whereby two CSTRs are connected. Assuming that $C_t$ denotes the concentration of tracer, derive the transfer functions $G_A(s)$ and $G_B(s)$ that relate the output concentrations from CSTR A and CSTR B to the input concentration of tracer $C_{t,0}$. You may assume constant volume for both CSTRs and assume that $\beta = \frac{\dot{Q}_B}{\dot{Q}_A}$, where $\dot{Q}$ represents volumetric flow rate.**

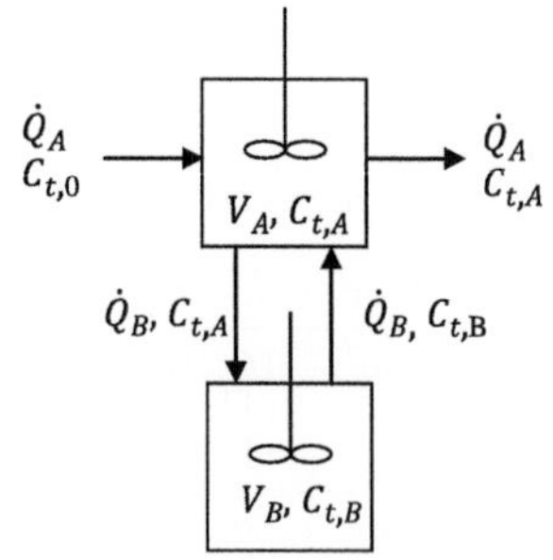


## Solution 24

***Worked Solution***

The mass balance of tracer in CSTR $A$ is as follows.

$$\frac{dN_{t,A}}{dt} = F_{\text{in},A} - F_{\text{out},A}$$

$$V_A \frac{dC_{t,A}}{dt} = \dot{Q}_A C_{t,0} + \dot{Q}_B C_{t,B} - \dot{Q}_A C_{t,A} - \dot{Q}_B C_{t,A}$$

Let $\tau_A = V_A/\dot{Q}_A$, therefore we have

$$\tau_A \frac{dC_{t,A}}{dt} = C_{t,0} + \frac{\dot{Q}_B}{\dot{Q}_A} C_{t,B} - \left(1 + \frac{\dot{Q}_B}{\dot{Q}_A}\right) C_{t,A}$$

Given that $\beta = \frac{\dot{Q}_B}{\dot{Q}_A}$, and take Laplace transform of the equation to obtain the following assuming $C_{t,A} = 0$ at $t = 0$.

$$\tau_A s \bar{C}_{t,A} = \bar{C}_{t,0} + \beta \bar{C}_{t,B} - (1 + \beta) \bar{C}_{t,A}$$

Similarly for CSTR B, we can perform a mass balance for the tracer and obtain the following

$$\frac{dN_{t,B}}{dt} = F_{\text{in},B} - F_{\text{out},B}$$

$$V_B \frac{dC_{t,B}}{dt} = \dot{Q}_B C_{t,A} - \dot{Q}_B C_{t,B}$$

$$\tau_B \frac{dC_{t,B}}{dt} = C_{t,A} - C_{t,B}$$

Taking Laplace transform of the equation assuming $C_{t,\ B} = 0$ at $t = 0$, we have

$$\tau_B s \bar{C}_{t,B} = \bar{C}_{t,A} - \bar{C}_{t,B}$$

$$\bar{C}_{t,B} = \frac{\bar{C}_{t,A}}{1 + \tau_B s}$$

Substituting this result into the mass balance equation for CSTR $A$, we have

$$\tau_A s \bar{C}_{t,A} = \bar{C}_{t,0} + \beta \left( \frac{\bar{C}_{t,A}}{1 + \tau_B s} \right) - (1 + \beta) \bar{C}_{t,A}$$

$$\bar{C}_{t,A} \left[ \tau_A s + 1 + \beta - \frac{\beta}{1 + \tau_B s} \right] = \bar{C}_{t,0}$$

$$G_A(s) = \frac{\bar{C}_{t,A}}{\bar{C}_{t,0}} = \frac{1 + \tau_B s}{(1 + \tau_B s)(1 + \tau_A s) + \beta \tau_B s}$$

$$G_B(s) = \frac{\bar{C}_{t,B}}{\bar{C}_{t,0}} = \frac{1}{(1 + \tau_B s)(1 + \tau_A s) + \beta \tau_B s}$$

## Problem 25

**Consider the following CSTR of volume $V = 1\ \text{m}^3$ in which a rapid and irreversible reaction occurs between $A$ and $B$. The inlet volumetric flow rate of $A$, $\dot{Q}_{A0} = 0.01\ \text{m}^3/s$ and that of $B$ is $\dot{Q}_{B0} = 0.001\ \text{m}^3/s$. One mole of $A$ reacts with one mole of $B$ to produce one mole of water.**

(a) **Show that the transfer functions relating the outlet concentrations of $A$ and $B$ to their inlet concentrations are given by, assuming that $C_A = C_B = 0$ at $t = 0$:**

$$\bar{C}_A = \frac{1}{100s + 1.1} \bar{C}_{A0} - \frac{0.1}{100s + 1.1} \bar{C}_{B0}$$

$$\bar{C}_B = \frac{-1}{100s + 1.1} \bar{C}_{A0} + \frac{0.1}{100s + 1.1} \bar{C}_{B0}$$

(b) **The system is at an initial steady state with $C_{A0} = 200$ mol/m$^3$, $C_{B0} = 1900$ mol/m$^3$. At $t = 0$, the inlet concentration of $A$ drops to 180 mol/m$^3$. Determine the outlet concentrations of $A$ and $B$ at $t = 0$ $s$, 40$s$, and 90$s$.**

## Solution 25

### *Worked Solution*

(a) First recall that in a well-mixed CSTR, we assume that outlet concentration is equivalent to concentration in the tank. To find the transfer function for outlet concentration of $A$, we assume first the scenario that $A$ is in large excess of $B$. In this scenario, $B$ is fully reacted and $C_B = 0$ in the tank and in the outlet. Setting this constraint also allows us to deduce that the rate at which A disappears due to reaction is equivalent to the rate at which $B$ enters the tank (rapid and irreversible reaction). Hence, consider the illustration below, which we can use to perform a mass balance for $A$.

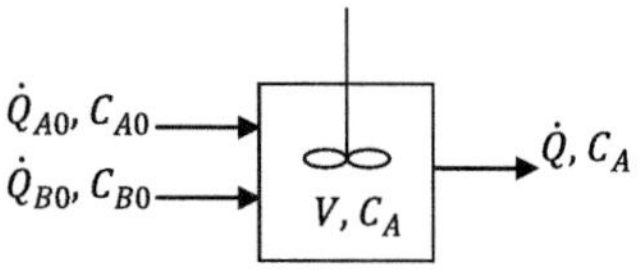


It is given that $V = 1$ m$^3$, $\dot{Q}_{A0} = 0.01$ m$^3$/s, and $\dot{Q}_{B0} = 0.001$ m$^3$/s. We can deduce that the outlet volumetric flow rate $\dot{Q} = \dot{Q}_{A0} + \dot{Q}_{B0} = 0.011$ m$^3$/s.

$$\frac{dN_A}{dt} = F_{\text{in},A} - F_{\text{out},A} - \text{reacted}$$

$$V\frac{dC_A}{dt} = \dot{Q}_{A0}C_{A0} - \dot{Q}C_A - \dot{Q}_{B0}C_{B0}$$

$$1\frac{dC_A}{dt} = 0.01C_{A0} - 0.011C_A - 0.001C_{B0}$$

We can take Laplace transform to convert the differential equation into a normal equation:

$$(s + 0.011)\bar{C}_A = 0.01\bar{C}_{A0} - 0.001\bar{C}_{B0}$$

$$\bar{C}_A = \frac{1}{100s + 1.1}\bar{C}_{A0} - \frac{0.1}{100s + 1.1}\bar{C}_{B0}$$

Similarly we can do a mass balance for species $B$ to find the transfer function for outlet concentration of $B$. Now we assume the scenario where $B$ is in large excess of $A$. In this scenario, $A$ is fully reacted and $C_A = 0$ in the tank and in the outlet.

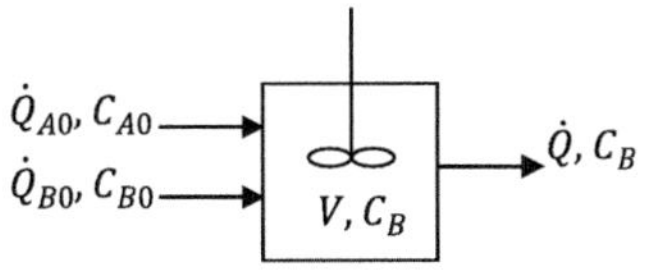


$$\frac{dN_B}{dt} = F_{\text{in},B} - F_{\text{out},B} - \text{reacted}$$

$$V\frac{dC_B}{dt} = \dot{Q}_{B0}C_{B0} - \dot{Q}C_B - \dot{Q}_{A0}C_{A0}$$

$$1\frac{dC_B}{dt} = 0.001C_{B0} - 0.011C_B - 0.01C_{A0}$$

We can take Laplace transform to convert the differential equation into a normal equation:

$$(s + 0.011)\bar{C}_B = -0.01\bar{C}_{A0} + 0.001\bar{C}_{B0}$$

$$\bar{C}_B = \frac{-1}{100s + 1.1}\bar{C}_{A0} + \frac{0.1}{100s + 1.1}\bar{C}_{B0}$$

(b) We are told that the initial steady state inlet concentrations are $C_{A0} = 200$ mol/m$^3$ and $C_{B0} = 1900$ mol/m$^3$. From this information, together with the known volumetric flow rates, $\dot{Q}_{A0} = 0.01$ m$^3$/s and $\dot{Q}_{B0} = 0.001$ m$^3$/s, we can find out if $A$ or $B$ is in excess in order to apply the appropriate transfer function found in part a.

$$\left(C_{A0}\dot{Q}_{A0}\right)_{ss} = 200 \times 0.01 = 2 \text{ mol/s}$$

$$\left(C_{B0}\dot{Q}_{B0}\right)_{ss} = 1900 \times 0.001 = 1.9 \text{ mol/s}$$

We note that $A$ is in excess, since $\left(C_{A0}\dot{Q}_{A0}\right)_{ss} > \left(C_{B0}\dot{Q}_{B0}\right)_{ss}$ at initial steady state when $t < 0$. The starting concentration of $A$ in the tank at $t = 0$, i.e., $C_A|_{t=0}$ will be the excess amount of $A$ left after the reaction between $A$ and $B$ during the initial steady state period. We can find $C_A|_{t=0}$ as follows:

$$\dot{Q}C_A|_{t=0} = \left(C_{A0}\dot{Q}_{A0}\right)_{ss} - \left(C_{B0}\dot{Q}_{B0}\right)_{ss}$$

$$0.011C_A|_{t=0} = 2 - 1.9$$

$$C_A|_{t=0} = \frac{0.1}{0.011} = 9.09 \text{ mol/m}^3$$

At $t = 0$, there is a step change (decrease) in $C_{A0}$ from 200 to 180 mol/m$^3$. This deviation of $-20$ mol/m$^3$can be expressed in terms of a unit step function, which in Laplace transform can be written as $-\frac{20}{s}$.

Note that $C_A$ can be expressed as the sum of the steady state value (remains constant), $C_{A,\,ss}$ and the deviation portion (or non-steady state portion) which we can denote as $C_A{}'$. So $C_A(t) = C_{A,\,ss} + C_A{}'$, where $C_{A,\,ss} = C_A|_{t=0} = 9.09$ mol/m$^3$.

We can use the transfer function found in part a (for the case where A is in excess) to find out the deviatoric concentration $C_A{}'$.

From part a, the relevant transfer function is:

$$\bar{C}_A = \frac{1}{100s + 1.1}\bar{C}_{A0} - \frac{0.1}{100s + 1.1}\bar{C}_{B0}$$

No change was introduced to inlet concentration of B, so $\bar{C}_{B0}{}' = 0$. As for $\bar{C}_{A0}{}'$, the change was $-\frac{20}{s}$; therefore

$$\bar{C}_A{}' = \frac{1}{100s + 1.1}\left(-\frac{20}{s}\right)$$

The typical way to simplify a Laplace expression so that we can convert easily back into original variable of time $t$ is to express in terms of partial fractions.

$$\bar{C}_A' = -0.2\left[\frac{1}{s(s + 0.011)}\right] = -0.2\left[\frac{A}{s + 0.011} + \frac{B}{s}\right]$$

Using the cover up rule we have

$$A = \frac{1}{-0.011}$$

$$B = \frac{1}{0.011}$$

$$\bar{C}_A{}' = -\frac{0.2}{0.011}\left[\frac{1}{s} - \frac{1}{s + 0.011}\right]$$

Taking the inverse Laplace transform, we have

$$C_A' = -\frac{0.2}{0.011}\left(1 - e^{-0.011t}\right)$$

$$C_A(t) = C_{A,ss} + C_A' = \frac{0.1}{0.011} - \frac{0.2}{0.011}\left(1 - e^{-0.011t}\right)$$

$$C_A(t) = \frac{0.1}{0.011}\left[1 - 2\left(1 - e^{-0.011t}\right)\right] = \frac{0.1}{0.011}\left[2e^{-0.011t} - 1\right]$$

We can find the time when $C_A = 0$,

$$0 = \frac{0.1}{0.011}\left[2e^{-0.011t} - 1\right]$$

$$t = \frac{1}{-0.011}\ln\frac{1}{2} = 63\ \mathrm{s}$$

When $t = 0$ s, $C_A = C_{A,ss} = \frac{0.1}{0.011} = 9.09\ \mathrm{mol/m^3}$, $C_B = 0$
When $t = 40$ s, $C_A = \frac{0.1}{0.011}\left[2e^{-0.011(40)} - 1\right] = 2.62\ \mathrm{mol/m^3}$, $C_B = 0$
When $t = 90$ s, note that we have already passed the time $t = 63$ s when $A$ is fully depleted. When this happens, we have $B$ starting to increase from zero to non-zero and become in excess as time progresses. So the following transfer function comes into play.

$$\bar{C}_B{}' = \frac{-1}{100s + 1.1}\bar{C}_{A0}{}' + \frac{0.1}{100s + 1.1}\bar{C}_{B0}{}'$$

At $t = 63$ s, $A$ just reaches the point when it is no longer in excess (i.e., $C_A = 0$). All of $A$ that enters the tank disappears upon full reaction with inflowing $B$ (new steady state portion). On the other hand, $B$ starts to go from zero to non-zero at this time, as it starts to accumulate (deviatoric or non-steady state portion) in excess of the amount that is reacted away due to inflowing $B >$ $B$ reacted with $A$.

So from $t = 63$ s, there is a step increase in B denoted by $\bar{C}_{B0}{}'$. To find out $\bar{C}_{B0}{}'$, consider that $C_{A0} = 180\ \mathrm{mol/m^3}$ and since it enters at a flow rate that is 10 times greater, it is diluted ten times. The matching concentration of $B$ for reaction with $A$ is ten times greater, at 1800 mol/m$^3$. We know that $C_{B0} = 1900\ \mathrm{mol/m^3}$, so the step change in B is $C'_{B0} = +100\ \mathrm{mol/m^3}$(Laplace form is $\bar{C}'_{B0} = \frac{100}{s}$).

$$\bar{C}'_B = \frac{0.1}{100s + 1.1}\left(\frac{100}{s}\right) = \frac{0.1}{s(s + 0.011)} = \frac{A}{s} + \frac{B}{s + 0.011}$$

$$A = \frac{0.1}{0.011}$$

$$B = \frac{0.1}{-0.011}$$

$$\bar{C}'_B = \frac{0.1}{0.011}\left(\frac{1}{s} - \frac{1}{s + 0.011}\right)$$

Taking inverse Laplace transform, we have

$$C'_B = \frac{0.1}{0.011}\left(1 - e^{-0.011t}\right)$$

So for $t > 63$ s,

$$C_B(t) = C_{B,ss} + C'_B = 0 + \frac{0.1}{0.011}\left(1 - e^{-0.011t}\right)$$

At $t = 90$ s,

$$C_B = \frac{0.1}{0.011}\left(1 - e^{-0.011(90)}\right) = 5.71\ \text{mol/m}^3$$

## Problem 26

**Consider the following transfer function that describes a process that has as its input, a unit step function.**

$$\mathbf{G(s) = \frac{1}{s^2 + \lambda s + 1}}$$

(a) **What values of λ would provide an oscillatory response as its output?**
(b) **What values of λ would give rise to an unstable output?**

## Solution 26

***Worked Solution***

(a) For a unit step input, the Laplace form is $\frac{1}{s}$. Therefore, since transfer function $G(s)$ is defined as shown below, where $\bar{y}(s)$ is the output response and $\bar{x}(s)$ the input,

$$G(s) = \frac{\bar{y}(s)}{\bar{x}(s)}$$

$$\bar{y}(s) = \left(\frac{1}{s}\right)\frac{1}{s^2 + \lambda s + 1}$$

We can find the characteristic roots of the polynomial $s^2 + \lambda s + 1$,

$$s^2 + \lambda s + 1 = 0$$

$$s = \frac{1}{2}\left[-\lambda \pm \sqrt{\lambda^2 - 4}\right]$$

In order for the solution to be a decaying oscillation, it has to have characteristic roots of the complex form $a \pm bi$ because this will give rise to solution with a decaying portion given by $e^{at}$ (for $a < 0$) and an oscillatory portion given by ($c_1 \cos bt + c_2 \sin bt$). The solution takes the form of the following:

$$e^{at}(c_1 \cos bt + c_2 \sin bt), \ a < 0$$

Therefore the conditions for us to have a decaying oscillation is that $b > 0$ so that we will retain the imaginary part, and $a < 0$ to have a negative real part.

$$s = \frac{1}{2}\left[-\lambda \pm \sqrt{\lambda^2 - 4}\right]$$

For $b > 0$, we need $\lambda^2 - 4 < 0$, $-2 < \lambda < 2$. Then we need to consider further that we also require $a < 0$, which means that we need $0 < \lambda < 2$.

(b) For an unstable output, we just require that in the solution, the exponential part does not decay, instead it unstably "explodes" to large values. This requires $a > 0$ in the term $e^{at}$. Therefore, $\lambda < 0$.

## Problem 27

**Consider the following CSTR in which a second-order irreversible reaction takes place. The inlet concentration of reactant $A$, $C_{A0} = 1800$ mol/m$^3$. The mean residence time $\tau = 660$ s and the reaction rate constant $k = 0.0001$ m$^3$s/mol.**

(a) **Determine the concentration of $A$ in the tank at steady state, $C_{A,ss}$**
(b) **Find the transfer function $G(s)$ that describes deviations from the steady state.**
(c) **Steady state is reached at $t < 0$ when the inlet concentration of $A$, $C_{A0} = 1800$ mol/m$^3$. At $t = 0$, $C_{A0}$ is instantly stepped up to 2000 mol/m$^3$. Determine $C_A$ at $t = 25$ s and at large times.**
(d) **Find the steady state concentration of $A$ after the step change described in part c.**

## Solution 27

***Worked Solution***

(a) We have illustrated our system as shown below:

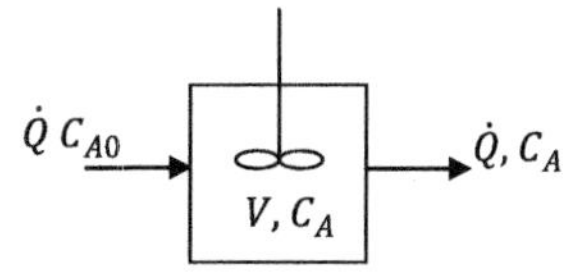

The mass balance for $A$ can be expressed as follows with a second-order reaction,

$$\frac{dN_A}{dt} = F_{A0} - F_A - kVC_A{}^2$$

Assuming a constant volume $V$ and volumetric flow rate, and we know that $\tau = V/\dot{Q}$

$$V\frac{dC_A}{dt} = \dot{Q}(C_{A0} - C_A) - kVC_A{}^2$$

$$\tau\frac{dC_A}{dt} = C_{A0} - C_A - k\tau C_A{}^2$$

At steady state, $\frac{dC_A}{dt} = 0$, and substituting the values of $\tau = 660$ s and $k = 0.0001$ m$^3$s/mol

$$0 = C_{A0} - C_{A,ss} - k\tau C_{A,ss}{}^2$$

$$0.066C_{A,ss}{}^2 + C_{A,ss} - 1800 = 0$$

$$C_{A,ss} = \frac{-1 \pm \sqrt{1 - 4(0.066)(-1800)}}{2(0.066)} = 158\ \text{mol/m}^3\ \text{(only take positive root)}$$

(b) To find the transfer function that describes the deviatoric part of concentration, let us first split the total concentration of $A$ into the steady state portion and the deviation portion.

$$C_A = C_{A,ss} + C_A{}'$$

$$C_{A0} = C_{A0,ss} + C_{A0}{}'$$

Substituting this expression into our differential equation of mass balance, we have the following. We note that since $C_{A,ss}$ is a constant, its derivative is zero.

$$\tau\frac{d(C_{A,ss} + C_A{}')}{dt} = (C_{A0,ss} + C_{A0}{}') - (C_{A,ss} + C_A{}') - k\tau(C_{A,ss} + C_A{}')^2$$

$$\tau\frac{dC_A{}'}{dt} = C_{A0,ss} - C_{A,ss} - k\tau C_{A,ss}{}^2 - 2k\tau C_{A,ss}C'_A - k\tau C_A{}'^2 + C'_{A0} - C_A{}'$$

Extracting only the deviation terms from this equation, and taking Laplace transform, we have the following. It is noted that $C'_A = 0$ at $t = 0$. Also note that the square of the deviation, i.e., $C_A{}'^2$, is similar to the variance concept in statistics, and this value remains invariant with the time parameter; hence, it does not appear in the equation below which only captures deviation terms that change with time.

$$\tau \frac{dC_A{}'}{dt} = -2k\tau C_{A,ss} C_A' - k\tau C_A{}'^2 + C_{A0}' - C_A{}'$$

$$\tau s \bar{C}_A' = -2k\tau C_{A,ss} \bar{C}_A' + \bar{C}_{A0}' - \bar{C}_A'$$

$$G(s) = \frac{\bar{C}_A'}{\bar{C}'_{A0}} = \frac{1}{\tau s + 1 + 2k\tau C_{A,ss}} = \frac{1}{660s + 1 + 2(660)(0.0001)(158)}$$

$$G(s) = \frac{1}{660s + 22}$$

(c) We note that there is a step change in the inlet concentration of +200 mol/m$^3$ at $t = 0$. In Laplace form, this is equivalent to $\bar{C}_{A0}' = \frac{200}{s}$.

$$\bar{C}_A' = \frac{1}{660s + 22}\left(\frac{200}{s}\right) = \frac{A}{s} + \frac{B}{660s + 22}$$

$$A = \frac{200}{22}$$

$$B = \frac{200}{-\left(\frac{22}{660}\right)}$$

$$\bar{C}_A' = \frac{200}{22}\left(\frac{1}{s} - \frac{660}{660s + 22}\right) = \frac{100}{11}\left(\frac{1}{s} - \frac{1}{s + 0.033}\right)$$

Taking inverse Laplace, we have the deviatoric concentration profile of $A$ in time $t$

$$C_A' = \frac{100}{11}\left(1 - e^{-0.033t}\right)$$

At $t = 0$, $C_A' = 0$
At $t = 25$ s,

$$C_A' = \frac{100}{11}\left(1 - e^{-0.033(25)}\right) = 5.1 \text{ mol/m}^3$$

$$C_A = C_{A,ss} + C_A' = 158 + 5.1 = 163 \text{ mol/m}^3$$

At $t \to \infty$,

$$C_A' = \frac{100}{11} = 9.09 \text{ mol/m}^3$$

$$C_A = 158 + 9.09 = 167 \text{ mol/m}^3$$

(d) The steady state concentration of $A$ after the step change is the concentration at large times $t \to \infty$. Hence, it is the result found in part c for $t \to \infty$.

$$C_{A,ss} = 167 \text{ mol/m}^3$$

## Problem 28

**Consider a system comprising of two CSTRs and a PFR with the following transfer function.**

$$\mathbf{G(s) = \frac{e^{-3s}}{(1 + 3s)(1 + 2s)}}$$

(a) **Determine the phase angle and amplitude ratio at an angular frequency $\omega = 0.15$ rad/s.**
(b) **Determine the exit concentration profile of an inert solute that enters the system with a function of $f(t) = 1.5 + 0.3 \cos 0.1t$.**
(c) **Assuming that the solute undergoes a first-order reaction with reaction rate constant $k = 0.025 \text{ s}^{-1}$, what would be the amplitude ratio, phase angle, and exit concentration?**

## Solution 28

*Worked Solution*

(a) To determine phase angle and amplitude ratio of the transfer function $G(s)$, we first substitute the Laplace variable $s$ with $i\omega$.

$$G(i\omega) = \frac{e^{-3i\omega}}{(1 + 3i\omega)(1 + 2i\omega)}$$

We can visualize the transfer function $G$ as analagous to a complex number with its phase angle (coefficient of $i$ in the exponential term) equivalent to the argument $\theta$ of the complex number, and its amplitude ratio equivalent to the modulus of the complex number.

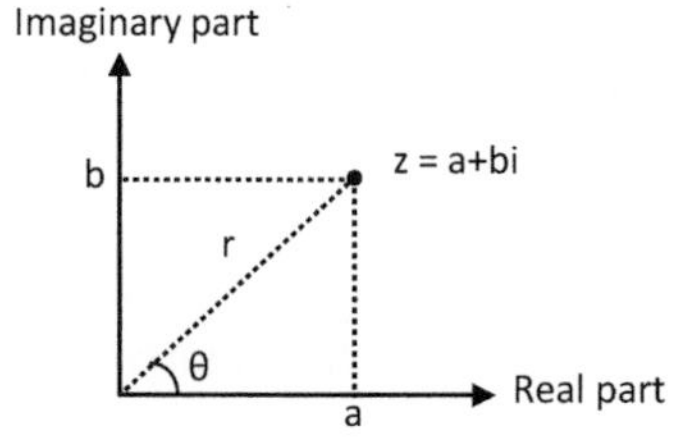

The complex number $z$ can be written in its polar form as follows. The polar form is convenient for working with products or quotients as the indices simply add up (or subtract) in a product (or quotient):

$$z = re^{i\theta}$$

$$|z| = r = \sqrt{a^2 + b^2}, \ \arg(z) = \theta = \tan^{-1}\frac{b}{a}$$

For example, the product of two complex numbers will be as shown:

$$z_1 z_2 = r_1 r_2 e^{i(\theta_1 + \theta_2)}$$

Therefore, when we have a transfer function that comprises of a combination of different reactor systems, we can also find the combined argument or amplitudes easily in the polar form.

$$G(i\omega) = \frac{e^{-3i\omega}}{(1 + 3i\omega)(1 + 2i\omega)}$$

$$|G| = \frac{1}{\sqrt{1^2 + (3\omega)^2}\sqrt{1^2 + (2\omega)^2}} = \frac{1}{\sqrt{(1 + 9\omega^2)(1 + 4\omega^2)}}$$

Substituting the value of $\omega = 0.15$rad/s, we have the amplitude ratio as follows.

$$|G| = \frac{1}{\sqrt{\left(1 + 9(0.15)^2\right)\left(1 + 4(0.15)^2\right)}} = 0.87$$

The argument can be obtained as follows. Consider first that the transfer function comprises the three complex numbers as shown below.

$$G(i\omega) = \frac{z_1}{z_2 z_3}$$

$$\arg(z_1) = -3\omega, \ \arg(z_2) = \tan^{-1}(3\omega), \ \arg(z_3) = \tan^{-1}(2\omega)$$

$$\arg(G) = \arg(z_1) - \arg(z_2) - \arg(z_3) = -3\omega - \tan^{-1}(3\omega) - \tan^{-1}(2\omega)$$
$$= -1.16$$

(b) To determine the exit concentration profile given the input function, $f(t) = 1.5 + 0.3 \cos 0.1t$, we note that the physical significance of the transfer function is to alter the amplitude of the input function by an amount given by the amplitude ratio |$G$|, and to shift the input function by a phase angle of arg($G$). Therefore, the exit concentration profile with time $t$ is given by

$$\begin{aligned}\text{Exit Concentration} &= 1.5 + (0.87)0.3\cos(0.1t - 1.16) \\ &= 1.5 + 0.26\cos(0.1t - 1.16)\end{aligned}$$

There is no change to the constant term of "1.5" in the input function because $G(i\omega = 0) = 1$. Therefore $1.5 \times 1 = 1.5$ (no change).

(c) This time we have a reaction. To visualize how the presence of a reaction affects the form of the transfer function, let us consider the mass balance equation of a simple CSTR example, where we have a constant volume $V$ and volumetric flow rate $\dot{Q}$. Let the mean residence time be denoted by $\tau$, and $C_A = 0$ at $t = 0$.

Let us first find the transfer function when there is no reaction:

$$V\frac{dC_A}{dt} = \dot{Q}(C_{A0} - C_A)$$

$$\tau\frac{dC_A}{dt} = C_{A0} - C_A$$

The Laplace form of this equation is $\tau s\bar{C}_A = \bar{C}_{A0} - \bar{C}_A$. Therefore the transfer function is

$$G(s) = \frac{1}{\tau s + 1}$$

Let us now find the transfer function when there is a first-order reaction:

$$V\frac{dC_A}{dt} = \dot{Q}(C_{A0} - C_A) - kVC_A$$

$$\tau\frac{dC_A}{dt} = C_{A0} - C_A - k\tau C_A$$

$$\tau s\bar{C}_A = \bar{C}_{A0} - \bar{C}_A - k\tau\bar{C}_A$$

$$G(s) = \frac{1}{\tau s + 1 + k\tau} = \frac{1}{\tau(s + k) + 1}$$

Therefore, we can deduce that the transfer function variable is shifted by k units.

$$G(s) \rightarrow G(s + k)$$

So we can now return to our problem, and we can apply this conclusion to our earlier result in part b, for the case where there is a first-order reaction with reaction rate constant $k = 0.025\ \text{s}^{-1}$.

$$G(s+k) = \frac{e^{-3(s+k)}}{(1+3(s+k))(1+2(s+k))} = \frac{e^{-3s}e^{-3k}}{(1+3s+3k)(1+2s+2k)}$$

$$G(i\omega) = \frac{e^{-3i\omega}e^{-3(0.025)}}{(1+3(0.025)+3i\omega)(1+2(0.025)+2i\omega)}$$

$$= \frac{e^{-3i\omega}e^{-0.075}}{(1.075+3i\omega)(1.05+2i\omega)}$$

Therefore the amplitude ratio is given by:

$$|G| = \frac{e^{-0.075}}{\sqrt{1.075^2+(3\omega)^2}\sqrt{1.05^2+(2\omega)^2}} = \frac{e^{-0.075}}{\sqrt{1.075^2+9(0.15)^2}\sqrt{1.05^2+4(0.15)^2}}$$

$$|G| = 0.73$$

The phase angle can be found as follows, similar to part b,

$$\arg(G) = -3\omega - \tan^{-1}\left(\frac{3\omega}{1.075}\right) - \tan^{-1}\left(\frac{2\omega}{1.05}\right) = -1.12$$

Note that this time, the value of $G(i\omega = 0)$ is different with and without reaction. Therefore, this has an effect on the constant term of the input function.

$$f(t) = 1.5 + 0.3\cos 0.1t$$

$$G(i\omega = 0) = \frac{e^{-0.075}}{(1.075)(1.05)} = 0.82$$

Therefore, putting all our findings together, the exit concentration is as follows when there is a first-order reaction.

$$\begin{aligned}\text{Exit Concentration} &= 1.5(0.82) + (0.73)0.3\cos(0.1t - 1.12)\\ &= 1.23 + 0.22\cos(0.1t - 1.12)\end{aligned}$$

## Problem 29

**Consider a cylindrical holding tank of radius $R = 0.5$ m. A fluid flows into the tank with an inlet flow rate $v_0$ maintained at 1.5 m$^3$/h, while its exit flow rate follows a sinusoidal function with amplitude of 0.12 m$^3$/h and angular frequency $\omega = 0.0087$ rad/s.**

(a) **Determine the amplitude of oscillations of the fluid level in the tank.**
(b) **Determine the deviation from the mean fluid level at the maximum outlet flow rate.**

## Solution 29

### *Worked Solution*

(a) We have the following simple sketch of the holding tank. This is an example with varying volume.

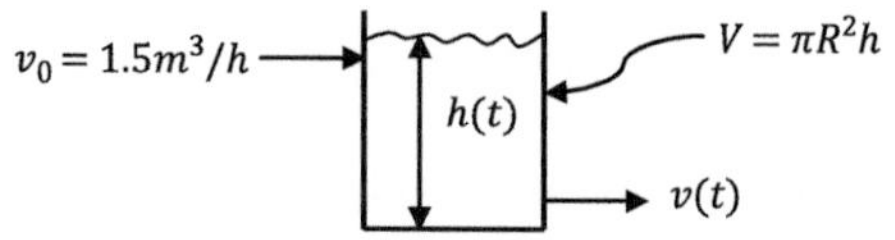


We can start with a mass balance of fluid in the tank,

$$\frac{dV}{dt} = v_0 - v$$

We note that we will need to find out the oscillations of the fluid level; therefore, we should try to relate volume to the height of the fluid level measured from the base of the tank as follows, where $A = \pi(0.5^2)$.

$$A\frac{dh}{dt} = v_0 - v$$

We can convert this ODE into a linear equation by taking Laplace transform, noting that $h = 0$ when $t = 0$.

$$As\bar{h} = \bar{v}_0 - \bar{v}$$

Considering only the deviation terms that change with time, we can find the transfer function $G(s)$ that outputs the height of fluid level (deviation) from an input of the outlet volumetric flow rate (deviation).

$$As\bar{h} = -\bar{v}$$

$$G(s) = \frac{\bar{h}}{\bar{v}} = -\frac{1}{As}$$

To find the amplitude ratio and argument of the transfer function, we can substitute the Laplace variable $s$ with $i\omega$. The amplitude ratio is given by $|G|$ and the argument of the input function is shifted by the phase angle or argument of the transfer function $\arg(G)$.

$$G(i\omega) = -\frac{1}{Ai\omega} = \frac{i^2}{Ai\omega} = \frac{1}{A\omega}i \text{ (no real part)}$$

$$|G| = \frac{1}{A\omega} = \frac{1}{0.25\pi(0.0087)} = 146$$

From the information provided in this problem, we can deduce that the sinusoidal variation of outlet flow rate can be expressed as a cosine function as shown below. However, do note that it is also valid to express *v*(*t*) as a sine function since it is also a sinusoidal function. The analysis shown here remains relevant.

$$v(t) = 0.12\cos(0.0087t)$$

Therefore the amplitude of oscillations of the fluid level in the tank is $0.12 \times 146 = 17.5$ m.

(b)

$$arg(G) = \pi/2$$

Applying the amplitude ratio and phase shift obtained from the transfer function, we can obtain the height profile as shown below.

$$h(t) = (0.12 \times 146)\cos(0.0087t + \pi/2)$$

This means that the height profile is shifted by $\pi/2$ from the outlet flowrate profile. Therefore, when the outlet flow rate is at its maximum, the height of fluid level in the tank is at its mean level.

## Multiple Integrals

**It is useful to be familiar with solving multiple integrals as they are used for many physical problems that are multidimensional. Below is an illustration of the basic concept of integration in more than one dimension.**

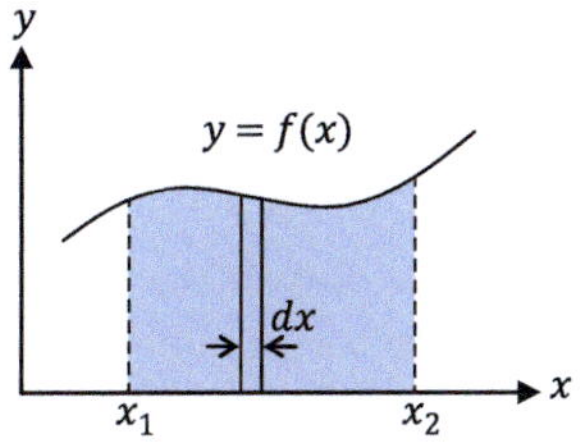

$$\int_{x_1}^{x_2} ydx = \text{shaded area}$$

**Consider a narrow rectangular strip under the curve, with a small width of *dx*. The area of this narrow strip below the curve is *ydx*. To find the total shaded area under the curve between $x_1$ and $x_2$, it requires summing the areas of an infinite number of these narrow rectangular strips below the curve, where *dx* is infinitesimally small or tends to zero. Integration is therefore a process of summing *ydx* over the defined range from $x_1$ to $x_2$, whereby the value of *y* (or height of rectangle) follows the variation as defined by the expression for $f(x)$ in $y = f(x)$.**

**Problem 30**

**Using multiple integrals, show that the area of the polygon below is 4.75.**

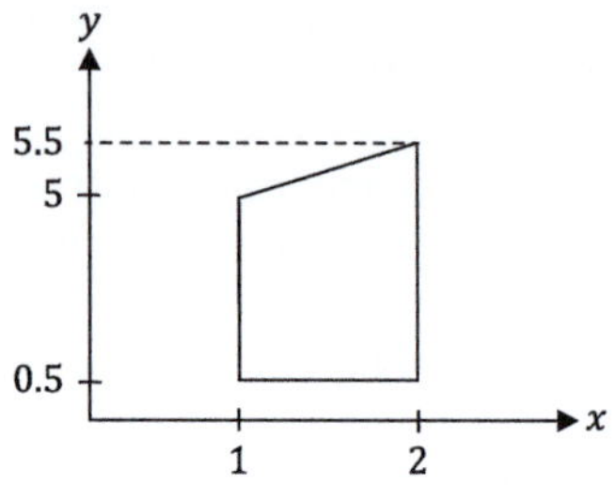


**Solution 30**

***Worked Solution***

We can visualize the area of this polygon, in two steps. Geometrically, integrals refer to the area under the curve, within the limits of integration. Therefore, we can find the area contained within the polygon by subtracting $A_2$ from $A_1$.

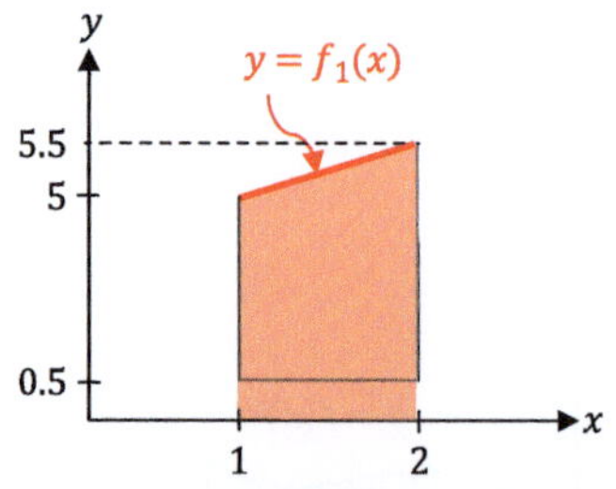


Step 1: Integrate from $x = 1$ to $x = 2$ over the function $y = f_1(x)$.

The expression for $y = f_1(x)$ is the equation of the red line as shown above. We can find this expression, since we know the coordinates that define the line, i.e., $(1, 5)$ and $(2, 5.5)$.

$$\frac{y - y_1}{x - x_1} = \frac{y_2 - y_1}{x_2 - x_1} = \frac{1}{2} \rightarrow y = f_1(x) = \frac{x}{2} + \frac{9}{2}$$

$$\text{Area } A_1 = \int_1^2 f_1(x)dx = \int_1^2 \left(\frac{x}{2} + \frac{9}{2}\right)dx = \left[\frac{x^2}{4} + \frac{9}{2}x\right]_1^2 = \frac{21}{4}$$

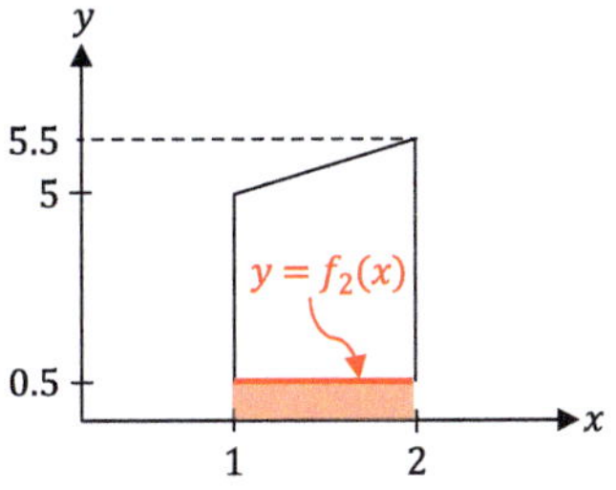


Step 2: Integrate from $x = 1$ to $x = 2$ over the function $y = f_2(x)$

The expression for $y = f_2(x)$ is the equation of the red line as shown above. We can find this expression easily since it is a horizontal line.

$$y = f_2(x) = \frac{1}{2}$$

$$\text{Area } A_2 = \int_1^2 f_2(x)dx = \int_1^2 \frac{1}{2}dx = \left[\frac{1}{2}x\right]_1^2 = \frac{1}{2}$$

$$\text{Area of polygon} = A_1 - A_2 = \frac{21}{4} - \frac{1}{2} = 4.75$$

We can verify our result by applying the simple formula for the area of a trapezium since our polygon is in fact a trapezium.

$$\text{Area of trapezium} = \frac{1}{2}(2 - 1)[(5.5 - 0.5) + (5 - 0.5)] = 4.75$$

**Problem 31**

**Using multiple integrals, find the area of the semicircle as shown below.**

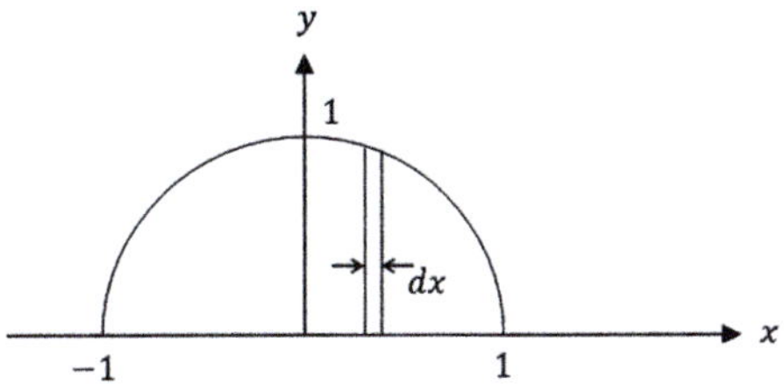

## Solution 31

***Worked Solution***

Let us first define the equation of the curve that encloses the semicircle as $y = f(x)$. Since the radius of the semicircle is 1, anywhere on the curve follows the equation $x^2 + y^2 = 1$. The limits of integration are $-1 \leq x \leq 1$. The limits of $y$ should be obtained from the equation of the curve, as we note that the heights of the narrow strips of rectangles will change with $x$, according to the function of $y = f(x)$. Since $x^2 + y^2 = 1$, we can deduce that the limits of $y$ are $0 \leq y \leq \sqrt{1 - x^2}$.

$$\text{Area of semicircle} = \int_{-1}^{1} y dx = \int_{-1}^{1} \sqrt{1 - x^2} dx$$

In order to solve this integral more easily, we can apply a substitution where $\theta$ is the angle measured from the $x$ axis. Then $x = \cos\theta$, $dx = -\sin\theta d\theta$, and $x = [-1, 1]$ can be changed to $\theta = [\pi, 0]$. The trigonometric identity $1 - \cos^2\theta = \sin^2\theta$ and double angle formula $\sin^2\theta = \frac{1}{2} - \frac{\cos 2\theta}{2}$ help to simplify the equation.

$$\int_{-1}^{1} \sqrt{1 - x^2} dx = \int_{\pi}^{0} \sqrt{1 - \cos^2\theta} (-\sin\theta) d\theta = \int_{\pi}^{0} \sin\theta(-\sin\theta) d\theta$$

$$= \int_{0}^{\pi} \frac{1}{2} - \frac{\cos 2\theta}{2} d\theta = \left[\frac{1}{2}\theta - \frac{1}{4}\sin 2\theta\right]_{0}^{\pi} = \frac{\pi}{2}$$

A quick check with the formula for area of a semicircle shows that the result is consistent.

$$\text{Area of semicircle} = \frac{1}{2}\pi(1)^2 = \frac{\pi}{2}$$

## Problem 32

**Find the shaded area in the diagram below via integration.**

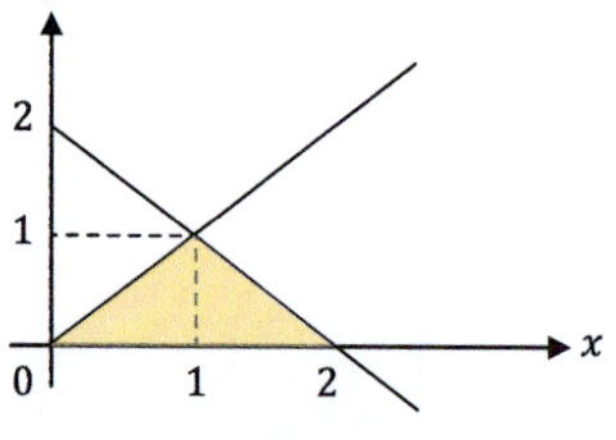

## Solution 32

### *Worked Solution*

In order to find the area of the shaded triangle using integration, we can first define the function that describes the height of the narrow rectangular strips with width of $dx$.

For $0 \leq x \leq 1$, $y = x$, and for $1 \leq x \leq 2$, $y = -x + 2$

Therefore we can construct the multiple integral as follows:

$$\text{Area} = \int_0^1 x dx + \int_1^2 (-x + 2) dx = \left[\frac{x^2}{2}\right]_0^1 + \left[2x - \frac{x^2}{2}\right]_1^2 = \frac{1}{2} + \frac{1}{2} = 1$$

We can verify this result using the formula for area of a triangle, which is consistent with the result above.

$$\text{Area} = \frac{1}{2}(2)(1) = 1$$

## Problem 33

**Evaluate the multiple integral and show that it is equal to $\sqrt{\pi}/4$.**

$$I = \int_0^\infty dx \int_0^\infty dy \frac{yx^2}{x^2 + y^2} e^{-(x^2+y^2)}$$

## Solution 33

### *Worked Solution*

One trick to simplifying integrals that contain the expression $x^2 + y^2$ is to convert to polar coordinates whereby $x^2 + y^2 = r^2$. $\theta$ is measured from the $x$-axis. $x = r \cos \theta$ and $y = r \sin \theta$.

For the limits of the integrals, we can convert the quadrant represented by $x = [0, \infty]$ and $y = [0, \infty]$ into $r = [0, \infty]$ and $\theta = \left[0, \frac{\pi}{2}\right]$.

Finally we need to be cautious when converting the area integral $\int_0^\infty dx \int_0^\infty dy$ into polar form. Note that the differential element is now the small square element of sides $dr$ and $rd\theta$.

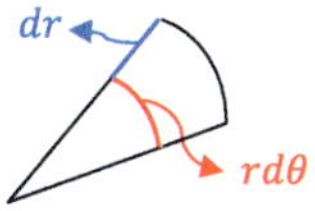

Therefore, the integral can be re-expressed as follows:

$$I = \int_{r=0}^{\infty}\int_{\theta=0}^{\frac{\pi}{2}} \frac{r\sin\theta r^2\cos^2\theta}{r^2} e^{-r^2} r d\theta dr = \int_{r=0}^{\infty} r^2 e^{-r^2} dr \int_{\theta=0}^{\frac{\pi}{2}} \sin\theta\cos^2\theta d\theta$$

The integral can be determined in two parts, with respect to $r$ and with respect to $\theta$. The integral with respect to $\theta$ can be done using integration by parts.

$$\int_{\theta=0}^{\frac{\pi}{2}} \sin\theta\cos^2\theta d\theta = \left[-\frac{\cos^3\theta}{3}\right]_0^{\frac{\pi}{2}} = \frac{1}{3}$$

$$\int_{r=0}^{\infty} r^2 e^{-r^2} dr = \int_{r=0}^{\infty} (r) r e^{-r^2} dr = \left[r\left(-\frac{1}{2}\right)e^{-r^2}\right]_0^{\infty} - \int_0^{\infty}\left(-\frac{1}{2}\right)e^{-r^2} dr$$

$$= 0 + \frac{1}{2}\int_0^{\infty} e^{-r^2} dr$$

From here, we note that the function $e^{-r^2}$ is even. Therefore, we can convert the limits of the integral this way, and make use of the result $\int_{-\infty}^{\infty} e^{-r^2} dr = \sqrt{\pi}$.

$$\int_{r=0}^{\infty} r^2 e^{-r^2} dr = \frac{1}{2}\int_{-\infty}^{\infty} \frac{1}{2} e^{-r^2} dr = \frac{\sqrt{\pi}}{4}$$

Combining our results, we have the value of the integral as shown below.

$$I = \frac{\sqrt{\pi}}{12}$$

## Problem 34

**Determine the integral as shown below.**

$$\int \tanh xy \,\mathrm{sech}\, xy \, dy$$

**Use your result to find the value of the following integral, where $\beta > 1$.**

$$I = \int_0^{\infty} \frac{1}{x}(\mathrm{sech}\, x - \mathrm{sech}\, \beta x) dx$$

## Solution 34

***Worked Solution***

$$\int \tanh xy \,\mathrm{sech}\, xy \, dy = \int \frac{\sinh xy}{\cosh^2 xy} dy = -\frac{1}{x\cosh xy} + \text{constant} = -\frac{\mathrm{sech}\, xy}{x} + \text{constant}$$

We can turn this indefinite integral into a definite integral by imposing limits for $y = [1, \beta]$.

$$\int_1^{\beta} \tanh xy \operatorname{sech} xy\, dy = -\frac{\operatorname{sech}\beta x}{x} + \frac{\operatorname{sech} x}{x}$$

Substituting this result into our integral, we have

$$I = \int_0^{\infty} \frac{1}{x}(\operatorname{sech} x - \operatorname{sech}\beta x)dx = \int_{x=0}^{\infty} \int_{y=1}^{\beta} \tanh xy \operatorname{sech} xy\, dy dx$$

We can interchange the order of integration since it does not matter which comes first.

$$I = \int_{y=1}^{\beta} \left[\int_{x=0}^{\infty} \tanh xy \operatorname{sech} xy\, dx\right] dy = \int_{y=1}^{\beta} \left[-\frac{1}{y}\operatorname{sech} xy\right]_{x=0}^{\infty} dy$$

From the graph of $\cosh xy$, we know that when $x = 0$, $\cosh xy = 1$ and when $x = \infty$, $\cosh xy = \infty$ and $\operatorname{sech} xy = 0$. Therefore,

$$I = \int_{y=1}^{\beta} \left(\frac{1}{y}\right) dy = [\ln y]_1^{\beta} = \ln \beta$$

## Fourier Series

**Fourier series are useful in representing a function as a sum of simpler sinusoidal functions. It decomposes any periodic function (which may represent a periodic signal) into a sum of oscillating functions, namely sines and cosines. These sine and cosine waves can also be expressed in complex exponentials. It is worth noting that sines and cosines are unique in that their amplitudes vary between (1, $-1$) and can output values from $-\infty$ to $+\infty$. They also follow a range of identities and formulae which help to solve problems due to their versatility modeling practical phenomena such as vibrations and waves, and in transforming from one form to another mathematically.**

**Problem 35**

**Explain what is meant by a Fourier series expansion, and describe its characteristics using a suitable example. Show how Fourier coefficients may be determined.**

## Solution 35

***Worked Solution***

If a function $f(x)$ is periodic with period $2L$, then $f(x)$ may be written as a Fourier series. The terms $a_0$, $a_n$, and $b_n$ are Fourier coefficients.

$$f(x + 2L) = f(x)$$

$$f(x) = \frac{1}{2}a_0 + \sum_{n=1}^{\infty}\left(a_n \cos\frac{n\pi x}{L} + b_n \sin\frac{n\pi x}{L}\right)$$

The individual terms (also called basis functions) that make up the Fourier series expansion have certain properties:

- Each is a periodic function on its own with the same period $2L$. Their periodicity can be expressed as follows:

$$\cos\frac{n\pi x}{L} = \cos\frac{n\pi(x + 2L)}{L}$$

- These basis functions are 1 (for the first term "$\frac{1}{2}a_0$"), $\cos\frac{\pi x}{L}$, $\cos\frac{2\pi x}{L}$, $\cos\frac{n\pi x}{L}$, $\sin\frac{\pi x}{L}$, $\sin\frac{2\pi x}{L}$, and $\sin\frac{n\pi x}{L}$ where $n$ is a positive integer.
- We may first observe if the function $f(x)$ is an even or odd function. If it is a straightforward case, then we can immediately simplify the Fourier series expansion to the following forms:

$$f(x) = \frac{1}{2}a_0 + \sum_{n=1}^{\infty} a_n \cos\frac{n\pi x}{L} \quad \text{for even function}$$

$$f(x) = \frac{1}{2}a_0 + \sum_{n=1}^{\infty} b_n \sin\frac{n\pi x}{L} \quad \text{for odd function}$$

- The basis functions are orthogonal over any interval of length $2L$. And this orthogonality means that their dot products are zero as shown below, where $m$ and $n$ are positive integers and $m \neq n$. This is a useful result especially for determining Fourier coefficients, $a_0$, $a_n$, and $b_n$.

$$\int_{-L}^{L} \cos\frac{n\pi x}{L} \cos\frac{m\pi x}{L} dx = 0 \quad (1)$$

$$\int_{-L}^{L} \cos\frac{n\pi x}{L} \sin\frac{m\pi x}{L} dx = 0 \quad (2)$$

$$\int_{-L}^{L} \sin\frac{n\pi x}{L} \sin\frac{m\pi x}{L} dx = 0 \tag{3}$$

$$\int_{-L}^{L} 1.\cos\frac{m\pi x}{L} dx = 0 \tag{4}$$

$$\int_{-L}^{L} 1.\sin\frac{m\pi x}{L} dx = 0 \tag{5}$$

- It is also useful to note the following useful results of the squares of the basis functions, where $n \neq 0$.

$$\int_{-L}^{L} \cos^2\frac{n\pi x}{L} dx = L \tag{6}$$

$$\int_{-L}^{L} \sin^2\frac{n\pi x}{L} dx = L \tag{7}$$

$$\int_{-L}^{L} 1^2 dx = 2L \tag{8}$$

To evaluate Fourier coefficients, we can integrate from –$L$ to $L$. We have below the Fourier series expansion restated, and different integrals performed as follows.

$$f(x) = \frac{1}{2}a_0 + \sum_{n=1}^{\infty}\left(a_n \cos\frac{n\pi x}{L} + b_n \sin\frac{n\pi x}{L}\right)$$

To find out $a_0$, we can integrate $f(x)$:

$$\int_{-L}^{L} f(x)dx = \int_{-L}^{L} \frac{1}{2}a_0 dx + \sum_{n=1}^{\infty} a_n \int_{-L}^{L} \cos\frac{n\pi x}{L} dx + \sum_{n=1}^{\infty} b_n \int_{-L}^{L} \sin\frac{n\pi x}{L} dx$$

Applying results (4) and (5) of the orthogonality property, we can eliminate the cos and sin terms. Therefore, we have

$$\int_{-L}^{L} f(x)dx = \int_{-L}^{L} \frac{1}{2}a_0 dx + 0 + 0 = a_0 L$$

$$a_0 = \frac{1}{L}\int_{-L}^{L} f(x)dx \tag{9}$$

To find out $a_n$, we can integrate the product of $f(x)$ and $\cos\frac{n\pi x}{L}$:

$$\int_{-L}^{L} f(x)\cos\frac{n\pi x}{L}dx = \int_{-L}^{L} \frac{1}{2}a_0\cos\frac{n\pi x}{L}dx + \sum_{n=1}^{\infty} a_n \int_{-L}^{L} \cos^2\frac{n\pi x}{L}dx + \sum_{n=1}^{\infty} b_n \int_{-L}^{L} \sin\frac{n\pi x}{L}\cos\frac{n\pi x}{L}dx$$

Applying results (4–6), we can simplify to

$$\int_{-L}^{L} f(x)\cos\frac{n\pi x}{L}dx = 0 + a_n L + \sum_{n=1}^{\infty} b_n \int_{-L}^{L} \frac{1}{2}\sin\frac{2n\pi x}{L}dx$$

$$\int_{-L}^{L} f(x)\cos\frac{n\pi x}{L}dx = 0 + a_n L + 0$$

$$a_n = \frac{1}{L}\int_{-L}^{L} f(x)\cos\frac{n\pi x}{L}dx \tag{10}$$

To find out $b_n$, we can integrate the product of $f(x)$ and $\sin\frac{n\pi x}{L}$:

$$\int_{-L}^{L} f(x)\sin\frac{n\pi x}{L}dx = \int_{-L}^{L} \frac{1}{2}a_0\sin\frac{n\pi x}{L}dx + \sum_{n=1}^{\infty} a_n \int_{-L}^{L} \cos\frac{n\pi x}{L}\sin\frac{n\pi x}{L}dx + \sum_{n=1}^{\infty} b_n \int_{-L}^{L} \sin^2\frac{n\pi x}{L}dx$$

Applying results (4), (5), and (7), we can simplify to

$$\int_{-L}^{L} f(x)\sin\frac{n\pi x}{L}dx = 0 + \sum_{n=1}^{\infty} a_n \int_{-L}^{L} \frac{1}{2}\sin\frac{2n\pi x}{L}dx + b_n L$$

$$\int_{-L}^{L} f(x)\sin\frac{n\pi x}{L}dx = 0 + 0 + b_n L$$

$$b_n = \frac{1}{L}\int_{-L}^{L} f(x)\sin\frac{n\pi x}{L}dx \tag{11}$$

## Problem 36

**Find the Fourier series expansion for the function shown below.**

$$f(x) = \begin{cases} 2 + \frac{x}{\pi}, & -\pi \le x \le 0 \\ 2 - \frac{x}{\pi}, & 0 \le x \le \pi \end{cases}$$

## Solution 36

### *Worked Solution*

We can observe that this is an even function from a simple plot. Therefore the Fourier series expansion will not have sine terms.

$$f(x) = \frac{1}{2}a_0 + \sum_{n=1}^{\infty} a_n \cos\frac{n\pi x}{L}$$

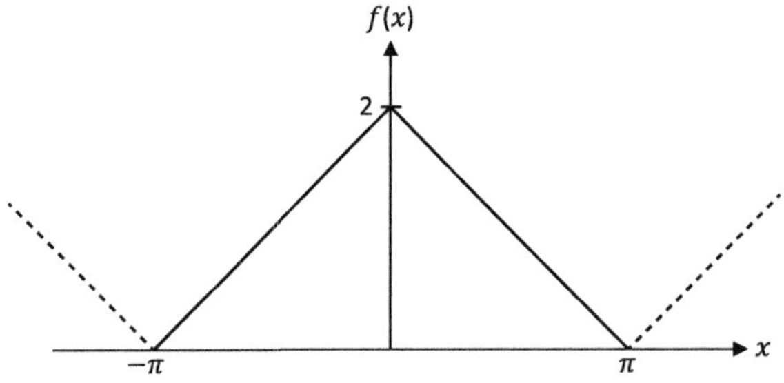


The period of this function is $2\pi$, therefore $L = \pi$

$$a_0 = \frac{1}{L}\int_{-L}^{L} f(x)dx = \frac{1}{\pi}\left[\int_{-\pi}^{0} 2 + \frac{x}{\pi}dx + \int_{0}^{\pi} 2 - \frac{x}{\pi}dx\right]$$

$$= \frac{1}{\pi}\left(\left[2x + \frac{x^2}{2\pi}\right]_{-\pi}^{0} + \left[2x - \frac{x^2}{2\pi}\right]_{0}^{\pi}\right)$$

$$= \frac{1}{\pi}\left(2\pi - \frac{\pi}{2} + 2\pi - \frac{\pi}{2}\right)$$

$$= \frac{1}{\pi}(4\pi - \pi)$$

$$a_0 = 3$$

Now we can find the Fourier coefficient $a_n$ for the cosine term.

$$a_n = \frac{1}{L}\int_{-L}^{L} f(x)\cos\frac{n\pi x}{L}\,dx = \frac{1}{\pi}\left[\int_{-\pi}^{0}\left(2 + \frac{x}{\pi}\right)\cos\,nx\,dx + \int_{0}^{\pi}\left(2 - \frac{x}{\pi}\right)\cos\,nx\,dx\right]$$

$$\pi a_n = \int_{-\pi}^{0}\left(2 + \frac{x}{\pi}\right)\cos\,nx\,dx + \int_{0}^{\pi}\left(2 - \frac{x}{\pi}\right)\cos\,nx\,dx$$

$$= \int_{-\pi}^{0} 2\cos\ nx\,dx + \int_{-\pi}^{0} \frac{x}{\pi}\cos\ nx\,dx + \int_{0}^{\pi} 2\cos\ nx\,dx - \int_{0}^{\pi} \frac{x}{\pi}\cos\ nx\,dx$$

$$= \int_{-\pi}^{0} \frac{x}{\pi}\cos\ nx\,dx - \int_{0}^{\pi} \frac{x}{\pi}\cos\ nx\,dx$$

We can integrate by parts, recall the following

$$\int_{a}^{b} udv = [uv]_a^b - \int_{a}^{b} vdu$$

$$\int_{-\pi}^{0} \frac{x}{\pi}\cos\ nx\,dx = \left[\frac{x}{\pi}\frac{\sin\ nx}{n}\right]_{-\pi}^{0} - \int_{-\pi}^{0} \frac{1}{\pi}\frac{\sin\ nx}{n}dx$$

$$= 0 + \left[\frac{1}{\pi}\frac{\cos\ nx}{n^2}\right]_{-\pi}^{0} = \frac{1}{\pi}\frac{1}{n^2} - \frac{1}{\pi}\frac{\cos\ (-n\pi)}{n^2}$$

$$= \frac{1}{\pi n^2}[1 - \cos\ (-n\pi)\ ] = \frac{1}{\pi n^2}(1 - \cos\ n\pi\ )$$

Similarly, we can integrate the other term using integration by parts to get the following.

$$\int_{0}^{\pi} \frac{x}{\pi}\cos\ nx\,dx = \frac{1}{\pi n^2}(\cos\ n\pi - 1)$$

Putting your results together, we have

$$\pi a_n = \frac{1}{\pi n^2}(1 - \cos\ n\pi\ ) - \frac{1}{\pi n^2}(\cos\ n\pi - 1)$$

$$a_n = \frac{2}{\pi^2 n^2}[1 - \cos\ n\pi\ ]$$

This can be expressed in terms of odd or even integers of $n$,

$$a_n = \begin{cases} \dfrac{4}{\pi^2 n^2}, & \text{for } n \text{ is odd} \\ 0\ , & \text{for } n \text{ is even} \end{cases}$$

The final expression for the Fourier series expansion of the function $f(x)$ is,

$$f(x) = \frac{1}{2}a_0 + \sum_{n=1}^{\infty} a_n \cos\frac{n\pi x}{L}$$

$$= \frac{3}{2} + \sum_{n=1}^{\infty} \frac{4}{\pi^2 n^2}\cos\ nx, \quad \text{where } n \text{ is odd}$$

Another way to express an odd positive integer is $2m - 1$, where $m$ is 1 to $\infty$, to match the limits of the summation.

$$f(x) = \frac{3}{2} + \sum_{n=1}^{\infty} \frac{4}{\pi^2 (2m-1)^2} \cos\left[(2m-1)x\right]$$

We know that $f(0) = 2$,

$$2 = \frac{3}{2} + \sum_{n=1}^{\infty} \frac{4}{\pi^2 (2m-1)^2}$$

$$\frac{\pi^2}{8} = \sum_{n=1}^{\infty} \frac{1}{(2m-1)^2}$$

## Problem 37

**Consider the following function with a period of $2\pi$, therefore $L = \pi$.**

(a) **Find the Fourier series expansion for this function.**

$$f(x) = \begin{cases} -\sin 2x, & -\pi \le x \le 0 \\ \sin 2x, & 0 \le x \le \pi \end{cases}$$

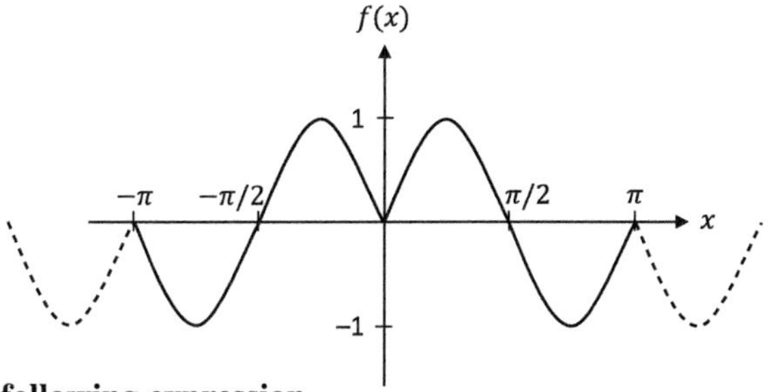


(b) **Show the following expression.**

$$\frac{1}{3} = \sum_{n=1}^{\infty} \frac{1}{(2m+1)^2 - 4}$$

## Solution 37

***Worked Solution***

(a) We can observe that this is an even function from the plot. Therefore the Fourier series expansion will not have sine terms. The general form of the Fourier series expansion without sine terms is as follows.

$$f(x) = \frac{1}{2}a_0 + \sum_{n=1}^{\infty} a_n \cos \frac{n\pi x}{L}$$

The period of this function is $2\pi$; therefore $L = \pi$

$$a_0 = \frac{1}{L}\int_{-L}^{L} f(x)dx = \frac{1}{\pi}\left[\int_{-\pi}^{0} -\sin\ 2x\,dx + \int_{0}^{\pi} \sin\ 2x\,dx\right]$$

$$= \frac{1}{\pi}\left(\left[\frac{\cos\ 2x}{2}\right]_{-\pi}^{0} + \left[-\frac{\cos\ 2x}{2}\right]_{0}^{\pi}\right)$$

$$= \frac{1}{\pi}\left[\frac{1}{2} - \frac{1}{2} + \left(-\frac{1}{2}\right) - \left(-\frac{1}{2}\right)\right] = 0$$

$$a_0 = 0$$

Now we can find the Fourier coefficient $a_n$ for the cosine term.

$$a_n = \frac{1}{L}\int_{-L}^{L} f(x)\cos\frac{n\pi x}{L}\,dx = \frac{1}{\pi}\left[\int_{-\pi}^{0} (-\sin\ 2x)\cos\ nx\,dx + \int_{0}^{\pi} (\sin\ 2x)\cos\ nx\,dx\right]$$

$$\pi a_n = \int_{-\pi}^{0} (-\sin\ 2x)\cos\ nx\,dx + \int_{0}^{\pi} (\sin\ 2x)\cos\ nx\,dx$$

We can integrate by recalling the trigonometric addition formula

$$\sin\ (A + B) = \sin\ A\ \cos\ B + \cos\ A\ \sin\ A$$

($\sin 2x \cos nx$) can be expressed as a single trigonometric function,

$$\frac{1}{2}\sin\ (2x + nx) = \frac{1}{2}\sin\ 2x\ \cos\ nx + \frac{1}{2}\cos\ 2x\ \sin\ nx$$

$$\frac{1}{2}\sin\ (2x - nx) = \frac{1}{2}\sin\ 2x\ \cos\ nx - \frac{1}{2}\cos\ 2x\ \sin\ nx$$

$$\sin\ 2x\ \cos\ nx = \frac{1}{2}\sin\ (2x + nx) + \frac{1}{2}\sin\ (2x - nx)$$

$$= \frac{1}{2}[\sin\ (2 + n)\,x + \sin\ (2 - n)x\,]$$

Similarly we can express $-\sin\ 2x\ \cos\ nx = -\frac{1}{2}[\sin\ (2 + n)\,x + sin\ (2 - n)x]$. Putting our results together, we have

$$\pi a_n = \int_{-\pi}^{0} (-\sin\ 2x)\cos\ nx\,dx + \int_{0}^{\pi} (\sin\ 2x)\cos\ nx\,dx$$

$$= \int_{-\pi}^{0} -\frac{1}{2}[\sin (2+n)\, x + \sin (2-n)x\,]dx$$
$$+ \int_{0}^{\pi} \frac{1}{2}[\sin (2+n)\, x + \sin (2-n)x\,]dx$$

$$= \frac{1}{2}\left\{\int_{-\pi}^{0} [-\sin (2+n)\, x - \sin (2-n)x\,]dx + \int_{0}^{\pi} [\sin (2+n)\, x + \sin (2-n)x\,]dx\right\}$$

$$= \frac{1}{2}\left\{\left[\frac{\cos (2+n)x}{(2+n)} + \frac{\cos (2-n)x}{(2-n)}\right]_{-\pi}^{0} + \left[-\frac{\cos (2+n)x}{(2+n)} - \frac{\cos (2-n)x}{(2-n)}\right]_{0}^{\pi}\right\}$$

$$= \frac{1}{2}\left\{\left[\frac{\cos (2+n)x}{(2+n)} + \frac{\cos (2-n)x}{(2-n)}\right]_{-\pi}^{0} + \left[-\frac{\cos (2+n)x}{(2+n)} - \frac{\cos (2-n)x}{(2-n)}\right]_{0}^{\pi}\right\}$$

$$a_n = \frac{1}{2\pi}\left\{\frac{2}{2+n} + \frac{2}{2-n} - \frac{2\cos (2+n)\pi}{(2+n)} - \frac{2\cos (2-n)\pi}{(2-n)}\right\}$$
$$= \frac{1}{\pi}\left\{\frac{4}{4-n^2} - \frac{\cos (2+n)\pi}{(2+n)} - \frac{\cos (2-n)\pi}{(2-n)}\right\}$$

This can be expressed in terms of odd or even integers of $n$,
When $n$ is even,

$$a_n = 0$$

When $n$ is odd,

$$a_n = \frac{8}{\pi(4-n^2)}$$

Combining our results,

$$f(x) = \sum_{m=0}^{\infty} \frac{8}{\pi\left(4-(2m+1)^2\right)} \cos (2m+1)x$$

(b) We know that $f(0) = 0$

$$0 = \sum_{m=0}^{\infty} \frac{8}{\pi\left(4-(2m+1)^2\right)}$$

$$0 = \frac{1}{3} + \sum_{m=1}^{\infty} \frac{1}{\left(4-(2m+1)^2\right)}$$

$$\frac{1}{3} = \sum_{m=1}^{\infty} \frac{1}{(2m+1)^2 - 4}$$

## Problem 38

**Consider the following sawtooth function. Evaluate the Fourier coefficients in the form of the Fourier series expansion as shown below**

$$f(x) = \frac{1}{2}a_0 + \sum_{n=1}^{\infty}\left(a_n \cos\frac{n\pi x}{L} + b_n \sin\frac{n\pi x}{L}\right)$$

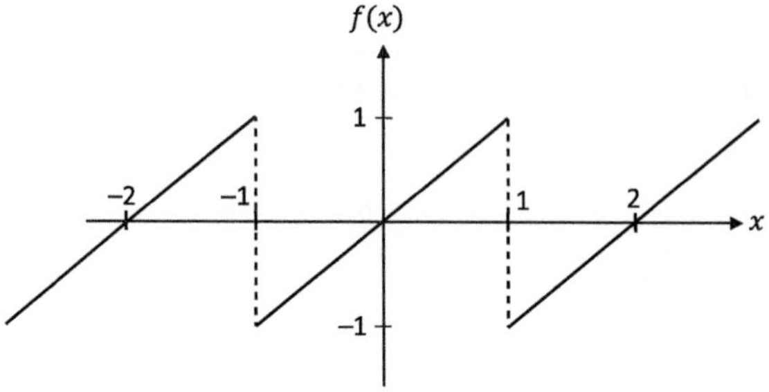


## Solution 38

***Worked Solution***

We can observe that this is an odd function from the plot. Therefore the Fourier series expansion will not have cosine terms and $a_n = 0$. The general form of the Fourier series expansion without cosine terms is as follows.

$$f(x) = \frac{1}{2}a_0 + \sum_{n=1}^{\infty} b_n \sin\frac{n\pi x}{L}$$

The function can be expressed as $f(x) = x$ with period equivalent to 2; therefore $L = 1$

$$a_0 = \int_{-1}^{1} f(x)dx = \int_{-1}^{1} xdx$$

$$a_0 = 0$$

Next we can find the Fourier coefficient $b_n$ for the sine term.

$$b_n = \int_{-1}^{1} x \sin n\pi x \, dx = \left[ (x) \frac{\cos n\pi x}{-n\pi} \right]_{-1}^{1} - \int_{-1}^{1} \left( \frac{\cos n\pi x}{-n\pi} \right) dx$$

$$= \frac{\cos n\pi}{-n\pi} - \frac{\cos(-n\pi)}{n\pi} + \left[ \frac{\sin n\pi x}{n^2\pi^2} \right]_{-1}^{1}$$

$$= \frac{2\cos n\pi}{-n\pi} + 0$$

We can observe that when $n$ is even, $\cos n\pi = 1$, and when $n$ is odd, $\cos n\pi = -1$. Therefore we can express in this form,

$$b_n = \frac{2(-1)^n}{-n\pi}$$

Putting the results together,

$$f(x) = \frac{2}{\pi} \sum_{n=1}^{\infty} \frac{(-1)^{n+1} \sin n\pi x}{n}$$

## Problem 39

**Consider the following function that has a plot as shown below.**

(a) **Derive the equation that describes this function.**
(b) **Express the function in the form of a Fourier series expansion, and determine the Fourier coefficients.**

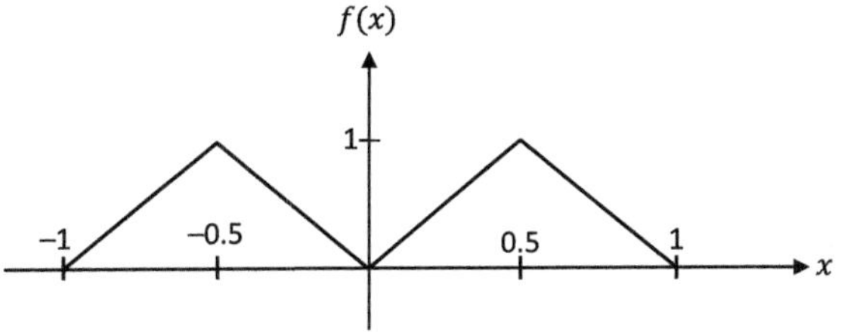


## Solution 39

***Worked Solution***

(a) The function has a period of 1, and therefore $L = 0.5$, where the reference frame for the period is $[-L, L]$ for the Fourier series of the general form as shown below. The function has a shape that matches the expression $f(x) = |2x|$.

$$f(x) = \frac{1}{2} a_0 + \sum_{n=1}^{\infty} \left( a_n \cos \frac{n\pi x}{L} + b_n \sin \frac{n\pi x}{L} \right)$$

(b) We may observe that this is an even function, hence there will not be sine terms and we can start from the following simplified form.

$$f(x) = \frac{1}{2}a_0 + \sum_{n=1}^{\infty} a_n \cos\frac{n\pi x}{L}$$

To find the Fourier coefficient $a_0$,

$$a_0 = \int_{-0.5}^{0.5} |2x| dx = \int_{-0.5}^{0} -2x dx + \int_{0}^{0.5} 2x dx$$

$$= \left[-x^2\right]_{-0.5}^{0} + \left[x^2\right]_{0}^{0.5} = 0.5$$

In fact, we can observe from the graph that the integral is simply the area under the graph from −0.5 to 0.5. And from geometry, we can deduce directly that the integral is equivalent to 0.5.

$$a_n = \frac{1}{L}\int_{-L}^{L} f(x) \cos\frac{n\pi x}{L} dx$$

$$= \frac{1}{0.5}\int_{-0.5}^{0.5} |2x| \cos\,(2n\pi x) dx$$

$$= \frac{1}{0.5}\left[\int_{-0.5}^{0} -2x \cos\,(2n\pi x) dx + \int_{0}^{0.5} 2x \cos\,(2n\pi x) dx\right]$$

We can simplify the equation using integration by parts.

$$\int_{-0.5}^{0} -2x \cos\,(2n\pi x) dx = \left[-2x\frac{\sin\,(2n\pi x)}{2n\pi}\right]_{-0.5}^{0} - \int_{-0.5}^{0} \frac{-2\sin\,(2n\pi x)}{2n\pi} dx$$

$$= 0 - \left[\frac{\cos\,(2n\pi x)}{2n^2\pi^2}\right]_{-0.5}^{0} = -\frac{1}{2n^2\pi^2} + \frac{\cos\,(-n\pi)}{2n^2\pi^2}$$

$$= -\frac{1}{2n^2\pi^2} + \frac{\cos\, n\pi}{2n^2\pi^2} = \frac{1}{2n^2\pi^2}[\cos\, n\pi - 1]$$

We can also simplify the other integral using integration by parts and arrive at

$$\int_{0}^{0.5} 2x \cos\,(2n\pi x) dx = \left[2x\frac{\sin\,(2n\pi x)}{2n\pi}\right]_{0}^{0.5} - \int_{0}^{0.5} \frac{2\sin\,(2n\pi x)}{2n\pi} dx$$

$$= \frac{\sin\,(n\pi)}{2n\pi} + \left[\frac{\cos\,(2n\pi x)}{2n^2\pi^2}\right]_{0}^{0.5} = \frac{\sin\,(n\pi)}{2n\pi} + \left[\frac{\cos\,(2n\pi x)}{2n^2\pi^2}\right]_{0}^{0.5}$$

$$= 0 + \frac{\cos\,(n\pi)}{2n^2\pi^2} - \frac{1}{2n^2\pi^2} = \frac{1}{2n^2\pi^2}[\cos\, n\pi - 1]$$

Therefore, the Fourier coefficient $a_n$ can be expressed as

$$a_n = \frac{1}{0.5}\left[\frac{1}{2n^2\pi^2}[\cos\ n\pi - 1] + \frac{1}{2n^2\pi^2}[\cos\ n\pi - 1]\right]$$
$$= \frac{2}{n^2\pi^2}[\cos\ n\pi - 1]$$

We note that when $n$ is even, $cos\ n\pi = 1$, and when $n$ is odd, $cos\ n\pi = -1$. Therefore we can express $cos\ n\pi = (-1)^n$.

$$a_n = \frac{2}{n^2\pi^2}[(-1)^n - 1]$$

We note that when $n$ is even, $a_n = 0$. Cosine terms only present for $n$ is odd and $a_n = \frac{-4}{n^2\pi^2}$. For a range of summation from 1 to $\infty$, we can re-express $n = 2m - 1$, to cover the range of positive and odd integers.

Now we can put our results together, back in the form of Fourier expansion as shown below.

$$f(x) = 1 + \sum_{n=1}^{\infty} a_n \cos\frac{n\pi x}{L}$$

$$f(x) = 1 + \sum_{n=1}^{\infty} \frac{-4}{n^2\pi^2} \cos\frac{n\pi x}{L}$$

$$f(x) = 1 + \sum_{m=1}^{\infty} \frac{-4}{(2m-1)^2\pi^2} \cos\frac{m\pi x}{0.5}$$

$$f(x) = 1 - \frac{4}{\pi^2}\sum_{m=1}^{\infty} \frac{\cos[2(2m-1)\pi x]}{(2m-1)^2}$$

## Problem 40

**Consider the following step function.**

$$f(x) = \begin{cases} -1, & -\pi \leq x \leq 0 \\ 1, & 0 \leq x \leq \pi \end{cases}$$

(a) **Express this function in terms of its Fourier series expansion.**

(b) **Using your results in (a), find the Fourier series expansion for the function shown in the plot below.**

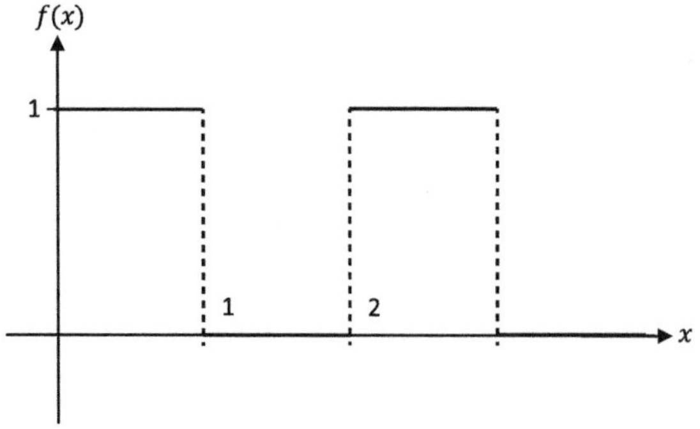


## Solution 40

***Worked Solution***

(a) The function provided in part a can be shown in the following plot.

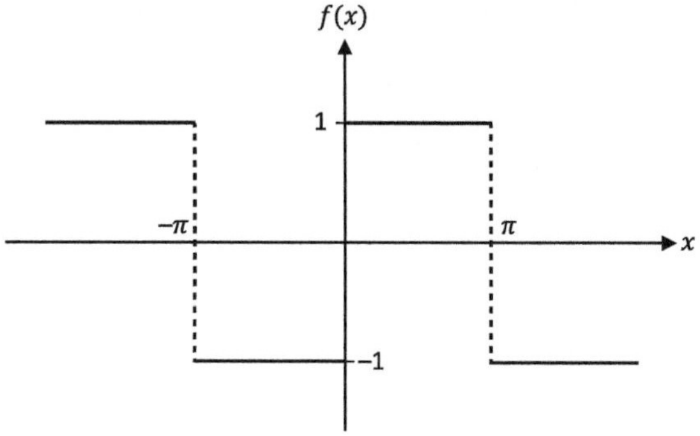


We can observe that this is an odd function from the plot. Therefore the Fourier series expansion will not have cosine terms and $a_n = 0$ in the general form of the Fourier series expansion as shown below

$$f(x) = \frac{1}{2}a_0 + \sum_{n=1}^{\infty}\left(a_n \cos\frac{n\pi x}{L} + b_n \sin\frac{n\pi x}{L}\right)$$

The Fourier series expression without cosine terms becomes

$$f(x) = \frac{1}{2}a_0 + \sum_{n=1}^{\infty} b_n \sin\frac{n\pi x}{L}$$

Our function can be expressed as $f(x) = x$ with period equivalent to $2\pi$; therefore $L = \pi$

$$a_0 = \int_{-\pi}^{\pi} f(x)dx = \int_{-\pi}^{0} -1dx + \int_{0}^{\pi} 1dx$$

$$a_0 = 0$$

Next we can find the Fourier coefficient $b_n$ for the sine term.

$$b_n = \frac{1}{\pi}\int_{-\pi}^{\pi} f(x)\sin\frac{n\pi x}{\pi}dx = \frac{1}{\pi}\int_{-\pi}^{0} -\sin nxdx + \frac{1}{\pi}\int_{0}^{\pi}\sin nxdx$$

$$= \frac{1}{\pi}\left[\frac{\cos nx}{n}\right]_{-\pi}^{0} + \frac{1}{\pi}\left[\frac{-\cos nx}{n}\right]_{0}^{\pi} = \frac{2}{n\pi}(1 - \cos n\pi)$$

We can observe that when $n$ is even, $cos\ n\pi = 1$, then $b_n = 0$. So $b_n$ is only applicable when $n$ is odd, $cos\ n\pi = -1$, and

$$b_n = \frac{4}{n\pi}$$

Putting the results together we have the following, where we have created the term $(2m - 1)$ to represent odd numbers.

$$f(x) = \sum_{m=1}^{\infty}\frac{4}{(2m-1)\pi}\sin\left[(2m-1)x\right]$$

(b) Note that the function is neither even nor odd. But we can use our result in part a to arrive at the Fourier series equivalent for the function in part b, via some simple graphical transformations. We will first vertically scale the graph from an earlier amplitude of $\pm 1$ to $\pm 0.5$. This is shown below as we go from the orange to black line. In terms of the equation for the function, this means going from $f_1(x)$ to $f_2(x)$.

$$f_1(x) = \sum_{m=1}^{\infty}\frac{4}{(2m-1)\pi}\sin\left[(2m-1)x\right]$$

$$f_2(x) = \frac{1}{2}\sum_{m=1}^{\infty}\frac{4}{(2m-1)\pi}\sin\left[(2m-1)x\right]$$

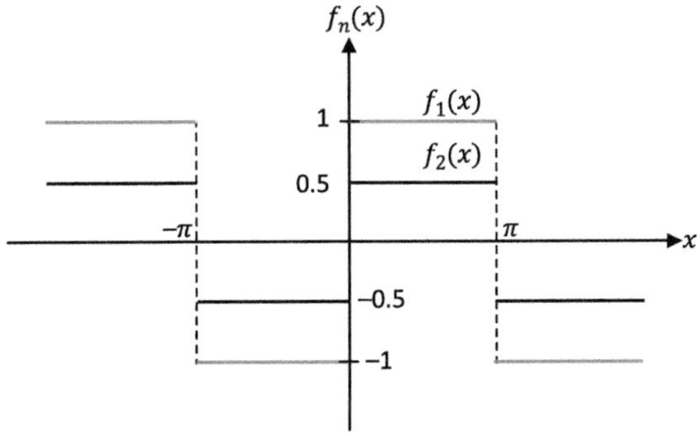

Next, we move the graph up by 0.5 units from the black to blue line. This means going from $f_2(x)$ to $f_3(x)$.

$$f_2(x) = \frac{1}{2}\sum_{m=1}^{\infty} \frac{4}{(2m-1)\pi} \sin[(2m-1)x]$$

$$f_3(x) = \frac{1}{2}\sum_{m=1}^{\infty} \frac{4}{(2m-1)\pi} \sin[(2m-1)x] + \frac{1}{2}$$

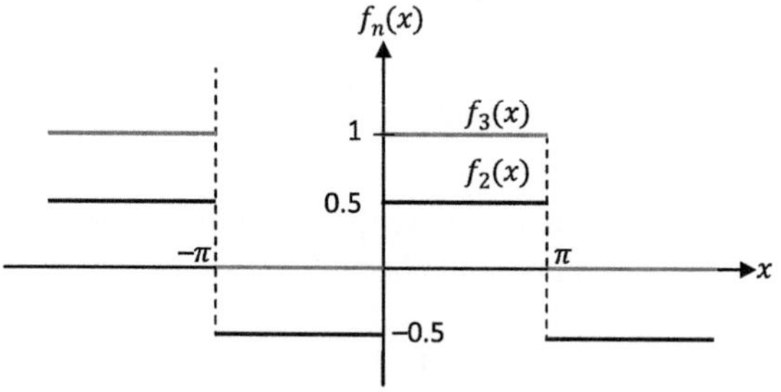


Finally, we scale the $x$-variable by a factor of $1/\pi$, going from the blue to red line. This means $x = \pi z$ as we go from $f_3(x)$ to $f_4(z)$.

$$f_3(x) = \frac{1}{2}\sum_{m=1}^{\infty} \frac{4}{(2m-1)\pi} \sin[(2m-1)x] + \frac{1}{2}$$

$$f_4(z) = \frac{1}{2}\sum_{m=1}^{\infty} \frac{4}{(2m-1)\pi} \sin[(2m-1)\pi z] + \frac{1}{2}$$

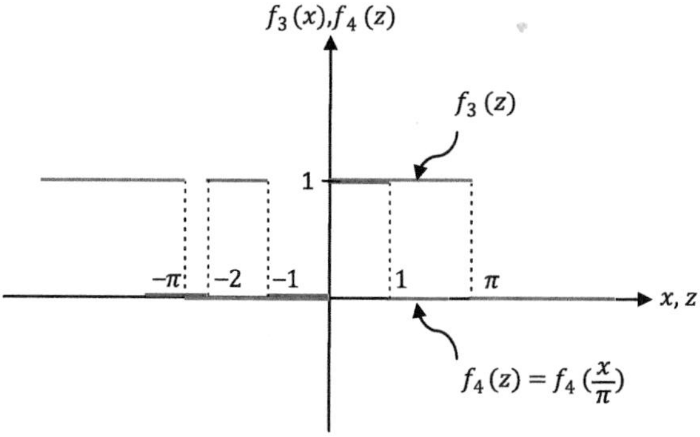


## Statistics

**Statistics is an increasingly important area of study, especially applicable in the field of engineering as it enables the analysis of collected data and measurements. Statistical methods help make sense of numerical data through**

**comparisons and identification of patterns, which facilitate the interpretation and drawing of useful conclusions that apply to a cause. In a pilot plant or production facility for example, the ability to draw relationships between inputs and outputs will allow engineers to design and operate processes better.**

## Problem 41

**A machine was considered to be in good operating condition if it produced metal rods with a mean length of 20.00 cm. The standard deviation of the rod length was stated as 0.8 cm.**

(a) **Assuming a sample of 60 rods had an average length of 20.15 cm, conduct a hypothesis test to determine if the machine can be considered to be in good operating condition. You may assume a significance level of 5%.**
(b) **Discuss why errors may occur in statistical results and explain the difference between a Type I and Type II error using a simple example.**
(c) **If in reality, the production equipment produced rods with a mean length of 20.31 cm, what is the probability of a Type II error occurring? Assume that the Type I error probability is still 5%.**

## Solution 41

*Worked Solution*

(a) We need to conduct a two-tailed hypothesis test with the null hypothesis $H_0$ and alternative hypothesis $H_1$ as follows:

$$H_0 : \mu = 20.00 \text{ cm}$$

$$H_1 : \mu \neq 20.00 \text{ cm}$$

Next we can calculate the test statistic, $z_{calc}$

$$z_{calc} = \frac{20.15 - 20.00}{0.8/\sqrt{60}} \approx 1.452$$

Applying a two-tailed test with a significance level, $\alpha = 0.05$, we can determine the required value of $\Phi$ from statistical tables (refer to normal distribution function in data booklet),

$$\Phi(z) = 1 - \frac{\alpha}{2} = 1 - 0.025 = 0.975 \rightarrow z_{crit} = 1.96$$

Since $z_{calc} < z_{crit}$, we accept the null hypothesis $H_0$ and conclude that the machine is working well as desired.

(b) Statistics do not always accurately reflect reality since we are only taking measurements of a sample size out of an entire population of data. It therefore follows that the decisions we make based on statistical analysis of a sample set of data can therefore either be right (i.e., appropriately reflects reality) or turn out to be an error.

The above can be illustrated using a simple example of a two-tailed test on a normal distribution with mean value $\mu$ as follows. Over here, we establish a null hypothesis $H_0$ and an alternative hypothesis $H_1$, and $k$ is a constant.

$$H_0 : \mu = k$$
$$H_1 : \mu \neq k$$

In reality, the null hypothesis can be true or false. Separately, our decision can be to accept or reject the null hypothesis. In this way, there are four possible decisions, whereby two would be errors.

1. We accept the null hypothesis, and in reality the null hypothesis is true. $\rightarrow$ **Right** decision
2. We reject the null hypothesis, and in reality the null hypothesis is false $\rightarrow$ **Right** decision
3. We accept the null hypothesis, and in reality the null hypothesis is false $\rightarrow$ **Type II Error**
4. We reject the null hypothesis, and in reality the null hypothesis is true $\rightarrow$ **Type I Error**

(c) A Type II error occurs when we accept the null hypothesis, and in reality the null hypothesis is false. For this error to occur, the average rod length, denoted here as $\bar{x}$, will fall within the range as shown.

$$\begin{aligned}\bar{x} &\in \left[20.00 \pm z_{\text{crit}}\left(\frac{0.8}{\sqrt{60}}\right)\right] \\ &= \left[20.00 \pm 1.96\left(\frac{0.8}{\sqrt{60}}\right)\right] \\ &\approx [19.798, 20.202]\end{aligned}$$

The test statistic $z_{\text{calc}}$ can be calculated as follows, and we can determine the required value of $\Phi$ from statistical tables (refer to normal distribution function in data booklet)

$$z_{\text{calc}} = \frac{|20.202 - 20.31|}{0.8/\sqrt{60}} = 1.05$$
$$F(z_{\text{calc}}) = F(1.05) \approx 0.853$$
$$\beta = 1 - 0.853 = 0.147 \approx 15\%$$

The probability of a Type II error occurring (i.e., accept $H_0$ even though $H_0$ is false and $H_1$ is true) is therefore 15%.

## Problem 42

**A fertilizer company has just launched a new product and claims that it is superior to their previous product in terms of helping farmers achieve greater agricultural yields. 11 farmers volunteered on a random basis to try out the new fertilizer product on 0.5 hectare of land, and the yield from this plot was compared to another 0.5 hectare of land that used the previous fertilizer product. Data on agricultural yields for both plots were collected as shown below. It may be assumed that the two populations of data (new and previous fertilizer) are independent.**

| Farmer | Yield from new fertilizer | Yield from previous fertilizer |
|---|---|---|
| 1 | 33 | 26 |
| 2 | 43 | 23 |
| 3 | 31 | 43 |
| 4 | 30 | 38 |
| 5 | 47 | 20 |
| 6 | 36 | 21 |
| 7 | 33 | 38 |
| 8 | 40 | 31 |
| 9 | 42 | 26 |
| 10 | 66 | 69 |
| 11 | 22 | 37 |

**The owner of the fertilizer company wants to decide if the new fertilizer improves yield using the data collected, and he wants to be 95% sure that his decision is right.**

(a) **Describe the steps in performing a hypothesis test in the context of data collected for the two fertilizer products, and comment on the difference between a parametric and a non-parametric test.**
(b) **Develop a parametric test to decide if the yield has improved using the new fertilizer and confirm whether this improvement can be claimed at 95% level of confidence.**

## Solution 42

*Worked Solution*

(a) A statistical test consists of the following key steps:

i. Determine the parameter of interest
ii. List down assumptions (if any) on the population

iii. Establish the null hypothesis $H_0$, and the alternative hypothesis $H_1$
iv. Define a level of significance
v. State a test statistic
vi. Identify the rejection region
vii. Calculate derived quantities from the sample data
viii. From the results of step vii, decide if the null hypothesis should be accepted or rejected.

A parametric test is relevant when the population can be appropriately assumed to be normally distributed. Conversely, a non-parametric test does not have this assumption. Non-parametric tests are often suited for smaller sample sizes, or when the population deviates significantly enough from a normal distribution.

(b) Let us denote the data set for the new and previous fertilizers by X and Y, respectively. Before we perform a parametric test, we should make clear our assumptions as follows:

- X and Y are independent populations
- X and Y are normally distributed with mean values that we can denote as $\mu_1$ and $\mu_2$, respectively.

We will now establish our null and alternative hypothesis. In this test, we want to know if $\mu_1 > \mu_2$ since we want to determine if the new fertilizer improves yield. We may then state the null hypothesis as the case where the new fertilizer is just as good as the previous fertilizer, while the alternative hypothesis is such that the new fertilizer is better than the previous fertilizer at improving yields. The null hypothesis $H_0$ and alternative hypothesis $H_1$ are therefore stated as follows:

$$H_0 : \mu_1 = \mu_2$$

$$H_1 : \mu_1 > \mu_2$$

Next, we note that we want to have 95% level of confidence, which translates to a 5% level of significance α.

$$\alpha = 1 - 0.95 = 0.05$$

We will now determine the appropriate test statistic. In this problem, we have two independent populations with their respective means, and we wish to compare the two mean values. We note further that the variance (or standard deviation) for the two populations are unknown, and the sample size is relatively small. Hence we use the t-distribution and the corresponding test function T as follows, where $\bar{X}$ and $\bar{Y}$ are the empirical means for the new and previous fertilizers, respectively, $S_1$ and $S_2$ are the respective standard deviations, and $n_1$ and $n_2$ are the respective sample sizes:

$$T = \frac{(\bar{X} - \bar{Y}) - (\mu_1 - \mu_2)}{\sqrt{\frac{S_1^2}{n_1} + \frac{S_2^2}{n_2}}}$$

We next determine the degrees of freedom, $v$.

$$n_1 = n_2 = 11 = n$$

$$v = n - 1 = 11 - 1 = 10$$

At $\alpha = 0.05$, $v = 10$, we can refer to the t-distribution in the data booklet to obtain the test statistic $t_{0.95}$. Note that since we are performing a one-tailed test, we need to be conscious that the statistical tables provided in data booklets tabulate t-values for two-tailed tests; hence, for a 5% significance level, we need to refer to the corresponding P-value where P[%] = 5 × 2 = 10. The t-distribution is symmetrical such that $P(-t_{\alpha/2} \leq T \leq t_{\alpha/2}) = P(T \leq t_\alpha)$

$$t_{0.95} = 1.81$$

The rejection region is given by $t > t_{0.95}$ or $t > 1.81$, where $t = \frac{\bar{X} - \bar{Y}}{\sqrt{\frac{S_1^2}{n_1} + \frac{S_2^2}{n_2}}}$.

From the data provided, we can determine some derived quantities from the sample as shown below.

$$\bar{x} = \frac{33 + 43 + \ldots + 22}{11} \approx 38.455$$

$$\bar{y} = \frac{26 + 23 + \ldots + 37}{11} \approx 33.818$$

In this case, since the sample size is small, it is preferred to use an unbiased estimator for variance, i.e., we use n − 1 instead of n to calculate the unbiased sample variance $S^2$ (or empirical variance in table below). n − 1 is also the number of degrees of freedom. Note that as the sample size increases, the amount of bias decreases, and the value of n − 1 becomes closer to n.

$$S^2 = \frac{1}{n-1} \sum_{i=1}^{n} (x_i - \bar{x})^2$$

$$S_1^2 = \frac{(33 - 38.455)^2 + (43 - 38.455)^2 + \ldots + (22 - 38.455)^2}{11 - 1}$$

$$S_2^2 = \frac{(26 - 33.818)^2 + (23 - 33.818)^2 + \ldots + (37 - 33.818)^2}{11 - 1}$$

| Sample | $n$ | Empirical mean $\bar{X}$, $\bar{Y}$ | Empirical standard deviation $S_1$, $S_2$ | Empirical variance $S_1^2$, $S_2^2$ |
|---|---|---|---|---|
| $X$ | 11 | 38.455 | 11.536 | 133.073 |
| $Y$ | 11 | 33.818 | 14.034 | 196.964 |

From the table above, we can calculate the test statistic

$$t = \frac{\bar{X} - \bar{Y}}{\sqrt{\frac{S_1^2}{n_1} + \frac{S_2^2}{n_2}}} = \frac{38.455 - 33.818}{\sqrt{\frac{133.073+196.964}{11}}} = 0.847$$

Recall that the rejection region is given by $t > t_{0.95} = 1.81$. Since $0.847 < 1.81$, $t < t_{0.95}$ and we conclude that the null hypothesis $H_0$ is accepted, since the test statistic does not fall within the rejection region.

## Problem 43

**A particular production process is required to produce circular tubes whereby the radius of the circular cross section is 12.00 cm and the standard deviation is 0.1 cm. Assuming a sample size of 60 tubes with an average radius of 12.02 cm,**

(a) **State the null and alternative hypotheses.**
(b) **Determine if the production requirement is met for the said sample, using a 5% level of significance.**
(c) **Discuss briefly what is meant by Type I and Type II errors. If the average radius of the population from which the sample was taken was 12.01 cm, calculate the probability of a Type II error occurring in part b.**
(d) **If the average radius of the population is 12.01 cm (as per part c), what is the desired sample size in order to reduce the Type II error to half of the value calculated in part c?**

## Solution 43

*Worked Solution*

(a) The null hypothesis $H_0$ and alternative hypothesis $H_1$ are as follows:

$$H_0 : \mu = 12.00 \text{ cm}$$

$$H_1 : \mu \neq 12.00 \text{ cm}$$

(b) First, we have to find out the test statistic, $z_{\text{calc}}$

$$z_{\text{calc}} = \frac{12.02 - 12.00}{0.1/\sqrt{60}} = 1.549$$

Applying a two-tailed test with a significance level, $\alpha = 0.05$, we can determine the required value of $\Phi$ from statistical tables (refer to normal distribution function in data booklet),

$$\Phi(z) = 1 - \frac{\alpha}{2} = 1 - 0.025 = 0.975 \rightarrow z_{crit} = 1.96$$

Since $z_{calc} < z_{crit}$, we accept the null hypothesis $H_0$ and conclude that the production requirement is met.

(c) A Type I error occurs when we decide to accept the alternative hypothesis (or reject the null hypothesis) even though in reality the null hypothesis is true. On the other hand, a Type II error occurs when we decide to accept the null hypothesis (or reject the alternative hypothesis) even though in reality the null hypothesis is false.

For a Type II error to occur in part b, we accept the null hypothesis even though it is false. For this error to occur, the average radius, denoted here as $\bar{x}$, will fall within the range as shown.

$$\bar{x} \in \left[12.00 \pm z_{crit}\left(\frac{0.1}{\sqrt{60}}\right)\right]$$
$$= \left[12.00 \pm 1.96\left(\frac{0.1}{\sqrt{60}}\right)\right]$$
$$\approx [11.975, 12.025]$$

The probability of a Type II error occurring in part b is therefore 0.87.

$$\beta = \Phi\left(\frac{12.025 - 12.01}{0.1/\sqrt{60}}\right) \approx \Phi(1.162) \approx 0.87$$

(d) In order to reduce the Type II error by half, we need a new value of $\beta_1 = \frac{\beta}{2} = \frac{0.87}{2} \approx 0.43$. Working backwards, we can determine the required sample size $n$ that satisfies the required test statistic $z_1$.

$$\beta_1 = 0.43 = \Phi\left(\frac{\left(12.00 + 1.96\left(\frac{0.1}{\sqrt{n}}\right)\right) - 12.01}{0.1/\sqrt{n}}\right) = \Phi(z_1)$$

$$z_1 = 1.96 - \left(\frac{0.01}{0.1}\right)\sqrt{n}$$

$$\sqrt{n} = -10(z_1 - 1.96)$$

From statistical tables, we can determine the value of $z_1$ at $\Phi(z_1) = 0.43$. Note that since $0.43 < 0.50$, and the statistical table values for $\Phi(z)$ starts from 0.500 onwards, hence, we first calculate

$$1 - 0.43 = 0.57$$

At $\Phi(z) = 0.57$, $z = 0.18$. Therefore, it can be deduced that $z_1 = -0.18$. Putting this value back into the earlier expression, we have

$$\sqrt{n} = -10(-0.18 - 1.96)$$

$$n = 458$$

## Problem 44

**Sarah manages a greenhouse and periodically measures the heights of her 20 cucumber plants. After a period of growth, the mean height of a sample was found to be 75 cm and the standard deviation of height for this sample is 5.2 cm. Using a confidence level of 95% for the mean height, and noting that a random sample of 20 was taken, determine the confidence interval. You may assume that plant height is normally distributed in the population.**

## Solution 44

### *Worked Solution*

A confidence interval represents a range of values that contains a population parameter. If we draw random samples from a population multiple times, a certain percentage of the confidence interval will contain the population mean. This percentage is also known as the confidence level. The confidence level is also equal to $1 - \alpha$, where $\alpha$ is the corresponding significance level.

In this example, there is no information given about whether the population mean $\mu$ is greater or less than the empirical mean; therefore, we conduct a two-tailed test (as opposed to one-tailed).

We have a sample whereby the size $n = 20$. The mean height derived from the random sample taken is also known as the empirical mean, $\bar{x}$. The test statistic $z$ can be calculated as follows, where $\sigma$ refers to standard deviation and $\mu$ refers to population mean.

$$z = \frac{|\bar{x} - \mu|}{\sigma/\sqrt{n}}$$

At 95% confidence level, $\alpha = 0.05$ and $z_{1-\alpha/2} = z_{0.975} = 1.96$. Substituting into the equation, we have

$$1.96 = \frac{|75 - \mu|}{5.2/\sqrt{20}}$$

$$|75 - \mu| \approx 2.3$$

The interval is therefore $75 \pm 2.3$.

## Problem 45

**A company produced ropes that had to comply with regulatory guidelines on breaking strength. The breaking strength of 8 ropes manufactured by this company was found to have a mean value of 7400N and a standard deviation of 130N after performing some tests. However, the company owner claimed that the mean breaking strength of their ropes is 7800N. Using significance levels of 5 and 1%, determine if the owner's claim is true.**

## Solution 45

***Worked Solution***

We can first establish the null hypothesis $H_0$ and alternative hypothesis $H_1$ where the null hypothesis represents the company owner's claim and the alternative hypothesis represents the case whereby the breaking strength has been over-represented:

$$H_0 : \mu = 7800 \text{ N}$$

$$H_1 : \mu < 7800 \text{ N}$$

We now observe that the sample size is relatively small at $n = 8$; therefore, we perform a one-tailed test on a $t$-distribution to better represent this population. The $t$-distribution describes the frequency distribution of small samples taken from a normal distribution. Therefore, it follows that as we increase our sample size, the $t$-distribution will become more and more like a normal distribution.

The test statistic $t_{\text{crit}}$ may be calculated as follows, where $S$ is the standard deviation, $n$ is the sample size, $\bar{x}$ is the empirical mean obtained from the random sample taken, and $\mu$ is the population mean:

$$t_{\text{crit}} = \frac{|\bar{x} - \mu|}{S/\sqrt{n}} = \frac{|7400 - 7800|}{130/\sqrt{8}} = 8.7$$

As we are performing a one-tailed test, we need to consider that the statistical tables provided in data booklets tabulate $t$-values for two-tailed tests, hence for a 5% significance level, we need to refer to the corresponding $P$-value where $P[\%] = 5 \times 2 = 10$. In the same way, at 1% significance level, we determine $P$ to be $P[\%] = 2$.

For a $t$-distribution, we also need to determine the degrees of freedom $v$ which is defined as $v = n - 1$. In this case,

$$v = 8 - 1 = 7$$

The corresponding $t$-values can be found from statistical tables as follows:

$$t_{5\%,7} = 1.89$$

$$t_{1\%,7} = 3.00$$

Since $t_{crit} > t_{5\%,7}$ and $t_{crit} > t_{1\%,7}$, the null hypothesis $H_0$ is rejected at both significance levels, and we conclude that the owner's claim is not supported.

## Problem 46

**(a) Consider a normally distributed population with a known mean $\mu_0$ and unknown variance $\sigma^2$, from which we obtain a set of $n$ independent and identical random variables $X_i$ where $i$ takes values from 1 to $n$, i.e., $X_1$, $X_2, \ldots, X_n$. For a significance level [%] in the range $0 \leq \alpha \leq 1$, derive an expression for the confidence interval for the variance $\sigma^2$ and specify its corresponding confidence level.**

**(b) Data in the following table ($x_1, x_2, \ldots, x_9$) was collected after some measurements were taken. Assuming a normal distribution of the form $\mathcal{N}(3.5, \sigma^2)$, find the empirical confidence interval at 98% for the variance based on the measurements obtained.**

| $j$ | 1 | 2 | 3 | 4 | 5 | 6 | 7 | 8 | 9 |
|---|---|---|---|---|---|---|---|---|---|
| $x_j$ | 4.14 | 3.91 | 1.39 | 2.55 | 1.57 | 3.35 | 4.54 | 2.37 | 4.89 |

**(c) Comment on what it means to do a point estimation as opposed to an interval estimation in statistical analysis.**

## Solution 46

### *Worked Solution*

(a) This problem introduces the concept of a chi-squared distribution, often denoted as $\chi_n^2$ at $n$ degrees of freedom. This is a useful distribution to note in inferential statistics and is often applied in hypothesis testing, determining confidence intervals and estimating parameters (e.g., variance) of a normally distributed population. Assume we have independent and identical random variables denoted by $X_i$ which follow the normal distribution of known mean $\mu_0$ but unknown variance $\sigma^2$ as given in the problem, i.e., $\mathcal{N}(\mu_0, \sigma^2)$, then the random variable $Z$ as shown below will follow the chi-squared distribution of $\chi_n^2$.

$$Z \sim \chi_n^{\ 2}$$

$$Z = \sum_{i=1}^{n} \left( \frac{X_i - \mu_0}{\sigma} \right)^2$$

We can now express the confidence level probability $P$, in relation to the significance level $\alpha$ as follows

$$1 - \alpha = P\left(\chi^2_{n,\alpha/2} \leq Z \leq \chi^2_{n,1-\alpha/2}\right)$$

$$= P\left( \frac{\sum_{i=1}^{n} (X_i - \mu_0)^2}{\chi^2_{n,1-\alpha/2}} \leq \sigma^2 \leq \frac{\sum_{i=1}^{n} (X_i - \mu_0)^2}{\chi^2_{n,\alpha/2}} \right)$$

The confidence interval for the unknown variance $\sigma^2$ can then be expressed as follows at a confidence level of $1 - \alpha$.

$$\sigma^2 \in \left[ \frac{\sum_{i=1}^{n} (X_i - \mu_0)^2}{\chi^2_{n,1-\alpha/2}}, \frac{\sum_{i=1}^{n} (X_i - \mu_0)^2}{\chi^2_{n,\alpha/2}} \right]$$

(b) We need to refer to the chi-squared statistical table which can be found in data booklets to determine the values we need for our confidence interval at 98% and degrees of freedom $n = 9$. Note that we need to consider that the statistical table tabulates values for two-tailed tests, hence for a significance level $\alpha[\%] = 2$, we need to refer to the corresponding $P$-value at $\frac{\alpha}{2} = 1\%$ which means a $P =$ value of 99%.

$$\chi^2_{n,1-\alpha/2} = \chi^2_{9,99}(P = 1) = 21.67$$

$$\chi^2_{n,\alpha/2} = \chi^2_{9,1}(P = 99) = 2.09$$

From the expression for confidence interval as earlier determined in part a, we can compute the upper and lower bounds of the confidence interval from the table of values provided.

$$\left[ \frac{\sum_{i=1}^{n} (X_i - \mu_0)^2}{\chi^2_{n,1-\alpha/2}}, \frac{\sum_{i=1}^{n} (X_i - \mu_0)^2}{\chi^2_{n,\alpha/2}} \right]$$

$$= \left[ \frac{(X_1 - 3.5)^2 + (X_2 - 3.5)^2 + \ldots + (X_9 - 3.5)^2}{21.67}, \frac{(X_1 - 3.5)^2 + (X_2 - 3.5)^2 + \ldots + (X_9 - 3.5)^2}{2.09} \right]$$

$$= \left[\frac{(4.19-3.5)^2+(3.91-3.5)^2+\ldots+(4.89-3.5)^2}{21.67}, \frac{(4.19-3.5)^2+(3.91-3.5)^2+\ldots+(4.89-3.5)^2}{2.09}\right]$$

$$= [0.648, 6.72]$$

The empirical confidence interval is therefore [0.648, 6.72].

(c) A point estimation is used to estimate unknown parameters of a population. Point estimation methods can be parametric or non-parametric. For parametric methods, a probability model is assumed for the population and hence the quantity to be estimated, denoted by $\vartheta$ for example, is also assumed to follow a specific distribution (e.g., normal distribution). Conversely, non-parametric techniques do not have this assumption. In general for point estimation, we define a test statistic $T$ which is a function of a sample (e.g., of sample size n) of random variables $X_1, X_2, \ldots, X_n$.

Interval estimation is used when we want to find out how accurate a certain point estimation is at a defined confidence level. The underlying probability model used for the point estimation is then assumed in determining a confidence interval for the parameter $\vartheta$. The confidence level denoted by $p$ is related to the significance level $\alpha$, by $p = 1 - \alpha$. A confidence level of 95% would correspond to a 5% significance level, whereby $p = 0.95$, $\alpha = 0.05$. Within the lower and upper bound values of a 95% confidence interval, the parameter $\vartheta$ can be found with a certainty of 95%.

## Problem 47

**Consider two populations denoted by $X$ and $Y$ which are normal distributions with known variances $\sigma_X^2 = 2.6$ and $\sigma_Y^2 = 2.8$. In order to test if the population means are the same, a sample was taken from each population. The sample size from population $X$ was $n_X = 7$ and the empirical sample mean $\bar{x} = 19.5$. The sample size from population $Y$ was $n_Y = 9$ and the empirical sample mean $\bar{y} = 16.5$.**

**Determine and compute the suitable test statistic and decide at 99% confidence level if we should accept or reject the hypothesis that the populations means for the two populations are the same.**

## Solution 47

### *Worked Solution*

We first identify the parameter of interest, which in this problem is the population mean. The null hypothesis $H_0$ and alternative hypothesis $H_1$ can therefore be stated as shown below.

$$H_0 : \mu_X = \mu_Y$$

$$H_1 : \mu_X \neq \mu_Y$$

Notice that this is a two-tailed test since no information is given about whether the mean of population $X$ is greater or less than the mean of population $Y$. We can now state a suitable test statistic |z| for this problem as follows,

$$|z| = \frac{|\bar{x} - \bar{y}|}{\sqrt{\frac{\sigma_X^2}{n_X} + \frac{\sigma_Y^2}{n_Y}}}$$

Given that the confidence level is 99%, i.e., significance level $\alpha = 1\%$, we can establish the rejection region as follows:

$$\frac{|\bar{x} - \bar{y}|}{\sqrt{\frac{\sigma_X^2}{n_X} + \frac{\sigma_Y^2}{n_Y}}} > \left|z_{1-\alpha/2}\right|$$

$$\frac{|\bar{x} - \bar{y}|}{\sqrt{\frac{\sigma_X^2}{n_X} + \frac{\sigma_Y^2}{n_Y}}} > |z_{99.5}|$$

From the statistical table for normal distribution which can be found in data booklets, we note that at $\alpha = 1\%$, $z_{99.5} \approx 2.58$. Therefore, we can compute the test statistic using sample data as follows:

$$|z| = \frac{|19.5 - 16.5|}{\sqrt{\frac{2.6}{7} + \frac{2.8}{9}}} = 3.6$$

$$3.6 > 2.58$$

$$|z| > |z_{99.5}|$$

Since the value of the test statistic falls within the rejection region, we conclude that null hypothesis is rejected and conclude from the test that the population means for the two populations are not the same at a 1% significance level.

## Eigenfunctions and Eigenvalues

**The concept of eigenfunctions and eigenvalues is widely applied in science and engineering, and is thus unsurprisingly a common topic in engineering mathematics. The use of eigenvalues (as well as eigenvectors) is useful in making sense of large systems which can be mathematically described by large sets of equations (e.g., system of differential equations). This application in linear algebra enables engineers to analyze the stability linear systems, and hence design effective control methods.**

## Problem 48

**The canonical form of matrices is useful in the geometric interpretation of functions. Given a 2 × 2 symmetric matrix $Q = \begin{bmatrix} 6 & 2 \\ 2 & -3 \end{bmatrix}$,**

**(a) Derive the canonical form for matrix $Q$ as shown below, whereby $R$ is a 2 × 2 matrix containing unit eigenvectors ($e_1$ and $e_2$) of matrix $Q$.**

$$Q = R\Lambda R^T, \text{whereby } R = [e_1 \ e_2]$$

**(b) Discuss the implications of eigenvalues $\lambda_i$ on the geometric properties of the function below assuming that eigenvalues are positive non-zero values, $\lambda_i > 0$.**

$$q(x) = x^T Q x = 1$$

**The standard form of an ellipse is also given as follows, where the midpoint is at origin, and assuming that $a > b$, then the horizontal major axis lies along the $x$-axis and the minor axis is the vertical $y$-axis.**

$$\frac{x^2}{a^2} + \frac{y^2}{b^2} = 1$$

**(c) Comment also on the other possible values that eigenvalues can take.**

## Solution 48

*Worked Solution*

(a) A symmetric matrix $Q$ is one whereby it is equivalent to its transpose.

$$Q = Q^T$$

Let us consider an example of a symmetric 2 × 2 square matrix as follows:

$$Q = \begin{bmatrix} 6 & 2 \\ 2 & -3 \end{bmatrix} = Q^T$$

The eigenvalue (a scalar quantity) $\lambda$ of a symmetric matrix $Q$ is related to the 2 × 1 eigenvector $\mathbf{e}$ as follows where $\mathbf{e} = \begin{bmatrix} e_a \\ e_b \end{bmatrix}$ and $\mathbf{I} = \begin{bmatrix} 1 & 0 \\ 0 & 1 \end{bmatrix}$ is an identity matrix.

$$\boldsymbol{Q}\mathbf{e} = \lambda\mathbf{e} \quad (1)$$

$$\boldsymbol{Q}\mathbf{e} - \lambda\mathbf{e} = 0$$

$$(\boldsymbol{Q} - \lambda\boldsymbol{I})\mathbf{e} = 0 \quad (2)$$

In order for Eq. (2) to have meaningful solutions for **e**, the condition for the determinant of matrix ($\boldsymbol{Q} - \lambda\boldsymbol{I}$) has to be satisfied as follows:

$$\det(\boldsymbol{Q} - \lambda\boldsymbol{I}) = 0$$

Substituting values for $\boldsymbol{Q}$,

$$\boldsymbol{Q} = \begin{bmatrix} 6 & 2 \\ 2 & -3 \end{bmatrix}$$

$$\boldsymbol{Q} - \lambda\boldsymbol{I} = \begin{bmatrix} 6 & 2 \\ 2 & -3 \end{bmatrix} - \begin{bmatrix} \lambda & 0 \\ 0 & \lambda \end{bmatrix} = \begin{bmatrix} 6-\lambda & 2 \\ 2 & -3-\lambda \end{bmatrix}$$

$$\det(\boldsymbol{Q} - \lambda\boldsymbol{I}) = |\boldsymbol{Q} - \lambda\boldsymbol{I}| = 0$$

$$(6-\lambda)(-3-\lambda) - (2)(2) = 0$$

$$\lambda^2 - 3\lambda - 22 = 0$$

$$\lambda = \frac{3 \pm \sqrt{9 - 4(-22)}}{2}$$

$$\lambda_1 = 6.4 \text{ and } \lambda_2 = -3.4$$

Solving the matrix equation yields two real roots for the eigenvalues of our matrix. Note that the eigenvalues of a symmetric and real matrix will also be real. Substituting the eigenvalues back into our matrix Eq. (1), we arrive at eigenvectors ($\mathbf{e}_1, \mathbf{e}_2$) that belong to the eigenvalues ($\lambda_1, \lambda_2$) as follows where the equations are valid.

$$\boldsymbol{Q}\mathbf{e} = \lambda\mathbf{e}$$

$$\boldsymbol{Q}\mathbf{e}_1 = \lambda_1\mathbf{e}_1 \text{ and } \boldsymbol{Q}\mathbf{e}_2 = \lambda_2\mathbf{e}_2 \quad (3)$$

By definition, the eigenvectors ($\mathbf{e}_1, \mathbf{e}_2$) belonging to the eigenvalues ($\lambda_1, \lambda_2$) are orthogonal. This means that their dot product is equal to zero.

$$\mathbf{e}_1\mathbf{e}_2 = 0 \quad (4)$$

We are told that the eigenvectors are unit vectors (which is usually the case); therefore, the following is also true about the eigenvectors.

$$\mathbf{e}_1\mathbf{e}_1 = \mathbf{e}_2\mathbf{e}_2 = 1 \tag{5}$$

We can now define the 2 × 2 matrix $\boldsymbol{R}$ containing eigenvectors of matrix $\boldsymbol{Q}$ where $\mathbf{e_1}$ and $\mathbf{e_2}$ are column vectors in the expression written below. In our specific example, as we have

$$\boldsymbol{R} = [\mathbf{e}_1 \quad \mathbf{e}_2]$$

Recall our results from (3), we substitute the above into our matrix equation to obtain

$$\boldsymbol{QR} = \boldsymbol{Q}[\mathbf{e}_1 \quad \mathbf{e}_2]$$

$$\boldsymbol{QR} = [\lambda_1\mathbf{e}_1 \quad \lambda_2\mathbf{e}_2]$$

After multiplying the left-hand side of both sides of this equation by $\boldsymbol{R}^T$, we have

$$\boldsymbol{R}^T\boldsymbol{QR} = \boldsymbol{R}^T[\lambda_1\mathbf{e}_1 \quad \lambda_2\mathbf{e}_2] = \begin{bmatrix} \mathbf{e}_1 \\ \mathbf{e}_2 \end{bmatrix}[\lambda_1\mathbf{e}_1 \quad \lambda_2\mathbf{e}_2]$$

Using results from (4) and (5), we simplify the expression to arrive at a matrix identity that represents the diagonalization of our earlier matrix $\boldsymbol{Q}$. This new matrix identity which we denote here as $\boldsymbol{\Lambda}$ contains eigenvalues along its diagonal and zeroes everywhere else.

$$\boldsymbol{R}^T\boldsymbol{QR} = \begin{bmatrix} \lambda_1 & 0 \\ 0 & \lambda_2 \end{bmatrix} = \boldsymbol{\Lambda}$$

It can be observed from earlier results (4) and (5) that matrix $\boldsymbol{X}$ is orthogonal. And it follows that for an orthogonal matrix, $\boldsymbol{R}^{-1} = \boldsymbol{R}^T$ and $\boldsymbol{RR}^T = \boldsymbol{R}^T\boldsymbol{R} = \mathbf{I}$. Substituting $\boldsymbol{R}^T$ with $\boldsymbol{R}^{-1}$, then multiplying $\boldsymbol{R}$ on the left-hand side of both sides of the equation, and finally multiplying $\boldsymbol{R}^{-1}$ on the right-hand side of both sides of the equation, we can single out matrix $\boldsymbol{Q}$ in its canonical form.

$$\mathbf{R^{-1}QR = \Lambda}$$

$$\mathbf{RR^{-1}QR = R\Lambda}$$

$$\mathbf{QR = R\Lambda}$$

$$\mathbf{QRR^{-1} = R\Lambda R^{-1}}$$

$$\mathbf{Q = R\Lambda R^{-1} = R\Lambda R^{T}}$$

The canonical form is useful in interpreting the geometric implications of eigenvalues.

(b) Let us consider the function which we wish to analyze the geometry for. In this case, we assume eigenvalues are positive and non-zero only ($\lambda_i > 0$):

$$q(x) = \boldsymbol{x}^T \boldsymbol{Q} \boldsymbol{x} = 1$$

whereby

$$x = \begin{bmatrix} x_1 \\ x_2 \end{bmatrix} \quad \text{and} \quad x^T = \begin{bmatrix} x_1 & x_2 \end{bmatrix}$$

Using the canonical form from part a, we substitute for $\boldsymbol{Q}$ using $\boldsymbol{Q} = \boldsymbol{R\Lambda R}^T$

$$1 = \boldsymbol{x}^T \left(\boldsymbol{R\Lambda R}^T\right)\boldsymbol{x}$$

We can perform the following groupings to simplify our expression, $\boldsymbol{x}^T\boldsymbol{R} = \mathbf{x}'^T = \begin{bmatrix} x'_1 & x'_2 \end{bmatrix}$ and $\boldsymbol{R}^T\boldsymbol{x} = \boldsymbol{x}' = \begin{bmatrix} x'_1 \\ x'_2 \end{bmatrix}$

$$1 = \boldsymbol{x}'^T \boldsymbol{\Lambda} \boldsymbol{x}'$$

We know from earlier that $\boldsymbol{\Lambda} = \begin{bmatrix} \lambda_1 & 0 \\ 0 & \lambda_2 \end{bmatrix}$, substituting this result, we get

$$1 = \begin{bmatrix} x'_1 & x'_2 \end{bmatrix} \begin{bmatrix} \lambda_1 & 0 \\ 0 & \lambda_2 \end{bmatrix} \begin{bmatrix} x'_1 \\ x'_2 \end{bmatrix}$$

$$1 = \lambda_1 x'_1 2 + \lambda_2 x'_2 2$$

We can express this equation in the standard form equation for an ellipse, where $z_1$ is the horizontal major axis,

$$1 = \frac{x'_1 2}{\left(\frac{1}{\sqrt{\lambda_1}}\right)^2} + \frac{x'_2 2}{\left(\frac{1}{\sqrt{\lambda_2}}\right)^2}$$

Assuming that $\lambda_1 < \lambda_2$, i.e., $\frac{1}{\sqrt{\lambda_1}} > \frac{1}{\sqrt{\lambda_2}}$, we have the following ellipse, where the eigenvalues relate to key geometric parameters as shown

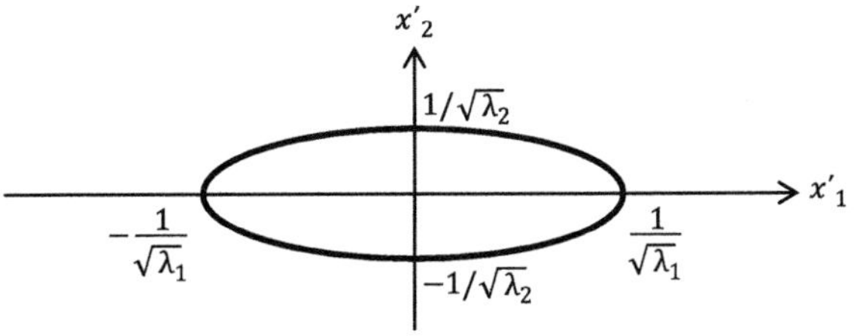

We may observe that the eigenvalue scales the ellipse along the major and minor axes, which are in this case of $\lambda_1 < \lambda_2$, the $x_1'$ and $x_2'$ axes, respectively. Notice also that the eigenvectors are the $x_1'$ and $x_2'$ axes, hence eigenvectors are useful in telling us the directions of the principal axes for the elliptical contours.

(c) In general, there are various possible values that eigenvalues can take and they are useful in giving us some idea about geometric properties:

$$\lambda_i > 0 \rightarrow A \text{ is a positive definite matrix} \rightarrow \text{convex quadratic function}$$
$$\lambda_i < 0 \rightarrow A \text{ is negative definite matrix} \rightarrow \text{concave quadratic function}$$

$$\lambda_i = 0 \rightarrow A \text{ is a singular matrix} \rightarrow \text{linear function}$$
$$\lambda_i > 0 \text{ and } \lambda_i < 0, \rightarrow A \text{ is indefinite}$$

## Problem 49

**Eigenvalues are useful in helping us identify the maximum and minimum points of functions. Given the function below,**

$$\boldsymbol{f(x,y) = 50\left(y - x^2\right)^2 + (4 - x)^2}$$

(a) **Find the gradient and Hessian matrix of the function $f(x, y)$.**
(b) **Calculate all the stationary points for this function and comment if they are global or local and maximum or minimum points.**

## Solution 49

***Worked Solution***

(a)

$$f(x,y) = 50\left(y - x^2\right)^2 + (4 - x)^2$$

We can first derive the gradient expressions.

$$\frac{\partial f}{\partial x} = 2(50)(-2x)\left(y - x^2\right) + 2(-1)(4 - x)$$

$$\frac{\partial f}{\partial y} = 2(50)\left(y - x^2\right)$$

In order to find stationary points, we need to set the gradients to zero.

$$\frac{\partial f}{\partial x} = 0$$

$$2(50)(-2x)\left(y-x^2\right)+2(-1)(4-x)=0$$

$$-200xy+200x^3-8+2x=0 \tag{1}$$

$$\frac{\partial f}{\partial y}=0$$

$$2(50)\left(y-x^2\right)=0$$

$$100y-100x^2=0$$

$$y=x^2 \tag{2}$$

Substitute Eq. (2) into (1)

$$-200x^3+200x^3-8+2x=0$$

$$x=4$$

$$y=16$$

The stationary point is (4, 16).
Now we can compute the Hessian matrix, which is defined as follows

$$\mathcal{H}(x,y)=\begin{bmatrix} \dfrac{\partial^2 f}{\partial x^2} & \dfrac{\partial^2 f}{\partial x \partial y} \\ \dfrac{\partial^2 f}{\partial y \partial x} & \dfrac{\partial^2 f}{\partial y^2} \end{bmatrix}$$

$$\frac{\partial f}{\partial x}=-200xy+200x^3-8+2x$$

$$\frac{\partial^2 f}{\partial x^2}=-200y+600x^2+2$$

$$\frac{\partial^2 f}{\partial x \partial y}=-200x$$

$$\frac{\partial f}{\partial y}=100y-100x^2$$

$$\frac{\partial^2 f}{\partial y^2}=100$$

$$\frac{\partial^2 f}{\partial y \partial x}=-200x$$

$$\mathcal{H} = \begin{bmatrix} -200y + 600x^2 + 2 & -200x \\ -200x & 100 \end{bmatrix}$$

Substituting the stationary point coordinates into the Hessian matrix, we have

$$\mathcal{H} = \begin{bmatrix} -200(16) + 600(4)^2 + 2 & -200(4) \\ -200(4) & 100 \end{bmatrix}$$
$$= \begin{bmatrix} 6402 & -800 \\ -800 & 100 \end{bmatrix}$$

To find the corresponding eigenvalues $\lambda$, we need to set the condition

$$\det[\mathcal{H} - \lambda I] = 0$$
$$\det\begin{bmatrix} 6402 - \lambda & -800 \\ -800 & 100 - \lambda \end{bmatrix} = 0$$
$$(6402 - \lambda)(100 - \lambda) - 800^2 = 0$$
$$640200 - 6402\lambda - 100\lambda + \lambda^2 - 640000 = 0$$
$$\lambda^2 - 6502\lambda + 200 = 0$$
$$\lambda = \frac{6502 \pm \sqrt{6502^2 - 4(200)}}{2}$$
$$\lambda_1 = 6502$$
$$\lambda_2 = 0.0308$$

Both eigenvalues are positive and non-zero, hence the function $f(x, y)$ is locally convex at the stationary point. This stationary point is also the local minima. Since there are no other stationary points, this local minima is also the global minima.

## Problem 50

**Using eigenvalue analysis, find the minimum surface area of a rectangular box that has a volume of eight cubic units. Identify if the minimum value obtained represents a local or global minima. The definition of the Hessian matrix is given below.**

$$\mathcal{H}(x, y) = \begin{bmatrix} \frac{\partial}{\partial x} & \frac{\partial}{\partial y} \end{bmatrix}^T \begin{bmatrix} \frac{\partial f}{\partial x} & \frac{\partial f}{\partial y} \end{bmatrix}$$

## Solution 50

***Worked Solution***

First, we should list down all constraints for the problem. We have an equality constraint for the volume of the box which can be expressed as shown below.

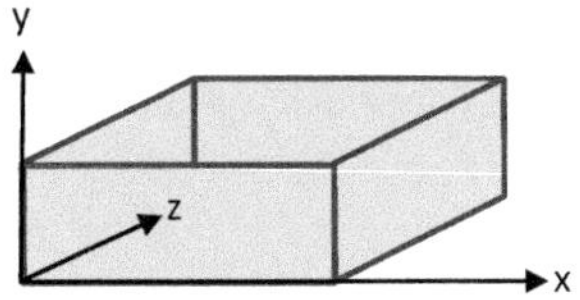


$$\text{Volume} = xyz = 8$$

The surface area needs to be minimized; hence, we set up an objective function $f(x, y, z)$ as follows:

$$f(x, y, z) = 2xy + 2xz + 2yz$$

We now substitute the equality constraint into the objective function

$$f(x, y, z) = 2xy + 2x\left(\frac{8}{xy}\right) + 2y\left(\frac{8}{xy}\right) = 2xy + \frac{16}{y} + \frac{16}{x}$$

A minimum point is also a stationary point; hence, we have the conditions

$$\frac{\partial f}{\partial x} = 0, \quad \frac{\partial f}{\partial y} = 0$$

$$\frac{\partial f}{\partial x} = 2y - \frac{16}{x^2} = 0$$

$$y = \frac{8}{x^2} \tag{1}$$

$$\frac{\partial f}{\partial y} = 2x - \frac{16}{y^2} = 0$$

$$x = \frac{8}{y^2} \tag{2}$$

Substituting Eq. (2) into (1), we have

$$y = \frac{8}{\left(\frac{8}{y^2}\right)^2} = \frac{y^4}{8}$$

$$y\left(1-\frac{y^3}{8}\right)=0 \rightarrow y=0 \text{ or } \sqrt[3]{8}=2$$

$$x=\frac{8}{y^2}=\infty(\text{rejected}) \text{ or } 2$$

We have one stationary point at (2, 2). Now we can formulate the Hessian matrix $\mathcal{H}(x, y)$ in order to find out the nature of this point. By definition, $\mathcal{H}(x, y)$ can be expressed as follows,

$$\mathcal{H}(x,y)=\begin{bmatrix}\frac{\partial}{\partial x} & \frac{\partial}{\partial y}\end{bmatrix}^T \begin{bmatrix}\frac{\partial f}{\partial x} & \frac{\partial f}{\partial y}\end{bmatrix}=\begin{bmatrix}\frac{\partial^2 f}{\partial x^2} & \frac{\partial^2 f}{\partial x \partial y} \\ \frac{\partial^2 f}{\partial y \partial x} & \frac{\partial^2 f}{\partial y^2}\end{bmatrix}$$

Evaluating the differentials using the values found earlier, we obtain the following expression. We then substitute the value of $x$ and $y$ for the stationary point as follows.

$$\mathcal{H}(x,y)=\begin{bmatrix}\frac{32}{x^3} & 2 \\ 2 & \frac{32}{y^3}\end{bmatrix}=\begin{bmatrix}4 & 2 \\ 2 & 4\end{bmatrix}$$

To find the eigenvalues at this stationary point, we set the following condition,

$$\det[\mathcal{H}-\lambda I]=0$$

$$\det\begin{bmatrix}4-\lambda & 2 \\ 2 & 4-\lambda\end{bmatrix}=0$$

$$(4-\lambda)^2-2^2=0$$

$$16-8\lambda+\lambda^2-4=0$$

$$\lambda^2-8\lambda+12=0$$

$$(\lambda-2)(\lambda-6)=0$$

$$\lambda_1=2$$

$$\lambda_2=6$$

Both eigenvalues are positive and non-zero, hence the function $f(x, y)$ is locally convex at the stationary point. This stationary point is thus proven to be a local minima. Since there are no other valid stationary points (the other solution for stationary point condition was rejected since the coordinate value was $\infty$), this local minima is also the global minima.

## Problem 51

**Consider the quadratic function below, whereby $A = \begin{pmatrix} 6 & 8 \\ 8 & -6 \end{pmatrix}$ and $x = \begin{pmatrix} x_1 \\ x_2 \end{pmatrix}$:**

$$\mathbf{10 = x^T A x}$$

(a) **Determine the eigenvalues $\lambda_1$ and $\lambda_2$ for matrix $A$ and sketch the geometry described by the function.**

(b) **Determine the unit eigenvectors $\hat{e}_1$ and $\hat{e}_2$ for matrix $A$, and demonstrate diagonalization as shown below, where $R = (\hat{e}_1 \quad \hat{e}_2)$. Show that matrix $R$ is orthogonal.**

$$\mathbf{R^T A R = \Lambda = \begin{bmatrix} \lambda_1 & 0 \\ 0 & \lambda_2 \end{bmatrix}}$$

## Solution 51

*Worked Solution*

(a) From the function provided, we can substitute the expressions for matrix $\boldsymbol{A}$ and vectors of $\boldsymbol{x}$.

$$10 = \boldsymbol{x}^T \boldsymbol{A} \boldsymbol{x}$$

$$10 = (x_1 \quad x_2)\begin{pmatrix} 6 & 8 \\ 8 & -6 \end{pmatrix}\begin{pmatrix} x_1 \\ x_2 \end{pmatrix} = 6{x_1}^2 + 16x_1x_2 - 6{x_2}^2$$

To compute the eigenvalues $\lambda$ of matrix $\boldsymbol{A}$, we set the condition as follows

$$\det(A - \lambda I) = 0$$

$$\det\begin{pmatrix} 6-\lambda & 8 \\ 8 & -6-\lambda \end{pmatrix} = 0$$

$$(6-\lambda)(-6-\lambda) - 64 = 0$$

$$-36 + \lambda^2 - 64 = 0$$

$$\lambda^2 = 100$$

$$\lambda_1 = 10$$

$$\lambda_2 = -10$$

Using the eigenvalues, we can define a new matrix containing eigenvalues along its diagonal, and rewrite the function equation as follows.

$$10 = \boldsymbol{x}'^T \boldsymbol{\Lambda} \boldsymbol{x}' = \begin{pmatrix} x_1' & x_2' \end{pmatrix} \begin{pmatrix} 10 & 0 \\ 0 & -10 \end{pmatrix} \begin{pmatrix} x_1' \\ x_2' \end{pmatrix}$$

What we have done here is effectively to convert from the previous coordinate system of $x_1$ and $x_2$ axes to a new coordinate system of $x_1'$ and $x_2'$. It is then clear that the eigenvalues scale the shape of the function along the eigenvectors which form the principal axes of the contours.

Dividing both sides of the equation by 10, we obtain the equation in the form of a rectangular hyperbola.

$$1 = x_1'^2 - x_2'^2$$

We can sketch the hyperbola as follows where the asymptotes are shown in dotted lines, and graphs are symmetrical about the both axes.

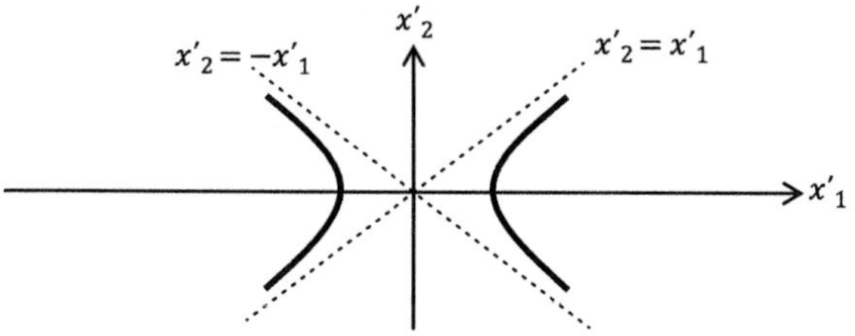


(b) We can also compute the eigenvectors of matrix $\mathbf{A}$, which we denote here as $\mathbf{e_1}$ and $\mathbf{e_2}$ where unknown values $a$, $b$, $c$, and $d$ are to be determined.

$$\mathbf{e}_1 = \begin{pmatrix} a \\ b \end{pmatrix} \quad \text{for} \quad \lambda_1 = 10$$

$$\mathbf{e}_2 = \begin{pmatrix} c \\ d \end{pmatrix} \quad \text{for} \quad \lambda_2 = -10$$

The eigenvalues scale the function in the direction of the eigenvectors; hence, the following relationship can be established

$$\mathbf{Ae_1} = \lambda_1 \mathbf{e_1}$$

$$\begin{pmatrix} 6 & 8 \\ 8 & -6 \end{pmatrix} \begin{pmatrix} a \\ b \end{pmatrix} = 10 \begin{pmatrix} a \\ b \end{pmatrix}$$

This gives us two equations as follows, which both produce the result $2b = a$.

$$6a + 8b = 10a$$

$$8a - 6b = 10b$$

$$\therefore 2b = a$$

We can use this result to rewrite the expression for eigenvector $\mathbf{e_1}$ where $k$ is a scalar quantity.

$$\mathbf{e_1} = \begin{pmatrix} a \\ b \end{pmatrix} = k\begin{pmatrix} 2 \\ 1 \end{pmatrix}$$

We can choose to make the eigenvector a unit vector by using a suitable value of $k$.

$$\hat{\mathbf{e}}_1 = \frac{1}{\sqrt{5}}\begin{pmatrix} 2 \\ 1 \end{pmatrix}$$

Similarly, we can find the other unit eigenvector $\hat{\mathbf{e}}_2$.

$$\boldsymbol{A}\mathbf{e_2} = \lambda_2\mathbf{e_2}$$

$$\begin{pmatrix} 6 & 8 \\ 8 & -6 \end{pmatrix}\begin{pmatrix} c \\ d \end{pmatrix} = -10\begin{pmatrix} c \\ d \end{pmatrix}$$

This gives us two equations as follows, which both produce the result $-2c = d$.

$$6c + 8d = -10c$$

$$8c - 6d = -10d$$

$$\therefore -2c = d$$

Like before, we can use this result to rewrite the expression for eigenvector $\mathbf{e_2}$ where $h$ is a scalar quantity and choose to make the eigenvector a unit vector by using a suitable value of $h$.

$$\mathbf{e_2} = \begin{pmatrix} c \\ d \end{pmatrix} = h\begin{pmatrix} -1 \\ 2 \end{pmatrix}$$

$$\hat{\mathbf{e}}_2 = \frac{1}{\sqrt{5}}\begin{pmatrix} -1 \\ 2 \end{pmatrix}$$

We can now construct a matrix of eigenvectors denoted by $\boldsymbol{R}$, and prove that it is orthogonal.

$$\boldsymbol{R} = \begin{pmatrix} \hat{\mathbf{e}}_1 & \hat{\mathbf{e}}_2 \end{pmatrix} = \frac{1}{\sqrt{5}}\begin{pmatrix} 2 & -1 \\ 1 & 2 \end{pmatrix}$$

We can demonstrate diagonalization of eigenvalues by showing the following is true.

$$\boldsymbol{R}^T\boldsymbol{A}\boldsymbol{R} = \boldsymbol{\Lambda}$$

$$\begin{aligned}
\boldsymbol{R}^T\boldsymbol{A}\boldsymbol{R} &= \frac{1}{\sqrt{5}}\begin{pmatrix} 2 & 1 \\ -1 & 2 \end{pmatrix}\begin{pmatrix} 6 & 8 \\ 8 & -6 \end{pmatrix}\frac{1}{\sqrt{5}}\begin{pmatrix} 2 & -1 \\ 1 & 2 \end{pmatrix} \\
&= \frac{1}{5}\begin{pmatrix} 20 & 10 \\ 10 & -20 \end{pmatrix}\begin{pmatrix} 2 & -1 \\ 1 & 2 \end{pmatrix} \\
&= \frac{1}{5}\begin{pmatrix} 50 & 0 \\ 0 & -50 \end{pmatrix} \\
&= \begin{pmatrix} 10 & 0 \\ 0 & -10 \end{pmatrix} \\
&= \begin{pmatrix} \lambda_1 & 0 \\ 0 & \lambda_2 \end{pmatrix} = \boldsymbol{\Lambda}
\end{aligned}$$

To prove that matrix $\boldsymbol{R}$ is orthogonal, we need to show that

$$\boldsymbol{R}^T = \boldsymbol{R}^{-1}$$

$$\boldsymbol{R} = \frac{1}{\sqrt{5}}\begin{pmatrix} 2 & -1 \\ 1 & 2 \end{pmatrix}$$

$$\boldsymbol{R}^{-1} = \frac{1}{\sqrt{5}}\begin{pmatrix} 2 & 1 \\ -1 & 2 \end{pmatrix} = \boldsymbol{R}^T$$

# Thermodynamics

## Problem 1

**Calculate the fugacity and fugacity coefficient of steam at 2 MPa and 700 °C.**

## Solution 1

***Worked Solution***

- - - - - - - - - - - - - - - - - - - - - - - - - - - - - - - - - - - - - - - - - - -

***Background Concepts***

Chemical potential is used to provide criteria for chemical equilibrium of species $i$ between phases $\alpha$ and $\beta$ in a multiphase system, whereby $\mu_i^\alpha = \mu_i^\beta$. Recall that chemical potential is simply partial molar Gibbs free energy defined by:

$$\mu_i = \left(\frac{\partial G}{\partial n_i}\right)_{T,P,n_{j\neq i}}$$

Chemical potential is a derived thermodynamic property, as opposed to measurable ones like temperature and pressure. However, there are limitations to the use of chemical potential, and fugacity was further derived from chemical potential as a better measure for equilibrium.

Fugacity has units of pressure, and it applies to gases, liquids, and solids. It can be understood as "corrected pressure" or the tendency for a substance to "flee" from the phase it is in. The higher the fugacity, the higher the partial pressure of ideal gas required to prevent the material from fleeing its phase and entering into the ideal gas. The fugacity of a component (either pure or in a mixture) is the partial pressure of an ideal gas of the same species which would be in equilibrium with that component.



X. W. Ng, *Engineering Problems for Undergraduate Students*,
https://doi.org/10.1007/978-3-030-13856-1_2

It also follows that as we tend towards real gases, where intermolecular forces become significant, fugacity decreases as gas molecules have less tendency to escape since they are held by those forces, and the fugacity coefficient (a fraction between 0 and 1) is a lower value to reflect a more "corrected pressure".

Fugacity plays the same role in real gases as partial pressure plays in ideal gases. As such, fugacity coefficient compares fugacity to partial pressure if the gas was ideal:

$$\hat{\varphi}_i = \frac{\hat{f}_i}{y_i P_{\text{sys}}}$$

The "hat" notation denotes the contribution of species $i$ to the mixture, as opposed to a pure species fugacity $f_i$ or a total solution fugacity $f$.

- - - - - - - - - - - - - - - - - - - - - - - - - - - - - - - - - - - - - - - - - - - - - - - -

So back to our problem which asked about fugacity. Fugacity is defined as follows:

$$\mu_i - \mu_i^0 = RT \ln\left(\frac{\hat{f}_i}{\hat{f}_i^0}\right)$$

Begin with the expression for fugacity of pure species $i$ using the ideal gas reference state.

$$g_i - g_i^0 = RT \ln\left(\frac{f_i^v}{y_i P_{\text{low}}}\right)$$

At $T = 700$ °C or 973.15 K, $g_i$ and $f_i^v$ are at 2 MPa. Therefore,

$$y_i = 1, \quad y_i P_{\text{low}} = 10 \text{ kPa}$$

We can refer to the steam table since $T$ and $P$ are defined, to find the enthalpy, $h$ and entropy, $s$ to calculate $g$. In the case of low pressure of 10 kPa, we find from the steam table that water is in the superheated vapor phase. Therefore,

$$g_i^0 = h_i^0 - Ts_i^0 = 3928.7 - 973.15(10.403) = -6195 \text{ kJ/K}$$

As for $P = 2$ MPa, the pure species $g_i$ is found as follows

$$g_i = 3917.5 - 973.15(7.9487) = -3818 \text{ kJ/K}$$

Combining our results, we have

$$g_i - g_i^0 = RT \ln\left(\frac{f_i^v}{10\ \text{kPa}}\right)$$

$$f_i^v = 13.4\ \text{kPa}, \quad P_i = P = 2\ \text{MPa}$$

$$\varphi_i = 0.0067$$

## Problem 2

**Consider the equation of state below and develop an expression for the fugacity and fugacity coefficient of a pure species.**

$$\frac{\boldsymbol{RT}}{\boldsymbol{v-b}} - \frac{\boldsymbol{a}}{\boldsymbol{Tv^2}}$$

## Solution 2

***Worked Solution***

Notice that equations of state (EOS) are often more easily expressed with $P$ as the subject of the formula. It is relatively more difficult to extract $v$ from the equation due to power relationships. This is the reason behind thinking about integrals with respect to $P$ instead of $v$.

Recall the definition of fugacity in integral form:

$$\mu_i - \mu_i^0 = \int_{P_{\text{low}}}^{P} \overline{V}_i dP = RT \ln\left(\frac{f_i^v}{y_i P_{\text{low}}}\right)$$

Since we are dealing with pure species, we can use $g_i$ instead of $\mu_i$ and $y_i P_{\text{low}} = P_{\text{low}}$. We can determine $\frac{dP}{dV}$ from the EOS.

$$\frac{dP}{dV} = \left[\frac{-RT}{(v-b)^2} + \frac{2a}{Tv^3}\right]$$

$$dP = \left[\frac{-RT}{(v-b)^2} + \frac{2a}{Tv^3}\right] dV$$

$$\ln\left(\frac{f_i^v}{P_{\text{low}}}\right) = \int_{P_{\text{low}}}^{P} \overline{V}_i \left[\frac{-RT}{(v-b)^2} + \frac{2a}{Tv^3}\right] dV = \int_{\frac{RT}{P_{\text{low}}}}^{v} \left[\frac{-RTv}{(v-b)^2} + \frac{2a}{Tv^2}\right] dV$$

Note that $v = v_i$ for pure species. For integrating, we can convert into partial fractions

$$\frac{v}{(v-b)^2} = \frac{1}{v-b} + \frac{b}{(v-b)^2}$$

Therefore the expression becomes

$$RT\ln\left(\frac{f_i^v}{P_{\text{low}}}\right) = -RT\int_{\frac{RT}{P_{\text{low}}}}^{v}\left[\frac{v}{(v-b)^2} - \frac{2a}{RT^2v^2}\right]dV$$

$$= -RT\int_{\frac{RT}{P_{\text{low}}}}^{v}\left[\frac{1}{v-b} + \frac{b}{(v-b)^2} - \frac{2a}{RT^2v^2}\right]dV$$

$$\ln\left(\frac{f_i^v}{P_{\text{low}}}\right) = -\int_{\frac{RT}{P_{\text{low}}}}^{v}\left[\frac{1}{v-b} + \frac{b}{(v-b)^2} - \frac{2a}{RT^2v^2}\right]dV$$

$$= \left[-\ln(v-b) + \left(\frac{b}{v-b}\right) - \frac{2a}{RT^2v}\right]_{\frac{RT}{P_{\text{low}}}}^{v}$$

$$\ln f_i^v - \ln P_{\text{low}} = -\ln(v-b) + \left(\frac{b}{v-b}\right) - \frac{2a}{RT^2v} + \ln\left(\frac{RT}{P_{\text{low}}} - b\right) - \left(\frac{b}{\frac{RT}{P_{\text{low}}} - b}\right) + \frac{2a}{RT^2\frac{RT}{P_{\text{low}}}}$$

For ideal gas, assume $\frac{RT}{P_{\text{low}}} \gg b$

$$\ln f_i^v - \ln P_{\text{low}} = -\ln(v-b) + \left(\frac{b}{v-b}\right) - \frac{2a}{RT^2v} + \ln\left(\frac{RT}{P_{\text{low}}}\right) - \left(\frac{b}{\frac{RT}{P_{\text{low}}}}\right) + \frac{2a}{RT^2\frac{RT}{P_{\text{low}}}}$$

$$\ln f_i^v = -\ln(v-b) + \left(\frac{b}{v-b}\right) - \frac{2a}{RT^2v} + \ln(RT) - \left(\frac{b}{\frac{RT}{P_{\text{low}}}}\right) + \frac{2a}{RT^2\frac{RT}{P_{\text{low}}}}$$

For ideal gas, we can also assume $P_{\text{low}} = 0$

$$\ln f_i^v = \ln\left(\frac{RT}{v-b}\right) + \left(\frac{b}{v-b}\right) - \frac{2a}{RT^2 v}$$

$$f_i^v\left(\frac{v-b}{RT}\right) = \exp\left(\frac{b}{v-b} - \frac{2a}{RT^2 v}\right)$$

Finally, we can determine fugacity coefficient since by definition, $f_i^v = \varphi_i P_{sys}$

$$f_i^v = \left(\frac{RT}{v-b}\right)\exp\left(\frac{b}{v-b} - \frac{2a}{RT^2 v}\right)$$

$$\varphi_i = \frac{\mathrm{RT}}{P(v-b)}\exp\left(\frac{b}{v-b} - \frac{2a}{RT^2 v}\right)$$

## Problem 3

**Data is available for a substance at 35 °C. Describe how one can use the data provided to obtain the fugacity and fugacity coefficient of the substance at 60 bar and 35 °C. [Actual calculations not required.]**

| $P$ [bar] | $v$ [m$^3$/mol] |
|---|---|
| 2.5 | 0.0075 |
| 6.0 | 0.0035 |
| 15.5 | 0.0016 |
| ... | ... |
| 60.0 | 0.00048 |

## Solution 3

***Worked Solution***

In this problem, we know the values of $T$, $P$, and $v$. Gas constant $R$ is a known value. We can make use of the equation of state. Note that $z = 1$ for an ideal gas.

$$z = \frac{Pv}{RT}$$

$$v = \frac{zRT}{P}$$

Recall definition of fugacity in integral form, and definition of fugacity coefficient.

$$\int_{P_{\text{low}}}^{P} \overline{V} dP = RT \ln\left(\frac{f_i^v}{P_{\text{low}}}\right) = RT \ln\left(\frac{\varphi_i P}{P_{\text{low}}}\right)$$

$$\int_{P_{\text{low}}}^{P} \frac{zRT}{P} dP = RT[\ln(\varphi_i) + \ln(P) - \ln(P_{\text{low}})]$$

$$\int_{P_{\text{low}}}^{P} \frac{z}{P} dP = \ln \varphi_i + \int_{P_{\text{low}}}^{P} \frac{1}{P} dP$$

$$\int_{P_{\text{low}}}^{P} \frac{z-1}{P} dP = \ln \varphi_i$$

We can find $\varphi_i$ if we can work out the integral. Recall the trapezium rule as shown below where $h = \frac{x_n - x_0}{n}$.

$$\int_{x_0}^{x_n} f(x)dx = \frac{1}{2}h[y_0 + y_n + 2(y_1 + y_2 + \ldots + y_{n-1})]$$

We can tabulate corresponding values of $\frac{z-1}{P}$ and $P$. Using trapezium rule, we can then find out $\ln \varphi_i$. Fugacity at the specified 60 bar and 35 °C can be determined from the known $\varphi_i$ and $P$, using

$$f_i^v = \varphi_i P_{\text{sys}}$$

Note that this is true for this case of a pure substance. In a mixture, it is worth noting that the expression is more accurately expressed as

$$f_i^v = \varphi_i P_i = \varphi_i y_i P_{\text{sys}}$$

**Problem 4**

**Consider a ternary system of methane (i), ethane (j), and propane (k) at 25 °C and 15 bar. Assume this system can be represented by the virial equation truncated at the second term.**

$$z = 1 + \frac{B_{\text{mix}}}{v}$$

**At 25 °C, the second virial coefficients [cm$^3$/mol] are given by:**

| | |
|---|---|
| $B_{ii}$ | −42 |
| $B_{jj}$ | −185 |
| $B_{kk}$ | −399 |
| $B_{ij}$ | −93 |
| $B_{ik}$ | −139 |
| $B_{jk}$ | −274 |

(a) **Develop an expression for the fugacity coefficient of methane in the mixture.**
(b) **Estimate the fugacity and the fugacity coefficient of methane for a mixture with 20% methane, 30% ethane, and 50% propane in mole percent.**

## Solution 4

*Worked Solution*

(a) Virial equation states the following where $B$ constants are given. This is a gas mixture by observation of the substances at the specified $T$ and $P$.

$$z = 1 + \frac{B_{\text{mix}}}{v}$$

Recall that the virial equation of state is a power series expansion for the compressibility factor z in terms of density $\rho$ (also possible in terms of pressure). Note that $\rho = \frac{1}{v}$.

$$z = \frac{Pv}{RT} = 1 + B\rho + C\rho^2 + D\rho^3 + \ldots$$

Whether to use the power series expression in terms of $\rho$, $v$, or $P$ depends on the question. We shall use the expression that is useful for our integration based on known parameters as described in the problem statement.

Another question is how accurate we should be, i.e., how many terms to consider. In general, for low pressures, the equation can be up to second term (i.e., linear). For moderate to higher pressures, quadratic and cubic expressions can be considered.

Finally, it is worth understanding that virial coefficients provide an indication of the intermolecular forces between molecules. By definition, $f_i^v = \varphi_i P_i = \varphi_i y_i P$ and $\int_{P_{\text{low}}}^{P} \overline{V}_i dP = RT \ln \left( \frac{f_i^v}{y_i P_{\text{low}}} \right)$. At this point, we recall a useful mathematical relation, the cyclic rule as follows:

$$-1 = \left(\frac{\partial x}{\partial z}\right)_y \left(\frac{\partial y}{\partial x}\right)_z \left(\frac{\partial z}{\partial y}\right)_x$$

$$\left(\frac{\partial V}{\partial n_i}\right)_{T,P} \left(\frac{\partial P}{\partial V}\right)_{T,n_i} \left(\frac{\partial n_i}{\partial P}\right)_{T,V} = -1$$

When the composition of species $i$ remains unchanged (which is the case unless there are inflows or outflows to the system), then $\left(\frac{\partial P}{\partial V}\right)_{T,n_i} = \frac{dP}{dV}$ and we get the correlation $\left(\frac{\partial V}{\partial n_i}\right)_{T,P} dP = -\left(\frac{\partial P}{\partial n_i}\right)_{T,V} dV$.

$$RT \ln\left(\frac{f_i^v}{P_i}\right) = -\int_{\frac{nRT}{P_{\text{low}}}}^{V} \left(\frac{\partial P}{\partial n_i}\right)_{T,V,n_{j,k\neq i}} dV$$

We can derive an expression for the integral using the Equation of State,

$$z = 1 + \frac{B_{\text{mix}}}{v} = \frac{Pv}{RT}$$

$$P = \frac{RT}{v}\left(1 + \frac{B_{\text{mix}}}{v}\right) = RT\left(\frac{1}{v} + \frac{B_{\text{mix}}}{v^2}\right)$$

Since $n_T = n_i + n_j + n_k$ and $v = \frac{V}{n_T} = \frac{V}{n_i+n_j+n_k}$,

$$P = RT\left(\frac{1}{v} + \frac{B_{\text{mix}}}{v^2}\right) = RT\left(\frac{n_i + n_j + n_k}{V} + \frac{B_{\text{mix}}}{V^2}\right)$$

Now we recall mixing rules which describe how two bodies interact. There are three types of two-body interactions in this example, like interactions denoted by $B_{ii}$, $B_{jj}$, and $B_{kk}$, as well as unlike interactions denoted by $B_{ij}$, $B_{jk}$, and $B_{ik}$. Note that $B_{\text{mix}}$ is the parameter for the entire mixture while the $B$ values in the table refer to specific binary interactions.

For like interactions such as between molecules $i$, a molecule $i$ must find another molecule $i$, and this will occur in proportion to the mole fraction of the first molecule $i$, multiplied by the mole fraction of the second molecule $i$ (probability rule), giving us the term "$y_i^2$". As for unlike interactions such as that between species $i$ and $j$, the same concept applies, giving rise to a term "$y_iy_j$".

However one should note that each molecule type should be accounted for, i.e., the scenario where molecule $i$ finds molecule $j$ and where molecule $j$ finds molecule $i$. Each of these scenarios is different and should be considered. Incidentally, both cases give us the same coefficient of "$y_iy_j$" (since $y_iy_j = y_jy_i$);

hence, assuming that $B_{ij}$ is equal to $B_{ji}$ as is the case in this example, we see the term "$2y_iy_jB_{ij}$". Therefore we arrive at the following:

$$P = RT\left(\frac{n_i + n_j + n_k}{V} + \frac{B_{\text{mix}}}{V^2}\right)$$

$$= RT\left(\frac{n_i + n_j + n_k}{V} + \frac{{n_i}^2 B_{ii} + {n_j}^2 B_{jj} + {n_k}^2 B_{kk} + 2n_in_jB_{ij} + 2n_in_kB_{ik} + 2n_jn_kB_{jk}}{V^2}\right)$$

$$\left(\frac{\partial P}{\partial n_i}\right)_{T,V,n_{j,k\neq i}} = RT\left(\frac{1}{V} + \frac{2n_iB_{ii} + 2n_jB_{ij} + 2n_kB_{ik}}{V^2}\right)$$

$$RT\ln\left(\frac{f_i^v}{P_{\text{low}}}\right) = -\int_{\frac{nRT}{P_{\text{low}}}}^{V} RT\left(\frac{1}{V} + \frac{2n_iB_{ii} + 2n_jB_{ij} + 2n_kB_{ik}}{V^2}\right)dV$$

$$\ln\left(\frac{f_i^v}{P_{\text{low}}}\right) = \left[-\ln V + \frac{2n_iB_{ii} + 2n_jB_{ij} + 2n_kB_{ik}}{V}\right]_{\frac{nRT}{P_{\text{low}}}}^{V}$$

$$\ln\left(\frac{f_i^v}{P_{\text{low}}}\right) = -\ln V + \frac{2n_iB_{ii} + 2n_jB_{ij} + 2n_kB_{ik}}{V} + \ln\frac{nRT}{P_{\text{low}}} - \frac{(2n_iB_{ii} + 2n_jB_{ij} + 2n_kB_{ik})P_{\text{low}}}{nRT}$$

Assume $P_{\text{low}}$ is a small value

$$\ln\left(\frac{f_i^v}{y_iP}\right) = \frac{2n_iB_{ii} + 2n_jB_{ij} + 2n_kB_{ik}}{V} + \ln\frac{nRT}{P_{\text{low}}V}$$

$$\ln\varphi_i = \frac{2n_iB_{ii} + 2n_jB_{ij} + 2n_kB_{ik}}{V} + \ln\left(\frac{1}{z}\right)$$

$$\varphi_i = \left(\frac{1}{z}\right)\exp\left[\frac{2n_iB_{ii} + 2n_jB_{ij} + 2n_kB_{ik}}{V}\right]$$

$$f_i^v = \left(\frac{y_iP}{z}\right)\exp\left[\frac{2n_iB_{ii} + 2n_jB_{ij} + 2n_kB_{ik}}{V}\right]$$

(b) If given mole percent, we will know $y_i$, $y_j$, *and* $y_k$.

$$B_{mix} = {y_i}^2 B_{ii} + {y_j}^2 B_{jj} + {y_k}^2 B_{kk} + 2y_iy_jB_{ij} + 2y_iy_kB_{ik} + 2y_jy_kB_{jk}$$

We know $y_i = 0.2$, $y_j = 0.3$, and $y_k = 0.5$,

$$B_{\text{mix}} = 0.04(-42) + 0.09(-185) + 0.25(-399) + 2(0.06)(-93) + 2(0.1)(-139) + 2(0.15)(-274)$$

$$= -2.39 \times 10^{-4}\ \text{m}^3/\text{mol}$$

$$z = 1 + \frac{B_{\text{mix}}}{v} = \frac{Pv}{RT}$$

$$1 + \frac{-2.39 \times 10^{-4}}{v} = \frac{(1{,}500{,}000\ \text{Pa})v}{\left(\frac{8.314\ \text{J}}{\text{mol} \cdot \text{K}}\right)(298.15\ \text{K})}$$

Solve for $v$, then $z$. Afterwhich $\varphi_i$ and $f_i^v$ can be found.

$$\varphi_i = \left(\frac{1}{z}\right)\exp\left[\frac{2n_iB_{ii} + 2n_jB_{ij} + 2n_kB_{ik}}{V}\right] = \left(\frac{1}{z}\right)\exp\left[\frac{2y_iB_{ii} + 2y_jB_{ij} + 2y_kB_{ik}}{v}\right]$$

$$f_i^v = \varphi_i y_i P$$

**Problem 5**

**Show that**

$$\left(\frac{\partial g_i}{\partial P}\right)_T = v_i = RT\left(\frac{\partial(\ln f_i)}{\partial P}\right)_T$$

**Solution 5**

***Worked Solution***

Gibbs free energy can be written as $dg_i = -s_i dT + v_i dP$. Therefore

$$\left(\frac{\partial g_i}{\partial P}\right)_T = \left(\frac{-s_i dT + v_i dP}{\partial P}\right)_T = v_i$$

From first principles, fugacity of a pure gas $g_i = g_i^0 + RT\ln\left(\frac{f_i^v}{P_{\text{low}}}\right) = g_i^0 + RT\ln f_i^v - RT\ln P_{\text{low}}$. So we can differentiate with respect to $P$, holding $T$ constant. Note that $P_{\text{low}}$ is a reference state low value for pressure where ideal gas is assumed, and it is a constant.

$$\left(\frac{\partial g_i}{\partial P}\right)_T = \left(\frac{\partial\left(g_i^0 + RT\ln f_i^v - RT\ln P_{\text{low}}\right)}{\partial P}\right)_T = RT\left(\frac{\partial \ln f_i^v}{\partial P}\right)_T$$

$$\left(\frac{\partial g_i}{\partial P}\right)_T = v_i = RT\left(\frac{\partial \ln f_i^v}{\partial P}\right)_T$$

## Problem 6

**Find fugacity of pure liquid acetone at 100 bar and 382 K. The molar volume of the liquid $v_i$ is 73.4 cm$^3$/mol. You may assume molar volume does not change with pressure.**

## Solution 6

### *Worked Solution*

Liquid fugacity is expressed as follows:

$$f_i = P^{\text{vap}} = P^{\text{sat}}\exp\left[\frac{(P-P^{\text{sat}})v^2}{RT}\right]$$

If we assume the vapor phase is an ideal gas (e.g., at low system pressure $P$), $(P - P^{\text{sat}}) \approx 0$, then

$$f_i = P^{\text{sat}}$$

To derive the above expression, we may first consider the pure liquid at a convenient pressure, such as $P^{\text{sat}}$ (Lewis/Randall reference state), then adjust to the stated system pressure, $P$.

$$\left(\frac{\partial g_i}{\partial P}\right)_T = v_i^l = RT\left(\frac{\partial \ln f_i^l}{\partial P}\right)_T$$

At system pressure $P, f_i^l$ can be found first:

$$v_i^l = RT\left(\frac{\partial \ln f_i^l}{\partial P}\right)_T$$

$$\left(\frac{v_i^l}{RT}\right)dP = d\ln f_i^l$$

$$\left(\frac{v_i^l}{RT}\right)\int_{P^{\text{sat}}}^{P} dP = \int_{f_i^{\text{sat}}}^{f_i} d\ln f_i^l$$

$$\ln\frac{f_i^l}{f_{i\text{sat}}^l} = \left(\frac{v_i^l}{RT}\right)(P - P^{\text{sat}})$$

Separately, we find that at saturated conditions, $f_{i\text{sat}}^l = f_{i\text{sat}}^v = \varphi_i^{\text{sat}}P_i^{\text{sat}}$

$$f_i^l = f_{i\text{sat}}^l\exp\left[\left(\frac{v_i^l}{RT}\right)(P-P^{\text{sat}})\right] = \varphi_i^{\text{sat}}P_i^{\text{sat}}\exp\left[\left(\frac{v_i^l}{RT}\right)(P-P^{\text{sat}})\right]$$

Note that $v_i^l$ is that of the liquid. Since liquids are taken to be incompressible, it is a constant.

How do we find saturated pressure vales from temperature, for a known substance? Saturated pressures are commonly reported in terms of the Antoine equation where values of $A$, $B$, and $C$ can be found from the data booklet at $T = 382$ K.

$$\ln P_i^{\text{sat}} = A - \frac{B}{C + T}$$

From published data, we can obtain $P_i^{\text{sat}}$ for acetone, which is about 4.64 bar. Note that this is greater than atmospheric pressure 1 bar, and hence $\varphi_i^{\text{sat}} \neq 1$.

In order to find the actual value of $\varphi_i^{\text{sat}}$, we need to use the correlation:

$$\log\varphi_i^{sat} = \log\varphi^{(0)} + \omega\log\varphi^{(1)}$$

To use this correlation, we first find reduced temperature and pressure, $T_r$ and $P_r$, where $T_r = T/T_{\text{critical}}$ and $P_r = P/P_{\text{critical}}$. Critical point values and $\omega$ values can be obtained from the data booklet. $T$ and $P$ are as defined in the problem statement, $T = 382$ K and $P_i^{\text{sat}} = 4.64$ bar. After $T_r$ and $P_r$ are obtained, we can find $\log\varphi^{(0)}$ and $\log\varphi^{(1)}$ using the data booklet. Finally we will obtain $\varphi_i^{\text{sat}} = 0.904$.

Substituting all values into our earlier expression, we obtain $f_i^l$ at 100 bar and 382 K.

$$f_i^l = \varphi_i^{\text{sat}} P_i^{\text{sat}} \exp\left[\left(\frac{v_i^l}{RT}\right)(P - P^{\text{sat}})\right]$$

$$f_i^l = 0.904(4.64)\exp\left[\left(\frac{7.34 \times 10^{-5}}{8.314 \times 382}\right)(100 - 4.64)\right] = 5.23 \text{ bar}$$

## Problem 7

**There is a binary mixture containing species $a$ and $b$ at $T = 300$ K, and $P = 20$ kPa. A table showing $\hat{f}_a^l$ values at different mole fractions $x_a$ is given. Use Henry's law as the reference state for species $a$ and the Lewis/Randall rule for species $b$.**

| $\hat{f}_a^l$/kPa | $x_a$ |
|---|---|
| 0 | 0 |
| 2 | 0.1 |
| 8 | 0.3 |
| 12 | 0.4 |
| 26 | 0.7 |
| 40 | 1 |

(a) **What is the Henry's law constant, $H_a$ for species $a$?**
(b) **What is the activity coefficient for species $a$ at $x_a = 0.4$? At $x_a = 0.8$?**
(c) **Is the activity coefficient for species $b$ at $x_a = 0.4$ greater than or less than 1?**
(d) **Is the $a$–$b$ interaction stronger than the pure species interactions?**
(e) **Consider the vapor phase to be ideal. What is the vapor phase mole fraction of $a$ in equilibrium with 40% liquid $a$?**

## Solution 7

***Worked Solution***

(a) First, we can draw a best fit curve for the data points provided. The Henry's law reference state can be understood as a pure fluid in which the dominant intermolecular interactions are unlike interactions (i.e., $a$–$b$ interactions). Mathematically, a straight line tangent to the curve at $x_a = 0$. Then the Henry's constant will be the fugacity value on this straight line, when $x_a = 1$. Since $\hat{f}_a^{\text{ideal}} = (1)H_a$.

There are two ways to approximate an ideal solution, using Henry's or Lewis/Randall (LR) reference states. If Henry's law is used as the reference state to describe species $a$, then $\hat{f}_a^{\text{ideal}} = x_a H_a$, where $H_a$ is the value of $\hat{f}_a^l$ at $x_a = 1$, and the reference state can be approximated as a point on a straight line tangent to the curve where $a$–$b$ interactions dominate, i.e., at $x_a = 0$. If the Lewis/Randall reference state is used, then $\hat{f}_a^{\text{ideal}} = x_a f_a$, where $f_a$ is the value of $\hat{f}_a^l$ at $x_a = 1$, and the reference state can be approximated as a point on a straight line tangent to the curve where $a$–$a$ interactions dominate, i.e., at $x_a = 1$. Since $\hat{f}_a^{\text{ideal}} = (1)f_a$.

Graphically, both are straight line approximations (ideal solution) using different points to lie tangent to the actual curve. By drawing a straight line tangent to the curve at $x_a = 0$, we can read off the value of $\hat{f}_a^l$ when $x_a = 1$, to obtain the value of $H_a = 19.5$ kPa.

(b) By definition, activity coefficient $\gamma_i = \frac{\hat{f}_i^l}{\hat{f}_i^{\text{ideal}}} = \frac{\hat{f}_i^l}{x_i f_i^0}$. Since species $a$ is to be described by Henry's reference state,

$$\gamma_a = \frac{\hat{f}_a^l}{x_a H_a}$$

At $x_a = 0.4$ and $x_a = 0.8$, we can read off graphically, the corresponding values of $\hat{f}_a^l$. Following this approach, we have $\gamma_a$ ($x_a = 0.4$) $= 1.54$ and $\gamma_a$ ($x_a = 0.8$) $= 1.99$.

(c) The Gibbs–Duhem (GD) equation is useful in relating properties of *different species* in a mixture. These properties may be partial molar volume, enthalpy, entropy, internal energy, Gibbs free energy, chemical potential, etc. The general form of the GD equation is as follows at constant $T$ and $P$, where K denotes the partial molar property.

$$0 = \sum n_i dK_i$$

When we know the activity coefficient of one species, we can use that to relate to the activity coefficient of another species.

$$0 = \sum n_i d\mu_i$$

We can express the above in terms of fugacity, which helps us relate to activity coefficient.

$$\mu_i - \mu_i^0 = RT \ln\left(\frac{\hat{f}_i}{\hat{f}_i^0}\right)$$

$$d\mu_i = RT\, d\ln\hat{f}_i$$

$$\sum n_i d\mu_i = RT \sum n_i d\ln\hat{f}_i$$

Fugacity can now be expressed in terms of activity coefficient:

$$\hat{f}_i = \gamma_i x_i \hat{f}_i^0$$

$$\sum n_i d\ln\hat{f}_i = \sum n_i d\ln\left(\gamma_i x_i \hat{f}_i^0\right)$$

$$\sum n_i d\ln\left(\gamma_i x_i \hat{f}_i^0\right) = \sum n_i d\ln\gamma_i + \sum n_i d\ln x_i + \sum n_i d\ln\hat{f}_i^0 = 0$$

$$\sum n_i d\ln x_i = \sum n_i\left(\frac{1}{x_i}\right)dx_i = \sum (nx_i)\left(\frac{1}{x_i}\right)dx_i = n\sum dx_i$$

Since $\sum x_i = 1$, $\sum dx_i = 0$, this gives us

$$\sum n_i d\ln\gamma_i + \sum n_i d\ln\hat{f}_i^0 = 0$$

And knowing that $\hat{f}_i^0$ is a reference state fugacity value and hence a constant, we have $d\ln\hat{f}_i^0 = 0$.

$$\sum n_i d\ln\gamma_i = 0$$

$$\sum nx_i d\ln\gamma_i = 0$$

$$\sum x_i d\ln\gamma_i = 0$$

Assuming we have only two species $a$ and $b$ in the mixture, then:

$$x_a d\ln\gamma_a + x_b d\ln\gamma_b = 0$$

$$x_a\left(\frac{d\ln\gamma_a}{dx_a}\right) + x_b\left(\frac{d\ln\gamma_b}{dx_a}\right) = 0$$

Let us return to the question, which asks if $\gamma_b$ at $x_a = 0.4$ is greater or less than 1. Recall that activity coefficient represents the ratio of the actual fugacity against the reference state fugacity (which itself is a choice). Graphically, if the actual curve lies *above* the reference state straight line approximation (tangent at $x_a = 0$ for Henry's and tangent at $x_a = 1$ for LR), then $\gamma_a > 1$. And likewise, if the actual curve lies *below* the reference state straight line approximation, then $\gamma_a < 1$.

The question required that species $a$ follows Henry's reference state and species $b$ follows LR reference state. So as $x_a$ tends to zero, $\gamma_a$ tends to one (or $\ln\gamma_a = 0$). We noticed from part (b) that as $x_a$ increases from 0.4 to 0.8, $\gamma_a$ increases where both values are greater than 1 (or $\ln\gamma_a > 0$). So we can observe from gradient analysis that at $x_a = 0.4$, $\frac{d\ln\gamma_a}{dx_a}$ is a positive value. At this point from sign analysis, we know that $\frac{d\ln\gamma_b}{dx_a}$ has to be a negative value to satisfy the equation. As $x_a$ tends to zero, $\gamma_b$ tends to 1 (or $\ln\gamma_b = 0$). $\frac{d\ln\gamma_b}{dx_a}$ is a negative value, which means that as $x_a$ increases, $\ln\gamma_b$ decreases from zero to negative values, so $\gamma_b < 1$.

(d) *a–b* interactions are stronger than *b–b* (or pure species) interactions because $\gamma_b < 1$, using the Lewis Randall reference state. The "tendency to escape" is less when we introduce *a–b* interactions to a pure *b*. As for $\gamma_a > 1$, using the Henry's reference state, it means that the "tendency to escape" is greater when we become more pure in *a*, which implies that *a–a* interactions are weaker than *a–b* interactions. So we can deduce that *a–b* interactions are stronger than pure species interactions.

(e) At equilibrium, $\hat{f}_a^v = \hat{f}_a^l$. For ideal gas, $\hat{f}_a^v = y_a P$. $y_a = \frac{\hat{f}_a^l}{P}$. We can find out the value of $\hat{f}_a^l$ at $x_a = 0.4$. We also know from the problem statement that $P = 20$ kPa. So $y_a = 0.6$.

Be careful to be clear about liquid versus vapor mole fractions. They are different. Also, be sure to check that the equilibrium equation is consistent for the specific species, i.e., do not mistakenly equate this way, $\hat{f}_a^v = \hat{f}_b^l$ if there are two species for example.

## Problem 8

**Derive an expression that associates the entropy change of an ideal gas between ($T_1$, $P_1$) and ($T_2$, $P_2$).**

## Solution 8

### *Worked Solution*

We can break down the path into two steps, first by keeping $T$ constant and changing $P$ from $P_1$ to $P_2$. Then by keeping $P$ constant at $P_2$ and changing $T$ from $T_1$ to $T_2$.

*Step 1: Isothermal*

The change in internal energy of an ideal gas is zero.

$$du = \delta q_{\text{rev}} + \delta w_{\text{rev}} = 0$$

$$\delta q_{\text{rev}} = -\delta w_{\text{rev}}$$

By definition, we apply the concept of work done as a product of force and distance moved by the force, so it follows that

$$W = -P\int dV$$

$$-\delta w_{\text{rev}} = Pdv$$

And for an ideal gas,

$$\delta q_{\text{rev}} = Pdv$$

By definition, entropy can be expressed as heat absorbed during a reversible process.

$$ds = \frac{\delta q_{\text{rev}}}{T}$$

$$\Delta s = \int \frac{\delta q_{\text{rev}}}{T}$$

So for our Step 1 process for an ideal gas,

$$Pv = RT$$

$$\Delta s_{\text{step1}} = \int \frac{\delta q_{\text{rev}}}{T} = \int \frac{Pdv}{T} = \int_{v_1}^{RT_1/P_2} \frac{R}{v}dv = R\ln \frac{RT_1/P_2}{v_1} = R\ln\left(\frac{P_1}{P_2}\right) = -R\ln\left(\frac{P_2}{P_1}\right)$$

*Step 2: Isobaric*

$$du = \delta q_{\text{rev}} + \delta w_{\text{rev}} = \delta q_{\text{rev}} - Pdv$$

There is a mathematical "trick" that is useful for us to introduce enthalpy to the expression, where $h = u + Pv$. We know that $dP = 0$ for isobaric process, so adding this dummy term is inconsequential though it is useful mathematically.

$$du + Pdv + vdP = \delta q_{\text{rev}} + 0$$

$$d(u + Pv) = dh = \delta q_{\text{rev}}$$

$$\Delta s_{\text{step2}} = \int \frac{\delta q_{\text{rev}}}{T} = \int \frac{dh}{T} = \int \frac{C_p dT}{T}$$

So summing both steps, we have the following entropy change for going from $(T_1, P_1)$ to $(T_2, P_2)$.

$$\Delta s = -R \ln\left(\frac{P_2}{P_1}\right) + \int_{T_1}^{T_2} \frac{C_p}{T} dT$$

**Problem 9**

**Calculate the thermodynamic property changes of mixing for a binary ideal gas mixture, under isothermal conditions. The pressures of each gas species drop from their pure species pressures with mixing as they isothermally expand to take up the space of the container.**

**Solution 9**

***Worked Solution***

Let us start with entropy change of mixing for a binary mixture of species $a$ and $b$,

$$\Delta s_{\text{mix}} = y_a(\bar{S}_a - s_a) + y_b(\bar{S}_b - s_b)$$

$$\Delta s_{\text{mix}} = y_a\left(-R \ln\left(\frac{P_a}{P}\right)\right) + y_b\left(-R \ln\left(\frac{P_b}{P}\right)\right)$$

$$\Delta s_{\text{mix}} = y_a\left(-R \ln\left(\frac{y_a P}{P}\right)\right) + y_b\left(-R \ln\left(\frac{y_b P}{P}\right)\right) = y_a(-R \ln y_a) + y_b(-R \ln y_b)$$

$$\Delta s_{\text{mix}}^{\text{IG}} = -R \sum_{i=1}^{m} y_i \ln y_i$$

The other properties below are zero for ideal gas.

$$\Delta h_{\text{mix}}^{\text{IG}} = 0$$

$$\Delta u_{\text{mix}}^{\text{IG}} = 0$$

$$\Delta v_{\text{mix}}^{\text{IG}} = 0$$

And finally for the Gibbs free energy of mixing, we have

$$\Delta g_{\text{mix}}^{\text{IG}} = \Delta h_{\text{mix}}^{\text{IG}} - T\Delta s_{\text{mix}}^{\text{IG}} = RT\sum_{i=1}^{m} y_i \ln y_i$$

**Problem 10**

**Derive an expression for the pressure dependence of partial molar Gibbs free energy, $\bar{G}_i$. Show that the temperature dependence of excess Gibbs free energy can be expressed as shown below.**

$$\left(\frac{\partial\left(\frac{g^E}{T}\right)}{\partial T}\right)_{P,n_i} = \frac{-h^E}{T^2}$$

**Solution 10**

***Worked Solution***

The pressure dependence of partial molar Gibbs free energy is more straightforward.

$$d\bar{G}_i = -\bar{S}_i dT + \overline{V}_i dP + \sum_{i=1}^{m} \bar{G}_i dn_i$$

At constant temperature and constant composition, $dT = 0$ and $dn_i = 0$.

$$\left(\frac{\partial \bar{G}_i}{\partial P}\right)_{P,n_i} = \overline{V}_i$$

To express the temperature dependence of excess Gibbs free energy change $g^E$ we start with the following equation. [Note that Gibbs free energy of mixing, $\Delta g_{\text{mix}} = g^E$.]

$$dg^E = -s^E dT + v^E dP + \sum_{i=1}^{m} g^E dn_i$$

At constant pressure and constant composition, $dP = 0$ and $dn_i = 0$.

$$\left(\frac{\partial g^E}{\partial T}\right)_{P,n_i} = -s^E$$

We can take partial derivative of $\left(\frac{\partial g^E}{\partial T}\right)_{P,n_i}$ with respect to $T$,

$$\left(\frac{\partial\left(\frac{g^E}{T}\right)}{\partial T}\right)_{P,n_i} = \frac{T\partial g^E - g^E\partial T}{T^2} = \frac{1}{T}\left(\frac{\partial g^E}{\partial T}\right)_{P,n_i} - \frac{g^E}{T^2} = -\frac{s^E}{T} - \frac{g^E}{T^2} = \frac{-s^E T - g^E}{T^2}$$

$$g^E = h^E - Ts^E$$

$$\left(\frac{\partial\left(\frac{g^E}{T}\right)}{\partial T}\right)_{P,n_i} = \frac{-h^E}{T^2}$$

## Problem 11

**Consider a binary liquid mixture of species *a* and *b*. The activity coefficients for this mixture are adequately described by the two-suffix Margules equation.**

(a) **If $\Delta h_{\mathrm{mix}} = 0$, what can you say about the two-suffix Margules parameter, $A$?**
(b) **If $\Delta v_{\mathrm{mix}} = 0$, what can you say about the two-suffix Margules parameter, $A$?**

## Solution 11

### *Worked Solution*

(a) An "excess property" describes the difference between the actual thermodynamic property of a real mixture versus the value it would have if it was an ideal solution at the same temperature, pressure, and composition. Hence by definition, excess property for an ideal mixture is zero.

We may define the ideal solution reference state as Lewis Randall, Henry, or the ideal gas. An ideal solution is where all the mixing rules are the same as for an ideal gas, where there are no intermolecular forces.

Even though excess property is by definition zero, the property of mixing may or may not be zero, for example, $\Delta h_{\mathrm{mix}}^{\mathrm{IG}} = \Delta u_{\mathrm{mix}}^{\mathrm{IG}} = \Delta v_{\mathrm{mix}}^{\mathrm{IG}} = 0$, but note that even for ideal mixture, the property change of mixing is non-zero for $\Delta g_{\mathrm{mix}}^{\mathrm{IG}}$ and $\Delta s_{\mathrm{mix}}^{\mathrm{IG}}$.

If $\Delta h_{\text{mix}} = 0$, it means that there is no difference between the actual enthalpy of the real mixture relative to if the solution was ideal, hence it means that $h^E = \Delta h_{\text{mix}} = 0$.

Recall that the temperature dependence of excess Gibbs free energy can be written as follows.

$$\left(\frac{\partial\left(\frac{g^E}{T}\right)}{\partial T}\right)_{P,n_i} = \frac{-h^E}{T^2}$$

If $h^E = 0$, then $\left(\frac{\partial\left(\frac{g^E}{T}\right)}{\partial T}\right)_{P,n_i} = 0$, which means that $\frac{g^E}{T}$ is independent of $T$. Since the two-suffix Margules equation describes $g^E = Ax_a x_b$, therefore $\frac{Ax_a x_b}{T}$ is independent of $T$ at constant pressure and composition. It then follows that $\frac{A}{T}$ is also independent of $T$. Mathematically, this means that $\frac{A}{T} =$ constant, or $A{\sim}T$.

(b) If $\Delta v_{\text{mix}} = v^E = 0$, then since the two-suffix Margules equation describes $g^E = Ax_a x_b$, and the pressure dependence of $g^E$ can be expressed as $\left(\frac{\partial g^E}{\partial P}\right)_{P,n_i} = -v^E$, then $\left(\frac{\partial Ax_a x_b}{\partial P}\right)_{P,n_i} = 0$. This means that $A$ is independent of $P$, at constant temperature and composition.

## Problem 12

**Below is a plot showing how activity coefficients of a mixture containing species *a* and *b* behave, relative to the mole fraction of *a*, $x_a$, at a defined temperature.**

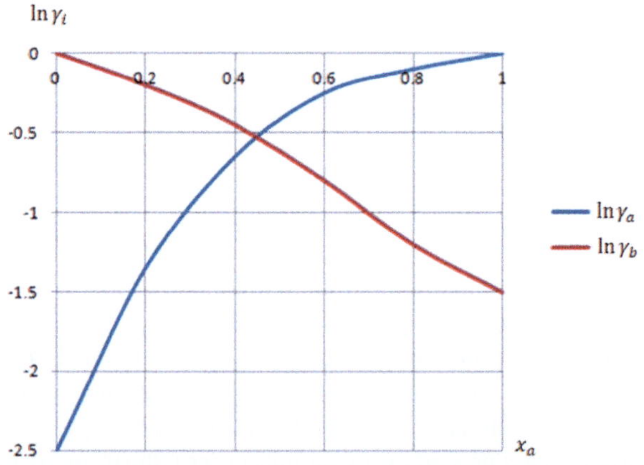

(a) **What is the reference state for each species?**
(b) **Show that the Gibbs–Duhem equation is satisfied at a mole fraction $x_a = 0.4$.**
(c) **Come up with an appropriate model for $g^E$ for this system and find the values of the model parameters.**
(d) **Is it possible for species *a* and *b* to separate into two liquid phases? Explain.**

## Solution 12

### *Worked Solution*

(a) There are a few common reference states. For example, Henry's reference state, Lewis Randall state, or ideal gas. We can observe from the graph that both species $a$ and $b$ are based on the Lewis Randall reference states because as the species compositions approach pure, activity coefficients tend to 1 which is the case for the reference state. When $x_a \to 1$, i.e., pure $a$, then ln $\gamma_a \to 1$; when $x_b \to 1$, i.e., pure $b$, then ln $\gamma_b \to 1$ at $x_a \to 0$.
(b) The graphical relevance of the Gibbs–Duhem equation is commonly in the clever use of the mathematical definition of tangent to a curve that is plotted for natural log of activity coefficient vs. species composition. The Gibbs–Duhem equation is as follows:

$$x_a\left(\frac{d\ln\gamma_a}{dx_a}\right) + (x_b)\left(\frac{d\ln\gamma_b}{dx_a}\right) = 0$$

We can observe that $\left(\frac{d\ln\gamma_a}{dx_a}\right)$ and $\left(\frac{d\ln\gamma_b}{dx_a}\right)$ describe gradients of tangents to the curves as plotted. So all we need to do is specify the point of interest to take tangent from. The problem statement states $x_a = 0.4$; hence, we draw tangents to the two curves at the point $x_a = 0.4$, and the gradients of the tangents will provide values for $\left(\frac{d\ln\gamma_a}{dx_a}\right)$ and $\left(\frac{d\ln\gamma_b}{dx_a}\right)$. Numerically, we substitute the values of the tangent gradients, together with $x_a = 0.4$ and $x_b = 0.6$, to show that the equation is satisfied.
(c) Before we tackle this question, we need to understand more about some simple models that describe non-ideal mixtures. One such example is the Margules equation, which may exist in the two-suffix form or three-suffix form.

The two-suffix Margules equation is in the form $g^E = Ax_ax_b$. This model is suitable if the following conditions are satisfied. We assume for simplicity of discussion, a binary solution of species $a$ and $b$ with the Lewis Randall reference state used. Then we note that for pure $a$ and pure $b$, each will be an ideal solution as defined by the choice of reference state. Then $g^E = 0$ when $x_a = 1$ and $x_b = 1$.

Another useful observation is that the two-suffix Margules equation is symmetric between $x_a$ and $x_b$, since $g^E$ behaves similarly relative to each. From deduction, we can see that in this problem, activity coefficients behave in different manners with respect to composition; this implies that $g^E$ will also behave dissimilarly with respect to $x_a$ and $x_b$.

There is also a three-suffix Margules equation which may be used as an asymmetric model. The way that this equation breaks the symmetry is by an additional term as shown in $g^E = x_a x_b[A + B(x_a - x_b)]$. In this equation, $x_a$ and $x_b$ are no longer interchangeable to give the same value of $g^E$.

In order to find the model parameters $A$ and $B$, we need to express excess Gibbs free energy in differential form.

$$n_T g^E = G^E$$

$$\bar{G}_b^E = \left(\frac{\partial G^E}{\partial n_b}\right)_{T,P,n_a} = \left(\frac{\partial n_T g^E}{\partial n_b}\right)_{T,P,n_a} = \left(\frac{\partial (n_a + n_b) g^E}{\partial n_b}\right)_{T,P,n_a}$$

$$\bar{G}_b^E = (n_a + n_b)\left(\frac{\partial g^E}{\partial n_b}\right)_{T,P,n_a} + g^E\left(\frac{\partial (n_a + n_b)}{\partial n_b}\right)_{T,P,n_a}$$

$$= (n_a + n_b)\left(\frac{\partial g^E}{\partial n_b}\right)_{T,P,n_a} + g^E$$

Using the chain rule, we can express $\left(\frac{\partial g^E}{\partial n_b}\right)_{T,P,n_a}$ in terms of the parameters $A$ and $B$ of the three-suffix Margules equation,

$$\left(\frac{\partial g^E}{\partial n_b}\right)_{T,P,n_a} = \left(\frac{\partial g^E}{\partial x_a}\right)_{T,P,n_a}\left(\frac{\partial x_a}{\partial n_b}\right)_{T,P,n_a}$$

Since $g^E = x_a x_b[A + B(x_a - x_b)]$, and $x_b = 1 - x_a$. Therefore we can express $g^E$ as follows.

$$g^E = x_a(1 - x_a)[A + B(x_a - 1 + x_a)] = \left(x_a - {x_a}^2\right)(A + 2Bx_a - B)$$

$$= (A - B)x_a + (3B - A){x_a}^2 - 2B{x_a}^3$$

$$\left(\frac{\partial g^E}{\partial x_a}\right)_{T,P,n_a} = (A - B) + (6B - 2A)x_a - 6B{x_a}^2$$

$$\left(\frac{\partial x_a}{\partial n_b}\right)_{T,P,n_a} = \left(\frac{\partial (n_a/(n_a + n_b))}{\partial n_b}\right)_{T,P,n_a} = \frac{-n_a}{(n_a + n_b)^2} = \frac{-x_a}{n_a + n_b}$$

$$\left(\frac{\partial g^E}{\partial n_b}\right)_{T,P,n_a} = \left[(A - B) + (6B - 2A)x_a - 6B{x_a}^2\right]\left(\frac{-x_a}{n_a + n_b}\right)$$

Putting our results together, we have

$$\bar{G}_b^E = (n_a + n_b)\left(\frac{\partial g^E}{\partial n_b}\right)_{T,P,n_a} + g^E$$

$$= (n_a + n_b)\left[(A - B) + (6B - 2A)x_a - 6B{x_a}^2\right]\left(\frac{-x_a}{n_a + n_b}\right) + (A - B)x_a$$

$$+ (3B - A){x_a}^2 - 2B{x_a}^3$$

$$= 4Bx_a^3 + (A - 3B)x_a^2$$

We can also express partial molar excess Gibbs free energy in terms of activity coefficient which helps us relate to the data obtained from the graph,

$$\bar{G}_b^E = \bar{G}_b - \bar{G}_b^{\text{ideal}} = \mu_b - \mu_b^{\text{ideal}}$$

$$\bar{G}_b^E = RT \ln\left(\frac{\bar{f}_b}{f_b^{\text{ideal}}}\right) = RT \ln \gamma_b$$

Combining both expressions, we have

$$RT \ln \gamma_b = 4Bx_a^3 + (A - 3B)x_a^2$$

We know from earlier that both species follow the Lewis Randall reference state. For species $a$, when $x_b = 0$, $x_a = 1$, $\ln\gamma_b = -1.5$. At infinite dilution of species $b$, we have the following

$$RT \ln \gamma_b^\infty = B + A = -1.5$$

Similarly, when $x_a = 0$, $x_b = 1$, $\ln\gamma_a = -2.5$. Repeat the above process for species $a$, and we will arrive at two equations with two unknowns, i.e., $A$ and $B$, which we can use to solve for $A$ and $B$.

$$RT \ln \gamma_a = -4Bx_b^3 + (A + 3B)x_b^2$$

$$RT \ln \gamma_b^\infty = A - B = -2.5$$

$$A = -4988.4, \ B = 1247.1$$

(d) The natural logs of both activity coefficients are negative, this means that for both species, the activity coefficients are less than 1. Conceptually, this means that the actual fugacities are lower than that if ideal solution (Lewis Randall pure species reference state). This also means that the "tendency to escape" is lower when in mixture, and the $a$–$b$ intermolecular forces are stronger than $a$–$a$ and $b$–$b$. The two species are unlikely to separate.

We can also observe that the excess Gibbs free energy of mixing has a negative value.

$$g^E = x_a x_b[A + B(x_a - x_b)] = x_a x_b[-4988.4 + 1247.1(x_a - x_b)] < 0$$

$$g^E = \Delta g_{\text{mix}} - RT\sum x_i \ln x_i$$

Since the term "$-RT\sum x_i \ln x_i$" can only be positive or zero, and we observed that is $g^E$ negative, it means that $\Delta g_{\text{mix}}$ must be negative; hence, it is thermodynamically favorable to mix.

## Problem 13

**Substance (*a*) and substance (*b*) form two partially miscible liquid phases. At 220 °C and 1 atm, the compositions of the two phases are given by $x_a^\alpha$=0.9 and $x_a^\beta$=0.2. Estimate the parameters in the three-suffix Margules equation.**

## Solution 13

### *Worked Solution*

At equilibrium, the fugacities in each liquid phase must be equal for each species. So for species $a$, we have

$$\hat{f}_a^{l,\alpha} = \hat{f}_a^{l,\beta}$$

It follows that for a specified ideal reference state for species $a$ (which cancels off on both sides since $\hat{f}_a^{0,\alpha} = \hat{f}_a^{0,\beta} = f_a^0 \text{ } where \text{ } f_a^0 = f_a$ for Lewis Randall, or $H_a$ for Henry's law). We can express equilibrium of species $a$ as follows:

$$\gamma_a^\alpha x_a^\alpha f_a^{0,\alpha} = \gamma_a^\beta x_a^\beta f_a^{0,\beta}$$

$$\gamma_a^\alpha x_a^\alpha = \gamma_a^\beta x_a^\beta$$

Similarly for species $b$,

$$\hat{f}_b^{l,\alpha} = \hat{f}_b^{l,\beta}$$

$$\gamma_b^\alpha x_b^\alpha = \gamma_b^\beta x_b^\beta$$

Given that in phase $\alpha$, species $a$ mole fraction $x_a^\alpha = 0.9$, therefore species $b$ mole fraction will be $x_b^\alpha = 1 - 0.9 = 0.1$.

Similarly in phase $\beta$, species $a$ mole fraction $x_a^\beta = 0.2$, therefore species $b$ mole fraction will be $x_b^\beta = 1 - 0.2 = 0.8$.

Substitute the values for $x_a^\alpha, x_b^\alpha$ and $x_a^\beta, x_b^\beta$ to obtain two independent equations for the remaining four unknowns, $\gamma_a^\alpha, \gamma_a^\beta$ and $\gamma_b^\alpha, \gamma_b^\beta$. Then from the three-suffix Margules equation, $g^E = x_a x_b[A + B(x_a - x_b)]$, we can derive another 2 equations, which will help us solve for the activity coefficients.

$$RT \ln \gamma_a = -4Bx_b^3 + (A + 3B)x_b^2$$

$$RT \ln \gamma_b = 4Bx_a^3 + (A - 3B)x_a^2$$

## Problem 14

**The Gibbs–Duhem equation can be used to derive a thermodynamic consistency test that can be used to evaluate experimental data. This test may be expressed in the form $\int_0^1 \ln\left(\frac{\gamma_a}{\gamma_b}\right) dx_a = 0$, assuming constant *T* and *P*, a mixture containing species *a* and *b* with each using the Lewis Randall reference state. Show the derivation of this test equation, using the Gibbs–Duhem equation.**

## Solution 14

***Worked Solution***

By definition, excess Gibbs free energy measures the deviation of the Gibbs free energy of a real mixture from the ideal reference state. In general, for a species *i*,

$$\bar{G}_i^E = \bar{G}_i - \bar{G}_i^{\text{ideal}} = \mu_i - \mu_i^{\text{ideal}} = RT \ln\left(\frac{\bar{f}_i}{f_i^{\text{ideal}}}\right) = RT \ln \gamma_i$$

For a whole mixture considering all species present, the excess Gibbs free energy $g^E$ can be expressed as follows,

$$g^E = \sum x_i \bar{G}_i^E = RT \sum x_i \ln \gamma_i$$

For a binary mixture containing species *a* and *b*,

$$g^E = RT(x_a \ln \gamma_a + x_b \ln \gamma_b) = RT(x_a \ln \gamma_a + (1 - x_a) \ln \gamma_b)$$

Differentiating with respect to $x_a$, we get

$$\frac{dg^E}{dx_a} = RT\left(x_a \frac{d \ln \gamma_a}{dx_a} + \ln \gamma_a + x_b \frac{d \ln \gamma_b}{dx_a} - \ln \gamma_b\right)$$

We observe that this expression can be simplified by substituting the Gibbs–Duhem equation as shown below.

$$x_a\left(\frac{d \ln \gamma_a}{dx_a}\right) + x_b\left(\frac{d \ln \gamma_b}{dx_a}\right) = 0$$

$$\frac{dg^E}{dx_a} = RT(\ln \gamma_a - \ln \gamma_b) = RT \ln \left(\frac{\gamma_a}{\gamma_b}\right)$$

We can integrate the differential equation from $x_a = 0$ to $x_a = 1$.

$$\int_{x_a=0}^{x_a=1} dg^E = RT \int_0^1 \ln \left(\frac{\gamma_a}{\gamma_b}\right) dx_a$$

Observe that $g^E$ is zero when the mixture contains only a pure species (whether $a$ or $b$), since the ideal reference state is described by the Lewis Randall state.

$$\int_0^1 \ln \left(\frac{\gamma_a}{\gamma_b}\right) dx_a = 0$$

This thermodynamic consistency test is useful when we plot a graph of $\ln \left(\frac{\gamma_a}{\gamma_b}\right)$ against $x_a$. Then the area under the curve should sum to zero (where areas above the horizontal axis are positive in value, and areas below the horizontal axis are negative in value).

## Problem 15

**Show the derivation of the improved test equation for thermodynamic consistency (as shown below) which can be used even when *T* is not constant. This equation is able to account for the temperature dependence of the activity coefficient. Constant *P* may be assumed.**

$$\int_0^1 \ln \left(\frac{\gamma_a}{\gamma_b}\right) dx_a = \int_{T_{\text{pure } b}}^{T_{\text{pure } a}} \frac{\Delta h_{\text{mix}}}{RT^2} dT = -\int_{(1/T)_{\text{pure } b}}^{(1/T)_{\text{pure } a}} \frac{\Delta h_{\text{mix}}}{R} d\left(\frac{1}{T}\right)$$

## Solution 15

***Worked Solution***

---

***Background Concepts***

Before we proceed, it is timely to be reminded of the following concepts:

- Thermodynamic properties can be either intensive or extensive. Extensive properties depend on the size of the system, while intensive properties do not. Hence, extensive properties are additive, while intensive properties are not. Common examples of intensive properties are $T$ and $P$.

- For a system containing a pure substance, its thermodynamic state and hence all of its intensive thermodynamic properties can be determined from two independent intensive properties. Once these two independent intensive properties are specified (e.g., $T$ and $P$), then
  - The rest of the intensive properties (e.g., molar enthalpy $h$) are also constrained, and are hence dependent intensive properties.
  - The extensive property of the pure substance can also be constrained if the size of the system is specified (e.g., total solution enthalpy $H$ is constrained if $T$, $P$, and n are specified). For a mixture containing multiple species, this concept is extrapolated to the sum over all species and the following makes sense (where $K$ hypothetically denotes a thermodynamic property which could be V, H, U, etc.:

$$dK = \left(\frac{\partial K}{\partial T}\right)_{P,n_i} dT + \left(\frac{\partial K}{\partial P}\right)_{T,n_i} dP + \sum_i \left(\frac{\partial K}{\partial n_i}\right)_{T,P,n_j} dn_i$$

- Going back to the definition of a partial molar property:

$$\bar{K}_i = \left(\frac{\partial K}{\partial n_i}\right)_{T,P,n_j}$$

- - - - - - - - - - - - - - - - - - - - - - - - - - - - - - - - - - - - - - - - - - - - - -

Let us now return to the question, where we can first express $dK$ in terms of partial molar $K$.

$$dK = \left(\frac{\partial K}{\partial T}\right)_{P,n_i} dT + \left(\frac{\partial K}{\partial P}\right)_{T,n_i} dP + \sum_i \bar{K}_i dn_i$$

We know that $K = n_T k$ and $n_i = x_i n_T$, so $dK = n_T dk + k dn_T$ and $dn_i = x_i dn_T + n_T dx_i$.

$$\left(\frac{\partial (n_T k)}{\partial T}\right)_{P,n_i} dT + \left(\frac{\partial (n_T k)}{\partial P}\right)_{T,n_i} dP + \sum_i \bar{K}_i d(x_i n_T) = n_T dk + k dn_T$$

$$n_T \left(\frac{\partial k}{\partial T}\right)_{P,n_i} dT + n_T \left(\frac{\partial k}{\partial P}\right)_{T,n_i} dP + \sum_i \bar{K}_i x_i dn_T + \sum_i \bar{K}_i n_T dx_i = n_T dk + k dn_T$$

Let us collect the terms for $n_T$ and $dn_T$ and move all terms to one side of the equation,

$$\left[dk - \left(\frac{\partial k}{\partial T}\right)_{P,n_i} dT - \left(\frac{\partial k}{\partial P}\right)_{T,n_i} dP - \sum_i \bar{K}_i dx_i\right] n_T + \left[k - \sum_i \bar{K}_i x_i\right] dn_T = 0$$

We may deduce that the total size (i.e., no of moles) of the system denoted by $n_T$ should not be affected by how the total number of moles changes. These are independent variables. Hence it follows that the terms in brackets should be zero respectively for the equation to hold.

$$dk - \left(\frac{\partial k}{\partial T}\right)_{P,n_i} dT - \left(\frac{\partial k}{\partial P}\right)_{T,n_i} dP - \sum_i \bar{K}_i dx_i = 0$$

$$k - \sum_i \bar{K}_i x_i = 0$$

Now we can figure out which property is useful in helping us solve this problem. It would be excess Gibbs free energy. Recall that partial molar excess Gibbs free energy is a partial molar property, and so is $\ln\gamma_i$ as follows from the definition of $\bar{G}_i^E = RT\ln\gamma_i$.

For the total solution, molar excess Gibbs free energy is denoted by $g^E$, so substituting this property into the first equation, we obtain:

$$dg^E - \left(\frac{\partial g^E}{\partial T}\right)_{P,n_i} dT - \left(\frac{\partial g^E}{\partial P}\right)_{T,n_i} dP - \sum_i \bar{G}_i^E dx_i = 0$$

$$d\left(\frac{g^E}{RT}\right) - \left(\frac{\partial\left(\frac{g^E}{RT}\right)}{\partial T}\right)_{P,n_i} dT - \left(\frac{\partial\left(\frac{g^E}{RT}\right)}{\partial P}\right)_{T,n_i} dP - \sum_i \left(\frac{\bar{G}_i^E}{RT}\right) dx_i = 0$$

We know from our earlier derivation that

$$\left(\frac{\partial\left(\frac{g^E}{RT}\right)}{\partial T}\right)_{P,n_i} = \frac{-h^E}{RT^2} = \frac{-\Delta h_{\text{mix}}}{RT^2}$$

$$\left(\frac{\partial\left(\frac{g^E}{RT}\right)}{\partial P}\right)_{T,n_i} = v^E = \Delta v_{\text{mix}}$$

Substituting back into the equation,

$$d\left(\frac{g^E}{RT}\right) + \frac{\Delta h_{\text{mix}}}{RT^2} dT - \frac{\Delta v_{\text{mix}}}{RT} dP - \sum_i \ln\gamma_i dx_i = 0$$

For constant $P$, $dP = 0$

$$d\left(\frac{g^E}{RT}\right) = -\frac{\Delta h_{\text{mix}}}{RT^2}dT + \sum_i \ln\gamma_i dx_i = -\frac{\Delta h_{\text{mix}}}{RT^2}dT + \ln\gamma_a dx_a + \ln\gamma_b dx_b$$

$$= -\frac{\Delta h_{\text{mix}}}{RT^2}dT + \ln\gamma_a dx_a + \ln\gamma_b d(1 - x_a) = -\frac{\Delta h_{\text{mix}}}{RT^2}dT + \ln\gamma_a dx_a - \ln\gamma_b dx_a$$

$$= -\frac{\Delta h_{\text{mix}}}{RT^2}dT + \ln\left(\frac{\gamma_a}{\gamma_b}\right)dx_a$$

Integrate both sides, we have the following.

$$\int d\left(\frac{g^E}{RT}\right) = -\int \frac{\Delta h_{\text{mix}}}{RT^2}dT + \int \ln\left(\frac{\gamma_a}{\gamma_b}\right)dx_a$$

We know that for the two limits of integration, i.e., pure $a$ and pure $b$, excess Gibbs free energy is zero as that is the ideal reference state.

$$0 = -\int_{T_{\text{pure } b}}^{T_{\text{pure } a}} \frac{\Delta h_{\text{mix}}}{RT^2}dT + \int_0^1 \ln\left(\frac{\gamma_a}{\gamma_b}\right)dx_a$$

$$\int_{T_{\text{pure } b}}^{T_{\text{pure } a}} \frac{\Delta h_{\text{mix}}}{RT^2}dT = \int_0^1 \ln\left(\frac{\gamma_a}{\gamma_b}\right)dx_a$$

We can also show that $d\left(\frac{1}{T}\right) = -\frac{dT}{T^2}$. Therefore we obtain the following.

$$\int_0^1 \ln\left(\frac{\gamma_a}{\gamma_b}\right)dx_a = \int_{T_{\text{pure } b}}^{T_{\text{pure } a}} \frac{\Delta h_{\text{mix}}}{RT^2}dT = -\int_{T_{\text{pure } b}}^{T_{\text{pure } a}} \frac{\Delta h_{\text{mix}}}{R}d\left(\frac{1}{T}\right)$$

## Problem 16

(a) **Express $RT \ln \varphi_i$ in terms of a deviation function of g from ideal gas behavior.**
(b) **Derive an expression to evaluate the deviation function for different specified temperatures and pressures.**
(c) **What criterion applies for fugacity coefficient at vapor–liquid equilibrium?**

(d) **If the pure fluid obeys the Equation of State $z=\frac{PV}{RT}=1+AP+BP^2$, would you recommend using the fugacity coefficient as a means of predicting the vapor pressure for this fluid? $A = 10$ bar$^{-1}$,$B = 0.2$ bar$^{-2}$.**

## Solution 16

***Worked Solution***

(a)

---

***Background Concepts***

The fugacity coefficient $\varphi_i$ of a pure fluid helps to measure its deviation from ideal gas behavior. We know that $\varphi_i = f_i/P$ , so when $P$ tends to zero (or when $V$ tends to infinity), we tend towards ideal gas behavior where $\varphi_i = 1$. Fugacity is also correlated to Gibbs free energy or chemical potential of pure $i$ by the equation $g_i = \mu_i = RT \ln f_i + h(T)$. Recall that chemical potential is simply partial molar Gibbs free energy $\mu_i = \left(\frac{\partial G}{\partial n_i}\right)_{T,P,n_{j\neq i}} = \bar{G}_i = g_i$ (for pure $i$).

---

The deviation function may be expressed mathematically as shown, where $g_i^0$ is the ideal gas reference state at low $P$.

$$g_i^E = g_i(T,P) - g_i^0(T,P)$$

We can define a new function $h(T)$ in the expression $g_i = RT \ln f_i + h(T)$, which is a function of temperature obtained when we integrate at constant temperature from $P_{\text{low}}$ to $P$. So we arrive at $h(T) = \mu_i^0 - RT \ln f_i^0 = g_i^0 - RT \ln P_{low}$.

We know that $dg_i = -s_i dT + v_i dP$; hence $\left(\frac{\partial g_i}{\partial P}\right)_T = v_i$. For ideal gas, $v_i = \frac{RT}{P_{\text{low}}}$, while for a real gas, $v_i = \frac{RT}{P}$, where the measured pressure $P$ for a real gas is effectively a "corrected pressure" which is also called fugacity $f_i$.

At constant $T$,

$$dg_i = v_i dP$$

$$dg_i = \frac{RT}{P} dP = RTd \ln P$$

$$g_i - g_i^0 = RT \ln f_i - RT \ln P_{\text{low}}$$

Since $g_i = RT \ln f_i + h(T)$, we can observe that $h(T) = g_i^0 - RT \ln P_{\text{low}}$. Now we know that by definition $\varphi_i = f_i/P$, where $P$ here refers to the ideal gas pressure, or $P_{\text{low}}$, we continue from the above where

$$g_i(T,P) - g_i^0(T,P) = RT\ln f_i - RT\ln P_{\text{low}} = g_i^E$$

$$g_i^E = RT\ln\frac{f_i}{P_{\text{low}}} = RT\ln\varphi_i$$

(b)

$$dg_i^E = -s_i^E dT + v_i^E dP$$

At a particular isotherm $T_1$,

$$g_i^E = \int_0^{P_1} v_i^E dP = \int_0^{P_1}\left(v_i - \frac{RT_1}{P}\right)dP$$

A series of $g_i^E$ at specified temperatures and pressures can be found by substituting the values for ($T_1$, $P_1$).

(c) At vapor–liquid equilibrium (VLE) ,

$$\mu_i^V = \mu_i^L$$

$$\mu_i^V - \mu_i^L = 0 = RT\ln\frac{f_i^V}{f_i^L} = RT\ln\frac{\varphi_i^V P}{\varphi_i^L P_{\text{sat}}}$$

Because we have a pure substance, therefore $P = P_{\text{sat}}$ and the expression simplifies to

$$0 = RT\ln\frac{\varphi_i^V}{\varphi_i^L}$$

$$\varphi_i^V = \varphi_i^L$$

Note that fugacity is not limited to gases, and fugacity coefficient is similarly not limited. Fugacity coefficients may apply to liquids and solids as well.

(d) For this equation of state (EOS), it is a function of a power series expansion of pressure; hence it is volume explicit. Volume explicit EOS will not yield multiple roots for volume at a given $P$. This means that we are unable to obtain two distinct values of volume for a given $P$. In our case, at $P = P_{\text{sat}}$, we should have $v^V$ and $v^L$ that both satisfy the EOS to account for the two phase behavior at VLE. Hence this equation is not suitable for our case.

Mathematically we can show that the EOS is unsuitable to model for the vapor phase as follows:

$$g_i^E = g_i(T,P) - g_i^0(T,P) = RT\ln\varphi_i^V = \int_0^P\left(v_i - \frac{RT}{P}\right)dP$$

$$RT \ln \varphi_i^V = \int_0^P \frac{RT}{P}\left(1 + AP + BP^2 - 1\right)dP = \int_0^P (ART + BRTP)dP$$

$$\ln \varphi_i^V = \int_0^P (A + BP)dP = AP + BP^2$$

If we assume ideal gas, $\varphi_i^V = 1, AP + BP^2 = 0$, we have two roots for $P$ which are negative in value or zero, respectively, which has no physical meaning.

However, this EOS can make sense for the liquid phase. Assuming at VLE, $\varphi_i^V = \varphi_i^L$ as derived earlier since this is a pure substance VLE, then $\ln \varphi_i^V \neq 0$ and the values for $P$ may be physically meaningful. Hence, this EOS only applies to the liquid phase but not the vapor phase.

## Problem 17

**A new process was developed for synthesizing a new fertilizer. One of the process steps involves adding a strong acid to a neutral aqueous solution. Heat is produced from this mixing, and if it is too much, other by-reactions will become significant and affect product yield adversely. Hence, it is important to control the heat produced, to maintain isothermal conditions. You are required to find the heat load on your cooling system, assuming a base case of 1 kg/s flow rate of aqueous solution fed to the reactor, to which the acid is added. The only data you have is a couple of test results from adding different volumes of pure acid into 3 ml water and measurements of the heat produced while maintaining the solution at isothermal conditions. The results are shown below:**

**Heat produced when acid is added to 3 ml water under isothermal conditions at 25 °C**

| Volume of acid added/ml | Heat produced/J |
|---|---|
| 0.2 | −300 |
| 0.5 | −600 |
| 1.0 | −1170 |
| 1.3 | −1440 |

**Heat produced when water is added to 3 ml acid under isothermal conditions at 25 °C**

| Volume of water added/ml | Heat produced/J |
|---|---|
| 0.10 | −200 |
| 0.17 | −320 |
| 0.25 | −452 |
| 0.33 | −594 |

**Given that the molar volume of water is 0.018 L/mol and that of acid is 0.054 L/mol.**

(a) **Discuss how the data can be related to partial molar enthalpies. Estimate the infinite dilution partial molar enthalpies for both the acid and water. You may assume that the molar enthalpy of a pure species is 0.**
(b) **Which of the two models for partial molar enthalpies suggested below would be better for your analysis?**
(c) **Estimate values for $A_{ab}$ and $A_{ba}$ and discuss their significance.**
(d) **Determine the heating/cooling load in Watts on the system needed to maintain isothermal conditions as a function of the amount of acid added to the aqueous reagent feed.**

## Solution 17

***Worked Solution***

(a) You can calculate the columns in italics (shown below) from the data provided.

| Volume of acid added/liters | Heat produced/J | *Moles of acid added* | *Moles of water in 3 ml* | $\Delta h_{mix}$*/J per mol of solution* |
|---|---|---|---|---|
| 0.0002 | −300 | *0.0037 (=0.0002/0.054)* | *0.17 (=0.003/0.018)* | *−1730 (=−300/(0.17+0.0037))* |
| 0.0005 | −600 | ... | *0.17* | ... |
| 0.0010 | −1170 | ... | *0.17* | ... |
| 0.0013 | −1440 | ... | *0.17* | ... |

| Volume of water added/liters | Heat produced/J | *Moles of water added* | *Moles of acid in 3 ml* | $\Delta h_{mix}$*/J per mol of solution* |
|---|---|---|---|---|
| 0.00010 | −200 | *0.0056 (=0.00010/0.018)* | *0.056 (=0.003/0.054)* | *−3250 (=−200/(0.0056+0.056))* |
| 0.00017 | −320 | ... | *0.056* | ... |
| 0.00025 | −452 | ... | *0.056* | ... |
| 0.00033 | −594 | ... | *0.056* | ... |

The enthalpy change of mixing is defined as follows.

$$\Delta h_{\text{mix}} = h - (x_{\text{water}} h_{\text{water}} + x_{\text{acid}} h_{\text{acid}})$$

$$h = x_{\text{water}} \bar{H}_{\text{water}} + x_{\text{acid}} \bar{H}_{\text{acid}}$$

For pure substance, question defined that we may assume the pure species molar enthalpy is 0. Hence $h_{\text{water}} = 0$; $h_{\text{acid}} = 0$.

$$\Delta h_{\text{mix}} = h - (0)$$

$$\Delta h_{\text{mix}} = x_{\text{water}}\bar{H}_{\text{water}} + x_{\text{acid}}\bar{H}_{\text{acid}}$$

This is the equation relating the data to the partial molar enthalpies.

Now, we need to estimate the infinite dilution partial molar enthalpies for both the acid and water. To do so, we can first plot $\Delta H_{\text{mix}}$ against $x_{\text{acid}}$ for values of $x_{\text{acid}}$ from 0 to 1. We then linearly fit two straight lines that are tangents to the plot at $x_{\text{acid}} = 0$ and at $x_{\text{acid}} = 1$. Note that the general equation for a straight line is $Y = mX+C$.

For both straight line tangents, the straight line equation takes on the following form whereby the gradient m $= \bar{H}_{\text{acid}}$ and vertical axis intercept $C = \bar{H}_{\text{water}}$. Also notice that when no acid is added to water, $Y = \Delta H_{\text{mix}} = 0$:

$$Y = mX + C$$

$$\Delta h_{\text{mix}} = \bar{H}_{\text{acid}}x_{\text{acid}} + x_{\text{water}}\bar{H}_{\text{water}}$$

The linearly fit tangent at $x_{\text{acid}} = 0$ gives $Y = (-62{,}500)X + 0$. Hence, $\bar{H}^{\infty}_{\text{acid}} = -62,500$ J/mol (i.e., the $m$ value).

The linearly fit tangent at $x_{\text{acid}} = 1$ gives $Y = (32{,}000)X - 32{,}100$. Hence, $\bar{H}^{\infty}_{\text{water}} = -32,100$ J/mol (i.e., the $C$ value).

(b) Model 1:

$$\bar{H}_a = [A_{ab} + 2(A_{ba} - A_{ab})x_a x_b]$$

$$\bar{H}_b = [A_{ba} + 2(A_{ab} - A_{ba})x_a x_b]$$

Model 2:

$$\bar{H}_a = x_b^2[A_{ab} + 2(A_{ba} - A_{ab})x_a]$$

$$\bar{H}_b = x_a^2[A_{ba} + 2(A_{ab} - A_{ba})x_b]$$

Any partial molar property should satisfy the Gibbs–Duhem (GD) equation, as shown below:

$$x_a\frac{d\bar{H}_a}{dx_a} + x_b\frac{d\bar{H}_b}{dx_a} = 0$$

For model 1, we may derive the expressions to test the Gibbs–Duhem validity,

$$\bar{H}_a = [A_{ab} + 2(A_{ba} - A_{ab})x_a x_b]$$

$$\bar{H}_b = [A_{ba} + 2(A_{ab} - A_{ba})x_a x_b]$$

$$x_a \frac{d\bar{H}_a}{dx_a} = 2(A_{ba} - A_{ab})x_a x_b - 2(A_{ba} - A_{ab}){x_a}^2$$

$$x_b \frac{d\bar{H}_b}{dx_a} = 2(A_{ab} - A_{ba}){x_b}^2 - 2(A_{ab} - A_{ba})x_a x_b$$

Combining the two results into the GD equation,

$$2(A_{ba} - A_{ab})x_a x_b - 2(A_{ba} - A_{ab}){x_a}^2 + 2(A_{ab} - A_{ba}){x_b}^2 - 2(A_{ab} - A_{ba})x_a x_b = 0$$

$$(A_{ba} - A_{ab})(x_a - x_b)^2 = 0$$

We obtain the results $A_{ba} = A_{ab}$ or $x_a = x_b$. Model 1 does not work because if $A_{ba} = A_{ab}$, then $\bar{H}_a = A_{ab}$ and $\bar{H}_b = A_{ba}$ which means $\bar{H}_a$ and $\bar{H}_b$ should be constants that are independent of mole fractions of $a$ and $b$. However, this is false. Also $x_a = x_b$ need not always be true.

For model 2, we may derive the expressions to test the Gibbs–Duhem validity,

$$\bar{H}_a = x_b^2[A_{ab} + 2(A_{ba} - A_{ab})x_a]$$

$$\bar{H}_b = x_a^2[A_{ba} + 2(A_{ab} - A_{ba})x_b]$$

$$x_a \frac{d\bar{H}_a}{dx_a} = -2x_a x_b A_{ab} - 4(A_{ba} - A_{ab}){x_a}^2 x_b + 2(A_{ba} - A_{ab}){x_b}^2 x_a$$

$$x_b \frac{d\bar{H}_b}{dx_a} = 2x_a x_b A_{ba} + 4(A_{ab} - A_{ba}){x_b}^2 x_a - 2(A_{ab} - A_{ba}){x_a}^2 x_b$$

Combining the two results into the GD equation,

$$-2x_a x_b A_{ab} - 4(A_{ba} - A_{ab}){x_a}^2 x_b + 2(A_{ba} - A_{ab}){x_b}^2 x_a + 2x_a x_b A_{ba}$$
$$+4(A_{ab} - A_{ba}){x_b}^2 x_a - 2(A_{ab} - A_{ba}){x_a}^2 x_b = 0$$

$$(A_{ba} - A_{ab})x_a x_b(1 - x_a - x_b)^2 = 0$$

We obtain the results $A_{ba} = A_{ab}$ or $x_a = 0$ or $x_b = 0$ or $1 - x_a - x_b = 0$. Model 2 can work as it is true that $1 - x_a - x_b = 0$.

(c) Assume species $a$ is water, and $b$ is acid. Following from part a, we know that $\bar{H}^{\infty}_{\text{acid}} = \bar{H}^{\infty}_b = -62{,}500$ J/mol, and $\bar{H}^{\infty}_{\text{water}} = \bar{H}^{\infty}_a = -32{,}100$ J/mol. Substituting values into model 2,

For species $a$,

$$\bar{H}_a = x_b^2[A_{ab} + 2(A_{ba} - A_{ab})x_a]$$

$$\bar{H}^{\infty}_a = -32{,}100 = 1[A_{ab} + 0] = A_{ab}$$

For species $b$,

$$\bar{H}_b = x_a^2[A_{ba} + 2(A_{ab} - A_{ba})x_b]$$

$$\bar{H}_b^{\infty} = -62,500 = 1[A_{ba} + 0] = A_{ba}$$

The values of $A_{ab}$ and $A_{ba}$ are the infinite dilution partial molar enthalpies for species $a$ (i.e., pure $b$) and $b$ (i.e., pure $a$), respectively.

(d) The heat generated from mixing is defined as follows.

$$\Delta h_{\text{mix}} = x_{\text{water}}\bar{H}_{\text{water}} + x_{\text{acid}}\bar{H}_{\text{acid}}$$

Substituting model 2 expressions,

$$\bar{H}_a = x_b^2[A_{ab} + 2(A_{ba} - A_{ab})x_a]$$

$$\bar{H}_b = x_a^2[A_{ba} + 2(A_{ab} - A_{ba})x_b]$$

We get the following expression for the enthalpy change of mixing:

$$\Delta h_{\text{mix}} = x_a x_b^2[A_{ab} + 2(A_{ba} - A_{ab})x_a] + x_b x_a^2[A_{ba} + 2(A_{ab} - A_{ba})x_b]$$

We may specify values of $x_a$, and $x_b$ will then be constrained and specified by $x_b = 1 - x_a$. $A_{ab}$ and $A_{ba}$ are known from part c; therefore, we may plot a graph of $\Delta h_{\text{mix}}$ against $x_a$, which will give us the amount of heat generated as a function of mole fraction of acid.

## Problem 18

**The compressibility factor of a substance *a* at 450 K and 10 bar is measured to be 0.89. At what temperature and pressure would substance *b* have a similar value for the compressibility factor? Thermodynamic data for both compounds appears below.**

| | Substance $a$ | Substance $b$ |
|---|---|---|
| **Molar mass [g/mol]** | 74 | 84 |
| **Acentric factor [$\omega$]** | 0.281 | 0.280 |
| **Critical temperature $T_c$ [K]** | 470 | 500 |
| **Critical pressure $P_c$ [bar]** | 36 | 31 |
| **Critical compressibility [$z$]** | 0.263 | 0.265 |
| **Critical volume $v_c$ [$cm^3$/mol]** | 280 | 350 |
| **Normal boiling point [K]** | 308 | 340 |

## Solution 18

***Worked Solution***

Before we tackle this problem, recall the law of corresponding states which states that all fluids behave similarly (i.e., same compressibility factor) at the same reduced temperature and pressure. Reduced parameters can be found from $T_r = \frac{T}{T_c}$, $P_r = \frac{P}{P_c}$. Compressibility factor is a dimensionless grouping denoted by $z = \frac{Pv}{RT} = f(T_r, P_r)$. Note that acentric factor $\omega$ measures how non-spherical a molecule is.

$$T_{r,a} = \frac{T}{T_{c,a}} = \frac{450}{470} = 0.96$$

$$P_{r,a} = \frac{P}{P_{c,a}} = \frac{10}{36} = 0.28$$

At $z = 0.89$, we want substance $b$ to behave similarly. This means that the reduced $T$ and P will be the same as that for substance $a$.

$$T = T_{r,b} \times T_{c,b} = 0.96 \times 500 = 480 \text{ K}$$

$$P = P_{r,b} \times P_{c,b} = 0.28 \times 31 = 8.7 \text{ bar}$$

## Problem 19

**The partial molar property of a particular component may be defined as follows:**

$$\bar{K}_i = \left(\frac{\partial K}{\partial n_i}\right)_{T,P,n_j}$$

(a) **A common mistake is to assume the following, explain why this is a mistake.**

$$\bar{K}_i = \left(\frac{\partial k}{\partial x_i}\right)_{T,P,n_j}$$

**Show that for a mixture containing two species *a* and *b*,**

$$\bar{K}_a = k + (1 - x_a)\left(\frac{\partial k}{\partial x_a}\right)_{T,P,n_b}$$

(b) **The partial molar heat of mixing of water (species $a$) and a non-volatile substance (species $b$) has been found from the temperature dependence of water partial pressures. This data can be expressed as follows:**

$$\overline{\Delta H}_{mix,a} = b(1 - x_a)^2$$

**Use this to determine the heat of mixing per mole solution, $\Delta h_{\mathrm{mix}}$ as a function of composition, $x_a$.**

## Solution 19

***Worked Solution***

(a)

$$\bar{K}_i = \left(\frac{\partial K}{\partial n_i}\right)_{T,P,n_j}$$

For a binary mixture, we have the following for species $a$

$$\bar{K}_a = \left(\frac{\partial(n_T k)}{\partial n_a}\right)_{T,P,n_b} = n_T\left(\frac{\partial k}{\partial n_a}\right)_{T,P,n_b} + k\left(\frac{\partial n_T}{\partial n_a}\right)_{T,P,n_b}$$

We know that $n_T = n_a + n_b$, so $\left(\frac{\partial n_T}{\partial n_a}\right)_{T,P,n_b} = \left(\frac{\partial(n_a+n_b)}{\partial n_a}\right)_{T,P,n_b} = 1$. Therefore we have the following,

$$\bar{K}_a = n_T\left(\frac{\partial k}{\partial n_a}\right)_{T,P,n_b} + k$$

Now we think about how to convert $\left(\frac{\partial k}{\partial n_a}\right)_{T,P,n_b}$ into the form $\left(\frac{\partial k}{\partial x_a}\right)_{T,P,n_b}$.

$$dn_a = d(n_T x_a) = n_T dx_a + x_a dn_T$$

Differentiating with respect to $x_a$, the expression becomes

$$\frac{dn_a}{dx_a} = n_T + x_a\frac{dn_T}{dx_a} = n_T + x_a\frac{d(n_a+n_b)}{dx_a} = n_T + x_a\frac{dn_a}{dx_a} + x_a\frac{dn_b}{dx_a}$$

In our context, we are finding $\bar{K}_a$. So $n_b$ is constant and $dn_b = 0$.

$$\frac{dn_a}{dx_a} = n_T + x_a\frac{dn_a}{dx_a}$$

$$\frac{dn_a}{dx_a}(1 - x_a) = n_T$$

$$\frac{dn_a}{dx_a} = \frac{n_T}{1 - x_a}$$

$$dn_a = \frac{n_T}{x_b} dx_a$$

Substitute this result back into our earlier expression for $\bar{K}_a$, we get

$$\bar{K}_a = x_b \left( \frac{\partial k}{\partial x_a} \right)_{T,P,n_b} + k = k + (1 - x_a) \left( \frac{\partial k}{\partial x_a} \right)_{T,P,n_b}$$

(b)

$$\Delta h_{\text{mix}} = x_a \overline{\Delta H}_{\text{mix},a} + x_b \overline{\Delta H}_{\text{mix},b}$$

We can relate the partial molar property of one species to that of another using the Gibbs–Duhem equation. We are given $\overline{\Delta H}_{\text{mix},a}$, so we can use Gibbs–Duhem to obtain the expression for $\overline{\Delta H}_{\text{mix},b}$ as we go about determining $\Delta h_{\text{mix}}$.

$$x_a \left( \frac{d\overline{\Delta H}_{\text{mix},a}}{dx_a} \right) + x_b \left( \frac{d\overline{\Delta H}_{\text{mix},b}}{dx_a} \right) = 0$$

$$-2bx_a(1 - x_a) + (1 - x_a) \left( \frac{d\overline{\Delta H}_{\text{mix},b}}{dx_a} \right) = 0$$

$$-2bx_a + \left( \frac{d\overline{\Delta H}_{\text{mix},b}}{dx_a} \right) = 0$$

$$d\overline{\Delta H}_{\text{mix},b} = 2bx_a dx_a$$

$$\int_0^{\overline{\Delta H}_{\text{mix},b}} d\overline{\Delta H}_{\text{mix},b} = \int_0^1 2bx_a dx_a$$

$$\overline{\Delta H}_{\text{mix},b} = bx_a^2$$

Substituting this result, we can find $\Delta h_{\text{mix}}$

$$\Delta h_{\text{mix}} = x_a \overline{\Delta H}_{\text{mix},a} + x_b \overline{\Delta H}_{\text{mix},b} = bx_a(1 - x_a)^2 + bx_a^2(1 - x_a)$$
$$= b(1 - x_a)(x_a - x_a^2 + x_a^2) = bx_a(1 - x_a)$$

## Problem 20

**At 350 K, the enthalpy of a solution is given by:**

$$\boldsymbol{h[\mathrm{J/mol}] = 200x_a + 200x_b + 50x_a x_b}$$

**A mixer operating at steady state combines the two inflows to form a single outflow stream Z. The following data is available about the inflow conditions:**

| | Flowrate [mol/s] | $x_a$ | $x_b$ | Temperature [K] |
|---|---|---|---|---|
| **Inflow X** | 2.5 | 0.3 | 0.7 | 350 |
| **Inflow Y** | 1.5 | 0.8 | 0.2 | 350 |

(a) **Calculate the mole fraction of component a in the outflow stream.**
(b) **Find the rate of heat transfer in Watts required by the mixer if the outflow temperature is 350 K.**
(c) **Is the binary solution in this problem ideal or non-ideal? Explain your reasons.**
(d) **Calculate the value of the partial molar enthalpies of components *a* and *b* for a solution with $x_a = 0.5$.**
(e) **If components *a* and *b* are needed to mix in a lab beaker, would you recommend pouring pure *a* into a beaker which contains pure *b*, or pouring pure *b* into a beaker which contains pure *a*? Explain your reasons.**

## Solution 20

***Worked Solution***

(a) This is an open system with inflows and outflows. We may construct mass balances as follows:

| | Flowrate [mol/s] | $x_a$ | $x_b$ | Temperature [K] |
|---|---|---|---|---|
| **Inflow X** | 2.5 | 0.3 | 0.7 | 350 |
| **Inflow Y** | 1.5 | 0.8 | 0.2 | 350 |

Total mass: $\dot{n}_X + \dot{n}_Y = \dot{n}_Z \rightarrow 2.5 + 1.5 = 4.0$
Component balance: $x_{X,a}\dot{n}_X + x_{Y,a}\dot{n}_Y = x_{Z,a}\dot{n}_Z \rightarrow 0.3(2.5) + 0.8(1.5) = x_{Z,a}(4.0)$

$$x_{Z,a} = 0.49; \quad x_{Z,b} = 1 - 0.49 = 0.51$$

(b) We may construct an energy balance as follows:

$$\dot{n}_X h_X + \dot{n}_Y h_Y - \dot{n}_Z h_Z + \dot{Q} = 0$$

$$2.5h_X + 1.5h_Y - 4.0h_Z + \dot{Q} = 0$$

We are given $h = 200x_a + 200x_b + 50x_ax_b$, or $h = 200(x_a + x_b) + 50x_ax_b = 200 + 50x_ax_b$, so we may substitute into the energy balance:

$$h_X = 200 + 50(0.3)(0.7) = 210.5$$

$$h_Y = 200 + 50(0.8)(0.2) = 208$$

$$h_Z = 200 + 50(0.49)(0.51) = 212.5$$

$$\dot{Q} = -2.5h_X - 1.5h_Y + 4.0h_Z = 11.75W$$

(c) If the mixture was non-ideal, then $\Delta h_{\text{mix}} \neq 0$. Heat input is required to maintain the temperature at 350 K, this implies that the mixing causes temperature to drop (endothermic), and the mixture is therefore non-ideal. We can observe that in the equation $h = 200x_a + 200x_b + 50x_ax_b$ that when $x_a = 1$, $h = 200$ and when $x_b = 1$, $h = 200$. And as long as there is a mixture of $a$ and $b$, $h$ is lowered from the pure species $h = 200$ J/mol. Therefore $\Delta h_{\text{mix}} = h^E$ is negative.

(d) Given that $x_a = 0.5$, we know that $x_b = 1 - 0.5 = 0.5$. So we can find $\bar{H}_a$ and $\bar{H}_b$.

$$h = 200x_a + 200x_b + 50x_ax_b = 200\left(\frac{n_a}{n_T}\right) + 200\left(\frac{n_b}{n_T}\right) + 50\left(\frac{n_a}{n_T}\right)\left(\frac{n_b}{n_T}\right)$$

$$n_T h = H = 200n_a + 200n_b + 50\left(\frac{n_a n_b}{n_T}\right) = 200n_a + 200n_b + 50\left(\frac{n_a n_b}{n_a + n_b}\right)$$

$$\bar{H}_a = \left(\frac{\partial H}{\partial n_a}\right)_{T,P,n_b} = 200 + 50\left(\frac{n_b(n_a + n_b) - n_a n_b}{(n_a + n_b)^2}\right) = 200 + 50{x_b}^2$$

$$= 200 + 50(0.5)^2 = 212.5\ \text{J/mol}$$

$$\bar{H}_b = \left(\frac{\partial H}{\partial n_b}\right)_{T,P,n_a} = 200 + 50\left(\frac{n_a(n_a + n_b) - n_a n_b}{(n_a + n_b)^2}\right) = 200 + 50{x_a}^2$$

$$= 200 + 50(0.5)^2 = 212.5\ \text{J/mol}$$

(e) At 350 K, the heat released is symmetric with respect to $x_a$, this means that if we interchange species $a$ and $b$, the equation for $\Delta h_{\text{mix}}$ is the same. Species $a$ behaves in the same way in a given proportion of species $b$, as species $b$ does in that same proportion of species $a$. Hence, regardless whether we mix species $a$ into $b$, or vice versa, the final solution enthalpy is the same.

## Problem 21

**Indicate whether each of the following statements is true or false and provide suitable explanations and justifications for each.**

(a) $\sum \left(x_i d\hat{\varphi}_i\right) = 0$ **by the Gibbs–Duhem equation.**
(b) **For an ideal solution,** $S^E = -R \sum (x_i \ln x_i)$**, to accommodate the effects of mixing, and** $S \neq \sum (x_i S_i)$**.**
(c) **The modified Raoult's expression assumes that the interactions between molecules in the gas phase and between molecules in the liquid phase are both zero.**
(d) **At equilibrium, a flavoring agent dissolved in salad dressing (a phase of oil in a phase of water) would have the same molar concentration.**

## Solution 21

***Worked Solution***

(a) False because the Gibbs–Duhem equation only holds for partial molar properties. $\ln \hat{\varphi}_i$ is a partial molar property as it is related to excess Gibbs free energy, but $\hat{\varphi}_i$ is not itself a partial molar property.
(b) False. Any excess property for an ideal solution is zero, since excess properties measure the deviation from the ideal solution reference state. The expression $-R \sum (x_i \ln x_i)$ is the entropy of mixing from ideal mixing. It is not an excess property.
(c) False. The modified Raoult's expression assumes that intermolecular forces between gas molecules are zero for ideal gas. But it does not assume this is the case for liquid phase. For ideal solutions, there exists intermolecular forces although these forces are assumed to be the same between different molecules, i.e., like interactions (*a–a* or *b–b*) are the same as unlike interactions (*a–b* or *b–a*) in a binary mixture. For non-ideal solutions, the activity coefficient will account for any deviations from this assumption.
(d) False. At equilibrium, the fugacities of the specie (e.g., flavoring agent) in oil phase and water phase are equal. However the molar concentrations in each phase need not be equal.

$$\hat{f}_i^{\text{oil}} = \hat{f}_i^{\text{water}}$$

$$\gamma_i^{\text{oil}} x_i^{\text{oil}} P_i^{\text{sat, oil}} = \gamma_i^{\text{water}} x_i^{\text{water}} P_i^{\text{sat, water}}$$

From here, you may observe that $x_i^{\text{oil}}$ need not be equal to $x_i^{\text{water}}$.

## Problem 22

**At atmospheric pressure, substances *a* and *b* form an azeotrope that boils at 65 °C and has a mole fraction of substance *a* in the liquid of 0.3. The saturation vapor pressures of *a* and *b* at 65 °C are 1.3 and 0.97 atm, respectively.**

(a) **Calculate the activity coefficients of substance *a* and *b* at the azeotrope.**
(b) **Determine a value for excess Gibbs energy $g^E$ for the azeotrope in kcal/mol. If enthalpic factors dominated in influencing $g^E$ for this system, would the adiabatic mixing of pure *a* and pure *b* (each at 65 °C) to generate a solution at the composition of the azeotrope produce a solution that is hotter or cooler than 65 °C?**
(c) **Using earlier results, calculate the vapor composition over the liquid solution at 65 °C that contains 12 mol% of substance *a*. Consider the van Laar equations to describe this system as shown below:**

$$g^E = x_a x_b \left( \frac{AB}{Ax_a + Bx_b} \right)$$

$$RT \ln \gamma_a = A \left( \frac{Bx_b}{Ax_a + Bx_b} \right)^2$$

$$RT \ln \gamma_b = B \left( \frac{Ax_a}{Ax_a + Bx_b} \right)^2$$

**Also note the following result.**

$$\frac{\ln \gamma_b}{\ln \gamma_a} = \frac{A'_{ab}}{A'_{ba}} \left[ \frac{x_a}{x_b} \right]^2$$

(d) **Calculate the total pressure over the solution at the condition of part c.**

## Solution 22

*Worked Solution*

(a) At vapor–liquid equilibrium for our binary mixture,

$$\hat{\varphi}_a y_a P = \gamma_a x_a P_a^{\text{sat}}$$

$$\hat{\varphi}_b y_b P = \gamma_b x_b P_b^{\text{sat}}$$

The azeotrope is a point in the phase diagram where the Px and Py curves go through a maximum or minimum. At this point, the mole fractions of each species in the liquid phase are the same as that in the vapor phase, i.e., $x_a = y_a$; $x_b = y_b$. Azeotropes are typically observed in mixtures where the deviation from

Raoult's law is significant, i.e., the unlike interactions are very different from the like interactions. Moreover, the saturation pressures of the two species are close in value. Our system is at atmospheric pressure, which is considerably low and hence ideal gas may be assumed for the gas phase and $\hat{\varphi}_a = \hat{\varphi}_b = 1$.

The equations become simplified to:

$$P = \gamma_a P_a^{\text{sat}} \rightarrow \gamma_a = \frac{1}{1.3} = 0.77 \text{ bar}$$

$$P = \gamma_b P_b^{\text{sat}} \rightarrow \gamma_b = \frac{1}{0.97} = 1.03 \text{ bar}$$

(b)

$$g^E = RT \sum x_i \ln \gamma_i = RT(x_a \ln \gamma_a + x_b \ln \gamma_b)$$

$$g^E = \left(\frac{2 \text{ cal} \cdot \text{K}}{\text{mol}}\right)(273.15 + 65)(0.3 \ln (0.77 \text{ bar}) + 0.7 \ln (1.03 \text{ bar}))$$
$$= -0.04 \text{ kcal/mol}$$

Note that $g^E = h^E - Ts^E < 0$. If we assume enthalpic effects dominate, then we may ignore contributions from the term $Ts^E$ and assume that $h^E < 0$. For adiabatic mixing of pure components, $\Delta h_{\text{mix}} = h^E$ since $\Delta h_{\text{mix}}^{\text{ideal}} = 0$. If $\Delta h_{\text{mix}} < 0$, then mixing is exothermic and temperature will increase. If we want to maintain constant temperature, we need to remove heat from the system. Since the system is adiabatic, $Q = 0$; therefore, temperature will rise.

(c) Let us now evaluate values of $A'_{ab}$ and $A'_{ba}$ at the azeotrope:

$$\frac{\ln \gamma_b}{\ln \gamma_a} = \frac{A'_{ab}}{A'_{ba}} \left[\frac{x_a}{x_b}\right]^2$$

$$\frac{\ln 1.03}{\ln 0.77} = \frac{A'_{ab}}{A'_{ba}} \left[\frac{0.3}{0.7}\right]^2$$

$$\frac{A'_{ab}}{A'_{ba}} = -0.616$$

Using the Van Laar equations, we let $A$ be $A'_{ab}$ and $B$ be $A'_{ba}$.

$$RT \ln \gamma_a = A'_{ab} \left(\frac{A'_{ba} x_b}{A'_{ab} x_a + A'_{ba} x_b}\right)^2 = A'_{ab} \left(\frac{1}{\left(\frac{A'_{ab}}{A'_{ba}}\right)\left(\frac{x_a}{x_b}\right) + 1}\right)^2$$

$$= [(8.314)(273.15 + 65) \ln 0.77] \left[-0.616 \left(\frac{0.3}{0.7}\right) + 1\right]^2 = A'_{ab}$$
$$= -398$$

$$\frac{A'_{ab}}{A'_{ba}} = -0.616 \rightarrow A'_{ba} = \frac{-398}{-0.616} = 646$$

Now, we can calculate the vapor composition over the liquid solution at 65 °C that contains 12 mol% of substance $a$.

$$x_a = 0.12, \ x_b = 1 - 0.12 = 0.88$$

$$\ln \gamma_a = \frac{-398}{8.314(273.15 + 65)} \left( \frac{646(0.88)}{-398(0.12) + 646(0.88)} \right)^2 = -0.17$$

$$\gamma_a = 0.84$$

$$\ln \gamma_b = \frac{646}{8.314(273.15 + 65)} \left( \frac{-398(0.12)}{-398(0.12) + 646(0.88)} \right)^2 = 0.0019$$

$$\gamma_b = 1.00$$

(d) To find out the total pressure over the solution at the condition of part c, we can start with the VLE condition assuming ideal gas but non-ideal liquid.

$$y_a P = \gamma_a x_a P_a^{\text{sat}} = 0.84 \times 0.12 \times 1.3 = 0.13$$

$$y_b P = (1 - y_a) P = \gamma_b x_b P_b^{\text{sat}} = 1.00 \times 0.88 \times 0.97 = 0.85$$

Solving the above simultaneous equations, we can determine $y_a$ and $P$.

$$P = 0.98 \text{ atm}, \quad y_a = 0.13$$

## Problem 23

**A cylinder of substance *a* at 60 bar and 30 °C contains some substance *b*. Substance *a* is sparingly soluble in *b* under the stated conditions. Assume that $\ln \gamma_a = A x_b^2$. The saturation pressure of substance *a* at this temperature is 10 bar, and that of substance *b* is 0.03 bar. We are told that the critical pressure of substance *a* is 42 bar, and the critical temperature is 370 K.**

**(a) What are the fugacities of substance *a* in both phases?**
**(b) What is the value of Henry's law constant for substance *a* in *b* at this pressure, and express it as a function of *A* only.**

## Solution 23

***Worked Solution***

(a) At VLE, the fugacities of substance $a$ in the vapor and liquid phase are equal.

$$\hat{f}_a^L = \hat{f}_a^V$$

$$\gamma_a x_a P_a^{sat} = \hat{\varphi}_a y_a P$$

We are told that substance $a$ is only sparingly soluble in $b$; hence, liquid phase has negligible amounts of substance $a$ ($x_a \ll 1$). Also we notice that $P_a^{\text{sat}} \gg P_b^{\text{sat}}$, so vapor phase mostly consists of substance $a$. Hence, we can more easily find out fugacity from $\hat{f}_a^V$ instead of $\hat{f}_a^L$ since $x_a$ is unknown, but this simplifying assumption of $y_a \approx 1$ can help us solve for $\hat{f}_a^V$.

Note that in this problem, $P = 60$ bar which is high; hence, we cannot assume ideal gas and $\hat{\varphi}_a = 1$. Instead, we can try to determine $\hat{\varphi}_a$ using the generalized correlation which uses the following equation:

$$\log\varphi_i = \log\varphi^{(0)} + \omega\log\varphi^{(1)}$$

We can refer to the values of $\varphi^{(0)}$, $\varphi^{(1)}$, and $\omega$ after we have calculated the reduced temperature and pressure for our system. Then we can substitute these values into the above correlation to obtain $\hat{\varphi}_a$.

$$P_r = \frac{P}{P_c} = \frac{60}{42} = 1.4$$

$$T_r = \frac{T}{T_c} = \frac{273.15 + 30}{370} = 0.82$$

These reduced parameters give us $\varphi^{(0)} = 0.2$, $\varphi^{(1)} = 0.36$, and $\omega = 0.15$. So $\hat{\varphi}_a = 0.2$.

$$\hat{f}_a^V = 0.2 \times 1 \times 60 = 12 \text{ bar} = \hat{f}_a^L$$

(b) Henry's law constant for substance $a$ occurs at infinite dilution of a in the liquid phase, i.e., $x_a = 0$, $x_b = 1$.

$$\hat{f}_a^L = \gamma_a^\infty x_a P_a^{\text{sat}} = x_a H_a \rightarrow H_a = \gamma_a^\infty P_a^{\text{sat}}$$

$$\ln\gamma_a = Ax_b^2 \rightarrow \gamma_a^\infty = e^{Ax_b^2} = e^A$$

$$H_a = e^A\left(P_a^{\text{sat}}\right) = 10e^A \text{ bar}$$

## Problem 24

**What vapor pressure $P^V$ is at equilibrium with a droplet of radius $r$ at temperature $T$? You may assume ideal gas vapor phase. From the result, explain the phenomenon whereby small droplets may shrink further while large bubbles may continue growing.**

## Solution 24

***Worked Solution***

Let us consider a case of a pure liquid droplet surrounded by its pure vapor. Note that the special condition about a droplet is that there is surface tension, which makes the pressure in the liquid dissimilar to the pressure in the vapor. The pressure in the liquid droplet has to be higher in order for the droplet to be at equilibrium with the external vapor phase $\mu^L = \mu^V$.

$$\left(P^{\text{in}} - P^{\text{out}}\right)\left(\pi r^2\right) = 2\pi r\sigma$$

Note that the pressure difference acts on the projected cross-sectional area of the spherical bubble; hence, the area is that of the circular cross section at $\pi r^2$. This higher pressure acting from inside the bubble outwardly is balanced by the surface tension force pulling inwards to keep the spherical shape of the bubble. Note that surface tension acts along the perimeter; hence, that is the circumference of a circle at $2\pi r$. Back to the question, we require for VLE that chemical potentials are equal.

$$\mu^L = \mu^V$$

$$\mu^L = g^L = g^{\text{sat}}(T) + \int_{P^{\text{sat}}}^{P^L} v^L dP = g^{\text{sat}}(T) + v^L\left(P^L - P^{\text{sat}}\right)$$

Note that in the case of a liquid droplet, the pressure inside the liquid droplet $P^L$ is significantly high and so the assumption that $(P^L - P^{\text{sat}})$ is small enough to be neglected cannot be made. This is different for the case of a bubble in liquid phase.

$$\mu^V = g^V = g^{\text{sat}}(T) + \int_{P^{\text{sat}}}^{P^V} v^V dP$$

$$= g^{\text{sat}}(T) + \int_{P^{\text{sat}}}^{P^V} \frac{RT}{P} dP = g^{\text{sat}}(T) + RT\ln\frac{P^V}{P^{\text{sat}}}$$

Finally equating both results, we get

$$v^L\left(P^L - P^{\text{sat}}\right) = RT\ln\frac{P^V}{P^{\text{sat}}}$$

$$\left(P^L - P^V\right)\left(\pi r^2\right) = 2\pi r\sigma \rightarrow P^L = \frac{2\sigma}{r} + P^V$$

$$v^L\left(\frac{2\sigma}{r}+P^V-P^{\text{sat}}\right)=RT\ln\frac{P^V}{P^{\text{sat}}}$$

From the expression, we observe that smaller droplets have a higher chemical potential and so are in equilibrium with a higher vapor pressure. If there was a droplet just smaller than the equilibrium radius, the liquid phase will have a higher chemical potential than the vapor and material migrates from the liquid to the vapor. Hence the droplet shrinks further to try to equilibrate. Conversely, for droplets larger than the equilibrium radius, the reverse is true due to the similarly unstable equilibrium which makes the droplets grow further in size.

## Problem 25

**Explain why superheated liquids can be stable (i.e., not boil even though temperature is above boiling point) in the absence of nucleation sites.**

## Solution 25

### *Worked Solution*

Boiling is observed when bubbles first become stable. When superheated liquids are stable and do not boil even though temperature has exceeded boiling point, it means that bubbles cannot form stably because the moment small bubbles form, the direction of equilibrium is in shrinking the bubbles further till they disappear. For small bubbles, the surface tension is high, and this inward force, together with atmospheric pressure, cause bubbles to shrink. The force that is expanding it is vapor pressure from inside the bubble.

Boiling occurs if bubble formation and growth is stable, and this is when saturated vapor pressure is equal to atmospheric pressure. This explains why boiling temperature is related to atmospheric pressure. It is easier to form bubbles (i.e., require lower boiling temperature) when the external atmospheric pressure (that works to crush boiling bubbles) is lower.

$$P^{\text{in}}=P^{\text{out}}+\frac{2\sigma}{r}$$

$$P^V=P^L+\frac{2\sigma}{r}=P^{\text{atm}}+\frac{2\sigma}{r}$$

In the above equation, the left-hand side is the expanding force that promotes growth of the bubble, while the right-hand side is the crushing force that shrinks the bubble.

$$\mu^V=g^V=g^{\text{sat}}(T)+\int_{P^{\text{sat}}}^{P^V}v^V dP=g^{\text{sat}}(T)+\int_{P^{\text{sat}}}^{P^V}\frac{RT}{P}dP=g^{\text{sat}}(T)+RT\ln\frac{P^V}{P^{\text{sat}}}$$

$$\mu^L = g^L = g^{\mathrm{sat}}(T) + \int_{P^{\mathrm{sat}}}^{P^L} v^L dP = g^{\mathrm{sat}}(T) + v^L\left(P^L - P^{\mathrm{sat}}\right) \approx g^{\mathrm{sat}}(T)$$

Note that $(P^L - P^{\mathrm{sat}})$ may be assumed to be small, since $P^L = P^{\mathrm{atm}}$ and this value is not significant. Moreover, $v^L$ is small. Hence we can assume $g^L$ is not a function of pressure. This means that $\mu^V = \mu^L \rightarrow RT \ln \frac{P^V}{P^{\mathrm{sat}}} = 0$ and $P^V$(higher boiling $T$) $=$ $P^{\mathrm{sat}}$(normal boiling $T$). For boiling, the interface is not flat, $P^L = P^{\mathrm{atm}} = P^V + \frac{2\sigma}{r} = P^{\mathrm{sat}}(\text{normal boiling } T) + \frac{2\sigma}{r}$.

At normal boiling point (with presence of nucleation sites) we know that $P^{\mathrm{sat}}$(normal boiling $T$) $= P^{\mathrm{atm}}$.

In the absence of nucleation sites at atmospheric pressure, bubbles are exceptionally small and this makes $\frac{2\sigma}{r}$ larger, and hence superheat needs to be applied for boiling to occur at a $T > T_{\text{normal bp}}$. It follows that $P_{\mathrm{sat}}(T) > P_{\mathrm{sat}}(T_{\text{normal bp}}) = P_{\mathrm{atm}}$. Using the Clausius–Clapeyron equation, we can calculate the superheat required to be applied.

$$\frac{d \ln P^{\mathrm{sat}}}{dT} = \frac{\Delta H_{\mathrm{vap}}}{RT^2}$$

$$\ln\left(\frac{P_1}{P_2}\right) = -\frac{\Delta H_{\mathrm{vap}}}{R}\left(\frac{1}{T_1} - \frac{1}{T_2}\right)$$

$$\ln\left(\frac{P_{\mathrm{sat}}(T)}{P_{\mathrm{atm}}}\right) = -\frac{\Delta H_{\mathrm{vap}}}{R}\left(\frac{1}{T} - \frac{1}{T_{\text{normal bp}}}\right)$$

## Problem 26

**(a) Starting from the second law, show that the Gibbs energy of a closed system is minimized at equilibrium if temperature and pressure are constant.**

**(b) Show that for a pure substance with coexisting vapor and liquid phases at equilibrium, the molar Gibbs energy of the vapor is equal to the molar Gibbs energy of the liquid.**

**(c) Show that for a binary mixture with coexisting vapor and liquid phases at equilibrium, the chemical potential of each component in the vapor phase is equal to the chemical potential of the same component in the liquid phase.**

## Solution 26

***Worked Solution***

(a) The second law states that $dS_{\mathrm{uni}} \geq 0$, where it is equal to zero for a reversible process. For a closed system at constant $T$, it is at thermal equilibrium with the universe. For a process occurring inside the system, there is a transfer of heat $dq$ from the universe into the system. Since we have thermal equilibrium, outside the system we have $dS_{\text{surr}} = -\frac{dQ}{T}$. This equation means that the heat added to the system must equate the heat lost from the surroundings, to remain at isothermal conditions.

The second law also states that $dS_{\text{sys}} \geq \frac{dQ}{T}$ in order for $dS_{\text{uni}} \geq 0$. For a closed system, $dU = dQ + dW = dQ + d(W_{\text{shaft}} + W_{\text{flow work}})$, where $dW$ is the work done on the system. Also for a closed system, there is no inflow/outflow of internal energy $U$ that is contributing to $dU$.

$W_{\text{flow work}}$ is $PV$ work and consists only of $PdV$ because external pressure is constant for closed system. $dU = dQ + dW_{\text{shaft}} - PdV$. Assume work is done by the system in an expansion of volume; hence, $dV$ is positive and there is a negative sign before the term $PdV$.

In general, $G = H - TS = (U + PV) - TS$. For an open system,

$$\left.\frac{dU}{dt}\right|_{\text{sys}} = \sum \dot{n}_{\text{in}} h_{\text{in}} + \dot{n}_{\text{out}} h_{\text{out}} + \dot{Q} + \dot{W}_{\text{shaft}}$$

For a closed system, $dG_{\text{sys}} = dU + PdV + VdP - TdS - SdT$. Under isothermal conditions ($dT = 0$) and for a closed system ($dP = 0$), $dG_{\text{sys}} = dU + PdV - TdS$.

$$dG_{\text{sys}} = (dQ + dW_{\text{shaft}} - PdV) + PdV - TdS$$

If no shaft work is done on system, $dW_{\text{shaft}} = 0$.

So it follows that from the second law, we can deduce that $TdS_{\text{sys}} \geq dQ$ in the equation $dG_{\text{sys}} = dQ - TdS$, and therefore $dG_{\text{sys}} \leq 0$ at constant $T$ and $P$ inside a closed system without shaft work.

(b) Pure substance at VLE at constant $T$ and $P$,

$$g^V = g^L$$

$dG^V = d(n^V g^V) = g^V dn^V$ since $g^V$ is a function of $T$ and $P$ only for pure substance. In this case it is a constant that can be taken out from the differential.

$$dG = dG^L + dG^V = g^L dn^L + g^V dn^V$$

$$dn^V = -dn^L$$

At equilibrium, $dG = 0$ and so

$$dG = \left(g^L - g^V\right) dn^L = 0$$

Therefore it is shown that $g^L = g^V$. In addition, we may observe that in order for $G$ to be lowered, there can be a transfer of material from vapor to liquid phase ($dn^L > 0$ *and* $(g^L - g^V) < 0$) or liquid to vapor phase ($dn^L < 0$ *and* $(g^L - g^V) > 0$).

(c) Consider differential amount of component 1 moving from vapor to liquid phase at constant $T$ and $P$.

$$dG^L = \left.\frac{\partial G^L}{\partial n_1^L}\right|_{T,P,n_2^L} dn_1^L = \mu_1^L dn_1^L$$

$$dG^V = \left.\frac{\partial G^V}{\partial n_1^V}\right|_{T,P,n_2^V} dn_1^V = \mu_1^V dn_1^V$$

$$dn_1^V = -dn_1^L$$

$$dG = dG^L + dG^V = \mu_1^L dn_1^L + \mu_1^V dn_1^V = \left(\mu_1^L - \mu_1^V\right) dn_1^L$$

At equilibrium, $dG = 0$, so $\mu_1^L = \mu_1^V$. The same applies for component 2.

## Problem 27

**Derive the Poynting correction factor, and use it to calculate the vapor pressure exerted by liquid water at 25 °C in a cylinder of argon at 190 bar.**

## Solution 27

***Worked Solution***

We have a VLE consisting of a pure liquid in equilibrium with a vapor phase mixture. The Poynting correction is used to account for pressure dependence of pure species fugacity in liquid phase (pressure of pure liquid is also $P^{sys} = P$), by adjusting for the significantly higher system pressure relative to the specie's saturation pressure, i.e., where $\left(P - P_i^{sat}\right)$ becomes significant.

The Poynting correction term is the exponential term in the general expression shown below:

$$f_a^L = \varphi_a^{sat} P_a^{sat} \exp\left[\int_{P_a^{sat}}^{P} \left(\frac{v_a^L}{RT}\right) dP\right] = \varphi_a^{sat} P_a^{sat} \exp\left[\frac{v_a^L}{RT}\left(P - P_a^{sat}\right)\right]$$

$$\hat{\varphi}_a^V = \frac{\hat{f}_a^V}{P_a} = \frac{\hat{f}_a^V}{y_a P}, \quad \mu_a - \mu_a^0 = RT \ln\left(\frac{\hat{f}_a}{\hat{f}_a^0}\right)$$

If saturation pressure $P_a^{sat}$ of the species is low, the saturation fugacity coefficient $\varphi_a^{sat} = 1$. If ideal mixing is assumed for vapor phase mixture, i.e., like interactions are the same as unlike interactions, then $\hat{\varphi}_a^V = 1$ and $y_a P = P_a$.

For VLE, where component $a$ is the pure liquid,

$$\mu_a^L = \mu_a^V$$

For pure species $a$ in liquid phase, $\mu_a^L = g_a^L(T, P)$, and at constant $T$ we have

$$dg = vdP - sdT = vdP$$

$$g_a^L(T, P) = g_a^L(T, P_a^{\text{sat}}) + \int_{P_a^{\text{sat}}}^{P} v_a^L dP = g_a^L(T, P_a^{\text{sat}}) + v_a^L(P - P_a^{\text{sat}})$$

For vapor phase, there is a mixture,

$$\mu_a^V(T, P) = g_a^V(T, P) + RT \ln\left(\frac{\hat{f}_a}{\hat{f}_a^0}\right) = g_a^V(T, P) + RT \ln\left(\frac{\hat{\varphi}_a^V y_a P}{P}\right)$$

If we assume an ideal mixture in vapor phase, $\hat{\varphi}_a^V = 1$

$$\mu_a^V(T, P) = g_a^V(T, P) + RT \ln y_a$$

If we adjust for pressure from saturation pressure to system pressure,

$$\mu_a^V(T, P) = g_a^V(T, P_a^{\text{sat}}) + \int_{P_a^{\text{sat}}}^{P} v_a^V dP + RT \ln y_a$$

Assume that the pure species $a$ in vapor phase behaves as ideal gas for the pressure range,

$$\mu_a^V(T,P) = g_a^V(T,P_a^{\text{sat}}) + \int_{P_a^{sat}}^{P} \frac{RT}{P} dP + RT \ln y_a = g_a^V(T,P_a^{\text{sat}}) + RT \ln \frac{P}{P_a^{\text{sat}}} + RT \ln y_a$$

$$= g_a^V(T, P_a^{\text{sat}}) + RT \ln \frac{P_a}{P_a^{\text{sat}}}$$

For VLE, we have the following equation relating the liquid and vapor phases.

$$g_a^L(T, P_a^{\text{sat}}) + v_a^L(P - P_a^{\text{sat}}) = g_a^V(T, P_a^{\text{sat}}) + RT \ln \frac{P_a}{P_a^{\text{sat}}}$$

For pure species at saturation conditions, $g_a^L(T, P_a^{\text{sat}}) = g_a^V(T, P_a^{\text{sat}})$

$$v_a^L\left(P - P_a^{\text{sat}}\right) = RT\ln\frac{P_a}{P_a^{\text{sat}}}$$

$$P_a = P_a^{\text{sat}}\exp\left[\frac{v_a^L\left(P - P_a^{\text{sat}}\right)}{RT}\right]$$

When we compare this to the general form of equation,

$$f_a^L = \varphi_a^{\text{sat}}P_a^{\text{sat}}\exp\left[\frac{v_a^L}{RT}\left(P - P_a^{\text{sat}}\right)\right]$$

We may observe that in this case, after accounting for all the assumptions, $f_a^L = P_a$. Note that $f_a^L \neq P$ because vapor phase is a mixture and not pure, and moreover the definition of fugacity of a component (pure or in a mixture) is the partial pressure of an ideal gas of the same species which would be in equilibrium with that component. Also note that $f_a^L \neq P_a^{\text{sat}}$ because the VLE is not occurring at saturation conditions.

In order to calculate the vapor pressure exerted by liquid water at 25 °C in a cylinder of argon at 190 bar, we can use the expression derived earlier, where species $a$ is water and $P$ is 190 bar.

$$P_a = P_a^{\text{sat}}\exp\left[\frac{v_a^L\left(P - P_a^{\text{sat}}\right)}{RT}\right]$$

## Problem 28

**Explain the meanings of the following terms:**

(a) **Chemical potential**
(b) **Fugacity**
(c) **Partial fugacity**
(d) **Activity coefficient**

## Solution 28

***Worked Solution***

(a) Chemical potential is the partial molar Gibbs energy, which serves as the driving force for the movement of a species between phases, whether in a chemical reaction, by diffusion or other means, etc.

$$\mu_i = \left(\frac{\partial G}{\partial n_i}\right)_{T,P,n_j}$$

In an open system where the total number of moles in the system may change,

$$dG = -SdT + VdP + \sum_i \mu_i dn_i$$

(b) Fugacity of a pure substance (solid/liquid/vapor) is the vapor pressure of an ideal gas (or partial pressure) of that same substance which will be in equilibrium with that pure substance. Or in other words, the pressure of a real gas of that same substance in equilibrium with that pure substance after multiplied by the fugacity coefficient $\varphi_i$.

$$g_i(T,P) = g_i^0(T,P^0) + RT\ln\left(\frac{f_i}{f_i^0}\right)$$

$g_i(T, P)$ is the pure real gas $i$, note that species $i$ can be solid/liquid/gas, as this term is a "dummy" that is created whereby $g_i(T,P) = f_i^V$ such that how it relates to the actual phase is through the VLE equation, $f_i^V = f_i^S$ (or $f_i^L$). $g_i^0(T,P^0)$ is the same pure gas at a chosen $P^0$ such that the species behaves as an ideal gas which is a conveniently chosen reference state. Finally, note that $f_i$ is the pure real gas fugacity, so it is obtained after correcting system pressure in $f_i = \varphi_i P$.

(c) Partial fugacity of a component in a real mixture, $\hat{f}_i^L$, is the pressure of a hypothetical gas (ideal mixing rules) of the same component which would be in equilibrium with the component $i$ in the mixture.

$$\hat{f}_i^L = \hat{f}_i^V = \hat{\varphi}_i^V y_i P$$

The fugacity of this hypothetical ideal gas of species $i$ features in the expression as follows:

$$\mu_i(T,P) = g_i^0(T,P^0) + RT\ln\left(\frac{\hat{f}_i^V}{f_i^0}\right)$$

$\mu_i(T, P)$ is for component $i$ in the real liquid mixture, $g_i^0(T,P^0)$ is a pure ideal gas of component $i$ at $T$ and a $P^0$ where it behaves ideally and refers to a hypothetical reference state. $\hat{f}_i^V$ is the partial fugacity of real gas $i$ in an ideal mixture whereby $\hat{f}_i^V = \hat{\varphi}_i^V y_i P$ and finally, $f_i^0 = P^0$ is the fugacity of a pure ideal gas of component $i$, chosen as the reference state.

(d) Activity coefficient $\gamma_i$ is a measure of the deviation from ideal mixture of a component in a real mixture. When $\gamma_i > 1$, the mixing is thermodynamically unfavorable, and conversely when $\gamma_i < 1$ the mixing is thermodynamically favorable. The activity coefficient features in the expression as follows:

$$\mu_i^L(T,P,x_i) = g_i^L(T,P,x_i = 1) + RT\ln(\gamma_i x_i)$$

$\mu_i(T, P, x_i)$ refers to that of component $i$ in a real mixture, $g_i(T, P, x_i = 1)$ refers to that of pure $i$, and $RT$ ln $(\gamma_i x_i)$ is the chemical potential of mixing which is a sum of two subcomponents ($\Delta g_{\text{mix}} = \Delta g_{\text{ideal mix}} + \Delta g^E$). We can interchange $x_i$ with $y_i$ for a gas mixture.

## Problem 29

**The two-constant Margules equation for the excess Gibbs free energy of a binary mixture is given below. Derive equations for the activity coefficients $\gamma_a$ and $\gamma_b$.**

$$\boldsymbol{g^E = x_a x_b [A + B(x_a - x_b)]}$$

## Solution 29

***Worked Solution***

$$g_a^E = \mu_a^E = \left.\frac{\partial G^E}{\partial n_a}\right|_{T,P,n_b} = \left.\frac{\partial (n_T g^E)}{\partial n_a}\right|_{T,P,n_b} = g^E + n_T \left.\frac{\partial g^E}{\partial n_a}\right|_{T,P,n_b}$$
$$= g^E + n_T \left.\frac{\partial g^E}{\partial n_a}\right|_{T,P,n_b}$$

Now we think about how to convert $\left(\frac{\partial g^E}{\partial n_a}\right)_{T,P,n_b}$ into the form $\left(\frac{\partial g^E}{\partial x_a}\right)_{T,P,n_b}$

$$dn_a = d(n_T x_a) = n_T dx_a + x_a dn_T$$

Differentiating the above with respect to $x_a$, we have

$$\frac{dn_a}{dx_a} = n_T + x_a \frac{dn_T}{dx_a} = n_T + x_a \frac{d(n_a + n_b)}{dx_a} = n_T + x_a \frac{dn_a}{dx_a} + x_a \frac{dn_b}{dx_a}$$

In our context, $n_b$ is constant; hence $dn_b = 0$

$$\frac{dn_a}{dx_a} = n_T + x_a \frac{dn_a}{dx_a}$$
$$\frac{dn_a}{dx_a}(1 - x_a) = n_T$$
$$\frac{dn_a}{dx_a} = \frac{n_T}{1 - x_a}$$
$$dn_a = \left(\frac{n_T}{x_b}\right) dx_a$$

Returning to our earlier expression for excess Gibbs free energy, and substituting the above results, we have

$$g_a^E = g^E + n_T \left.\frac{\partial g^E}{\partial n_a}\right|_{T,P,n_b} = g^E + x_b \left.\frac{\partial g^E}{\partial x_a}\right|_{T,P,n_b}$$

$$= x_a x_b [A + B(x_a - x_b)] + x_b \left[(A - B) + (6B - 2A)x_a - 6B{x_a}^2\right]$$

$$= Ax_a x_b + B{x_a}^2 x_b - Bx_a {x_b}^2 + Ax_b - Bx_b + 6Bx_a x_b - 2Ax_a x_b - 6B{x_a}^2 x_b$$

$$= -B(1 - x_b){x_b}^2 + (A - B)x_b + (6B - A)(1 - x_b)x_b - 5B\left(1 - 2x_b + {x_b}^2\right)x_b$$

$$= -B{x_b}^2 + B{x_b}^3 + Ax_b - Bx_b + 6Bx_b - Ax_b - 6B{x_b}^2 + A{x_b}^2 - 5Bx_b + 10B{x_b}^2 - 5B{x_b}^3$$

$$= (A + 3B){x_b}^2 - 4B{x_b}^3$$

Since by definition, $g_a^E = RT \ln \gamma_a$,

$$RT \ln \gamma_a = (A + 3B){x_b}^2 - 4B{x_b}^3$$

As for species $b$,

$$g_b^E = g^E + n_T \left.\frac{\partial g^E}{\partial n_b}\right|_{T,P,n_a} = g^E + x_a \left.\frac{\partial g^E}{\partial x_b}\right|_{T,P,n_a} = g^E - x_a \left.\frac{\partial g^E}{\partial x_a}\right|_{T,P,n_a}$$

In the same way, we may find the following for species $b$,

$$g_b^E = x_a x_b [A + B(x_a - x_b)] - x_a \left[(A - B) + (6B - 2A)x_a - 6B{x_a}^2\right]$$

$$= x_a(1 - x_a)[A + B(2x_a - 1)] - (A - B)x_a - (6B - 2A){x_a}^2 + 6B{x_a}^3$$

$$= Ax_a - A{x_a}^2 + 2B{x_a}^2 - 2B{x_a}^3 - Bx_a + B{x_a}^2 - Ax_a + Bx_a - 6B{x_a}^2 + 2A{x_a}^2 + 6B{x_a}^3$$

$$= (A - 3B){x_a}^2 + 4B{x_a}^3$$

Since by definition, $g_b^E = RT \ln \gamma_b$,

$$RT \ln \gamma_b = (A - 3B){x_a}^2 + 4B{x_a}^3$$

## Problem 30

**If the activity coefficients for species *a* in a mixture is given by the expression below, derive an expression for $\ln\gamma_b$.**

$$\ln \gamma_a = a{x_b}^2 + b{x_b}^3 + c{x_b}^4$$

## Solution 30

***Worked Solution***

From Gibbs–Duhem, we know that

$$\sum_i x_i d\ln\gamma_i = 0$$

$$x_a\frac{d\ln\gamma_a}{dx_b} + x_b\frac{d\ln\gamma_b}{dx_b} = 0$$

$$x_a\left(2ax_b + 3bx_b^2 + 4cx_b^3\right) + x_b\frac{d\ln\gamma_b}{dx_b} = 0$$

$$(1 - x_b)\left(2ax_b + 3bx_b^2 + 4cx_b^3\right) + x_b\frac{d\ln\gamma_b}{dx_b} = 0$$

$$-2ax_b - 3bx_b^2 - 4cx_b^3 + 2ax_b^2 + 3bx_b^3 + 4cx_b^4 = x_b\frac{d\ln\gamma_b}{dx_b}$$

$$\frac{d\ln\gamma_b}{dx_b} = -2a + (2a - 3b)x_b + (3b - 4c)x_b^2 + 4cx_b^3$$

$$\ln\gamma_b = -2ax_b + ax_b^2 - \frac{3}{2}x_b^2 + bx_b^3 - \frac{4}{3}cx_b^3 + cx_b^4 + \text{constant}$$

To find the constant, we can use the fact that when $x_b = 1$, $\gamma_b = 1$

$$0 = a - 2a + b - \frac{3}{2}b - \frac{4}{3}c + c + \text{constant}$$

$$\text{constant} = a + \frac{b}{2} + \frac{c}{3}$$

Substituting the constant, and knowing that $x_b = 1 - x_a$, we get

$$\ln\gamma_b = ax_a^2 + bx_a^2\left(\frac{3}{2} - x_a\right) + cx_a^2\left(x_a^2 - \frac{8}{3}x_a + 2\right)$$

## Problem 31

**A mixture of species 1 and 2 is in vapor–liquid equilibrium at 1.01 bar and 110 °C. The liquid phase is 15 mol % of species 2 and the vapor phase is 75 mol % of species 2. Assume that at 110 °C, saturation pressures of species 1 and 2 are 0.17 bar and 1.41 bar, respectively. Find out the parameters of the Van Laar equation, $\alpha$ and $\beta$ for this system. Note that the Van Laar equations are as follows:**

$$\ln \gamma_1 = \frac{\alpha}{\left(1 + \frac{\alpha x_1}{\beta x_2}\right)^2}$$

$$\ln \gamma_2 = \frac{\beta}{\left(1 + \frac{\beta x_2}{\alpha x_1}\right)^2}$$

## Solution 31

### *Worked Solution*

We know that in the liquid phase, $x_2 = 0.15$, $x_1 = 0.85$ and in the vapor phase, $y_2 = 0.75$, $y_1 = 0.25$.

For VLE, we have the balance of chemical potentials in the liquid and vapor phases.

$$\gamma_1 x_1 P_1^{\text{sat}} = y_1 P$$

$$\gamma_1 = \frac{y_1 P}{x_1 P_1^{\text{sat}}} = \frac{0.25 \times 1.01}{0.85 \times 0.17} = 1.75$$

Similarly for species 2, we have

$$\gamma_2 = \frac{y_2 P}{x_2 P_2^{sat}} = \frac{0.75 \times 1.01}{0.15 \times 1.41} = 3.58$$

Now we can find simplified expressions from the Van Laar equations,

$$\frac{\ln \gamma_1}{\ln \gamma_2} = \frac{\alpha\left(1 + \frac{\beta x_2}{\alpha x_1}\right)^2}{\beta\left(1 + \frac{\alpha x_1}{\beta x_2}\right)^2} = \frac{\alpha}{\beta}\left(\frac{\frac{(\alpha x_1 + \beta x_2)^2}{\alpha^2 {x_1}^2}}{\frac{(\beta x_2 + \alpha x_1)^2}{\beta^2 {x_2}^2}}\right) = \frac{\alpha}{\beta}\left(\frac{\beta^2 {x_2}^2}{\alpha^2 {x_1}^2}\right) = \frac{\beta {x_2}^2}{\alpha {x_1}^2}$$

$$\frac{\alpha x_1}{\beta x_2} = \left(\frac{x_2}{x_1}\right)\frac{\ln \gamma_2}{\ln \gamma_1}$$

$$\ln \gamma_1 = \frac{\alpha}{\left(1 + \frac{\alpha x_1}{\beta x_2}\right)^2}$$

$$\alpha = \ln \gamma_1 \left(1 + \frac{x_2 \ln \gamma_2}{x_1 \ln \gamma_1}\right)^2 = \ln 1.75\left(1 + \frac{0.15 \ln 3.58}{0.85 \ln 1.75}\right)^2 = 1.1$$

Similarly, we have the following for species *b*,

$$\frac{\beta x_2}{\alpha x_1} = \left(\frac{x_1}{x_2}\right)\frac{\ln \gamma_1}{\ln \gamma_2}$$

$$\ln \gamma_2 = \frac{\beta}{\left(1 + \frac{\beta x_2}{\alpha x_1}\right)^2}$$

$$\beta = \ln \gamma_2 \left(1 + \frac{x_1 \ln \gamma_1}{x_2 \ln \gamma_2}\right)^2 = \ln 3.58\left(1 + \frac{0.85 \ln 1.75}{0.15 \ln 3.58}\right)^2 = 15.5$$

## Problem 32

**Seawater is approximately 4%wt of sodium chloride. Estimate its freezing point. You are provided with the enthalpy of fusion of water = 6 kJ/mol.**

## Solution 32

***Worked Solution***

In analyzing freezing point depression, we may consider a solid–liquid equilibrium system. In a system that consists of a pure species, the temperature at which boiling occurs (at a specified pressure) is fixed and termed the boiling point at atmospheric pressure of 1 atm. Consider the scenario where we now add a solute *b* into a liquid solvent *a*, we will then notice that the temperature for this mixture to boil will be higher. This occurrence is known as boiling point elevation.

Similarly if we add a small amount of solute *b* into liquid solvent *a*, liquid a will freeze into solid a at a lower temperature than the freezing point of pure liquid *a*. This is known as freezing point depression.

We can apply the concepts of phase equilibria to explain the above observations.

*Freezing point depression (solid–liquid equilibrium)*

In general, for solid–liquid equilibrium, the following holds:

$$\hat{f}_i^s = \hat{f}_i^l$$

Assume that solid phase (species *a*) is pure, and the liquid phase consists of solvent *a* and solute *b*, where the Lewis Randall reference state is used for the liquid phase, then for species *a*, we have

$$f_a^s = x_a \gamma_a f_a^l$$

$$x_a \gamma_a = \frac{f_a^s}{f_a^l}$$

Separately, we know that by definition $d\mu = RTd \ln f$, so for a pure substance that is melting,

$$\Delta g_{\text{fus}} = g_a^l - g_a^s = RT \ln \left( \frac{f_a^l}{f_a^s} \right) = -RT \ln (x_a \gamma_a)$$

It is more common that we have values of enthalpy and entropy of fusion at the melting point (at atmospheric pressure) from data booklets, so we may express $\Delta g_{\text{fus}}$ in terms of the thermodynamic properties of enthalpy and entropy, and also adjust these properties to our desired temperature from their values at the normal melting point.

$$-\ln (x_a \gamma_a) = \frac{\Delta g_{\text{fus}}}{RT} = \frac{\Delta h_{\text{fus}}}{RT} - \frac{\Delta s_{\text{fus}}}{R}$$

We typically know $\Delta h_{\text{fus}}$ at the melting point, $T_m$. We can find out $\Delta h_{\text{fus}}$ at any temperature $T$ as follows. Let us assume $T < T_m$ for a case of freezing point depression:

$$\Delta h_{\text{fus}}(T) = \Delta h_{\text{fus}}(T_m) + \int_T^{T_m} c_p^s dT + \int_{T_m}^{T} c_p^l dT$$

Similarly for entropy of fusion,

$$\Delta s_{\text{fus}}(T) = \Delta s_{\text{fus}}(T_m) + \int_T^{T_m} \frac{c_p^s}{T} dT + \int_{T_m}^{T} \frac{c_p^l}{T} dT$$

We know that at the melting point, $T_m$

$$\Delta g_{\text{fus}} = \Delta h_{\text{fus}} - T_m \Delta s_{\text{fus}} = 0$$

$$\Delta s_{\text{fus}} = \frac{\Delta h_{\text{fus}}}{T_m}$$

Substituting into the earlier expression, we get

$$\Delta s_{\text{fus}}(T) = \frac{\Delta h_{\text{fus}}}{T_m} + \int_T^{T_m} \frac{c_p^s}{T} dT + \int_{T_m}^{T} \frac{c_p^l}{T} dT$$

Finally we can relate activity coefficient and molar composition of the species to the enthalpy of fusion (at any desired temperature that may not be the normal melting point) and heat capacities.

$$-\ln(x_a\gamma_a) = \frac{\Delta h_{\text{fus}}(T_m) + \int_T^{T_m} c_p^s dT + \int_{T_m}^{T} c_p^l dT}{RT} - \frac{\frac{\Delta h_{\text{fus}}(T_m)}{T_m} + \int_T^{T_m} \frac{c_p^s}{T} dT + \int_{T_m}^{T} \frac{c_p^l}{T} dT}{R}$$

To simplify the expression further, we may define a term $\Delta c_p^{sl} = c_p^l - c_p^s$, which may then be assumed constant for simplifying the integration.

$$-\ln(x_a\gamma_a) = \frac{\Delta h_{\text{fus}}(T_m)}{R}\left(\frac{1}{T} - \frac{1}{T_m}\right) - \frac{1}{R}\int_{T_m}^{T} \frac{\Delta c_p^{sl}}{T} dT + \frac{1}{RT}\int_{T_m}^{T} \Delta c_p^{sl} dT$$

$$\ln(x_a\gamma_a) = -\frac{\Delta h_{\text{fus}}(T_m)}{R}\left(\frac{1}{T} - \frac{1}{T_m}\right) - \frac{\Delta c_p^{sl}}{R}\ln\left(\frac{T}{T_m}\right) + \frac{\Delta c_p^{sl}}{RT}(T - T_m)$$

$$\ln(x_a\gamma_a) = -\frac{\Delta h_{\text{fus}}(T_m)}{R}\left(\frac{1}{T} - \frac{1}{T_m}\right) + \frac{\Delta c_p^{sl}}{R}\left(1 - \frac{T_m}{T} - \ln\left(\frac{T}{T_m}\right)\right)$$

If solute $b$ is dilute in solvent $a$, the liquid mixture can be assumed as ideal (close to pure $a$) where $\gamma_a = 1$, and the freezing point depression is small where $T_m \approx T$. Then the following holds,

$$\ln(x_a) = -\frac{\Delta h_{\text{fus}}(T_m)}{R}\left(\frac{1}{T} - \frac{1}{T_m}\right) = \ln(1 - x_b)$$

Using the power series expansion for the ln function, we know that if $x_b$ is small, then we can further approximate $\ln(1 - x_b) \approx -x_b$, so now we have the following expression for the freezing point depression:

$$-\frac{\Delta h_{\text{fus}}(T_m)}{R}\left(\frac{1}{T} - \frac{1}{T_m}\right) = -x_b$$

$$-\frac{\Delta h_{\text{fus}}(T_m)}{R}\left(\frac{T_m - T}{TT_m}\right) = -x_b$$

$$\left(\frac{T_m - T}{TT_m}\right) = \frac{R}{\Delta h_{\text{fus}}(T_m)} x_b$$

Assume that the freezing point depression is small so $T_m \approx T$, and $TT_m \approx T_m{}^2$

$$T_m - T \approx \frac{RT_m{}^2}{\Delta h_{\text{fus}}(T_m)} x_b$$

If we have a significant amount of solute $b$ in solvent $a$, then liquid mixture a is no longer ideal, and $\gamma_a \neq 1$, then

$$T_m - T \approx \frac{RT_m{}^2}{\Delta h_{\text{fus}}(T_m)} \gamma_a x_b$$

$$\gamma_a = \frac{\Delta h_{\text{fus}}(T_m)(T_m - T)}{x_b R T_m{}^2}$$

*Boiling point elevation (vapor–liquid equilibrium)*

We can now consider the other similar phenomenon of boiling point elevation, and so we start with the vapor–liquid equilibrium condition:

$$\hat{f}_i^v = \hat{f}_i^l$$

Assume that vapor phase (i.e., species $a$) is pure, and the liquid phase consists of species $a$ and $b$ (solute), where the Lewis Randall reference state is used for the liquid phase, then for species $a$, we get

$$f_a^v = x_a \gamma_a f_a^l$$

$$x_a \gamma_a = \frac{f_a^v}{f_a^l}$$

Separately, we know that by definition $d\mu = RTd \ln f$, so for a pure substance that is vaporizing,

$$\Delta g_{\text{vap}} = g_a^v - g_a^l = RT \ln \left( \frac{f_a^v}{f_a^l} \right) = RT \ln (x_a \gamma_a)$$

Similar to the derivation for freezing point depression, we have

$$\ln (x_a \gamma_a) = \frac{\Delta g_{\text{vap}}}{RT} = \frac{\Delta h_{\text{vap}}}{RT} - \frac{\Delta s_{\text{vap}}}{R}$$

We typically know $\Delta h_{\text{vap}}$ at the boiling point, $T_b$. We can find out $\Delta h_{\text{vap}}$ at any temperature $T$ as follows. Let us assume $T > T_b$ for a case of boiling point elevation:

$$\Delta h_{\text{vap}}(T) = \Delta h_{\text{vap}}(T_b) + \int_{T_b}^{T} c_p^l dT + \int_{T}^{T_b} c_p^v dT$$

Similarly for entropy of vaporization,

$$\Delta s_{\text{vap}}(T) = \Delta s_{\text{vap}}(T_b) + \int_{T_b}^{T} \frac{c_p^l}{T} dT + \int_{T}^{T_b} \frac{c_p^v}{T} dT$$

We know that at the boiling point, $T_b$

$$\Delta g_{\text{vap}} = \Delta h_{\text{vap}} - T_b \Delta s_{\text{vap}} = 0$$

$$\Delta s_{\text{vap}} = \frac{\Delta h_{\text{vap}}}{T_b}$$

Substituting into the earlier expression, we get

$$\Delta s_{\text{vap}}(T) = \frac{\Delta h_{\text{vap}}}{T_b} + \int_{T_b}^{T} \frac{c_p^l}{T} dT + \int_{T}^{T_b} \frac{c_p^v}{T} dT$$

Finally we can relate activity coefficient and molar composition of the species to the enthalpy of vaporization (at any desired temperature that may not be the normal boiling point) and heat capacities.

$$\ln(x_a \gamma_a) = \frac{\Delta h_{\text{vap}}(T_b) + \int_{T_b}^{T} c_p^l dT + \int_{T}^{T_b} c_p^v dT}{RT} - \frac{\frac{\Delta h_{\text{vap}}}{T_b} + \int_{T_b}^{T} \frac{c_p^l}{T} dT + \int_{T}^{T_b} \frac{c_p^v}{T} dT}{R}$$

To simplify the expression further, we may define a term $\Delta c_p^{lv} = c_p^v - c_p^l$, which may then be assumed constant for simplifying the integration.

$$\ln(x_a \gamma_a) = \frac{\Delta h_{\text{vap}}(T_b)}{R}\left(\frac{1}{T} - \frac{1}{T_b}\right) + \frac{1}{R}\int_{T_b}^{T} \frac{\Delta c_p^{lv}}{T} dT - \frac{1}{RT}\int_{T_b}^{T} \Delta c_p^{lv} dT$$

$$\ln(x_a \gamma_a) = \frac{\Delta h_{\text{vap}}(T_b)}{R}\left(\frac{1}{T} - \frac{1}{T_b}\right) + \frac{\Delta c_p^{lv}}{R} \ln\left(\frac{T}{T_b}\right) - \frac{\Delta c_p^{lv}}{RT}(T - T_b)$$

$$= \frac{\Delta h_{\text{vap}}(T_b)}{R}\left(\frac{1}{T} - \frac{1}{T_b}\right) - \frac{\Delta c_p^{lv}}{R}\left(1 - \frac{T_b}{T} - \ln\left(\frac{T}{T_b}\right)\right)$$

If solute $b$ is dilute in solvent $a$, the liquid mixture can be assumed as ideal (close to pure $a$) where $\gamma_a = 1$, and the boiling point elevation is small where $T_b \approx T$. Then the following holds,

$$\ln(x_a) = \frac{\Delta h_{\text{vap}}(T_b)}{R}\left(\frac{1}{T} - \frac{1}{T_b}\right) = \ln(1 - x_b)$$

Using the power series expansion for the ln function, we know that if $x_b$ is small, then we can further approximate $\ln(1 - x_b) \approx -x_b$, so now we have the following expression for the boiling point elevation:

$$\frac{\Delta h_{\text{vap}}(T_b)}{R}\left(\frac{1}{T}-\frac{1}{T_b}\right)=-x_b$$

$$\frac{\Delta h_{\text{vap}}(T_b)}{R}\left(\frac{T_b-T}{TT_b}\right)=-x_b$$

$$\left(\frac{T-T_b}{TT_b}\right)=\frac{R}{\Delta h_{\text{vap}}(T_b)}x_b$$

Assume that the boiling point elevation is small so $T_b \approx T$, and $TT_b \approx {T_b}^2$

$$T-T_b \approx \frac{R{T_b}^2}{\Delta h_{\text{vap}}(T_b)}x_b$$

If we have a significant amount of solute $b$ in solvent $a$, then liquid mixture a is no longer ideal, and $\gamma_a \neq 1$, then

$$T-T_b \approx \frac{R{T_b}^2}{\Delta h_{\text{vap}}(T_b)}\gamma_a x_b$$

$$\gamma_a=\frac{\Delta h_{\text{vap}}(T_b)(T-T_b)}{x_b R{T_b}^2}$$

So finally, returning to the problem statement, we note that the solute in this case is sodium chloride, whereby one molecule of NaCl gives two solute ions ($Na^+$ and $Cl^-$). We also know that $\Delta h_{\text{fus}}(T_m)$ of water is 6 kJ/mol, and $T_m = 273$ K. The weight percent of water is 96 % = 100 % − 4 % .

$$x_a = x_{\text{water}} = \frac{\frac{96}{MW_{\text{water}}}}{\frac{96}{MW_{\text{water}}}+2\left(\frac{4}{MW_{\text{NaCl}}}\right)} = 0.974$$

$$x_b = x_{\text{ion}} = 1-0.974 = 0.026$$

From the value of $x_b$, we confirm that solute is dilute enough for the solution to be assumed as an ideal solution, and so the freezing point depression is assumed small, and it follows that

$$T_m - T \approx \frac{R{T_m}^2}{\Delta h_{\text{fus}}(T_m)}x_b = \frac{8.314\times 273^2}{6000}(0.026) = 2.7\ \text{K}$$

$$T = 0^{\circ}\text{C} - 2.7^{\circ}\text{C} = -2.7^{\circ}\text{C}$$

Hence we find that the depressed freezing point is 2.7°C.

## Problem 33

**Estimate how much salt you would need to add to 100 g of water to keep its boiling point at 100 °C at 200 m above sea level.**

## Solution 33

***Worked Solution***

We will assume salt is an involatile solute such that the vapor phase is pure while the liquid phase is a mixture of salt and water.

For the liquid phase, we have the following expression where we assume water is species *a* and salt is species *b*, and the pure liquid Gibbs free energy is adjusted for a mixture. We also assume that liquid is incompressible and that $v_l$ is small; hence, pressure dependence of $g_a^l$ is assumed negligible and so $g_a^l(T, P_{\text{atm}}) \approx g_a^l(T, P)$.

$$\mu_a^l(T, P) = g_a^l(T, P) + RT \ln \gamma_a x_a = g_a^l(T, P_{\text{atm}}) + RT \ln \gamma_a x_a$$

As for the vapor phase, we have pure vapor Gibbs free energy adjusted for pressure at $P$

$$\mu_a^v(T, P) = g_a^v(T, P_{\text{atm}}) + RT \ln \left(\frac{P}{P_{\text{atm}}}\right)$$

Now we try to relate the height above sea level to pressure, knowing that density (in kg/m$^3$) is related to molar volume by $\frac{\rho}{MW} = \frac{1}{v}$.

$$\frac{dP}{dh} = -\rho g = -\left(\frac{P}{RT}\right)(MW_{\text{air}})g$$

$$\frac{dP}{P} = -\frac{(MW_{\text{air}})g}{RT}dh$$

We know that $h = 200$,

$$\ln \frac{P}{P_{\text{atm}}} = -1.2 \times 10^{-4}(200) = -0.024$$

At vapor–liquid equilibrium during boiling,

$$\mu_a^v(T, P) = \mu_a^l(T, P)$$

$$g_a^v(T, P_{\text{atm}}) + RT \ln \left(\frac{P}{P_{\text{atm}}}\right) = g_a^l(T, P_{\text{atm}}) + RT \ln \gamma_a x_a$$

We know further that at $P_{\text{atm}}$, $g_a^v(T, P_{\text{atm}}) = g_a^l(T, P_{\text{atm}})$; therefore

$$\ln\left(\frac{P}{P_{\text{atm}}}\right) = \ln \gamma_a x_a = -0.024$$

If we assume that the solute is dilute, then the mixture is close to ideal solution consisting mainly of water, and $\gamma_a \approx 1$. It also follows then that $x_b$ is small.

$$\ln x_a = \ln(1 - x_b) \approx -x_b = -0.024$$

If we assume that in 100 g of salt solution we have $M$ grams of salt (or NaCl), then,

$$x_b = \frac{\frac{M}{MW_{\text{NaCl}}} \times 2}{\left(\frac{M}{MW_{\text{NaCl}}} \times 2\right) + \left(\frac{100-M}{MW_{\text{water}}}\right)} = 0.024$$

$$M = 3.89$$

We can then solve for the only unknown $M = 3.89$ g.

## Problem 34

**Pure species $b$ freezes from a liquid mixture of $a$ and $b$ consisting of 52 vol% species $a$ at $-40\ ^\circ$C. Construct a one-constant Margules model for the binary liquid mixture and estimate the eutectic point. How would you improve the accuracy of this estimate? You are provided with the following data.**

**Melting point of species $a$, $T_{m,a} = -16\ ^\circ$C**
**Melting point of species $b$, $T_{m,b} = 0\ ^\circ$C**
**Density of species $a$, $\rho_a = 1.1$ g/cm$^3$**
**Density of species $b$, $\rho_b = 1$ g/cm$^3$**
**Molecular Weight of species $a$, $MW_a = 62$**
**Molecular Weight of species $b$, $MW_b = 18$**
**Enthalpy of fusion of species $a$, $\Delta h_{\text{fus},a}(T_{m,a}) = 11{,}000$ J/mol**
**Enthalpy of fusion of species $b$, $\Delta h_{\text{fus},b}(T_{m,b}) = 6000$ J/mol**

## Solution 34

*Worked Solution*

Eutectic point is the lowest possible temperature at which we have only liquid. Recall that when we have a mixture containing two species $a$ and $b$, due to the concept of freezing point depression, we can have a single liquid phase at a lower temperature than the freezing point of either pure species $a$ or pure species $b$.

First we find out the mole fraction of species $a$ from the volume percent composition, using density and molecular weight information provided.

$$x_a = \frac{\frac{52\times\rho_a}{MW_a}}{\frac{52\times\rho_a}{MW_a} + \frac{(100-52)\times\rho_b}{MW_b}} = 0.26$$

From earlier derivation, we obtained the following,

$$\ln(x_i\gamma_i) = -\frac{\Delta h_{\text{fus}}(T_m)}{R}\left(\frac{1}{T} - \frac{1}{T_m}\right) + \frac{\Delta c_p^{sl}}{R}\left(1 - \frac{T_m}{T} - \ln\left(\frac{T}{T_m}\right)\right)$$

We can apply the above expression for species $b$, and assume that the freezing point depression is small. Then the expression simplifies to the following. Note that freezing point depression phenomenon means that $T_m > T$.

$$\ln(x_b\gamma_b) = -\frac{\Delta h_{\text{fus},b}(T_{m,b})}{R}\left(\frac{1}{T} - \frac{1}{T_{m,b}}\right)$$

$$\ln((1 - x_a)\gamma_b) = -\frac{\Delta h_{\text{fus},b}(T_{m,b})}{R}\left(\frac{1}{T} - \frac{1}{T_{m,b}}\right)$$

$$\ln((1 - 0.26)\gamma_b) = -\frac{6000}{8.314}\left(\frac{1}{(-40 + 273)} - \frac{1}{273}\right)$$

$$\ln\gamma_b = -0.15$$

Using the one-constant Margules equation,

$$RT\ln\gamma_b = Ax_a^2$$

$$A = \frac{8.314 \times (-40 + 273) \times (-0.15)}{0.26^2} = -4300\,\text{J/mol}$$

At the eutectic point, solutes $a$ and $b$ both exist in the liquid phase and exert their effects of freezing point depressions; hence, we have the two equations below whereby $T$ is the eutectic temperature for both equations.

$$\ln(x_a\gamma_a) = -\frac{\Delta h_{\text{fus},a}(T_{m,a})}{R}\left(\frac{1}{T} - \frac{1}{T_{m,a}}\right)$$

$$\ln(x_b\gamma_b) = -\frac{\Delta h_{\text{fus},b}(T_{m,b})}{R}\left(\frac{1}{T} - \frac{1}{T_{m,b}}\right)$$

The easiest way to solve the above simultaneous equations is by equating the common temperature term, $^1/_T$.

$$\frac{1}{T_{m,a}} - \frac{R(\ln(x_a\gamma_a))}{\Delta h_{\text{fus},a}(T_{m,a})} = \frac{1}{T_{m,b}} - \frac{R(\ln(x_b\gamma_b))}{\Delta h_{\text{fus},b}(T_{m,b})}$$

We know $T_{m,\,a} = -16 + 273 = 257$ K, $T_{m,\,b} = 273$ K, $A = -4300$ J/mol, $\Delta h_{\text{fus},\,a}(T_{m,\,a}) = 11{,}000$ J/mol, $\Delta h_{fus,\,b}(T_{m,\,b}) = 6000$ J/mol. So, we can express $\ln(x_a\gamma_a)$ in terms of a single unknown, $x_a$.

$$\ln(x_a\gamma_a) = \ln x_a + \ln\gamma_a = \ln x_a + \left(\frac{A}{RT}\right)x_b^{\,2} = \ln x_a + \left(\frac{A}{RT}\right)(1 - x_a)^2$$

Similarly we can express $\ln(x_b\gamma_b)$ in terms of a single unknown, $x_a$.

$$\ln(x_b\gamma_b) = \ln x_b + \ln\gamma_b = \ln(1 - x_a) + \left(\frac{A}{RT}\right)x_a^{\,2}$$

We can solve the equation to find the unknown $x_a$. Once this is done, we then substitute back into either one of the two earlier equations to find the eutectic temperature $T$.

$$\ln(x_a\gamma_a) = -\frac{\Delta h_{\text{fus},a}(T_{m,a})}{R}\left(\frac{1}{T} - \frac{1}{T_{m,a}}\right)$$

$$\ln(x_b\gamma_b) = -\frac{\Delta h_{\text{fus},b}(T_{m,b})}{R}\left(\frac{1}{T} - \frac{1}{T_{m,b}}\right)$$

We should find $T = 140$ K. We can find $T$ through an iterative trial and error process, by first assuming a eutectic $T$ for a specified $x_b$ say, for example, $T = 220$ K and $x_b = 0.6$, we then find $\gamma_b$ from the one-constant Margules equation and back substitute $\gamma_b$ and $x_b$ into the equation of freezing point depression, to find out if the value of $T$ matches the initial guess. If the value of $T$ shows to be less than the initial guess of 220 K, then we may lower our estimate and continue the iterative process to arrive at the correct value of $T$ that satisfies both equations.

This method can be improved as the one-constant Margules equation is a simplified model that starts to break down when the deviation from ideal behavior is significant. We can improve on the model by getting more data points and obtaining a best fit over a range of mole compositions.

We can also check the temperature dependence of the activity coefficient model parameter, $A$. For this, we can obtain a separate set of data points at different temperatures. In this solution, we have assumed that $A$ is independent of temperature.

## Problem 35

**Describe the conditions required for a bubble containing water vapor to be in equilibrium with pure liquid water. Calculate the equilibrium temperature for a bubble of diameter 0.08 μm at atmospheric pressure. Is this equilibrium stable? It is given that the surface tension of water $\sigma = 0.07\ \mathrm{Nm}^{-1}$. The enthalpy of evaporation of water is 40,600 J/mol.**

## Solution 35

***Worked Solution***

For equilibrium in this case there are three equations that hold. Thermal equilibrium whereby $T^L = T^V$, mechanical equilibrium whereby $P^V = P^L + \frac{2\sigma}{r}$, and chemical equilibrium whereby $\mu^L = \mu^V = g^L$.

For the liquid phase, $\mu^L = g^{\mathrm{sat}}(T)$. And for the vapor phase, $\mu^V = g^{\mathrm{sat}}(T) + RT\ln\left(\frac{P^V}{P^{\mathrm{sat}}}\right)$.

In order for chemical equilibrium to hold, $RT\ln\left(\frac{P^V}{P^{\mathrm{sat}}}\right) = 0$ and $P^V = P^{\mathrm{sat}}(T)$. From the mechanical equilibrium equation, we can equate to the chemical equilibrium result as follows,

$$P^V = P^L + \frac{2\sigma}{r} = P^{\mathrm{atm}} + \frac{2\sigma}{r} = P^{\mathrm{sat}}(T)$$

We may recall at this point that $P^{\mathrm{sat}}$ is correlated to enthalpy via the Clausius–Clapeyron equation as follows,

$$\frac{d\ln P^{\mathrm{sat}}}{dT} = \frac{\Delta h^{\mathrm{vap}}}{RT^2}$$

$$\ln\left(\frac{P^{\mathrm{sat}}(T)}{P^{\mathrm{atm}}}\right) = \frac{\Delta h^{\mathrm{vap}}}{R}\left(\frac{1}{T_{b,\mathrm{atm}}} - \frac{1}{T}\right)$$

$$\ln\left(\frac{P^V(T)}{P^{\mathrm{atm}}}\right) = \ln\left(\frac{P^{\mathrm{atm}} + \frac{2\sigma}{r}}{P^{\mathrm{atm}}}\right) = \ln\left(1 + \frac{2\sigma}{P^{\mathrm{atm}}(r)}\right)$$

We know that $r = 0.08 \times 0.5 \times 10^{-3}$ m, $\sigma = 0.07\ \mathrm{Nm}^{-1}$, $\Delta h^{\mathrm{vap}} = 40{,}600$ J/mol, $T_{b,\,\mathrm{atm}} = 100^{\circ}\mathrm{C} = 373$ K. Hence we may solve the equation to find that $T = 500$ K.

We know that this value of $T$ is much greater than the normal boiling point of 373 K. This means we require a large amount of superheat to allow this bubble to form spontaneously in the absence of nucleation sites. The smaller the size of the bubble, the greater the surface tension and the higher the pressure in the vapor phase within the bubble. This makes $\mu^V$ higher, and this provides a driving force for water to migrate from the vapor phase to the liquid phase in order to reduce Gibbs free energy to reach equilibrium. This would cause the bubble to reduce in size further and it is therefore an unstable equilibrium. An unstable equilibrium can be seen as a maxima point at which $\mu^V = \mu^L$; however, any deviation away from this point reduces total Gibbs free energy of the system.

## Problem 36

**A particular substance (1) is dissolved in a solvent (2) at 25 °C at a partial pressure of 1 atm. The vapor phase is assumed as an ideal gas. The liquid fugacity of substance (1) at 25 °C is 900 bar.**

(a) **Find the mole fraction of substance 1 present in the liquid phase assuming the two species 1 and 2 form an ideal solution.**
(b) **The activity coefficient of substance 1 is modeled by the one-constant Margules equation in the form $\ln\gamma_1 = 0.53(1 - x_1)^2$. Recalculate the mole fraction of substance 1 in the liquid phase.**
(c) **What is the Henry's constant for this system? Find an expression for the activity coefficient of substance 1 using the Henry's constant as the reference state.**

## Solution 36

*Worked Solution*

(a) For vapor–liquid equilibrium of this mixture, we have

$$\hat{f}_1^V = \hat{f}_1^L$$

We are told that the liquid phase is an ideal mixture; hence $\gamma_1 = 1$

$$\hat{f}_1^L = \gamma_1 x_1 f_1^L = x_1 f_1^L$$

We are also told that the vapor phase is ideal gas; hence

$$\hat{f}_1^V = y_1 P = P_1$$

We are required to find $x_1$, so we need to solve this equation

$$x_1 f_1^L = P_1$$

$$x_1 = \frac{1}{900} = 1.1 \times 10^{-3}$$

(b) We can obtain an expression for activity coefficient from the Margules model provided

$$\ln \gamma_1 = 0.53(1 - x_1)^2$$

$$\gamma_1 = \exp\left[0.53(1 - x_1)^2\right]$$

We then revisit the vapor–liquid equilibrium condition

$$\gamma_1 x_1 f_1^L = P_1$$

$$\left(\exp\left[0.53(1 - x_1)^2\right]\right)x_1(900) = 1$$

We can use an iterative method to find $x_1$, the initial estimate can be made by assuming $x_1$ is small, and hence $\gamma_1 \approx \exp[0.53(1)^2] = 1.7$.

$$x_1 = \frac{1}{900 \times 1.7} = 6.5 \times 10^{-4}$$

We then substitute this value of $x_1$ back into the equilibrium equation to see if it balances the right-hand side value of 1. It happens that in this case the first guess works, which means that our assumption that $x_1$ was sufficiently small enough to simplify the exponential term was fine. Hence $x_1 = 6.5 \times 10^{-4}$.

(c) If we use Henry's reference state for substance 1, then the value of Henry's constant is obtained by extending the tangent at infinite dilution point ($x_1 \to 0$), to the point when $x_1 = 1$. At this extrapolated point, the value of fugacity is the Henry's constant.

$$H = \lim_{x_1 \to 0} \left(\gamma_1 f_1^L\right) = \gamma_1^{\infty}(900) = \left(900\exp\left[0.53(1 - x_1)^2\right]\right)$$

Since substance 1 is at infinite dilution, $1 - x_1 \approx 1$, so we can find Henry's constant to be

$$H = 900\exp[0.53] = 1500$$

Now we try to derive an expression for the activity coefficient under the Henry's reference state,

$$\gamma_1^{\text{Henry}} x_1 H = \hat{f}_1^L = \gamma_1^{LR} x_1 f_1^L$$

$$\gamma_1^{\text{Henry}} H = \gamma_1 f_1^L$$

$$\gamma_1^{\text{Henry}} = \frac{\gamma_1 f_1^L}{H} = \frac{\left(\exp\left[0.53(1 - x_1)^2\right]\right)(900)}{1500}$$

## Problem 37

**A given system consists of two components 1 and 2 in VLE. The solution is not ideal with the activity coefficients given by $RT \ln \gamma_1 = Ax_2^3$ where $A$ is not a function of temperature. You are told that $P_1^{\text{sat}} = 0.5$ bar and $P_2^{\text{sat}} = 0.8$ bar at 298 K. If you mix equimolar portions of 1 and 2, the heat evolved is 800 J/mol solution. Use this information to determine the total pressure and vapor phase compositions at $T = 298$ K.**

## Solution 37

***Worked Solution***

To relate the activity coefficient of component 1 to that of component 2, recall the Gibbs–Duhem equation

$$x_1 d \ln \gamma_1 + x_2 d \ln \gamma_2 = 0$$

$$\frac{d \ln \gamma_2}{dx_2} = -\frac{x_1 d \ln \gamma_1}{x_2 dx_2}$$

We are given that

$$RT \ln \gamma_1 = Ax_2^3$$

$$\ln \gamma_1 = \frac{A}{RT} x_2^3$$

$$\frac{d \ln \gamma_1}{dx_2} = \frac{3A}{RT} x_2^2$$

Substituting back into the Gibbs–Duhem expression, we have

$$\frac{d \ln \gamma_2}{dx_2} = -\frac{x_1 d \ln \gamma_1}{x_2 dx_2} = -\frac{x_1}{x_2} \frac{3A}{RT} x_2^2 = -\frac{3A}{RT}(1 - x_2)x_2$$

$$d \ln \gamma_2 = -\frac{3A}{RT}(1 - x_2)x_2 dx_2$$

In order to solve this differential equation, we need to have boundary conditions that would set the limits of the integral. We know that as $\gamma_2 \to 1$ as $x_2 \to 1$, $\ln\gamma_2 \to 0$. This assumes the Lewis Randall reference state where ideality is reached when there is pure species.

$$\int_0^{\ln \gamma_2} d \ln \gamma_2 = \int_1^{x_2} -\frac{3A}{RT}(1 - x_2)x_2 dx_2$$

$$\ln \gamma_2 = \frac{3A}{RT}\int_1^{x_2}\left(x_2^2 - x_2\right)dx_2 = \frac{3A}{RT}\left[\frac{x_2^3}{3} - \frac{x_2^2}{2}\right]_1^{x_2}$$

$$= \frac{3A}{RT}\left[\frac{2x_2^3 - 3x_2^2 - 2 + 3}{6}\right] = \frac{A}{2RT}\left[2x_2^3 - 3x_2^2 + 1\right]$$

The problem statement also mentions about heat evolved during mixing of different species. This brings to mind the enthalpy of mixing, which then relates to Gibbs free energy. We can try to correlate Gibbs free energy and activity coefficient using the definition of excess Gibbs free energy.

$$\frac{g^E}{RT} = \sum x_i \ln \gamma_i = x_1 \ln \gamma_1 + x_2 \ln \gamma_2$$

$$\frac{g^E}{RT} = x_1 \frac{A}{RT} x_2^3 + x_2 \frac{A}{2RT}\left[2x_2^3 - 3x_2^2 + 1\right]$$

$$= \frac{A}{RT}x_2^3 - \frac{A}{RT}x_2^5 + \frac{A}{RT}x_2^5 - \frac{3A}{2RT}x_2^3 + x_2\frac{A}{2RT} = \frac{A}{2RT}\left[x_2 - x_2^3\right]$$

Now using Gibbs–Helmholtz equation to relate excess Gibbs free energy to enthalpy,

$$-\frac{h^E}{T^2} = \frac{\partial\left(\frac{g^E}{T}\right)}{\partial T} = \frac{\partial\left(\frac{A}{2T}\left[x_2 - x_2^3\right]\right)}{\partial T}$$

$$-\frac{h^E}{T^2} = -\frac{A}{2T^2}\left[x_2 - x_2^3\right]$$

$$A = \frac{2h^E}{x_2 - x_2^3}$$

$$h^E = \Delta h_{\text{mix}} = -800\ \text{J/mol}$$

So we can find $A$ as we know that the mole compositions would be $x_1 = x_2 = 0.5$ for equimolar mixing.

$$A = \frac{2h^E}{x_2 - x_2^3} = \frac{2 \times (-800)}{0.5 - 0.5^3} = -4300\ \text{J/mol}$$

We need to find out the total pressure and vapor phase compositions at $T = 298$ K.

$$y_1 P = x_1 \gamma_1 P_1^{\text{sat}}$$

$$y_2P = x_2\gamma_2P_2^{\text{sat}}$$

$$P = y_1P + y_2P = x_1\gamma_1P_1^{\text{sat}} + x_2\gamma_2P_2^{\text{sat}}$$

To solve for pressure, we need to calculate $\gamma_1$and $\gamma_2$

$$\ln\gamma_1 = \frac{A}{RT}{x_2}^3 = -\frac{4300}{8.314 \times 298}(0.5)^3$$

$$\gamma_1 = 0.8$$

$$\ln\gamma_2 = \frac{A}{2RT}\left[2{x_2}^3 - 3{x_2}^2 + 1\right] = -\frac{4300}{2 \times 8.314 \times 298}\left[2(0.5)^3 - 3(0.5)^2 + 1\right] = -0.4$$

$$\gamma_2 = 0.7$$

So we can now solve for total pressure.

$$P = (0.5)(0.8)(0.5) + (0.5)(0.7)(0.8) = 0.48$$

$$y_1 = \frac{x_1\gamma_1P_1^{\text{sat}}}{P} = \frac{(0.5)(0.8)(0.5)}{0.48} = 0.4$$

$$y_2 = 1 - 0.4 = 0.6$$

## Problem 38

**One mole of a gas is isothermally and reversibly compressed from a starting pressure of $P_1$ to a final pressure $P_2$. We may assume that this process is modeled by the equation of state $z = 1 + B'P$, where $B' < 0$.**

(a) **Identify the equation of state and comment.**
(b) **Find an expression for the work done in this process.**

## Solution 38

***Worked Solution***

(a) This equation takes on the form of the Virial Equation of State which is truncated and volume-explicit. A volume explicit equation of state refers to one where volume $v$ is not raised to higher powers and can be singly isolated as the subject of the formula. We may observe this as the left-hand side of the equation is $z$ which contains the variable $v$ since $z = \frac{Pv}{RT}$. In this equation, the variable that is raised to higher powers in a power series expansion is pressure $P$. However, this expression is truncated (i.e., simplified), as the higher order terms in $P$ are assumed insignificant and ignored here. Hence only the first two terms remain (i.e., zeroth and first orders of $P$)

(b)
$$z = 1 + B'P = \frac{Pv}{RT}$$

$$v = \frac{RT}{P}(1 + B'P) = \frac{RT}{P} + B'RT$$

Assuming constant $T$, since isothermal conditions, we have

$$dv = d\left(\frac{RT}{P} + B'RT\right) = -\frac{RT}{P^2}dP$$

We may derive an expression for work done for a reversible process. In a reversible process, we are only slightly out of equilibrium, reversibility is characterized by the fact that we are able to reverse the process (i.e., to expand back if we are compressing) by adding or removing infinitesimal amounts of differential mass. This reversal will have no net effect on the surroundings, whereby the same amount of work will be involved in each direction. The reversible work sets the limit of work done, i.e., the max work that can be done or min work that is needed to put in. Only in a reversible process, can we substitute system pressure with external pressure.

In this case, since our compression process is reversible, we have $P = P_{ext}$ and so

$$w = -\int P_{ext}dv = -\int Pdv = \int P\left(\frac{RT}{P^2}\right)dP = \int_{P_1}^{P_2} \frac{RT}{P}dP = RT\ln\left(\frac{P_2}{P_1}\right)$$

We may notice that since the system is compressed, $P_2 > P_1$ and hence $w$ is a positive value, and work is required to be done.

## Problem 39

**An unknown pure substance is found to have a vapor pressure of 2 bar at 320 K with a saturated liquid having a volume of $1 \times 10^{-4}$ m³/mol.**

(a) **At 1 bar and 320 K, what phase(s) of the unknown substance will be present?**
(b) **Comment on whether the critical temperature of this substance will be greater or less than 320 K and whether the critical pressure will be above or below 2 bar?**
(c) **Find out the number of moles of this pure substance at 2 bar and 320 K in a 10 m³ tank which is 70 vol% occupied by the saturated liquid and the rest of the volume is saturated gas.**
(d) **Find out the total reversible work needed to compress 100% saturated vapor to the final state as described in part c.**

## Solution 39

### *Worked Solution*

(a) We know that $P_{vap} = 2$ bar and $P_{sys} = 1$ bar.

Since $P_{vap} > P_{sys}$, we may deduce that at the same temperature of 320 K, the substance will exist as pure vapor for this said system.

(b) Let us recall some key features about the critical point. We know that in a Pv diagram, the critical point is located at the peak of the liquid–vapor dome, whereby $v_l = v_v$. At this point the temperature, $T_c$ and pressure $P_c$ are constrained. The critical point represents the point at which the liquid and vapor phases are no longer distinct and distinguishable from one another. At temperatures above $T_c$, we have a supercritical fluid that displays some vapor-like characteristics and some liquid-like characteristics.

At the system conditions, we have vapor–liquid equilibrium. This state exists as a point within the vapor–liquid dome. As we raise temperature and pressure, we will reach critical point. As such, we expect the critical temperature to be greater than 320 K and the critical pressure to be greater than 2 bar.

(c) We are told that $v_l = 1 \times 10^{-4}$ m$^3$/mol; therefore

$$v_v = \frac{RT}{P} = \frac{8.314 \times 320}{2 \times 10^5} = 0.01 \text{ m}^3/\text{mol}$$

We may convert volume percent into mole fraction

$$V_v = 0.3V_{\text{tank}} = 3 \text{ m}^3$$

$$V_l = 0.7V_{\text{tank}} = 7 \text{ m}^3$$

We can now find number of moles

$$N = N_l + N_v = \frac{7}{1 \times 10^{-4}} + \frac{3}{0.01} = 70,300 \text{ mol}$$

(d) For a reversible process, we may express work done as follows

$$W = -\int P_{\text{ext}} dV = -\int P dV$$

To reversibly compress from pure saturated vapor to only 30 vol% saturated vapor, we need to find out the molar volumes at the two different end points. Note that for a VLE, the molar volume will consider molar contributions of the liquid and vapor molar volumes, to make up the total mixture molar volume that matches the same total number of moles (mass is conserved in the system).

$$v_i = v_v^{\text{sat}} = 0.01 \text{ m}^3/\text{mol}$$

$$v_f = \frac{N_l}{N} v_l^{\text{sat}} + \frac{N_v}{N} v_v^{\text{sat}} = \frac{\frac{7}{1\times10^{-4}}}{70,300}\left(1\times10^{-4}\right) + \frac{\frac{3}{0.01}}{70,300}(0.01) = 1.4\times10^{-4}$$

$$W = -NP\left(v_f - v_i\right) = -(70,300)\left(2\times10^5\right)\left(1.4\times10^{-4} - 0.01\right) = 140 \text{ MJ}$$

**Problem 40**

**A heater is used to raise the temperature of 10 mol of gas isobarically at 47 bar from an initial temperature of 30 °C and to a final temperature of 230 °C. Ideal gas is assumed and the heat capacity is provided as follows,**

$$\frac{C_p^{ig}}{R} = 4.8 + 0.003T$$

(a) **Find the energy requirement for the heater assuming ideal gas.**
(b) **If the equation of state for the gas is given by $z = 1 + \frac{aP}{\sqrt{T}}$, where $a = -0.09 \text{ K}^{0.5}/\text{bar}$, comment on the intermolecular forces determining the behavior of the gas in this process.**
(c) **Find out the energy requirement of the heater if we do not assume ideal gas.**

**Solution 40**

***Worked Solution***

(a) For isobaric process in a closed system, $\Delta H = Q$. We also know that by definition,

$$c_p = \left(\frac{\partial h}{\partial T}\right)_P$$

So assuming ideal gas, we have the following

$$\Delta H = n\int_{T_i}^{T_f} c_p dT = nR\int_{T_i}^{T_f} (4.8 + (0.003)T)dT$$

$$\Delta H = nR\left[4.8T + \frac{(0.003)T^2}{2}\right]_{303}^{503}$$

$$Q = \Delta H = (10)(8.314)\left[4.8(200) + \frac{(0.003)\left(503^2 - 303^2\right)}{2}\right] = 100 \text{ kJ}$$

Since the value of $\Delta H > 0$, heat is required to be added to the system.

(b) We are given the equation of state for a real gas where $a = -0.09\ \mathrm{K}^{0.5}/\mathrm{bar}$.

$$z = \frac{Pv}{RT} = 1 + \frac{aP}{\sqrt{T}}$$

$$v = \frac{RT}{P} + \frac{RT}{P}\left(\frac{aP}{\sqrt{T}}\right) = \frac{RT}{P} + aR\sqrt{T}$$

$$v^{ig} = \frac{RT}{P}$$

By comparing the expressions of molar volume between ideal gas and real gas, we can observe that the additional term of $aR\sqrt{T}$ is the difference, and it is a negative value. Hence attractive forces dominate as $v^{\mathrm{ig}} > v^{\mathrm{real}}$.

(c) When we have a real gas, we may break down the pathway into three steps, where one of the steps will assume ideal gas, and the other two steps are to bring the state from real gas to ideal gas and vice versa.

We have this pathway to achieve: $H_i^{\mathrm{real}}(T_i, P) \to H_f^{\mathrm{real}}(T_f, P)$ and we may do this via three steps:

1. Isothermal Step 1: $H_i^{\mathrm{real}}(T_i, \boldsymbol{P}) \to H_1^{\mathrm{ig}}(T_i, \boldsymbol{P}^*)$
2. Isobaric Step 2: $H_1^{\mathrm{ig}}(\boldsymbol{T}_i, P^*) \to H_2^{\mathrm{ig}}(\boldsymbol{T}_f, P^*)$
3. Isothermal Step 3: $H_2^{\mathrm{ig}}(T_f, \boldsymbol{P}^*) \to H_f^{\mathrm{real}}(T_f, \boldsymbol{P})$

For Step 1,

$$\Delta h = \int_P^{P^*} \left(\frac{\partial h}{\partial P}\right) dP$$

Now the trick is to try to re-express $\frac{\partial h}{\partial P}$ into a form that is in terms of $v$ and $T$ since this would allow us to make use of the equation of state.

---

***Background Concepts***

It is now useful to recap on some thermodynamic relations. From the first and second laws of thermodynamics, we know that

$$du = \delta q + \delta w_{\mathrm{rev}}$$

From here, we can obtain the following fundamental property relations.

$$du = Tds - Pdv$$

$$dh = Tds + vdP$$

$$dg = -sdT + vdP$$

We can also derive the following:

$$\left(\frac{\partial u}{\partial s}\right)_v = T, \quad \left(\frac{\partial u}{\partial v}\right)_s = -P$$

$$\left(\frac{\partial h}{\partial s}\right)_P = T, \quad \left(\frac{\partial h}{\partial P}\right)_s = v$$

$$\left(\frac{\partial g}{\partial T}\right)_P = -s, \quad \left(\frac{\partial g}{\partial P}\right)_T = v$$

Also via Maxwell relations, which states that the order of partial differentiation does not matter, we have

$$\left(\frac{\partial T}{\partial v}\right)_s = -\left(\frac{\partial P}{\partial s}\right)_v, \quad \left(\frac{\partial T}{\partial P}\right)_s = \left(\frac{\partial v}{\partial s}\right)_P$$

$$\left(\frac{\partial s}{\partial v}\right)_T = \left(\frac{\partial P}{\partial T}\right)_v, \quad -\left(\frac{\partial s}{\partial P}\right)_T = \left(\frac{\partial v}{\partial T}\right)_P$$

---

So returning to the question, for step 1, we have

$$dh = Tds + vdP$$

$$\frac{dh}{dP} = \frac{T_i ds}{dP} + v = v - T_i\left(\frac{dv}{dT}\right)_P$$

$$\Delta h = \int_P^{P^*}\left[v - T_i\left(\frac{dv}{dT}\right)_P\right]dP$$

In this case, we have a real gas equation of state,

$$v = \frac{RT}{P} + aR\sqrt{T}$$

$$\Delta H_{\text{step1}} = n\Delta h = n\int_P^{P^*}\left[\frac{RT_i}{P} + aR\sqrt{T_i} - T_i\left(\frac{R}{P} + \frac{1}{2}aRT_i^{-\frac{1}{2}}\right)\right]dP$$

$$\Delta H_{\text{step1}} = nR\int_P^{P^*}\left[T_iP^{-1} + aT_i^{\frac{1}{2}} - T_iP^{-1} - \frac{1}{2}aT_i^{\frac{1}{2}}\right]dP = nR\int_P^{P^*}\left[\frac{1}{2}aT_i^{\frac{1}{2}}\right]dP$$

$$\Delta H_{\text{step1}} = naR\left[\frac{1}{2}T_i^{\frac{1}{2}}\right](P^* - P)$$

Step 3 is similar to step 1, so we can derive a similar expression for step 3 here.

$$\Delta H_{\text{step3}} = n\Delta h = n\int_{P^*}^{P}\left[\frac{RT_f}{P} + aR\sqrt{T_f} - T_f\left(\frac{R}{P} + \frac{1}{2}aRT_f^{-\frac{1}{2}}\right)\right]dP$$

$$\Delta H_{\text{step3}} = naR\left[\frac{1}{2}T_f^{\frac{1}{2}}\right](P - P^*)$$

Step 2 is a process that may assume ideal gas, which you have solved in part a

$$\Delta H_{\text{step2}} = n\Delta h = n\int_{T_i}^{T_f} c_p dT = 100 \text{ kJ}$$

The total heat required is as follows

$$\Delta H = naR\left[\frac{1}{2}T_i^{\frac{1}{2}}\right](P^* - P) + 100 \text{ kJ} - naR\left[\frac{1}{2}T_f^{\frac{1}{2}}\right](P^* - P)$$

$$\Delta H = \frac{naR}{2}\left(T_i^{\frac{1}{2}} - T_f^{\frac{1}{2}}\right)(P^* - P) + 100 \text{ kJ}$$

In order to achieve ideal gas state, $P^* \to 0$. We observe that less heat is required if we do not assume ideal gas.

$$\Delta H = -\frac{naR}{2}\left(T_i^{\frac{1}{2}} - T_f^{\frac{1}{2}}\right)P + 100 \text{ kJ}$$

$$Q = \Delta H = -\frac{10(-0.09)(8.314)}{2}\left(303^{\frac{1}{2}} - 503^{\frac{1}{2}}\right)(47) + 100{,}000 = 99 \text{ kJ}$$

## Problem 41

**One mole of a non-ideal gas is compressed isothermally of 350 K. The initial pressure is 1 bar at a saturation pressure of 20 bar. The final state is 25% saturated liquid. Over this range, the vapor phase follows the virial equation of state truncated to one term $z = 1 + B'P$, where $B' = -0.03 \text{ bar}^{-1}$ and is independent of pressure. The heat of vaporization at 20 bar is 600 kJ/mol.**

(a) **What is the enthalpy change for the compression?**
(b) **What is the entropy change for the compression?**

## Solution 41

### *Worked Solution*

(a) First we notice that after the non-ideal gas is compressed, liquid is formed from gas which means that phase change (condensation) occurred. Therefore, we may analyze the overall process as comprising of a compression part and a condensation part.

$$\Delta H = \Delta H_{\text{comp}} + \Delta H_{\text{cond}}$$

$$\Delta H = \left(\frac{\partial H}{\partial T}\right)_P dT + \left(\frac{\partial H}{\partial P}\right)_T dP + \Delta H_{\text{cond}}$$

$$\Delta H = C_p dT + \left[v - T\left(\frac{dv}{dT}\right)_P\right] dP + \Delta H_{\text{cond}}$$

We know that the equation of state is $z = 1 + B'P$,

$$z = \frac{Pv}{RT} = 1 + B'P$$

$$v = \frac{RT}{P} + RTB'$$

$$\Delta H = C_p dT + \left[\frac{RT}{P} + RTB' - T\left(\frac{R}{P} + RB'\right)\right] dP + \Delta H_{\text{cond}} = C_p dT + \Delta H_{\text{cond}}$$

For isothermal process, $dT = 0$

$$\Delta H = \Delta H_{\text{cond}} = -n\Delta h_{\text{vap}} = -0.25 \times 600 = -150 \text{ kJ/mol}$$

The change in enthalpy for the process is 150 kJ/mol.

(b)

$$\Delta S = \Delta S_{\text{comp}} + \Delta S_{\text{cond}}$$

$$\Delta S = \left(\frac{\partial S}{\partial T}\right)_P dT + \left(\frac{\partial S}{\partial P}\right)_T dP + \Delta S_{\text{cond}}$$

For an isothermal process,

$$\Delta S = \left(\frac{\partial S}{\partial P}\right)_T dP + \frac{Q_{\text{cond}}}{T}$$

We found from part a, $Q_{\text{cond}} = -150$ kJ/mol. We have also been told that temperature $T = 350$ K, $B' = -0.03 \text{ bar}^{-1}$ and $n = 1$.

$$\left(\frac{\partial S}{\partial P}\right)_T = -\left(\frac{\partial V}{\partial T}\right)_P$$

$$v = \frac{RT}{P} + RTB'$$

$$\left(\frac{\partial S}{\partial P}\right)_T = -\left(\frac{\partial\left(\frac{nRT}{P} + nRTB'\right)}{\partial T}\right)_P = -R\left(\frac{1}{P} + B'\right)$$

$$\Delta S = \int_1^{20}\left[-R\left(\frac{1}{P}+B'\right)\right]dP + \left(\frac{-150{,}000}{350}\right) = -R\ln\left(\frac{20}{1}\right) - B'R(20-1) + \left(\frac{-150{,}000}{350}\right)$$

$$= -8.314\ln\left(\frac{20}{1}\right) - 0.03(8.314)(20-1) + \left(\frac{-150{,}000}{350}\right) = -20 - 430 = -450\,\mathrm{J/mol/K}$$

## Problem 42

**A certain solid a is heated to a high temperature $T_H$ at a high pressure $P_H$ to form another solid *b*. Once *b* is formed, it would be quickly cooled and de-pressured to prevent a from reforming. You may assume that molar entropy and density are independent of temperature and pressure. Find an equation that relates $P_H$ as a function of $T_H$ when therxe is equilibrium between *a* and *b*. You may use the following data:**

| At 298 K and 1 bar | Substance *a* | Substance *b* |
|---|---|---|
| **Molar Gibbs Free Energy, $g_0$ (J/mol)** | 0 | 3000 |
| **Molar Entropy (J/mol K)** | 6 | 2 |
| **Density (kg/m$^3$)** | 2000 | 3500 |
| **Molecular Weight** | 12 | 12 |

## Solution 42

***Worked Solution***

At solid–solid equilibrium, $\Delta G = 0$ or in other words $G_a = G_b$

$$g_a = g_{0,a} + \int_{P_0}^{P_H}\left(\frac{\partial g}{\partial P}\right)_T dP + \int_{T_0}^{T_H}\left(\frac{\partial g}{\partial T}\right)_P dT$$

$$g_b = g_{0,b} + \int_{P_0}^{P_H}\left(\frac{\partial g}{\partial P}\right)_T dP + \int_{T_0}^{T_H}\left(\frac{\partial g}{\partial T}\right)_P dT$$

Recall thermodynamic relations,

$$\left(\frac{\partial g}{\partial T}\right)_P = -s, \quad \left(\frac{\partial g}{\partial P}\right)_T = v$$

$$g_a = g_{0,a} + \int_{P_0}^{P_H} vdP - \int_{T_0}^{T_H} sdT = 0 + \left[\frac{12 \times 10^{-3}}{2000}\right](P_H - 1)(10^5) - 6(T_H - 298)$$

$$g_b = 3000 + \left[\frac{12 \times 10^{-3}}{3500}\right](P_H - 1)(10^5) - 2(T_H - 298)$$

We can now derive an equation that relates $P_H$ to $T_H$

$$\left[\frac{12 \times 10^{-3}}{2000}\right](P_H - 1)(10^5) - 6(T_H - 298)$$
$$= 3000 + \left[\frac{12 \times 10^{-3}}{3500}\right](P_H - 1)(10^5) - 2(T_H - 298)$$
$$0.6P_H - 0.6 - 6T_H + 1788 = 3000 + 0.34P_H - 0.34 - 2T_H + 596$$
$$1800 = 0.26P_H - 4T_H$$

# Separation Processes

## Problem 1

**You are provided with experimental data for vapor–liquid equilibrium (VLE) of a binary liquid mixture of acetone and water at 1 bar.**

| T/°C | $x_{a,actual}$ | $y_{a,actual}$ |
|---|---|---|
| 100 | 0 | 0 |
| 84.75 | 0.02 | 0.4451 |
| 75.13 | 0.05 | 0.634 |
| 68.19 | 0.1 | 0.7384 |
| 65.02 | 0.15 | 0.7813 |
| 63.39 | 0.2 | 0.8047 |
| 61.45 | 0.3 | 0.8295 |
| 60.39 | 0.4 | 0.8426 |
| 59.91 | 0.5 | 0.8518 |
| 59.55 | 0.6 | 0.8634 |
| 58.79 | 0.7 | 0.8791 |
| 58.07 | 0.8 | 0.9017 |
| 57.07 | 0.9 | 0.9317 |
| 54.14 | 1 | 1 |

| T/°C | $P_a^{vap}$/bar | $P_w^{vap}$/bar |
|---|---|---|
| 15 | 0.1933 | 0.0171 |
| 20 | 0.2437 | 0.0233 |
| 30 | 0.3769 | 0.0424 |
| 40 | 0.5633 | 0.0737 |
| 50 | 0.8171 | 0.1233 |
| 60 | 1.154 | 0.1992 |
| 70 | 1.591 | 0.3116 |

(continued)



X. W. Ng, *Engineering Problems for Undergraduate Students*,
https://doi.org/10.1007/978-3-030-13856-1_3

| T/°C | $P_a^{vap}$/bar | $P_w^{vap}$/bar |
|---|---|---|
| 80 | 2.148 | 0.4734 |
| 90 | 2.843 | 0.701 |
| 100 | 3.697 | 1.013 |

(a) **Comment on whether the mixture is ideal.**
(b) **Determine the bubble point and dew point temperatures for equal mole fractions of acetone and water at 1 bar.**
(c) **If a liquid mixture at 52 °C containing acetone of mole fraction at 0.4 is heated at constant pressure of 1 bar until vapor begins to form, what will be the composition of this vapor?**

## Solution 1

### *Worked Solution*

(a) Below is a simple illustration of our system.

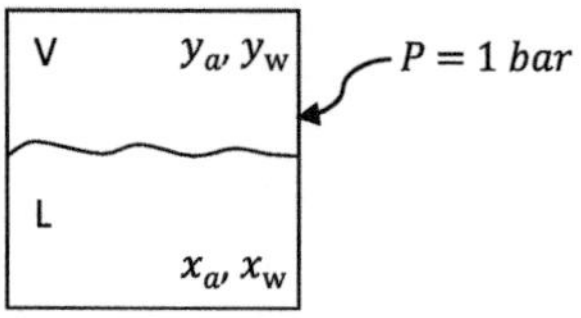


We have an ideal mixture when the VLE data follows Raoult's law, which states that for a particular species $i$ in the mixture, the following equation holds.

$$y_i P = x_i P_i^{vap}$$

$y_i$ is the mole fraction of species $i$ in the vapor phase, $P$ is the pressure of the system or total pressure, $x_i$ is the mole fraction of species $i$ in the liquid phase, and $P_i^{vap}$ is the vapor pressure of species $i$ which is also its pressure when it is in vapor–liquid equilibrium in a pure component system.

The derivation of Raoult's law can be followed as follows:

In order for equilibrium, chemical potentials of species $i$ are equivalent in the vapor and liquid phases.

$$\mu_i^L = \mu_i^V$$

$$\mu_{i,\text{pure}}^L + \text{RT}\ln x_i = \mu_{i,\text{pure,ref}}^V + \text{RT}\ln\frac{f_i}{P_{\text{pure,ref}}} \quad (1)$$

Equation (1) above represents the balance of chemical potentials under the actual system conditions. The term $\text{RT}\ln x_i$ accounts for the deviation of chemical potential of species $i$ in a liquid mixture from pure liquid $i$. On the right-hand side of the equation, $\mathrm{RT}\ln\frac{f_{i,\mathrm{pure}}}{P_{\mathrm{pure,ref}}}$ accounts for the deviation from pure gas $i$ under reference state conditions and actual conditions.

$$\mu^{L}_{i,\mathrm{pure}} = \mu^{V}_{i,\mathrm{pure,ref}} + \mathrm{RT}\ln\frac{f_{i,\mathrm{pure}}}{P_{\mathrm{pure,ref}}} \tag{2}$$

Equation (2) above represents the balance of chemical potentials under a hypothetical state where there is pure $i$ in VLE. We can subtract Eq. (2) from Eq. (1) and obtain

$$\mathrm{RT}\ln x_i = \mathrm{RT}\ln\frac{f_i}{f_{i,\mathrm{pure}}}$$

$$x_i = \frac{f_i}{f_{i,\mathrm{pure}}}$$

We can assume ideal gas, which means $f$ can be replaced by pressure $P$. $P_i$ represents the partial pressure of component $i$ in the vapor phase of the mixture. $P_{i,\mathrm{pure}}$ is the partial pressure of component $i$ in a pure vapor phase of $i$, so this pressure is in fact the vapor pressure of $i$ (at one bar in our case).

$$x_i = \frac{P_i}{P_{i,\mathrm{pure}}} = \frac{y_i P}{P_{i,\mathrm{pure}}}$$

If the mixture was ideal, then $x_i P_{i,\mathrm{pure}} = y_i P$ or $x_i P_i^{\mathrm{vap}} = y_i P$. We are given data for the corresponding $x_i$, $y_i$, $P_i^{\mathrm{vap}}$ for different values of $T$. For our binary mixture, we have acetone and water, and $P = 1$ bar.

$$x_a P_a^{\mathrm{vap}} = y_a(1) = y_a$$

$$x_w P_w^{\mathrm{vap}} = y_w(1) = y_w$$

We know that $y_a + y_w = 1$ and $x_a + x_w = 1$. Combining the equations, we obtain

$$x_a P_a^{\mathrm{vap}} + x_w P_w^{\mathrm{vap}} = y_a(1) + y_w(1) = 1$$

$$x_a P_a^{\mathrm{vap}} + (1 - x_a) P_w^{\mathrm{vap}} = 1$$

$$x_a = x_{a,\mathrm{Raoult}} = \frac{1 - P_w^{\mathrm{vap}}}{P_a^{\mathrm{vap}} - P_w^{\mathrm{vap}}};\ y_{a,\mathrm{Raoult}} = x_{a,\mathrm{Raoult}} P_a^{\mathrm{vap}}$$

If Raoult's law is satisfied, then the experimental data of $x_{a,\mathrm{actual}}$ and $y_{a,\mathrm{actual}}$ should be consistent with values of $x_a$ calculated using Raoult's law above (i.e., the values of $x_{a,\mathrm{Raoult}}$ and $y_{a,\mathrm{Raoult}}$ shown in table below) using values of $P_w^{\mathrm{vap}}$, $P_a^{\mathrm{vap}}$ at corresponding temperatures.

| T/°C | $P_a^{\text{vap}}$/bar | $P_w^{\text{vap}}$/bar | $x_{a,\text{Raoult}}$ | $y_{a,\text{Raoult}}$ |
|---|---|---|---|---|
| 15 | 0.1933 | 0.0171 | 5.57832 | 1.078289 |
| 20 | 0.2437 | 0.0233 | 4.431488 | 1.079954 |
| 30 | 0.3769 | 0.0424 | 2.86278 | 1.078982 |
| 40 | 0.5633 | 0.0737 | 1.891953 | 1.065737 |
| 50 | 0.8171 | 0.1233 | 1.263621 | 1.032504 |
| 60 | 1.154 | 0.1992 | 0.83871 | 0.967871 |
| 70 | 1.591 | 0.3116 | 0.538065 | 0.856061 |
| 80 | 2.148 | 0.4734 | 0.314463 | 0.675467 |
| 90 | 2.843 | 0.701 | 0.139589 | 0.396852 |
| 100 | 3.697 | 1.013 | −0.00484 | −0.01791 |

Below are the plots generated from the actual experimental data (red and blue) and the data calculated assuming Raoult's law is satisfied (purple and green). We can see that the experimental data and the Raoult's law model do not fit; hence the mixture is not ideal.

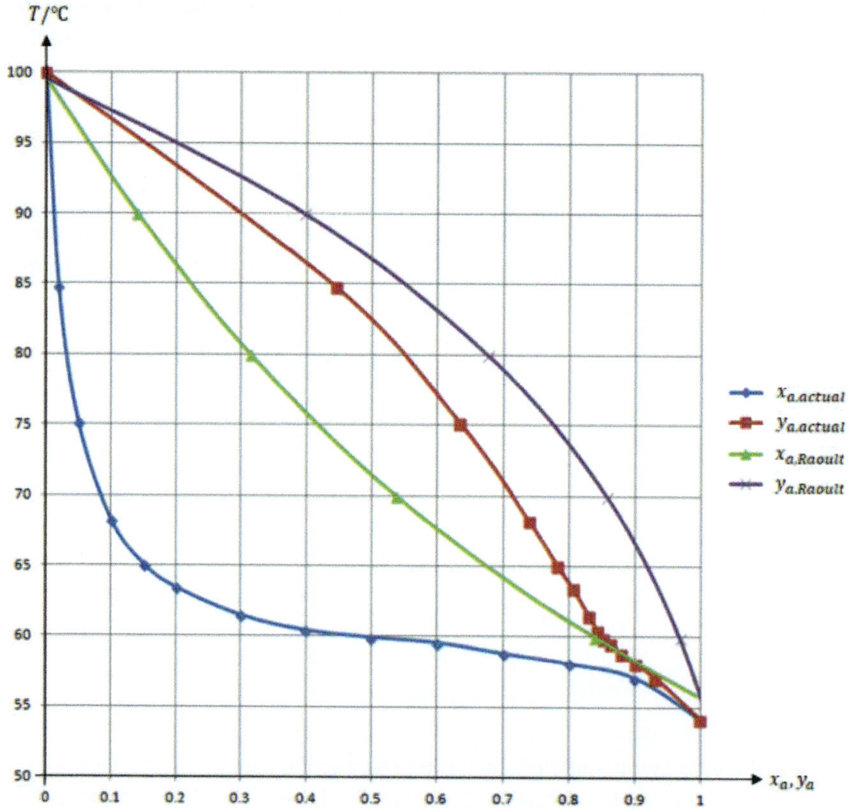

(b) Let us first recall some basic concepts about phase diagrams.

---

***Background Concepts***

There are two common types of phase diagrams, *Pxy* (at constant *T*) and *Txy* (at constant *P*) diagrams.

In a Txy diagram, bubble point occurs at the temperature (at specified *P*) when the first bubble of vapor is formed when we heat a liquid mixture. In a Pxy diagram, bubble point occurs at the pressure (at specified *T*) when the first bubble forms when we reduce pressure. Conversely, dew point occurs at the temperature (at specified *P*) when the first liquid droplet forms when we cool a liquid mixture. In a *Pxy* diagram, it occurs at the pressure (at specified *T*) when the first droplet forms when we raise pressure.

For a pure component, the bubble point and the dew point temperatures are the same, i.e., the boiling point. Graphically, these occur at the extreme ends of the *Pxy* or *Txy* diagrams when the two curves meet.

In a *Txy* or *Pxy* diagram, we are only analyzing one of the species in the mixture, e.g., species a. In order to find out the mole fractions for the other species in a binary mixture, we can apply the fact that mole compositions sum to 1 ($x_b = 1 - x_a$; $y_b = 1 - y_a$, where $x_a$ and $y_a$ may, for example, be obtained from the *Pxy*/*Txy* graphs for species a). When we have a mixture at VLE, we can find the percentage of liquid and vapor for the species by applying the lever rule, whereby *a* and *b* are as shown in the 2-phase region in the *Pxy* diagram below for species a.

$$\%L = \frac{q}{p+q}, \quad \%V = \frac{p}{p+q}$$

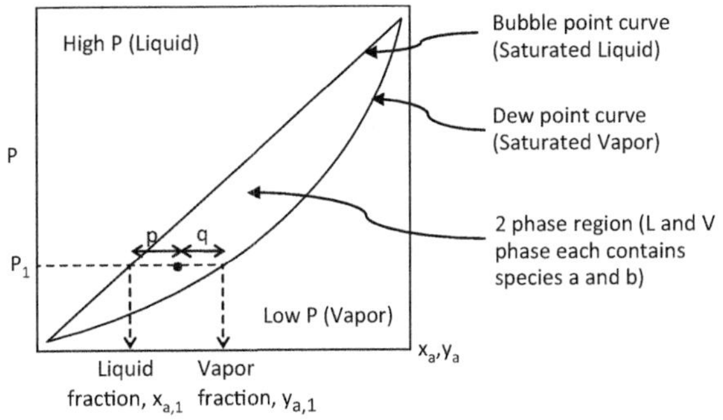


---

Returning to our problem, we have earlier plotted the *Txy* diagram for acetone at 1 bar. To determine the bubble point for equal mole fractions of acetone and water, we can find the value of *T* when $x_a = 0.5$ on the saturated liquid line from the graph $T = 60°\text{C}$.

Similarly, dew point is the value of $T$ when $x_a = 0.5$ on the saturated vapor line. Note that the saturated liquid line lies below the saturated vapor line on a $Txy$ diagram, as it makes sense that when we reduce temperature, we go from vapor to liquid $T = 82°C$.

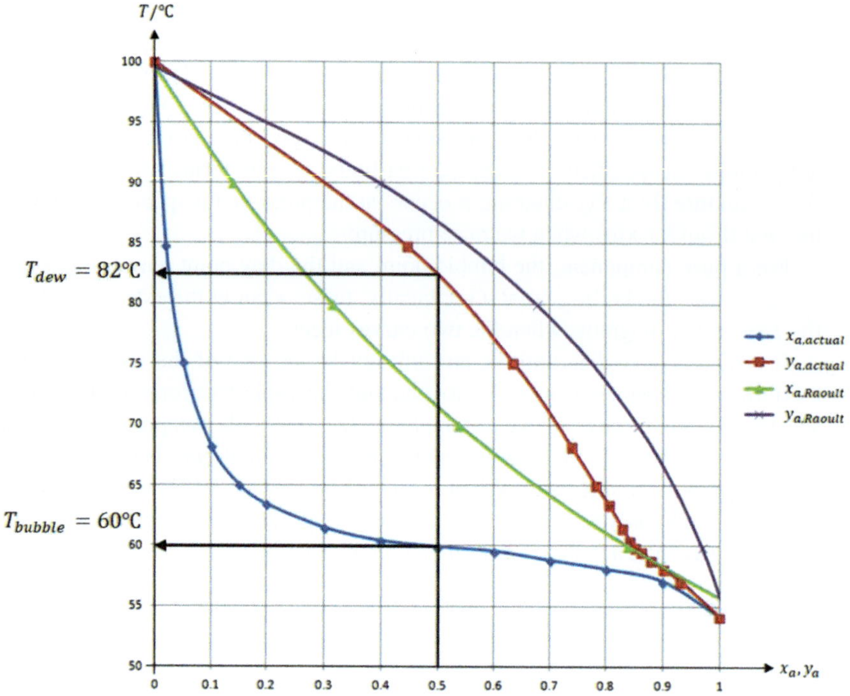


(c) We can analyze this process via the trajectory as shown in the graph below.

**Step 1:** We start from $x_a = 0.4$ and initial temperature of mixture at 52°C. When we heat the mixture, we travel vertically upwards, until we hit the bubble point/saturated liquid curve (blue line). At this point, first vapor bubble is formed, $T \approx 61°C$ from graph.

**Step 2:** At this bubble point temperature, we move horizontally to the right (keeping $T$ constant at 61°C) until we hit the dew point/saturated vapor curve (red line).

**Step 3:** To obtain the corresponding vapor fraction of acetone $y_a$ at the bubble point $T = 61°C$, we read off the value on the horizontal axis corresponding to this $T$ value on the graph. $y_a \approx 0.84$ from the graph, and the vapor fraction of water can be found $y_w = 1 - 0.84 = 0.16$.

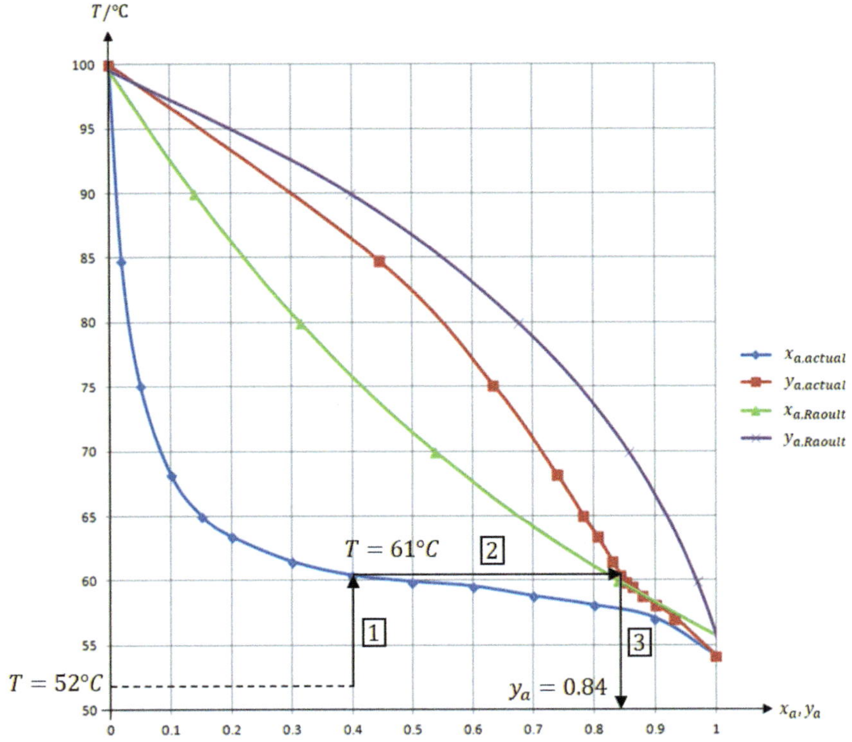


## Problem 2

**We have a mixture of four components 1, 2, 3, and 4, whereby ideal mixture can be assumed for the vapor and liquid phases. The vapor pressures of the components follow the Antoine equation $\ln(P[\text{bar}]) = A - \frac{B}{T[K]}$, whereby the Antoine constants $A$ and $B$ are provided below together with the mixture composition. The mixture is passed through a flash unit at steady state. The outlet mixture is at vapor–liquid equilibrium at a pressure of 5 bar.**

| Component | Feed composition, $z_i$ | $A$ | $B$ |
|---|---|---|---|
| 1 | 0.2 | 9.5 | 2200 |
| 2 | 0.1 | 10.2 | 3000 |
| 3 | 0.2 | 10.1 | 3400 |
| 4 | 0.5 | 10.5 | 3900 |

**Describe the method to determine the temperature range whereby the exit mixture is a liquid–vapor mixture. [Actual calculations are not required.]**

## Solution 2

### *Worked Solution*

Let us recall some basic concepts about the flash distillation unit.

---

### *Background Concepts*

Flash distillation is a common separation process. A feed mixture is partially vaporized in a flash unit (or "drum"), and exiting the flash drum are two streams, a vapor stream enriched in more volatile components from the feed, and a liquid stream enriched in the less volatile components.

The feed mixture is pressurized and heated, then passed through a nozzle into the flash unit. Due to a sudden large drop in pressure, part of the feed mixture vaporizes and the resulting vapor is then taken off overhead forming the outlet vapor stream, while the liquid portion drains to the bottom forming the outlet liquid stream.

This process is called "flash" due to the rapid process of vaporization when the feed enters the flash unit. Within the flash unit, vapor–liquid equilibrium may be assumed.

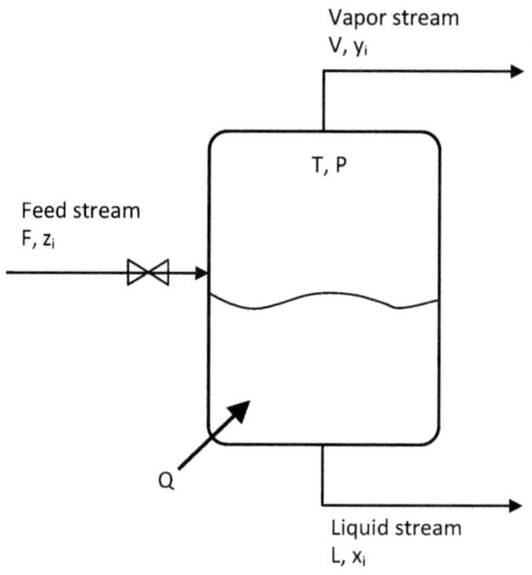


---

Returning to the problem, the temperature range for the outlet to contain both liquid and vapor phases is bounded by the temperature limits of a vapor–liquid equilibrium (or two phase region) to exist at that specific pressure. It follows that the lower limit and upper limit of this temperature range will be the bubble point and dew point, respectively, since these are the points when the vapor–liquid equilibrium (or two-phase region) first begins/ends.

Let us perform mass balances around the flash unit as shown below.

$$F = V + L \tag{1}$$

$$Fz_i = Vy_i + Lx_i \tag{2}$$

Let the vapor fraction be $\psi = V/F$; therefore, combining Eqs. (1) and (2)

$$z_i = \psi y_i + (1 - \psi)x_i$$

$$x_i = \frac{z_i}{\psi \frac{y_i}{x_i} + (1 - \psi)}$$

The equilibrium constant is defined as $K_i = y_i/x_i$; therefore

$$x_i = \frac{z_i}{\psi K_i + 1 - \psi}$$

Two other equations we know are

$$\sum_i x_i = 1 \tag{3}$$

$$\sum_i y_i = 1 \tag{4}$$

Therefore, combining Eqs. (3) and (4) gives us the following useful equation for analyzing feed data.

$$\sum_i \frac{z_i}{\psi K_i + 1 - \psi} = 1 = \sum_i y_i$$

$$\sum_i \frac{z_i}{\psi K_i + 1 - \psi} = \sum_i K_i x_i = \sum_i \frac{K_i z_i}{\psi K_i + 1 - \psi}$$

$$\sum_i \frac{K_i z_i}{\psi K_i + 1 - \psi} - \sum_i \frac{z_i}{\psi K_i + 1 - \psi} = 0$$

This equation is also called the Rachford–Rice equation and is useful for solving problems related to flash separation operations. The above equation can be rewritten as follows:

$$f(\psi) = \sum_i \frac{z_i(K_i - 1)}{\psi K_i + 1 - \psi} = 0$$

The roots of the Rachford–Rice equation can tell us the compositions of the resulting flash distillation fractions, as well as the respective liquid and/or vapor fractions exiting the flash unit. The variables $z_i$, $K_i$, $\psi$ can all be determined easily. In this problem, $z_i$ values are provided in the feed data. $\psi$ is defined by our need to find the temperature limits at bubble point and dew point, whereby $\psi = 0$ and 1, respectively. In order to determine $K_i$, we can apply Raoult's law assuming ideal mixture for both the liquid and vapor phases. Vapor and liquid phase fugacities for a particular species are equivalent under VLE,

$$f_i^L = f_i^V$$

$$y_i P = x_i P_i^V$$

$$\frac{y_i}{x_i} = \frac{P}{P_i^V} = K_i$$

We know the value of $P = 5$ bar from the problem statement. We can find vapor pressure $P_i^V$ at a specified temperature using the Antoine equation: $\ln(P[\text{bar}]) = A - \frac{B}{T[K]}$. This means we can find a series of $K_i$ values for a series of specified $T$ values.

Therefore, in order to find the limits of the temperature range, we can follow these steps:

1. Specify a trial value for $T$ (we will be iterating this value until our conditions are satisfied).
2. Find $P_i^V$ at this $T$ value using the Antoine equation. Then find $K_i$.
3. To determine the lower temperature limit, specify $\psi = 0$, and substitute $z_i$ and $K_i$ value (calculated in step 2) into the expression $\frac{z_i(K_i-1)}{\psi K_i+1-\psi}$ and sum over all components.
4. The summation result should be zero to fulfill mass balance. Iterate the value of $T$ (from Step 1) until this condition is met.

Using this method, we will find that $T = 340$ K for bubble point and $T = 416$ K for dew point. Thus, the temperature range for the vapor and liquid phases to exist in the outlet is [340 K, 416 K].

## Problem 3

**A binary mixture containing components 1 and 2 is fed into a continuous distillation column with a total condenser and partial reboiler. The saturated liquid feed contains component 1 at a mole fraction of 0.55, and enters the column at a feed rate of 120 mol/h. The distillate and bottoms streams contain 96 mol% and 4 mol% of component 1, respectively. The reflux ratio is 2.3, and equilibrium may be assumed at the distillation stages. VLE data is provided for component 1 as shown in the graph below. You may assume efficiency of 1.0.**

(a) **Find the flow rates for the distillate and bottoms streams.**
(b) **Find the rate at which vapor needs to be produced in the reboiler.**
(c) **The feed is to be separated into two equivalent substreams, whereby one of them will enter the column at the third stage from the bottom (where partial reboiler is defined as stage 1) and the other substream will enter the column at the sixth stage. Find the total number of stages needed to achieve the required separation.**
(d) **If all the feed entered the column at one stage, which stage should it enter to ensure the shortest length of column required.**

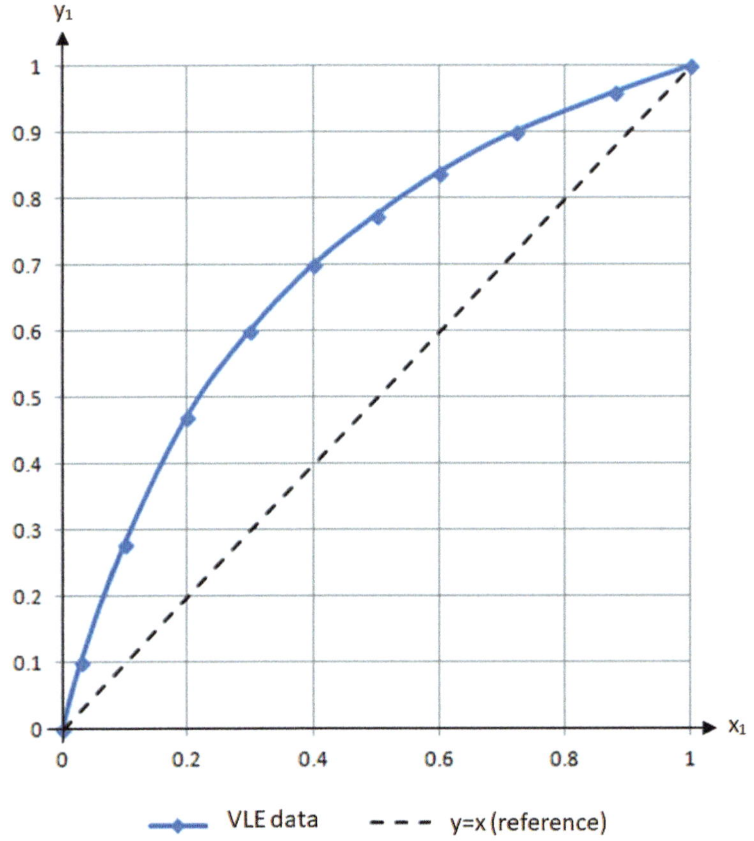

## Solution 3

### *Worked Solution*

(a) In this problem, we are introduced to the reflux ratio, $R$. When vapor stream exits the top of the distillation column, a fraction of it is fed back into the column, and this is the reflux stream $L_T$. The ratio between the flow rate of this reflux stream to the distillate stream (collected in the receiver) is defined as the reflux ratio, $R = \frac{L_T}{D} = \frac{V_T - D}{D}$.

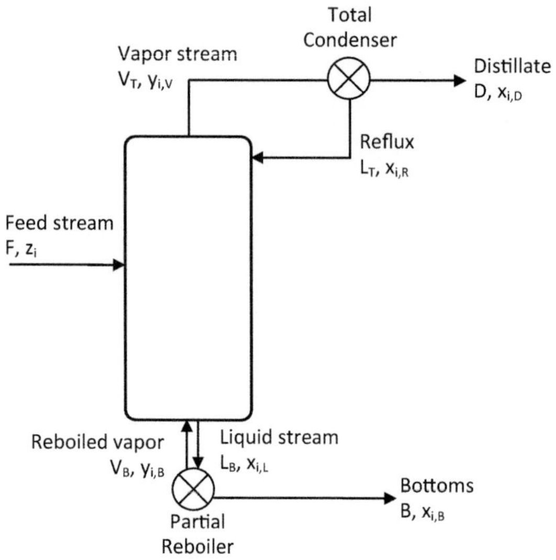


To find the flow rates for distillate and bottoms streams, we can first construct a mass balance for total flow rate and for component 1,

$$F = D + B$$

$$Fz_1 = Dx_{1,D} + Bx_{1,B} = Dx_{1,D} + (F - D)x_{1,B}$$

Substituting the values provided in the problem,

$$(120)(0.55) = D(0.96) + (120 - D)(0.04)$$

$$D = 66.5 \text{ mol/h}; \; B = 120 - 66.5 = 53.5 \text{ mol/h}$$

(b) The rate at which vapor needs to be produced in the reboiler is $V_B$ as shown in the diagram, and this rate has to balance the rate at which vapor leaves the top of the column (i.e., $V_T$), since the feed is a saturated liquid.

$$L_T = \mathrm{RD} = 2.3(66.5) = 153\ \mathrm{mol/h}$$

$$V_B = V_T = L_T + D = 153 + 66.5 = 219.5\ \mathrm{mol/h}$$

(c) Since the feed is separated into two equivalent substreams, it means each will be at a flow rate of

$$F_1 = F_2 = \frac{120}{2} = 60\ \mathrm{mol/h}$$

The separation of the feed into two at different stages separates the column into 3 parts, where each part follows each respective operating line (bottom operating line (BOL), intermediate operating line (IOL), and top operating line (TOL)).

For the BOL (<stage 3), we have the following mass balance:

$$V_B + B = L_B$$

$$V_B y_{1,B} = L_B x_{1,L} - B x_{1,B}$$

$$y_{1,B} = \frac{L_B}{V_B} x_{1,L} - \frac{B}{V_B} x_{1,B} = \frac{53.5 + 219.5}{219.5} x_{1,L} - \frac{53.5}{219.5}(0.04) = 1.2 x_{1,L} - 0.0097$$

$$\text{Equation of BOL}: y_1 = 1.2 x_1 - 0.0097$$

For the IOL (stage 3–6), we have the following mass balance. Note that $V_I = V_T$ for saturated liquid feed:

$$F_1 + V_I = L_I + D$$

$$60 + 219.5 = L_I + 66.5 \rightarrow L_I = 213\ \mathrm{mol/h}$$

$$F_1 z_1 + V_I y_{1,I} = L_I x_{1,I} + D x_{1,D}$$

$$(60)(0.55) + (219.5) y_{1,I} = (213) x_{1,I} + (66.5)(0.96)$$

$$y_{1,I} = 0.97 x_{1,I} + 0.14$$

$$\text{Equation of IOL}: y_1 = 0.97 x_1 + 0.14$$

For the TOL (>stage 6), we have the following mass balance:

$$V_T = L_T + D$$

$$V_T y_{1,V} = L_T x_{1,R} + D x_{1,D}$$

$$y_{1,V} = \frac{L_T}{V_T} x_{1,R} + \frac{D}{V_T} x_{1,D} = \frac{153}{219.5} x_{1,R} + \frac{66.5}{219.5}(0.96) = 0.7x_{1,R} + 0.29$$

$$\text{Equation of TOL} : y_1 = 0.7x_1 + 0.29$$

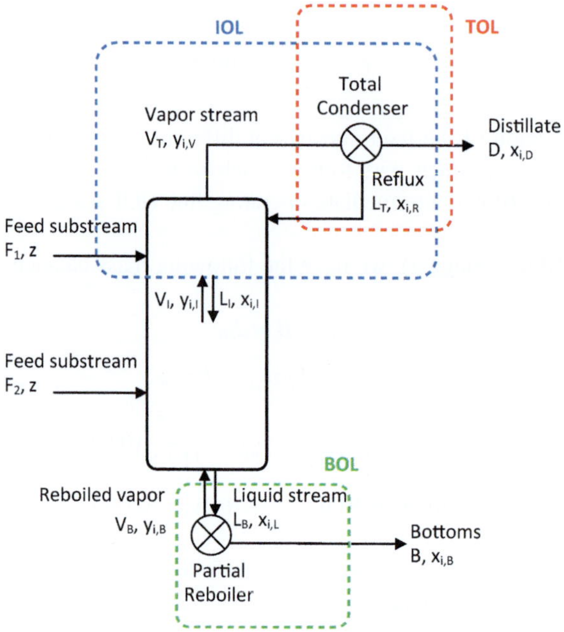


Now that we have come up with the three equations for the three operating lines, i.e., BOL, IOL, and TOL, we can plot them on our VLE data chart, and using the McCabe–Thiele method of "stepping up the stages". "Steps" begin from the bottoms composition, then rise up to the VLE curve and travels horizontally to the relevant operating line. It then continues stepping up to the VLE curve and horizontally to the operating line, in a general rightwards direction to reach the desired distillate composition. We can find that the required number of stages for the desired separation is 7.

Note that we step off the BOL, until we reach stage 3 when the IOL takes effect until stage 6 (feed substream entrance affects operating line equation). Then TOL is used for stage 7. The start and end points are defined by our feed composition and desired separation purity. Note that the operating lines intersect at the liquid mole fraction of $x_1 = 0.55$ which is consistent with the saturated liquid feed composition.

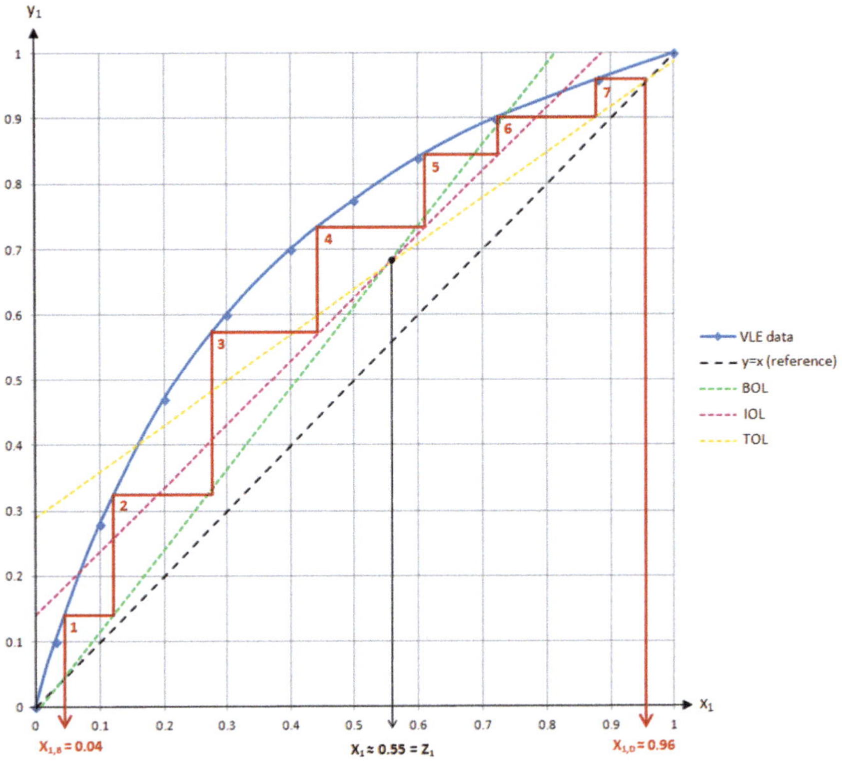


(d) Now we have the scenario whereby all the feed entered the column at one stage. Therefore, we will only require the TOL and BOL. Recall from part c:

$$\text{Equation of TOL}: y_1 = 0.7x_1 + 0.29$$

$$\text{Equation of BOL}: y_1 = 1.2x_1 - 0.0097$$

We can work backwards to find the intersection point between the two operating lines, and observe at which "step" (i.e., stage #) of the McCabe Thiele step-up sketch this intersection occurs. This intersection reflects the entry point of the feed for the minimum number of stages. The stage # is 4 from the bottom.

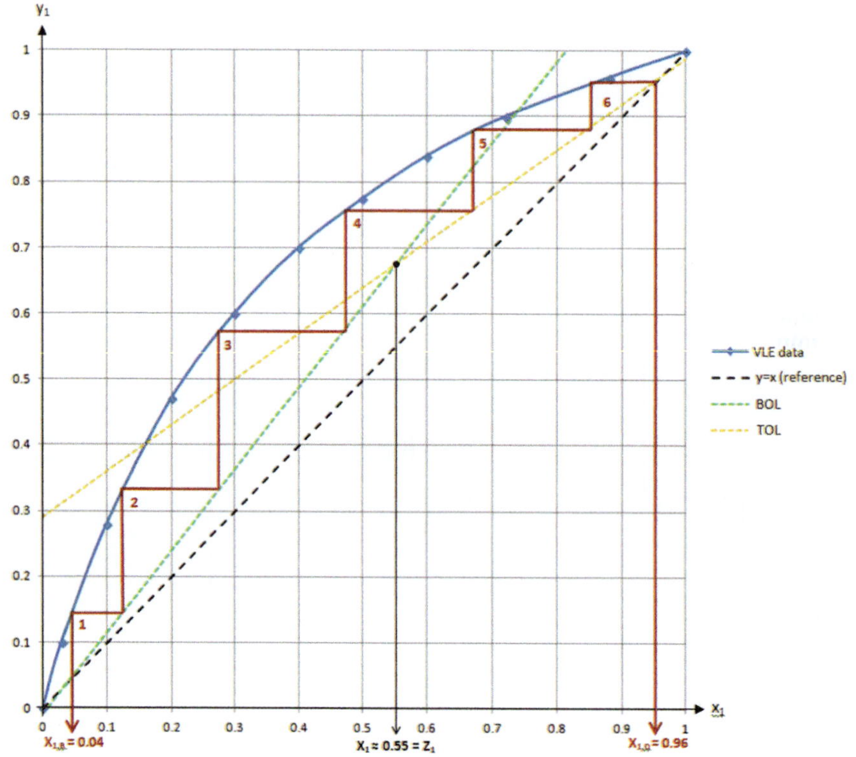


## Problem 4

**A distillation column is fed two feed streams containing components *P* and *Q*. The relative volatility *α* of *P* to *Q* is 2.3, and it is assumed to be independent of temperature. Properties of the two feed streams 1 and 2 are given below:**

| Feed stream | Phase | Mole fraction of *P* | Flow rate, *F* [mol/h] |
|---|---|---|---|
| 1 | Saturated liquid | 0.6 | 120 |
| 2 | Saturated liquid | 0.4 | 120 |

**If the required distillate and bottoms compositions are 99 mol% and 1 mol% of component *P*, respectively, consider the following two scenarios for the feed stream configuration:**

***Scenario 1:***

**The two feed streams are mixed together before entering the column as a saturated liquid at the optimal location. The column has a partial reboiler and a total condenser.**

(a) **Given the following equation for relative volatility $\alpha$ where $x_p$ and $y_p$ are the liquid and vapor phase mole fractions of *P*, respectively, determine the minimum reflux ratio and the minimum number of stages.**

$$y_p = \frac{\alpha x_p}{1 + (\alpha - 1)x_p}$$

(b) **If the column operates at 1.2 times the minimum reflux ratio, how many theoretical stages are needed?**

(c) **Determine the optimal feed location.**

***Scenario 2:***

**The two feed streams enter the column separately, at their respective optimal locations.**

(d) **If the column operates at the same reflux ratio as in part b, determine the number of theoretical stages needed.**

(e) **Determine the optimal feed location.**

**Solution 4**

*Worked Solution*

(a) We can construct a simple diagram to better understand the problem. For scenario 1, we have two feed streams 1 and 2 which are premixed before entering the column.

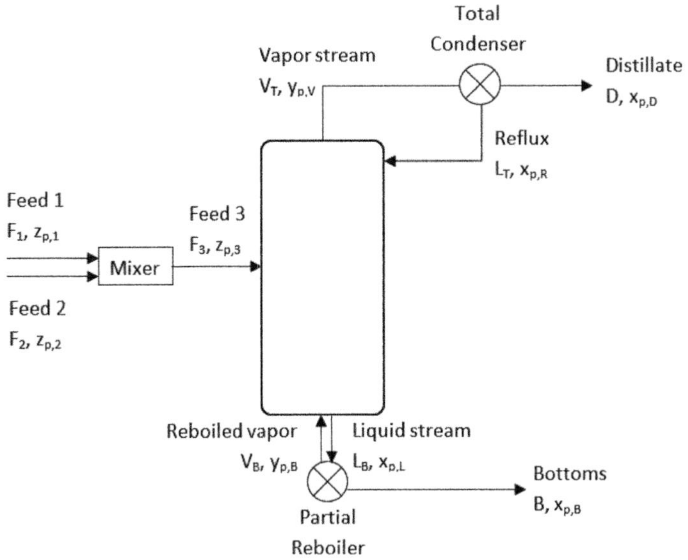

We can give information about feed streams 1 and 2; therefore, we can perform a simple mass balance around the mixer to use the given information to deduce the properties of feed stream 3. $F_n$ refers to the flow rate of feed stream $n$ and $z_{p,n}$ refers to the mole fraction of component $P$ in feed stream $n$.

$$F_1 + F_2 = F_3$$

$$z_{p,1}F_1 + z_{p,2}F_2 = z_{p,3}F_3$$

We know that $F_1 = F_2 = 120$ mol/h; therefore

$$F_3 = 120 + 120 = 240 \text{ mol/h}$$

$$z_{p,1}F_1 + z_{p,2}F_2 = 0.6(120) + 0.4(120) = z_{p,3}(240)$$

$$z_{p,3} = 0.5$$

In order to find the minimum number of stages and minimum reflux ratio, we can construct a McCabe–Thiele diagram. We are told that the relative volatility $\alpha$ of $P$ to $Q$ is 2.3 and this information allows us to draw the phase equilibrium line in our diagram using the expression for relative volatility below.

$$y_p = \frac{\alpha x_p}{1 + (\alpha - 1)x_p}$$

In order to draw the phase equilibrium line, we can tabulate a series of data points using an arbitrary set of values between 0 and 1 for $x_p$ as shown below.

| $x_p$ | $y_p$ |
|---|---|
| 0 | $y_p = \frac{2.3(0)}{1+(2.3-1)(0)} = 0$ |
| 0.1 | 0.204 |
| 0.2 | 0.365 |
| 0.3 | 0.496 |
| 0.4 | 0.605 |
| 0.5 | 0.697 |
| 0.6 | 0.775 |
| 0.7 | 0.843 |
| 0.8 | 0.902 |
| 0.9 | 0.954 |
| 1 | 1 |

The equilibrium line is plotted in orange and annotated as Vapor–Liquid Equilibrium (VLE) data.

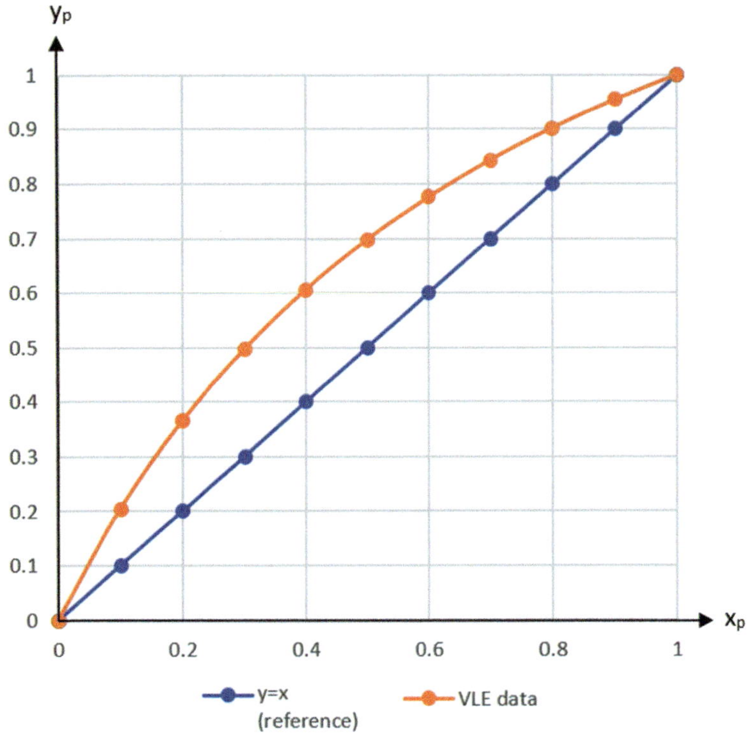


We can now add our feed line to the diagram. Since we were told that the feed stream after mixing, i.e., $F_3$, is still a saturated liquid, this feed line will be a vertical line. We also found earlier that the composition of feed stream $F_3$ is 0.5. Therefore, we can draw the feed line as shown in purple below.

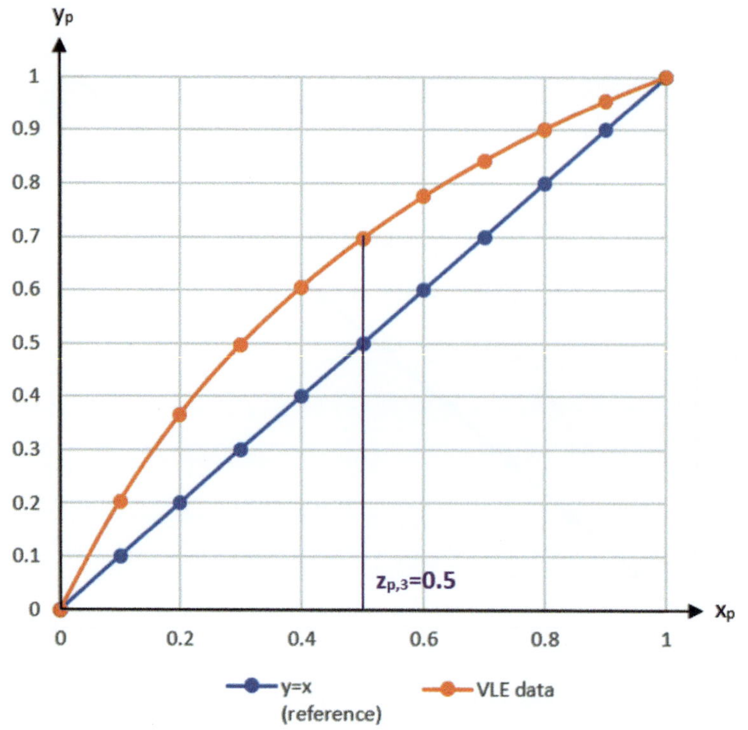


Next, we need to draw operating lines for the distillation at minimum reflux ratio. The slope of the Upper Operating Line (UOL which refers to section of column above feed) can be expressed in terms of reflux ratio $R$ as shown below. As R decreases, $1/R$ increases, and the slope of the UOL decreases. Therefore at minimum $R$, we draw the UOL with the smallest gradient possible.

$$\text{Slope of UOL} = \frac{R}{R+1} = \frac{1}{1+1/R} \tag{1}$$

Given that the distillate composition $x_{p,D}$ has to be 0.99 (shown by yellow line), the point (0.99,0.99) becomes specified as one end point for our UOL. Starting from this point, we can only have a UOL with minimum slope when the UOL intersects the feed line at the phase equilibrium line (note that operating lines cannot cross the phase equilibrium line since that is thermodynamically impossible). This point of intersection can also be called the pinch point and it occurs at $x_p = 0.5$ where the saturated liquid feed (therefore $z_p = x_p$) enters the column. The UOL is added to the diagram as described in red below.

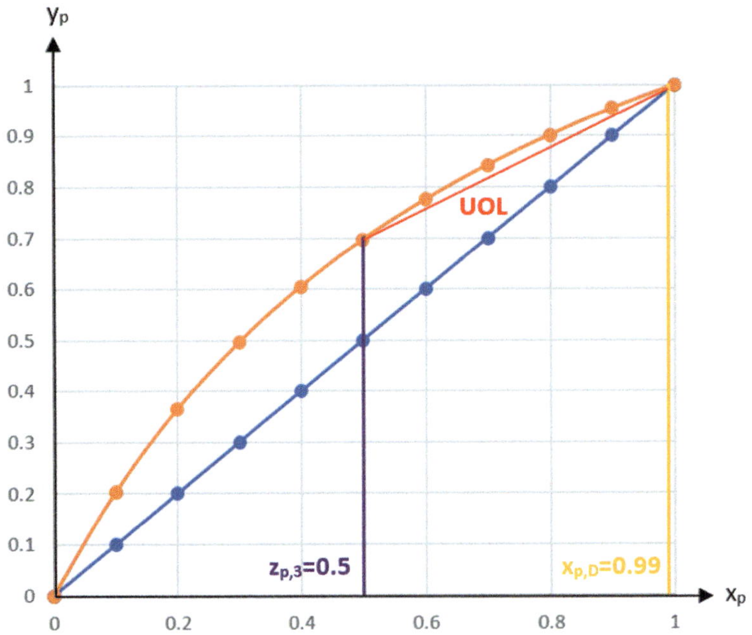


Now we can draw the Lower Operating Line (LOL) which refers to the section of the column below the feed location. The LOL is defined by the bottoms composition; therefore, it has to start from the point (0.01, 0.01), and this line has to intersect the UOL at the feed location.

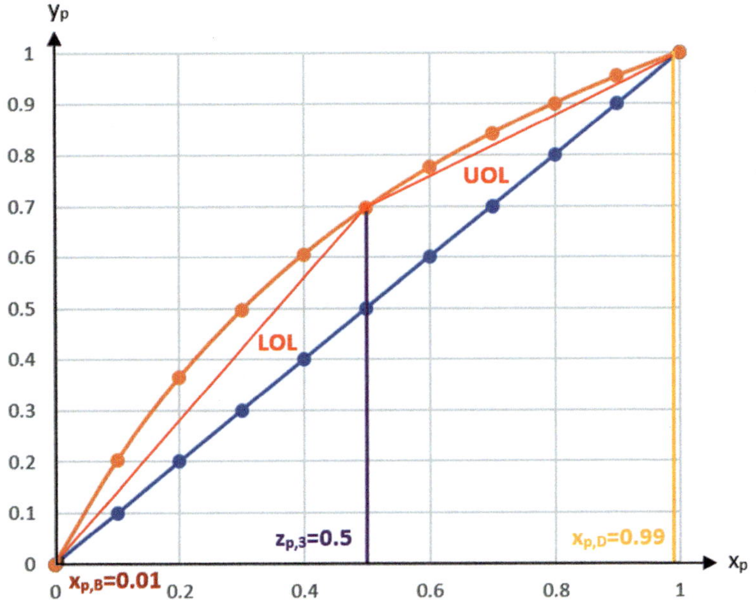

Going back to Eq. (1), we can find the minimum reflux ratio $R_m$ by calculating the slope of the UOL from our plot.

$$\text{Slope of UOL} = \frac{R_m}{R_m + 1} = 0.60$$

$$R_m = 1.5$$

The minimum reflux ratio is found to be 1.5. This occurs when we have an infinite number of stages. Conversely if we want the minimum number of stages, we set the condition whereby the reflux ratio is maximum, i.e., $R \to \infty$. The slope of the UOL is therefore equal to 1 as shown below, which is the same as the line $y = x$.

$$\text{Slope of UOL} = \lim_{R\to\infty} \frac{R}{R+1} = \lim_{R\to\infty} \frac{1}{1 + 1/R} = 1$$

To find the minimum number of stages, we note that our operating line ($y = x$) is still defined by the bottoms composition specified by the point (0.01, 0.01), feed composition specified by $x_p = 0.5$ and distillate composition specified by the point (0.99, 0.99). Using a McCabe Thiele construction, which consists of a series of steps as shown in red, between the operating line and the phase equilibrium line (annotated as Vapor–Liquid Equilibrium (VLE) data), we find that the number of steps is 11. This means that the minimum number of stages is 11, or 10 stages with the reboiler.

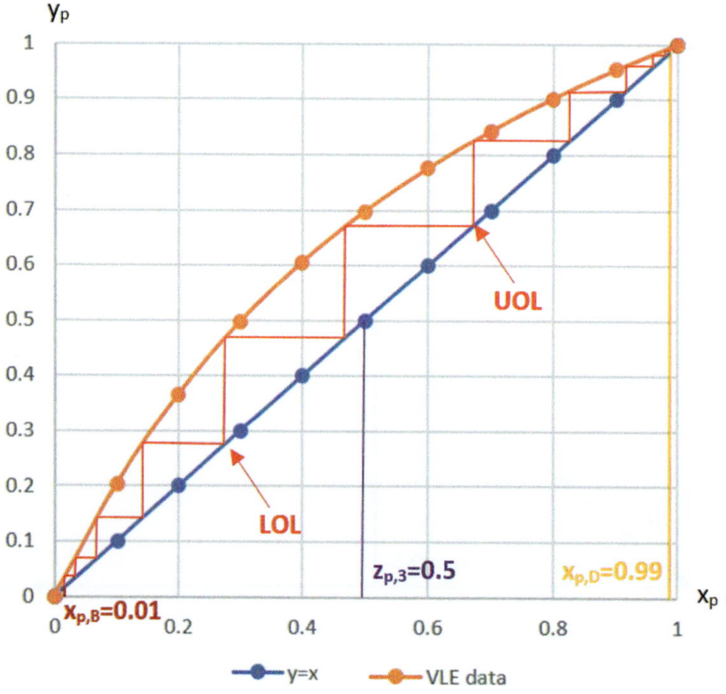

(b) Now we wish to find the number of stages needed for an operating reflux ratio of 1.2 times the minimum.

$$R = 1.2(R_m) = 1.2(1.5) = 1.8$$

The new slope for the upper operating line (UOL) is therefore found as follows. We know the distillate composition which fixes one end of the UOL. Using a slope of 0.643, we can therefore draw the UOL until it intersects the feed line at $x_p = 0.5$. This intersection point will fix one end of the lower operating line (LOL), while the other end of the LOL is fixed at the bottoms composition.

$$\frac{R}{R+1} = \frac{1.8}{1.8+1} = 0.643$$

With the two operating lines drawn, we can now find the number of stages using the same technique of "stepping up" using the McCabe Thiele construction as shown below.

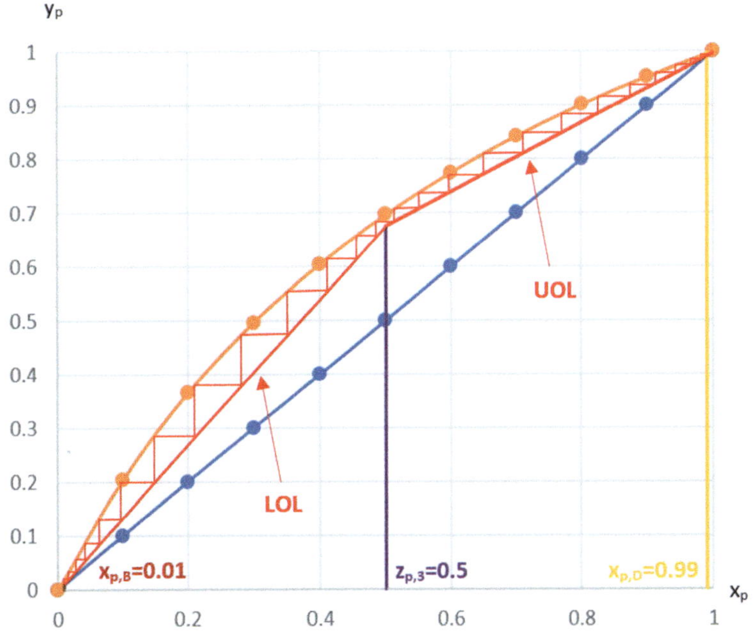


The total number of stages needed for the required separation is 25 stages or 24 stages with a reboiler.

(c) From the McCabe Thiele construction, we can read off the graph where the "step" crosses the feed line to find that the optimal feed location to be the 12th stage above the reboiler.

(d) We now have two separate feed locations at the same reflux ratio of 1.8. To better visualize this setup, we illustrate the diagram below.

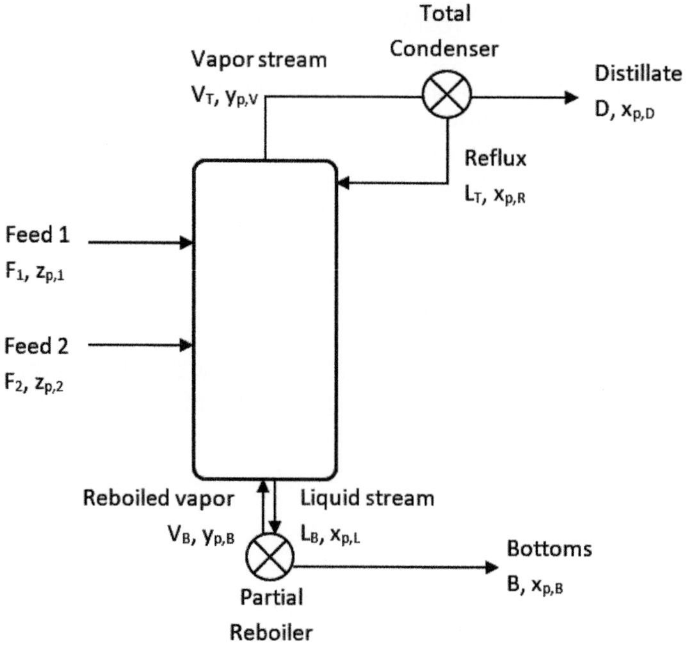


The McCabe Thiele construction will now be adjusted to include two feed lines, which are both vertical lines as we are told that the two feed streams are saturated liquids (and therefore $z_p = x_p$).

This time, we are constrained by the same bottoms and distillate compositions specified by the points (0.01,0.01) and (0.99,0.99), respectively, as well as the two feed compositions specified by $x_{p,1} = 0.6$ and $x_{p,2} = 0.4$ as given in the problem.

With these constraints, we note that we will now have three sections to the operating line as shown below:

- Upper Operating line (UOL) for the section between the upper feed stream and the top of the column
- Intermediate Operating Line (IOL) for the section between the upper and lower feed streams
- Lower Operating Line (LOL) for the section between the lower feed stream and the bottom of the column.

Since the reflux ratio is the same as in part b, the slopes of the UOL and LOL are also the same as in part b. The slope of the IOL will be defined automatically as it is drawn by joining the LOL and UOL at the specified feed compositions.

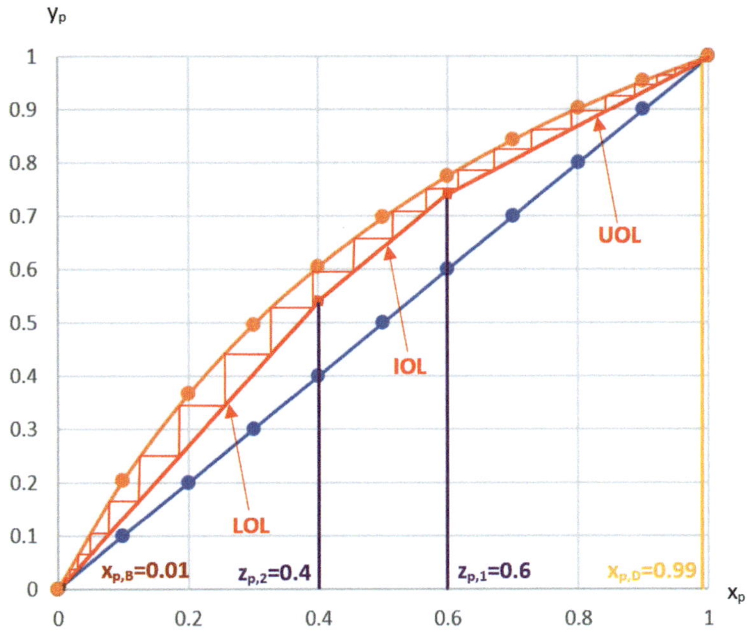


Counting the number of steps, there are 23 theoretical stages in total, or 22 stages with the reboiler.

(e) Feed stream 2 should be added at the 9th stage above the reboiler and feed stream 1 should be added at the 12th stage above the reboiler.

**Problem 5**

**A saturated liquid feed contains a mixture of components 1 and 2. We are required to separate them by a continuous steady-state distillation process comprising of a total condenser and partial reboiler. The mole fraction of 1 in the feed is 0.5 and the relative volatility of 1 to 2, $\alpha = 2.5$. The column has a very large number of equilibrium stages (approaching infinity) and the column will be run essentially at the minimum reflux ratio. The maximum rate at which vapor can be generated in the reboiler is 110 mol/h.**

(a) **Show that the following is true for vapor–liquid equilibrium, where vapor and liquid mole fractions are denoted as $y_1$ and $x_1$ respectively, and relative volatility is a constant value denoted as $\alpha_{12}$.**

$$y_1 = \frac{\alpha_{12} x_1}{1 + (\alpha_{12} - 1) x_1}$$

(b) **If the feed rate to the column is 110 mol/h, and the distillate and bottom product flow rates are equal, find the concentrations of the distillate and bottoms product streams.**

(c) **We need to produce distillate and bottom products of concentration 0.90 and 0.1, respectively. What is the fraction of feed that must be taken as distillate product? Given the limit on reboiler vapor rate remains the same, what is the rate at which the feed needs to be processed.**

(d) **We need to make the product in b, at a feed processing rate of 110 mol/h. To accomplish this, given the limit on reboiler vapor rate, it is proposed to vaporize some feed prior to entering the column. What fraction of feed should be pre-vaporized to achieve our desired result?**

## Solution 5

*Worked Solution*

(a) Relative volatility $\alpha$ measures the difference in volatility (and hence their boiling points) between two components, therefore it is also an indication of how easy or difficult it will be to separate the components from each other. For this problem, we define $\alpha_{12}$ as the relative volatility of component 1 with respect to 2 as follows. We may also rewrite the quantity $y_i/x_i$ in terms of equilibrium constant $K_i$ whereby $K_i = y_i/x_i$.

$$\alpha_{12} = \frac{y_1/x_1}{y_2/x_2} = \frac{K_1}{K_2}$$

Since we have a binary system comprising of components 1 and 2 only, the mole fractions of both components in the liquid and vapor phases have to sum to one; therefore we arrive at the following

$$\alpha_{12} = \frac{y_1/x_1}{(1-y_1)/(1-x_1)}$$

Rearranging the above equation, we can express in terms of $y_1$

$$(1-y_1)\alpha_{12}x_1 = y_1(1-x_1)$$

$$\alpha_{12}x_1 = y_1 - x_1y_1 + \alpha_{12}x_1y_1 = y_1(1 - x_1 + \alpha_{12}x_1)$$

$$y_1 = \frac{\alpha_{12}x_1}{(1 - x_1 + \alpha_{12}x_1)} = \frac{\alpha_{12}x_1}{1 + (\alpha_{12} - 1)x_1}$$

(b) Note that this time we do not have VLE data provided in terms of a data table or chart, but we can find the equation that defines the VLE curve and draw it ourselves, by using the relative volatility $\alpha$. From results of part a, we have the following equilibrium relationship between vapor and liquid mole fractions, denoted as $y$ and $x$, respectively, at constant relative volatility, $\alpha$.

$$y = \frac{\alpha x}{1 + (\alpha - 1)x} = \frac{5x}{2 + 3x}$$

This equation can be used to draw the VLE curve as shown in blue below.

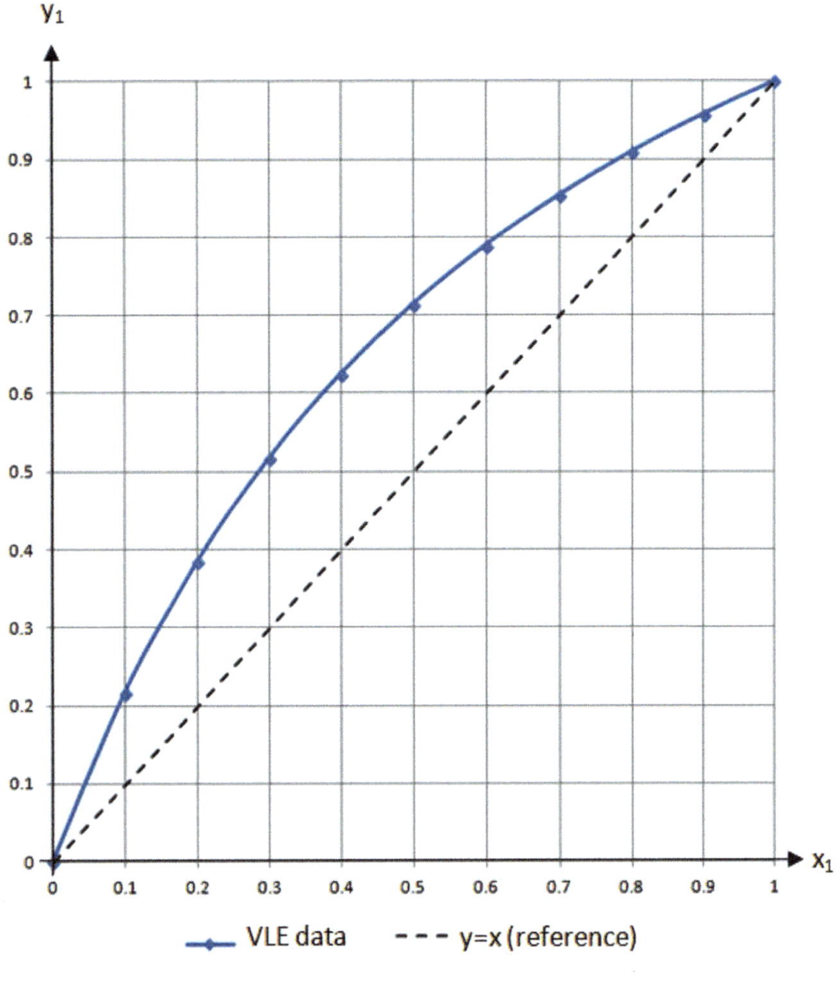


For a saturated liquid feed, a mass balance around the feed plate will give us the following two equations for the rising vapor and falling liquid streams, respectively. Maximum rate at which vapor can be generated in the reboiler is 110 mol/h; therefore $V_B = V_T = 110$ mol/h.

$$V_B = V_T = 110$$

$$F + L_T = L_B$$

We know that $F = 110$ mol/h and $D = B$. Therefore we have the following overall mass balances.

$$F = D + B \rightarrow D = B = \frac{110}{2} = 55 \text{ mol/h}$$

$$Fz_1 = Dx_{1,D} + Bx_{1,B}$$

To find $x_{1,D}$ and $x_{1,B}$, we need to determine the top operating line (TOL) and bottom operating line (BOL). The TOL will apply for all the stages above the feed and the BOL will apply for the stages below the feed.

We note that there is a large number of stages, i.e., a minimum reflux ratio and the top and bottom operating lines will intersect at the VLE curve, which coincides with the feed position. From the VLE curve, we can find that when $z_1 = x_1 = 0.5$ (saturated liquid feed), $y_1 = 0.7143$.

To find the equations of the operating lines, we construct the mass balances as shown in dotted lines below.

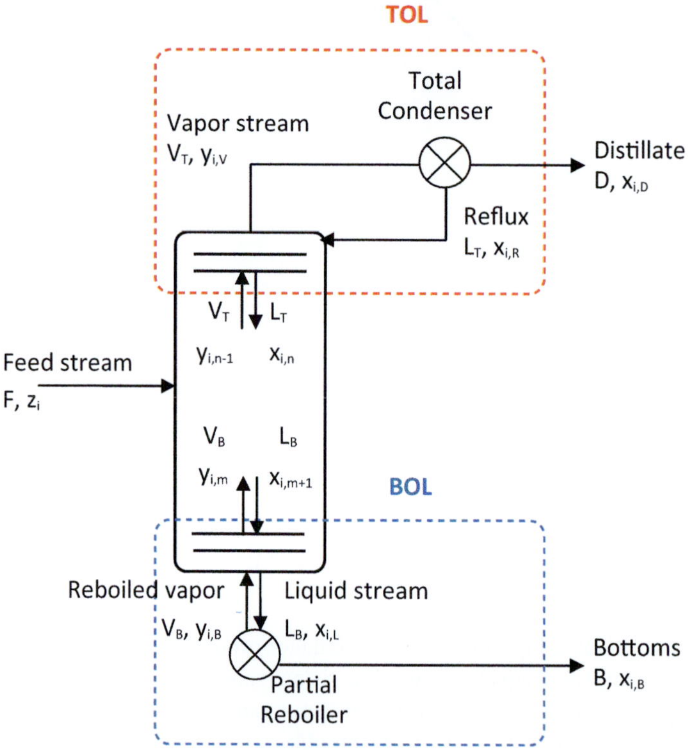


For the TOL, we have

$$V_T = D + L_T$$

$$110 = 55 + L_T \rightarrow L_T = 55 \text{ mol/h}$$

$$V_T y_{1,n-1} = D x_{1,D} + L_T x_{1,n}$$

$$110\left(y_{1,n-1}\right) = 55(x_{1,D}) + 55(x_{1,n})$$

We can find $x_{1,D}$ by substituting $x_{1,n} = 0.5$ and $y_{1,n-1} = 0.7143$ into the above TOL equation

$$x_{1,D} = \frac{110(0.7143) - 55(0.5)}{55} = 0.9286$$

For the BOL, we have

$$V_B + B = L_B$$

$$110 + 55 = L_B \rightarrow L_B = 165 \text{ mol/h}$$

$$V_B y_{1,m} + B x_{1,B} = L_B x_{1,m+1}$$

$$110\left(y_{1,m}\right) + 55(x_{1,B}) = 165(x_{1,m+1})$$

We can find $x_{1,B}$ by substituting $x_{1,m+1} = 0.5$ and $y_{1,m} = 0.7143$ into the above BOL equation

$$x_{1,B} = \frac{165(0.5) - 110(0.7143)}{55} = 0.0714$$

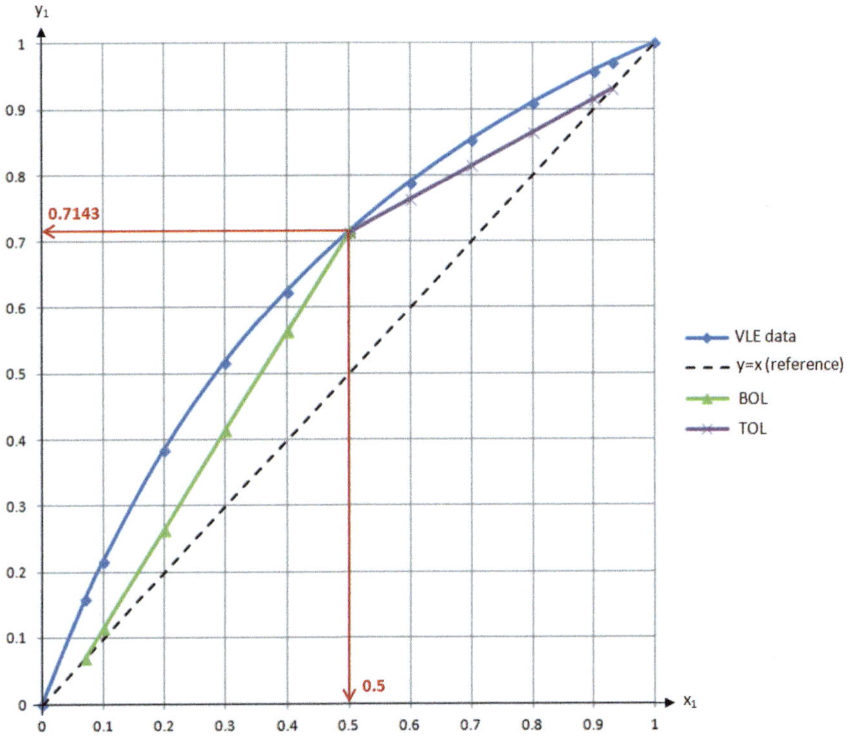

(c) Now we are given specified values for $x_{1,B} = 0.1$ and $x_{1,D} = 0.9$, so we have the following mass balance.

$$F = D + B$$

$$Fx_1 = Dx_{1,D} + Bx_{1,B}$$

$$F(0.5) = D(0.9) + (F - D)(0.1)$$

$$0.5 = 0.9\left(\frac{D}{F}\right) + 0.1\left(1 - \frac{D}{F}\right)$$

$$\frac{D}{F} = 0.5$$

Using the BOL, we substitute the value of $x_{1,B} = 0.1$, the given reboiler vapor rate of $V_B = 110$ mol/h and $x_{1,m+1} = 0.5$ and $y_{1,m} = 0.7143$ into the equation.

$$V_B y_{1,m} + Bx_{1,B} = L_B x_{1,m+1}$$

$$y_{1,m} + 0.1\left(\frac{L_B}{V_B}\right) - 0.1 = \left(\frac{L_B}{V_B}\right) x_{1,m+1}$$

$$\frac{L_B}{V_B} = \frac{0.7143 - 0.1}{0.5 - 0.1} = 1.53575$$

$$L_B = 110(1.53575) = 168.93$$

$$B = 168.93 - 110 = 58.93$$

Fraction of feed taken as distillate is

$$\frac{B}{F} = 1 - \frac{D}{F} = 1 - 0.5 = 0.5$$

$$F = \frac{58.93}{0.5} = 117.86 \text{ mol/h}$$

(d) Now that some of the feed is pre-vaporized, we need to note that $V_B \neq V_T$. Let us look at the mass balance around the feed plate, whereby a fraction of feed, $\lambda$ is now vaporized, and $(1 - \lambda)$ remains liquid.

$$V_B + \lambda F = V_T$$

$$(1 - \lambda)F + L_T = L_B$$

We have $F = 110$ and the same reboiler rate $V_B = 110$. From part b, we know that $\frac{D}{F} = 0.5$ still holds if the desired distillate and bottoms compositions are the same. So

$$D = \left(\frac{D}{F}\right)F = 0.5(110) = 55$$

$$B = F - D = 110 - 55 = 55$$

$$L_B = V_B + B = 110 + 55 = 165$$

$$\frac{L_B}{V_B} = \frac{165}{110} = 1.5$$

The pre-vaporization of feed means the BOL equation will change since $\left(\frac{L_B}{V_B}\right)$ is now changed to 1.5 instead. The equation of the BOL is:

$$V_B y_{1,m} + Bx_{1,B} = L_B x_{1,m+1}$$

$$y_{1,m} + \left(\frac{L_B - V_B}{V_B}\right)0.1 = \left(\frac{L_B}{V_B}\right)x_{1,m+1}$$

$$y_{1,m} = 1.5x_{1,m+1} - 0.05$$

We can now construct this new BOL line, and find the new intersection with the VLE curve from the graph, which is

$$x_{1,m+1} = 0.52$$

$$y_{1,m} = 0.73$$

The slope of the feed line (or $q$-line) is by definition whereby $q$ denotes the mole fraction of liquid in the feed. The feed line is drawn from the reference line of $y = x$ to the point where the BOL intersects with the VLE curve (which is the feed position for infinite number of stages)

$$\text{Slope} = \frac{q}{q-1} = \frac{1-\lambda}{-\lambda} = \frac{0.6 - 0.73}{0.6 - 0.52} = -1.625$$

$$\lambda = 0.38$$

The mole fraction of feed to pre-vaporize is 0.38, and this equates to an amount of feed to pre-vaporize of $\lambda F = 0.38(110) = 42$ mol/h.

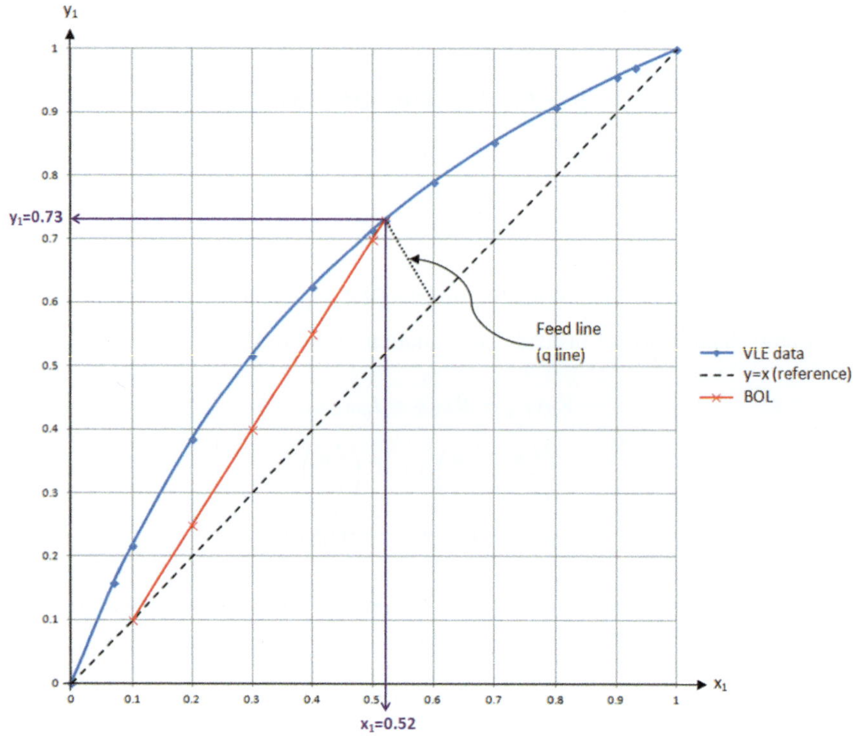


## Problem 6

**Consider a continuous distillation of two saturated liquid feed streams *A* and *B*, each containing components 1 and 2. Each of the two streams has a flow rate of 120 mol/h, and the molar compositions are shown below. We are required to obtain a distillate of 98.5 mol% of component 1, and a bottoms product of 1.5 mol% component 1. The relative volatility of 1 to 2 is 1.8.**

| | $x_1$ | $F$ [mol/h] |
|---|---|---|
| Feed *A* | 0.25 | 120 |
| Feed *B* | 0.65 | 120 |

**If the two feed streams are mixed together before entering the column at the optimal location,**

(a) **Find the minimum reflux ratio and the minimum number of stages required.**
(b) **If the operating reflux ratio is 1.2 times the minimum reflux ratio, determine the number of stages required. Also, identify the location that the feed should enter the column.**

**If the two streams entered the column separately, each at its optimal location,**

(c) **Find the number of stages required and locations for the feeds to enter the column if the reflux ratio is the same as that in part b.**

## Solution 6

***Worked Solution***

(a) Given that the relative volatility of component 1 to 2 is 1.8, it means that component 1 is more volatile. So it would be a good choice to construct the McCabe–Thiele diagram (VLE graph of $y_1$ against $x_1$) for component 1.

For the case when the streams are mixed before entering the column as shown in the diagram below, we can construct mass balance equations.

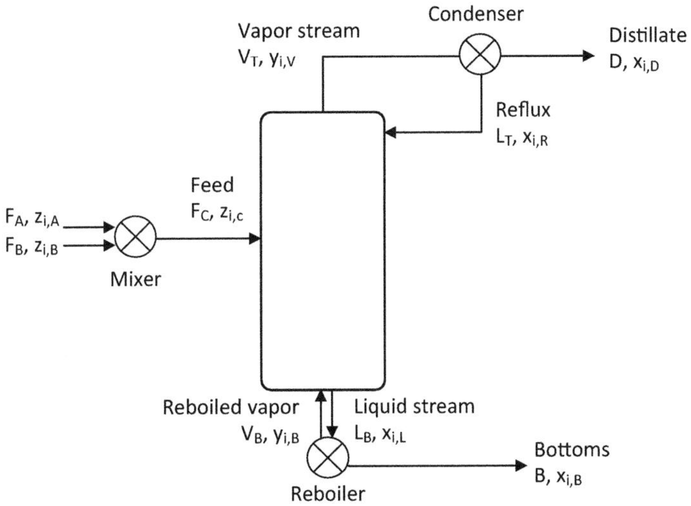


$$F_A + F_B = F_C$$

$$F_A z_{1,A} + F_B z_{1,B} = F_C z_{1,C}$$

Substituting known values, we have

$$F_C = 120 + 120 = 240 \text{ mol/h}$$

$$120(0.25) + 120(0.65) = 240(z_{1,C})$$

$$z_{1,C} = 0.45$$

Using the relative volatility $\alpha = 1.8$, we can establish the following equilibrium relationship between vapor and liquid mole fractions for component 1.

$$y_1 = \frac{\alpha x_1}{1 + (\alpha - 1)x_1} = \frac{1.8x_1}{1 + 0.8x_1}$$

To find the minimum reflux ratio, we need to draw the operating lines on the McCabe–Thiele diagram. This can be deduced from the following points:

- The operating lines will intersect at the feed position and the equilibrium line.
- Since the feed is saturated liquid, the feed line (also known as $q$-line) will be vertical at the composition $z_{1,C} = x_1 = 0.45$.
- The operating lines before and after the feed position are constrained by the bottoms composition and distillate composition requirements of $x_{1,B} = 0.015$ ($y_1 = 0.015$) and $x_{1,D} = 0.985$ ($y_1 = 0.985$).

We can draw the McCabe–Thiele diagram as shown below from the points above.

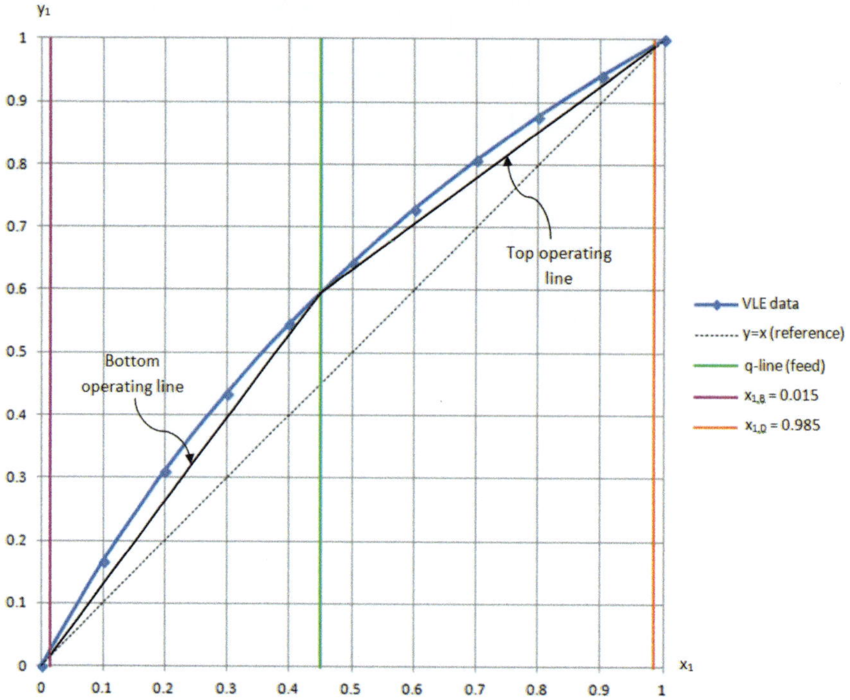


To find out the minimum reflux ratio, we note the important feature about this graph that the slope of the top operating line (for rectifying section) is as follows, where $R$ is the reflux ratio

$$\text{Slope of TOL} = \frac{R}{R+1}$$

We can find either by observing from the graph or by substituting $x_1 = 0.45$ into the equation $y_1 = \frac{1.8x_1}{1+0.8x_1}$ that the intersection point is at $y_1 = 0.6$. Therefore the slope of the TOL is as follows. A zoom-in of the graph is shown on the next page.

$$\frac{0.6 - 0.985}{0.45 - 0.985} = \frac{R}{R+1}$$

$$R = R_{\min} = 2.66$$

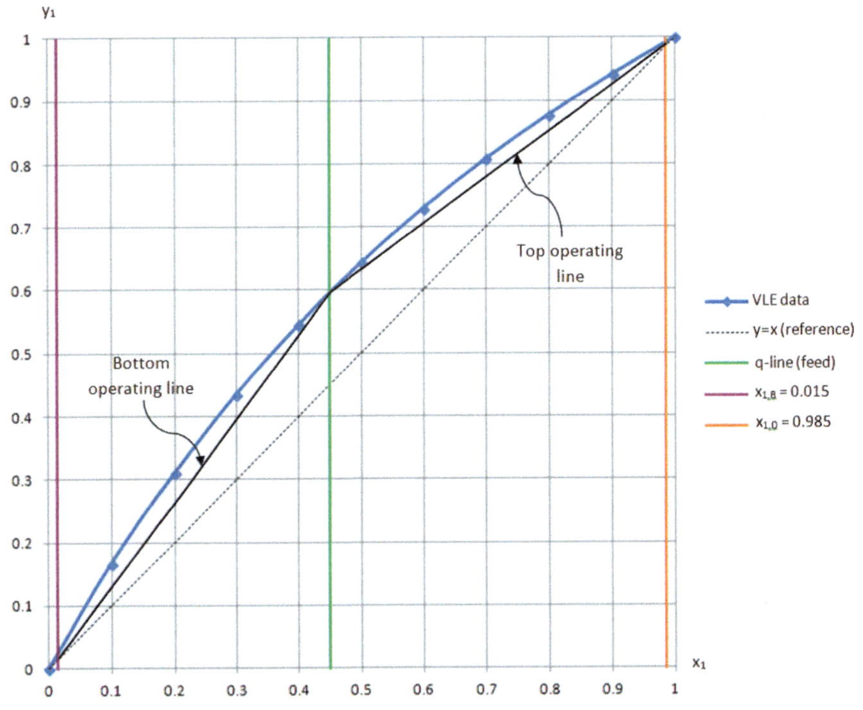


The minimum number of stages required occurs when reflux ratio is maximum, i.e., $R \to \infty$

$$\lim_{R\to\infty} \frac{R}{R+1} = \lim_{R\to\infty} \frac{1}{1+1/R} = 1$$

This means that the slope of the TOL is 1, which coincides with the line of $y = x$. The feed position becomes (0.45, 0.45), and this constrains the BOL to also be the line $y = x$. We can find the minimum number of stages by finding the number of steps required to go from (0.015, 0.015) to (0.985, 0.985). From the construction, we can see that the equilibrium line is hit 14 times; therefore, the minimum number of stages is 14.

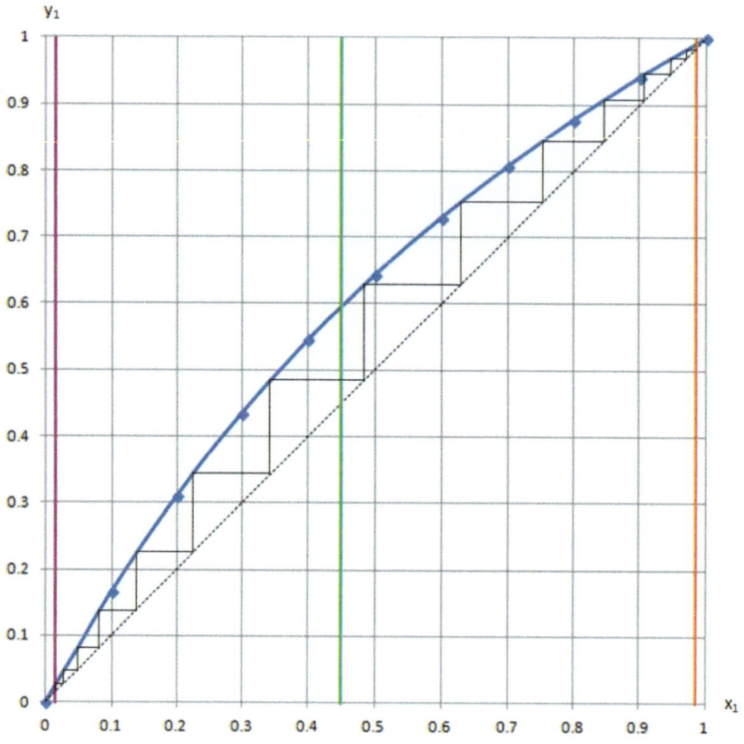


(b) If the reflux ratio is 1.2 times the minimum reflux ratio, then

$$R = 1.2R_{\text{min}} = 1.2(2.66) = 3.192$$

To determine the number of stages required, we only need to know the new slope of the top operating line, since it is already defined at one point, (0.985, 0.985).

$$\text{Slope of TOL} = \frac{R}{R+1} = \frac{3.192}{4.192} = 0.76145$$

Therefore the equation of the top operating line is:

$$\frac{y_1 - 0.985}{x_1 - 0.985} = 0.76145$$

We can find the feed locus which is at (0.45, 0.5776) which is the intersection between the TOL and the $q$-line (feed). We can also construct the bottom operating line since it is defined by the point at the feed locus, i.e., (0.45, 0.5776) and the point that defines the bottoms composition of $x_{1,B} = 0.015$.

With the two operating lines drawn, we can step up the space between the operating lines and the VLE curve to find the number of stages required for this new value of reflux ratio $R$. Note that we start from $x_{1,\,B} = 0.015$ and step off the bottom operating line until the feed locus is just crossed, then we switch to stepping off the top operating line. The optimal stage for the feed to enter is the stage number just before the feed locus is crossed.

From the McCabe–Thiele construction shown below, the feed position is at stage 14. There is a total of 28 stages.

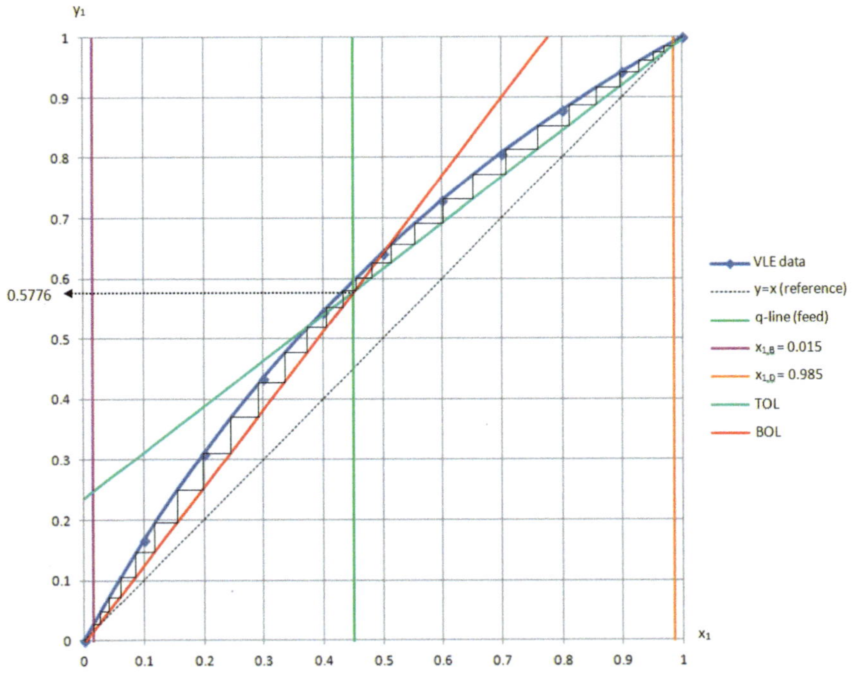

(c) Now we have two entry points for the feed. This means that we have a third operating line (which we term here as the Intermediate Operating Line (IOL)) that occurs between the two feed stages; therefore, we will now have 2 feed loci to our McCabe–Thiele diagram, at $x_1 = 0.25$ and 0.65, instead of the previous case of one feed at 0.45 only. Then we construct the steps, starting from the bottoms composition on the BOL to the IOL (after passing first feed stage) and then to the TOL (after passing second feed stage), until we reach the distillate composition. We observe that the first feed stage occurs optimally at stage 10, and the second feed stage occurs optimally at stage 14. The total number of stages is 24.

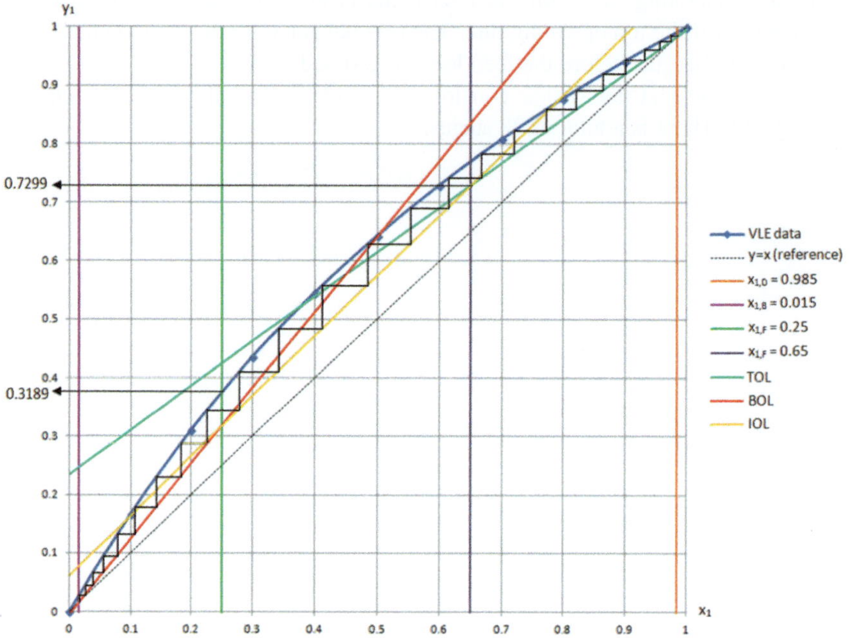


## Problem 7

**The equilibrium data for ethanol–water is provided below at 1 atm, where component 1 is ethanol.**

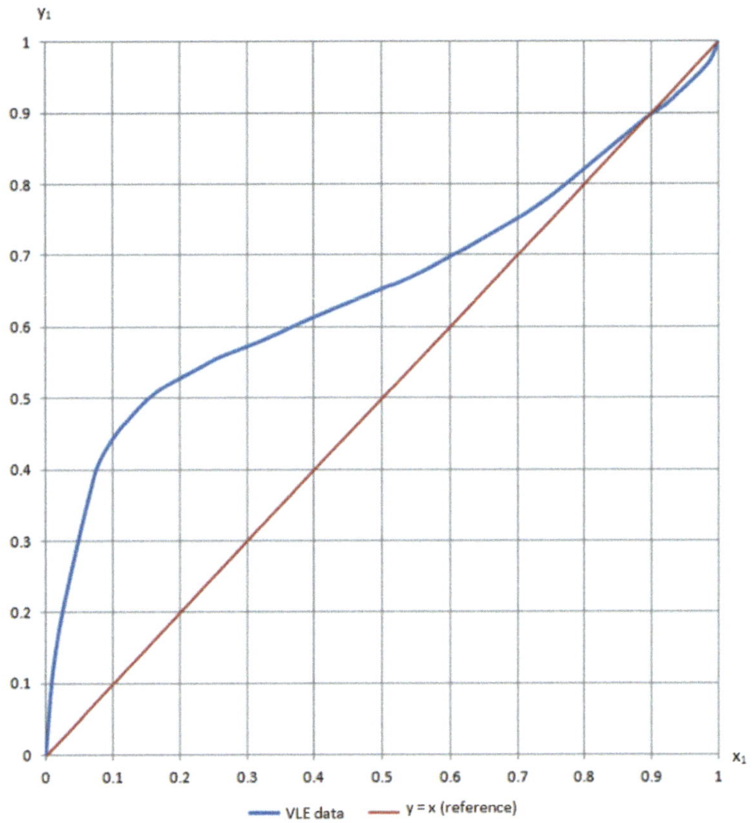


(a) **Determine the maximum and minimum ethanol concentrations that can be achieved in the distillate and bottoms products at 1 atm.**

(b) **If the feed stream is a saturated liquid with the mole fraction of ethanol = 0.25, and the required ethanol mole fractions in the distillate and bottoms are 0.85 and 0.05, respectively. Find the flow rates of the product streams for a feed flow rate of 1 mol/h. Find the minimum reflux ratio and minimum number of stages for this separation.**

## Solution 7

### *Worked Solution*

(a) We are required to find the maximum and minimum ethanol concentrations of ethanol that can be obtained in the distillate and bottoms streams at 1 atm.

Note that the shape of the VLE data shows an intersection with the line $y = x$ at $x_1 = 0.9$. This is an azeotrope, and it represents the maximum ethanol concentration that can be achieved. The minimum ethanol concentration is $x_1 = 0$.

(b) We can perform the following mass balances

$$F = D + B$$

$$Fx_1 = Dx_{1,D} + Bx_{1,B}$$
$$1(0.25) = D(0.85) + B(0.05)$$
$$B = 0.75\ \text{mol/h},\ D = 0.25\ \text{mol/h}$$

It is useful to note that the slope of the top operating line is defined as follows.

$$\text{Slope of TOL} = \frac{R}{R+1}$$

Normally, to find the minimum reflux ratio, we draw the top operating line such that it joins the distillate point (on $y = x$ line) with the feed locus (on the VLE curve). However, in this case, connecting these 2 points will violate VLE equilibrium as the line crosses the VLE curve such that there is a section of the TOL that lies above the VLE curve, which is not possible.

Therefore, we start with the distillate point, and adjust the slope of the TOL until it is just tangential to the VLE without crossing above it, and this line hits the VLE curve at the other point (near to the bottoms) at around (0.055, 0.33).

$$\frac{0.85 - 0.33}{0.85 - 0.055} = \frac{R_{\min}}{R_{\min} + 1}$$
$$R_{\min} = 1.89$$

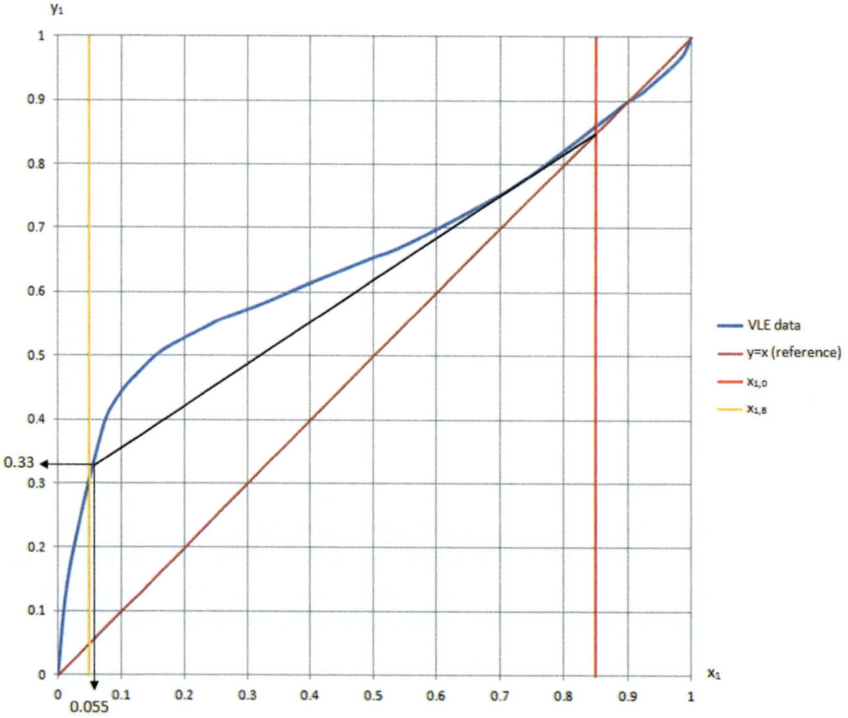

To find the minimum number of stages, the top and bottom operating lines coincide with $y = x$. The number of steps is 8, from the graphical construction as shown below. Hence $N_{min} = 8$.

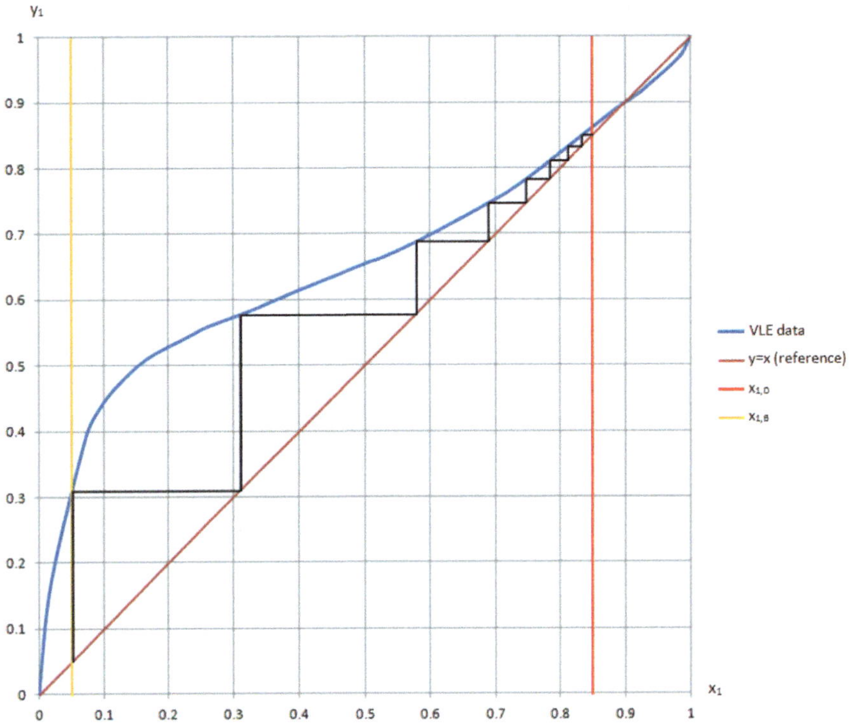


## Problem 8

**Consider flash unit with a single equilibrium stage. There are 3 components in the feed as shown below**

| Component | Feed composition, $z_i$ | $K_i$ Case 1 | Case 2 | Case 3 | Case 4 | Case 5 |
|---|---|---|---|---|---|---|
| 1 | 0.55 | 2.40 | 3.30 | 1.00 | $\infty$ | $\infty$ |
| 2 | 0.15 | 1.00 | 2.20 | 0.450 | 4.10 | 3.10 |
| 3 | 0.30 | 0.483 | 1.00 | 0.120 | 0.720 | 0.00 |

**Under equilibrium conditions, the Rachford–Rice equation below holds, where $V$ and $F$ are the vapor and feed flow rates.**

$$\sum_i \frac{z_i(1-K_i)}{\frac{V}{F}(K_i-1)+1}=0$$

**Identify the cases whereby both liquid and vapor streams leave the flash unit and find the vapor fraction in the feed if this is the case. Identify also the cases whereby only a single phase leaves the flash unit, and specify its phase.**

## Solution 8

### *Worked Solution*

We are given the following equation, whereby we have defined the vapor fraction in the feed as $\beta = V/F$. Note that $K_i = y_i/x_i$ at equilibrium.

$$F(\beta) = \sum_i \frac{z_i(1-K_i)}{\beta(K_i-1)+1} = 0$$

It helps to understand the derivation of the above RR equation by revisiting its derivation which originates from the following four key equations:

$$F = V + L \quad (1)$$

$$Fz_i = Vy_i + Lx_i \quad (2)$$

$$\sum_i x_i = 1 \quad (3)$$

$$\sum_i y_i = 1 \quad (4)$$

When a mixture is at its bubble point,

$$\sum_i K_i x_i = 1$$

And when a mixture is at its dew point,

$$\sum_i \frac{y_i}{K_i} = 1$$

When both vapor and liquid streams exist, it means that there is a valid solution for $\beta$ that is between 0 and 1 for the Rachford–Rice equation.

Conversely, if there is only a single phase, either liquid-only or vapor-only, the following cases are possible:

- Superheated vapor—all $K_i$ values of all components in the mixture are $\geq 1$.
- Subcooled liquid—all $K_i$ values of all components in the mixture are $\leq 1$.

- The $K_i$ values of some components are $\geq 1$ and some are $\leq 1$. This means that the overall mixture is either:
  - Liquid only, and below bubble point if $F(\beta = 0) > 0$
  - Liquid only at bubble point if $F(\beta = 0) = 0$
  - Vapor only, and above dew point if $F(\beta = 1) < 0$
  - Vapor only at dew point if $F(\beta = 1) = 0$

  Now let us analyze the 5 cases that we have been given.

*Case 1:*

Some $K_i$ values are $\geq 1$ and some are $\leq 1$, so we need to determine $F(\beta = 0)$ and $F(\beta = 1)$ to find out more.

$$F(0) = \sum_i z_i(1 - K_i) = 0.55(1 - 2.4) + 0.15(1 - 1) + 0.30(1 - 0.483) = -0.615 < 0$$

$$F(1) = \sum_i \frac{z_i(1 - K_i)}{K_i} = \frac{0.55(1 - 2.4)}{2.4} + \frac{0.15(1 - 1)}{1} + \frac{0.30(1 - 0.483)}{0.483} = 0.0003 \approx 0$$

This means that the mixture is a single phase vapor at dew point.

*Case 2:*

Observe that all $K_i$ values are $\geq 1$, the mixture is a single phase vapor (superheated) above dew point.

$$F(1) = \sum_i \frac{z_i(1 - K_i)}{K_i} = \frac{0.55(1 - 3.3)}{3.3} + \frac{0.15(1 - 2.2)}{2.2} + \frac{0.30(1 - 1)}{1} = -0.465 < 0$$

*Case 3:*

Observe that all $K_i$ values are $\leq 1$, the mixture is a single phase liquid (subcooled) below bubble point.

$$F(0) = \sum_i z_i(1 - K_i) = 0.55(1 - 1) + 0.15(1 - 0.45) + 0.30(1 - 0.12)$$
$$= 0.3465 > 0$$

*Case 4:*

Some $K_i$ values are $\geq 1$ and some are $\leq 1$, so we need to determine $F(\beta = 0)$ and $F(\beta = 1)$ to find out more.

We have $K_1 = \infty$, it means all of component 1 is in the vapor phase. Also, we need to use the L'Hospital's rule to evaluate the value of the function.

$$\lim_{x\to\infty}\frac{f(x)}{g(x)}=\lim_{x\to\infty}\frac{f'(x)}{g'(x)}$$

$$\lim_{K_i\to\infty}\frac{z_i(1-K_i)}{\beta(K_i-1)+1}=\lim_{K_i\to\infty}\frac{z_i-K_iz_i}{\beta K_i-\beta+1}=\lim_{K_i\to\infty}\frac{\frac{\partial}{\partial K_i}(z_i-K_iz_i)}{\frac{\partial}{\partial K_i}(\beta K_i-\beta+1)}=-\frac{z_i}{\beta}$$

$$F(\beta)=\sum_i\frac{z_i(1-K_i)}{\beta(K_i-1)+1}=-\frac{z_1}{\beta}+\frac{z_2(1-K_2)}{\beta(K_2-1)+1}+\frac{z_3(1-K_3)}{\beta(K_3-1)+1}$$

$$F(\beta)=-\frac{0.55}{\beta}+\frac{0.15(1-4.1)}{\beta(4.1-1)+1}+\frac{0.3(1-0.72)}{\beta(0.72-1)+1}$$

When we solve this equation for $F(\beta)=0$, we find that $\beta=2.48$ which is not physically meaningful since $0\le\beta\le1$. And so this mixture cannot be fully liquid.

Now, we test the value of $\beta=1$.

$$F(\beta=1)=-\frac{0.55}{1}+\frac{0.15(1-4.1)}{1(4.1-1)+1}+\frac{0.3(1-0.72)}{1(0.72-1)+1}=-0.547<0$$

This result confirms that there is a single vapor phase which is above dew point.

*Case 5:*

We have $K_3=0$, it means that all of component 3 is in the liquid phase. All of component 1 is in the vapor phase since $K_1=\infty$. This already tells us that there are both liquid and vapor phases, and so there must be VLE. In this case, then $F(\beta)=0$ can be solved to find $\beta$.

$$F(\beta)=\sum_i\frac{z_i(1-K_i)}{\beta(K_i-1)+1}=-\frac{z_1}{\beta}+\frac{z_2(1-K_2)}{\beta(K_2-1)+1}+\frac{z_3(1-K_3)}{\beta(K_3-1)+1}$$

$$F(\beta)=0=-\frac{0.55}{\beta}+\frac{0.15(1-3.1)}{\beta(3.1-1)+1}+\frac{0.3(1-0)}{\beta(0-1)+1}$$

$$\beta=0.68$$

## Problem 9

**Consider the following components in a feed mixture at equilibrium at 95°C and 12 bar. Ideal mixture can be assumed for both the vapor and liquid phases. The vapor pressures of the components follow the following correlation.**

$$\ln(P^{vap}[\text{bar}])=A-B/(T[K])$$

| Component | $z$ | $A$ | $B$ |
|---|---|---|---|
| 1 | 0.20 | 9.623 | 2240 |
| 2 | 0.45 | 9.915 | 2598 |
| 3 | 0.35 | 10.053 | 3099 |

(a) **Find out whether the mixture is a subcooled liquid, a liquid–vapor mixture, or a superheated vapor.**
(b) **In order to bring the mixture to its bubble point as a saturated liquid, what is the pressure required for the temperature at 95°C.**
(c) **If pressure is maintained at 12 bar, what temperature must the mixture be to reach the bubble point.**

## Solution 9

***Worked Solution***

(a) Let us determine the $K$ values of each component at the operating conditions 120°C and 12 bar.

$$\ln\left(P^{\text{vap}}[\text{bar}]\right) = A - B/(T[K])$$

$$P_i^{\text{vap}} = \exp\left(A_i - \frac{B_i}{T}\right)$$

Under vapor–liquid equilibrium, chemical potentials in each phase are equivalent. Assuming ideal gas for the vapor phase and ideal mixtures in both the vapor and liquid phases, then

$$\mu_i^L = \mu_i^V$$

$$\mu_{i,\text{pure}}^L + \text{RT}\ln x_i = \mu_{i,\text{pure,ref}}^V + \text{RT}\ln\frac{f_i}{P_{\text{pure,ref}}}$$

$$\left(\mu_{i,\text{pure,ref}}^V + \text{RT}\ln\frac{f_{i,\text{pure}}}{P_{\text{pure,ref}}}\right) + \text{RT}\ln x_i = \mu_{i,\text{pure,ref}}^V + \text{RT}\ln\frac{f_i}{P_{\text{pure,ref}}}$$

$$\text{RT}\ln x_i = \text{RT}\ln\frac{f_i}{f_{i,\text{pure}}}$$

$$x_i = \frac{f_i}{f_{i,\text{pure}}}$$

$$x_i = \frac{P_i}{P_{i,\text{pure}}} \quad \text{for ideal gas}$$

$$x_i = \frac{y_i P}{P_{i,\text{pure}}} \text{ for ideal mixing in vapor phase}$$

In fact, $P_{i,\text{pure}} = P_i^{\text{vap}}$; therefore,

$$x_i = \frac{y_i P}{P_i^{\text{vap}}}$$

$$K_i = \frac{y_i}{x_i} = \frac{P_i^{\text{vap}}}{P} = \frac{\exp\left(A_i - \frac{B_i}{T}\right)}{P}$$

| Component | $z_i$ | $A$ | $B$ | $P_i^{\text{vap}}$[bar] | $K_i$ | $z_i K_i$ | $z_i/K_i$ |
|---|---|---|---|---|---|---|---|
| 1 | 0.20 | 9.623 | 2240 | 34.331 | 2.8609 | 0.57218 | 0.069908 |
| 2 | 0.45 | 9.915 | 2598 | 17.378 | 1.4482 | 0.65168 | 0.31073 |
| 3 | 0.35 | 10.053 | 3099 | 5.1131 | 0.42609 | 0.14913 | 0.82142 |
| | | | | | | 1.37 | 1.20 |

If all $K_i > 1$, the mixture will be a superheated vapor. Conversely if all $K_i < 1$, the mixture will be a subcooled liquid. In our case, we have some $K_i$ values greater than 1 and some less than 1.

Consider the Rachford–Rice equation for vapor–liquid equilibrium, where $\beta = V/F$.

$$\sum_i \frac{z_i(K_i - 1)}{\beta(K_i - 1) + 1} = 0$$

It is worth noting first that at bubble point, $\beta = 0$ at $F = 0$ and so

$$\sum_i z_i(K_i - 1) = 0$$

$$\sum_i z_i K_i = 1$$

For our mixture, the value of $\sum_i z_i K_i = 1.37 > 1$. This means that our operating temperature of 95°C is above the bubble point.

Separately, we note that at dew point, $\beta = 1$ at $F = 0$ and so

$$\sum_i \frac{z_i(K_i - 1)}{K_i} = 0$$

$$\sum_i \frac{z_i}{K_i} = 1$$

For our mixture, the value of $\sum_i z_i/K_i = 1.20 > 1$. This means that our operating temperature of 95°C is below the dew point. We can deduce that we have a vapor–liquid mixture and can find the value of $\beta$ for it such that $0 < \beta < 1$.

$$\sum_i \frac{z_i(K_i - 1)}{\beta(K_i - 1) + 1} = 0$$

$$\frac{0.20(2.8609 - 1)}{\beta(2.8609 - 1) + 1} + \frac{0.45(1.4482 - 1)}{\beta(1.4482 - 1) + 1} + \frac{0.35(0.42609 - 1)}{\beta(0.42609 - 1) + 1} = 0$$

$$\beta = 0.66$$

(b) We note from part a that we have a mixture that is above bubble point and below dew point. In order to bring a mixture that is above its bubble point to its bubble point without changing temperature, we can increase pressure to greater than 12 bar.

$$\sum_i z_i(K_i - 1) = \sum_i z_i\left(\frac{P_i^{\text{vap}}}{P} - 1\right)$$

$$= \frac{1}{P}\sum_i z_i P_i^{\text{vap}} - \sum_i z_i = \frac{1}{P}\sum_i z_i P_i^{\text{vap}} - 1 = 0$$

$$\sum_i z_i P_i^{\text{vap}} = P$$

$$P = \sum_i z_i\left[\exp\left(A_i - \frac{B_i}{95 + 273}\right)\right] = 16.5 \text{ bar}$$

(c) Similar to part b, in order to bring a mixture that is above its bubble point to its bubble point without changing pressure, we can decrease temperature to lower than 95°C.

$$\sum_i z_i(K_i - 1) = 0$$

$$\sum_i z_i\left(\frac{P_i^{\text{vap}}}{P} - 1\right) = \sum_i z_i\left(\frac{\left[\exp\left(A_i - \frac{B_i}{T}\right)\right]}{12} - 1\right) = 0$$

$$T = 78.6^{\circ}\text{C}$$

## Problem 10

**Consider a mixture containing components 1 and 2 which are to be separated by continuous distillation. The mixture is at bubble point and contains 25 mol% of component 1 and 75 mol% component 2. The relative volatility of 1 to 2 is 2.8. The feed flow rate is 150 mol/h and the separation objective is to obtain a**

**distillate product of 97 mol% of component 1 and which contains 0.6 of the amount of component 1 in the feed.**

(a) **What is the minimum reflux ratio and minimum number of stages required?**

(b) **If there was a partial reboiler and total condenser for the column, and the operating reflux ratio was set at 1.25 of the minimum, how many stages are required? Determine the optimal location for the feed.**

(c) **Assuming that we have a test column that is fitted with a partial reboiler and total condenser. It comes with 10 plates above the reboiler and operates at reflux ratio of 3.8. The feed is to enter at stage 5 (where stage 1 is above the reboiler). The reboiler has a vaporization rating of 120 mol/h.**

i. **Is this column able to achieve the desired separation objective, and comment on the adequacy of the reboiler provided. Can the feed stream enter directly into the reboiler (instead of at stage 5)?**

ii. **What is the minimum number of stages for this test column, and identify the optimal feed location.**

## Solution 10

*Worked Solution*

(a) Let us illustrate a diagram for our system.

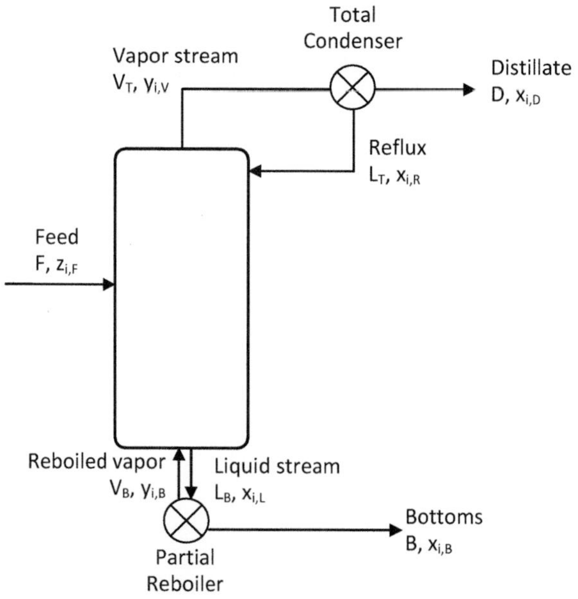

Using the relative volatility $\alpha = 2.8$, we can establish the following equilibrium relationship between vapor and liquid mole fractions for component 1, which we can use to construct the VLE curve.

$$y_1 = \frac{\alpha x_1}{1 + (\alpha - 1)x_1} = \frac{2.8x_1}{1 + 1.8x_1}$$

We can perform the mass balances as shown below based on the information provided.

$$F = D + B$$

$$Fz_{1,F} = Dx_{1,D} + Bx_{1,B}$$

$$0.6 = \frac{Dx_{1,D}}{Fz_{1,F}} = \frac{D(0.97)}{(150)(0.25)}$$

$$D = 23.196 \text{ mol/h}$$

$$B = F - D = 150 - 23.196 = 126.804 \text{ mol/h}$$

$$x_{1,B} = \frac{Fz_{1,F} - Dx_{1,D}}{B} = \frac{(150)(0.25) - (23.196)(0.97)}{126.804} = 0.1183$$

For minimum reflux ratio $R$, the top operating line (TOL) intersects the feed $q$-line at the VLE curve. We know that the feed is a saturated liquid at bubble point; hence the $q$-line is vertical.

As a rule of thumb, it is useful to note the following characteristics of the $q$-line for different feed types.

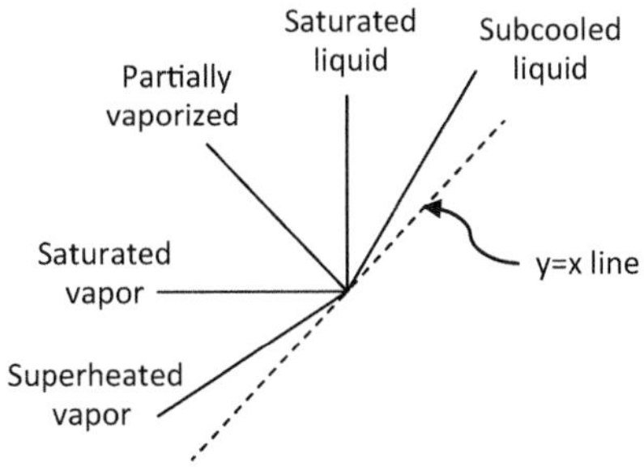


Therefore we can construct the McCabe Thiele diagram for the case of minimum reflux ratio. Note that the slope of the TOL is related to the reflux ratio as follows:

$$\text{Slope of TOL} = \frac{R_{\min}}{R_{\min}+1} = \frac{0.97-0.483}{0.97-0.25} = 0.6767$$

$$R_{\min} = 2.09$$

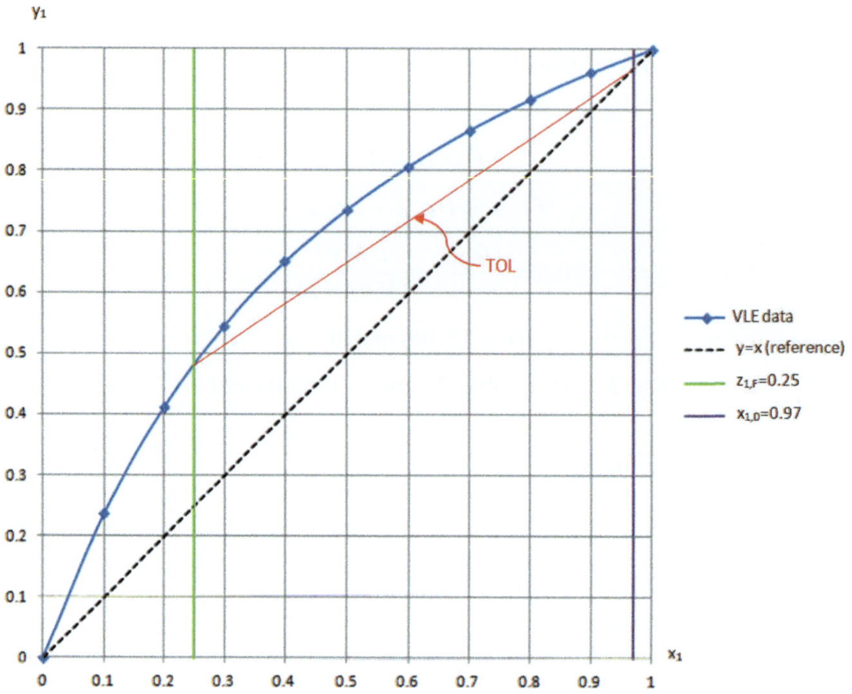


The minimum number of stages occurs when our operating lines coincide with the line $y = x$. There are 6 equilibrium stages from the graphical construction and this number includes the partial reboiler. Therefore there are 5 stages and 1 partial reboiler.

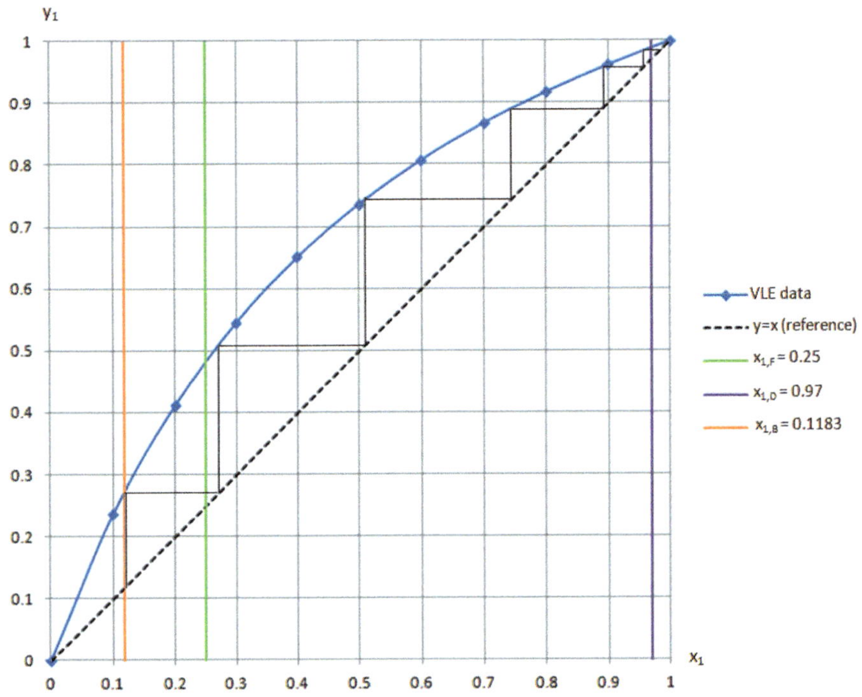


(b) Now that we have an operating reflux ratio which is 1.25 times of the minimum, we have a new slope for the top operating line = 0.723. Since the TOL is also defined at the point, (0.97, 0.97), we can construct the TOL line.

$$R = 1.25(2.09) = 2.61$$

$$\text{Slope of TOL} = \frac{2.61}{1 + 2.61} = 0.723$$

From the diagram, we observe that the TOL intersects the $q$-line (feed) at $x = 0.25$, $y = 0.449$. We can draw the bottom operating line now as it passes through this feed locus at (0.25, 0.449) and the point (0.1183, 0.1183) defined by the bottoms composition.

Once we have both operating lines, we can draw "steps" on the McCabe–Thiele diagram to find out the number of stages required at this new reflux ratio. We find that there are 11 equilibrium stages (i.e., 10 stages and 1 reboiler stage), where the optimal feed location is at stage 3 (where stage 1 is after the reboiler) where we step over from lower operating line to top operating line.

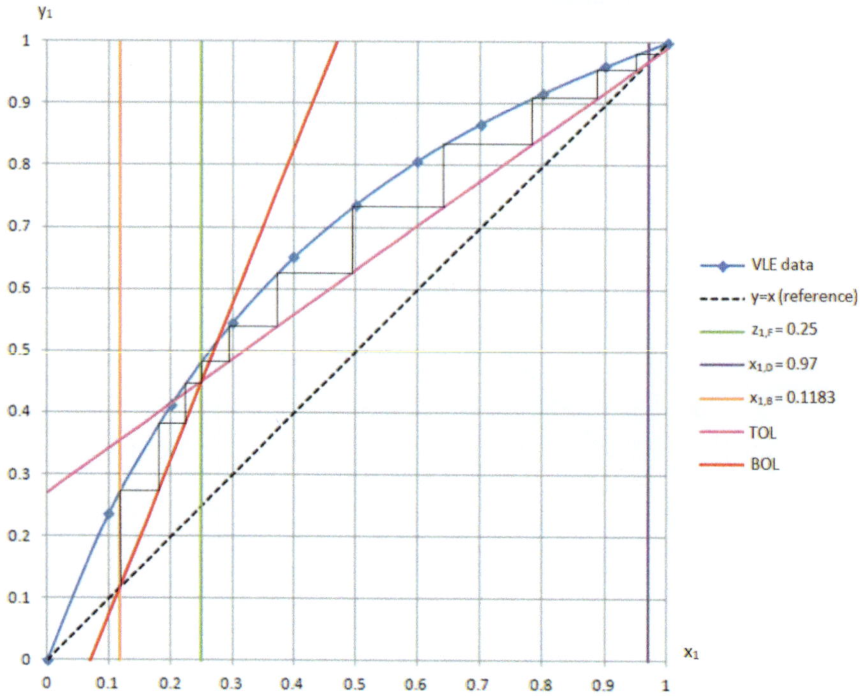


(c) (i) Now that $R = 3.8$. We have a new slope for the TOL.

$$\text{Slope of TOL} = \frac{3.8}{1 + 3.8} = 0.79167$$

We still want to retain the column from part b, which consists of 11 equilibrium stages (1 reboiler + 10 stages). But this time, we are asked to introduce the feed at plate 5 as one of the options. We note from the graph that this design will be able to give us a distillate product that is purer than the required $x_{1,D} = 0.97$.

Now we check if the reboiler has adequate capacity. The BOL is already defined based on the specifications in the problem. It has a slope as follows from the graphical construction.

$$\text{Slope of BOL} = \frac{0.1183 - 0.4}{0.1183 - 0.25} = 2.139$$

The equation of the BOL can also be obtained from a mass balance at the bottom of the column,

$$V_B y_{1,m} + Bx_{1,B} = L_B x_{1,m+1}$$

$$y_{1,m} = \left(\frac{L_B}{V_B}\right)x_{1,m+1} - \left(\frac{B}{V_B}\right)x_{1,B}$$

$$\text{Slope of BOL} = \frac{L_B}{V_B} = \frac{F + \text{RD}}{V_B} = \frac{150 + 3.8(23.196)}{V_B} = \frac{238.14}{V_B}$$

Equating both expressions for slope of BOL,

$$\frac{238.14}{V_B} = 2.139$$

$$V_B = 111.3 \text{ mol/h}$$

We are given a reboiler that has $V_B = 120$ mol/h which is greater than 111.3 mol/h. Therefore the reboiler has adequate capacity. We also observe that it is not possible for the feed to enter directly into the reboiler as that would require stepping from the BOL to TOL at the first step.

(ii) From graphical construction (black line), we can determine the minimum number of equilibrium stages = 8 (7 stages + reboiler), with the optimal feed location at plate 2 (above reboiler).

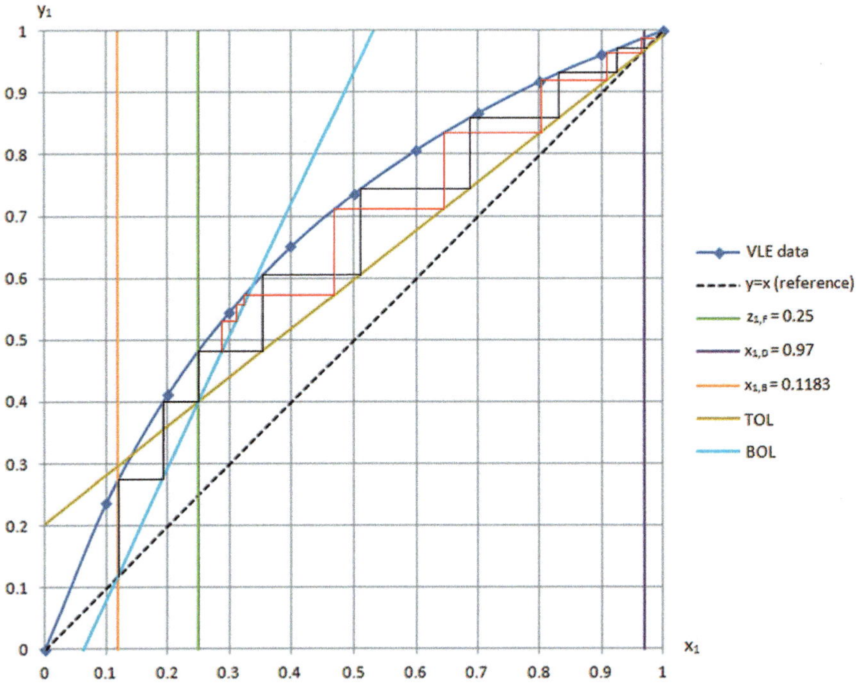

## Problem 11

**Consider a flash unit that is used to process two hydrocarbon streams *A* and *B*. The flash operation is performed at 90°C and 1.4 bar. Each stream contains three components 1, 2, and 3 as shown below:**

| | Mole fraction, $z_i$ | | | | Vapor pressure [bar] | | |
|---|---|---|---|---|---|---|---|
| Feed stream | 1 | 2 | 3 | Flow rate [mol/h] | 1 | 2 | 3 |
| *A* | 0.52 | 0.28 | 0.20 | 110 | 2.10 | 1.32 | 0.62 |
| *B* | 0.45 | 0.45 | 0.10 | 110 | | | |

**You may assume that there is an ideal gas vapor phase and both the liquid and vapor phases have ideal mixing.**

(a) **Comment on whether both vapor and liquid phases are present in the exit stream, and find the fraction of vapor flow rate to feed flow rate.**
(b) **Determine the vapor and liquid flow rates at the exit streams and their molar compositions.**
(c) **If the flash unit is maintained at 90°C, what is the pressure required to ensure only liquid at its bubble point leaves the unit?**

## Solution 11

***Worked Solution***

(a) Let us illustrate our system in a simple diagram.

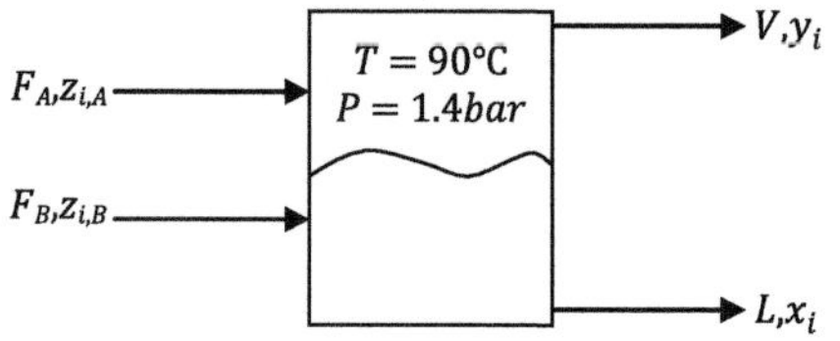


We know that by definition, $K_i = y_i/x_i$ at vapor–liquid equilibrium. Also, since we have ideal mixing in both vapor and liquid phases, and an ideal gas in the vapor phase, it follows that

$$y_i P = x_i P_i^{\text{sat}}$$

$$K_i = \frac{P_i^{\text{sat}}}{P}$$

| Feed stream | Mole fraction, $z_i$ | | | Flow rate [mol/h] | Vapor pressure [bar] | | | $K_i$ | | |
|---|---|---|---|---|---|---|---|---|---|---|
| | 1 | 2 | 3 | | 1 | 2 | 3 | 1 | 2 | 3 |
| $A$ | 0.52 | 0.28 | 0.20 | 110 | 2.10 | 1.32 | 0.62 | $2.10/1.4 = 1.5$ | 0.943 | 0.443 |
| $B$ | 0.45 | 0.45 | 0.10 | 110 | | | | | | |

We can calculate the effective single feed composition for the two feed streams.

$$\bar{z}_1 = \left(\frac{F_A}{F_A + F_B}\right)0.52 + \left(\frac{F_B}{F_A + F_B}\right)0.45 = 0.485$$

$$\bar{z}_2 = \left(\frac{F_A}{F_A + F_B}\right)0.28 + \left(\frac{F_B}{F_A + F_B}\right)0.45 = 0.365$$

$$\bar{z}_3 = \left(\frac{F_A}{F_A + F_B}\right)0.20 + \left(\frac{F_B}{F_A + F_B}\right)0.10 = 0.15$$

For equilibrium, we can apply the Rachford Rice equation where $\beta = V/F$ and denotes the ratio of vapor flow rate to feed flow rate.

$$F = \sum_i \frac{z_i(1 - K_i)}{(1 - \beta) + \beta K_i} = 0$$

Let us check the values of the function $F$ assuming $\beta = 0$ (bubble point) and $\beta = 1$ (dew point). At bubble point, $\beta = 0$ at $F = 0$ and so

$$\sum_i z_i(1 - K_i) = 0$$

$$\sum_i z_i K_i = 1$$

$$0.485(1.5) + 0.365(0.943) + 0.15(0.443) = 1.14$$

For our mixture, the value of $\sum_i z_i K_i = 1.14 > 1$. This means that our operating temperature of 90°C is above the bubble point. Separately, we know that at dew point, $\beta = 1$ at $F = 0$ and so

$$\sum_i \frac{z_i(1 - K_i)}{K_i} = 0$$

$$\sum_i \frac{z_i}{K_i} = 1$$

$$\frac{0.485}{1.5} + \frac{0.365}{0.943} + \frac{0.15}{0.443} = 1.05$$

For our mixture, the value of $\sum_i \frac{z_i}{K_i} = 1.05 > 1$. This means that our operating temperature of 90°C is below the dew point. Therefore, we have a mixture between bubble point and dew point, which means that we have both vapor and liquid phases in the exit stream. To find the fraction of vapor, we can solve for $\beta$ which is the fraction of vapor to feed flow rate.

$$\frac{0.485(1-1.5)}{(1-\beta)+1.5\beta} + \frac{0.365(1-0.943)}{(1-\beta)+0.943\beta} + \frac{0.15(1-0.443)}{(1-\beta)+0.443\beta} = 0$$

$$\beta = 0.801$$

(b) Since the total feed flow rate entering the unit is

$$F = F_A + F_B = 110 + 110 = 220 \text{ mol/h}$$

$$V = 0.801(220) = 176 \text{ mol/h}$$

$$L = 220 - 176 = 43.8 \text{ mol/h}$$

To find the molar compositions, we can use the mass balance equation

$$Fz_i = Vy_i + Lx_i$$

$$(220)0.485 = (176)K_1x_1 + (43.8)x_1 = (176)(1.5)x_1 + (43.8)x_1$$

$$x_1 = 0.347, \ y_1 = 1.5(0.347) = 0.520$$

$$(220)0.365 = (176)K_2x_2 + (43.8)x_2 = (176)(0.943)x_2 + (43.8)x_2$$

$$x_2 = 0.383, \ y_2 = 0.943(0.383) = 0.361$$

$$(220)0.15 = (176)K_3x_3 + (43.8)x_3 = (176)(0.443)x_3 + (43.8)x_3$$

$$x_3 = 0.271, \ y_3 = 0.443(0.271) = 0.120$$

(c) If temperature is held constant, the way to make the 2-phase mixture reach its bubble point (single liquid phase) is to increase pressure to a value greater than 1.4 bar. At bubble point, the following is true at $\beta = 0$.

$$\sum_i z_i(1 - K_i) = 0$$

$$\sum_i z_iK_i = \sum_i z_i\left(\frac{P_i^{\text{sat}}}{P}\right) = 1$$

$$0.485\left(\frac{2.1}{P}\right)+0.365\left(\frac{1.32}{P}\right)+0.15\left(\frac{0.62}{P}\right)=1$$

$$P=1.59\ \text{bar}$$

## Problem 12

**Consider distillation of a binary mixture containing $A$ and $B$. The relative volatility of $A$ to $B$ is 2.4 and mole fraction of $A$ in the feed is 0.6. The distillate composition needs to be 80 mol% of $A$. Determine if this can be achieved for the following scenarios, and determine the ratio of distillate to feed flow rate where the desired separation is possible.**

(a) **The feed mixture enters directly into a partial reboiler as a saturated liquid at bubble point. The reboiler is fitted to a total condenser with a reflux ratio of $R = 2.3$. Vapor–liquid equilibrium can be assumed in the partial reboiler.**
(b) **The feed mixture enters a distillation column comprising of one plate and one total condenser. The feed is a saturated liquid at bubble point and reflux ratio $R = 2.3$.**
(c) **Total reflux is used for the conditions in part b.**
(d) **Reflux ratio $R = 0$ for the conditions in part b.**
(e) **The saturated liquid feed enters the column into the single plate for the conditions in part b.**
(f) **A partial condenser is used instead of total condenser for the same conditions as in part a. Liquid returns to reboiler and vapor is removed as distillate. The liquid and vapor leaving the condenser may be assumed to be in equilibrium with each other.**

## Solution 12

*Worked Solution*

(a) Let us sketch a diagram for our setup.

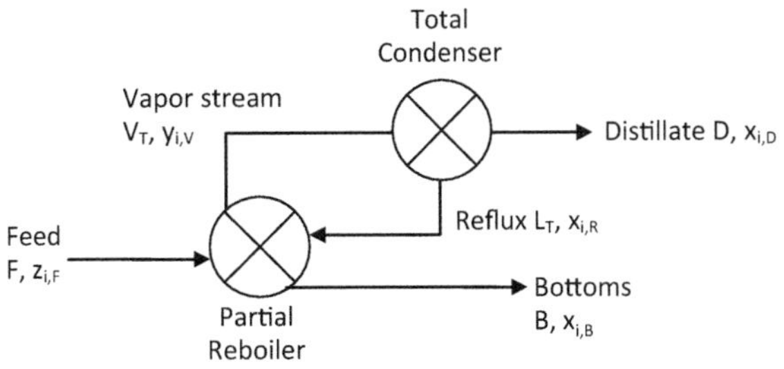

We know that relative volatility $\alpha = 2.4$; therefore, we can draw the VLE curve using the following equilibrium relationship.

$$y_A = \frac{\alpha x_A}{1 + (\alpha - 1)x_A} = \frac{2.4x_A}{1 + 1.4x_A}$$

At the reboiler, we have equilibrium; therefore we can find $x_{A,B}$

$$y_{A,V} = x_{A,D} = 0.8 = \frac{2.4x_{A,B}}{1 + 1.4x_{A,B}}$$

$$x_{A,B} = 0.625 > z_{A,F} = 0.6$$

This means that the bottoms product has a higher composition of component $A$ than in the feed, which is not possible. Therefore, this setup does not give us our desired product.

(b) Now we have a setup that includes one stage in the column.

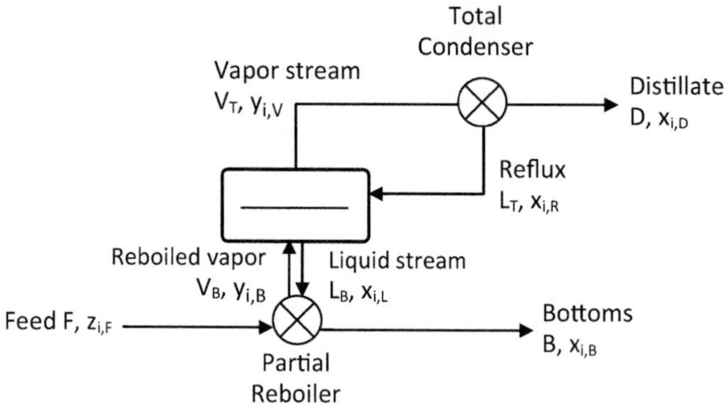


We know that reflux ratio is 2.4, and a mass balance around the total condenser gives us the following.

$$\mathrm{RD} = L_T$$

$$V_T = D + L_T = D(1 + R) = 3.4D$$

We have equilibrium reached at the stage in the column, and we know that $y_{A,V} = x_{A,D} = 0.8$; therefore

$$y_{A,V} = \frac{2.4x_{A,L}}{1 + 1.4x_{A,L}}$$

$$0.8 = \frac{2.4x_{A,L}}{1 + 1.4x_{A,L}}$$

$$x_{A,L} = 0.625$$

A mass balance around the first stage and total condenser gives us

$$V_B y_{A,B} = Dx_{A,D} + L_B x_{A,L}$$

$$y_{A,B} = \frac{D}{V_B}(0.8) + \frac{L_B}{V_B}(0.625)$$

The slope of the top operating line is defined by the reflux ratio as follows. This coincides with the gradient term $\frac{L_B}{V_B}$ of the above equation for the TOL. Also we know that $V_B = V_T$

$$\text{Slope of TOL} = \frac{R}{1+R} = \frac{2.3}{3.3} = 0.69697 = \frac{L_B}{V_B}$$

$$y_{A,B} = \frac{D}{V_T}(0.8) + 0.69697(0.625) = \frac{1}{3.4}(0.8) + 0.69697(0.625) = 0.671$$

$$y_{A,B} = \frac{2.4x_{A,B}}{1 + 1.4x_{A,B}} = 0.671$$

$$x_{A,B} = 0.460 < z_{A,F} = 0.6$$

Therefore, it is possible to produce the required distillate product. To find the required flow rate (as a ratio of distillate to feed flow rate) to achieve this, we can perform the following mass balances.

$$F = D + B$$

$$Fz_{A,F} = Dx_{A,D} + Bx_{A,B}$$

$$z_{A,F} = \frac{D}{F}x_{A,D} + \left(1 - \frac{D}{F}\right)x_{A,B}$$

$$\frac{D}{F} = \frac{z_{A,F} - x_{A,B}}{x_{A,D} - x_{A,B}} = \frac{0.6 - 0.46}{0.8 - 0.46} = 0.412$$

We note that we will produce a distillate stream at flow rate of 41.2 moles per 100 moles of feed.

(c) Now we have total reflux, i.e., $R \to \infty$, which also means that all of the liquid obtained from the total condenser is returned to the column, and there is no outlet distillate stream. This means that it is not possible to produce a distillate stream of 80 mol% $A$ as required.

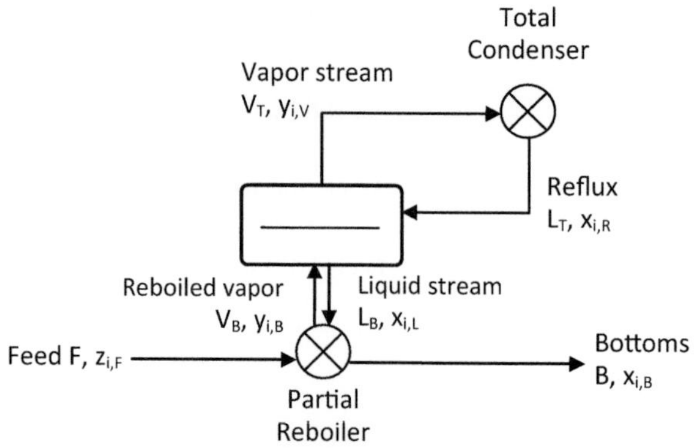


(d) For reflux ratio $R = 0$, there is no liquid return into the column. Therefore, there is effectively only one equilibrium stage at the reboiler itself, and the stage in the distillation column does not serve as an equilibrium stage. This setup is the same as in part a, which we have earlier shown to be unfeasible to produce our desired distillate product.

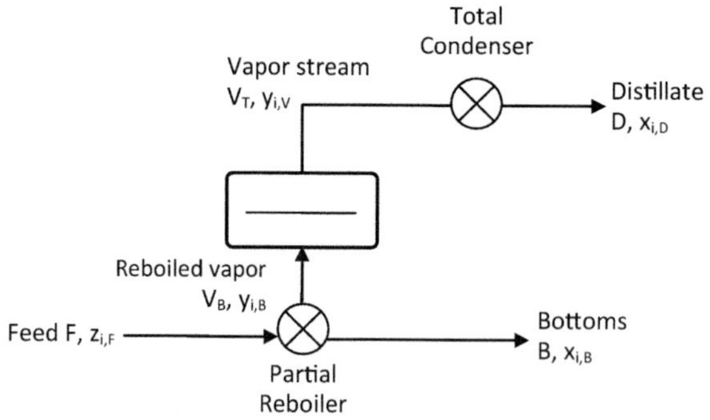


(e) Now we have the saturated liquid feed entering the column into the plate.

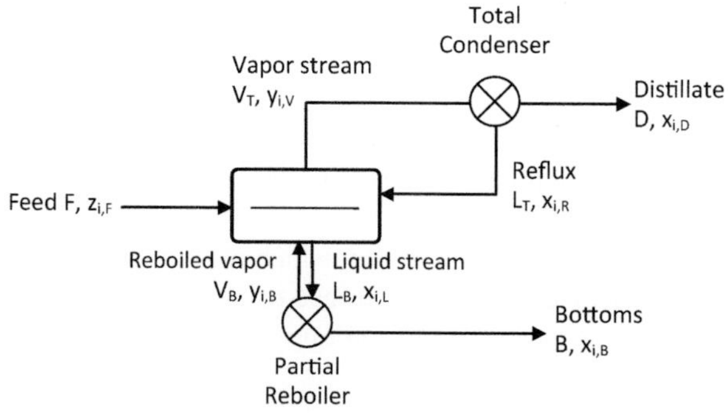

This gives us a new mass balance around the total condenser.

$$V_T = D + L_T = D(1 + R) = 3.3D$$

We also know from a mass balance around the single stage,

$$V_T = V_B$$

$$L_B = F + L_T = F + \mathrm{RD}$$

Finally, we know that $y_{A,V}$ and $x_{A,L}$ are related by the equilibrium equation. $y_{A,V}$ is already constrained by the required distillate purity, i.e., $y_{A,V} = x_{A,D} = 0.8$.

$$y_{A,V} = \frac{2.4x_{A,L}}{1 + 1.4x_{A,L}} = 0.8$$

$$x_{A,L} = 0.625$$

We can now derive a bottom operating line, which applies for the partial reboiler as an equilibrium stage. This can be done by doing a mass balance around the partial reboiler and substituting earlier results.

$$L_B x_{A,L} = Bx_{A,B} + V_B y_{A,B}$$

$$(F + \mathrm{RD})x_{A,L} = (F - D)x_{A,B} + V_T y_{A,B}$$

$$(F + \mathrm{RD})x_{A,L} = (F - D)x_{A,B} + (3.3D)y_{A,B}$$

$$\left(1 + 2.3\frac{D}{F}\right)x_{A,L} = \left(1 - \frac{D}{F}\right)x_{A,B} + 3.3\left(\frac{D}{F}\right)y_{A,B}$$

$$\left(1 + 2.3\frac{D}{F}\right)(0.625) = \left(1 - \frac{D}{F}\right)x_{A,B} + 3.3\left(\frac{D}{F}\right)\left(\frac{2.4x_{A,B}}{1 + 1.4x_{A,B}}\right) \quad (1)$$

An overall mass balance gives us one more equation required to solve for $x_{A,B}$ and $\frac{D}{F}$.

$$Fz_{A,F} = Dx_{A,D} + Bx_{A,B} = Dx_{A,D} + (F - D)x_{A,B}$$

$$z_{A,F} = \left(\frac{D}{F}\right)x_{A,D} + \left(1 - \frac{D}{F}\right)x_{A,B}$$

$$0.6 = \left(\frac{D}{F}\right)0.8 + \left(1 - \frac{D}{F}\right)x_{A,B} \quad (2)$$

Solving the simultaneous Eqs. (1) and (2) gives us

$$x_{A,B} = 0.49, \quad \frac{D}{F} = 0.35$$

(f) A partial condenser is used instead of total condenser for the same conditions as in part a.

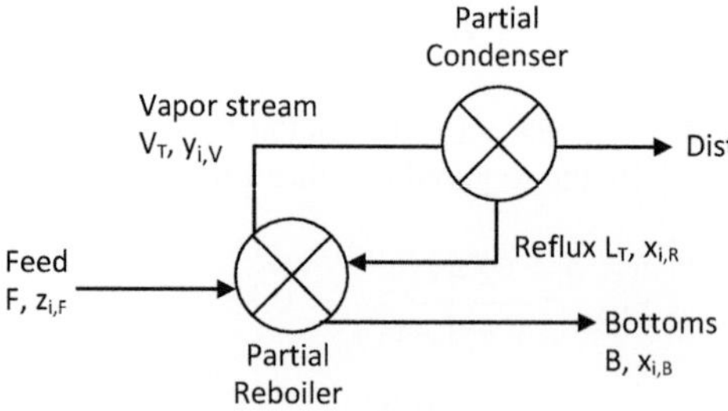


In this case, we have $x_{A,R}$ in equilibrium with $y_{A,D}$ as the partial condenser effectively becomes an equilibrium stage and this scenario is equivalent to part b.

$$y_{A,D} = 0.8 = \frac{2.4x_{A,R}}{1 + 1.4x_{A,R}}$$

$$x_{A,R} = 0.625$$

Performing a mass balance around the partial condenser, we have

$$L_T + D = V_T = D(R+1) = 3.3D$$

$$V_T y_{A,V} = L_T x_{A,R} + D y_{A,D}$$

$$y_{A,V} = \frac{L_T}{V_T}(0.625) + \frac{D}{V_T}(0.8) = \frac{2.3}{3.3}(0.625) + \frac{1}{3.3}(0.8) = 0.678$$

$$y_{A,V} = \frac{2.4x_{A,B}}{1 + 1.4x_{A,B}} = 0.678$$

$$x_{A,B} = 0.468$$

We can also specify the required *D*/*F* by performing an overall mass balance

$$F z_{A,F} = D y_{A,D} + B x_{A,B}$$

$$0.6 = \frac{D}{F}(0.8) + \left(1 - \frac{D}{F}\right)(0.468)$$

$$\frac{D}{F} = 0.40$$

Therefore, we have shown that it is possible to produce the required distillate at a flow rate of 40 mol per 100 mol of feed.

## Problem 13

**Consider a feed mixture comprising of three hydrocarbons 1, 2, and 3, which is flashed at 1.8 bar and 93 °C. The molar compositions and *K* values are shown below:**

| | Hydrocarbon 1 | Hydrocarbon 2 | Hydrocarbon 3 |
|---|---|---|---|
| $z_i$ | 0.25 | 0.55 | 0.20 |
| $K_i$ at 93 °C, 1.8 bar | 2.3 | 0.95 | 0.42 |
| $K_i$ at 25 °C, 1.8 bar | 0.32 | 0.10 | 0.028 |

(a) **Comment on whether 93 °C falls between the bubble and dew point temperatures of the feed at 1.8 bar. Calculate the fraction of feed that is vaporized, and the composition of the product streams.**
(b) **If a liquid of the same composition as the feed were stored under a nitrogen blanket at 25 °C and 1.8 bar, what would be the partial pressure of nitrogen and of the hydrocarbons in the vapor phase.**
(c) **If 0.8 kmol of that same feed as part a were mixed with 0.2 kmol of $H_2O$ at 1.8 bar, describe the method to find the bubble point of the resulting stream (actual calculations are not necessary) if you are provided with the steam table and K value nomogram for hydrocarbons?**
(d) **Further from part c, explain how the dew point temperature can be found, and how you would determine if the hydrocarbon phase or water phase condenses first.**

## Solution 13

*Worked Solution*

(a) At bubble point, the first bubble of vapor is formed from a saturated liquid. The composition of vapor in the bubble must satisfy

$$\sum_i y_i = 1$$

By definition, $K_i = y_i/x_i$. Since there is VLE and the liquid composition is known (and is not altered much by the first bubble), the bubble point condition becomes

$$\sum_i K_i x_i = 1$$

| | Hydrocarbon 1 | Hydrocarbon 2 | Hydrocarbon 3 | Sum |
|---|---|---|---|---|
| $z_i$ | 0.25 | 0.55 | 0.20 | |
| $K_i$ at 93 °C | 2.3 | 0.95 | 0.42 | |
| $K_i x_i$ (to find bubble pt, $z_i = x_i$) | 0.575 | 0.5225 | 0.084 | 1.18 > 1 |

The result of $\sum_i K_i x_i > 1$ means that the temperature of 93 °C is greater than the bubble point temperature. Since $K_i$ represents equilibrium for an endothermic (vaporization) process, therefore $K_i \sim T$. And it follows that a greater value of $K_i x_i$ than 1 (condition for bubble point temperature) means that the operating temperature of 93 °C is greater than the bubble point temperature.

At dew point, the first droplet of liquid is formed from a superheated vapor. The composition of liquid in the droplet must satisfy

$$\sum_i x_i = 1$$

Similarly, by definition, $K_i = y_i / x_i$, and since there is VLE with a known vapor composition (not altered much by the first droplet formation), the dew point condition is

$$\sum_i \frac{y_i}{K_i} = 1$$

| | Hydrocarbon 1 | Hydrocarbon 2 | Hydrocarbon 3 | Sum |
|---|---|---|---|---|
| $z_i$ | 0.25 | 0.55 | 0.20 | |
| $K_i$ at 93 °C | 2.3 | 0.95 | 0.42 | |
| $\frac{y_i}{K_i}$ (to find dew pt, $z_i = y_i$) | 0.109 | 0.579 | 0.476 | 1.16 > 1 |

The result of $\frac{y_i}{K_i} > 1$ means that the temperature of 93 °C is lower than the dew point temperature. Since $K_i$ represents equilibrium for an endothermic (vaporization) process, therefore $K_i \sim T$. And it follows that a greater value of $\frac{y_i}{K_i}$ than 1 (for dew point) means $K_i$ is smaller than is required for dew point, and hence the temperature of 93 °C is lower than the dew point temperature.

Let us consider our system in the simple diagram below.

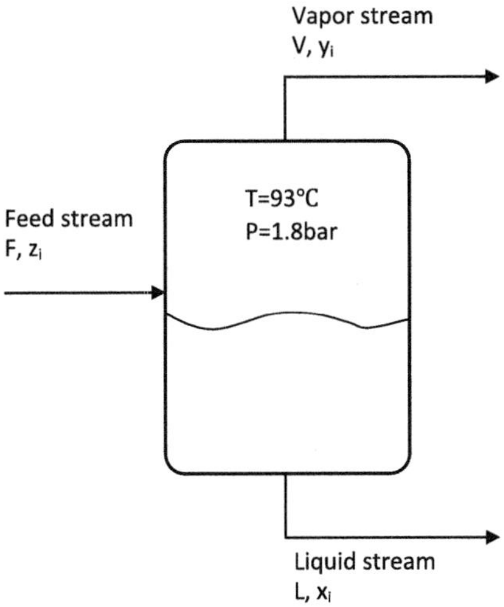


To find the fraction of feed that is evaporated, we can begin with mass balance equations.

$$F = V + L$$

$$Fz_i = Vy_i + Lx_i$$

$$Fz_i = V(K_i x_i) + (F - V)x_i = x_i(VK_i + F - V)$$

$$x_i = \frac{Fz_i}{VK_i + F - V} = \frac{z_i}{\left(\frac{V}{F}\right)K_i + 1 - \left(\frac{V}{F}\right)} = \frac{z_i}{\frac{V}{F}(K_i - 1) + 1}$$

We know that $\sum_i x_i = \sum_i y_i = 1$ is always true. Therefore,

$$\sum_i \frac{z_i}{\frac{V}{F}(K_i - 1) + 1} - \sum_i y_i = 0$$

$$\sum_i \frac{z_i}{\frac{V}{F}(K_i - 1) + 1} - \sum_i K_i x_i = \sum_i \frac{z_i}{\frac{V}{F}(K_i - 1) + 1} - \sum_i K_i \left(\frac{z_i}{\frac{V}{F}}(K_i - 1) + 1\right) = 0$$

$$\sum_i \frac{z_i(1 - K_i)}{\frac{V}{F}(K_i - 1) + 1} = 0$$

$$\frac{0.25(1 - 2.3)}{\frac{V}{F}(2.3 - 1) + 1} + \frac{0.55(1 - 0.95)}{\frac{V}{F}(0.95 - 1) + 1} + \frac{0.2(1 - 0.42)}{\frac{V}{F}(0.42 - 1) + 1} = 0$$

$$\frac{V}{F} = 0.52$$

Therefore the fraction of feed that is evaporated is 0.52 and the compositions of the product streams at 93 °C and 1.8 bar are as follows.

Liquid stream:

| | Hydrocarbon 1 | Hydrocarbon 2 | Hydrocarbon 3 |
|---|---|---|---|
| $z_i$ | 0.25 | 0.55 | 0.20 |
| $K_i$ | 2.3 | 0.95 | 0.42 |
| $x_i = \frac{z_i}{0.52(K_i-1)+1}$ | 0.15 | 0.56 | 0.29 |

Vapor stream:

| | Hydrocarbon 1 | Hydrocarbon 2 | Hydrocarbon 3 |
|---|---|---|---|
| $z_i$ | 0.25 | 0.55 | 0.20 |
| $K_i$ | 2.3 | 0.95 | 0.42 |
| $y_i = \frac{K_i z_i}{0.52(K_i-1)+1}$ | 0.34 | 0.54 | 0.12 |

(b) Nitrogen blankets are used as they are inert and also do not dissolve readily into the liquid phase of the mixture. Therefore, we can assume that nitrogen only features in the vapor composition and not the liquid composition.

$$P_{\text{total}} = P_1 + P_2 + P_3 + P_{N2} = P_{\text{HC}} + P_{N2}$$

We have the following liquid feed composition, which we can use to find the vapor molar composition $y_i = K_i x_i$.

| | Hydrocarbon 1 | Hydrocarbon 2 | Hydrocarbon 3 | Nitrogen |
|---|---|---|---|---|
| $x_i$ | 0.25 | 0.55 | 0.20 | 0 |
| $K_i$ at 25 °C, 1.8 bar | 0.32 | 0.10 | 0.028 | – |
| $y_i = K_i x_i$ | 0.08 | 0.055 | 0.0056 | 0.86 |

$$\sum_i y_i = 1$$

$$y_{N2} = 1 - 0.08 - 0.055 - 0.0056 = 0.86$$

$$P_{N2} = y_{N2} P_{\text{total}} = 0.86(1.8) = 1.55 \text{ bar}$$

$$P_{\text{HC}} = 0.14(1.8) = 0.25 \text{ bar}$$

(c) We assume that water and hydrocarbon phases do not mix. Hence the liquid composition we are considering here is for the hydrocarbon phase. To determine bubble point, we need the condition below to be met for the hydrocarbon phase at the bubble point temperature. $K_i$ values for hydrocarbons are defined at a specific temperature and pressure from a nomogram.

$$\sum_i K_i x_i^{\text{HC}} = \sum_i y_i^{\text{HC}} = 1$$

A nomogram is also called a DePriester chart which looks like the following. On the left-hand side, we have a vertical axis of pressure values, and on the right-hand side, we have a vertical axis of temperature values. The $K_i$ value of a hydrocarbon can be found by drawing a straight line connecting the relevant temperature and pressure values, and reading the $K_i$ value corresponding to the specific hydrocarbon of interest. The example below is shown for methane at $T = 60°F$, $P = 100$ psia.

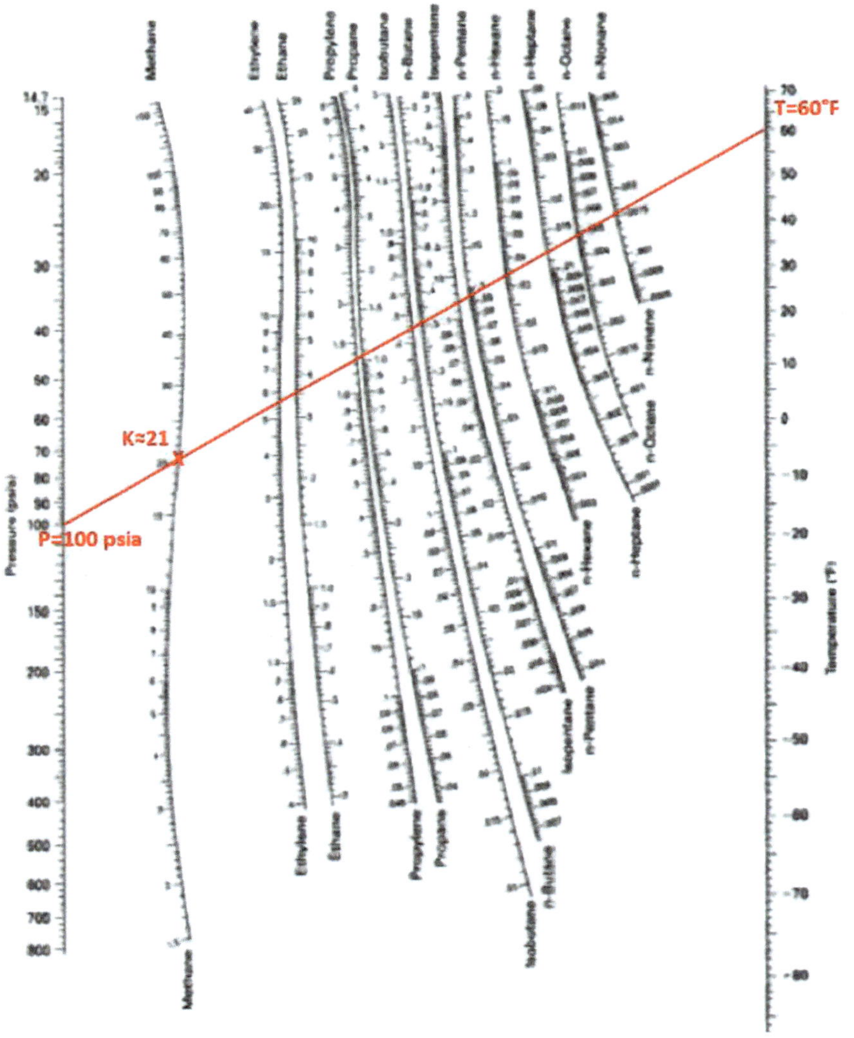


Also at the bubble point, the following pressure condition will be fulfilled whereby $P_{H_2O}$ is the partial pressure of water and $P_{HC}$ is the partial pressure of the hydrocarbons. Note that since the liquid water phase is immiscible with the hydrocarbon phase, it is separately exerting its full saturated vapor pressure. The saturated vapor pressure of water can be found from steam tables for a specified temperature.

$$P_{HC} + P_{H_2O} = P_{total} = 1.8 \text{ bar}$$

We know that $P = 1.8$ bar. Therefore, we need to follow the below steps:

1. Guess a bubble point temperature, $T_{guess}$.
2. From steam tables, determine the corresponding $P_{H_2O}$ at $T_{guess}$.
3. We can then determine $P_{HC}$ since $P_{HC} = 1.8$ bar – $P_{H_2O}$ (from step 2).
4. With knowledge of $P_{HC}$ and $T_{guess}$, we can find $K_i$ values for the hydrocarbons.
5. We know that the hydrocarbon phase feed has the same molar composition as in part a shown below. Therefore we can find $K_i x_i^{HC}$ using the $K_i$ values obtained from step 4. If the result of $\sum_i K_i x_i^{HC} > 1$, then we need to reduce $T_{guess}$ and vice versa. We iterate until we obtain the bubble point temperature when $\sum_i K_i x_i^{HC} = 1$

| | Hydrocarbon 1 | Hydrocarbon 2 | Hydrocarbon 3 |
|---|---|---|---|
| $x_i^{HC}$ | 0.25 | 0.55 | 0.20 |

6. We can then find the corresponding $P_{H_2O}$ at that temperature from steam tables. We can then obtain $P_{HC} = 1.8$ bar – $P_{H_2O}$. The partial pressures of each hydrocarbon component can be found using $y_i^{HC} P_{HC} = P_i^{HC}$, where $y_i^{HC} = K_i x_i^{HC}$.

   To find dew point, the condition to be met is as follows.

$$\sum_i \frac{y_i^{HC}}{K_i} = 1$$

(d) For dew point calculation, we similarly guess a temperature value and find $K_i$ at that temperature for the hydrocarbons. We then calculate the value of $\sum_i \frac{y_i^{HC}}{K_i}$ using the feed composition below, and iterate until we get the dew point temperature when $\sum_i \frac{y_i^{HC}}{K_i} = 1$.

| | Hydrocarbon 1 | Hydrocarbon 2 | Hydrocarbon 3 |
|---|---|---|---|
| $y_i^{HC}$ | 0.25 | 0.55 | 0.20 |

This time, since we start with a vapor phase as we go towards dew point, the vapor phase starts off with miscible vapors of water and hydrocarbons, and the vapor compositions and partial pressures will therefore be as follows:

$$y_i^{HC} = 0.8,\ y_i^{H_2O} = 0.2$$

$$P_{H_2O} = y_i^{H_2O} P_{total} = 0.2(1.8\text{ bar}) = 0.36\text{ bar}$$

$$P_{HC} = 1.8\text{ bar} - P_{H_2O} = 1.44\text{ bar}$$

If the hydrocarbon phase condenses first, it means that at the dew point temperature, the saturated vapor pressure of water (which can be found from steam tables) should be >0.36 bar.

## Problem 14

**A mixture of 70 kmol of hydrocarbon *A*, 5 kmol of water, and 10 kmol of nitrogen occupies a container of volume 90 m$^3$ at 130 °C. Find the number of phases present and determine the pressure in the container and the composition in each phase. List any other assumptions made.**

**You are given the saturated vapor pressure of hydrocarbon *A* at 130 °C = 5.2 bar, liquid density = 750 kg/m$^3$, and molecular weight = 84 kg/kmol.**

## Solution 14

### *Worked Solution*

First, we will assume that the vapor phase is an ideal gas, i.e., intermolecular interactions are negligible. Also, we may assume that nitrogen gas is immiscible in the liquid phase. The liquid phases include a liquid hydrocarbon phase and liquid water phase which are immiscible with each other.

To find the number of phases present, we can first guess that there is a single vapor phase (and no liquid phase), at a total system pressure denoted as $P_2$. For 1 mol of an ideal gas, we can assume that it occupies a volume of 24 dm$^3$ at room temperature and pressure of 298 K and 1 bar. This is equivalent to 24 m$^3$ for 1 kmol. Using the ideal gas relationship, we have

$$\frac{P_1 V_1}{T_1} = \frac{P_2 V_2}{T_2}$$

Therefore we can find the volume of 1 kmol of vapor at 130 °C (or 403.15 K),

$$V_2 = \frac{P_1 V_1 T_2}{P_2 T_1} = \frac{1(24)(403.15)}{P_2(298)} = \frac{32.5}{P_2}$$

If the total volume of the system is 90 m$^3$, then $V_2 = V_{\text{total}}$, we can find the total system pressure $P_2 = P_{\text{total}}$ as follows, where $n_T$ denotes the total number of kmols of gas.

$$\frac{32.5}{P_{\text{total}}}(n_T) = 90$$

$$P_{\text{total}} = \frac{32.5(70 + 5 + 10)}{90} = 30.7 \text{ bar}$$

If our assumption was correct, then we can calculate the partial pressures of the hydrocarbon and water phases. For water, we can find the saturated vapor pressure at 130 °C using the steam table.

$$P_{\text{H}_2\text{O}} = y_{\text{H}_2\text{O}} P_{\text{total}} = \frac{5}{85}(30.7) = 1.8 \text{ bar} < \text{saturated vap pr at } 2.7 \text{ bar}$$

This means that water still exists as a vapor, and there is no liquid phase for water.

$$P_{\text{HC}} = y_{\text{HC}} P_{\text{total}} = \frac{70}{85}(30.7) = 25.3 \text{ bar} \gg \text{saturated vap pr at } 5.2 \text{ bar}$$

This means that some of the hydrocarbon has already condensed into liquid, i.e., there is a vapor phase and a liquid phase for the hydrocarbon. It follows that we can now guess that we have 1 vapor phase (containing both water and hydrocarbon) and 1 liquid phase (containing hydrocarbon only). Therefore, the total number of kmols in the vapor phase is

$$n_T^{\text{vap}} = n_{\text{HC}}^{\text{vap}} + 5 + 10 = n_{\text{HC}}^{\text{vap}} + 15$$

Since the liquid phase is pure hydrocarbon, the partial pressure of hydrocarbon is also the saturated vapor pressure of hydrocarbon, 5.2 bar, at the same temperature of 130 °C.

$$P_{\text{HC}} = \frac{n_{\text{HC}}^{\text{vap}}}{n_{\text{HC}}^{\text{vap}} + 15} P_{\text{total}} = 5.2 \text{ bar}$$

$$P_{\text{total}} = \frac{5.2\left(n_{\text{HC}}^{\text{vap}} + 15\right)}{n_{\text{HC}}^{\text{vap}}} \quad (1)$$

$$V_{\text{total}} = V_{\text{vapor phase}} + V_{\text{liquid phase}} = \frac{32.5}{P_{\text{total}}}\left(n_T^{\text{vap}}\right) + \frac{\left(70 - n_{\text{HC}}^{\text{vap}}\right)(84)}{750}$$

$$90 = \frac{32.5}{P_{\text{total}}}\left(n_{\text{HC}}^{\text{vap}} + 15\right) + \frac{\left(70 - n_{\text{HC}}^{\text{vap}}\right)(84)}{750} \qquad (2)$$

We can solve the pair of simultaneous Eqs. (1) and (2), to find

$$P_{\text{total}} = 11.03 \text{ bar}$$

$$n_{\text{HC}}^{\text{vap}} = 13.4 \text{ kmol}$$

We can now check if our assumption of phases was correct.

$$P_{\text{H}_2\text{O}} = y_{\text{H}_2\text{O}} P_{\text{total}} = \frac{5}{13.4 + 15}(11.03) = 1.94 \text{ bar} < \text{saturated vap pr at 2.7 bar}$$

This means that water still exists as a vapor, and there is no liquid phase for water.

$$V_{\text{liquid phase}} = \frac{(70 - 13.4)(84)}{750} = 6.3 \text{ m}^3$$

The volume of liquid phase is 0 m$^3$ < 6.3 m$^3$ < 90 m$^3$, which is reasonable. The vapor phase compositions are therefore,

$$y_{\text{HC}} = \frac{13.4}{13.4 + 15} = 0.47, \ P_{\text{HC}} = 0.47(11.03) = 5.2 \text{ bar} = \text{saturated vap pr}$$

$$y_{\text{H}_2\text{O}} = \frac{5}{13.4 + 15} = 0.18$$

$$y_{\text{HC}} = \frac{10}{13.4 + 15} = 0.35$$

## Problem 15

**Consider 1 kmol of a hydrocarbon feed mixture with the following composition, which is flashed at 60 °C, 8 bar. Find the quantity and composition of the final liquid.**

| Component | $z_i$ | $K_i$ at 60 °C, 8 bar |
|---|---|---|
| 1 | 29.90 | 20.9 |
| 2 | 7.05 | 4.5 |
| 3 | 5.13 | 1.7 |
| 4 | 5.05 | 0.57 |
| 5 | 4.80 | 0.31 |
| 6 | 4.32 | 0.081 |
| 7 | 4.71 | 0.025 |
| 8 | 4.68 | 0.011 |
| Involatile bottoms | 34.36 | 0.000 |

## Solution 15

### *Worked Solution*

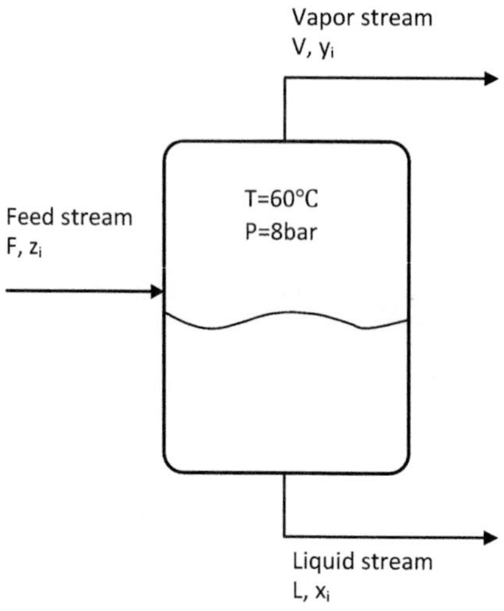


$$F = V + L$$

$$Fz_i = Vy_i + Lx_i$$

$$Fz_i = V(K_i x_i) + (F - V)x_i = x_i(VK_i + F - V)$$

$$x_i = \frac{Fz_i}{VK_i + F - V} = \frac{z_i}{\frac{V}{F}(K_i - 1) + 1}$$

We know that $\sum_i x_i = 1$ is always true. Therefore,

$$\sum_i \frac{z_i}{\frac{V}{F}(K_i - 1) + 1} = 1$$

We know the values of $z_i$ and $K_i$; therefore, we can start with an initial guess for $\frac{V}{F}$.

| Component | $z_i$ | $K_i$ at 60 °C, 8 bar | $z_i$ in mole fraction | $x_i = \frac{z_i}{\frac{V}{F}(K_i-1)+1}$ |
|---|---|---|---|---|
| 1 | 29.90 | 20.9 | 0.299 | 0.035 |
| 2 | 7.05 | 4.5 | 0.0705 | 0.030 |
| 3 | 5.13 | 1.7 | 0.0513 | 0.040 |
| 4 | 5.05 | 0.57 | 0.0505 | 0.060 |
| 5 | 4.80 | 0.31 | 0.048 | 0.065 |

(continued)

| Component | $z_i$ | $K_i$ at 60 °C, 8 bar | $z_i$ in mole fraction | $x_i = \frac{z_i}{\frac{V}{F}(K_i-1)+1}$ |
|---|---|---|---|---|
| 6 | 4.32 | 0.081 | 0.0432 | 0.070 |
| 7 | 4.71 | 0.025 | 0.0471 | 0.075 |
| 8 | 4.68 | 0.011 | 0.0468 | 0.075 |
| Involatile bottoms | 34.36 | 0.000 | 0.3436 | 0.55 |
| | | | | $\sum_i x_i$ when $\frac{V}{F}$ is 0.38 = 1.00 |

$$\frac{V}{F} = 0.38$$

Given that $F = 1$ kmol, $V = 0.38$ kmol and therefore the quantity of resulting liquid is as follows with the mole fractions $x_i$ as shown in the table above.

$$L = 1 - 0.38 = 0.62 \text{ kmol}$$

## Problem 16

**Hydrocarbon $A$ is to be separated from hydrocarbon $B$ by distillation, with a partial condenser at the top of the column. The separation requirement is that the vapor distillate should contain mole fraction of $A$, $y_{A,D} = 0.95$. The outlet temperature from the partial condenser is 45°C. The $K$ values for $A$ and $B$ at 45°C and various pressures are given below.**

| Pressure/bar | $K_A$ | $K_B$ |
|---|---|---|
| 10 | 1.33 | 0.25 |
| 11 | 1.21 | 0.23 |
| 12 | 1.10 | 0.20 |
| 13 | 1.015 | 0.185 |

(a) **Find the pressure at the top of the column and the composition of the liquid reflux. You may assume negligible pressure drop between the condenser and the top of the column.**

(b) **A liquid bottoms product from a distillation column has the following component flow rates and $K$ values at 710 kPa and varying temperatures. If the pressure at the bottom of the column is 710 kPa, find the temperature of the bottoms product.**

| Component | Flow rate [kmol/h] | $K_i$ at 710 kPa, 63 °C | $K_i$ at 710 kPa, 68 °C | $K_i$ at 710 kPa, 73 °C |
|---|---|---|---|---|
| 1 | 10 | 1.08 | 1.35 | 1.69 |
| 2 | 230 | 0.9 | 1.1 | 1.35 |
| 3 | 24 | 0.35 | 0.5 | 0.72 |
| 4 | 15 | 0.27 | 0.43 | 0.71 |

(c) **The stream in part b is subsequently flashed to 120 kPa and 15°C. Determine if the feed remains liquid? Describe briefly how you would find out the enthalpy required to be added or removed in this process if the identities of the components are known hydrocarbons.**

| Component | $K_i$ at 120 kPa, 15 °C |
|---|---|
| 1 | 2.05 |
| 2 | 1.54 |
| 3 | 0.48 |
| 4 | 0.38 |

## Solution 16

### *Worked Solution*

(a) Let us illustrate a simple diagram for our setup.

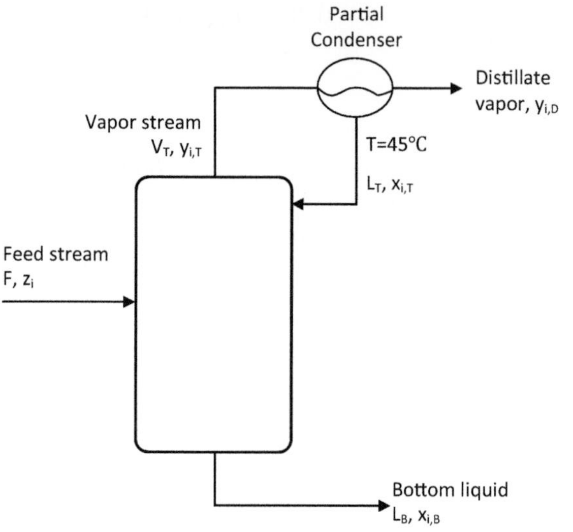


We note that the separation objective in the vapor distillate is $y_{A,D} = 0.95$, and hence $y_{B,D} = 0.05$. In the partial condenser, the temperature will be at dew point; therefore the following is satisfied. When $\sum_i \frac{y_i}{K_i} > 1$, we need to increase $K_i$ which will be at a lower pressure if temperature remains the same.

$$\sum_i x_i = \sum_i \frac{y_i}{K_i} = 1$$

| Component | Mole fraction $y_{i,D}$ | $K_i$ at 45 °C, 12 bar | $x_i = \frac{y_i}{K_i(12\text{ bar})}$ | $K_i$ at 45 °C, 11 bar | $x_i = \frac{y_i}{K_i(11\text{ bar})}$ |
|---|---|---|---|---|---|
| $A$ | 0.95 | 1.10 | $\frac{0.95}{1.1} = 0.864$ | 1.21 | 0.785 |
| $B$ | 0.05 | 0.20 | 0.250 | 0.23 | 0.217 |
| | | | $\sum_i x_i = 1.11$ | | $\sum_i x_i = 1.00$ |

Therefore the pressure at the top of the column is 11 bar, where the dew point condition is met in the condenser.

The composition of the liquid reflux is given by $x_i = \frac{y_i}{K_i(11\text{ bar})}$, and is therefore, $x_{A,T} = 0.785$ and $x_{B,T} = 0.217$ as calculated in the table above.

(b) We have a liquid bottoms product, which means that we have a partial reboiler at the bottom, since it has to generate vapor that feeds back into the column, and produce a liquid bottom stream that is removed as product.

In the partial reboiler, the mixture is at bubble point temperature and hence

$$\sum_i y_i = \sum_i K_i x_i = 1$$

We can find the mole fractions from the flow rates,

| Component | Flow rate [kmol/h] | Mole fraction $x_{i,B}$ |
|---|---|---|
| 1 | 10 | $\frac{10}{279} = 0.036$ |
| 2 | 230 | 0.824 |
| 3 | 24 | 0.086 |
| 4 | 15 | 0.054 |
| | $\sum_i L_{i,B} = 279$ | $\sum_i x_{i,B} = 1$ |

| Component | Mole fraction $x_{i,B}$ | $K_i$ at 63 °C | $y_i = K_i x_i$ at 63 °C | $K_i$ at 68 °C | $y_i = K_i x_i$ at 68 °C | $K_i$ at 73 °C | $y_i = K_i x_i$ at 73 °C |
|---|---|---|---|---|---|---|---|
| 1 | 0.036 | 1.08 | 0.04 | 1.35 | 0.05 | 1.69 | 0.06 |
| 2 | 0.824 | 0.9 | 0.74 | 1.1 | 0.91 | 1.35 | 1.11 |
| 3 | 0.086 | 0.35 | 0.03 | 0.5 | 0.04 | 0.72 | 0.06 |
| 4 | 0.054 | 0.27 | 0.02 | 0.43 | 0.02 | 0.71 | 0.04 |
| | | | $\sum_i y_i = 0.83$ | | $\sum_i y_i = 1.02$ | | $\sum_i y_i = 1.27$ |

We can see that the bubble point condition is met at a temperature of approximately 68 °C. Therefore the temperature of the bottoms product is 68 °C.

(c) The liquid bottom stream in part b becomes the feed to the flash unit for this part. Therefore the molar composition of the feed, $z_i$, is as shown below added to the table of $K$ values provided.

| Component | $K_i$ at 120 kPa, 15 °C | $z_i$ | $K_i z_i$ | $\frac{z_i}{K_i}$ |
|---|---|---|---|---|
| 1 | 2.05 | 0.036 | 0.07 | 0.02 |
| 2 | 1.54 | 0.824 | 1.27 | 0.54 |
| 3 | 0.48 | 0.086 | 0.04 | 0.18 |
| 4 | 0.38 | 0.054 | 0.02 | 0.14 |
| | | | $\sum_i K_i z_i = 1.4 > 1$ | $\sum_i \frac{z_i}{K_i} = 0.88 < 1$ |

Since $\sum_i K_i z_i > 1$, it means that the temperature of 15 °C at 120 kPa is above the bubble point. Therefore some of the feed is vaporized. In order to find out if all of the liquid is vaporized, we need to check against dew point.

For dew point, $\sum_i \frac{z_i}{K_i} = 1$. Since we have $\sum_i \frac{z_i}{K_i} < 1$, it means that our $K$ values are higher than if they were at dew point temperature. This means that the temperature of 15 °C is higher than the dew point temperature at 120 kPa. Therefore, we can conclude that all of the liquid in the feed is vaporized, and none remains as liquid.

The enthalpy required to be added to the flash unit, $\dot{q}$, is found from an energy balance around the unit, where $h_{\text{in}}$ denotes the enthalpy of the saturated liquid feed at 68 °C, $h_{\text{out}}$ denotes the enthalpy of the exit vapor stream at 15 °C and 120 kPa. Enthalpies can be found from data booklets for known hydrocarbons based on specified operating conditions such as temperature and pressure.

$$h_{\text{in}} - h_{\text{out}} + \dot{q} = 0$$

## Problem 17

**A hydrocarbon feed has the following composition.**

| Component | 1 | 2 | 3 | 4 | Involatiles |
|---|---|---|---|---|---|
| Mole fraction | 30 | 29 | 20 | 11 | 10 |

**A distillation column operating at 1 bar is used to separate the components, consisting of a total condenser and a partial reboiler. The separation objective is to have a distillate of $x_{3,D} = 0.02$ and bottoms product with $x_{2,B} = 0.006$.**

(a) **Determine the molar composition of the product streams.**

(b) **Check if the temperatures at the top plate and the reboiler are set appropriately at 25 °C and 89 °C, respectively. You are provided with the following $K$ values**

| Component | 1 | 2 | 3 | 4 | Involatiles |
|---|---|---|---|---|---|
| $K$ at 1 bar, 89 °C | – | 3.7 | 1.65 | 0.76 | – |
| $K$ at 1 bar, 25 °C | 2.3 | 0.70 | 0.2 | – | – |

(c) **By using the Fenske equation, find the number of theoretical plates needed for total reflux.**

## Solution 17

### *Worked Solution*

(a) Before we solve this problem, let us revisit some basic concepts in multicomponent distillation of volatiles.

We may encounter the terms, "Light Non-Key (LNK)," "Light Key (LK)," "Heavy Key (HK)," and "Heavy Non-Key (HNK) ," representing 4 components of adjacent volatilities with the most volatile to least volatile in the order LNK → LK → HK → HNK.

- The feed entry point separates the column into a top rectifying section and a bottom stripping section.
- LNK is the most volatile, and its concentration decreases down the column (pink line).
- HNK is the least volatile, and its concentration increases down the column (orange line).
- The intermediate species, such as HK and LK, experience maxima in their concentrations, where this maxima occurs at the top section for the LK (green line), and the bottom section for the HK (yellow line).

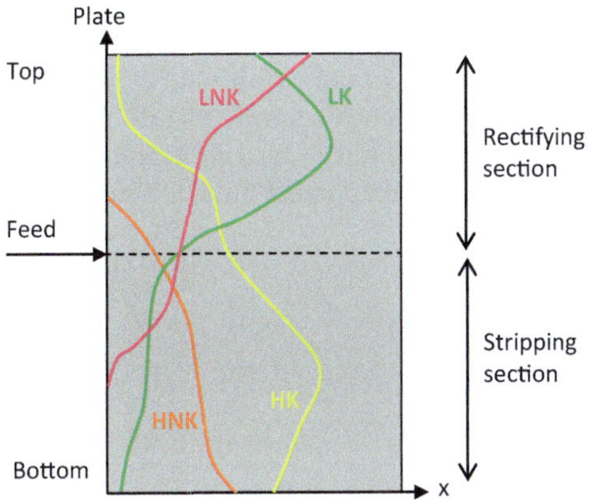


In this problem, we have component 2 as our LK, and component 3 as our HK. Assuming 100 kmol of feed we have the following, where $P$ and $Q$ are amounts of LK and HK in the bottoms and tops products, respectively:

| Component | 1 | 2 (LK) | 3 (HK) | 4 | Involatiles | Total/kmol |
|---|---|---|---|---|---|---|
| Mole fraction | 0.3 | 0.29 | 0.2 | 0.11 | 0.10 | 1 |
| Feed/kmol | 30 | 29 | 20 | 11 | 10 | 100 |
| Tops/kmol | 30 | $29 - P = \mathbf{28.76}$ | $Q = \mathbf{1.2}$ | 0 | 0 | $59 - P + Q = 59.96$ |
| Bottoms/kmol | 0 | $P = \mathbf{0.24}$ | $20 - Q = \mathbf{18.8}$ | 11 | 10 | $41 + P - Q = 40.04$ |

The required LK concentration in the bottoms product is

$$\frac{P}{41 + P - Q} = 0.006$$

The required HK concentration in the tops product is

$$\frac{Q}{59 - P + Q} = 0.02$$

Solving the simultaneous equations, we have $P = 0.24$ and $Q = 1.2$.

(b) The mole fractions in the tops product is

| Component | 1 | 2 (LK) | 3 (HK) | 4 | Involatiles | Total/ kmol |
|---|---|---|---|---|---|---|
| Mole fraction in tops, $x_{\text{tops}}$ | 30/ 59.96 = 0.500 | 0.48 | 0.02 | 0 | 0 | 1 |
| Mole fraction in bottoms, $x_{\text{bottoms}}$ | 0 | 0.006 | 0.469 | 0.275 | 0.25 | 1 |

Since we have a total condenser, $x_{\text{tops}} = y_T$, where $y_T$ is the vapor mole fraction exiting the top plate. To find out if the top plate temperature should be 25 °C, we need to verify if it coincides with the dew point temperature. For dew point, the following condition holds. From the calculations in the table below, it is shown that the dew point condition is fulfilled, hence the top plate temperature is correctly set at 25 °C.

$$\sum_i \frac{y_{T,i}}{K_i} = 1$$

| Component | 1 | 2 (LK) | 3 (HK) | 4 | Involatiles | |
|---|---|---|---|---|---|---|
| Mole fraction in tops, $y_T$ | 0.500 | 0.48 | 0.02 | 0 | 0 | |
| $K$ at 1 bar, 25 °C | 2.3 | 0.70 | 0.2 | – | – | |
| $\frac{y_{T,i}}{K_i}$ | 0.5/2.3 = 0.22 | 0.69 | 0.1 | – | – | $\sum_i \frac{y_{T,i}}{K_i} \approx 1$ |

Similarly, for the reboiler, we need to check if the temperature of 89 °C is the bubble point temperature for the liquid bottoms composition. For bubble point, the following condition holds; therefore, the temperature of the reboiler is correctly set at 89 °C.

$$\sum_i K_i x_i = 1$$

| Component | 1 | 2 (LK) | 3 (HK) | 4 | Involatiles | |
|---|---|---|---|---|---|---|
| Mole fraction in Bottoms, $x_B$ | 0 | 0.006 | 0.469 | 0.275 | 0.25 | |
| $K$ at 1 bar, 89 °C | – | 3.7 | 1.65 | 0.76 | – | |
| $K_ix_i$ | – | 3.7 (0.006) = 0.022 | 0.77 | 0.21 | – | $\sum_i K_i x_i \approx 1$ |

(c) The Fenske equation has the general form as follows for a binary mixture, where $N$ denotes the number of theoretical plates where the reboiler is considered a plate, $\alpha$ is the relative volatility of the more volatile component to less volatile component, and $x_D$ and $x_B$ are the mole fractions of the more volatile component in the tops and bottoms product streams, respectively.

$$N = \frac{\log\left[\left(\frac{x_D}{1-x_D}\right)\left(\frac{1-x_B}{x_B}\right)\right]}{\log\alpha}$$

When we have a multicomponent mixture, we can express the Fenske equation in the form below, where the more volatile and less volatile components are the LK and HK, respectively. When we have total reflux, $N$ is the minimum, i.e., $N_m$.

$$N = \frac{\log\left[\left(\frac{x_{\mathrm{LK}}}{x_{\mathrm{HK}}}\right)_D\left(\frac{x_{\mathrm{HK}}}{x_{\mathrm{LK}}}\right)_B\right]}{\log\alpha_{\mathrm{avg}}}$$

When $\alpha$ is not a constant across the height of column, a mean value can be used whereby $\alpha_{\mathrm{avg}}$ is averaged across the height of column using $\alpha_{\mathrm{avg}} = \sqrt{(\alpha_{\mathrm{LK-HK},D})(\alpha_{\mathrm{LK-HK},B})}$.

$$\alpha_{\mathrm{LK-HK}} = \frac{\left(\frac{y_{\mathrm{LK}}}{x_{\mathrm{LK}}}\right)}{\left(\frac{y_{\mathrm{HK}}}{x_{\mathrm{HK}}}\right)} = \frac{K_{\mathrm{LK}}}{K_{\mathrm{HK}}}$$

At the top of the column at 25 °C,

$$\alpha_{\mathrm{LK-HK},D} = \left.\frac{K_{\mathrm{LK}}}{K_{\mathrm{HK}}}\right|_{25^\circ\mathrm{C}} = \frac{0.7}{0.2} = 3.5$$

At the bottom of the column at 89 °C

$$\alpha_{\mathrm{LK-HK},D} = \left.\frac{K_{\mathrm{LK}}}{K_{\mathrm{HK}}}\right|_{89^\circ\mathrm{C}} = \frac{3.7}{1.65} = 2.24$$

Therefore, the average relative volatility can be found as follows.

$$\alpha_{avg} = \sqrt{(\alpha_{LK-HK,D})(\alpha_{LK-HK,B})} = \sqrt{(3.5)(2.24)} = 2.8$$

Under total reflux, the minimum number of theoretical plates is 6, since the value of $N_m$ includes the partial reboiler as one of the stages.

$$N_m = \frac{\log\left[\left(\frac{0.48}{0.02}\right)\left(\frac{0.469}{0.006}\right)\right]}{\log 2.8}$$

$$2.8^{N_m} = \left(\frac{0.48}{0.02}\right)\left(\frac{0.469}{0.006}\right) = 1876$$

$$N_m = \frac{\log 1876}{\log 2.8} = 7.3 \approx 7$$

## Problem 18

**The Fenske's equation is given as follows. $\alpha_{ij}$ is the relative volatility of component $i$ to component $j$.**

$$N_m = \frac{\log\left[\frac{\left(\frac{x_{i,D}}{x_{j,D}}\right)}{\left(\frac{x_{i,B}}{x_{j,B}}\right)}\right]}{\log \alpha_{ij}}$$

**Consider a liquid feed at its bubble point into a distillation column with the following composition.**

| Component | 1 | 2 | 3 | 4 | 5 | 6 |
|---|---|---|---|---|---|---|
| Mole fraction | 0.14 | 0.11 | 0.22 | 0.28 | 0.06 | 0.19 |
| Relative volatility $\alpha_{i4}$ | 3.21 | 2.20 | 1.83 | 1.00 | 0.612 | 0.399 |

(a) **If we require the recovery of component 3 in the distillate to be 98.5% of its amount in the feed and the recovery of component 4 in the bottoms stream to be 97.5% of its amount in the feed, what is the minimum number of equilibrium stages required to achieve these targets.**
(b) **Find the compositions of the distillate and bottoms streams assuming that the light non-keys end up in the distillate stream only, and the heavy non-keys end up in the bottoms stream only. Determine also, the distribution of the non-keys.**
(c) **Given Underwood's equations as follows, determine the minimum reflux ratio $R_m$ and value of $\theta$.**

$$\sum_i \frac{\alpha_i x_{iF}}{\alpha_i - \theta} = 1 - q$$

$$\sum_i \frac{\alpha_i x_{iD}}{\alpha_i - \theta} = 1 + R_m$$

(d) **If the actual reflux ratio $R = 1.2R_m$, find $N$ using an appropriate equation from Gilliland's correlation as shown below. Comment on other methods to determine $N$ using this correlation.**

**For** $\left(\frac{R-R_m}{R+1}\right) \in [0, 0.01]$,

$$\frac{N - N_m}{N+1} = 1.0 - 18.57\left(\frac{R - R_m}{R+1}\right)$$

**For** $\left(\frac{R-R_m}{R+1}\right) \in [0.01, 0.9]$,

$$\frac{N - N_m}{N+1} = 0.5458 - 0.5914\left(\frac{R - R_m}{R+1}\right) + \frac{0.002743}{\left(\frac{R-R_m}{R+1}\right)}$$

**For** $\left(\frac{R-R_m}{R+1}\right) \in [0.9, 1]$,

$$\frac{N - N_m}{N+1} = 0.1659 - 0.1659\left(\frac{R - R_m}{R+1}\right)$$

## Solution 18

***Worked Solution***

(a) Let us assume that we have 100 kmol of feed for easy computation. Applying the below principles, we are able to derive the values in bold in the table below.

- From the separation objectives, we can define component 3 as the Light Key (LK) and component 4 as the Heavy Key (HK). The implication of this is that:

  *Distillate*

  - There will be none of components 5 and 6 (i.e., heavier than HK) in the distillate stream.
  - Components 1 and 2 (i.e., lighter than LK) will have no change in amount from the feed.

  *Bottoms*

  - There will be none of components 1 and 2 (i.e., lighter than LK) in the bottoms stream.

  - Components 5 and 6 (i.e., heavier than HK) will have no change in amount from the feed.

- The requirement of $x_{3,D} = 0.985$ and $x_{4,B} = 0.975$ are used to compute $n_i$ in the distillate and bottoms streams, respectively.
- Finally for the same component, we note that mass balance has to apply; therefore $n_{i,\text{feed}} = n_{i,D} + n_{i,B}$.

| Component $i$ | 1 | 2 | 3 (LK) | 4 (HK) | 5 | 6 | Sum |
|---|---|---|---|---|---|---|---|
| Mole fraction in feed, $z_i$ | 0.14 | 0.11 | 0.22 | 0.28 | 0.06 | 0.19 | $\sum_i z_i=1$ |
| **Amount in feed, $n_{i,\text{feed}}$/ kmol** | **0.14(100) = 14** | **11** | **22** | **28** | **6** | **19** | $\sum_i n_{i,\text{feed}}=100$ |
| Relative volatility, $\alpha_{i4}$ | 3.21 | 2.20 | 1.83 | 1.0 | 0.612 | 0.399 | |
| **Amount in distillate, $n_{i,D}$/ kmol** | **14** | **11** | **0.985 (22) = 21.7** | **28 – 27.3 = 0.7** | **0** | **0** | $\sum_i n_{i,D} \cong 47.4$ |
| **Amount in bottoms, $n_{i,B}$/ kmol** | **0** | **0** | **22 – 21.7 = 0.3** | **0.975(28) = 27.3** | **6** | **19** | $\sum_i n_{i,B} \cong 52.6$ |
| **Mole fraction in distillate, $x_{i,D}$** | **14/47.4 = 0.30** | **0.23** | **0.46** | **0.015** | **0** | **0** | $\sum_i x_{i,D} \cong 1$ |
| **Mole fraction in bottoms, $x_{i,B}$** | **0** | **0** | **0.3/52.6 = 4.0057** | **0.52** | **0.11** | **0.36** | $\sum_i x_{i,B} \cong 1$ |

We are given the following Fenske's equation, which we can apply to components 3 and 4, i.e., the LK and HK.

$$N_m = \frac{\log\left[\dfrac{\left(\dfrac{x_{i,D}}{x_{j,D}}\right)}{\left(\dfrac{x_{i,B}}{x_{j,B}}\right)}\right]}{\log\alpha_{ij}} = \frac{\log\left[\dfrac{x_{3,D}x_{4,B}}{x_{3,B}x_{4,D}}\right]}{\log\alpha_{34}}$$

We are told that recovery of component 3 in the distillate is 98.5% of its amount in the feed.

$$0.985 = \frac{Dx_{3,D}}{Fz_3} \qquad (1)$$

$$1 - 0.985 = \frac{Fz_3 - Dx_{3,D}}{Fz_3} = \frac{Bx_{3,B}}{Fz_3} \tag{2}$$

Equation (2) divided by Eq. (1) gives:

$$\frac{Dx_{3,D}}{Bx_{3,B}} = \frac{0.985}{1 - 0.985} \tag{3}$$

Similarly, we are told that the recovery of component 4 in the bottoms stream is 97.5% of its amount in the feed.

$$0.975 = \frac{Bx_{4,B}}{Fz_4}$$

$$1 - 0.975 = \frac{Fz_4 - Bx_{4,B}}{Fz_4} = \frac{Dx_{4,D}}{Fz_4}$$

$$\frac{Bx_{4,B}}{Dx_{4,D}} = \frac{0.975}{1 - 0.975} \tag{4}$$

The product of Eqs. (3) and (4) gives

$$\left(\frac{Dx_{3,D}}{Bx_{3,B}}\right)\left(\frac{Bx_{4,B}}{Dx_{4,D}}\right) = \left(\frac{0.985}{1 - 0.985}\right)\left(\frac{0.975}{1 - 0.975}\right)$$

$$N_m = \frac{\log\left[\frac{x_{3,D}x_{4,B}}{x_{3,B}x_{4,D}}\right]}{\log\alpha_{34}} = \frac{\log\left[\left(\frac{0.985}{1-0.985}\right)\left(\frac{0.975}{1-0.975}\right)\right]}{\log 1.83}$$

$$N_m = 12.99 \approx 13$$

Therefore, the minimum number of equilibrium stages required is 13.

(b) Before we solve this problem, let us revisit some related concepts in multicomponent distillation of volatiles with non-keys.

Non-key profiles are characterized by long sections where mole fraction $x$ does not change much. Assume the case when we have an intermediate non-key component between the LK (green line) and HK (yellow line) as shown in the diagram below, its destination in the distillate or bottoms is unimportant. In section AB near the bottom of column, separation is between NK and HK, and the mole fraction of LK does not change much. In the remaining section BC, both LK/NK and LK/HK separation occurs. The mole fraction of NK can reach an appreciable value (purple line).

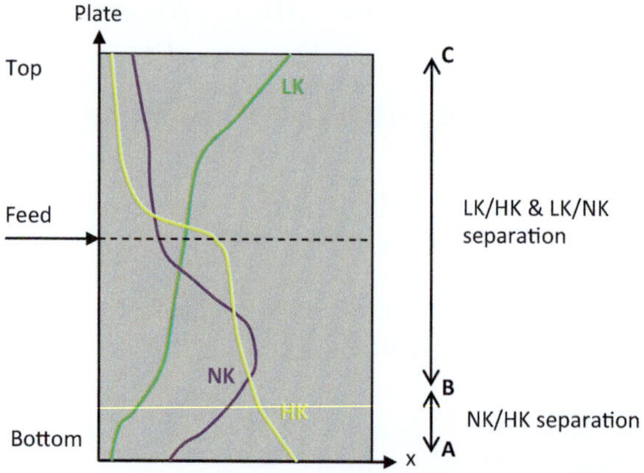


In our problem, we have a heavy non-key (less volatile than HK) and a light non-key (more volatile than LK). We can modify our Fenske equation, such that $\beta$ represents the recovery fraction of component $i$ in the distillate relative to its amount in the feed.

$$N_m = \frac{\log\left[\left(\frac{\beta_i}{1-\beta_i}\right)\left(\frac{0.975}{1-0.975}\right)\right]}{\log\alpha_{i4}} = 12.99 \cong 13$$

$$\beta_i = \frac{Dx_{i,D}}{Fz_i} = \frac{n_{i,D}}{n_{i,\text{feed}}}$$

$$\alpha_{i4}{}^{13} = \left(\frac{\beta_i}{1-\beta_i}\right)\left(\frac{0.975}{1-0.975}\right) = \left(\frac{\beta_i}{1-\beta_i}\right)(39)$$

$$\beta_i = \frac{\alpha_{i4}{}^{13}}{39+\alpha_{i4}{}^{13}}$$

We can determine the values of bold, as shown in the table below. The compositions of the distillate and bottoms streams, and the distribution of non-keys are evaluated. It can also be verified that the light non-keys end up mainly in the distillate and heavy non-keys end up mainly in the bottoms stream.

| Component $i$ | 1 (LNK) | 2 (LNK) | 3 (LK) | 4 (HK) | 5 (HNK) | 6 (HNK) | Sum |
|---|---|---|---|---|---|---|---|
| Relative volatility, $\alpha_{i4}$ | 3.21 | 2.20 | 1.83 | 1.0 | 0.612 | 0.399 | |
| $\beta_i$ | **$3.21^{13}/(39+3.21^{13}) =$ 1.00** | **0.999** | **0.985** | **0.025** | **4.33E-05** | **1.67E-07** | |
| Amount in feed, $n_{i,\text{feed}}$/kmol | 14 | 11 | 22 | 28 | 6 | 19 | $\sum_i n_{i,\text{feed}} = 100$ |

(continued)

| Component $i$ | 1 (LNK) | 2 (LNK) | 3 (LK) | 4 (HK) | 5 (HNK) | 6 (HNK) | Sum |
|---|---|---|---|---|---|---|---|
| **Amount in distillate, $n_{i,D}$/ kmol** | **1.00(14) = 14.0** | **11.0** | **21.7** | **0.7** | **2.60E-04** | **3.17E-06** | $\sum_i n_{i,D} \cong 47.4$ |
| **Amount in bottoms, $n_{i,B}$/ kmol** | **14 – 14.0 = 0.0** | **0.0** | **0.3** | **27.3** | **6.00** | **19.0** | $\sum_i n_{i,B} \cong 52.6$ |
| **Mole fraction in distillate, $x_{i,D}$** | **14.0/47.4 = 0.295** | **0.232** | **0.458** | **0.0148** | **5.49E-06** | **6.69E-08** | $\sum_i x_{i,D} \cong 1$ |
| **Mole fraction in bottoms, $x_{i,B}$** | **0.0/52.6 = 0** | **0.0** | **0.00570** | **0.519** | **0.114** | **0.361** | $\sum_i x_{i,B} \cong 1$ |

(c) The Underwood's equations can be obtained from the data booklet. They are also provided in our problem as shown:

$$\sum_i \frac{\alpha_i x_{iF}}{\alpha_i - \theta} = 1 - q$$

In distillation, $q$ denotes the fraction of liquid phase in the feed. Conversely, the fraction of vapor phase in the feed is $1 - q$. In our problem, we have a feed at bubble point; therefore $q = 1$.

$$\sum_i \frac{\alpha_i x_{iF}}{\alpha_i - \theta} = 0$$

$$\frac{\alpha_{14}x_{1,F}}{\alpha_{14} - \theta} + \frac{\alpha_{24}x_{2,F}}{\alpha_{24} - \theta} + \frac{\alpha_{34}x_{3,F}}{\alpha_{34} - \theta} + \frac{\alpha_{44}x_{4,F}}{\alpha_{44} - \theta} + \frac{\alpha_{54}x_{5,F}}{\alpha_{54} - \theta} + \frac{\alpha_{64}x_{6,F}}{\alpha_{64} - \theta} = 0$$

$$\frac{3.21(0.14)}{3.21 - \theta} + \frac{2.20(0.11)}{2.20 - \theta} + \frac{1.83(0.22)}{1.83 - \theta} + \frac{1.0(0.28)}{1.0 - \theta} + \frac{0.612(0.06)}{0.612 - \theta} + \frac{0.399(0.19)}{0.399 - \theta} = 0$$

$$\theta = 1.265$$

We can substitute this value of $\theta$ into the equation below to find $R_m$

$$\sum_i \frac{\alpha_i x_{iD}}{\alpha_i - \theta} = 1 + R_m$$

$$\frac{\alpha_{14}x_{1,D}}{\alpha_{14} - \theta} + \frac{\alpha_{24}x_{2,D}}{\alpha_{24} - \theta} + \frac{\alpha_{34}x_{3,D}}{\alpha_{34} - \theta} + \frac{\alpha_{44}x_{4,D}}{\alpha_{44} - \theta} + \frac{\alpha_{54}x_{5,D}}{\alpha_{54} - \theta} + \frac{\alpha_{64}x_{6,D}}{\alpha_{64} - \theta} = 1 + R_m$$

$$\frac{3.21(0.296)}{3.21-1.265}+\frac{2.20(0.232)}{2.20-1.265}+\frac{1.83(0.458)}{1.83-1.265}+\frac{1.0(0.0148)}{1.0-1.265}+\frac{0.612(5.52E-6)}{0.612-1.265}$$
$$+\frac{0.399(6.74E-8)}{0.399-1.265}=1+R_m$$

$$R_m = 1.459$$

(d) Gilliland established a method that empirically correlates the number of stages $N$ at a finite reflux ratio $R$ to the minimum number of stages $N_m$ (which occurs at total reflux) and the minimum reflux ratio $R_m$ (which occurs at infinite number of stages). To develop this correlation, a series of empirical calculations were made and a correlation was found between the function $\frac{N-N_m}{N+1}$ and $\frac{R-R_m}{R+1}$. The correlation can be presented in the form of equations as given in the problem, and the equation to use depends on the value of $\frac{R-R_m}{R+1}$.

We often use Gilliland's correlation to find out the number of stages, $N$, using values previously calculated from the Fenske and Underwood equations (i.e., $N_m$, $R$, and $R_m$). Note that the Gilliland correlation is empirical and useful for rough estimates, but does not provide exact solutions.

We can determine the specific Gilliland correlation equation to use by calculating the value as follows.

$$\frac{R-R_m}{R+1}=\frac{1.2(1.459)-1.459}{1.2(1.459)+1}\approx 0.1061$$

Therefore we use the following equation to determine $N$.

$$\frac{N-N_m}{N+1}=0.5458-0.5914\left(\frac{R-R_m}{R+1}\right)+\frac{0.002743}{\left(\frac{R-R_m}{R+1}\right)}$$

Given that the actual reflux ratio $R = 1.2R_m$, we compute the value of $N$ as follows

$$\frac{N-12.99}{N+1}=0.5458-0.5914(0.1061)+\frac{0.002743}{0.1061}$$

$$N \approx 28$$

The Gilliland correlation can also be presented in a plot where the value of $N$ can be read off as shown below.

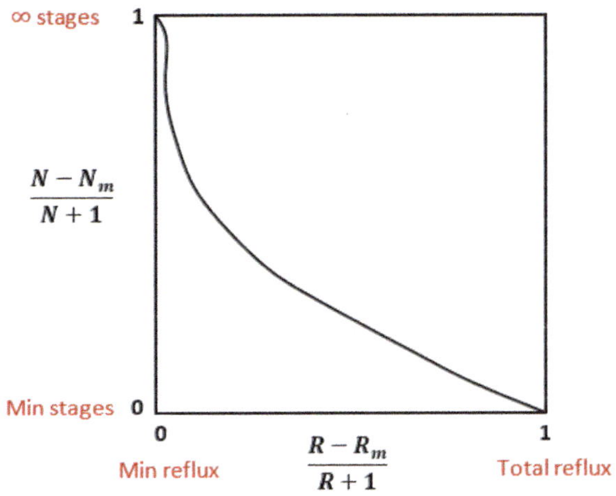


**Problem 19**

**In multicomponent distillation comprising more than 3 components, it is often required to use rigorous programming methods to perform rating calculations that help determine specific compositions of distillate and bottom streams. This is necessary for production plants to ensure that the distillation process meets product specification requirements. These rigorous programming methods require data inputs such as operating conditions (e.g., flow rates of inlet and outlet streams) as well as initial estimates for key distillation parameters (e.g., number of stages, reflux ratio). The programming method then performs subsequent iterations using this initial data set, to converge to specific output values of product stream compositions.**

**Shortcut methods have been developed to find estimates for these data inputs, and they include the combined use of the Fenske's equation, Underwood's equations, and Gilliland's correlation.**

(a) **Discuss the use of the alternative form of the Fenske's equation given below for multicomponent distillation calculations, where $N_m$ is the minimum number of stages at total reflux, $\alpha_{ij}$ is the relative volatility of a certain component $i$ to another component $j$. $(\alpha_{ij})_{\text{geom avg}}$ is the geometric average for $\alpha_{ij}$ and $x_{iD}$ and $x_{iB}$ refer to the mole fractions of component $i$ in the distillate and bottoms streams, respectively (same for component $j$). Comment also on the choice of components $i$ and $j$, as well as the use of a geometric average for $\alpha_{ij}$.**

$$N_m = \frac{\ln\left[\frac{x_{iD}/x_{jD}}{x_{iB}/x_{jB}}\right]}{\ln\left(\alpha_{ij}\right)_{\text{geom avg}}}$$

(b) **A saturated liquid feed with a flow rate of 100 kmol/h and the following composition enters a column operated at 1 atm. Using the Fenske's equation from part a, find the minimum number of stages to achieve a light key composition in the bottom stream $x_{iB} = 0.01$ and a heavy key composition in the distillate stream $x_{jD} = 0.015$. You may assume that $(\alpha_{ij})_{\text{geom avg}} = 2.4$.**

| Component | $i$ | $j$ | $k$ |
|---|---|---|---|
| Feed mole fraction | 0.4 | 0.3 | 0.3 |
| Number of carbon atoms in molecule | 6 | 7 | 8 |

(c) **Underwood's equations are useful in estimating the minimum reflux ratio $R_m$ for multicomponent distillations. Explain the terms used in Underwood's First and Second equations shown below and state any assumptions made when using these equations.**

$$\sum_i \frac{\alpha_i z_i}{\alpha_i - \theta} = 1 - q \qquad (1)$$

$$\sum_i \frac{\alpha_i x_{iD}}{\alpha_i - \theta} = 1 + R_m \qquad (2)$$

(d) **Using the distillation example in part b, and assuming that $(\alpha_{kj})_{\text{geom avg}} = 0.3$, find the minimum reflux ratio $R_m$ using Underwood's equations.**

(e) **Following from the example in part b, if the operating reflux ratio is 1.2 times of the minimum $R_m$, explain how Gilliland's empirical correlation in the graphical form as shown below (taken from "Separation Process Principles—Chemical and Biochemical Operations" by Seader, Henley and Roper) can be used to estimate the total number of equilibrium stages $N$ needed for the separation example in part b.**

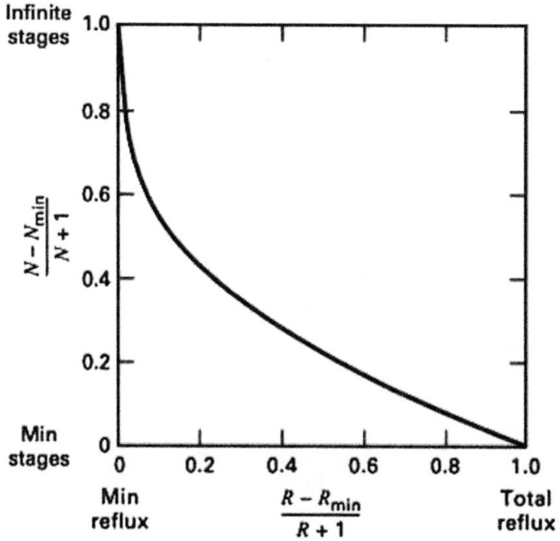

## Solution 19

### *Worked Solution*

(a) Shortcut methods have been developed to find estimates for the number of stages $N$ required for specified separation objectives. A common method is the combined use of the Fenske equation, Underwood's equations, and Gilliland's empirical correlation.

The Fenske equation is used in binary distillations to estimate the minimum number of stages at total reflux. An alternative form of the equation was also developed for multicomponent distillations to obtain estimates for the minimum number of stages at total reflux $N_m$ by defining a most volatile component as the "Light Key (LK)" and a least volatile component as the "Heavy Key (HK)".

$$N_m = \frac{\ln\left[\frac{x_{iD}/x_{jD}}{x_{iB}/x_{jB}}\right]}{\ln\left(\alpha_{ij}\right)_{\text{geom avg}}}$$

In the equation, the LK is component $i$ and the HK is component $j$. The choice of the LK and HK is usually done by identifying two components whose compositions in the outlet streams are good indicators of the effectiveness of the separation process, i.e., the separation achieves a good split between the two keys. These two key components are typically adjacent in order of volatility. The more volatile LK ends up mostly in the distillate stream, while the less volatile HK ends up mostly in the bottom stream. Components lighter than the LK are referred to as Light Light Keys (LLK) or Light Non-Keys (LNK), and end up almost exclusively in the distillate stream. Components heavier than the HK (i.e., Heavy Heavy Keys (HHK), or Heavy Non-Keys (HNK)) end up almost exclusively in the bottom stream.

$$\alpha_{ij} = K_i/K_j$$

$$\left(\alpha_{ij}\right)_{\text{geom avg}} = \sqrt{\alpha_{ijD}\alpha_{ijB}}$$

When we use the Fenske's equation, we assume a constant value for $\alpha_{ij}$. However, we note that volatilities are not constant along the column since temperature varies. Pressure also changes slightly along the column but this is a minor variation unless the column operates under significant vacuum. The use of a geometric average is an attempt to account for this temperature effect, although not fully, by taking the average value for $\alpha_{ij}$ using $\alpha_{ijD}$ where temperature $T_{\text{top}}$ occurs at the top end of the column (total condenser) and $\alpha_{ijB}$ where temperature $T_{\text{bottom}}$ occurs at the bottom end (partial reboiler) of the column.

For a known component, if we define a temperature and pressure, we can find the corresponding Vapor–Liquid equilibrium constant $K$ using diagrams such as the DePriester Chart (for simple hydrocarbons).

We can find $T_{top}$ (through a series of iterative guesses) such that the dew point condition is met at the total condenser. We do so by first assuming a guess value for $T_{top}$. Using this initial guess and the known operating pressure $P$, we can find $K$ from a data chart. Then, using known values for the distillate composition $y_{iD}$, we check if the condition is met. If the condition is not met, we adjust our guess value for $T_{top}$ until the condition is satisfied.

Dew Point condition at temperature $T_{top}$ and pressure $P$:

$$\sum_i x_{iD} = \sum_i \frac{y_{iD}}{K_{iD}} = 1$$

Note that since the top end is almost pure in the lighter component, it is reasonable to assume that temperature also remains fairly constant at $T_{top}$.

This method provides us with the values of $K_i$ and $K_j$ for the light and heavy keys, respectively, at the top of the column, which we can use to find $(\alpha_{ij})_{top}$.

$$(\alpha_{ij})_{top} = \left.\frac{K_i}{K_j}\right|_{T_{top}}$$

Similarly at the bottom of the column, we can find $T_{bottom}$ through a series of iterations such that the bubble point condition is met at the partial reboiler. Again, we first guess a value for $T_{bottom}$ at the known operating pressure $P$ to find the corresponding $K$ from a data chart. Using known values for the bottoms composition $x_{iB}$, we check if the bubble point condition is met. If the condition is not met, we adjust our guess value for $T_{bottom}$ until the condition is satisfied.

Bubble Point condition at temperature $T_{bottom}$ and pressure $P$:

$$\sum_i y_{iB} = \sum_i K_{iB} x_{iB} = 1$$

This gives us with the values of $K_i$ and $K_j$ for the light and heavy keys, respectively, at the bottom of the column, which we can use to find $(\alpha_{ij})_{bottom}$.

$$(\alpha_{ij})_{bottom} = \left.\frac{K_i}{K_j}\right|_{T_{bottom}}$$

With these two values $(\alpha_{ij})_{top}$ and $(\alpha_{ij})_{bottom}$, we can find a geometric average for $\alpha_{ij}$ as an attempt to account for the temperature variation along the column.

$$(\alpha_{ij})_{geom\ avg} = \sqrt{\alpha_{ijD}\alpha_{ijB}} = \sqrt{(\alpha_{ij})_{top}(\alpha_{ij})_{bottom}}$$

(b) We can illustrate our setup in the diagram below.

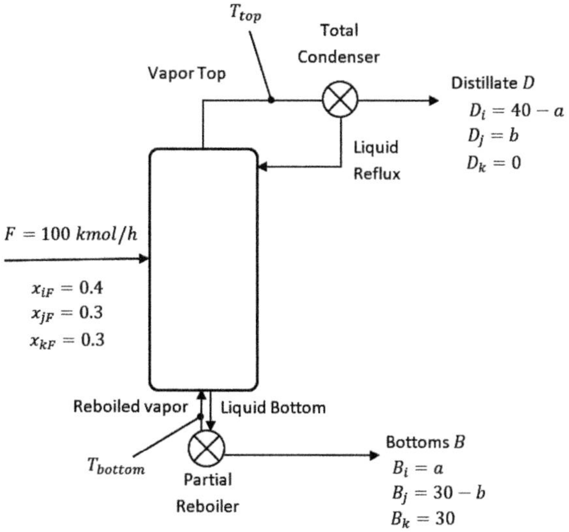


We note that the separation objectives are defined by components $i$ and $j$. From the table provided, we also note that component $j$ is a larger carbon chain than component $i$, so it is reasonable to identify component $i$ as less volatile than component $j$ and define our light and heavy keys as $i$ and $j$, respectively. It follows that component $k$ is the Heavy Non-Key. The LK component $i$ will be found mainly in the distillate while the HK component $j$ will be found mainly in the bottoms stream.

For ease of reference, let us denote the molar flow rate (in kmol/h) of LK component $i$ in the bottom stream as $a$, and the molar flow rate of HK component $j$ in the distillate as $b$.

We are told that $x_{iB} = 0.01$ and $x_{jD} = 0.015$. Since component $k$ is a HNK, it collects almost exclusively in the bottom stream and we can deduce that the distillate stream comprises mainly of $i$ and $j$; hence $x_{iD} \cong 1 - 0.015 = 0.985$ and $x_{kD} \cong 0$.

We can tabulate the respective compositions and flow rates as shown:

| | Component $i$ (LK) | Component $j$ (HK) | Component $k$ (HNK) |
|---|---|---|---|
| Feed flowrate, $F$ [kmol/h] | $F_i = 0.4(100) = 40$ | 30 | 30 |
| Bottom flowrate, $B$ [kmol/h] | $a$ | $30 - b$ | 30 |
| Distillate flowrate, $D$ [kmol/h] | $40 - a$ | $b$ | 0 |
| $x_B$ | $x_{iB} = \frac{a}{a+(30-b)+30} = 0.01$ | ? | ? |
| $x_D$ | $x_{iD} = \frac{40-a}{(40-a)+b} = 0.985$ | $x_{jD} = \frac{b}{(40-a)+b} = 0.015$ | 0 |

We can solve for $a$ and $b$, using the compositions $x_{iD}$ and $x_{iB}$ to form two simultaneous equations for the two unknowns.

$$x_{iD} = \frac{40 - a}{(40 - a) + b} = 0.985 \tag{1}$$

$$x_{iB} = \frac{a}{a + (30 - b) + 30} = 0.01 \tag{2}$$

From Eq. (1), we obtain an expression for $a$

$$0.6 = 0.015a + 0.985b$$

$$a = 40 - 65.667b \tag{3}$$

From Eq. (2), we have the following,

$$0.6 = 0.99a + 0.01b$$

Substituting Eq. (3) into the above result, we get

$$0.6 = 0.99(40 - 65.667b) + 0.01b$$

$$b = 0.6$$

$$a = 40 - 65.667b = 0.6$$

We can now use the values of $a$ and $b$ to find $x_{jB}$ and $x_{kB}$. These values are updated in the table below in bold.

| | Component $i$ (LK) | Component $j$ (HK) | Component $k$ (HNK) |
|---|---|---|---|
| Feed flowrate, $F$ [kmol/h] | 40 | 30 | 30 |
| Bottom flowrate, $B$ [kmol/h] | $a$ | $30 - b$ | 30 |
| Distillate flowrate, $D$ [kmol/h] | $40 - a$ | $b$ | 0 |
| $x_B$ | $x_{iB} = \frac{a}{a+(30-b)+30} = 0.01$ | $\boldsymbol{x_{jB} = \frac{30-b}{a+(30-b)+30} = 0.49}$ | $\boldsymbol{x_{kB} = \frac{30}{a+(30-b)+30} = 0.5}$ |
| $x_D$ | $x_{iD} = \frac{40-a}{(40-a)+b} = 0.985$ | $x_{jD} = \frac{b}{(40-a)+b} = 0.015$ | 0 |

Therefore, we can find $N_m$ as follows using values determined in the table above. Note that the partial reboiler is also an equilibrium stage; therefore, $N_m = 9.23$ is equivalent to 1 partial reboiler and 8.23 (round up to 9) stages in the column.

$$N_m = \frac{\ln\left[\frac{x_{iD}/x_{jD}}{x_{iB}/x_{jB}}\right]}{\ln\left(\alpha_{ij}\right)_{\text{geom avg}}} = \frac{\ln\left[\frac{0.985/0.015}{0.01/0.49}\right]}{\ln 2.4}$$

$$N_m = 9.23$$

(c) When reflux ratio is at its minimum for a binary distillation, we have a pinch point at the feed location. For multicomponent distillations, pinches also occur, but they are not necessarily at the feed location, and may appear in the rectification (above the feed) and stripping (below the feed) sections of the column.

The key assumptions made in Underwood's analysis include:

- Constant molar overflow (i.e., constant liquid and vapor molar flow rates in the rectifying section and stripping section are constant (but need not be the same flow rates in these two sections)).
- Constant relative volatility $\alpha_i$ between pinch points. If the values of $\alpha_i$ significantly differ in the rectifying and stripping sections of the column, you may use a geometric average as explained in part a.
- High recovery of keys (i.e., a sharp separation)
- No other components with $\alpha$ values close to that of the keys.

Underwood's First and Second equations are given as shown:

$$\sum_i \frac{\alpha_i z_i}{\alpha_i - \theta} = 1 - q \tag{1}$$

$$\sum_i \frac{\alpha_i x_{iD}}{\alpha_i - \theta} = 1 + R_m \tag{2}$$

$\alpha_i$ typically refers to the relative volatility of component $i$ with respect to the component identified as the heavy key. $z_i$ is the feed composition for component $i$. $q$ describes the nature of the feed, whereby $q = 1$ for a saturated liquid feed (i.e., at bubble point) and $q = 0$ for a saturated vapor feed (i.e., at dew point). $\theta$ is an unknown parameter to be determined.

Note that mathematically, there are multiple solutions for $\theta$ to Eq. (1). We solve Eq. (1) iteratively until we obtain the only valid solution, whereby $\theta$ lies between the relative volatilities for the light and heavy keys as shown below.

$$\alpha_{\text{HK}} < \theta < \alpha_{\text{LK}}$$

With $\theta$ value found from Eq. (1), we substitute into Eq. (2) to obtain $R_m$.

(d) We start with Underwood's First equation as given below

$$\sum_i \frac{\alpha_i z_i}{\alpha_i - \theta} = 1 - q \tag{1}$$

To use (1), we need to first find the relative volatilities for each component with respect to the heavy key $j$. It is given that $\alpha_{ij} = 2.4$ and $\alpha_{kj} = 0.3$. It is straightforward to note that $\alpha_{jj} = 1$. [Note that we are using the geometric average for $\alpha$, averaged between the top and bottom ends of the column to account partially for the temperature variation of $\alpha$ along the column.]

From the problem statement, we know that $z_i = 0.4$, $z_j = 0.3$, and $z_k = 0.3$. For a saturated liquid feed, $q = 1$.

$$\sum_i \frac{\alpha_i z_i}{\alpha_i - \theta} = \frac{\alpha_{ij} z_i}{\alpha_{ij} - \theta} + \frac{\alpha_{jj} z_j}{\alpha_{jj} - \theta} + \frac{\alpha_{kj} z_k}{\alpha_{kj} - \theta} = 1 - q$$

$$\frac{2.4(0.4)}{2.4 - \theta} + \frac{1(0.3)}{1 - \theta} + \frac{0.3(0.3)}{0.3 - \theta} = 1 - 1 = 0$$

$$\frac{0.96}{\theta - 2.4} + \frac{0.3}{\theta - 1} + \frac{0.09}{\theta - 0.3} = 0$$

$$\frac{0.96(\theta - 1)(\theta - 0.3) + 0.3(\theta - 2.4)(\theta - 0.3) + 0.09(\theta - 2.4)(\theta - 1)}{(\theta - 2.4)(\theta - 1)(\theta - 0.3)} = 0$$

$$\left(0.96\theta^2 - 1.248\theta + 0.288\right) + \left(0.3\theta^2 - 0.81\theta + 0.216\right) + \left(0.09\theta^2 - 0.306\theta + 0.216\right) = 0$$

$$1.35\theta^2 - 2.364\theta + 0.72 = 0$$

$$\theta = \frac{-(-2.364) \pm \sqrt{(-2.364)^2 - 4(1.35)(0.72)}}{2(1.35)} = 1.36 \text{ or } 0.39$$

The valid root is 1.36 since it meets the following condition.

$$\alpha_{\text{HK}} < \theta < \alpha_{\text{LK}}$$

$$\alpha_{jj} < \theta < \alpha_{ij}$$

$$1 < \theta < 2.4$$

Using this value of $\theta = 1.36$, and the distillate compositions found in part b, we substitute into Underwood's Second Eq. (2)

$$\sum_i \frac{\alpha_i x_{iD}}{\alpha_i - \theta} = 1 + R_m \tag{2}$$

$$R_m = -1 + \frac{\alpha_{ij} x_{iD}}{\alpha_{ij} - \theta} + \frac{\alpha_{jj} x_{jD}}{\alpha_{jj} - \theta} + \frac{\alpha_{kj} x_{kD}}{\alpha_{kj} - \theta}$$

$$R_m = -1 + \frac{2.4(0.985)}{2.4 - 1.36} + \frac{1(0.015)}{1 - 1.36} + \frac{0.3(0)}{0.3 - 1.36}$$

$$R_m = 1.23$$

(e) Given that the operating reflux ratio $R = 1.2R_m = 1.2(1.23) = 1.48$, we can look up Gilliland's correlation chart by calculating the required value on the horizontal axis as shown below

$$\frac{R - R_m}{R + 1} = \frac{1.48 - 1.23}{1.48 + 1} = 0.101$$

The corresponding value on the vertical axis can be read off Gilliland's correlation chart for $\frac{R-R_m}{R+1} = 0.101$

$$\frac{N - N_m}{N + 1} \approx 0.52$$

We obtained $N_m = 9.23$ from Fenske's equation in part b. Therefore we can calculate $N$

$$\frac{N - 9.23}{N + 1} = 0.52$$

$$N = 20.3 \approx 21$$

The total number of equilibrium stages required is 21, or 20 stages with 1 partial reboiler.

## Problem 20

**In distillation, we often use a $q$-line to find the locus of intersection of operating lines for the rectifying and stripping sections.**

$$y = \frac{q}{q - 1}x - \frac{z_F}{q - 1}$$

**Explain the $q$-line and show how the following equation is derived where $q$ is defined as the heat required to convert 1 mol of feed from its initial enthalpy to a saturated vapor, divided by the molal latent heat. Comment on the usefulness of this equation.**

## Solution 20

***Worked Solution***

The introduction of feed changes the slope of operating lines, by triggering a switch between the stripping section (below feed and closer to bottoms) and the rectifying section (above feed and closer to tops).

The diagram below shows the key vapor and liquid streams at the feed stage.

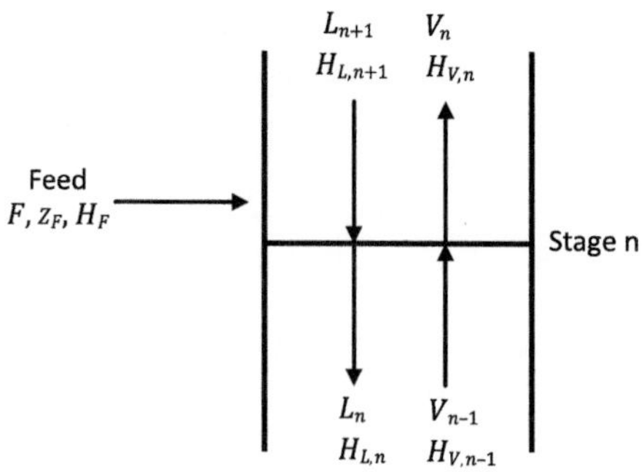


The quantities of the liquid and vapor streams change when the feed is introduced as it may consist of a liquid, vapor, or both. For example, if the feed is a saturated liquid, the liquid stream will exceed its initial flow rate by the amount of added feed. This can be shown in the following general mass balance equation.

$$F + L_{n+1} + V_{n-1} = V_n + L_n$$

$$\frac{V_{n-1} - V_n}{F} = \frac{L_n - L_{n+1}}{F} - 1$$

The energy balance is also shown below.

$$FH_F + L_{n+1}H_{L,n+1} + V_{n-1}H_{V,n-1} = V_nH_{V,n} + L_nH_{L,n}$$

The vapor and liquid inside the distillation column are saturated, and the molal enthalpies of all the saturated vapors and liquid over one tray are almost equivalent due to almost similar temperature and compositions. So we may assume that $H_{L,n+1} = H_{L,n} = H_L$ and $H_{V,n} = H_{V,n-1} = H_V$ and the equation simplifies to the following.

$$FH_F + (V_{n-1} - V_n)H_V = (L_n - L_{n+1})H_L$$

We can combine this energy balance with the earlier mass balance equation.

$$H_F + \left(\frac{V_{n-1} - V_n}{F}\right)H_V = \left(\frac{L_n - L_{n+1}}{F}\right)H_L$$

$$\frac{H_F}{H_V} + \left(\frac{L_n - L_{n+1}}{F}\right) - 1 = \left(\frac{L_n - L_{n+1}}{F}\right)\frac{H_L}{H_V}$$

$$\frac{L_n - L_{n+1}}{F} = \frac{1 - \frac{H_F}{H_V}}{1 - \frac{H_L}{H_V}} = \frac{H_V - H_F}{H_V - H_L} = q \qquad (1)$$

We can observe from the above that $q$ is the amount of heat energy required to convert 1 mole of feed from its initial enthalpy $H_F$ to the enthalpy of the saturated vapor $H_V$ at stage $n$, divided by the molal latent heat $(H_V - H_L)$.

The feed can take on a range of different conditions, from being a subcooled liquid, saturated liquid, partial liquid partial vapor, saturated vapor to a superheated vapor. These different types of feed can be characterized by their respective $q$ values.

$$V_{n-1} - V_n = F(q - 1) \tag{2}$$

The mass balances around the top rectifying section (excluding feed) and bottom stripping section (excluding feed) give:

$$yV_n = L_{n+1}x + Dx_D$$
$$yV_{n-1} + Bx_B = L_n x$$

Combining both equations give us

$$(V_{n-1} - V_n)y = (L_n - L_{n+1})x - (Bx_B + Dx_D) \tag{3}$$

An overall mass balance around the entire column is

$$Fz_F = Dx_D + Bx_B \tag{4}$$

Combining Eqs. (1), (2), (3), and (4), we can derive the given equation. Starting with Eq. (1):

$$(L_n - L_{n+1}) = Fq$$
$$(L_n - L_{n+1})x - (Bx_B + Dx_D) = Fqx - Dx_D - Bx_B$$

Substituting Eq. (3), we get

$$V_{n-1} - V_n = \frac{Fqx}{y} - \frac{Dx_D}{y} - \frac{Bx_B}{y}$$

Substituting Eqs. (2) and (4), we get

$$(q - 1)F = \frac{Fqx}{y} - \frac{Fz_F}{y}$$
$$(q - 1)y = qx - z_F$$
$$y = \frac{q}{q - 1}x - \frac{z_F}{q - 1}$$

One of the key features of this equation is the value of the slope $\frac{q}{q-1}$ which can help us determine the state of the feed (e.g., saturated liquid or saturated vapor)

| State of feed | $\frac{q}{q-1}$ | $q$ |
|---|---|---|
| Subcooled liquid | >1.0 | >1.0 |
| Saturated liquid | $\infty$ | 1.0 |
| Partial liquid partial vapor | For $F = V_F + L_F$, $\frac{q}{q-1} = \frac{L_F}{L_F - F}$ | $0 < q < 1.0$ |
| Saturated vapor | 0 | 0 |
| Superheated vapor | $0 < \frac{q}{q-1} < 1.0$ | < 0 |

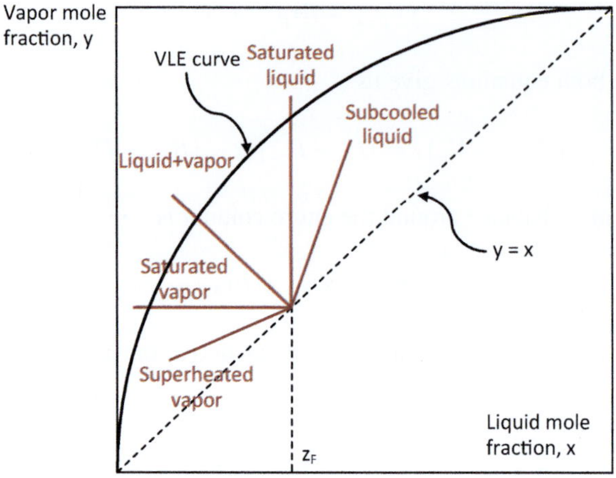


## Problem 21

**Reflux ratio is a common parameter encountered in distillation column design. It is a ratio between the amount of reflux that goes back into the column to the amount of reflux that is collected as distillate. Explain the terms minimum reflux and maximum reflux.**

## Solution 21

***Worked Solution***

Consider a simple column shown in the diagram where the reflux ratio is defined as follows.

$$R = \frac{L_T}{D}$$

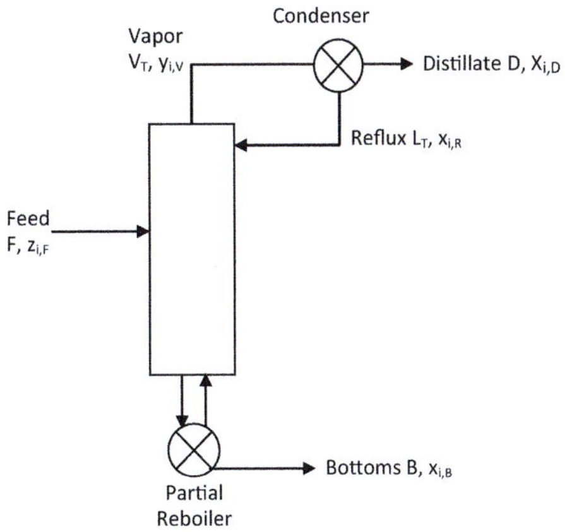


When there is total reflux, the reflux ratio $R = \infty$ and there is no distillate collected in the receiver as all of the material is returned back into the column. In order to achieve mass balance for the overall column, when there is no distillate collected, we will also need to reboil all of the bottoms product. Total reflux also means infinite reboiler heat and condenser cooling capacities.

At maximum or total reflux, $V_T = L_T$. In terms of the McCabe–Thiele construction, the operating lines of both the rectifying and stripping sections of the column coincide with the $y = x$ diagonal line. At this maximum reflux, the number of stages is the minimum. Below is a sample diagram showing distillation at maximum reflux, with the "steps" corresponding to the theoretical stages, for a hypothetical scenario with defined mole fractions of the feed, distillate and bottoms.

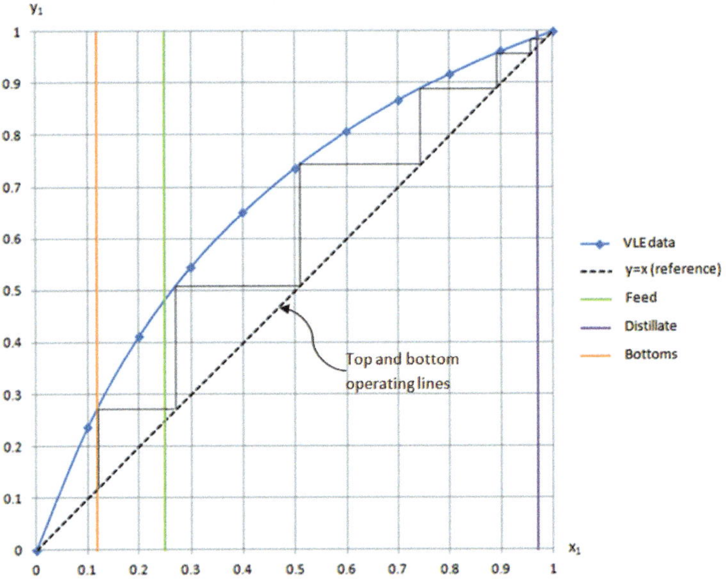

The operating lines move closer to the equilibrium curve as we decrease the reflux ratio. Consequently, at the minimum reflux ratio, we arrive at the maximum number of stages to achieve the desired separation. This also means minimum reboiler heat and condenser cooling capacities. The operating lines for the top rectifying section and bottom stripping section intersect with the feed line. The top operating line is further constrained by the required distillate composition, while the bottom operating line is constrained by the bottoms composition. Below is an example with a subcooled liquid feed, where we included operating lines at a minimum reflux ratio.

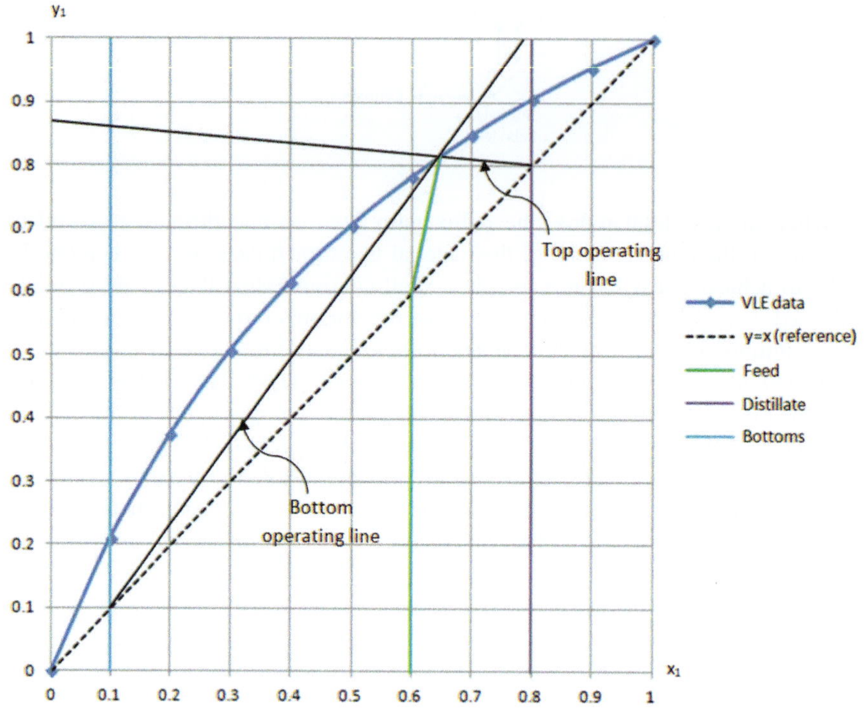


The minimum reflux ratio depends on the shape of the VLE curve, as its curvature can constrain its value. An example is shown below for a saturated liquid feed, where the slope of the top operating line cannot be any less steep, otherwise it will cross over the VLE curve above the point marked with a cross which is not physically possible.

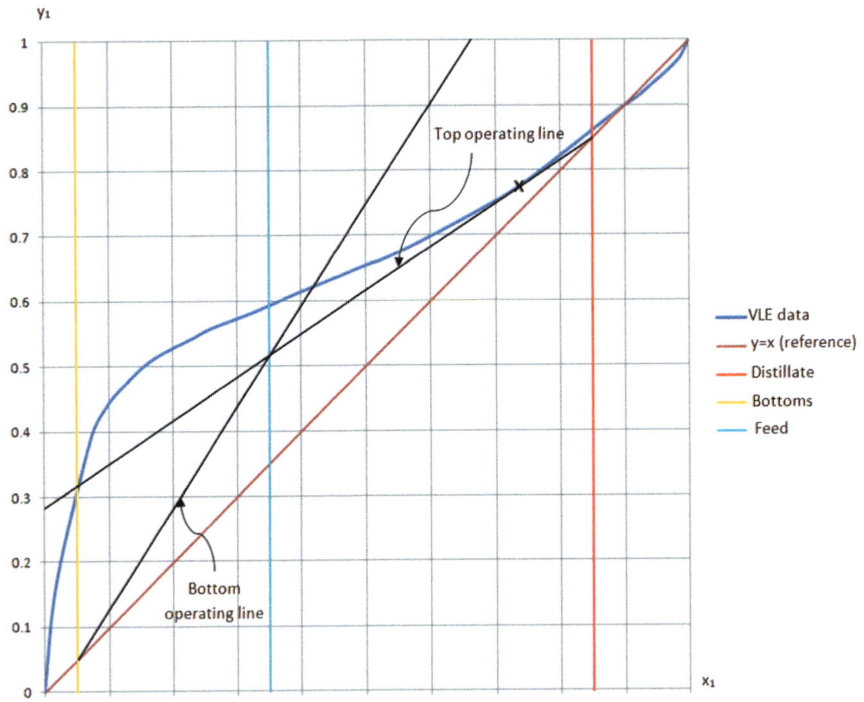


## Problem 22

**Consider a stripping column comprising countercurrent flows of water and clean air. Volatile hydrocarbon compounds are removed from the water source at 25 °C and 1 bar, with the purification objective to remove 99.95% of compound 1. The compounds present and their K values (molar basis) at 25 °C and 1 bar are shown below.**

| Compound | *K* | Concentration (mass basis) in water [kg/m$^3$] | Molecular weight [kg/mol] |
|---|---|---|---|
| 1 | 260 | 0.182 | 0.078 |
| 2 | 251 | 0.065 | 0.092 |
| 3 | 280 | 0.047 | 0.106 |

(a) **Determine the volumetric ratio of flow rate of air to water if the air flow rate is 3 times the minimum required to achieve the purification objective for compound 1.**

(b) **List the assumptions that allow multicomponent absorption processes to be modeled using the analytical Kremser–Souders–Brown procedure and the key results of this analysis.**

(c) **Assuming that our system is sufficiently dilute and may be modeled using the Kremser–Souders–Brown equation for a stripping column as shown below, find the number of plates required for an overall efficiency of 10%. $x_B^*$ denotes the hypothetical equilibrium value for liquid mole fraction at the bottom of the column, $L$ and $G$ denote the liquid and gas molar flow rates, respectively, and $K$ values (molar basis) are as provided in the above.**

$$\frac{x_T - x_B}{x_T - x_B^*} = \frac{\frac{1}{A} - \left(\frac{1}{A}\right)^{N+1}}{1 - \left(\frac{1}{A}\right)^{N+1}}, \quad \text{where } A = \frac{L}{GK}$$

(d) **Determine the concentrations (mass basis) of the compounds in the water at its exit from the column.**

## Solution 22

*Worked Solution*

(a) We can illustrate our system as shown.

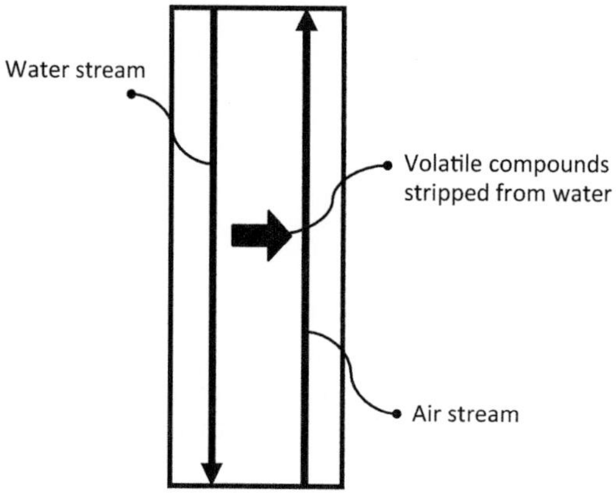


As the clean air stream rises, it is increasingly enriched with the volatile compounds from the water stream; hence, the driving force for compounds to move into the air stream is largest at the bottom part of the column. Conversely, at the top of the column, we can assume equilibrium between the vapor and liquid streams.

We are given $K$ values on a molar basis, which means that $K_i = y_i/x_i$, where $y_i$ and $x_i$ are the mole fractions of component $i$ in the vapor phase and liquid phase, respectively.

We can first convert the mass concentration to molar concentration.

$$\rho_1 = 0.182\left[\frac{\text{kg}}{\text{m}^3\ \text{water}}\right] = \frac{0.182}{0.078}\left[\frac{\text{mol}}{\text{m}^3\ \text{water}}\right] = 2.33\left[\frac{\text{mol}}{\text{m}^3\ \text{water}}\right]$$

We know that $\rho_{\text{water}} = 1000\ \text{kg/m}^3$ and molecular weight is 0.018 kg/mol. Therefore 1 $\text{m}^3$ of water has an equivalent molar amount of $\frac{1000}{0.018} = 5.55 \times 10^4$ mol. Therefore the mole fraction of compound 1 in water, $x_1$, is as follows.

$$x_1 = \frac{2.33}{5.55 \times 10^4} = 4.2 \times 10^{-5} = x_{1,\text{Top}}$$

$x_1$ calculated above represents incoming compound 1 mole fraction at the top of the column. Equilibrium is reached near the top, so the vapor mole fraction of compound 1 at the top is,

$$y_{1,\text{Top}} = K_1 x_{1,\text{Top}} = 260\left(4.2 \times 10^{-5}\right) = 0.0109$$

We require the following to be true based on the purification objective for compound 1, and also noting that incoming air at the bottom is clean.

$$\frac{x_{1,\text{Bottom}}}{x_{1,\text{Top}}} = \frac{1 - 0.9995}{1} = 0.0005$$

$$y_{1,\text{Bottom}} = 0$$

We can construct a mass balance for compound 1 around the stripping column, where $L$ and $G$ are the molar flow rates for the liquid and gas streams, respectively.

$$Lx_{1,\text{Top}} + Gy_{1,\text{Bottom}} = Lx_{1,\text{Bottom}} + Gy_{1,\text{Top}}$$

$$x_{1,\text{Top}} = x_{1,\text{Bottom}} + \frac{G}{L}(0.0109)$$

$$\frac{G}{L} = \frac{x_{1,\text{Top}} - x_{1,\text{Bottom}}}{0.0109} = \frac{\left(4.2 \times 10^{-5}\right)(1 - 0.0005)}{0.0109} = 3.85 \times 10^{-3} = \left.\frac{G}{L}\right|_{\text{min}}$$

The value of $\frac{G}{L}$ above is the minimum required to meet the purification objective. It is stated that the operating $G = 3G_{\text{min}}$. Therefore the operating $\frac{G}{L}$ is as follows, on a molar basis.

$$\left.\frac{G}{L}\right|_{\text{molar}} = 3\left(3.85 \times 10^{-3}\right) = 0.01155$$

We can convert the molar ratio into a volumetric ratio as follows:

$$\left.\frac{G}{L}\right|_{\text{volume}} = \frac{n_{\text{air}}/\rho_{m,\text{air}}}{n_{\text{water}}/\rho_{m,\text{water}}} = \frac{n_{\text{air}}}{n_{\text{water}}}\left(\frac{\rho_{m,\text{water}}}{\rho_{m,\text{air}}}\right) = \left.\frac{G}{L}\right|_{\text{molar}}\left(\frac{\rho_{m,\text{water}}}{\rho_{m,\text{air}}}\right)$$

We can find the molar density of air $\rho_{m,\text{ air}}$ [mol/m$^3$] using the ideal gas law

$$\rho_{m,\text{air}} = \frac{P}{\text{RT}} = \frac{10^5}{8.314(298)} = 40.36\ \text{mol/m}^3$$

$$\rho_{m,\text{water}} = \frac{\rho_{\text{water}}}{MW_{\text{water}}} = \frac{1000}{0.018}\ \text{mol/m}^3$$

Therefore, putting our results together, we have the operating volumetric ratio as follows:

$$\left.\frac{G}{L}\right|_{\text{volume}} = 0.01155\left(\frac{\frac{1000}{0.018}}{40.36}\right) = 15.9$$

(b) As a general overview, design calculations for multicomponent absorption processes can be performed by the McCabe–Thiele method or the analytical Kremser–Souders–Brown (KSB) procedure. In the KSB procedure, both the equilibrium and operating lines are assumed straight. This can be extended to multicomponent problems if the following are assumed:

- System is dilute—The total flow rates of the gas and liquid streams, $G$ and $L$, respectively, are constant over the column. Moreover, the partition of any solute between the two phases is independent of the other solutes present.
- There is a constant temperature and pressure from plate to plate. This means constant K values (linear equilibrium) throughout the column can be assumed and no complications from variations of pressure and/or significant heats of absorption.

In this case, a pseudo single-component calculation for each component can be done for the mixture. It is common to focus on a particular component, called the "key component" which is referenced to when separation objectives are specified. For example, the key component has to achieve a specified minimum level of removal from the liquid stream (absorbed into the gas phase) in a stripping column.

Analysis for this key component, for example, component $i$, can be done assuming it was a single component.

Operating Line from Mass Balance (purple outline): $y_{n,i} = \frac{L}{G}x_{n+1,i} - y_{T,i}$

Equilibrium Relationships (for all components): $y_{n,i} = K_i x_{n,i}$

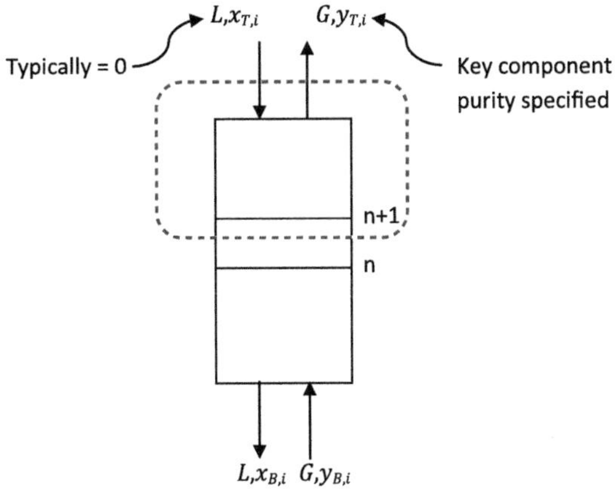


The purpose of the KSB analysis is to arrive at the number of stages $N$ required to meet the purification objective. The key results of this analysis are as follows where $A = \frac{L}{GK}$:

For stripping column:

$$\frac{x_T - x_B}{x_T - x_B^*} = \frac{\frac{1}{A} - \left(\frac{1}{A}\right)^{N+1}}{1 - \left(\frac{1}{A}\right)^{N+1}}$$

For rectifying column:

$$\frac{y_B - y_T}{y_B - y_T^*} = \frac{A - A^{N+1}}{1 - A^{N+1}}$$

(c) Using component 1, we can first calculate the value of $A$,

$$A_1 = \frac{L}{GK_1} = \frac{1}{0.01155(260)} = 0.33$$

Note that $x_B^*$ is the hypothetical equilibrium mole fraction at the bottom of the column; however, in reality, equilibrium is not reached at the bottom for a stripping column. This parameter is used in the KSB equation as a hypothetical reference state.

We note that clean air enters the bottom, hence $y_{1,B} = 0$ and given that by definition, $K_{1,B} = y_{1,B}/x_{1,B}$ at equilibrium, $x_{1,B} = 0$ at equilibrium, and $x_{1,B}^* = 0$.

Substituting values into the KSB equation, we have

$$\frac{x_T - x_B}{x_T - x_B^*} = \frac{\frac{1}{A} - \left(\frac{1}{A}\right)^{N+1}}{1 - \left(\frac{1}{A}\right)^{N+1}}$$

$$\frac{x_{1,T} - x_{1,B}}{x_{1,T} - 0} = 0.9995 = \frac{\frac{1}{0.33} - \left(\frac{1}{0.33}\right)^{N+1}}{1 - \left(\frac{1}{0.33}\right)^{N+1}}$$

$$N = 5.6$$

Given that the overall efficiency is 10%, the number of plates required will be 56.

$$N = \frac{5.6}{0.1} = 56$$

(d) Now that we have found the value of $N$, we can back-calculate the compositions of the other compounds using the KSB equation.

For compound 1, the concentration (mass basis) is more straightforward as it is defined by the purification objective.

$$\rho_1 = (1 - 0.9995)\rho_{1,\text{inlet}} = 0.0005(0.182) = 9.1 \times 10^{-5}\ \text{kg/m}^3$$

We can calculate the $A$ values for compounds 2 and 3,

$$A_2 = \frac{L}{GK_2} = \frac{1}{0.01155(251)} = 0.345$$

$$A_3 = \frac{L}{GK_3} = \frac{1}{0.01155(280)} = 0.309$$

Substituting into the KSB equation for compounds 2 and 3, we have

$$\frac{x_{2,T} - x_{2,B}}{x_{2,T} - 0} = \frac{\frac{1}{0.345} - \left(\frac{1}{0.345}\right)^{5.6+1}}{1 - \left(\frac{1}{0.345}\right)^{5.6+1}} = 0.99822$$

$$\rho_2 = (1 - 0.99822)\rho_{2,\text{inlet}} = (1 - 0.99822)0.065 = 1.16 \times 10^{-4}\ \text{kg/m}^3$$

$$\frac{x_{3,T} - x_{3,B}}{x_{3,T} - 0} = \frac{\frac{1}{0.309} - \left(\frac{1}{0.309}\right)^{5.6+1}}{1 - \left(\frac{1}{0.309}\right)^{5.6+1}} = 0.99898$$

$$\rho_3 = (1 - 0.99898)\rho_{3,\text{inlet}} = (1 - 0.99898)0.047 = 4.80 \times 10^{-5}\ \text{kg/m}^3$$

## Problem 23

(a) **Derive the following operating line equation for an arbitrary stage $n$ in an absorption column with $N$ number of stages in total ($1 \leq n \leq N$). Let $L'$ and $V'$ denote the molar flow rates of solute-free liquid absorbent and vapor streams, respectively. The mole fractions of solute to solute-free liquid absorbent and vapor stream are denoted as $x$ and $y$, respectively.**

$$y_{n+1} = x_n\left(\frac{L'}{V'}\right) + y_1 - x_0\left(\frac{L'}{V'}\right)$$

(b) **Plot operating lines for the following 3 cases for $L'$. Also include in your plot a sketch of the equilibrium curve and explain the relative positions of the plots.**

(i) $L' = L'_{\text{min}}$
(ii) $L'_{\text{min}} < L' < \infty$
(iii) $L' = \infty$

(c) **An air feed stream containing 1.8 vol% of impurity $Z$ is to be scrubbed with pure water in a column packed with ceramic rings at atmospheric conditions. The absorption requirement is to achieve an outlet concentration of solute $Z$, $y_{\text{out}} = 0.0045$. If the total vapor and liquid flow rates are $V = 0.065$ kmol/s and $L = 2.5$ kmol/s, respectively, and the equilibrium correlation between the vapor mole fraction ($y_{\text{eqm}}$) and liquid mole fraction ($x_{\text{eqm}}$) for solute $Z$ is as follows,**

$$y_{\text{eqm}} = 42x_{\text{eqm}}$$

- **Using results from part a, show that the required liquid stream flow rate $L$ needs to be twice the value of $L'_{\text{min}}$ for the desired separation in this problem.**
- **State any assumptions made in your solution to this problem.**

## Solution 23

*Worked Solution*

(a) Let us perform a general mass balance around a section of an absorber that has a total number of $N$ stages as shown in the diagram below. The mass balance denoted by dotted line starts from the top end of the tower (i.e., stage 1) and ends at an arbitrary equilibrium stage $n$ (where $n < N$). The mass balance equation gives us the operating line equation for absorption.

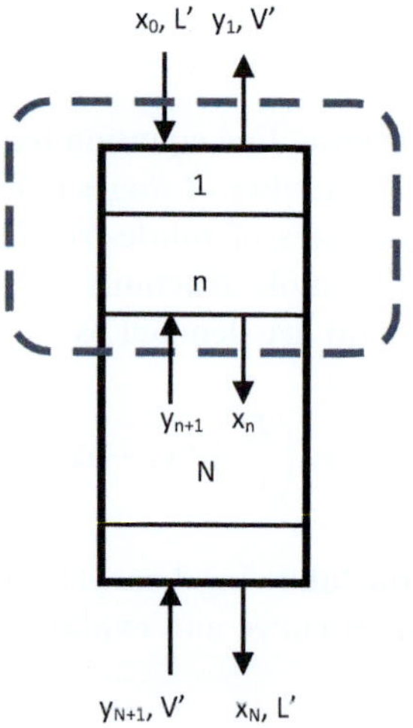


The mass balance equation can therefore be written as follows

$$x_0 L' + y_{n+1} V' = x_n L' + y_1 V'$$

$$y_{n+1} = x_n \left(\frac{L'}{V'}\right) + y_1 - x_0 \left(\frac{L'}{V'}\right)$$

(b) Note that by convention, the subscript for mole fraction $x$ (and similarly for $y$) refers to the stage from which the stream *leaves*; therefore, the inlet feed (liquid) mole fraction has a subscript 0 instead of 1 at the point on the graph that refers to stage 1. This convention also explains why the point corresponding to stage $N$ in the graph has a vapor mole fraction of $y_{N+1}$, where the subscript is $N + 1$ instead of $N$.

For absorption, the operating line is drawn *above* the equilibrium curve since for a specific solute concentration in the liquid absorbent stream, the solute concentration in the vapor stream is always greater than the equilibrium value, otherwise no absorption can take place. This difference in concentration between the liquid and vapor streams creates a driving force for mass transfer.

The two extreme cases whereby the liquid absorbent flow rate $L' = L'_{\text{min}}$ and $\infty$ set the limits of the straight line plot for this absorption. At $L'_{\text{min}}$, there is equilibrium between the vapor and liquid phases at stage $N$; hence, the line touches the equilibrium curve at the bottom of the column (or top right of the plot). This occurs when we have an infinite number of stages.

As $L'$ increases to a value $L'_{\text{min}} < L' < \infty$, the gradient of the operating line increases until it approaches a vertical line whereby $L' = \infty$ (number of stages required is zero).

We note that as $L'$ increases, the required number of stages decreases. This shows a trade-off between building a taller column (increased material cost) and increasing $L'$ (increased energy cost). Only one of the two factors can be cost-optimized at one time to achieve a desired separation.

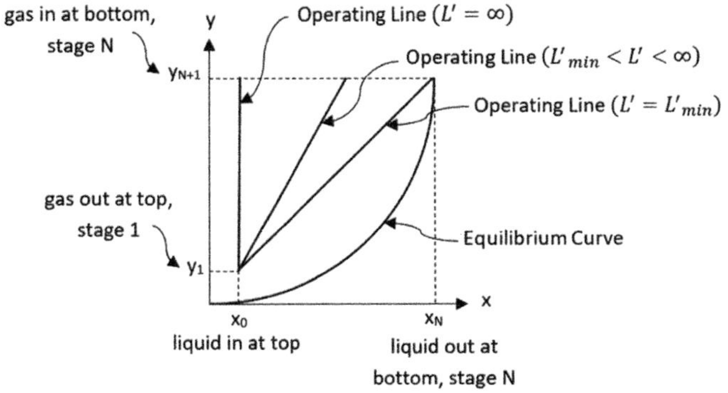


(c) (i) At $L' = L'_{\text{min}}$, we reach equilibrium between the liquid absorbent stream leaving the bottom of the column ($n = N$) and the vapor feed stream entering the bottom of the column. The corresponding solute concentrations in the liquid stream at stage $N$ and feed gas are therefore $x_N$ and $y_{N+1}$. Note that it takes an infinite number of stages to achieve this equilibrium.

An expression for $L'_{\text{min}}$ at stage $N$ can be obtained from the operating line equation derived in part a

$$y_{N+1} = x_N\left(\frac{L'_{\text{min}}}{V'}\right) + y_1 - x_0\left(\frac{L'_{\text{min}}}{V'}\right)$$

$$L'_{\text{min}} = \frac{V'\left(y_{N+1} - y_1\right)}{x_N - x_0}$$

We substitute the equilibrium condition at stage $N$, $x_N = y_{N+1}/K_N$ to obtain

$$L'_{\text{min}} = \frac{V'\left(y_{N+1} - y_1\right)}{\frac{y_{N+1}}{K_N} - x_0}$$

Since the liquid absorbent entering the column is pure water and has not yet absorbed any solute $Z$, $x_0 = 0$,

$$L'_{\text{min}} = \frac{V'\left(y_{N+1} - y_1\right)}{\frac{y_{N+1}}{K_N}} = V'\left(\frac{y_{N+1} - y_1}{y_{N+1}}\right)K_N \qquad (1)$$

We observe that the quantity in brackets, $\frac{y_{N+1}-y_1}{y_{N+1}}$, represents the fraction of solute Z absorbed by the column. We can calculate this fraction from the values given in the problem. We start with an inlet vapor feed containing air and 1.8 vol% of solute Z. We can compute $V'$, i.e., the solute-free molar flow rate for the vapor stream

$$V' = V(1 - 0.018) = 0.065(1 - 0.018) = 0.064\ \text{kmol/s}$$

The molar flow rate of solute Z at the inlet of the column is therefore

$$V_{Z,\,\text{inlet}} = V(0.018) = 0.065(0.018) = 0.0012\ \text{kmol/s}$$

We know that the mole fraction of solute Z in the vapor stream exiting the column, $y_{\text{out}} = 0.0045$. This means that the mole fraction of solute-free vapor in the vapor stream at the outlet of the column is $1 - y_{\text{out}} = 0.9955$.

The molar flow rate of solute Z at the outlet of the column is therefore

$$\frac{V_{Z,\,\text{outlet}}}{V'} = \frac{y_{\text{out}}}{1 - y_{\text{out}}} = \frac{0.0045}{0.9955}$$

$$V_{Z,\,\text{outlet}} = 0.064\left(\frac{0.0045}{0.9955}\right) = 0.00029\ \text{kmol/s}$$

Since the liquid absorbent entering the column was pure water that is solute-free, all of the solute Z that was removed from the vapor stream entered the liquid stream. Hence the molar flow rate of Z in the liquid stream at the outlet of the column, $L_{Z,\text{outlet}}$, can be calculated as follows

$$L_{Z,\text{outlet}} = V_{Z,\text{inlet}} - V_{Z,\text{outlet}} = 0.0012 - 0.00029 = 0.00091\ \text{kmol/s}$$

Fraction of solute Z absorbed can also be found,

$$\frac{L_{Z,\text{outlet}}}{V_{Z,\text{inlet}}} = \frac{0.00091}{0.0012} = 0.76 = \frac{y_{N+1} - y_1}{y_{N+1}}$$

We now return to Eq. (1) to substitute the fraction of solute absorbed into the expression. We note that $K_N = 42$ as given in the problem; therefore

$$L'_{\text{min}} = V'\left(\frac{y_{N+1} - y_1}{y_{N+1}}\right)K_N = 0.064(0.76)(42) = 1.22$$

Finally we arrive at the required condition for the liquid flow rate in order to achieve the absorption objective

$$\frac{L}{L'_{\mathrm{min}}} = \frac{2.5}{1.22} \approx 2$$

(ii) Assumptions and important points to note for absorption are as follows:

- Continuous and steady-state operation.
- Vapor–liquid phase equilibrium is assumed between the vapor and liquid streams leaving each tray (in countercurrent directions in this problem).
- Only solute $Z$ is transferred from one phase to another, no other mass transfers occur between the countercurrent streams.
- The values of $L'$ and $V'$ are assumed constant throughout the column.
  - We assume no vaporization of liquid absorbent into the vapor stream and no absorption of vapor from the vapor stream by the liquid.
  - The vapor feed mixture is dilute (in solute $Z$) such that the mass transfer of $Z$ as a result of absorption from the vapor to liquid stream does not alter the values of $L'$ and $V'$ significantly.

## Problem 24

**Absorption towers are often used in industrial plants to remove air pollutants from gaseous process streams before the "cleaned" air is released into the atmosphere. One example of such a pollutant is sulfur dioxide or $SO_2$, which is easily absorbed by water.**

**Consider the case where $SO_2$ is to be removed from an air–$SO_2$ mixture by contacting a feed gas with pure water in a packed tower at 30 °C and 1 atm. $SO_2$-free water enters the tower at the top with a flow rate of 420 kg/h. The liquid stream leaving the tower at the bottom contains 0.5 g $SO_2$ per 100 g water. As for the gas stream leaving the top of the tower, there is a partial pressure of $SO_2$ measuring 30 mmHg and a mole ratio of water to air ($SO_2$-free) equivalent to 20:1. You may assume that the system is dilute.**

(a) **Determine the percentage of $SO_2$ in the feed gas that is absorbed by the tower.**
(b) **Find the partial pressure of $SO_2$ in the feed gas stream.**

## Solution 24

***Worked Solution***

(a) We know that the molecular weight of water is 18, so we can convert the mass flow rate of the incoming liquid absorbent stream (i.e., pure water) at the top of the tower to molar flow rate as follows

$$L_{w,\text{top}} = \frac{420}{18} = 23.3\ \text{kmol/h}$$

We are also told that at the top of the tower, the mole ratio of water to air ($SO_2$-free) is 20:1. In this case, water refers to the liquid absorbent stream entering the top, while air refers to the vapor stream exiting the top. Using this information, we can calculate the molar flow rate of air ($SO_2$-free) leaving the tower at the top

$$V_{\text{air,top}} = \frac{L_{w,\text{top}}}{20} = 1.165\ \text{kmol/h}$$

We know that the tower is operating at 1 atm, which is equivalent to 760 mmHg. Since the partial pressure of $SO_2$ in the vapor stream at the top, $p_s$, is 30 mmHg, the resulting partial pressure of air ($SO_2$-free) in this vapor stream, $p_{\text{air}}$, can be found

$$p_{s,\text{top}} = 30\ \text{mmHg}$$

$$p_{\text{air,top}} = 760 - 30 = 730\ \text{mmHg}$$

The molar flow rate of $SO_2$ in the vapor stream exiting the top is therefore

$$V_{s,\text{top}} = V_{\text{air,top}}\left(\frac{p_{s,\text{top}}}{p_{\text{air,top}}}\right) = 1.165\left(\frac{30}{730}\right) = 0.0479\ \text{kmol/h}$$

We know that the molecular weight of $SO_2$ is 64 and the liquid stream leaving the tower at the bottom contains 0.5 g $SO_2$ per 100 g water. We also note that the system is dilute such that the amount of $SO_2$ absorbed into the liquid stream as the liquid flows down the tower has no impact on the mass flow rate of the liquid stream. Hence, the mass flow rate of water (i.e., 420 kg/h) entering the top of the tower is the same as that leaving the bottom of the tower (although enriched with $SO_2$). We can compute the molar flow rate of $SO_2$ in the effluent liquid stream at the bottom

$$L_{s,\text{bottom}} = \frac{0.5\left(\frac{420}{100}\right)}{64} = 0.0328\ \text{kmol/h}$$

We can perform a simple mass balance around the tower for $SO_2$ as illustrated below. This gives us the molar flow rate of $SO_2$ entering the tower in the feed gas at the bottom.

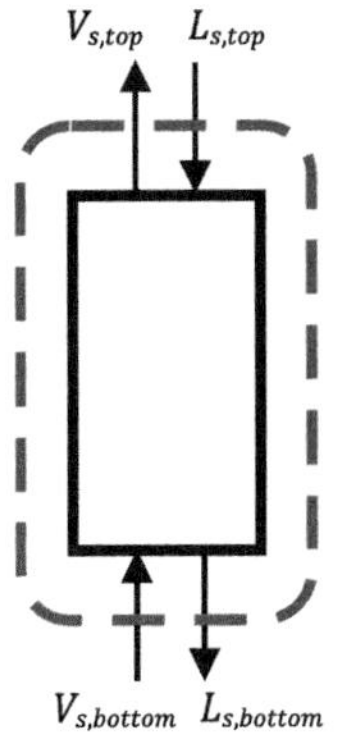


$$\text{In} - \text{Out} = 0$$

$$L_{s,\text{top}} + V_{s,\text{bottom}} - L_{s,\text{bottom}} - V_{s,\text{top}} = 0$$

We know that the liquid absorbent stream is pure water that is free of $SO_2$; therefore $L_{s,\text{ top}} = 0$. Substituting values found from earlier, we get

$$V_{s,\text{bottom}} = L_{s,\text{bottom}} + V_{s,\text{top}}$$

$$V_{s,\text{bottom}} = 0.0328 + 0.0479 = 0.0807\ \text{kmol/h}$$

The percentage of $SO_2$ absorbed is therefore

$$\frac{V_{s,\text{top}}}{V_{s,\text{bottom}}} \times 100\% = \frac{0.0479}{0.0807} \times 100\% \approx 59\%$$

(b) We note that air is just a carrier medium for the gaseous $SO_2$ in the vapor stream and is not affected by the absorption process; hence, the molar flow rate of air in the vapor stream is constant throughout the tower

$$V_{\text{air,bottom}} = V_{\text{air,top}} = 1.165\ \text{kmol/h}$$

The partial pressure of $SO_2$ in the vapor feed stream at the bottom is therefore

$$p_{s,\text{bottom}} = \frac{V_{s,\text{bottom}}}{V_{s,\text{bottom}} + V_{\text{air,bottom}}} = \frac{0.0807}{0.0807 + 1.165} = 0.065\ \text{atm}$$

## Problem 25

**Consider a column packed with rings that is used to strip a solute $Q$ from liquid water using air (gas stream) as illustrated below. $G$ and $L$ denote gas and liquid streams, and the mole fraction of solute $Q$ in the liquid stream entering the top is given as $x_{top} = 0.015$. The required mole fraction of $Q$ in the outlet liquid stream is $x_0 = 0.0015$. You may assume that the inlet gas stream is solute-free.**

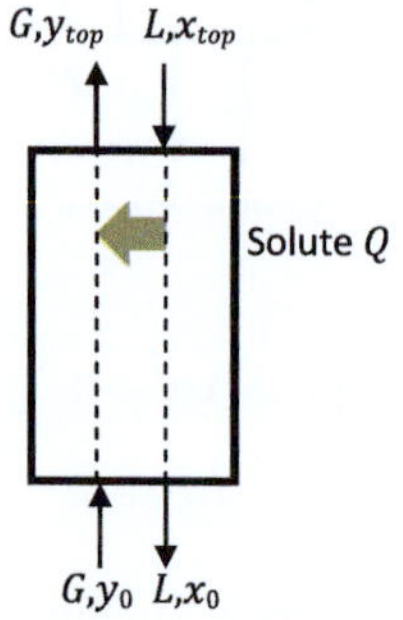


**It is given that at vapor–liquid equilibrium for solute $Q$, the following correlation between the mole fraction in the vapor (air) phase, $y$, and that in the liquid (water) phase holds true**

$$y_{eqm} = 8.2x_{eqm}$$

**The rate of mass transfer of solute $Q$ is also assumed to be controlled by the liquid side's resistance such that the overall mass-transfer coefficient based on the liquid phase, $K_x$, can be approximated as the liquid side coefficient, $k_L$, as shown below, where $a$ denotes the interfacial area per unit volume.**

$$K_x a \approx k_L a$$

**If the liquid stream enters at a flow rate per unit area of $L = 6000$ mol/(m$^2$ h), and the stripping column has a height $l = 30$ m and operates at 30 °C and 1 atm,**

(a) **Find the minimum flow rate of air, $G_{min}$.**
(b) **Sketch a graph of $y$ against $x$ for solute $Q$ for the scenario in part a. Include in your plot, the equilibrium curve.**
(c) **If the operating flow rate of air $G$ is 2.5 times that of $G_{min}$, determine the value of $y_{top}$. Plot the operating line for this scenario.**
(d) **Given the following expressions for the number of overall liquid phase transfer units, $N_{OL}$, and height of an overall transfer unit based on the liquid phase, $H_{OL}$, show that the value of $k_L a \approx 640$ mol/(m$^3$.h.$\Delta x$).**

$$N_{OL} = \int_{x_{\text{top}}}^{x_{\text{bottom}}} \frac{dx}{x_{\text{eqm}} - x}$$

$$H_{OL} = \frac{L}{K_x a}$$

(e) **A faulty valve was found to occur at a position measuring 10 m from the base of the tower, causing gas leakage such that the gas flow rate in the section of the column above the valve to be reduced to 50% of the value of $G$ (as found in part c). The gas flow rate in the bottom section of the column remains unaffected. If the mole fraction of solute $Q$ in the liquid stream leaving the tower is found to increase to $x_0 = 0.0018$, determine the mole fraction of $Q$ in the gas stream at the position of the valve, $y_{\text{valve position}}$. Plot the operating line for this faulty mode of operation. You may assume that the faulty valve does not affect the liquid stream.**

## Solution 25

### *Worked Solution*

(a) We will get the minimum flow rate of air if we set the condition whereby vapor–liquid equilibrium for solute $Q$ is just reached at the top of the column.

We can construct a mass balance for solute $Q$ as shown in dotted line below which cuts the column at an arbitrary position between the top and bottom of the column. This method is useful in deriving an operating equation for the stripping process in the column.

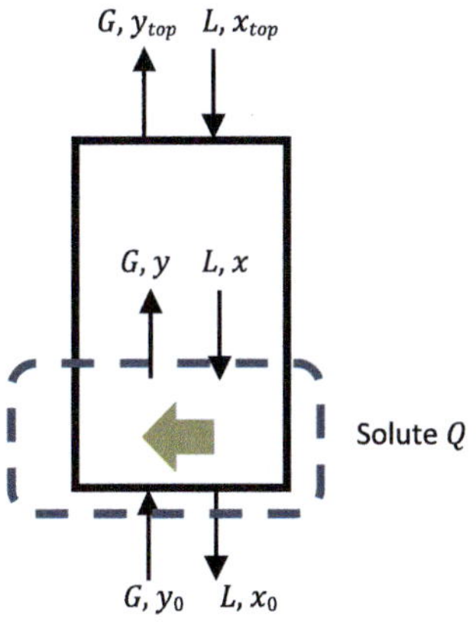


$$\text{In} - \text{Out} = 0$$

$$Lx + Gy_0 - Gy - Lx_0 = 0$$

We note that the inlet air is solute-free; this means $y_0 = 0$. Dividing by $G$ throughout, we get the operating line.

$$y = \frac{L}{G}(x - x_0)$$

At minimum $G$, equilibrium is just reached at the top of the tower. We are also told that $x_0 = 0.0015$ and $x_{top} = 0.015$. Using the equilibrium correlation provided, and the operating line equation (specified at the top of the column) found above, we have

$$y_{top} = 8.2x_{top} = 8.2(0.015) = 0.123$$

$$y_{top} = \frac{L}{G_{min}}(x_{top} - x_0)$$

$$\frac{L}{G_{min}} = \frac{y_{top}}{x_{top} - x_0} = \frac{0.123}{0.015 - 0.0015} = 9.1$$

The problem states that the liquid stream enters at a rate (per unit area) of 6000 mol/(m$^2$.h); hence substituting $L = 6000$, we obtain $G_{min} = 660$ mol/(m$^2$.h).

$$G_{min} = \frac{6000}{9.1} \approx 660$$

(b) Part b examines the graphical significance of the results in part a. Note that when equilibrium is reached at the top of the column, we may also refer to this point as the "pinch point" where the operating line intersects the equilibrium curve.

Since $\frac{L}{G_{min}}$ is equivalent to $\frac{L}{G}\big|_{max}$. The ratio $\frac{L}{G}$ represents the gradient of the operating line in the plot below. Hence the scenario in part a occurs when we have a pinch point at the top of the column and when the operating line is the steepest possible given the specified mole fractions at the top and bottom of the column.

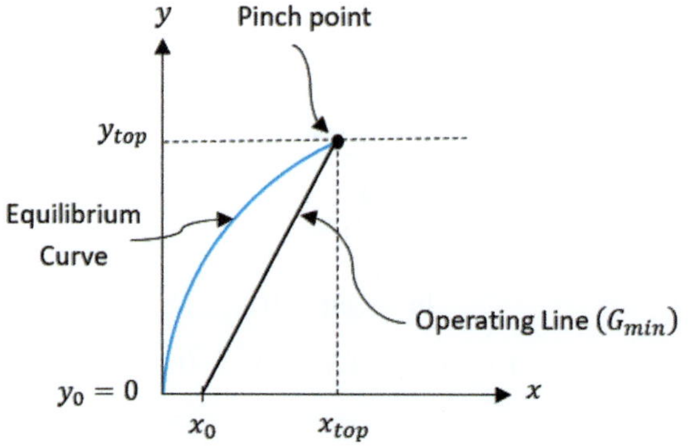

(c) We found the value of $G_{\mathrm{min}}$ in part a; therefore, we can compute the operating flow rate of air

$$G = 2.5 \times G_{\mathrm{min}} = 2.5 \times 660 = 1650$$

Performing a mass balance around the entire column, we have

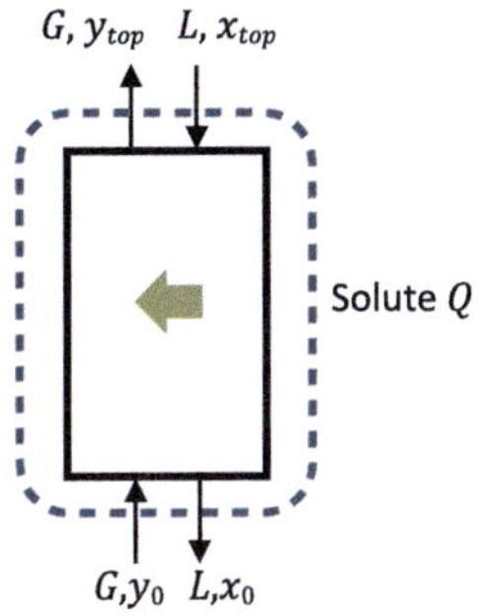


$$\mathrm{In} - \mathrm{Out} = 0$$

$$Lx_{\mathrm{top}} + Gy_0 - Gy_{\mathrm{top}} - Lx_0 = 0$$

Again, we note that the inlet air is solute-free; this means $y_0 = 0$. Dividing by $G$ throughout, we get

$$y_{\mathrm{top}} = \frac{L}{G}\left(x_{\mathrm{top}} - x_0\right) = \frac{6000}{1650}(0.015 - 0.0015) = 0.049$$

We can plot the operating line as shown below, where the driving force for mass transfer is due to the difference between equilibrium and operating conditions as shown in the red arrow.

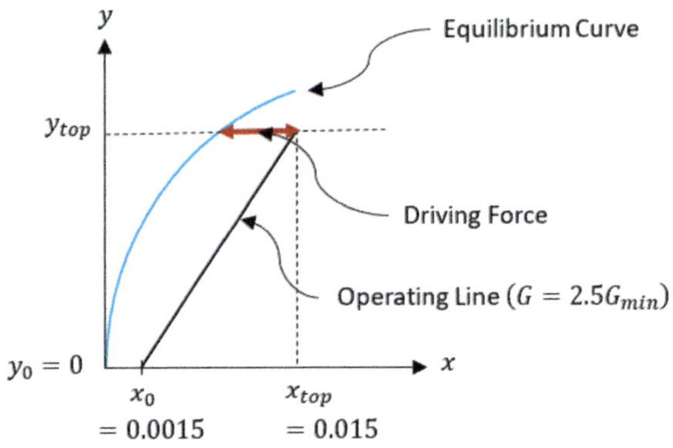

(d) We are given the following definition for $N_{OL}$ since solute $Q$ is stripped from the liquid phase (water) and is highly soluble in the air stream. The resistance to mass transfer is expected to be in the liquid film.

$$N_{OL} = \int_{x_{\text{top}}}^{x_{\text{bottom}}} \frac{dx}{x_{\text{eqm}} - x}$$

In simplifying this integral, we can use the logarithmic mean (LM) average as shown below, since the relationship is non-linear. Note that on the contrary, arithmetic average is better used for linear correlations.

$$N_{OL} = \int_{x_{\text{top}}}^{x_{\text{bottom}}} \frac{dx}{x_{\text{eqm}} - x} = \frac{x_{\text{bottom}} - x_{\text{top}}}{(x_{\text{eqm}} - x)_{LM}}$$

$$= \frac{x_{\text{bottom}} - x_{\text{top}}}{(x_{\text{eqm}} - x)_{\text{bottom}} - (x_{\text{eqm}} - x)_{\text{top}}} \ln\left[\left(\frac{(x_{\text{eqm}} - x)_{\text{bottom}}}{(x_{\text{eqm}} - x)_{\text{top}}}\right)\right]$$

To determine the value of $N_{OL}$, we need to find $x_{\text{eqm}}$ at the top and bottom of the column using the equilibrium relationship. We note that it is given in the problem that $y_0 = 0$ and $K = 8.2$, we also found the value of $y_{\text{top}}$ in part c.

$$(x_{\text{eqm}})_{\text{bottom}} = \frac{y_{\text{bottom}}}{K} = \frac{y_0}{K} = \frac{0}{8.2} = 0$$

$$(x_{\text{eqm}})_{\text{top}} = \frac{y_{\text{top}}}{K} = \frac{0.049}{8.2} = 0.006$$

Substituting these values back into the expression for $N_{OL}$, and noting from the problem that $x_{\text{bottom}} = x_0 = 0.0015$ and $x_{\text{top}} = 0.015$, we have

$$N_{OL} = \frac{0.0015 - 0.015}{(0 - 0.0015) - (0.006 - 0.015)} \ln\left(\frac{0 - 0.0015}{0.006 - 0.015}\right) = 3.2$$

Given that the packed column height is 30 m, we can determine $H_{OL}$

$$H_{OL} = \frac{l}{N_{OL}} = \frac{30}{3.2} = 9.4$$

Using the expression for $H_{OL}$ provided in the problem, we can find $k_L a$ as shown below

$$H_{OL} = \frac{L}{K_x a} \approx \frac{L}{k_L a} = \frac{6000}{k_L a} = 9.4$$

$$k_L a = 640 \text{ mol}/\left(\text{m}^3.\text{h}.\Delta x\right)$$

(e) We note from the following expression that the height of the column $l$ is proportional to $N_{OL}$. The value of $H_{OL}$ is dependent on the properties of the liquid stream, which is unaffected by the leaky valve and remains constant throughout the column.

$$l = H_{OL} N_{OL}$$

$$l \propto N_{OL}$$

Since the valve is located 10 m from the base of the tower which has a total height of 30 m, the value of $N_{OL,\text{bottom}}$ for the bottom section of the tower can be determined.

$$N_{OL,\text{bottom}} = \frac{N_{OL}}{3} = \frac{3.2}{3} \approx 1.1$$

$$N_{OL,\text{bottom}} = \frac{x_{\text{bottom}} - x_{\text{valve position}}}{\left(x_{\text{eqm}} - x\right)_{\text{bottom}} - \left(x_{\text{eqm}} - x\right)_{\text{valve position}}} \ln\left[\left(\frac{\left(x_{\text{eqm}} - x\right)_{\text{bottom}}}{\left(x_{\text{eqm}} - x\right)_{\text{valve position}}}\right)\right]$$

Equating the value of $N_{OL,\text{bottom}}$, and noting that this time, $x_{\text{bottom}} = x_0 = 0.0018$, we have

$$1.1 = \frac{0.0018 - x_{\text{valve position}}}{\left(x_{\text{eqm}} - 0.0018\right)_{\text{bottom}} - \left(x_{\text{eqm}} - x\right)_{\text{valve position}}} \ln\left[\left(\frac{\left(x_{\text{eqm}} - 0.0018\right)_{\text{bottom}}}{\left(x_{\text{eqm}} - x\right)_{\text{valve position}}}\right)\right] \quad (1)$$

We can find the corresponding $x_{\text{eqm}}$ at the two ends of the bottom section of the column using the equilibrium relationship. It is given that $y_0 = 0$ and $K = 8.2$; therefore Eq. (1) simplifies to the following

$$\left(x_{\text{eqm}}\right)_{\text{bottom}} = \frac{y_{\text{bottom}}}{K} = \frac{y_0}{K} = \frac{0}{8.2} = 0$$

$$1.1 = \frac{0.0018 - x_{\text{valve position}}}{-0.0018 - \left(x_{\text{eqm}} - x\right)_{\text{valve position}}} \ln\left[\left(\frac{-0.0018}{\left(x_{\text{eqm}} - x\right)_{\text{valve position}}}\right)\right] \quad (1)$$

Using the equilibrium relationship at the valve position, we can express $(x_{\text{eqm}})_{\text{valve position}}$ in terms of $y_{\text{valve position}}$

$$\left(x_{\text{eqm}}\right)_{\text{valve position}} = \frac{y_{\text{valve position}}}{K} = \frac{y_{\text{valve position}}}{8.2} \quad (2)$$

We also obtained a general operating line equation from part a which we can apply to the column section below the valve as shown. The gas stream flow rate is only reduced in the top section of the column, so $G_{\text{bottom}} = G$. This helps us to express $x_{\text{valve position}}$ in terms of $y_{\text{valve position}}$.

$$y = \frac{L}{G}(x - x_0) \rightarrow y_{\text{valve position}} = \frac{L}{G_{\text{bottom}}}\left(x_{\text{valve position}} - 0.0018\right)$$

$$x_{\text{valve position}} = \left(\frac{G_{\text{bottom}}}{L}\right)y_{\text{valve position}} + 0.0018$$

$$x_{\text{valve position}} = \left(\frac{1650}{6000}\right)y_{\text{valve position}} + 0.0018 \tag{3}$$

Substituting Eqs. (2) and (3) into Eq. (1), we obtain an equation that we can solve for $y_{\text{valve position}}$

$$1.1 = \frac{-\left(\frac{1650}{6000}\right)}{-\frac{1}{8.2} + \left(\frac{1650}{6000}\right)} \ln\left[\left(\frac{-0.0018}{\frac{y_{\text{valve position}}}{8.2} - \left(\frac{1650}{6000}\right)y_{\text{valve position}} - 0.0018}\right)\right]$$

$$= -1.8 \ln\left[\left(\frac{-0.0018}{-0.15y_{\text{valve position}} - 0.0018}\right)\right]$$

$$y_{\text{valve position}} = 0.0101$$

Now we can plot the operating line for the faulty mode of operation. To do so, we can analyze the top section (above the faulty valve) and bottom section (below the valve) separately.

Above the valve, the gas flow rate is reduced by 50%. We can perform a mass balance as follows following the control volume marked in dotted line.

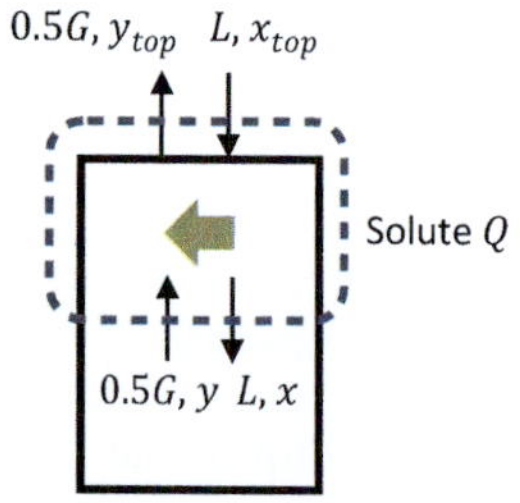


$$\text{In} - \text{Out} = 0$$

$$Lx_{top} + (0.5G)y - (0.5G)y_{top} - Lx = 0$$

$$y = \frac{L}{0.5G}\left(x - x_{top}\right) + y_{top}$$

$$y = \frac{6000}{0.5 \times 1650}\left(x - x_{top}\right) + y_{top} = 7.3\left(x - x_{top}\right) + y_{top}$$

The above equation can be referred to as the Upper Operating Line (UOL), and it is a straight line with a gradient (or slope) of 7.3.

Next, we move on to the section below the valve, to find a similar equation for the Lower Operating Line (LOL). Below the valve, the gas flow rate is not affected. We perform a mass balance as marked in dotted line.

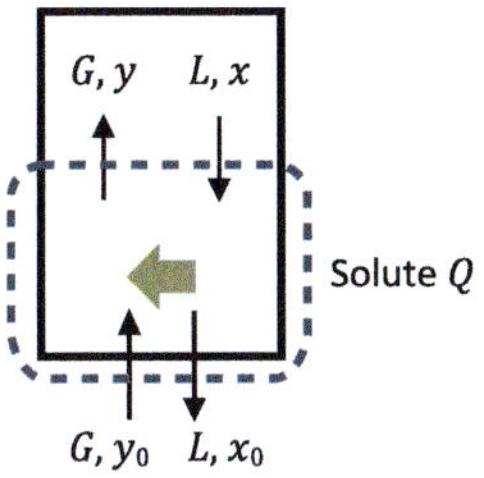


$$\text{In} - \text{Out} = 0$$

$$Lx + Gy_0 - Gy - Lx_0 = 0$$

It is given that $y_0 = 0$ and $x_0 = 0.0018$; therefore the equation simplifies to

$$y = \frac{L}{G}(x - x_0)$$

$$y = \frac{6000}{1650}(x - 0.0018) = 3.6(x - 0.0018)$$

The above equation describes the LOL, and it is a straight line with a gradient (or slope) of 3.6.

We observe that the slope of the UOL is steeper than that of the LOL. We also note key points in the plot, i.e., the mole fractions at the bottom end of the column ($x_0 = 0.0018$ and $y_0 = 0$) and the intersection point between the two operating lines where the faulty valve is located has a value for $y$ of $y_{\text{valve position}} = 0.0101$ as found earlier.

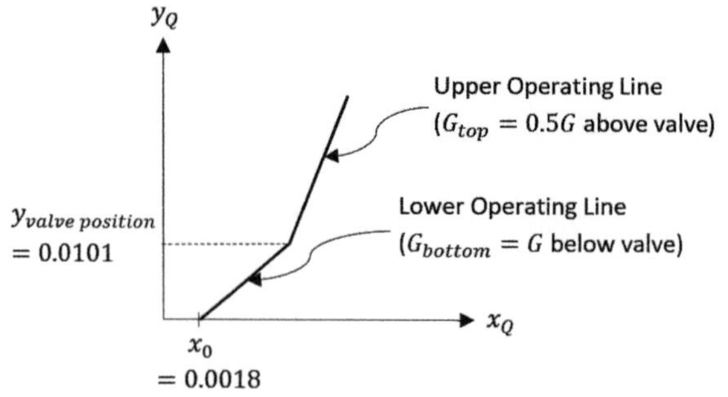


## Problem 26

**A stream of polluted water is processed in a stripping column with the aim of removing a volatile organic compound $A$ from the liquid water stream via contact with an air stream at 298 K and pressure of 1.8 bar. The inlet mole fraction of solute $A$ in the liquid stream is 15 ppm by weight, while the required outlet mole fraction of $A$ is 0.003 ppm. As for the air stream, it is assumed to enter the column solute-free. The volumetric mass-transfer coefficients for the liquid and gaseous phases for solute $A$ are given as $k_La = 0.061\ s^{-1}$ and $k_Ga = 0.84\ s^{-1}$, respectively. The equilibrium constant $K$ of solute $A$ is also given as $K = 149$.**

(a) **Discuss the use of a volumetric mass-transfer coefficient and explain what *a* means.**

(b) **Using an appropriate diagram, discuss the two-film mass transfer theory for the stripping column and explain how the following equation comes about, where *r* refers to the rate of mass transfer of solute $A$ across the liquid and vapor phases. State any assumptions clearly.**

$$r = k_GaC_G(y_b - y_i) = k_LaC_L(x_i - x_b) = K_GaC_G(y_b - y^*)$$

(c) **Determine the value of $K_Ga$ with units clearly indicated, if we base it on a**

   (i) **Concentration driving force**
   (ii) **Partial pressure driving force**

## Solution 26

*Worked Solution*

(a) Volumetric mass-transfer coefficients of the form $k_La$ for the liquid phase and $k_Ga$ for the vapor phase are useful for packed bed operations such as in a packed column. This is because the surface area for mass transfer in a packed bed is difficult to measure easily and hence determine accurately. Therefore it is convenient to adopt volumetric mass-transfer coefficients which use a volume (rather than area) basis. The additional parameter $a$ which is defined as the mass transfer area per unit volume of packed bed helps to convert from an area basis to a volume basis.

(b) Upon examination of the expression, we notice that it essentially equates the rate of mass transfer of solute $A$ in the liquid phase (as denoted by subscript "$L$") and the gas phase (as denoted by subscript "$G$").

This equation originates from Fick's law of diffusion, which helps explain mass transfer phenomena that are largely caused by Brownian motion of molecules. In this case, the mass transfer rate is determined by a chemical potential difference, which is translated into either a concentration or pressure gradient. This gradient drives the movement of molecules from an area of higher concentration to lower concentration. [Note that diffusion is opposed to bulk motion of fluids which can work against a concentration or pressure gradient due to convective effects.]

In this example of a stripping column, we have a gas (air) phase and a liquid (water) phase. Under the two-film theory, we assume there is a gas film and a liquid film on both sides of the gas–liquid interface. It is also assumed that the bulk liquid and bulk gas are perfectly mixed. Under steady-state conditions and assuming no chemical reactions occur, the rate of mass transfer in gas film equals that in the liquid film. This system is illustrated in the diagram below.

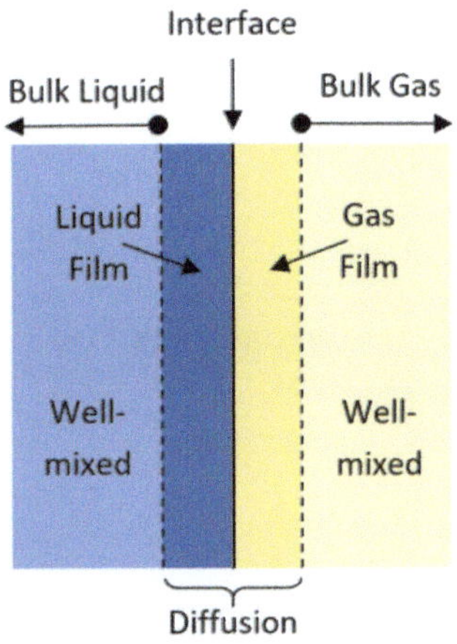


Using Fick's law, we arrive at the equation given:

$$r = k_G a C_G (y_b - y_i) = k_L a C_L (x_i - x_b)$$

In the expression above, $C_G$ and $C_L$ are the total gas and liquid phase concentrations, respectively. $r$ can be expressed in terms of the mole fraction driving force in either the liquid or gas phase, whereby the partial pressure driving force is used if the basis was the gas phase (note that $C_G$ is related to pressure via the ideal gas law where $C_G = P/\mathrm{RT}$), while the concentration driving force is used if the basis was the liquid phase.

The rate expression can be re-expressed in terms of an *overall* mass transfer coefficient $K_G$ which is defined in terms of an overall driving force. The reason for this is because solute compositions at the interface between the gas and liquid films are, in practice, difficult to determine. So we refer to an overall driving force, and we make use of the equilibrium constant $K$ for the solute to create an "imaginary" mole fraction, $y^*$ in this case, which is the gas phase mole fraction of $A$ in equilibrium with the mole fraction of $A$ in the bulk liquid, $x_b$. The two mole fractions are related in terms of $y^* = Kx_b$. In addition to using an overall mass transfer coefficient, we note that the controlling resistance to mass transfer is in the gas film (and not the liquid film); hence $K_G$ instead of $K_L$ is used.

$$r = K_G a C_G (y_b - y^*)$$

We can annotate the two-film theory diagram with mole fractions and concentrations for a stripping column example as shown below:

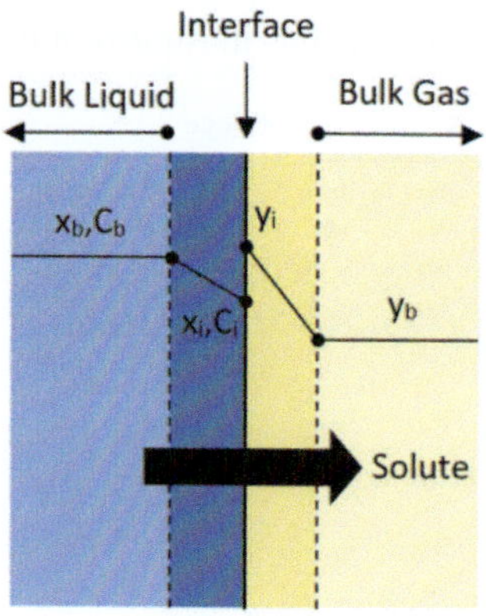


(c) (i) We are given the rate equation and phase equilibrium relationship as follows. We further note that we can express the mole fractions at the interface using the equilibrium constant $K$; therefore, we have a system of simultaneous equations as follows. Note that the rate equation in (1) is such that the overall volumetric mass transfer coefficient, $K_G a$, is based on a concentration driving force.

$$r = K_G a C_G (y_b - y^*) \tag{1}$$

$$y^* = Kx_b \tag{2}$$

$$y_i = Kx_i \tag{3}$$

Our objective is to determine $K_Ga$ so we can substitute Eqs. (2) and (3) into Eq. (1) in a way that helps simplify the expression and single out this parameter such that it can be solved using known values provided. One useful trick is to create the useful pairs of $(y_b - y_i)$ and $(y^* - y_i)$ which are quantities relative to the interface mole fraction $y_i$.

$$\frac{1}{K_GaC_G} = \frac{y_b - y^*}{r} = \frac{(y_b - y_i) - (y^* - y_i)}{r}$$

$$= \frac{y_b - y_i}{r} + \frac{y_i - y^*}{r}$$

$$= \frac{y_b - y_i}{r} + \frac{K(x_i - x_b)}{r}$$

We can now substitute the other expressions for mass transfer rate $r = k_GaC_G(y_b - y_i) = k_LaC_L(x_i - x_b)$ into the equation above. This substitution is useful since we are given values for the liquid and gas phase mass transfer coefficients; hence, we should try to incorporate these parameters into the expression.

$$\frac{y_b - y_i}{r} + \frac{K(x_i - x_b)}{r} = \frac{1}{k_GaC_G} + \frac{K}{k_LaC_L} = \frac{1}{K_GaC_G}$$

$$\frac{1}{K_Ga} = \frac{1}{k_Ga} + \frac{KC_G}{k_LaC_L}$$

Total gas concentration $C_G$ can be determined using the ideal gas law as follows

$$C_G = \frac{P}{\mathrm{RT}} = \frac{180,000}{8.314 \times 298} = 72.7\ \mathrm{mol/m^3}$$

Total gas concentration $C_L$ can be determined using the known properties of liquid water, which are the density $\rho_L$ and molecular weight $M_L$.

$$C_L = \frac{\rho_L}{M_L} = \frac{1,000,000}{18} = 5.56 \times 10^4\ \mathrm{mol/m^3}$$

Finally we can put together all known parameters and solve for $K_Ga$ for a concentration driving force.

$$K_Ga = \frac{1}{\frac{1}{k_Ga} + \frac{KC_G}{k_LaC_L}} = \frac{1}{\frac{1}{0.84} + \frac{149(72.7)}{0.061(5.56 \times 10^4)}} = 0.23\ \mathrm{s^{-1}}$$

(ii) For a partial pressure driving force, we have included the expression on the right with subscript "$p$" in the term $K_G|_p$ to indicate this, and the use of total pressure $P$ instead of total concentration $C_G$

$$r = K_G a C_G (y_b - y^*) = K_G|_p a P (y_b - y^*)$$

$$K_G|_p a = K_G a \frac{C_G}{P}$$

We can express $C_G$ in terms of pressure using the ideal gas law.

$$C_G = \frac{P}{\mathrm{RT}}$$

Substituting the expression for $C_G$ into the mass transfer rate equation, and using our earlier result for $K_G a$ from part c(i), we obtain the overall volumetric mass transfer coefficient based on a partial pressure driving force as follows.

$$K_G|_p a = K_G a \frac{C_G}{P} = \frac{K_G a}{\mathrm{RT}} = \frac{0.23}{8.314 \times 10^{-3} \times 298} = 0.093\ \mathrm{mol/(kPa.m^3.s)}$$

## Problem 27

**Show that for a binary gas mixture containing components $A$ and $B$, the if the permeate side pressure is assumed to be small relative to feed side pressure, the mole fraction of component $A$ on the permeate side, $y_A$ may be expressed as follows, where $x_A$ denotes the mole fraction of component $A$ on the feed side, and $\alpha$ represents selectivity ratio for membrane coefficients where $\alpha = K_{m,A}/K_{m,B}$. Describe any assumptions made in the analysis.**

$$y_A = \frac{\alpha x_A}{1 + (\alpha - 1) x_A}$$

## Solution 27

### *Worked Solution*

Before we solve this problem, let us revisit some basic concepts.

The general equation for flux of component $i$ across a membrane from feed side to permeate side is as follows, where $C_{f,i}$ and $C_{p,i}$ are the feed side and permeate side concentrations, respectively. In gas systems, we typically use partial pressures or mole fractions in place of concentration:

$$N_i'' = K_i (C_{f,i} - C_{p,i})$$

In gas separation, $K_i$ may be approximated to the membrane coefficient $K_{m,i}$ as it dominates over mass transfer effects on $K_i$, and it is further assumed that $K_{m,i}$ (different for different components) is based on partial pressures.

Let us now consider the following diagram for a general binary mixture containing components $A$ and $B$. $p_{f,\mathrm{A}}$ and $p_{f,\mathrm{B}}$ are the partial pressures of $A$ and $B$ on the feed side. $p_{p,\mathrm{A}}$ and $p_{p,\mathrm{B}}$ are the partial pressures of $A$ and $B$ on the permeate side. $P_f$ and $P_p$ are the total pressures on the feed and permeate sides, respectively.

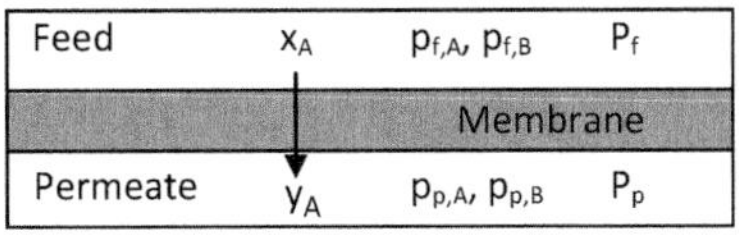


In general, analysis can be done as follows for component $A$:

$$N''_A = K_{m,A}\left(p_{f,A} - p_{p,A}\right)$$

$$N''_A = K_{m,A}\left(P_f x_A - P_p y_A\right)$$

Similarly, for component $B$:

$$N''_B = K_{m,B}\left(p_{f,B} - p_{p,B}\right)$$

$$N''_B = K_{m,B}\left(P_f(1 - x_A) - P_p(1 - y_A)\right)$$

Let the pressure ratio be $\beta = P_p/P_f$ where $0 < \beta < 1$, and selectivity ratio is defined as $\alpha = K_{m,A}/K_{m,B}$.

Substituting $\beta$ into the equation, we have

$$N''_A = K_{m,A}P_f(x_A - \beta y_A)$$

$$N''_B = K_{m,B}P_f((1 - x_A) - \beta(1 - y_A))$$

The amount of component $A$ on the permeate side can be expressed as

$$y_A = \frac{N''_A}{N''_A + N''_B} = \frac{K_{m,A}P_f(x_A - \beta y_A)}{K_{m,A}P_f(x_A - \beta y_A) + K_{m,B}P_f((1 - x_A) - \beta(1 - y_A))}$$

Substituting $\alpha$ into the equation, we have the following quadratic equation in $y_A$.

$$y_A = \frac{x_A - \beta y_A}{(x_A - \beta y_A) + (1/\alpha)((1 - x_A) - \beta(1 - y_A))}$$

One useful simplification is to consider when $\beta \rightarrow 0$, then the equation can be approximated to

$$y_A = \frac{\alpha x_A}{1 + (\alpha - 1)x_A}$$

## Problem 28

**An industrial gas stream containing 55% by volume of hydrogen, 40% of propane, 5% ethane needs to be separated into a pure hydrogen product stream and the remaining hydrocarbon stream using a membrane separation process. The membrane selectivity for hydrogen-propane is 120. The feed is at a pressure of 28 bar and the gas may be assumed as a pseudo-binary mixture.**

(a) **Determine the mole fraction of hydrogen in the permeate at the inlet if the permeate pressure is 2 bar, 7 bar, and 15 bar.**
(b) **Find the permeate pressure necessary to have an initial permeate concentration of 97% hydrogen.**
(c) **If the permeate pressure is 2 bar, and the gas mixture leaves the unit with a hydrogen concentration of 3% by volume, use the arithmetic average (across length of separator from inlet to outlet) of permeate mole fraction to determine the recovery $R$ of permeate and the fraction of hydrogen recovered in the permeate. $R$ is defined as follows whereby $Q_p$ is the volumetric flow rate of permeate (across membrane) and $Q_{in}$ is the volumetric flow rate of feed into the separator at the inlet (parallel to membrane).**

$$R = \frac{Q_p}{Q_{in}}$$

(d) **Using a logarithmic mean average for the volume fluxes at the two ends of the separation unit, find the surface area of membrane necessary to achieve the required purification, if the permeability of hydrogen is $4.5 \times 10^{-5}$ $Nm^3/m^2.s.bar$ and the feed flow rate is 0.15 $Norm^3/s$.**

## Solution 28

*Worked Solution*

(a) Let us illustrate our system as shown below.

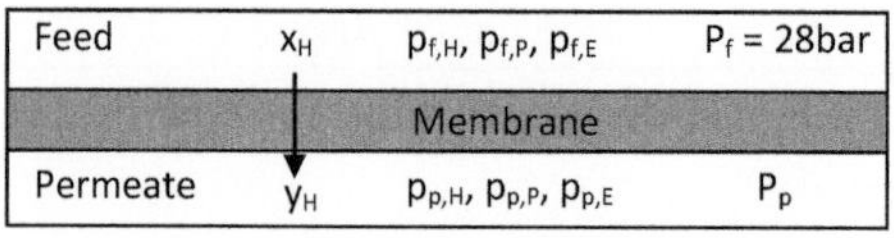

Assuming ideal gas behavior, the volume ratio is equivalent to molar ratio for a constant temperature and pressure. $PV = n\text{RT} \rightarrow V{\sim}n$. We have in this problem $x_H = 0.55$, $x_P = 0.40$, and $x_E = 0.05$. We can construct the volume flux equations for propane and hydrogen, assuming that we have a pseudo binary mixture of hydrogen and the main other component propane. Therefore, $x_P \approx 1 - x_H$ and $y_P \approx 1 - y_H$.

$$N''_H = K_{m,H}\left(p_{f,H} - p_{p,H}\right)$$

$$N''_H = K_{m,H}\left(P_f x_H - P_p y_H\right)$$

$$N''_P = K_{m,P}\left(p_{f,P} - p_{p,P}\right)$$

$$N''_P = K_{m,P}\left(P_f(1 - x_H) - P_p(1 - y_H)\right)$$

Let the pressure ratio be $P_p/P_f$, where $P_f = 28$ bar and selectivity ratio $\alpha = \frac{K_{m,H}}{K_{m,P}} = 120$. Substituting $\beta$ into the equation, we have

$$N''_H = K_{m,H}P_f(x_H - \beta y_H)$$

$$N''_P = K_{m,P}P_f((1 - x_H) - \beta(1 - y_H))$$

The amount of hydrogen on the permeate side can be expressed as

$$y_H = \frac{N''_H}{N''_H + N''_P} = \frac{K_{m,H}P_f(x_H - \beta y_H)}{K_{m,H}P_f(x_H - \beta y_H) + K_{m,P}P_f((1 - x_H) - \beta(1 - y_H))}$$

Substituting selectivity ratio $\alpha = \frac{K_{m,H}}{K_{m,P}} = 120$ into the equation, we have the following quadratic equation in $y_H$.

$$y_H = \frac{x_H - \beta y_H}{(x_H - \beta y_H) + (1/120)((1 - x_H) - \beta(1 - y_H))}$$

$$y_H = \frac{120(x_H - \beta y_H)}{120(x_H - \beta y_H) + (1 - x_H) - \beta(1 - y_H)}$$

We have the following permeate pressures, $P_p = 2$ bar, 7 bar, and 15 bar, to determine the corresponding $y_H$ values. This means $\beta = \frac{2}{28}, \frac{7}{28}$, and $\frac{15}{28}$.

$$y_H = 0.992, \text{ when } \beta = \frac{2}{28} = 0.0714$$

$$y_H = 0.988, \text{ when } \beta = \frac{7}{28} = 0.250$$

$$y_H = 0.935, \text{ when } \beta = \frac{15}{28} = 0.536$$

(b) Now we need to find $P_p$ to have an initial permeate concentration of 97% hydrogen, $y_H = 0.97$. We can substitute this back into the earlier quadratic equation in $y_H$ and solve for $\beta$.

$$0.97 = \frac{120(0.55 - 0.97\beta)}{120(0.55 - 0.97\beta) + (1 - 0.55) - \beta(1 - 0.97)}$$

$$\beta = 0.44$$

$$P_p = (28)\beta = 12.4 \text{ bar}$$

(c)

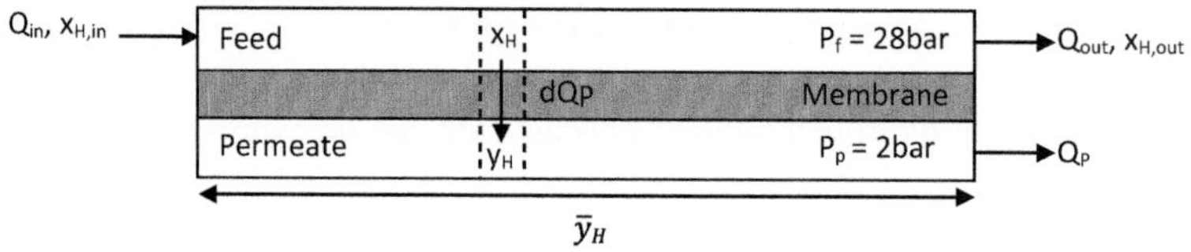


Note that the mole fractions of the gas mixture components vary along the length of membrane from inlet to outlet, therefore we can use an average such as the arithmetic or logarithmic mean average to approximate the permeate side mole fraction.

In this problem, we have $x_{H,\text{in}} = 0.55$ and $x_{H,\text{out}} = 0.03$.

$$\beta = \frac{P_p}{P_f} = \frac{2}{28} = 0.0714$$

$$y_{H,\text{in}} = \frac{120\left(0.55 - 0.0714y_{H,\text{in}}\right)}{120\left(0.55 - 0.0714y_{H,\text{in}}\right) + (1 - 0.55) - 0.0714\left(1 - y_{H,\text{in}}\right)}$$

$$y_{H,\text{in}} = 0.992$$

$$y_{H,\text{out}} = \frac{120\left(0.03 - 0.0714y_{H,\text{out}}\right)}{120\left(0.03 - 0.0714y_{H,\text{out}}\right) + (1 - 0.03) - 0.0714\left(1 - y_{H,\text{out}}\right)}$$

$$y_{H,\text{out}} = 0.359$$

$$\bar{y}_H = \frac{1}{2}(0.992 + 0.359) = 0.676$$

In order to relate to recovery $R$, we can construct a mass balance using volumetric flow rates

$$Q_p\bar{y}_H + Q_{\text{out}}x_{H,\text{out}} = Q_{\text{in}}x_{H,\text{in}}$$

$$\left(\frac{Q_p}{Q_{\text{in}}}\right)\bar{y}_H + \left(\frac{Q_{\text{out}}}{Q_{\text{in}}}\right)x_{H,\text{out}} = x_{H,\text{in}}$$

Since $Q_{\text{out}} = Q_{\text{in}} - Q_p$ and $R = \frac{Q_p}{Q_{\text{in}}}$, therefore,

$$R\bar{y}_H + (1 - R)x_{H,\text{out}} = x_{H,\text{in}}$$

$$R = \frac{x_{H,\text{in}} - x_{H,\text{out}}}{\bar{y}_H - x_{H,\text{out}}} = \frac{0.55 - 0.03}{0.676 - 0.03} = 0.806$$

The fraction of hydrogen recovered in the permeate is

$$\frac{Q_p\bar{y}_H}{Q_{\text{in}}x_{H,\text{in}}} = 0.806\left(\frac{0.676}{0.55}\right) = 0.945$$

(d) Note that permeability $K_{m,H}$ is expressed in units of $\text{Nm}^3/\text{m}^2$. s. bar whereby $\text{Nm}^3$ refers to volume measured at normal conditions of 1 bar and 25 °C.

$$N''_{H,\text{in}} = K_{m,H}P_f\left(x_{H,\text{in}} - \beta y_{H,\text{in}}\right) = \left(4.5 \times 10^{-5}\right)(28)(0.55 - (0.0714)0.992)$$

$$N''_{H,\text{in}} = 6.04 \times 10^{-4}\ \text{Nm}^3/\text{m}^2.\text{s}$$

$$N''_{H,\text{out}} = K_{m,H}P_f\left(x_{H,\text{out}} - \beta y_{H,\text{out}}\right) = \left(4.5 \times 10^{-5}\right)(28)(0.03 - (0.0714)0.359)$$

$$N''_{H,\text{out}} = 5.50 \times 10^{-6}\ \text{Nm}^3/\text{m}^2.\text{s}$$

Note that typically if $\frac{N''_{H,\text{in}}}{N''_{H,\text{out}}} \gg 2$, we can use a logarithmic mean to find an appropriate average. The logarithmic mean volume flux is calculated as follows:

$$\bar{N}''_H = \frac{N''_{H,\text{in}} - N''_{H,\text{out}}}{\ln\left(\frac{N''_{H,\text{in}}}{N''_{H,\text{out}}}\right)} = \frac{6.04 \times 10^{-4} - 5.50 \times 10^{-6}}{\ln\left(\frac{6.04 \times 10^{-4}}{5.50 \times 10^{-6}}\right)} = 1.27 \times 10^{-4}\ \text{Nm}^3/\text{m}^2.\text{s}$$

The area of membrane $A$ required is as follows, where

$$\bar{N}''_H = \frac{Q_P}{A}\bar{y}_H = \frac{RQ_{\text{in}}}{A}\bar{y}_H$$

$$1.27 \times 10^{-4} = \frac{0.806(0.15)}{A}0.676$$

$$A = 643.5 \text{ m}^2$$

## Problem 29

**An air separation unit is used to process a binary feed containing components $A$ and $B$, with 25% by volume of component $A$. The membrane has a selectivity ratio given by $\alpha = \frac{K_{m,A}}{K_{m,B}} = 4.6$ and the permeability of component $A$, $K_{m,A} = 1.6 \times 10^{-5}$ Nm$^3$/m$^2$.s.bar. The unit operates in a cocurrent flow mode, and the retentate contains 97.5% by volume of component $B$.**

**(a) Determine the permeate composition and the fraction of feed obtained as permeate if the pressure at feed side is 13 bar and the pressure at permeate side is 1 bar.**

**(b) Find the area of the membrane needed to treat a feed of 1.4 Nm$^3$/s using numerical integration with step size of 0.025 (you may use a spreadsheet).**

**You may use the following expression for local value of permeate mole fraction $y_i$ without derivation, where $\beta$ is the ratio of permeate side pressure to feed side pressure:**

$$y_i = \frac{x_i - \beta y_i}{x_i - \beta y_i + (1/\alpha)[(1 - x_i) - \beta(1 - y_i)]}$$

## Solution 29

### *Worked Solution*

(a) Let us illustrate our system as shown below.

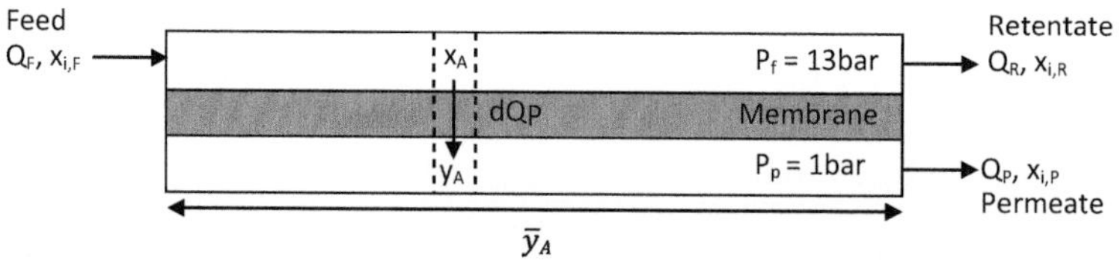


Let us establish some known values from the problem statement. $\alpha$ and $\beta$ are assumed constant along the length of the membrane.

$$\alpha = \frac{K_{m,A}}{K_{m,B}} = 4.6$$

$$\beta = \frac{P_p}{P_f} = \frac{1}{13} = 0.0769$$

We assume ideal gas at constant $P$ and $T$; therefore, volume fraction is the same as mole fraction.

$$x_{A,R} = 1 - 0.975 = 0.025, \; x_{A,F} = 0.25$$

We can find the local values of $y_A$ using the expression given:

$$y_{A,F} = \frac{x_{A,F} - \beta y_{A,F}}{x_{A,F} - \beta y_{A,F} + (1/\alpha)\left[(1 - x_{A,F}) - \beta\left(1 - y_{A,F}\right)\right]}$$

$$y_{A,F} = 0.5695$$

$$y_{A,R} = \frac{x_{A,R} - \beta y_{A,R}}{x_{A,R} - \beta y_{A,R} + (1/\alpha)\left[(1 - x_{A,R}) - \beta\left(1 - y_{A,R}\right)\right]}$$

$$y_{A,R} = 0.08562$$

(b) To find the area of membrane required using manual numerical integration, we can tabulate data on a spreadsheet. To help us get there, let us establish some principles:

- Interval size is given as 0.025. Therefore $\Delta x_A = 0.025$ as we step along the membrane from inlet (first row of table) to outlet (last row of table).
- We can find the corresponding local $y_A$ for each specified local $x_A$ using the equation provided

$$y_A = \frac{x_A - \beta y_A}{x_A - \beta y_A + (1/\alpha)[(1 - x_A) - \beta(1 - y_A)]}$$

- We can perform a mass balance to find differential expression to find recovery R.

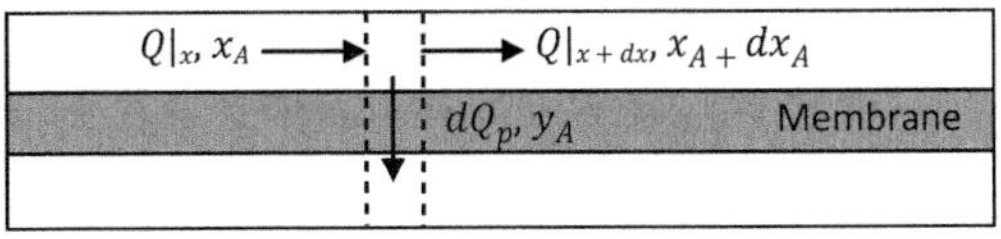

$$\left(Q|_{x}\right)x_A - \left(Q|_{x+dx}\right)(x_A + dx_A) - \left(dQ_p\right)y_A = 0$$

$$\left(Q|_{x}\right)x_A - \left(Q|_{x} + dQ\right)(x_A + dx_A) - \left(dQ_p\right)y_A = 0$$

$$[Q_F(1-R)]x_A - (Q_F(1-R) - Q_F dR)(x_A + dx_A) - (Q_F dR)y_A = 0$$

$$(1-R)x_A - (1-R-dR)(x_A + dx_A) - y_A dR = 0$$

Ignoring higher order terms (i.e., $dRdx_A$) due to their small value, we have

$$(1-R)dx_A - x_A dR + y_A dR = 0$$

$$\frac{dR}{1-R} = \frac{dx_A}{x_A - y_A}$$

$$-\ln(1 - R(x)) = \int_{x_{A,F}}^{x} \frac{dx_A}{x_A - y_A}$$

The above continuous integral can be discretized into the following form for an interval with interval end points from 1 to 2 (see diagram below).

$$-\ln\left(1 - R\left(x_A|_2\right)\right) = \sum_{\text{intervals}} \frac{\Delta x_A}{\bar{x}_A - \bar{y}_A} = \sum_{\text{intervals}} \frac{x_A|_2 - x_A|_1}{\left(\frac{x_A|_1 + x_A|_2}{2}\right) - \left(\frac{y_A|_1 + y_A|_2}{2}\right)}$$

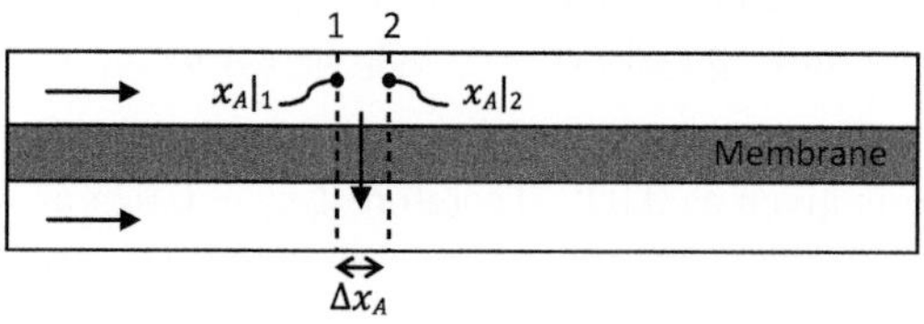


- From a differential element mass balance, we can deduce that differential volume flux is $d\left(N''_A A\right) = y_A dQ_p$, where $y_A$ and $N''_A$ are both local values. For a constant membrane coefficient and constant feed side and permeate side pressures, $N''_A$ is constant along the membrane length. Therefore the equation may be expressed as

$$N''_A dA = y_A dQ_p \rightarrow \bar{N}''_A dA = \bar{y}_A dQ_p$$

In a manual numerical integration using defined step sizes, we will approximate the local values $y_A$ and $N''_A$ using mean values over the interval. It therefore follows that the smaller the step, the closer we get to the local value.

$$A = \int_{\text{inlet}}^{\text{outlet}} dA = \int_{\text{inlet}}^{\text{outlet}} \frac{y_A}{N''_A} dQ_p$$

The above continuous integral can be discretized into the following form:

$$\sum\nolimits_{\text{inlet}}^{\text{outlet}} \Delta A = \sum\nolimits_{\text{inlet}}^{\text{outlet}} \frac{\bar{y}_A}{\bar{N}''_A} \Delta Q_p$$

We know further that we can define recovery as $R = Q_p/Q_F$. Therefore $dQ_p = Q_F dR$. Therefore we can express $dA$ in terms of $dR$:

$$dA = \frac{y_A}{N''_A} dQ_p = \frac{Q_F y_A}{N''_A} dR \rightarrow \frac{Q_F \bar{y}_A}{\bar{N}} {''}_A \Delta R$$

$$A = \int_{\text{inlet}}^{\text{outlet}} \frac{Q_F y_A}{N''_A} dR$$

Again, the above continuous integral can be discretized into the following form:

$$A = \sum\nolimits_{\text{inlet}}^{\text{outlet}} \frac{Q_F \bar{y}_A}{\bar{N}''_A} \Delta R$$

- We can find the mean volume flux over the interval, $\bar{N}''_A$. Considering an interval with interval end points from 1 to 2 (refer to earlier diagram above)

$$\bar{N}''_A\big|_2 = K_{m,A} P_f \left(\bar{x}_A - \beta \bar{y}_A\right)$$
$$= \left(1.6 \times 10^{-5}\right)(13)\left[\left(\frac{x_A|_1 + x_A|_2}{2}\right) - (0.0769)\left(\frac{y_A|_1 + y_A|_2}{2}\right)\right]$$

Applying the points described above, we obtain the data of values shown in the table below.

$$= \frac{0.569533 + 0.533775}{2}$$

$$= \frac{0.225 - 0.25}{\left(\frac{0.25 + 0.225}{2}\right) - \left(\frac{0.569533 + 0.533775}{2}\right)}$$

$$= 0.149929 - 0.076495$$

$$= 1 - e^{-0.162436}$$

$$= \frac{1.4(0.514225)}{3.59724E-05}(0.073434)$$

| Local $x_A$ | Local $y_A$ | Mean over interval $\bar{y}_A$ | $\frac{\Delta x_A}{\bar{x}_A - \bar{y}_A}$ | $\sum \frac{\Delta x_A}{\bar{x}_A - \bar{y}_A}$ | $R$ | $\Delta R$ | Mean vol flux $\bar{N}''_A$ | $\Delta A$ | $\sum A$ |
|---|---|---|---|---|---|---|---|---|---|
| 0.25 ($x_{A,F}$ at inlet) | 0.569533 | - | - | - | - | - | - | - | - |
| 0.225 | 0.533775 | 0.551654 | 0.079579 | 0.079579 | 0.076495 | 0.076495 | 4.05735E-05 | 1456.075 | 1456.075 |
| 0.2 | 0.494675 | 0.514225 | 0.082857 | 0.162436 | 0.149929 | 0.073434 | 3.59724E-05 | 1469.643 | 2925.718 |
| 0.175 | 0.45179 | 0.473233 | 0.087494 | 0.24993 | 0.221145 | 0.071215 | 3.14283E-05 | 1501.263 | 4426.981 |
| 0.15 | 0.404748 | 0.428269 | 0.094067 | 0.343997 | 0.291069 | 0.069924 | 2.69477E-05 | 1555.784 | 5982.764 |
| 0.125 | 0.352984 | 0.378866 | 0.103577 | 0.447574 | 0.360823 | 0.069754 | 2.25381E-05 | 1641.596 | 7624.36 |
| 0.1 | 0.295914 | 0.324449 | 0.117953 | 0.565527 | 0.431939 | 0.071116 | 1.82088E-05 | 1774.031 | 9398.391 |
| 0.075 | 0.232768 | 0.264341 | 0.14137 | 0.706896 | 0.506828 | 0.074889 | 1.39705E-05 | 1983.786 | 11382.18 |
| 0.05 | 0.16296 | 0.197864 | 0.184687 | 0.891584 | 0.589994 | 0.083166 | 9.83418E-06 | 2342.638 | 13724.81 |
| 0.025 ($x_{A,R}$ at outlet) | 0.085621 | 0.124291 | 0.288049 | 1.179633 | 0.692608 | 0.102614 | 5.81135E-06 | 3072.543 | 16797.36 |

$$= 0.079579 + 0.082857$$

$$= (1.6 \times 10^{-5})(13)\left[\left(\frac{0.25 + 0.225}{2}\right) - (0.0769)(0.551654)\right]$$

$$= 1456.075 + 1469.643$$

$$y_A = \frac{x_A - \beta y_A}{x_A - \beta y_A + (1/\alpha)[(1 - x_A) - \beta(1 - y_A)]}$$

Finally, we can determine the membrane area required, which is 16,797 $m^2$.

## Problem 30

**Key factors affecting the rate of transport of a component through a membrane include a driving force and a resistance to mass transfer.**

(a) **Comment on this statement, and explain the following general equation for overall permeability of a membrane separation system for component *i*.**

$$\frac{1}{K_{\text{overall},i}}=\frac{1}{k_{f,i}}+\frac{1}{K_{m,i}}+\frac{1}{k_{p,i}}$$

(b) **Explain the difference between membrane separation for liquid and gas phase systems and describe the phenomenon of concentration polarization. Show this expression, in the case of a steady state between back diffusion caused by concentration polarization and convective transport of solute across the membrane, where $J_v''$ is solvent volumetric flux, $k_{f,i}$ is the mass transfer coefficient on the feed side, and $C_{F,i}|_w$, $C_{F,i}$, and $C_{P,i}$ are the solute concentrations at the wall on the feed side, on the feed side, and on the permeate side.**

$$J_v''\left[\frac{\text{m}^3\ \text{solvent}}{\text{m}^2.\text{s}}\right]=k_{f,i}\ln\left(\frac{C_{F,i}\big|_w-C_{P,i}}{C_{F,i}-C_{P,i}}\right)$$

## Solution 30

### *Worked Solution*

(a) Driving force for mass transport through a membrane is analogous to the difference in fugacities of a component between each side of the membrane. This fugacity difference may be approximated as a concentration difference, $\Delta C_i$, or a difference in partial pressure for gases, $\Delta p_i$.

In liquid separation systems, we have a bulk flow of solvent through the membrane, driven by a pressure difference between each side of the membrane, i.e., the feed side (also called the retentate side at the downstream position), and the permeate side. Below is a simple diagram showing a cocurrent flow membrane separation system where the pressure difference is $P_f - P_p$, and the mole fractions of species $i$ on the feed side and permeate side are $x_i$ and $y_i$, respectively.

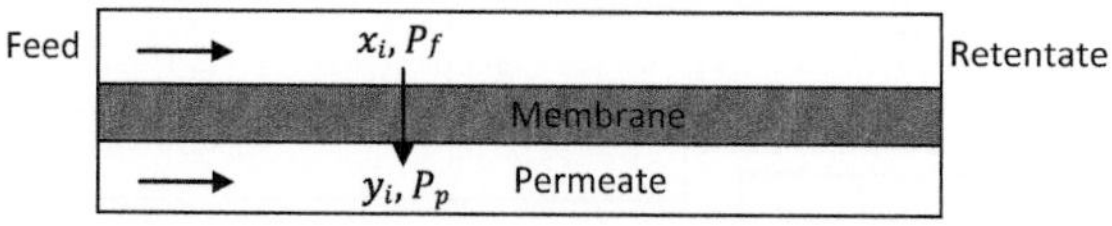

We may find this concept similar in the molar flux equation derived from Fick's law of diffusion as shown below for species $i$. $\mathcal{D}_i$ denotes the diffusion coefficient, $\nabla C_i$ is the concentration gradient, and $\delta_{BL} = x_2 - x_1$ is the boundary layer width over which mass transfer driving force exists.

Molar Flux in $x$ direction:

$$N''_{x,i} = -\mathcal{D}_i\left(\frac{dC_i}{dx}\right) = -\mathcal{D}_i\left(\frac{C_i|_{x_2} - C_i|_{x_1}}{x_2 - x_1}\right) = -\frac{\mathcal{D}_i}{\delta_{BL}}\left(C_i|_{x_2} - C_i|_{x_1}\right) = \frac{\mathcal{D}_i}{\delta_{BL}}\Delta C_i$$

In this equation, the term $\frac{\mathcal{D}_i}{\delta_{BL}}$ is analogous to the overall permeability or mass transfer coefficient in a membrane separation system.

The transport of component $A$ through a membrane in the direction shown in the diagram below can be modeled via a series of mass transfer resistances. $\delta_P$ and $\delta_F$ denote the mass transfer fluid films on the permeate side and feed side, respectively. The feed film resistance is characterized by the mass transfer coefficient, $k_f$, similarly the permeate side film resistance is characterized by the mass transfer coefficient $k_p$. The concentrations of solute at the walls of the membrane in the membrane are $C'_{F,w}$ and $C'_{P,w}$ on the feed side and permeate side, respectively. These corresponding wall concentrations in the free fluid are $C_{F,w}$ and $C_{P,w}$, respectively. In this diagram, $C'_w \neq C_w$ whereby the difference is measured by a solubility factor. When the solute dissolves fully in the membrane, then $C'_w = C_w$. The concentration gradient within the membrane is almost linear, indicating a constant value for membrane permeability, $K_m$.

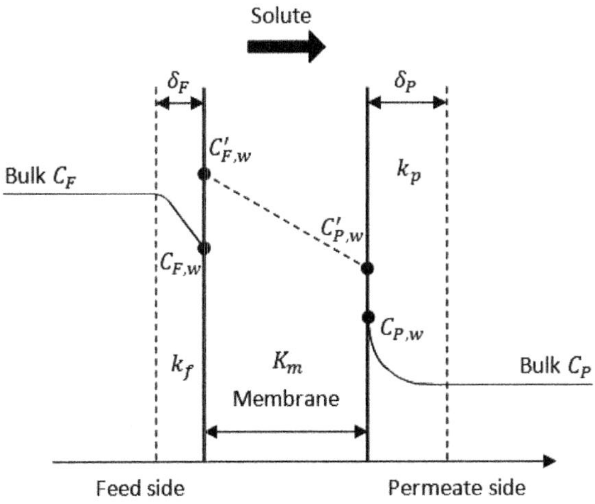


The flux of solute $i$ is given by $N''_i = K_{\text{overall},i}(C_{F,i} - C_{P,i})$, where $K_{\text{overall},i}$ denotes the overall permeability of the system. It is a sum of a series of resistances,

$$\frac{1}{K_{\text{overall},i}} = \frac{1}{k_{f,i}} + \frac{1}{K_{m,i}} + \frac{1}{k_{p,i}}$$

(b) In gas systems, it is typical for the membrane coefficient component to dominate (i.e., mass transfer effects are negligible when $k_{f,i}$ and $k_{p,i} \gg K_{m,i}$); hence the above may be simplified to:

$$\frac{1}{K_{\text{overall},i}} \approx \frac{1}{K_{m,i}}$$

In liquid systems, concentration polarization may occur, i.e., $k_{f,i}$ is small. Concentration polarization is a phenomenon whereby the flux through the membrane is controlled by the film mass transfer resistance on the feed side, rather than the resistance of the membrane, $K_{m,i}$.

$$\frac{1}{K_{\text{overall},i}} \approx \frac{1}{k_{f,i}}$$

Concentration polarization arises because a solute is selectively retained by the membrane, and accumulates at the wall of the membrane, generating a mass transfer resistance for the solute in its mass transport. The flux of solute and solvent towards the membrane convects solute to the wall of the membrane where it becomes "stranded" in the stagnant boundary layer (width of $\delta$) at the wall. At steady state, this is balanced by a back diffusion (in the reverse direction of desired mass transport direction, i.e., back into bulk feed fluid) of the solute down the concentration gradient created by this accumulation of solute at the membrane wall.

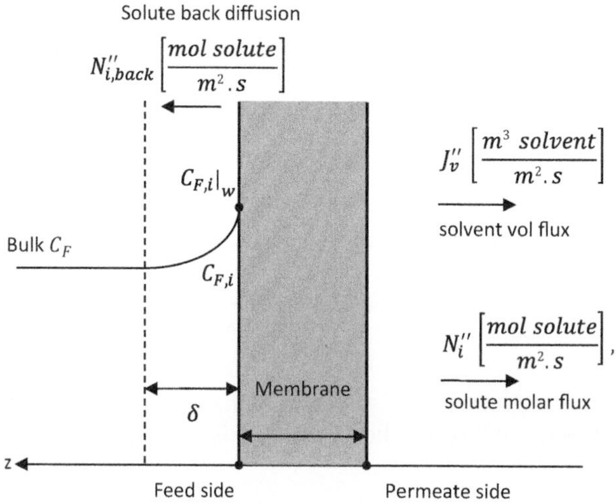

At steady state in the +$z$ direction, we have the molar fluxes on the permeate side equivalent to that on the feed side.

$$N_i''\big|_P = N_i''\big|_F$$

$$-J_v''C_{P,i} = -J_v''C_{F,i} + N_{i,\text{back}}''$$

$$-J_v''C_{P,i} = -J_v''C_{F,i} - \mathcal{D}_i\frac{dC_{F,i}}{dz}$$

$$J_v''(C_{F,i} - C_{P,i}) = -\mathcal{D}_i\frac{dC_{F,i}}{dz}$$

$$\int_0^\delta J_v''dz = \int_{C_{F,i}|_w}^{C_{F,i}} \frac{-\mathcal{D}_i}{C_{F,i} - C_{P,i}}dC_{F,i}$$

$$J_v''\delta = \mathcal{D}_i \ln\left(\frac{C_{F,i}|_w - C_{P,i}}{C_{F,i} - C_{P,i}}\right)$$

$$J_v'' = \frac{\mathcal{D}_i}{\delta}\ln\left(\frac{C_{F,i}|_w - C_{P,i}}{C_{F,i} - C_{P,i}}\right) = k_{f,i}\ln\left(\frac{C_{F,i}|_w - C_{P,i}}{C_{F,i} - C_{P,i}}\right)$$

The above equation describes the concentration polarization effect, where $k_{f,i}$ denotes the mass transfer coefficient across the stagnant film on the feed side. It is noted that the solvent flux, $J_v''$, depends greatly on the solution concentration profile. For a larger solvent flux, the higher the solute concentration due to wall accumulation, $C_{F,i}|_w$. To decrease back $C_{F,i}|_w$, $k_{f,i}$ needs to be increased by methods such as increasing the rate of bulk flow parallel to the surface of the membrane (tangential to surface of membrane).

This trend of an increasing $J_v''$ with increasing $C_{F,i}|_w$ as implied by the equation, however, is not indefinite in reality because of the following factors which work to decrease back the solvent flux as $C_{F,i}|_w$ increases.

- Increased local osmotic pressure will decrease the solvent flux.
- The formation of a solute gel (gel effect) which is a high viscosity structured liquid through which the solute cannot penetrate and poses resistance to the solvent flux. Then $C_{P,i} = 0$ and $C_{F,i}|_w = C_{F,\text{gel},i}$ and a limiting flux $J_v''|_{\text{limit}}$ is reached. We can increase solvent flux by increasing the pressure difference between feed and permeate side; however, this is only temporary before a thicker gel forms to decrease back the flux to a limiting value.

$$J_v''\big|_{\text{limit}} = k_{f,i}\ln\left(\frac{C_{F,\text{gel},i}}{C_{F,i}}\right)$$

## Problem 31

**A membrane separation unit used to remove salt from seawater contains hollow tubes of outer diameter $D_2 = 2 \times 10^{-4}$ m and inner diameter $D_1 = 1 \times 10^{-4}$ m. The volume flux of water $J_v''$ is $3.9 \times 10^{-6}$ m$^3$/m$^2$.s. The feed contains salt concentration of $C_{F,i} = 1.5 \times 10^{-4}$ mol/m$^3$ at room temperature of 25 °C, and flows perpendicular to the outside of the tubes at velocity of 0.006 m/s with an apparent solute retention ratio $R_{app} = 0.98$. Comment on whether concentration polarization is significant.**

**You may assume that the solute has the same properties as water, and $\mathcal{D}_{salt} = 1.5 \times 10^{-9}$ m$^2$/s. Note that flow that is perpendicular to a cylinder can be described by the correlation Sh = 0.6 Re$^{0.5}$Sc$^{0.3}$. The kinematic viscosity of water at 25 °C is $\nu = 8.93 \times 10^{-7}$ m$^2$/s. You may use the following expression without derivation, where $J_v''$ is solvent volumetric flux, $k_{f,i}$ is the mass transfer coefficient on the feed side, and $C_{F,i}|_w$, $C_{F,i}$, and $C_{P,i}$ are the solute concentrations at the wall on the feed side, on the feed side, and on the permeate side.**

$$J_v'' = k_{f,i} \ln\left(\frac{C_{F,i}|_w - C_{P,i}}{C_{F,i} - C_{P,i}}\right)$$

## Solution 31

### *Worked Solution*

Before we solve this problem, let us revisit some concepts. Apparent solute retention ratio, $R_{app}$ is defined as follows where $C_{i,P}$ and $C_{i,F}$ are the solute concentrations on the permeate side and feed side, respectively. This is an apparent value as the true value is based on wall concentration on the feed side. Note that solute retention ratio is the same as solute rejection ratio.

$$R_{app} = 1 - \frac{C_{P,i}}{C_{F,i}}, \quad R_{true} = 1 - \frac{C_{P,i}}{C_{F,i}|_{wall}}$$

In our system, we have a cross flow pattern across tubular membranes, which is common in commercial operations. This can help remove fouling, and facilitate semicontinuous operations instead of batch mode. Hollow tubular membranes are typically asymmetric membranes that are specially prepared such that the membrane is composite in nature, being supported by a highly porous layer capable of withstanding pressure drops while offering little resistance to flow. This supporting layer allows the membrane itself to remain thin (to maximize flux) while ensuring structural integrity. Some examples of membrane separation in commercial processes include desalination via reverse osmosis, ultrafiltration for macromolecular concentration, hydrogen recovery in gas separation, and ion separation in electrodialysis.

Looking at this problem, we can first calculate the dimensionless quantities that describe the flow.

The Reynolds number is commonly used in fluid analysis and represents the ratio of inertial forces to viscous forces. It can be used to determine if flow is laminar or turbulent.

$$\text{Re} = \frac{VD_2}{\nu} = \frac{(0.006)\left(2 \times 10^{-4}\right)}{8.93 \times 10^{-7}} = 1.34$$

The Schmidt number represents the ratio of momentum diffusivity (kinematic viscosity) and mass diffusivity, and is used for fluid flows where there are both momentum and mass diffusion convection processes.

$$\text{Sc} = \frac{\nu}{\mathcal{D}} = \frac{8.93 \times 10^{-7}}{1.5 \times 10^{-9}} = 595$$

The Sherwood number represents the ratio of the convective mass transfer to the rate of diffusive mass transport.

$$\text{Sh} = \frac{k_f D_2}{\mathcal{D}} = 0.6\text{Re}^{0.5}\text{Sc}^{0.3} = 0.6\left(1.34^{0.5}\right)\left(595^{0.3}\right) = 4.72$$

$$k_f = \frac{4.72\left(1.5 \times 10^{-9}\right)}{2 \times 10^{-4}} = 3.5 \times 10^{-5}\ m/s$$

We can find the solute concentration on the permeate side, $C_{P,i}$.

$$R_{\text{app}} = 0.98 = 1 - \frac{C_{P,i}}{C_{F,i}}$$

$$C_{P,i} = (1 - 0.98)\left(1.5 \times 10^{-4}\right) = 3 \times 10^{-6}\ \text{mol/m}^3$$

To find out whether concentration polarization is significant, we can determine the solute concentration at the wall on the feed side and compare it with the concentration in the free fluid on the feed side.

$$J''_v = k_{f,i} \ln\left(\frac{C_{F,i}|_w - C_{P,i}}{C_{F,i} - C_{P,i}}\right)$$

$$C_{F,i}|_w = (C_{F,i} - C_{P,i})\exp\left(\frac{J''_v}{k_{f,i}}\right) + C_{P,i}$$

$$= \left(1.5 \times 10^{-4} - 3 \times 10^{-6}\right)\exp\left(\frac{3.9 \times 10^{-6}}{3.5 \times 10^{-5}}\right) + 3 \times 10^{-6}$$

$$C_{F,i}\big|_w = 1.66 \times 10^{-4}\ \mathrm{mol/m^3}$$

$$\frac{C_{F,i}\big|_w - C_{F,i}}{C_{F,i}} = \frac{1.66 \times 10^{-4} - 1.5 \times 10^{-4}}{1.5 \times 10^{-4}} = 0.106 > 0.1$$

This means that the solute concentration at the wall due to concentration polarization just exceeds about 10% accumulation, which is just about significant.

We were given the apparent retention ratio, $R_{\mathrm{app}} = 0.98$. We can also find the true retention ratio to observe the difference. We find that $R_{\mathrm{true}}$ is slightly greater than $R_{\mathrm{app}}$.

$$R_{\mathrm{true}} = 1 - \frac{C_{P,i}}{C_{F,i}\big|_{\mathrm{wall}}} = 1 - \frac{3 \times 10^{-6}}{1.66 \times 10^{-4}} = 0.982$$

## Problem 32

**Determine the osmotic pressure difference between pure water and an aqueous solution containing the following molar compositions of an ideal solute at 298 K.**

(a) **1.5 mol%**
(b) **2.5 mol%**

## Solution 32

***Worked Solution***

(a) The system can be illustrated as shown below:

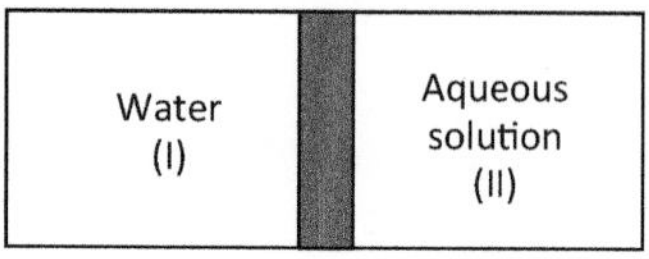


$$\mu_w^I = g_w\left(T, P^0\right) \tag{1}$$

$$\mu_w^{II} = g_w(T, P) + \mathrm{RT}\ln x_w$$

$$\mu_w^{II} = g_w\left(T, P^0\right) + \frac{\partial g}{\partial P}\bigg|_T \left(P - P^0\right) + \mathrm{RT}\ln x_w \tag{2}$$

The chemical potentials of water in ($I$) and ($II$) are equivalent at equilibrium. Also, we know that $dg = vdP - sdT = vdP$ or $\frac{\partial g}{\partial P}\big|_T = v$ at constant $T$. Therefore,

$$\mu_w^I = \mu_w^{II}$$

$$g_w(T,P^0) = g_w(T,P^0) + \left.\frac{\partial g}{\partial P}\right|_T (P - P^0) + \mathrm{RT}\ln x_w$$

$$0 = v(P - P^0) + \mathrm{RT}\ln x_w$$

$$\frac{-\mathrm{RT}\ln x_w}{v} = P - P^0 = \Delta\Pi$$

The density of water is 1000 kg/m$^3$, and its molecular weight is 0.018 kg/mol, hence the molar volume of water is

$$v = \frac{0.018}{1000}\ \mathrm{m^3/mol}$$

Therefore we can find the osmotic pressure differences at the two different solute molar compositions.

$$\Delta\Pi|_{1.5\ \mathrm{mol\%}} = \frac{-(8.314)(298)\ln(1 - 0.015)}{\left(\frac{0.018}{1000}\right)} = 2.08 \times 10^6\ \mathrm{Pa}$$

(b) The same method is used to find the osmotic pressure difference at 2.5 mol% of solute.

$$\Delta\Pi|_{2.5\ \mathrm{mol\%}} = \frac{-(8.314)(298)\ln(1 - 0.025)}{\left(\frac{0.018}{1000}\right)} = 3.48 \times 10^6\ \mathrm{Pa}$$

## Problem 33

**A reverse osmosis system is used to remove a solute $X$ from water containing 35 g/L of $X$. The pressure on the feed side is 65 barg while that at the permeate side is at atmospheric pressure. The osmotic pressure in the solution can be expressed as $\Pi$[bar] = 0.75$C$ [g/L], where $C$ refers to concentration of solute in the solution. The apparent solute rejection factor and permeability of the membrane are $R_{app} = 0.997$ and $K_m = 3.8 \times 10^{-3}$ m$^3$/m$^2$.h.bar. We are required to process a feed of flow rate $Q_f = 10$ m$^3$/day.**

(a) **Determine the maximum recovery $S_{max}$ if $S$ remains constant and concentration polarization effects may be ignored.**
(b) **If the operating value of recovery, $S = 0.5S_{max}$, find the membrane area required and the concentration of solute in the permeate. Arithmetic mean values may be used.**
(c) **If concentration polarization effects are significant, the expression below holds whereby $C_f|_w$ is the solute concentration at the membrane wall. Find $C_f|_w$ at the inlet if the mass transfer coefficient $k_f = 1.2 \times 10^{-4}$ m/s.**

$$J_v'' = k_f \ln\left(\frac{C_f|_w - C_p}{C_f - C_p}\right)$$

## Solution 33

### *Worked Solution*

(a) Let us start with an illustration of our reverse osmosis system. As the feed solution flows along membrane, water is removed from it and it becomes more concentrated in $X$ as it reaches the tail end of the feed side (i.e., retentate side). The permeate side will contain water with the desired purity.

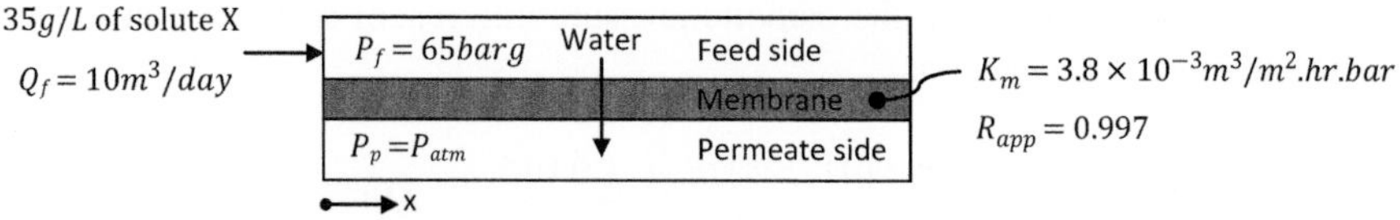


At maximum recovery $S$, osmotic pressure is equivalent to pressure difference across the membrane.

$$\Delta \Pi = \Delta P$$

$$\Pi_{f,w} - \Pi_p = P_f - P_p = (65 + P_{\text{atm}}) - P_{\text{atm}} = 65$$

When concentration polarization effects are ignored, then $C_{f,w} = C_f$ and $\Pi_{f,w} = \Pi_f$. Therefore,

$$\Pi_f - \Pi_p = 65 \tag{1}$$

We note that by definition, $R_{\text{app}} = 1 - C_p/C_f$ therefore $R_{\text{app}}C_f = C_f - C_p$

$$\Pi_f - \Pi_p = 0.75(C_f - C_p) = 0.75(R_{\text{app}}C_f) = 0.75(0.997)C_f$$

Substituting this result into Eq. (1), we have the following equation at maximum $S$.

$$0.75(0.997)C_f|_{S_{\max}} = 65$$

$$C_f|_{S_{\max}} = 86.927g/L$$

Recovery $S$ is the ratio of volumetric flow rate across the membrane to the feed side flow rate. We can perform a mass balance to find a differential expression for $S$.

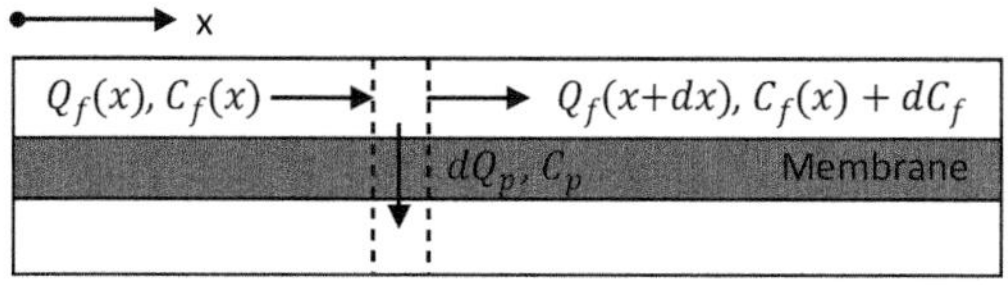


$$Q_f(x)C_f(x) - Q_f(x+dx)\left(C_f(x)+dC_f\right) - \left(dQ_p\right)C_p = 0$$
$$Q_f(x)C_f(x) - \left(Q_f(x)+dQ_f\right)\left(C_f(x)+dC_f\right) - \left(dQ_p\right)C_p = 0$$

Let $Q_f(x=0) = Q_{\text{in}}$ therefore $Q_f(x) = Q_{\text{in}}(1-S)$

$$[Q_{\text{in}}(1-S)]C_f(x) - (Q_{\text{in}}(1-S) - Q_{\text{in}}dS)\left(C_f(x)+dC_f\right) - (Q_{\text{in}}dS)C_p = 0$$
$$(1-S)C_f(x) - (1-S-dS)\left(C_f(x)+dC_f\right) - C_p dS = 0$$

Ignoring higher order terms (i.e., $dSdC_f$) due to their small value, we have

$$(1-S)dC_f - C_f dS + C_p dS = 0$$
$$\frac{dS}{1-S} = \frac{dC_f}{C_f - C_p}$$
$$-\ln(1-S(x)) = \int_{C_f(x=0)}^{C_f(x)} \frac{dC_f}{C_f - C_p}$$

In our problem, we know that $C_f(x=0) = 35$ g/L, and we found earlier that $C_f\big|_{S_{\max}} = 86.927\ \text{g/L}$; therefore,

$$-\ln(1-S_{\max}) = \int_{35\text{g/L}}^{86.927\text{g/L}} \frac{dC_f}{R_{\text{app}}C_f} = \frac{1}{0.997}\ln\left(\frac{86.927}{35}\right)$$
$$S_{\max} = 0.59846$$

(b) We are told that the operating recovery is $0.5S_{\max}$, and let us define the solute concentration at the outlet on the feed side be $C_{f,\text{out}}$

$$S = 0.5(0.59846) = 0.299$$
$$-\ln(1-0.299) = \frac{1}{0.997}\ln\left(\frac{C_{f,\text{out}}}{35}\right)$$

$$C_{f,\text{out}} = 49.875g/L$$

Therefore the concentration of solute in the permeate $C_{p,\text{ out}}$ is

$$C_{p,\text{out}} = C_{f,\text{out}}(1 - R) = 49.875(1 - 0.997) = 0.1496 \text{ g/L}$$
$$C_{p,\text{in}} = C_{f,\text{in}}(1 - R) = 35(1 - 0.997) = 0.105 \text{ g/L}$$

To find the area of membrane required, we can relate to the volume flux of solvent (i.e., water) across the membrane averaged over the length of the unit from inlet to outlet, $\bar{J}''_v$. We are also told that $Q_f = 10$ m$^3$/day.

$$\bar{J}''_v A = Q_p \quad (2)$$

$$Q_p = S(Q_f) = 0.299(10)\left(\frac{1}{24}\right) = 0.1246 \text{ m}^3/\text{h}$$

$$\bar{J}''_v = \frac{J''_v|_{\text{inlet}} + J''_v|_{\text{outlet}}}{2} = \frac{K_m\left[(\Delta P - \Delta\Pi)_{\text{inlet}} + (\Delta P - \Delta\Pi)_{\text{outlet}}\right]}{2}$$

$$\bar{J}''_v = \frac{3.8 \times 10^{-3}\left\{\left[60 - 0.75\left(C_{f,\text{in}} - C_{p,\text{in}}\right)\right] + \left[60 - 0.75\left(C_{f,\text{out}} - C_{p,\text{out}}\right)\right]\right\}}{2}$$

$$\bar{J}''_v = \frac{3.8 \times 10^{-3}\{[60 - 0.75(35 - 0.105)] + [60 - 0.75(49.875 - 0.1496)]\}}{2}$$
$$= 0.107 \text{ m}^3/\text{m}^2.\text{h}$$

Combining our results in Eq. (2), we have

$$0.107A = 0.1246$$
$$A = 1.16 \text{ m}^2$$

(c) If concentration polarization effects are significant, then to find $C_{f,\text{in}}|_w$, we can use the equation

$$J''_v\big|_{\text{in}} = k_f \ln\left(\frac{C_{f,\text{in}}|_w - C_{p,\text{in}}}{C_{f,\text{in}} - C_{p,\text{in}}}\right)$$

$$J''_v\big|_{\text{in}} = K_m(\Delta P - \Delta\Pi)_{\text{inlet}} = K_m\left[\Delta P - 0.75\left(C_{f,\text{in}}|_w - C_{p,\text{in}}\right)\right]$$

$$J''_v\big|_{\text{in}} = \frac{3.8 \times 10^{-3}}{3600}\left[60 - 0.75\left(R_{\text{app}}C_{f,\text{in}}|_w\right)\right]$$
$$= \frac{3.8 \times 10^{-3}}{3600}\left[60 - 0.75\left(0.997C_{f,\text{in}}|_w\right)\right]$$

Combining the expressions for solvent volume flux at the inlet, we have the following given that $k_f = 1.2 \times 10^{-4}$ m/s,

$$\frac{3.8 \times 10^{-3}}{3600}\left[60 - 0.75\left(0.997C_{f,\text{in}}\big|_w\right)\right] = \left(1.2 \times 10^{-4}\right)\ln\left(\frac{C_{f,\text{in}}\big|_w - C_{p,\text{in}}}{C_{f,\text{in}} - C_{p,\text{in}}}\right)$$

$$\frac{3.8 \times 10^{-3}}{3600}\left[60 - 0.75\left(0.997C_{f,\text{in}}\big|_w\right)\right] = \left(1.2 \times 10^{-4}\right)\ln\left[\frac{0.997C_{f,\text{in}}\big|_w}{35 - (1 - 0.997)C_{f,\text{in}}\big|_w}\right]$$

$$60 - 0.74775C_{f,\text{in}}\big|_w = 113.68\ln\left[\frac{0.997C_{f,\text{in}}\big|_w}{35 - 0.003C_{f,\text{in}}\big|_w}\right]$$

$$C_{f,\text{in}}\big|_w = 44.3 \text{ g/L}$$

## Problem 34

**An ultrafiltration unit uses a membrane to concentrate a solution from a feed of solute concentration 70 g/L to a product of 300 g/L. The solute has a rejection ratio of $R = 1$ by the membrane. At high solute concentrations exceeding 390 g/L, a gel layer forms. A single stage membrane is used with a recycle of retentate. The recycle stream flow rate $Q_R$ is much greater than the inlet feed flow rate at $Q_f = 60$ m$^3$/h. Changes in osmotic pressure and concentrations along the membrane are assumed negligible. Given that the membrane permeability to water is $8.2 \times 10^{-6}$ m/s.bar, and the mass transfer coefficient $k_f$ is $2.5 \times 10^{-5}(V^{0.68})$m/s.**

(a) **Determine the minimum membrane surface area required to process a process stream at $V = 1.5$ m/s and 4 m/s.**
(b) **If two units are placed in series, find the area required for the same velocities (as indicated in part a) of process streams passing through both units.**

## Solution 34

*Worked Solution*

(a) Let us illustrate a diagram for our ultrafiltration unit.

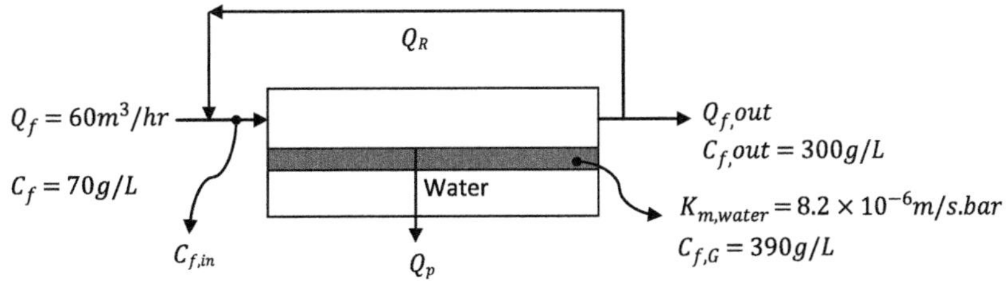

To relate to membrane area, we can start with the expression for volume flux of solvent (i.e., water in this case). Note that a constant mean flux $\bar{J}''_v$ across the length of membrane is used here, since we are told that variations in concentrations and osmotic pressure may be ignored along the membrane length. For the minimum area, we will require maximum flux, therefore

$$\bar{J}''_{v,\max} A_{\min} = Q_p$$

In order to find $Q_p$, we can construct a simple mass balance. Maximum volume flux occurs when we reach the limiting solute concentration on feed side at the membrane wall, $C_{f,\mathrm{wall}} = C_{f,G}$. We are told that $R = 1$. By definition, $RC_f = C_f - C_p$; therefore $C_p = 0$.

$$Q_f C_f = Q_p C_p + Q_{f,\mathrm{out}} C_{f,\mathrm{out}}$$

$$\frac{60}{3600}(70) = Q_p(0) + Q_{f,\mathrm{out}}(300)$$

$$Q_{f,\mathrm{out}} = 3.889 \times 10^{-3}\ \mathrm{m^3/s}$$

$$Q_p = Q_f - Q_{f,\mathrm{out}} = \frac{60}{3600} - 3.889 \times 10^{-3} = 0.01278\ \mathrm{m^3/s}$$

$$\bar{J}''_{v,\max} A_{\min} = 0.01278$$

We can also express $\bar{J}''_{v,\max}$ in terms of mass transfer coefficient $k_f$ and solute concentrations for the concentration polarization effect. Note that since $Q_R \gg Q_f$, therefore $C_{f,\mathrm{in}} \approx C_{f,\mathrm{out}} = 300$ g/L (and not $C_f = 60$ g/L).

$$\bar{J}''_{v,\max} = k_f \ln\left(\frac{C_{f,\mathrm{wall}} - C_p}{C_{f,\mathrm{in}} - C_p}\right) = \left(2.5 \times 10^{-5}\right)\left(V^{0.68}\right) \ln\left(\frac{C_{f,G} - 0}{C_{f,\mathrm{in}} - 0}\right)$$

Note that $k_f$ is a function of flow velocity; therefore, we have the following fluxes at each velocity

$$\bar{J}''_{v,\max}\Big|_{V=1.5\ \mathrm{m/s}} = \left(2.5 \times 10^{-5}\right)\left(1.5^{0.68}\right) \ln\left(\frac{390}{300}\right) = 8.641 \times 10^{-6}\ \mathrm{m^3/m^2.s}$$

$$A_{\min}\Big|_{V=1.5\ \mathrm{m/s}} = \frac{0.01278}{8.641 \times 10^{-6}} = 1479\ \mathrm{m^2}$$

$$\bar{J}''_{v,\max}\Big|_{V=4\ \mathrm{m/s}} = \left(2.5 \times 10^{-5}\right)\left(4^{0.68}\right) \ln\left(\frac{390}{300}\right) = 1.684 \times 10^{-5}\ \mathrm{m^3/m^2.s}$$

$$A_{\min}\Big|_{V=4\ \mathrm{m/s}} = \frac{0.01278}{1.684 \times 10^{-5}} = 759\ \mathrm{m^2}$$

(b)

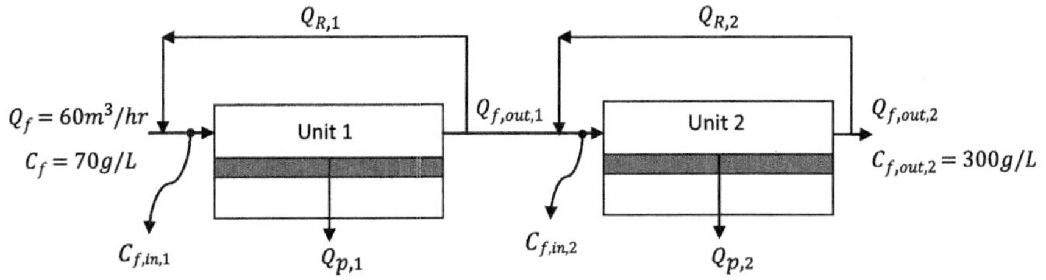


We know that $Q_f = 60$ m$^3$/h. From part a, we obtained $Q_{f,\text{out}} = 3.889 \times 10^{-3}$ m$^3$/s which will be equivalent to $Q_{f,\text{out},2}$ since we are constrained by the same feed entering the two-unit system.

$$Q_{f,\text{out},2} = 3.889 \times 10^{-3}\ \text{m}^3/\text{s}$$

Furthermore, since the two units are the same, we have $Q_{p,1} + Q_{p,2} = 0.01278$ m$^3$/s ($Q_p$ from part a) whereby $Q_{p,1} = Q_{p,2}$; therefore

$$Q_{p,1} = Q_{p,2} = 0.5(0.01278) = 6.39 \times 10^{-3}\ \text{m}^3/\text{s}$$

We can perform a mass balance around unit 1 to find $C_{f,\text{out},1}$

$$Q_f = Q_{p,1} + Q_{f,\text{out},1}$$

$$\frac{60}{3600} = 6.39 \times 10^{-3} + Q_{f,\text{out},1}$$

$$Q_{f,\text{out},1} = 0.010277\ \text{m}^3/\text{s}$$

$$Q_f C_f = Q_{p,1} C_{p,1} + Q_{f,\text{out},1} C_{f,\text{out},1} = Q_{p,1}(0) + Q_{f,\text{out},1} C_{f,\text{out},1}$$

$$C_{f,\text{out},1} = \frac{\frac{60}{3600}(70)}{0.010277} = 113.5\ \text{g/L}$$

We can also perform a mass balance around both units,

$$Q_f C_f = Q_{p,1} C_{p,1} + Q_{p,2} C_{p,2} + Q_{f,\text{out},2} C_{f,\text{out},2}$$

$$\frac{60}{3600}(70) = Q_{p,1}(0) + Q_{p,2}(0) + Q_{f,\text{out},2}(300)$$

$$Q_{f,\text{out},2} = 3.889 \times 10^{-3}\ \text{m}^3/\text{s}$$

Now we can find the membrane areas required for each unit, starting with unit 1.

$$\bar{J}''_{v,\max,1} A_{\min,1} = Q_{p,1} = 6.39 \times 10^{-3}\ \text{m}^3/\text{s}$$

Similar to part a, note that since $Q_{R,1}$ and $Q_{R,2}$ are large, therefore $C_{f,\text{in},1} \approx C_{f,\text{out},1}$ (and not $C_f$) and $C_{f,\text{in},2} \approx C_{f,\text{out},2}$ (and not $C_{f,\text{out},1}$)

$$\bar{J}''_{v,\max,1} = k_f \ln\left(\frac{C_{f,\text{wall}} - C_{p,1}}{C_{f,\text{in},1} - C_{p,1}}\right) = \left(2.5 \times 10^{-5}\right)\left(V^{0.68}\right) \ln\left(\frac{C_{f,G} - 0}{C_{f,\text{out},1} - 0}\right)$$

$$\bar{J}''_{v,\max,1}\big|_{V=1.5\ \text{m/s}} = \left(2.5 \times 10^{-5}\right)\left(1.5^{0.68}\right) \ln\left(\frac{390}{113.5}\right) = 4.0655 \times 10^{-5}\text{m}^3/\text{m}^2.\text{s}$$

$$A_{\min,1}\big|_{V=1.5\ \text{m/s}} = \frac{6.39 \times 10^{-3}}{4.0655 \times 10^{-5}} = 157\ \text{m}^2$$

$$\bar{J}''_{v,\max,1}\big|_{V=4\ \text{m/s}} = \left(2.5 \times 10^{-5}\right)\left(4^{0.68}\right) \ln\left(\frac{390}{113.5}\right) = 7.921 \times 10^{-5}\ \text{m}^3/\text{m}^2.\text{s}$$

$$A_{\min,1}\big|_{V=4\ \text{m/s}} = \frac{6.39 \times 10^{-3}}{7.921 \times 10^{-5}} = 80.67\ \text{m}^2$$

Repeating the steps for unit 2, we have

$$\bar{J}''_{v,\max,2} A_{\min,2} = Q_{p,2} = 6.39 \times 10^{-3}\ \text{m}^3/\text{s}$$

$$\bar{J}''_{v,\max,2} = k_f \ln\left(\frac{C_{f,\text{wall}} - C_{p,2}}{C_{f,\text{in},2} - C_{p,2}}\right) = \left(2.5 \times 10^{-5}\right)\left(V^{0.68}\right) \ln\left(\frac{C_{f,G} - 0}{C_{f,\text{out},2} - 0}\right)$$

$$\bar{J}''_{v,\max,2}\big|_{V=1.5\ \text{m/s}} = \left(2.5 \times 10^{-5}\right)\left(1.5^{0.68}\right) \ln\left(\frac{390}{300}\right) = 8.641 \times 10^{-6}\ \text{m}^3/\text{m}^2.\text{s}$$

$$A_{\min,2}\big|_{V=1.5\ \text{m/s}} = \frac{6.39 \times 10^{-3}}{8.641 \times 10^{-6}} = 739\ \text{m}^2$$

$$\bar{J}''_{v,\max,2}\big|_{V=4\ \text{m/s}} = \left(2.5 \times 10^{-5}\right)\left(4^{0.68}\right) \ln\left(\frac{390}{300}\right) = 1.684 \times 10^{-5}\ \text{m}^3/\text{m}^2.\text{s}$$

$$A_{\min,2}\big|_{V=4\ \text{m/s}} = \frac{6.39 \times 10^{-3}}{1.684 \times 10^{-5}} = 379.5\ \text{m}^2$$

Therefore the total areas required for the two flow velocities are

$$A_{\min}\big|_{V=1.5\ \text{m/s}} = 157 + 739 = 896\ \text{m}^2$$

$$A_{\min}\big|_{V=4\ \text{m/s}} = 80.67 + 379.5 = 460\ \text{m}^2$$

## Problem 35

**Consider a solution containing a solute $A$ in solvent $B$. Explain the origin of osmotic pressure $\Pi$ and how reverse osmosis can occur, using chemical potential and activity of solvent $B$ in your explanation. Show that a dilute and ideal solution has an osmotic pressure given by the following expression, if 1 mol of salt at concentration $C_A$ dissociates to give $\alpha$ mols of solute $A$, where $\alpha$ is a positive integer.**

$$\Pi = \alpha \mathrm{RTC_A}$$

## Solution 35

***Worked Solution***

It is a natural occurrence for a solvent to diffuse down its (i.e., solvent) concentration gradient, from a region of lower solute concentration to a region of higher solute concentration. When there is a membrane separating two regions of differing concentrations, this diffusion process occurs from one side to the other side of the membrane, as shown by the arrow in the diagram below left, where we assume that the membrane rejection for $A$ is $R = 1$ (i.e., permeable to $B$ but impermeable to $A$).

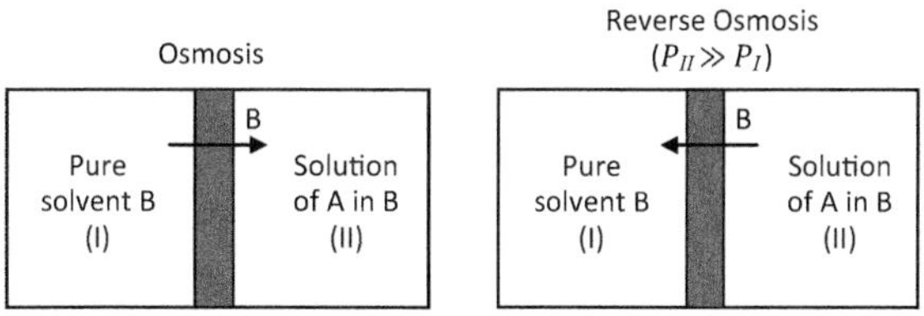


This diffusion process continues until equilibrium is reached when the activity of the solvent is the same on both sides. We can make this happen artificially by increasing pressure on the solution side (i.e., region II), and this increased pressure can cause solvent to move in the reverse direction from II to I (diagram above right). This is called reverse osmosis which is applied in commercial processes of ultrafiltration and hyperfiltration for macromolecular concentration and desalination of seawater.

The concentration of $B$ in region $I$ is higher than that in II. If the pressure of both regions were equivalent, the difference in chemical potential of $B$ will be as shown below.

$$\mu_B(P) = \overset{\circ}{\mu}_B + \mathrm{RT} \ln a_B + \int_{P^\circ}^{P} \left(\frac{\partial \mu_B}{\partial P}\right)_T dP$$

Activity of a species $a_i$ is, to a real solution, what "concentration" is to an ideal solution. Unlike ideal solutions, real solutions mean that interactions between species in the solution are significant. Activity is related to the mole fraction, $x_i$, via the activity coefficient, $\gamma_i$, and this is commonly expressed as $a_i = \gamma_i x_i$.

Recall the thermodynamic relationship as shown below where $\bar{G}_i$ is a partial molar gibbs free energy (contribution by species $i$ in a mixture).

$$dG = \left(\frac{\partial G}{\partial T}\right)_{P,n_i} dT + \left(\frac{\partial G}{\partial P}\right)_{T,n_i} dP + \sum_i \bar{G}_i dn_i$$

When there is no change in composition, $dn_i = 0$, the equation should match the following thermodynamic relationship $dG = -SdT + VdP$. Therefore for our mixture, and at equilibrium when $dG = 0$. Note also that an equivalent notation for $\bar{G}_i$ is $\mu_i$.

$$0 = -SdT + VdP + \sum_i \bar{G}_i dn_i$$

$$\left(\frac{\partial G}{\partial T}\right)_{P,n_i} = -S, \quad \left(\frac{\partial G}{\partial P}\right)_{T,n_i} = V$$

$$\left(\frac{\partial \mu_i}{\partial T}\right)_{P,n_i} = \left(\frac{\partial \bar{G}_i}{\partial T}\right)_{P,n_i} = -\bar{S}_i, \quad \left(\frac{\partial \mu_i}{\partial P}\right)_{T,n_i} = \left(\frac{\partial \bar{G}_i}{\partial P}\right)_{T,n_i} = \overline{V}_i$$

Therefore, going back to our earlier equation for our system, we can deduce that $\left(\frac{\partial \mu_B}{\partial P}\right)_T = v_B$, where $v_B$ is the molar volume of $B$ [m$^3$/mol]. Since $B$ is a liquid, it can be assumed constant.

$$\mu_B(P) = \mu_B^{\circ} + \mathrm{RT} \ln a_B + \int_{P^{\circ}}^{P} v_B dP$$

$$\text{Region } I: \quad \mu_{B,I}(P_I) = \mu_B^{\circ} + \mathrm{RT} \ln a_B + v_B\left(P_I - P^{\circ}\right)$$

$a_B = 1$ in region $I$ since it is pure solvent $B$. Therefore,

$$\mu_{B,I}(P_I) = \mu_B^{\circ} + v_B\left(P_I - P^{\circ}\right)$$

As for region $II$, we have the following where $a_{B,\,II} < 1$ and the term "RT ln $a_{B,II}$" is a negative value

$$\text{Region } II: \quad \mu_{B,II}(P_{II}) = \mu_B^{\circ} + \mathrm{RT} \ln a_{B,II} + v_B\left(P_{II} - P^{\circ}\right)$$

The above analysis is for the case where $P_I = P_{II}$, hence $\mu_{B,I} > \mu_{B,II}$ will cause the diffusion of solvent $B$ from $I$ to II to increase $\mu_{B,II}$ in a normal osmosis scenario.

This direction of transport of $B$ can be reversed if we apply pressure $\Pi$ to region II so that we increase back $\mu_{B,\,II}$ artificially until it reverses the inequality $\mu_{B,I} < \mu_{B,II}$. Let us denote this new increased pressure in region II as $P_{II,\mathrm{new}} = P_I + \Pi$.

$$\mu_{B,I}(P_I) = \mu_{B,II}(P_{II,\text{new}})$$

$$\mu_B^{\circ} + v_B\left(P_I - P^{\circ}\right) = \mu_B^{\circ} + \text{RTln}a_{B,II} + v_B\left(P_{II,\text{new}} - P^{\circ}\right)$$

$$0 = \text{RT}\ln a_{B,II} + v_B(\Pi)$$

$$\Pi = -\frac{\text{RT}\ln a_{B,II}}{v_B}$$

This increment in pressure $\Pi$ is known as osmotic pressure and drives the dilution of the solution.

Taking this into account, our expression for solvent volume flux becomes as shown, where $\Pi_{f,\text{wall}}$ is the osmotic pressure with respect to the concentration on feed side at the wall ($C_{f,\text{wall}}$), and $\Pi_p$ is with respect to the concentration on the permeate side ($C_p$). Note that if rejection factor of membrane $R = 1$, then $\Pi_p = 0$ since $C_p = 0$.

$$J_v'' = K_m(\Delta P - \Delta\Pi) = K_m\left[\Delta P - \left(\Pi_{f,\text{wall}} - \Pi_p\right)\right]$$

If we have a dilute solution, then $\gamma_{B,II} \approx 1$ as the solution behaves close to pure solvent.

$$a_{B,II} = \gamma_{B,II}x_{B,II} \approx 1(1 - x_{A,II}) = 1 - x_{A,II}$$

When both sides of the membrane are in equilibrium, then we have

$$\Pi = -\frac{\text{RT}\ln\left(1 - x_{A,II}\right)}{v_B}$$

Since the solution is dilute, $x_{A,II}$ is small, we know that the power series expansion of the function $\ln(1 - x) = -x - \frac{x^2}{2} - \frac{x^3}{3} + \ldots$ which can be approximated to $\ln(1 - x) \approx -x$ for small values of $x$. Applying this to our equation, we have the following, where $n_A$ and $n_B$ are the number of moles of $A$ and $B$ in the solution.

$$\Pi = \frac{\text{RT}}{v_B}x_{A,II} = \frac{\text{RT}}{v_B}\left(\frac{n_A}{n_A + n_B}\right)$$

For a dilute solution, we may approximate $n_A + n_B \approx n_B$ and $V_B \approx V$, where $V$ is the total volume of the solution. Finally we have a simplified expression

$$\Pi \approx \frac{\text{RT}}{v_B}\left(\frac{n_A}{n_B}\right) = \frac{\text{RT}n_A}{V_B} \approx \frac{\text{RT}n_A}{V}$$

For 1 mole of salt at concentration $C_A$ that dissociates to produce $\alpha$ moles of solutes (for example, 1 mole of NaCl gives 2 moles of solutes since it dissociates into $Na^+$ and $Cl^-$), we have $\frac{n_A}{V} = \alpha C_A$; therefore, we arrive at the expression:

$$\Pi = \alpha RTC_A$$

## Problem 36

**The drying of solids by thermal vaporization is usually the final step in industrial processing when we require liquid to be removed before the material is fed into a heated drier.**

(a) **Describe the main types of driers and comment on the differences between adiabatic driers and non-adiabatic driers.**
(b) **Derive general expressions for the rate of liquid removal from the material for adiabatic drying in a cross-circulation drier, considering both the constant rate period and the falling rate period.**

## Solution 36

### *Worked Solution*

(a) There are various types of driers to cater for different needs. Wet solids may exist in the form of crystals, granules, powders, or continuous sheets. The liquid to be removed may be on the surface of the solid, or inside the solid, or both. Driers may operate in continuous or batch mode, and may be with or without agitation of the solids, and be operated under vacuum for cases that require reduced drying temperatures.

The main types of driers are adiabatic (or direct) driers and non-adiabatic (or indirect) driers.

In direct driers, the wet solids are exposed directly to a heated gas. This can be done by passing hot gas over the solid surface in cross-circulation drying. Alternatively, the hot gas can be blown through a bed of solids in through circulation drying. In the latter, the solid bed may be a moving bed, whereby there is a downward movement of particles through a gas stream, or entrainment of particles in a high velocity gas stream.

In indirect driers, the only gas involved is the vaporized liquid. Heat is transferred to the wet solids via contact with a hot surface or via radiant/microwave energy. The solids may be spread across a heated surface or travel over the surface by a conveyor or agitator.

(b) We can obtain the rate of liquid removal/evaporation from wet solids via batch experiments. Let us consider an adiabatic drier, whereby a gas of constant temperature and humidity is passed over the solids, and the variation in the solids' liquid content is measured against time.

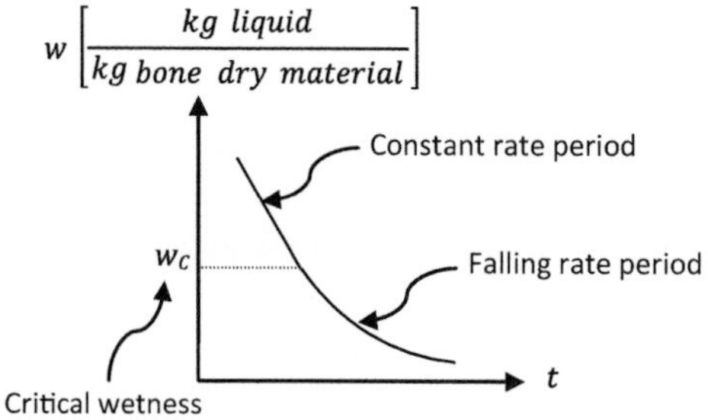


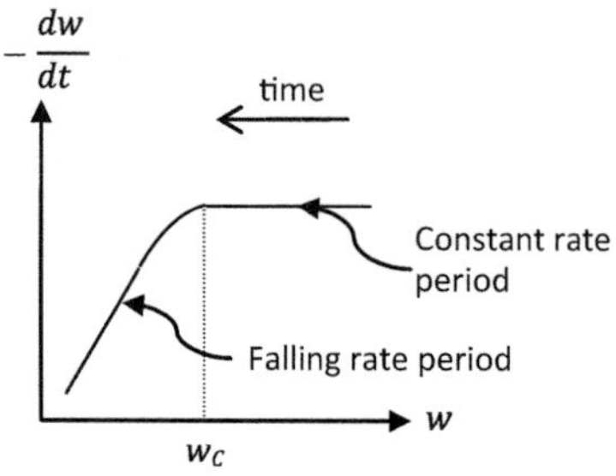

During the constant rate period, the liquid content is high ($w > w_c$) and the drying rate is independent of $w$. Drying rate is only a function of humidity driving force defined by $\Delta\mathcal{H} = \mathcal{H} - \mathcal{H}^*$, whereby $\mathcal{H}$ has units of mass of bone dry air per mass of bone dry material assuming the use of hot air for drying. During this phase, the surface of the solid is completely wet, and therefore the surface is at the wet bulb temperature $T_W$ of the gas.

$$-\frac{dw}{dt} = N''a = k(\mathcal{H} - \mathcal{H}^*)a$$

The units of $-\frac{dw}{dt}$ is $\left[\frac{\text{kg liquid removed}}{\text{kg bone dry material.s}}\right]$, the units of $a$ is $\left[\frac{\text{m}^2\text{ interfacial area}}{\text{kg bone dry material}}\right]$, and the units for $N''$ is $\left[\frac{\text{kg liquid removed}}{\text{m}^2\text{ interfacial area.s}}\right]$. $k$ is the mass transfer coefficient, and the value of $ka$ can be obtained experimentally, when we measure the drying rates for different values of imposed humidity difference, $\Delta\mathcal{H}$.

During the falling rate period, the liquid content has dropped to $w < w_c$. There is now insufficient liquid to completely wet the surface of the material, and the material may not be at its wet bulb temperature, $T_W$. The rate of mass transfer of liquid from the interiors of the solid to the surface is limiting and the drying rate becomes a function of both the humidity driving force and the liquid content or wetness $w$.

## Problem 37

**Consider the drying of a wet sheet that has a water content of 0.97 kg water/kg bone dry sheet, whereby the drying rates are given below:**

**$-\frac{dw}{dt} = 8.2\Delta\mathcal{H}\ [\text{h}^{-1}]$ in the constant rate period and**

**$-\frac{dw}{dt} = 0.4w\ [\text{h}^{-1}]$ in the falling rate period**

**We are provided with a countercurrent adiabatic drier and the required final product has to achieve $w = 0.09$ kg water/kg bone dry sheet. Air is supplied at 45°C which has a wet bulb temperature of 22°C. Determine the time required to dry the wet sheet as required, if the air flow rate is 1.3 times of the minimum.**

## Solution 37

***Worked Solution***

Let us illustrate our system.

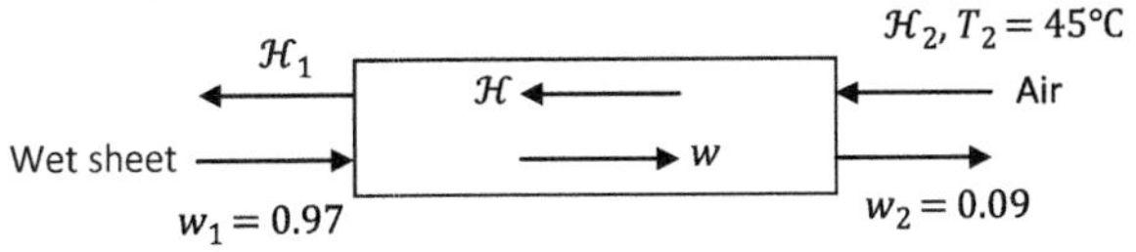

Let $m_{\text{air}}$ denote the air flow rate with units of mass of bone dry air per mass of bone dry sheet. A mass balance around the drier on water gives us the following, assuming 1 kg of bone dry sheet.

$$m_{\text{air}}(\mathcal{H}_1 - \mathcal{H}_2) = w_1 - w_2$$

There are two common assumptions in this case that we can make to simplify the analysis:

- The adiabatic saturation temperature $T_s$ remains constant as air moves through the dryer.
- For an air/water system, the wet bulb temperature $T_W$ remains constant and this is at the value that occurs at the inlet, which is 22 °C in this problem.
- For an air/water system, the specific enthalpy lines and the wet bulb temperature lines (diagonals running from left top to right bottom) coincide.
- At the left most part of the diagram, we have the bold line indicating the saturation wet bulb temperature where relative humidity is 100%.

We can find out the value of specific humidity $\mathcal{H}$ if have a specified value of temperature $T$ using a psychrometric chart. A simplified sketch of the air/water psychrometric chart is shown below. [Note that the vertical axis (on the right) is used for both specific humidity and specific enthalpy, but in this case, only specific humidity is shown for a simplified view.]

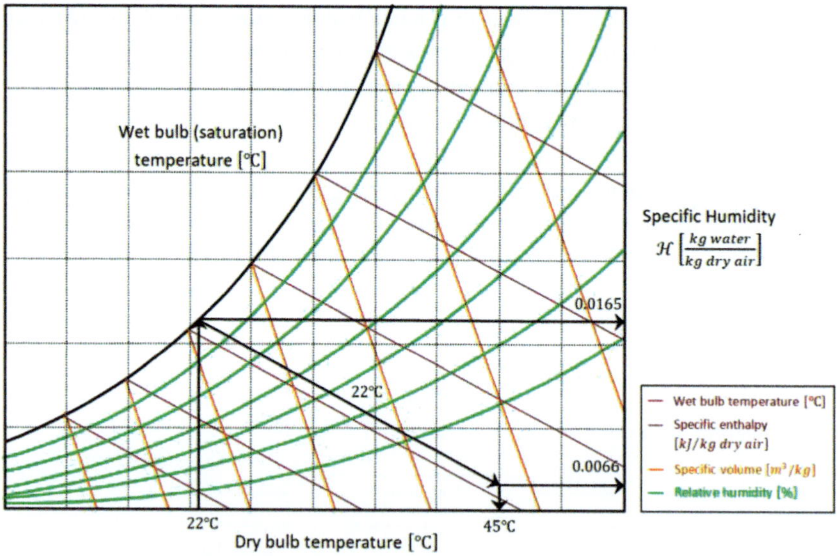

Take, for example, that we need to determine $\mathcal{H}_2$ at a temperature $T_2 = 45°\text{C}$ and wet bulb temperature of 22°C. We can first find the point on the wet bulb saturation curve at this wet bulb temperature of 22°C. We then follow the wet bulb temperature diagonal line down towards the right until it crosses the desired dry bulb temperature of $T_2 = 45°\text{C}$. At this point, we read off the specific humidity $\mathcal{H}_2$ from the vertical axis on the right.

$$\mathcal{H}_2 = 0.0066, T_2 = 45^\circ\text{C}, \qquad \mathcal{H}_W = 0.0165, T_W = 22^\circ\text{C}$$

The minimum air flow rate $m_{\text{min,air}}$ occurs when the air at the outlet has reached equilibrium with the incoming wet sheet. This means that saturation occurs at the outlet and $\mathcal{H}_1 = \mathcal{H}_W = 0.0165$.

$$m_{\text{min,air}}(\mathcal{H}_1 - \mathcal{H}_2) = w_1 - w_2$$

$$m_{\text{min,air}} = \frac{w_1 - w_2}{\mathcal{H}_1 - \mathcal{H}_2} = \frac{0.97 - 0.09}{0.0165 - 0.0066} = 88.9 \text{ kg BD air/kg BD sheet}$$

$$m_{\text{air}} = 1.3 m_{\text{min,air}} = 1.3(88.9) = 115.56$$

The operating line for this drier is shown below, where $\mathcal{H}$ and $w$ are values at any point within the drier. Also note that $\mathcal{H}_W$ is assumed constant, and we define $\Delta\mathcal{H} = \mathcal{H}_W - \mathcal{H}$.

$$m_{\text{air}}(\mathcal{H} - \mathcal{H}_2) = w - w_2$$

$$m_{\text{air}}(\Delta\mathcal{H}_2 - \Delta\mathcal{H}) = w - w_2$$

$$\Delta\mathcal{H} = (0.0165 - 0.0066) - \frac{w - 0.09}{115.56} = 0.01068 - 0.00865w$$

We can find the critical wetness which occurs at the transition point between the constant rate period and falling rate period.

$$-\frac{dw}{dt} = 8.2\Delta\mathcal{H} = 0.4w_c$$

$$8.2(0.01068 - 0.00865w_c) = 0.4w_c$$

$$w_c = 0.186 \text{ kg water/kg BD sheet}$$

The time required to dry the sheet to the required wetness can be found by summing the time spent in the constant rate period and falling rate periods.

$$-\frac{dw}{dt} = 8.2\Delta\mathcal{H} = 8.2(0.01068 - 0.00865w)$$

$$\frac{dw}{dt} - 0.07093w = -0.0876$$

To solve this differential equation, we can use an integrating factor of the form $e^{\int(-0.07093)dt} = e^{-0.07093t}$. $c$ is an integration constant.

$$we^{-0.07093t} = \int e^{-0.07093t}(-0.0876)dt$$

$$we^{-0.07093t} = \frac{e^{-0.07093t}}{-0.07093}(-0.0876) + c$$

$$w = \frac{0.0876}{0.07093} + ce^{0.07093t} = 1.235 + ce^{0.07093t}$$

When $t = 0$, $w = w_1 = 0.97$, so $c = -0.265$. Therefore the wetness profile for the constant rate period is as follows.

$$w = 1.235 - 0.265e^{0.07093t}$$

When $w = w_c = 0.186$, $t_{CRP} = 19.4$ h. We can now similarly solve the differential equation for the falling rate period.

$$-\frac{dw}{dt} = 0.4w$$

$$w = \frac{e^{-0.4t}}{0.4} = 2.5e^{-0.4t}$$

When $w = w_2 = 0.09$, $t = 8.31$ h. When $w = w_c = 0.186$, $t = 6.50$ h. $t_{FRP} = 8.31 - 6.50 = 1.81$ h

Total time spent to dry the wet cloth to required water content is 21.2 h.

$$t = t_{CRP} + t_{FRP} = 19.4 + 1.81 = 21.2 \text{ h}$$

## Problem 38

**A humidification unit is used to increase the moisture content of air. A feed stream of air at 18 °C and wet bulb temperature of 12.5 °C is heated to 48 °C and enters the unit. The outlet air is fully saturated. Determine the following:**

(a) **The dew point of the inlet air**
(b) **The humidity of the inlet air**
(c) **The percentage humidity of the inlet air**
(d) **The heat supplied per kg of dry air**
(e) **The water evaporated per kg of dry air**
(f) **The temperature of the exit air**

## Solution 38

### *Worked Solution*

(a) This problem is good for understanding the basic terms and concepts on humidity and drying. Below is a simple diagram showing our system.

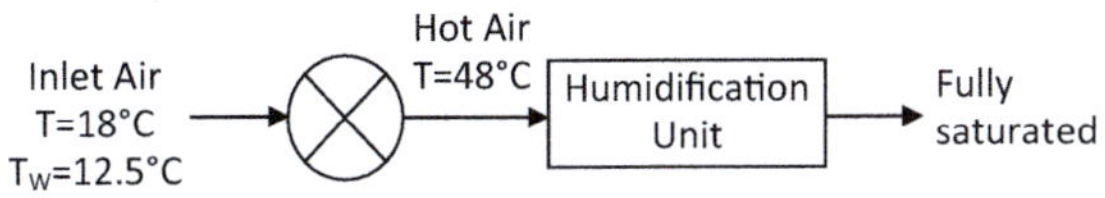


Dew point occurs at the temperature when saturation is reached. We can find the dew point by following these steps, and as shown by arrows in the psychrometric chart below.

1. On the horizontal axis of dry bulb temperature, draw a vertical line at the wet bulb temperature of 12.5 °C.
2. When this vertical line hits the saturation curve, follow the diagonal down towards the right until we reach the corresponding dry bulb temperature of 18 °C.
3. From the point identified from step 2, move horizontally to the left until we reach the saturation curve, and read off the temperature value at this point from the horizontal axis of dry bulb temperature. We find that dew point is 8 °C.

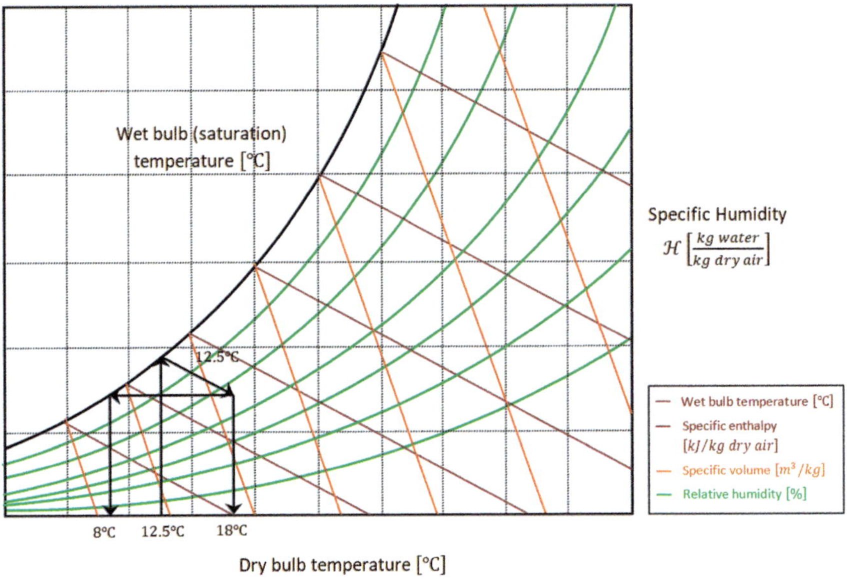

(b) The humidity of the inlet air at 18 °C at wet bulb temperature of 12.5 °C can be found by reading off the corresponding specific humidity $\mathcal{H}$ value from the vertical axis on the right. In this case, $\mathcal{H} = 0.0065$ kg water/kg dry air.

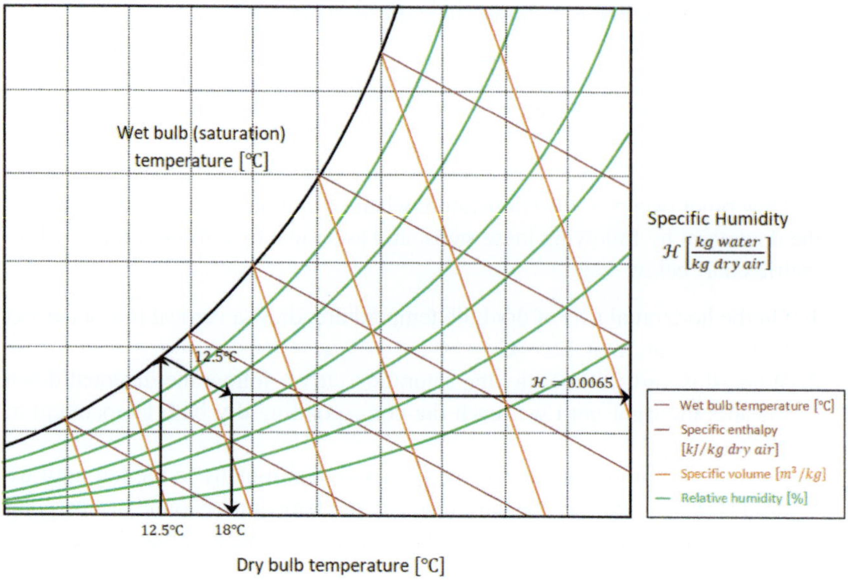


(c) The percentage of humidity is measured as relative humidity [%] and can be read off from the curved lines (indicated in green below). For our specific case of $T = 18$ °C at wet bulb temperature of $T_w = 12.5$ °C, the RH is 49%.

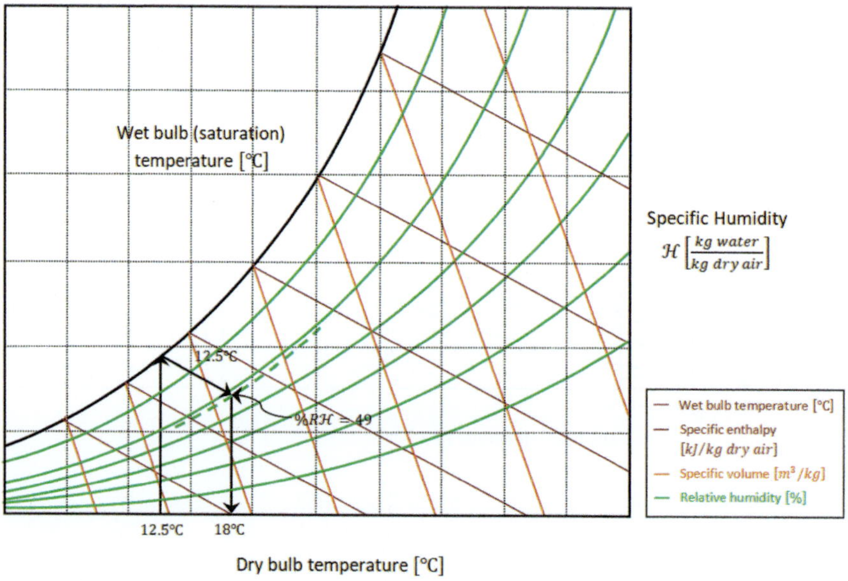

(d) The heat supplied per kg of dry air is related to enthalpy. We can find the specific enthalpies corresponding to the dry bulb temperature $T = 18$ °C and wet bulb temperature $T_w = 12.5$ °C, which are 35 and 65 kJ/kg, respectively. Therefore the enthalpy required to heat the air is $\Delta h = 65 - 35 = 30$ kJ/kg dry air.

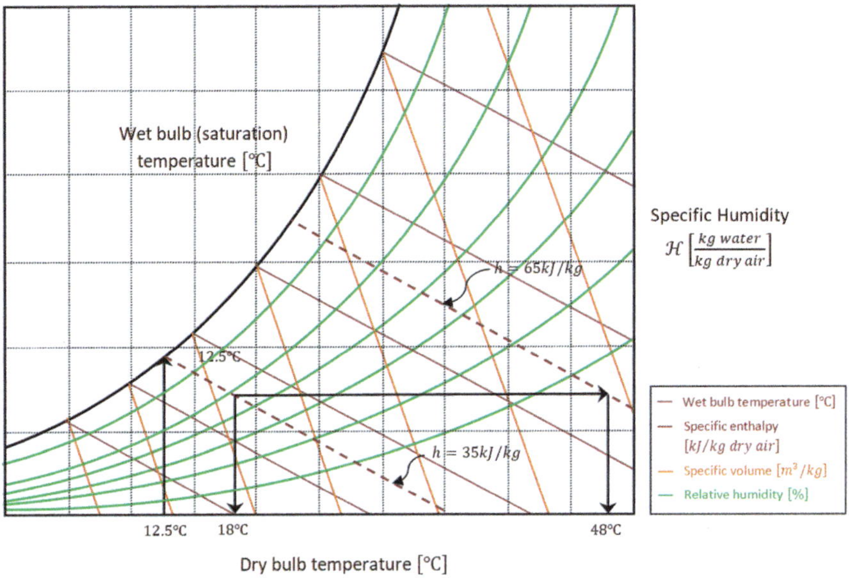


(e) To find amount of water removed per kg of dry air, we need to find the difference in specific humidities between the inlet and outlet. The specific humidity at the inlet is $\mathcal{H} = 0.0065$ kg water/dry air, which is at the point where $T = 48$ °C as shown in part b. We know that air at the outlet is fully saturated. To meet this condition, we move up the adiabatic saturation lines (diagonals in maroon) from the inlet point (i.e., at $T = 48$ °C and $\mathcal{H} = 0.0065$ kg water/dry air) until we hit the saturation curve (100% RH). This point has specific humidity, $\mathcal{H} = 0.0172$ kg water/dry air, and amount of water to be evaporated is $\Delta\mathcal{H} = 0.0172 - 0.0065 = 0.0107$ kg water/dry air.

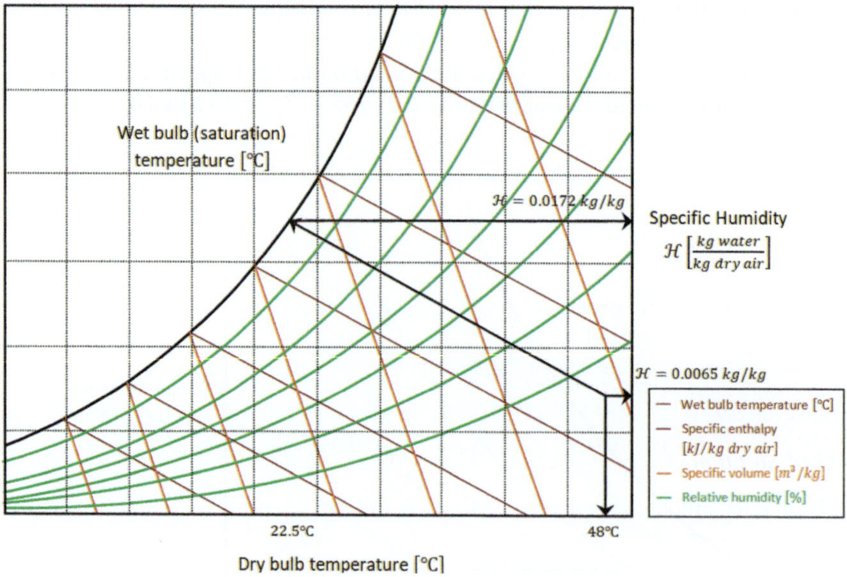


(f) The exit temperature is the dry bulb temperature that corresponds to the exit point, i.e., at $\mathcal{H} = 0.0172$ kg water/dry air. This value is $T = 22.5$ °C as read off from the horizontal axis.

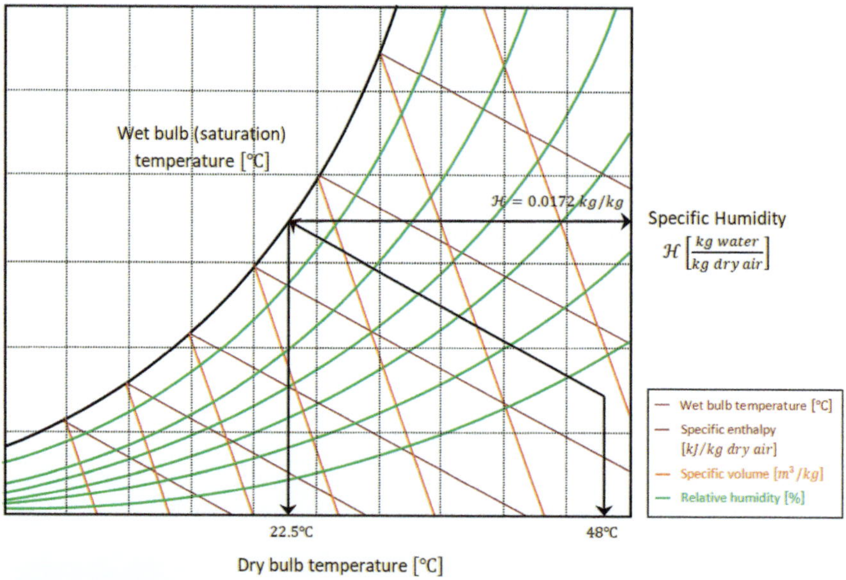


## Problem 39

**Consider 30,000 kg/h of dry air with a dry bulb temperature of 25 °C and wet bulb temperature of 15 °C. It is heated to 50 °C before passing over wet slabs. The slabs are to be dried from an initial water content of 0.85 kg water/kg bone dry slab to 0.02 kg water/kg bone dry slab.**

(a) **Determine the heat required to heat the air, and find the wet bulb temperature and dew point of the heated air.**
(b) **Determine the maximum mass of slab that can be dried if an adiabatic countercurrent drier is used.**
(c) **If the actual mass of slabs that is dried is 70% of the maximum, find the temperature and humidity of the exit air. Determine also the total drying time if the following apply to this scenario.**

$$-\frac{dw}{dt} = 12\Delta\mathcal{H}\left[\mathrm{h}^{-1}\right] \text{ in the constant rate period;}$$

$$-\frac{dw}{dt} = 0.5w\left[\mathrm{h}^{-1}\right] \text{ in the falling rate period.}$$

## Solution 39

*Worked Solution*

(a) To find the amount of heat required to heat the air, we consider enthalpy change from its initial temperature (dry bulb) of $T = 25$ °C to $T = 50$ °C. This can be found from the psychrometric chart as shown below. The amount of heat required is $\Delta h = 68 - 42 = 26$ kJ/kg dry air. The dry air flow rate is 30,000 kg/h; therefore, the power required [kW] is 217 kW. We can also find the specific humidity for inlet air which is $\mathcal{H} = 0.0066$ kg/kg.

$$\text{Power} = \frac{26(30,000)}{3600} = 217 \text{ kW}$$

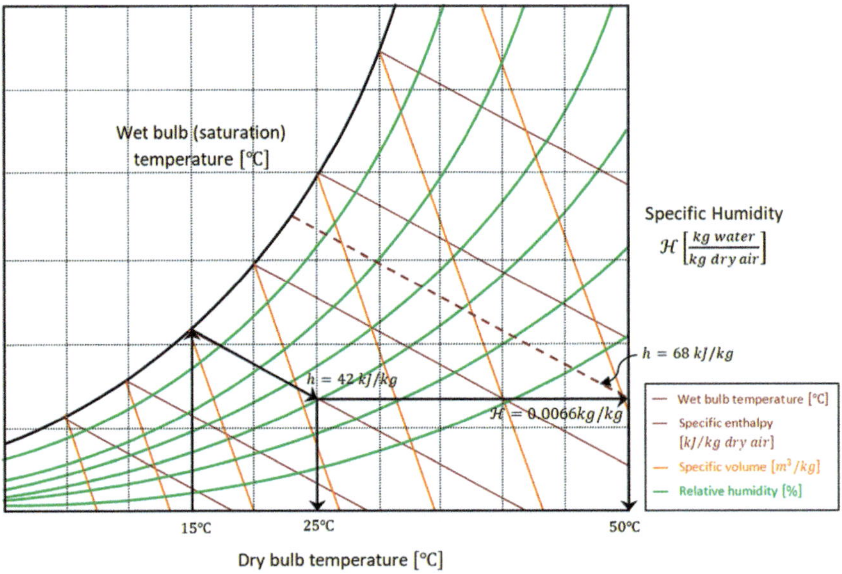

For the wet bulb temperature, we follow the line of constant wet bulb temperature going diagonally left-up, until we reach the saturation curve at 100% relative humidity. The wet bulb temperature is read from the horizontal axis, and has a value of 23 °C at a specific humidity of 0.0178 kg/kg. The dew point of the heated air can be found by following the horizontal line and going leftwards until we reach the saturation curve at 100% relative humidity. The dew point temperature is read from the horizontal axis and has a value of 8 °C.

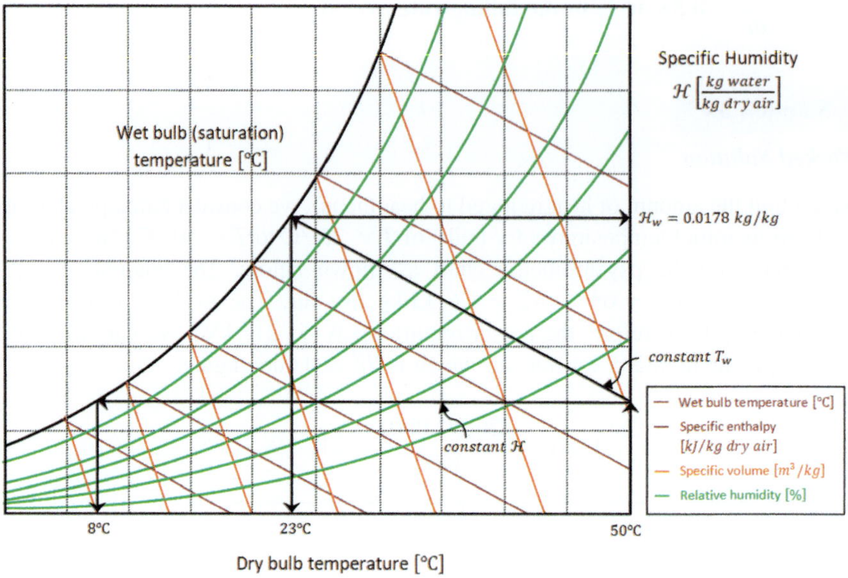


(b) Let us illustrate our drier operation as shown below.

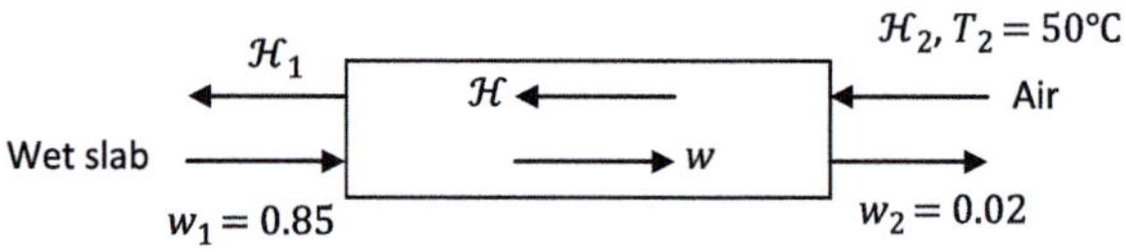


The maximum number of slabs that can be dried can be found by setting the condition whereby the air exiting the drier is at saturation, i.e., 100% relative humidity when $\mathcal{H}_1 = \mathcal{H}_w$. This scenario also corresponds to the minimum air flow rate.

We have found from part a that the saturation humidity $\mathcal{H}_w = 0.0178$ kg water/kg dry air and the specific humidity of the inlet air $\mathcal{H}_2 = 0.0066$ kg water/kg dry air.

$$\mathcal{H}_1 = 0.0178 \text{ kg water/kg dry air}, \ w_1 = 0.85 \text{ kg water/kg dry slab}$$
$$\mathcal{H}_2 = 0.0066 \text{ kg water/kg dry air}, \ w_2 = 0.02 \text{ kg water/kg dry slab}$$

A mass balance around the drier on water gives us the following, assuming 1 kg of bone dry slab, and where the air flow rate is $m_{\text{air}}$[kg dry air/kg dry slab].

$$m_{\text{air,min}}(\mathcal{H}_1 - \mathcal{H}_2) = w_1 - w_2$$
$$m_{\text{air,min}}(0.0178 - 0.0066) = 0.85 - 0.02$$
$$m_{\text{air,min}} = 74.1 \text{ kg/kg}$$

Maximum mass of slab = 1/74.1 = 0.0135 kg dry slab/kg dry air.

(c) Given that the actual mass of slabs dried is 70% of the maximum, we have actual mass of slabs = 0.009446 kg dry slab/kg dry air. Substituting this value into the mass balance, we can find the new value of $\mathcal{H}_1$,

$$\mathcal{H}_1 - \mathcal{H}_2 = 0.009446(w_1 - w_2)$$
$$\mathcal{H}_1 - 0.0066 = 0.009446(0.85 - 0.02)$$
$$\mathcal{H}_1 = 0.0144$$

The humidity of the exit air is 0.0144 kg water/kg dry air. From the psychrometric chart, we can find the temperature $T_1 = 31°\text{C}$.

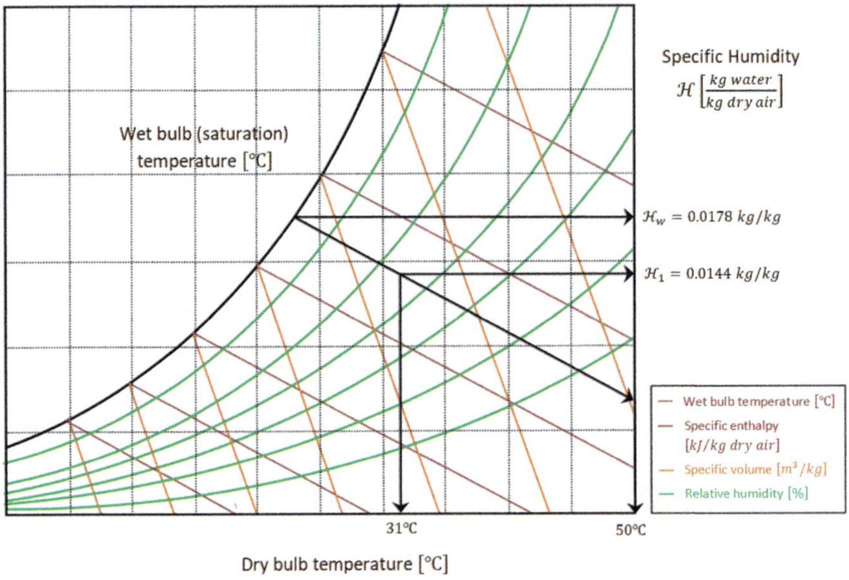

To find the total drying time, we can find the sum of the drying time in the constant rate period and falling rate period. The critical wetness $w_c$ needs to be determined to find out the transition point between the two periods. We found from earlier that $\mathcal{H}_w = 0.0178$.

At critical wetness,

$$12\Delta\mathcal{H} = 0.5w_c$$

$$12(\mathcal{H}_w - \mathcal{H}) = 12(0.0178 - \mathcal{H}) = 0.5w_c$$

The operating line is as follows, where the humidity and wetness at the air outlet is now a variable.

$$\mathcal{H} - \mathcal{H}_2 = 0.009446(w - w_2)$$

$$\mathcal{H} - 0.0066 = 0.009446(w - 0.02)$$

$$\mathcal{H} = 0.009446w + 0.00641$$

Combining the results, we have

$$12(0.0178 - 0.009446w_c - 0.00641) = 0.5w_c$$

$$w_c = 0.222$$

For the constant rate period, we have

$$-\frac{dw}{dt} = 12\Delta\mathcal{H} = 12(0.0178 - 0.009446w - 0.00641) = 0.137 - 0.113w$$

$$\int_{0.85}^{0.222} \frac{-1}{0.137 - 0.113w} dw = \int_0^{t_{\mathrm{CRP}}} dt$$

$$t_{\mathrm{CRP}} = \frac{1}{0.113} \ln\left[\frac{0.137 - 0.113(0.222)}{0.137 - 0.113(0.85)}\right] = 8.9 \text{ h}$$

For the falling rate period, we have

$$-\frac{dw}{dt} = 0.5w$$

$$\int_{0.222}^{0.02} \frac{-1}{0.5w} dw = \int_0^{t_{\mathrm{FRP}}} dt$$

$$t_{\mathrm{FRP}} = \frac{1}{0.5} \ln\left(\frac{0.222}{0.02}\right) = 4.8 \text{ h}$$

The total drying time is therefore

$$t = t_{\text{CRP}} + t_{\text{FRP}} = 8.9 + 4.8 = 13.7\ \text{h}$$

## Problem 40

**The drying of filter cakes can occur during a constant rate period and a falling rate period. Starting from the general equation for drying flux, $R''\left[\frac{\text{kg}}{\text{h.m}^2}\right]$ as shown below, where $X$ represents free moisture content [% dry basis], $m_s$ represents the mass of bone dry solid, and $A$ represents the surface area over which drying occurs.**

$$R'' = -\frac{m_s}{A}\frac{dX}{dt}$$

**Derive expressions for the following cases:**

(a) **Drying time in the phase of the constant rate period, if $X_1$ and $X_2$ represent the initial and final free moisture contents, respectively.**
(b) **Drying time in the phase of the falling rate period if the rate is a linear function of $X$ whereby $R'' = k$, $k > 0$ when $X = 0$. You may assume again that $X_1$ and $X_2$ represent the initial and final free moisture contents, respectively.**
(c) **Drying time for the falling rate period if the rate is a linear function of $X$ and $R'' = 0$ when $X = 0$. You may assume again that $X_1$ and $X_2$ represent the initial and final free moisture contents, respectively.**

## Solution 40

*Worked Solution*

(a) Before we derive the expressions, here are some key distinctions between the constant rate and falling rate periods.

The constant rate period occurs at the beginning of the drying process, when the surface of the solid is very wet and a continuous film exists on the drying surface. This water in the film is entirely unbound and acts as if the solid is not present. Therefore evaporation rate is independent of the solid and occurs from the free liquid surface. The constant rate period will last for as long as there is water supplied to the surface as fast as it is evaporated.

The constant rate period eventually transitions into the falling rate period when the free moisture content drops to a critical value $X_c$. There is now insufficient water to maintain a continuous wet film on the drying surface, and the wetted area decreases until completely dry.

The rate is constant, therefore $R''$ can be taken out of the integral as shown below. Note that $X_2 > X_c$.

$$\int_0^{t_{\text{CRP}}} dt = -\frac{m_s}{R''A}\int_{X_1}^{X_2} dX$$

$$t_{\text{CRP}} = \frac{m_s}{R''A}(X_1 - X_2)$$

(b) In the falling rate period, note that $X_1 < X_c$. If $R''$ is linear with respect to $X$ as described, then it can be expressed as follows where $a$ is the slope of this linear function.

$$R'' = aX + k$$

$$dR'' = adX$$

Therefore we can evaluate the integral, noting that $R''$ is no longer a constant with respect to $X$

$$\int_0^{t_{\text{FRP}}} dt = -\frac{m_s}{A}\int_{X_1}^{X_2} \frac{dX}{R''} = \frac{m_s}{A}\int_{X_2}^{X_1} \frac{dX}{R''} = \frac{m_s}{Aa}\int_{R_2''}^{R_1''} \frac{dR''}{R''}$$

$$t_{\text{FRP}} = \frac{m_s}{Aa}\ln\left(\frac{R_1''}{R_2''}\right)$$

We can derive an expression for $a$ as follows:

$$R_1'' = aX_1 + k, \quad R_2'' = aX_2 + k$$

$$a = \frac{R_1'' - R_2''}{X_1 - X_2}$$

Substituting this result into our earlier expression, we eliminate $a$ from the expression for $t_{\text{FRP}}$ as shown.

$$t_{\text{FRP}} = \frac{m_s(X_1 - X_2)}{A(R_1'' - R_2'')}\ln\left(\frac{R_1''}{R_2''}\right)$$

(c) This part is similar to part b, except that the linear function is now as follows, where $R'' = 0$ when $X = 0$.

$$R'' = aX$$

$$dR'' = adX$$

We evaluate the integral like in part b, again noting that $R''$ is no longer a constant with respect to $X$

$$\int_0^{t_{\text{FRP}}} dt = -\frac{m_s}{A}\int_{X_1}^{X_2}\frac{dX}{R''} = \frac{m_s}{A}\int_{X_2}^{X_1}\frac{dX}{R''} = \frac{m_s}{Aa}\int_{R_2''}^{R_1''}\frac{dR''}{R''}$$

$$t_{\text{FRP}} = \frac{m_s}{Aa}\ln\left(\frac{R_1''}{R_2''}\right)$$

This time, the expression for $a$ can be found by applying the fact that the falling rate period starts from the upper limit moisture content $X_c$ and falls to zero where $R''$ also becomes zero. This means that $R_1'' = R_c''$, $X_1 = X_c$ and $R_2'' = 0$, $X_2 = 0$. The expression for $a$ becomes simplified to:

$$a = \frac{R_c''}{X_c}$$

Substituting this result into our earlier expression, we eliminate $a$ from the expression for $t_{\text{FRP}}$ as shown.

$$t_{\text{FRP}} = \frac{m_s X_c}{AR_c''}\ln\left(\frac{R_c''}{R_2''}\right) = \frac{m_s X_c}{AR_c''}\ln\left(\frac{X_c}{X_2}\right)$$

For any $0 < R'' < R_c''$, $R'' = R_c''\left(\frac{X}{X_c}\right)$.

## Problem 41

**We are required to dry slabs of wet filter cake that fill rectangular trays measuring 90 cm (length) by 85 cm (width) by 2 cm (depth). Drying air is passed over the top surface of the filter cake, and drying may be assumed to occur at the top surface only. Drying air is maintained at 160 °C and 1 bar, with a wet bulb temperature of 60 °C and passes over the cake surface at a velocity of 4 m/s.**

**The specific humidity $\mathcal{H}$ of air at 160 °C can be obtained from a high humidity psychrometric chart and is given as 0.10 kg water/kg dry air. The heat of vaporization of water at 60 °C obtained from steam tables is 2400 kJ/kg.**

**The properties of the filter cake are as follows:**

- **Bone-dry density = 2500 kg/m$^3$**
- **Initial free moisture content, $X_1$ = 105% (dry basis)**
- **Critical free moisture content, $X_c$ = 65% (dry basis)**
- **Final free moisture content $X_f$ = 4% (dry basis)**

**The drying times in hours during the constant rate period (CRP) and falling rate period (FRP) may be modeled using the following equations:**

$$t_{\text{CRP}} = \frac{m_s}{R''A}(X_1 - X), \quad X_c < X < X_1$$

$$t_{\text{FRP}} = \frac{m_s X_c}{AR''_c}\ln\left(\frac{X_c}{X}\right), \quad 0 < X < X_c$$

**$R''\left[\frac{\text{kg}}{\text{h.m}^2}\right]$ denotes drying flux, $m_s$ denotes mass of bone dry solid, $A$ represents the surface area over which drying occurs, $X$ denotes free moisture content at time $t$ with $X_1$ and $X_c$ representing initial and critical values, respectively, $R''_c$ denotes drying flux at $X_c$.**

(a) **Determine the drying time in the constant rate period.**
(b) **Determine the drying time in the falling rate period.**

## Solution 41

### *Worked Solution*

(a) We can first illustrate our system as shown below. Note that we assume that all heat is used to evaporate liquid (i.e., latent heat), and not to raise the temperature of the evaporated moisture (i.e., sensible heat ignored).

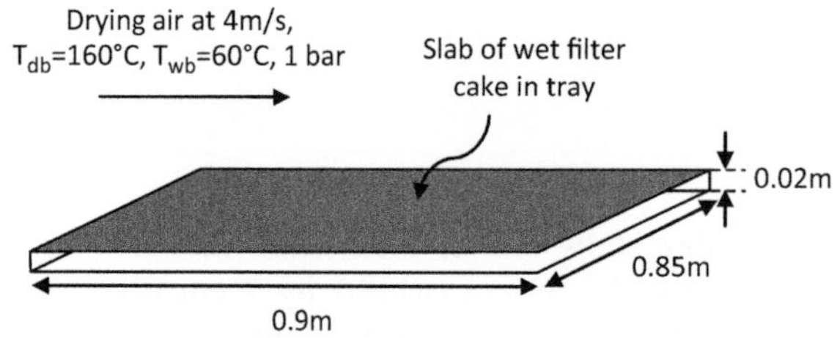


We can construct an energy balance as shown below

(Mass flux of water) × (heat absorbed per kg water evaporated) = (heat transfer coefficient) × ($\Delta T$)

$$(R'')\left(\Delta H_{\text{vap}}\big|_{T_{\text{solid surface}}}\right) = h(T_{\text{air}} - T_{\text{solid surface}})$$

$$R'' = \frac{h(T_{db} - T_{wb})}{\Delta H_{\text{vap}}\big|_{T_{wb}}} \tag{1}$$

For parallel flow of air to the solid surface, the heat transfer coefficient $h\left[\frac{W}{\text{m}^2K}\right]$ may be modeled by the following correlation in terms of $G\left[\frac{\text{kg}}{\text{m}^2\text{h}}\right]$, where $\rho$ and $u$ denote density $\left[\frac{\text{kg}}{\text{m}^3}\right]$ and velocity $\left[\frac{\text{m}}{\text{s}}\right]$ of air, respectively.

$$h = 0.0204G^{0.8},\ G = 3600\,\rho u$$

Assuming ideal gas behavior $P = \rho\text{RT}$ and molecular weight of air is 0.029 kg/mol.

$$\rho = \frac{101325(0.029)}{8.314(160 + 273)} = 0.8\ \text{kg/m}^3$$

However, this is the density of dry air, and we need to correct for moisture in air.

We know that in 1 kg of dry air, we have 0.10 kg water.

$$V_{\text{moist air}} = \frac{n_{\text{moist air}}\text{RT}}{P} = \frac{8.314(160 + 273)}{101325}\left[\frac{1}{0.029} + \frac{0.1}{0.018}\right] = 1.4$$

$$\rho_{\text{moist air}} = \frac{\rho}{1.4} = \frac{0.8}{1.4} = 0.6$$

Therefore, we can evaluate $h$ as follows,

$$h = 0.0204(9000)^{0.8} = 30,\ G = 3600(0.6)(4) = 9000$$

For CRP,

$$R''(A)(\Delta t) = m_s(\Delta X) \tag{2}$$

Combining Eqs. (1) and (2), we have

$$t_{\text{CRP}} = \frac{m_s \Delta X \Delta H_{\text{vap}}\big|_{T_{wb}}}{hA(T_{db} - T_{wb})}$$

The bone dry density of solid is 2500 kg/m$^3$. Tray volume is (0.9)(0.85)(0.02) = 0.015 m$^3$. Therefore, mass of solid in one tray $m_s$ is 2500 (0.015) = 38 kg. Drying area $A$ is tray area (0.9)(0.85) = 0.8 m$^2$.

$$t_{\text{CRP}} = \frac{(38)(1.05 - 0.65)\left(2400 \times 10^3\right)}{30(0.8)(160 - 60)} = 15{,}200\ \text{s} = 4.2\ \text{h}$$

(b) For FRP,

$$R''_c(A)(\Delta t) = m_s X_c \ln\left(\frac{X_c}{X_f}\right) \tag{3}$$

Combining Eqs. (1) and (3), we have

$$t_{\mathrm{FRP}} = \frac{m_s X_c \Delta H_{\mathrm{vap}}|_{T_{wb}}}{hA(T_{db} - T_{wb})} \ln\left(\frac{X_c}{X_f}\right) = \frac{(38)(0.65)(2400 \times 10^3)}{30(0.8)(160 - 60)} \ln\left(\frac{0.65}{0.04}\right) = 69000 \text{ s}$$

$$= 19 \text{ h}$$

**Problem 42**

**Consider a wet solid that takes 6 h to dry from 40 wt% to 8 wt% on a dry basis. The critical and equilibrium moisture contents are 17 wt% and 2 wt% on a dry basis, respectively. Find the drying time for the same wet solid from 50 wt% to 6 wt% on a dry basis. You may assume a linear falling rate period.**

**Solution 42**

***Worked Solution***

We know that the transition point from constant rate period to falling rate period is at the critical moisture content, $X_c$. Therefore, we have the total drying time as shown below.

$$t = t_{\mathrm{CRP}} + t_{\mathrm{FRP}} = \frac{m_s}{R_c''A}(X_1 - X_c) + \frac{m_s X_c}{R_c''A} \ln\left(\frac{X_c}{X_f}\right)$$

We know that it takes 6 h to reach $X_c = 0.17 - 0.02 = 0.15$ from an initial $X_1 = 0.40 - 0.02 = 0.38$. So we can use that to find $\frac{m_s}{R_c''A}$.

$$6 = \frac{m_s}{R_c''A}(0.38 - 0.15)$$

$$\frac{m_s}{R_c''A} = \frac{6}{0.38 - 0.15} = 26$$

Substituting this result back into our earlier equation, and with the new conditions of an initial $X_1 = 0.50 - 0.02 = 0.48$ and final $X_f = 0.06 - 0.02 = 0.04$, we can find the total drying time.

$$t = 26\left[(0.48 - 0.15) + 0.15 \ln\left(\frac{0.15}{0.04}\right)\right] = 13.7 \text{ h}$$

## Problem 43

**We have trays measuring 1.8 m by 1 m by 0.2 m which contain a wet solid of initial total moisture content of 120 wt% on a dry basis. The final total moisture content is 10 wt% dry basis. Drying air at 33 °C is blown over the wet solids at 1 bar and a relative humidity (RH) of 15%. The equilibrium moisture content is 3 wt% on a dry basis.**

**Data of total moisture content (dry basis) against time in minutes is provided below.**

| Time [min] | 0 | 100 | 200 | 300 | 400 | 500 | 600 | 700 | 800 | 900 | 950 | 1000 |
|---|---|---|---|---|---|---|---|---|---|---|---|---|
| Total moisture content, dry basis [%] | 120 | 93 | 65 | 44.5 | 32 | 23.5 | 17 | 13.5 | 11 | 10 | 9.1 | 8.7 |

**Find the drying time required for the test case of reducing from total moisture content of 110 wt% to 5 wt% if air of 52 °C and 20% relative humidity is used instead.**

**From data obtained from the psychrometric chart and steam table at 1 bar, the wet bulb temperatures are $T_{wb}$ ($T_{db}$ = 33 °C, RH = 15%) = 16 °C and $T_{wb}$ ($T_{db}$ = 52 °C, RH = 20%) = 29 °C, while the heat of vaporizations at these wet bulb temperatures are $\Delta H_{vap}(T_{wb}$ = 16°C) = 2800 kJ/kg and $\Delta H_{vap}(T_{wb}$ = 29°C) = 2400 kJ/kg.**

## Solution 43

*Worked Solution*

We can plot the data provided on a chart, and calculate the free moisture content which is the difference between total moisture content and equilibrium moisture content of 3 wt%.

| Time [min] | 0 | 100 | 200 | 300 | 400 | 500 | 600 | 700 | 800 | 900 | 950 | 1000 |
|---|---|---|---|---|---|---|---|---|---|---|---|---|
| Total moisture content, dry basis [%] | 120 | 93 | 65 | 44.5 | 32 | 23.5 | 17 | 13.5 | 11 | 10 | 9.1 | 8.7 |
| Free moisture content, dry basis [%] | 117 | 90 | 62 | 41.5 | 29 | 20.5 | 14 | 10.5 | 8 | 7 | 6.1 | 5.7 |

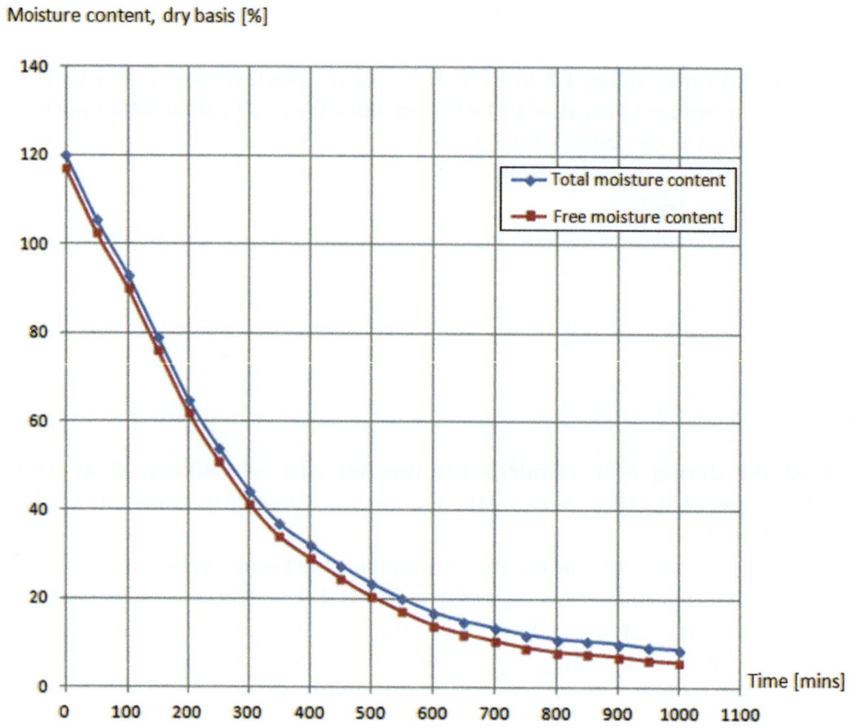


We can observe from the plot that the critical moisture content where the linear region (constant rate period) transitions into a non-linear region (falling rate period) is approximately at 210 min, where free moisture content is 59 wt% (or total moisture content of 62 wt%).

We can find the constant drying rate using the slope of the initial linear region. For CRP: $R'' = \frac{m_s \Delta X}{A.\Delta t}$, where $m_s$ denotes mass of bone dry solid.

$$\frac{R''A}{m_s} = \frac{\Delta X}{\Delta t} = \text{constant}$$

$$\frac{\Delta X}{\Delta t} = \frac{1.20 - 0.62}{210} = 0.00276 \text{ kg water/kg dry solid.min}$$

From an energy balance, we can relate drying rate to enthalpy of vaporization at the wet bulb temperature as shown below.

$$R'' = \frac{h(T_{db} - T_{wb})}{\Delta H_{\text{vap}}\big|_{T_{wb}}} = \frac{m_s \Delta X}{A \Delta t}$$

$$\frac{\Delta t(T_{db} - T_{wb})}{\Delta X \Delta H_{vap}|_{T_{wb}}} = \frac{m_s}{Ah} = \text{constant}$$

We can equate the constant between the two known cases:

$$\left[\frac{\Delta t(T_{db} - T_{wb})}{\Delta X \Delta H_{vap}|_{T_{wb}}}\right]_{\text{case 1}} = \left[\frac{\Delta t(T_{db} - T_{wb})}{\Delta X \Delta H_{vap}|_{T_{wb}}}\right]_{\text{case 2}}$$

$$\frac{210(33-16)}{(120-62)(2800\times 10^3)} = \frac{(\Delta t)_{\text{case 2}}(52-29)}{(110-62)(2400\times 10^3)}$$

$$(\Delta t)_{\text{case 2}} = 110 \text{ min}$$

Therefore, the drying time during the constant rate period is 110 min for the test case. Now we need to find the drying time during the falling rate period. For the falling rate period,

$$R_c''(A)(\Delta t) = m_s X_c \ln\left(\frac{X_c}{X_f}\right) \tag{1}$$

$$R'' = \frac{h(T_{db} - T_{wb})}{\Delta H_{vap}|_{T_{wb}}} \tag{2}$$

Combining Eqs. (1) and (2), we have

$$\Delta t = \frac{m_s X_c \Delta H_{vap}|_{T_{wb}}}{hA(T_{db} - T_{wb})} \ln\left(\frac{X_c}{X_f}\right)$$

$$\frac{(\Delta t)_{\text{case 1}}}{(\Delta t)_{\text{case 2}}} = \frac{\left[\frac{\Delta H_{vap}|_{T_{wb}}}{(T_{db} - T_{wb})} \ln\left(\frac{X_c}{X_f}\right)\right]_{\text{case 1}}}{\left[\frac{\Delta H_{vap}|_{T_{wb}}}{(T_{db} - T_{wb})} \ln\left(\frac{X_c}{X_f}\right)\right]_{\text{case 2}}}$$

$$\frac{900-210}{(\Delta t)_{\text{case 2}}} = \frac{\left[\frac{2800\times 10^3}{(33-16)} \ln\left(\frac{62}{10}\right)\right]}{\left[\frac{2400\times 10^3}{(52-29)} \ln\left(\frac{62}{5}\right)\right]}$$

$$(\Delta t)_{\text{case 2}} = 603 \text{ min}$$

Therefore the total drying time for the test case is

$$t_{\text{case 2}} = 110 + 603 = 713 \text{ min}$$

## Problem 44

**Explain the difference between bound moisture and unbound moisture in a wet solid with a simple sketch of total moisture content [kg water/kg dry solid] against relative humidity [%].**

## Solution 44

### *Worked Solution*

There are two main types of moisture in a wet solid. It is useful to know the difference in considering drying problems.

1. Bound moisture: moisture that is adsorbed into the material's cell walls, capillaries, or surfaces. Bound water in wet solid exerts a vapor pressure that is less than the partial pressure of water at the same temperature.
2. Free moisture: moisture in excess of the equilibrium moisture content. This excess, unbound water is held in the voids of the solid.

Let us visualize some of these concepts in the following diagrams (at constant temperature and pressure).

Equilibrium moisture content decreases with increasing temperature, and increases with increasing relative humidity. It varies with material type. The equilibrium content occurs at the point when the moisture content is at 100% relative humidity. Any moisture content above the equilibrium value will be free moisture. Different curves are shown for different materials.

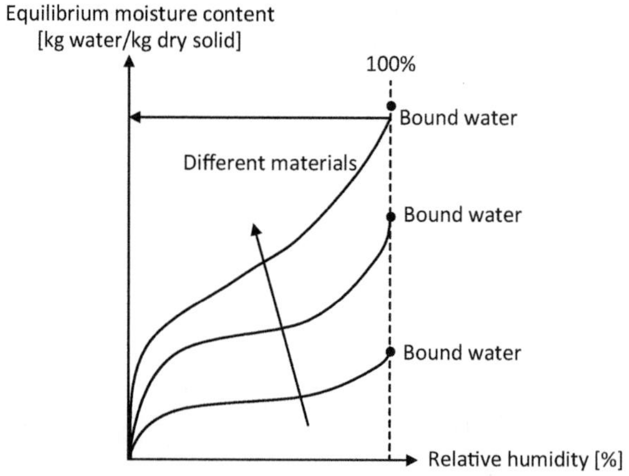

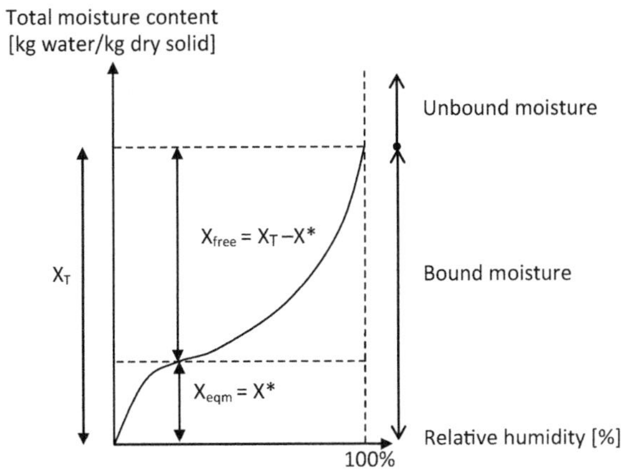


## Problem 45

**Describe the different phases in a drying process of a wet solid, and show how the drying flux varies with free moisture content and with time, highlighting any key features.**

## Solution 45

***Worked Solution***

At time $t = 0$, the initial free moisture content is indicated by point *A* (far right). In the beginning, the solid is usually at a temperature lower than its ultimate temperature and the evaporation rate will be higher due to a larger temperature difference. If the solid was hot to start with, the rate may start at point *A*'.

At point *B*, the surface temperature increases to an equilibrium value. From *B* to *C*, the phase of drying is known as the constant rate period (CRP). This phase is characterized by the fact that the solid surface is very wet and a continuous film exists on the drying surface. This water film is completely unbound and water behaves as if the solid was not there. Therefore, during the CRP, the evaporation rate is independent of the solid as evaporation occurs from the free liquid surface. The duration of the CRP depends on how long water can be supplied to the surface at least as fast as it evaporates.

At point *C* the CRP ends and we enter into the falling rate period (FRP). This occurs when the free moisture content reaches a critical value, typically denoted as $X_c$. From point *C* onwards, there is insufficient water to maintain a continuous film and the solid surface is no longer entirely wetted. The wetted area starts to decrease with time, until the solid is completely dry at point *D*.

From point $D$ onwards, we enter into a second falling rate period when the plane of evaporation recedes from surface inwards into the interior of the solid. Vaporized water moves through the solid into the air stream outside. At a given relative humidity, drying can only remove water until the free moisture content is zero, i.e., when total moisture content is equivalent to equilibrium moisture content.

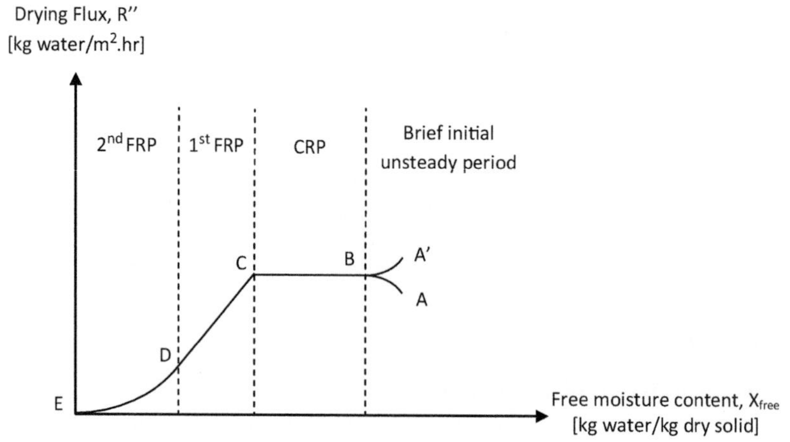


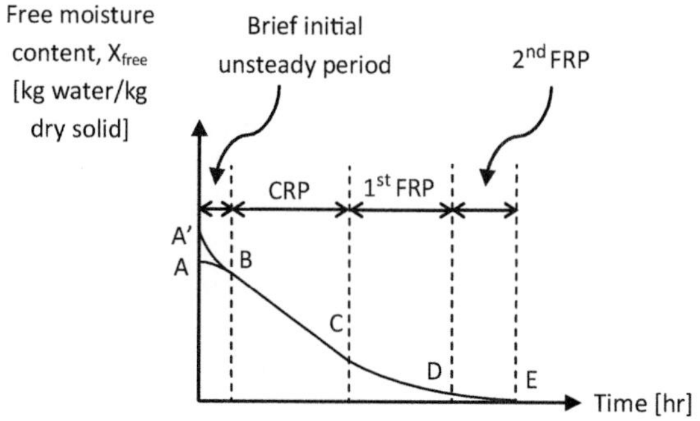


## Problem 46

**In a psychrometric chart, we can find diagonals running down towards the bottom right which indicate enthalpy values. Explain what these lines are, and show how they relate to physical properties of an air–water system. Also explain the terms adiabatic saturation temperature and adiabatic saturation humidity.**

## Solution 46

### *Worked Solution*

The enthalpy of a gas–vapor mixture comprises the enthalpy of the bone dry gas and the enthalpy of the vapor. [Note that over here, vapor refers to moisture, for example, water vapor in dry air]. We can therefore define enthalpy $h$ for a mixture of specific humidity $\mathcal{H}$ as shown below, where $\lambda$ is the enthalpy of vaporization (or latent heat of vaporization) at temperature $T_0$.

$$h = C_{p,\text{gas}}(T - T_0) + \mathcal{H}\left[C_{p,\text{gas}}(T - T_0) + \lambda\right]$$

Let us define a new variable, $s$, which simplifies the above equation. This is a useful quantity and is also known as humid heat, which is also the effective heat capacity of the gas–vapor mixture. Humid heat varies with specific humidity, $\mathcal{H}$.

$$h = s(T - T_0) + \mathcal{H}\lambda$$

$$s = C_{p,\text{gas}} + \mathcal{H}C_{p,\text{vapor}}$$

In a psychrometric chart, the enthalpy of a gas–vapor mixture can be read from lines of constant enthalpy. An example is shown below for an air–water mixture at 40 °C and specific humidity of 0.0345 kg/kg. The corresponding enthalpy line at this point is at 120 kJ/kg.

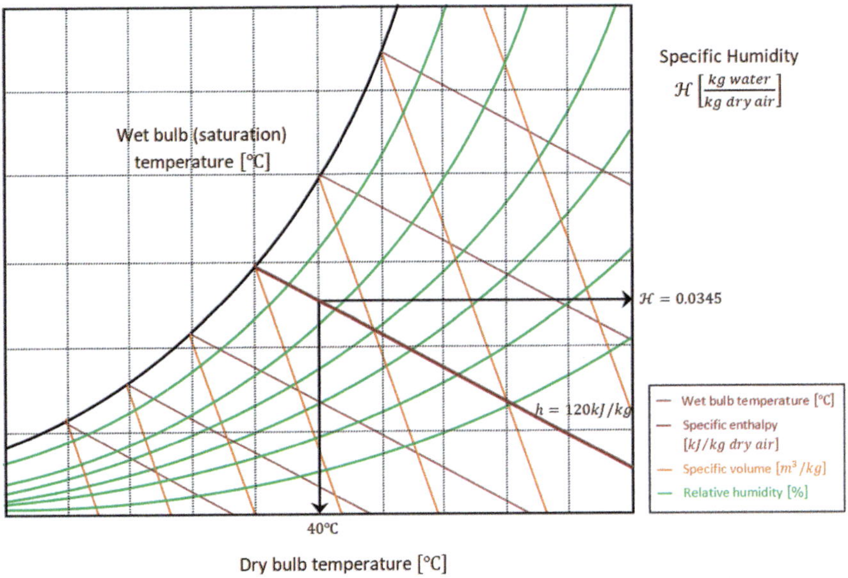

Lines of constant enthalpy are also known as adiabatic saturation lines. And this term can be better understood by considering the following scenario. We have mixing of a gas–vapor mixture with more liquid. The mixture is allowed to reach equilibrium under adiabatic conditions.

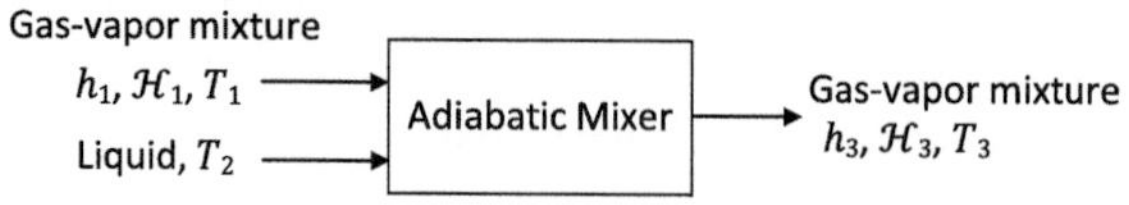


The mass of liquid added per kg of bone dry gas $= \mathcal{H}_3 - \mathcal{H}_1$. An energy balance can be done for the liquid, where $T_0$ denotes a reference temperature.

$$h_3 - h_1 = (\mathcal{H}_3 - \mathcal{H}_1) C_{P,\text{liq}}(T_2 - T_0)$$

The enthalpy of the liquid added, $h = s(T - T_0) - \lambda\mathcal{H}$. Therefore we can substitute this expression into the above equation to obtain

$$[s(T_3 - T_0) + \mathcal{H}_3\lambda] - [s(T_1 - T_0) + \mathcal{H}_1\lambda] = (\mathcal{H}_3 - \mathcal{H}_1) C_{P,\text{liq}}(T_2 - T_0)$$

$$s(T_3 - T_1) = -(\mathcal{H}_3 - \mathcal{H}_1)\left[\lambda + C_{P,\text{liq}}(T_2 - T_0)\right]$$

Note that the amount of sensible heat to raise temperature of the liquid from $T_0$ to $T_2$ is assumed negligible as compared to latent heat of vaporization $\lambda$. And we can derive the equation of the adiabatic saturation line as follows, which relates specific humidity $\mathcal{H}$ to temperature $T$. Note that the line has a slope of $-\frac{s}{\lambda}$.

$$\frac{\mathcal{H}_3 - \mathcal{H}_1}{T_3 - T_1} = -\frac{s}{\lambda}$$

By observing the form of the equation, we may note that as more liquid is added, humidity increases but temperature falls. This is because enthalpy is used to vaporize the added liquid adiabatically. As we add even more liquid, the mixture eventually becomes saturated and the temperature and humidity at this point are also known as the adiabatic saturation temperature $T^s$ and adiabatic saturation humidity $\mathcal{H}^s$.

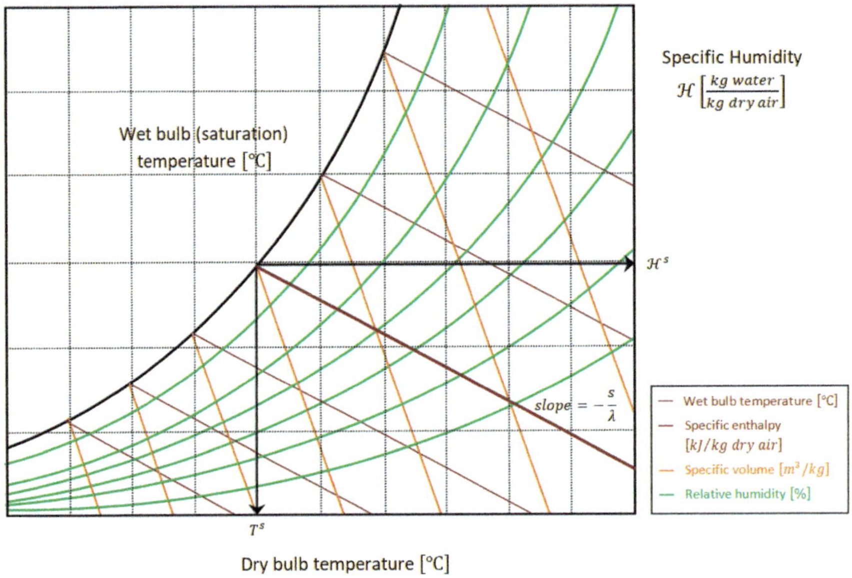


## Problem 47

**The understanding of crystallization processes, especially at the molecular level, is important in biological and industrial applications. Good control of the crystallization process is key in obtaining desired crystalline particles. Discuss the factors that influence the following aspects of crystallization:**

(a) **Crystal growth rate**
(b) **Particle size distribution**
(c) **Crystal shape**

## Solution 47

***Worked Solution***

(a) Crystallization involves the creation of a new daughter phase from a bulk parent phase and nucleation refers to the first step in the formation of this new phase through molecular self-assembly. The growth rate of crystals is largely dependent on nucleation conditions.

The nucleation step can be categorized into primary or secondary, and heterogeneous or homogeneous. Primary nucleation occurs without requiring pre-existing crystals to be present, while secondary nucleation produces new crystals from pre-existing crystals. Primary nucleation can be avoided by maintaining low levels of supersaturation. In industrial applications, secondary nucleation is the common mode of crystallization and can be enhanced by agitation. Pre-existing crystals in supersaturated solution can be initiated by fluid shear forces acting along crystal surfaces in a crystallizer, sweeping away nuclei as a result. Collisions between crystals, as well as with vessel surfaces or agitator blades also contribute to the generation of pre-existing crystal nuclei.

As for heterogeneous nucleation, it occurs when nucleation sites are on surfaces, while homogeneous nucleation occurs from the interior of a uniform substance. Heterogenous nucleation is more common and requires supersaturation. Homogeneous nucleation typically requires superheating or supercooling and occurs randomly and spontaneously.

Crystal growth is influenced by kinetic and thermodynamic factors.

*Thermodynamic Effects*

The total overall Gibbs free energy of formation of small crystal units in solution, $\Delta G_f$ is the sum of the surface free energy change, $\Delta G_s$ and the bulk free energy associated with the movement of small crystal units from solution into the bulk solid crystal phase (or lattice free energy), $\Delta G_b$. These units can be ions, atoms, or molecules.

$$\Delta G_f = \Delta G_s + \Delta G_b$$

$\Delta G_s$ arises due to the creation of an interface between the new solid phase and the solution phase. This quantity is always a positive value and works against crystallization as it increases the overall Gibbs free energy $\Delta G_f$. $\Delta G_s$ is representative of a surface tension, and is therefore proportional to the surface area of the new crystal forming.

$\Delta G_b$ is influenced by the degree of supersaturation $S$ of the solution which relates to the difference in chemical potential of the crystal $\mu_{\text{crystal}}$ and the solution of its ions $\mu_{\text{ions}}$. Using the example of a spherical crystal, $\Delta G_b$ can be expressed in terms $\Delta\mu$ as shown below, whereby $k$ is the Boltzmann constant, $T$ is temperature, $c$ is the concentration of ions in the solution, $c_{\text{eqm}}$ is the equilibrium concentration of ions in a saturated solution, $r$ is the radius of the crystal cluster being formed, and $a$ is the size of individual units making up the crystal cluster.

$$\Delta G_b = -\frac{4}{3}\pi\left(\frac{r}{a}\right)^3 \Delta\mu$$

$$\Delta\mu = \mu_{\text{ions}} - \mu_{\text{crystal}} = kT \ln S = kT \ln\left(\frac{c}{c_{\text{eqm}}}\right)$$

For $c > c_{\text{eqm}}$, $\mu_{\text{ions}} > \mu_{\text{crystal}}$, it then follows that $\Delta G_b$ is always negative. We can see that it is the competition between $\Delta G_s$ and $\Delta G_b$ that determines the overall thermodynamic conditions for crystal growth. To visualize this better, we can sketch the following diagram.

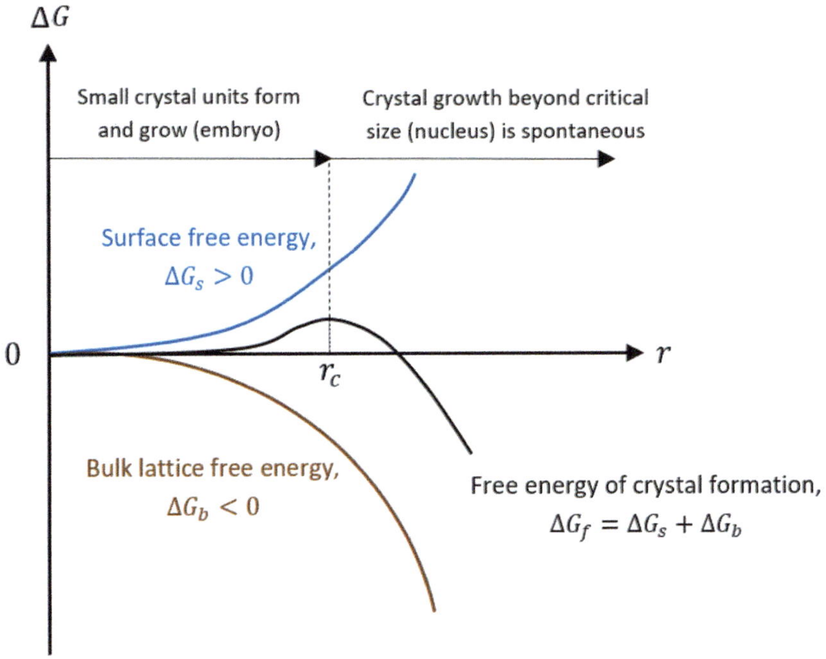


We note that the free energy of crystal formation $\Delta G_f$ increases initially; hence, this part of the process is thermodynamically challenging due to the need to overcome the dominating surface tension effect ($\Delta G_s > 0$) when crystal units are small. However, as the crystal grows further, the competing effect from bulk lattice free energy ($\Delta G_b > 0$) starts to dominate, and this makes it more thermodynamically favorable for crystal growth. There is a point in the process whereby the crystal size reaches a critical value, $r_c$. This is also the point when the crystal is referred to as a nucleus. Mathematically, this point is a maximum point defined by ($\frac{d\Delta G_f}{dr} = 0$), and a further increase in size from $r_c$ decreases overall Gibbs free energy and is therefore spontaneous.

*Kinetic Effects*

Kinetic factors refer more to mass transfer aspects, such as the adsorption rate of molecules into the crystal lattice, or diffusion to surfaces, etc.

Using the example of a pure ionic crystal with a cubic lattice structure, and assuming heterogenous nucleation, crystals grow as small crystal units deposit on a crystal surface. Prior to deposition, the units are transported to the surface, and are adsorbed. They may then diffuse to a site that is most energetically favorable. As the crystal grows, more surface sites are filled, and we eventually obtain a flat surface. Further growth results in the formation of new layers, and this layered structure gives rise to a two-dimensional island disc shape, characteristic of heterogenous nucleation. A high degree of supersaturation is kinetically favorable for this process. On the other hand, homogenous nucleation is more often characterized by three-dimensional shapes.

(b) The particle size distribution (PSD) can be considered at two levels, macroscopic and molecular. At the molecular level, PSD is influenced by the competition between different growth mechanisms, temperature effects, local concentrations, and presence of impurities in solution.

At the macroscopic level, we refer to aggregation and attrition of particles that commonly occur within industrial-scale mixers and growth chambers, and which affect crystal size. Intense mixing typically lead to smaller particles.

(c) The smallest repeating unit in a crystal is called a unit cell. Crystal shape is affected by the geometry and packing parameters of the unit cell. In reality however, we do not often have perfect shapes and/or repetitions of atoms in the crystal lattice due to the presence of impurities which distort crystal lattice structures.

The type of nucleation (e.g., homogeneous or heterogenous) and the relative rates of growth on the various crystal planes also influence overall shape. Other factors that affect crystal shape and morphology include temperature, the degree of supersaturation, and kinetic effects. Thermodynamic effects are relatively less prominent, and typically affect only smaller particle sizes.

## Problem 48

**(a) Explain briefly the use of seed crystals in industrial applications, and common methods to minimize the degree of nucleation.**

**(b) Consider the batch crystallization of a spherical hydrated salt ($X.nH_2O$) from a seeded aqueous solution at atmospheric pressure. A heat exchanger is used to maintain the operating temperature at 30 °C throughout the crystallization process. The crystallizer was initially filled with 2000 kg of saturated aqueous solution and inoculated with 1 kg of crystal seeds (hydrated salt $X.nH_2O$). The crystal seeds have radii measuring 20 μm and it is assumed that crystallization occurs predominantly from the seeds. The final desired size of the hydrated crystal salt is 350 μm in diameter. If the average crystal growth rate is $8 \times 10^{-8}$ m/s measured from the crystal plane and supersaturation level is 0.02 g crystal/g solution, determine the following:**

**(i) Final mass of hydrate crystals in the crystallizer.**
**(ii) Number of crystals in the seed inoculum.**
**(iii) Mass of water evaporated.**
**(iv) Time taken to reach the desired crystal size.**
**(v) Explain how supersaturation affects crystal size distribution in solution crystallization.**
**(vi) Given the following correlation for surface-averaged Sherwood number for mass transfer, determine if crystal growth is controlled by mass transfer or surface reactions.**

$$\mathbf{Sh = 2 + 0.6Re^{0.5}Sc^{0.3}}$$

**The following table is provided for reference.**

| | |
|---|---|
| Molecular weight of $X.\,nH_2O$ | 220 |
| Molecular weight of $X$ | 100 |
| Solubility of $X$ in aqueous solution [weight %] | 25 |
| Density of $X.\,nH_2O$ [kg/m$^3$] | 1700 |
| Density of aqueous solution [kg/m$^3$] | 1300 |
| Viscosity of aqueous solution [g/cm.s] | 0.07 |
| Diffusivity of $X$ in aqueous solution [cm$^2$/s] | $1.05 \times 10^{-5}$ |
| Solution velocity past crystal surface [m/s] | 0.04 |

## Solution 48

### *Worked Solution*

(a) Crystal growth is relatively easier compared to nucleation. Because nucleation is a random process that occurs at the molecular level, it is difficult to control and predict and this is a disadvantage for large-scale industrial processes requiring precise control and monitoring. It is challenging to detect crystal nuclei at atomic dimensions especially at initial stages of nucleation and a quantitative treatment for the nucleation of salts is therefore difficult. However, nucleation conditions have significant impact on the size and size distribution of crystal particles. In order to have reproducible results, efforts are made to get around the randomness of the nucleation phase, and one way is to use well-characterized seed crystals. A seed crystal is the first small piece of crystal that you use to grow into a larger crystal of desired specifications. For most electrolytes, there is a relatively well-defined critical supersaturation value, below which a stable supersaturated solution can be maintained for long periods without spontaneous crystallization occurring. By controlling experimental conditions within this range, the rate of growth of crystals is more reproducible and can be more easily studied from microscopic examinations and observed kinetics. In order to minimize primary nucleation events, we can alter supersaturation to sufficiently low levels. We can also adopt milder agitation to avoid secondary nucleation.

(b) (i) The initial crystal seeds have radii measuring 20 μm, this translates to a diameter of 40 μm $= 4 \times 10^{-5}$ m. The final desired crystal has a diameter of 350 μm $= 3.5 \times 10^{-4}$ m. We know the initial mass of crystal seeds, and we need to find the final mass of crystals after the crystallization process.

It helps to first find a correlation between mass and length scale (e.g., diameter) for a spherical particle. In general, the mass $M$ of a spherical particle is related to its diameter $D$, volume $V$, and density $\rho$ as follows:

$$M = \rho V = \rho\left(\frac{4}{3}\pi\left(\frac{D}{2}\right)^3\right) = \rho\left(\frac{\pi D^3}{6}\right)$$

$$M \sim D^3$$

$$\frac{M_1}{(D_1)^3} = \frac{M_2}{(D_2)^3}$$

We know that $M_1 = 1$ kg, $D_1 = 4 \times 10^{-5}$ m, and $D_2 = 3.5 \times 10^{-4}$ m; therefore, the final mass of the hydrate salt crystals is approximately 670 kg.

$$M_2 = M_1\left(\frac{D_2}{D_1}\right)^3 = 1\left(\frac{3.5 \times 10^{-4}}{4 \times 10^{-5}}\right)^3 = 670 \text{ kg}$$

(ii) Given the density of the hydrated salt crystal is 1700 kg/m$^3$, we can calculate the mass of a single crystal in the seed inoculum.

$$m_1 = \rho\left(\frac{\pi D_1{}^3}{6}\right) = 1700\left(\frac{\pi\left(4 \times 10^{-5}\right)^3}{6}\right) = 5.7 \times 10^{-11}\text{kg}$$

We can now find the number of crystals $n_{\text{crystal}}$ in the seed inoculum

$$n_{\text{crystal}} = \frac{M_1}{m_1} = \frac{1}{5.7 \times 10^{-11}} = 1.8 \times 10^{10}$$

(iii) To find the mass of water evaporated, we can start with a mass balance on water between the initial and final stages. The initial mass of water in the saturated solution $m_{w,\text{initial}}$ should be equivalent to the sum of (1) the mass of water left in the solution at the final stage $m_{w,\text{final soln}}$, (2) the mass of water entrapped in the newly formed hydrated crystals, $m_{w,n\text{H}_2\text{O}}$, and (3) the mass of water evaporated $m_{w,\text{evaporated}}$.

$$m_{w,\text{initial}} = m_{w,\text{final soln}} + m_{w,n\text{H}_2\text{O}} + m_{w,\text{evaporated}} \tag{1}$$

We need to determine $m_{w,\text{ evaporated}}$; therefore, we can focus on first determining $m_{w,\text{initial}}$, $m_{w,\text{final soln}}$, and $m_{w,n\text{H}_2\text{O}}$.

***To determine $m_{w,initial}$:***

We are told that the mass of the initial saturated solution is 2000 kg, and that solute *X* is at 25 wt% in solution. We can use this information to find out the initial mass of water and *X*, respectively, in the saturated solution.

$$m_{x,\text{initial}} = 0.25 \times 2000 = 500 \text{ kg}$$

$$m_{w,\text{initial}} = 2000 - 500 = 1500 \text{ kg}$$

***To determine*** $m_{w,nH_2O}$:

We know that we have 1 kg of crystal seeds added to the saturated solution at the initial stage, and from part a we note that the total final mass of hydrate crystals (seeds + newly formed) is 670 kg. The mass of crystals $m_{\text{new crystals}}$ that were newly formed is therefore

$$m_{\text{new crystals}} = 670 - 1 = 669 \text{ kg}$$

Given that the molecular weights of $X$ and $\boldsymbol{X.n\text{H}_2\text{O}}$ are 100 and 220, respectively, we can calculate the corresponding weight of the hydration portion (i.e., $\boldsymbol{n\text{H}_2\text{O}}$) of the crystal relative to one molar unit of crystal ($\boldsymbol{X.n\text{H}_2\text{O}}$) as follows

$$MW_{n\text{H}_2\text{O}} = MW_{\boldsymbol{X.n\text{H}_2\text{O}}} - MW_X = 220 - 100 = 120$$

We can now find the mass of water entrapped in the hydrated crystals that were newly formed, $m_{w,nH_2O}$. For one molar unit of crystal, we have

$$\frac{m_{w,n\text{H}_2\text{O}}}{MW_{n\text{H}_2\text{O}}} = \frac{m_{\text{new crystals}}}{MW_{\boldsymbol{X.n\text{H}_2\text{O}}}}$$

$$m_{w,n\text{H}_2\text{O}} = 120\left(\frac{669}{220}\right) = 365 \text{ kg}$$

***To determine*** $m_{w,final\ soln}$:

Finally let us find out the mass of water left in solution. We know from earlier that $m_{\text{new crystals}} = 669$ kg. We also found that within these newly formed crystals, the hydration mass is $m_{w,n\text{H}_2\text{O}} = 365$ kg. Therefore, we know that the portion of the newly formed crystals that comprises $X$ only has the following mass

$$m_{x,\text{new crystals}} = m_{\text{new crystals}} - m_{w,n\text{H}_2\text{O}}$$
$$= 669 - 365 = 304 \text{ kg}$$

The initial mass of $X$ in solution, $m_{x,\text{initial}}$, was found earlier to be 500 kg. Some of the $X$ went into the crystals as $m_{x,\text{new crystals}}$. Hence the remaining mass of $X$ in the final solution is

$$m_{x,f\text{inal soln}} = m_{x,\text{initial}} - m_{x,\text{new crystals}} = 500 - 304 = 196 \text{ kg}$$

We know that temperature and pressure are maintained throughout the process; hence, solubility of $X$ in the aqueous solution can also be reasonably assumed to be maintained at 25 wt% of X.

$$m_{w,f\text{inal soln}} = m_{x,f\text{inal soln}}\left(\frac{75}{25}\right) = 196\left(\frac{75}{25}\right) = 588 \text{ kg}$$

***Putting together our intermediate results into Eq. (1):***

$$m_{w,\text{initial}} = m_{w,f\text{inal soln}} + m_{w,n\text{H}_2\text{O}} + m_{w,\text{evaporated}}$$

$$m_{w,\text{evaporated}} = 1500 - 588 - 365 = 547 \text{ kg}$$

Hence the mass of water evaporated is 547 kg.

(iv) The growth rate from one plane of the crystal is $8 \times 10^{-8}$ m/s. Since we have a spherical crystal, this growth rate extends the diameter on both sides and the diameter therefore grows at twice the value of $8 \times 10^{-8}$ m/s, which is equivalent to $1.6 \times 10^{-7}$ m/s.

From earlier, we obtained the initial crystal seed diameter of 40 μm = $4 \times 10^{-5}$ m and final crystal diameter of 350 μm = $3.5 \times 10^{-4}$ m.

$$t = \frac{\left(3.5 \times 10^{-4}\right) - \left(4 \times 10^{-5}\right)}{1.6 \times 10^{-7}} = 1937.5 \text{ s}$$

The time taken is about 1937.5 s or 32 min.

(v) In solution crystallization, supersaturation is a key driving force and is equivalent to the difference between the actual concentration and the solubility concentration at a given temperature. Supersaturation is often controlled to avoid undesirable nucleation.

The solubility curve is an important aspect to consider when deciding the preferred method for crystallization. The effect of temperature on solubility varies between different types of compounds (e.g., organic vs. inorganic), and this determines the best method for controlling supersaturation levels. For compounds whereby solubilities decrease only slightly with large drops in temperature (e.g., systems of most inorganic salts in water), cooling the crystallizer unit by evaporation is preferred. As for compounds that have solubilities decrease significantly with decrease in temperature, crystallization by cooling (e.g., heat removal using heat exchanger) can be used. A general solubility curve is illustrated below where $\Delta c$ determines supersaturation levels. The solubilities of most inorganic compounds increase with temperature.

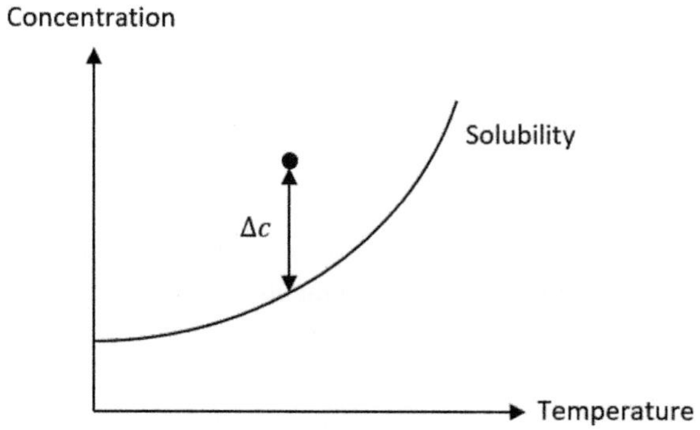

Supersaturation is a key driving force for crystal nucleation and growth. Nucleation refers to the creation of new crystal nuclei, while crystal growth refers to the enlargement of crystal size as solute is deposited from solution. Nucleation can be either primary (spontaneously occurs from solution) or secondary (pre-existing crystals). The importance of recognizing these two processes is in noting that the competition between these two processes at different supersaturation levels significantly influences the final crystal size distribution.

The rates of both processes $r$ can be denoted using a general form of the rate equation as shown below whereby $k$ is the growth (or nucleation) constant, $n$ is the growth (or nucleation) order:

$$r = k(\Delta c)^n$$

We can observe that supersaturation which is indicated by $\Delta c$ has an exponential correlation with the order of the reaction. Therefore, for systems whereby growth order is much smaller than nucleation order (e.g., organic systems), then at low supersaturation, crystal growth dominates as it is faster than nucleation and we have more large crystals. Conversely, at high supersaturation, nucleation is much faster than crystal growth and we end up with smaller crystals. Other than controlling supersaturation levels, mechanical agitation can also give rise to smaller crystals that are also higher in purity, more uniformly sized, and produced faster. Both supersaturation levels and uniformity of crystal sizes can be controlled by careful consideration of circulation between the crystallizing zone and the supersaturation zone. The use of vertical baffles promotes uniform mixing.

(vi) The surface-averaged Sherwood number is given in the problem as follows:

$$\text{Sh} = 2 + 0.6\text{Re}^{0.5}\text{Sc}^{0.3}$$

It is worth noting that this correlation is valid for the range of approximately $0 \leq \text{Re} < 200$, and $0 \leq \text{Sc} < 250$. At very low Re, Sh tends to a value of 2 and this considers the molecular diffusion from a sphere into a large volume of stagnant fluid.

From data booklets, we can find the definitions of the 3 dimensionless numbers, namely the Sherwood number (Sh), Reynolds number (Re), and Schmidt number (Sc).

***Sherwood Number***

The Sherwood number represents the ratio of convection rate to diffusion rate in mass transfer. Mathematically it can be expressed as follows whereby $k$ is the convective mass transfer coefficient, $\mathcal{D}$ is the mass diffusivity (or molecular diffusion coefficient of solute in solution), and $L$ is the characteristic length. In the case of a spherical particle in this problem, $L$ can be approximated as particle diameter $D_p$. In our problem, we note that $\mathcal{D} = 1.05 \times 10^{-5}$ cm²/s.

$$\mathrm{Sh} = \frac{k}{\mathcal{D}/L} = \frac{k}{\mathcal{D}/D_p} = \frac{kD_p}{1.05 \times 10^{-5}}$$

***Schmidt Number***

The Schmidt number Sc describes the ratio of momentum diffusivity to mass diffusivity. Mathematically, it is equivalent to the ratio of the kinematic viscosity of the fluid $\nu$ to diffusivity $\mathcal{D}$. $\nu$ can be further expressed in terms of dynamic viscosity $\mu$ and fluid density $\rho_l$. In our problem, we note that $\mu = 0.07$ g/(cm.s) and $\rho_l = 1300$ kg/m³ $= 1.3$ g/cm³.

$$\mathrm{Sc} = \frac{\nu}{\mathcal{D}} = \frac{\mu}{\mathcal{D}\rho_l} = \frac{0.07}{\left(1.05 \times 10^{-5}\right)1.3} \approx 5128$$

***Reynolds Number***

Lastly, the Reynolds number Re is useful in understanding flow patterns, and specifically it indicates whether fluid flow is steady or turbulent. Mathematically it is expressed as follows whereby $V$ is fluid velocity, $L$ is the characteristic length which is approximated as particle diameter $D_p$ in this problem, and $r$ is the crystal radius. In our problem, we note that $V = 0.04$ m/s $= 4$ cm/s:

$$\mathrm{Re} = \frac{VL}{\nu} = \frac{V\rho_l D_p}{\mu} = \frac{4(1.3)D_p}{0.07} \approx 74D_p = 148r$$

To solve the problem, let us first consider the case where mass transfer is the controlling mechanism. We can express the growth rate of the crystals as a differential equation with respect to time $t$ and in terms of a mass transfer coefficient and a concentration driving force. In the equation below, $n$ denotes the molar amount of hydrate crystals, $A_s$ refers to the surface area of the crystals, $k$ denotes the growth (assumed dominated by mass transfer) constant, and $\Delta c$ is a concentration difference that is correlated to supersaturation.

$$\frac{dn}{dt} = A_s k(\Delta c)$$

Since we have spherical crystals, we know that $A_s = 4\pi r^2$.

$$\frac{dn}{dt} = 4\pi r^2 k(\Delta c) \tag{1}$$

We can also relate $n$ to molar density $\rho_m$ of the crystal and this is useful since $\rho_m$ can be calculated easily from the given mass density ($\rho = 1.7$ g/cm$^3$) of the crystal.

$$\rho_m = \frac{\rho}{MW_{X.nH_2O}} = \frac{1.7}{220} = 0.0077 \text{ mol/cm}^3$$

The volume of a sphere is known to be Volume $= (4/3)\pi r^3$. Therefore, the number of moles of spherical crystal is written as

$$n = \frac{4}{3}\pi r^3 \rho_m$$

Substituting the above expression for $n$ into Eq. (1), we have

$$\frac{dn}{dt} = \frac{d}{dt}\left(\frac{4}{3}\pi r^3 \rho_m\right) = 4\pi r^2 k(\Delta c)$$

The left-hand side of the equation can be simplified as follows

$$\frac{d}{dt}\left(\frac{4}{3}\pi r^3 \rho_m\right) = \frac{4}{3}\pi \rho_m \frac{d(r^3)}{dt} = \frac{4}{3}\pi \rho_m (3r^2)\frac{dr}{dt} = 4\pi r^2 \rho_m \frac{dr}{dt}$$

Equating the above with the right-hand side of the equation, we now have a differential equation representative of the growth rate of the crystal in terms of its size (i.e., radius)

$$4\pi r^2 \rho_m \frac{dr}{dt} = 4\pi r^2 k(\Delta c)$$

$$\frac{dr}{dt} = \frac{k(\Delta c)}{\rho_m} = \frac{k(\Delta c)}{0.0077} \tag{2}$$

We can now make use of the given correlation for the Sherwood number in the problem and equate it to the definition of the Sherwood number.

$$\text{Sh} = \frac{kD_p}{1.05 \times 10^{-5}} = 2 + 0.6\text{Re}^{0.5}\text{Sc}^{0.3}$$

$$\frac{k(2r)}{1.05 \times 10^{-5}} = 2 + 0.6(148r)^{0.5}(5128)^{0.3}$$

The initial and final sizes of the crystal are $r_1 = 2 \times 10^{-3}$cm and $r_2 = 1.75 \times 10^{-2}$cm, respectively. We can therefore calculate the values of $k$ for the initial and final stages to examine the effect of mass transfer on crystal growth rate.

The initial and final values of mass transfer coefficient $k_1$ and $k_2$ can be found as follows

$$k_1 = \frac{(1.05 \times 10^{-5})\left(2 + 0.6(148r_1)^{0.5}(5128)^{0.3}\right)}{2r_1} = 0.016\ \text{cm/s}$$

$$k_2 = \frac{(1.05 \times 10^{-5})\left(2 + 0.6(148r_2)^{0.5}(5128)^{0.3}\right)}{2r_2} = 0.004\ \text{cm/s}$$

We observe that the mass transfer coefficient decreases as the crystal grows. Going back to our differential Eq. (2), we can examine how the different values of $k$ affect growth rate $\frac{dr}{dt}$.

$$\frac{dr}{dt} = \frac{k(\Delta c)}{0.0077}$$

The concentration driving force $\Delta c$ for the above equation is in molar units, and this can be obtained from the known value of supersaturation (mass basis) of 0.02 g $X.nH_2O$ per g of solution as follows.

$$\Delta c = \frac{0.02(\rho_l)}{MW_{X.nH_2O}} = \frac{0.02(1.3)}{220} = 0.00012\ \text{mol/cm}^3$$

Therefore the initial and final growth rates are:

$$\left.\frac{dr}{dt}\right|_1 = \frac{k_1(\Delta c)}{0.0077} = \frac{0.016(0.00012)}{0.0077} = 2.5 \times 10^{-4}\ \text{cm/s}$$

$$\left.\frac{dr}{dt}\right|_2 = \frac{k_2(\Delta c)}{0.0077} = \frac{0.004(0.00012)}{0.0077} = 6.2 \times 10^{-5}\ \text{cm/s}$$

We compare the above growth rates (assuming mass-transfer dominated) with the given observed average growth rate of $8 \times 10^{-6}$ cm/s and notice that even for the slower growth rate (assuming mass-transfer dominated) of $\left.\frac{dr}{dt}\right|_2 = 6.2 \times 10^{-5}$ cm/s, it is still much faster than the observed growth rate.

Therefore, we conclude that crystal growth in this case is not controlled by mass transfer, but by surface integration reactions instead.

## Problem 49

**Plot a graph showing the effect of temperature and solute concentration on solubility and explain the following phenomena and terms using the graph:**

(a) **No crystals are observed when a solution is undersaturated.**

(b) **Clouds at high altitude do not form ice although they are at temperatures below freezing point.**
(c) **"Supersaturation temperature difference," "supersaturation," and "relative supersaturation".**
(d) **Spontaneous nucleation can give rise to very small crystals under certain conditions, although these crystals are invisible to the naked eye (less than about 20 μm).**
(e) **"Metastable region" and "metastable limiting solubility".**

## Solution 49

***Worked Solution***

(a) The plot below is an example of a solubility curve, which correlates the concentration of solute with temperature. In most cases, solubility increases with increasing temperature. No crystals are observed when a solution is undersaturated; this can be represented by the point (marked in red) as shown in the plot below. As the point lies below the solubility curve, it is in the "undersaturated" region where crystals of all sizes dissolve. No crystals form unless we increase concentration of solute or reduce temperature (as shown in gray arrows) such that the point is sufficiently moved out of this "undersaturated" region.

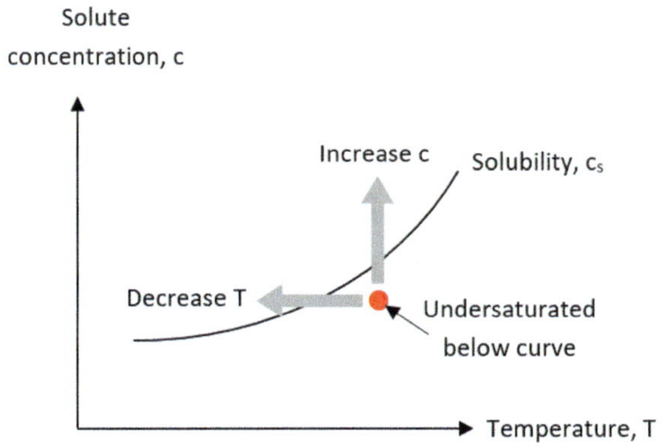


(b) Clouds are essentially a mass of tiny water droplets. These water droplets do no freeze into ice even though temperature may be below freezing point at high altitudes (assuming there are no pre-existing ice crystals that serve as seed crystals to trigger crystallization). This phenomenon can be explained in the diagram below, where we exist in the region as marked by red, below the metastable limit (represented by the c* curve) and the solubility curve. Along the solubility curve, $c = c_s$ and equilibrium exists between the crystal (visible) and saturated solution. At the point marked in red, $c_s < c < c^*$ and crystals can grow but no spontaneous nucleation occurs since $c < c^*$. In the absence of pre-existing seed crystals, if no nucleation occurs then no new crystals can form (Note: If pre-existing crystals are present, these crystals can grow).

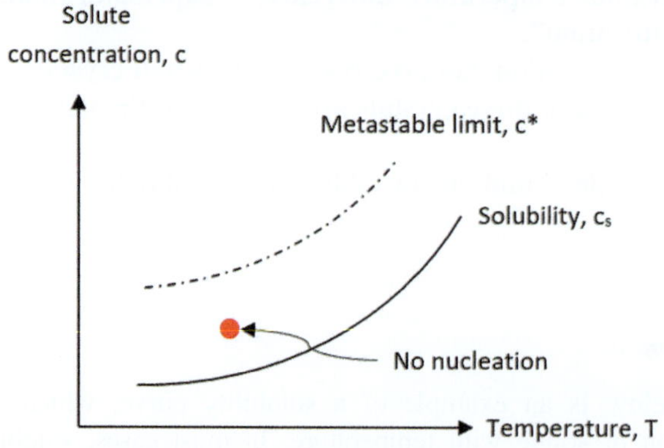


(c) Supersaturation temperature difference is shown by the horizontal gray arrow and represents the difference in temperature between a point above the solubility curve and a point on the solubility curve at constant solute concentration. Supersaturation $\Delta c$ is defined as the difference between actual solute concentration and solubility concentration. It can be represented by the vertical gray arrow between a point above the solubility curve and a point on the solubility curve at constant temperature.

$$\Delta c = c - c_s$$

Relative supersaturation $s$ is an alternate form of measuring the degree of supersaturation and is typically defined as follows, where we may find the use of a supersaturation ratio, $S = \frac{c}{c_s}$.

$$s = \frac{c - c_s}{c_s} = \frac{c}{c_s} - 1 = S - 1$$

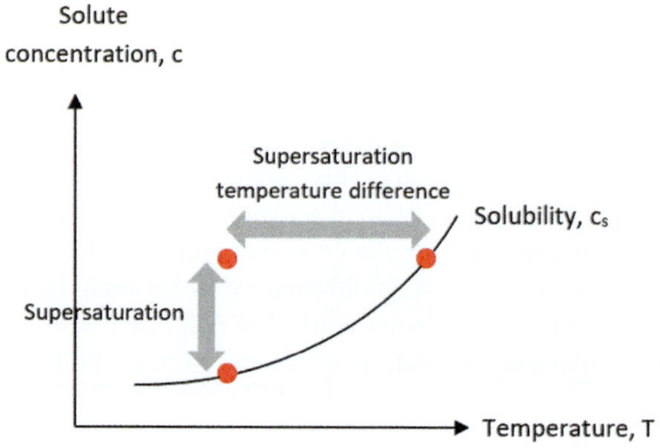

(d) Spontaneous nucleation can give rise to the creation of new and very small crystals (invisible to the naked eye), even if there are no pre-existing crystals present. This occurs when we are located on the metastable limit curve c* as marked in red below.

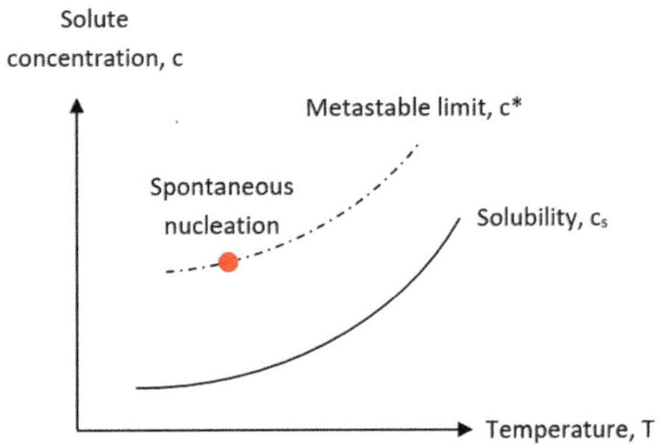


(e) The term metastable limiting solubility refers to the value of $c^*$, whereby the $c^*$ curve is simply a continuous plot of metastable liming solubility values over a temperature range. The metastable region is the region "sandwiched" between the metastable limit curve $c^*$ and the solubility curve $c_s$. We can also define a "limit of the supersaturation temperature difference" as shown below by the horizontal gray arrow, as well as a limiting supersaturation as shown by the vertical gray arrow ($\Delta c^* = c^* - c_s$).

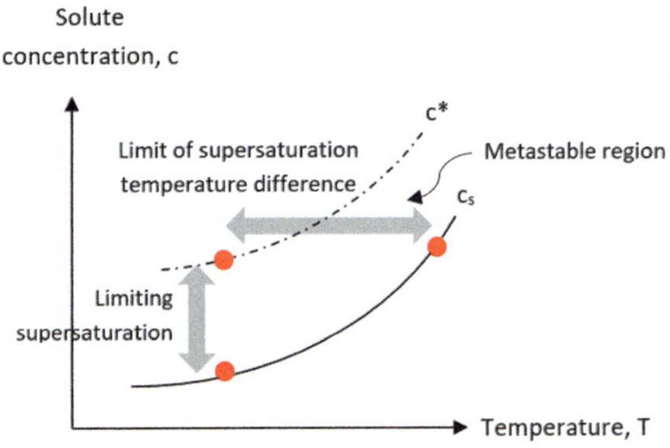


**Problem 50**

**Other than temperature, solubility is also influenced by crystal size. In general, there is greater solubility with smaller crystals. The equation below is used to model the relationship between crystal size and solubility of a particular**

**inorganic compound $A$ (assumed hydrate-free) in water at temperature $T$. In this equation, $c$ denotes actual concentration of solute $A$, $c_s$ denotes the solubility of hydrate-free $A$ at infinite dilution, $\sigma$ denotes interphase surface tension between solid and liquid phases, $\rho_m$ denotes molar density of crystals of $A$, $q$ denotes the number of ions dissociated from one solute molecule of $A$, $D_p$ refers to the diameter of the crystal particle, and $R$ refers to the gas constant.**

$$\ln\left(\frac{c}{c_s}\right) = \frac{4\sigma}{\rho_m qRTD_p}$$

**It is given that the molecular weight, density, and $q$ value for $A$ are 75, 2000 kg/m$^3$, and 2, respectively, and at a temperature of 30 °C, it was found that the values of $c_s$ and $\sigma$ for $A$ are 37 g/100 g $H_2O$ and 0.03 J/m$^2$, respectively.**

**Show how crystal size affects solubility of $A$ and comment on any key observations.**

## Solution 50

### *Worked Solution*

The equation that is given in this problem originates from the Kelvin equation, which is used to study solubility of small particles. In its general form, the Kelvin equation models the change in the vapor pressure of a liquid or the solubility of a solid due to the curvature of the interface between adjacent phases (e.g., liquid–vapor interface or solid–liquid interface). In this problem, we have a solid–liquid interface when solid crystals of $A$ dissolve in liquid water.

We are told that the solubility of $A$ at infinite dilution and at 30 °C is 37 g/100 g water on a hydrate-free basis. For systems comprising soluble organic or inorganic compounds in water, extensive data is available for solubility as a function of temperature. (Note: The heat of solution at infinite dilution at room temperature can also be found from such published data.)

The $q$ value of $A$ is given as 2. Solutes such as sodium chloride (NaCl) and potassium chloride (KCl) are examples of salts with this dissociation number since one molecule of NaCl (for example) dissociates into two ions $Na^+$ and $Cl^-$ in water.

We know that $R = 8314$ J/(kmol. K), and can substitute the given values for $A$ into the equation as follows:

$$\rho_m = \frac{\rho}{MW} = \frac{2000}{75} = 26.7\ \text{kmol/m}^3$$

$$\ln\left(\frac{c}{c_s}\right) = \frac{4\sigma}{\rho_m q\text{RT}D_p}$$

$$\ln\left(\frac{c}{37}\right) = \frac{4(0.03)}{(26.7)(2)(8314)(273+30)D_p}$$

We can obtain a function of actual solute concentration $c$ with respect to size of particle $D_p$.

$$c\ [g\ A/100\ g\ H_2O] = 37\exp\left(\frac{9.12 \times 10^{-10}}{D_p\ [m]}\right) = 37\exp\left(\frac{9.12 \times 10^{-4}}{D_p\ [\mu m]}\right)$$

After tabulating values of $c$ for an arbitrary range of $D_p$, we notice that the solubility of crystal $A$ increases as size starts to decrease from about 0.01 μm. However, beyond a size of about 0.1 μm, the size of crystal $A$ does not affect solubility much and solute concentration remains at around 37 g $A$ per 100 g water.

| $c$ [g $A$/100 g $H_2O$] | $D_p$ [μm] |
|---|---|
| 92.10396 | 0.001 |
| 67.95991 | 0.0015 |
| 53.28836 | 0.0025 |
| 44.40348 | 0.005 |
| 40.53306 | 0.01 |
| 38.37468 | 0.025 |
| 37.68107 | 0.05 |
| 37.33898 | 0.1 |
| 37.03376 | 1 |
| 37.00337 | 10 |
| 37.00034 | 100 |

A graph can be plotted to better visualize this trend, note that the horizontal axis is plotted on a log10 scale.

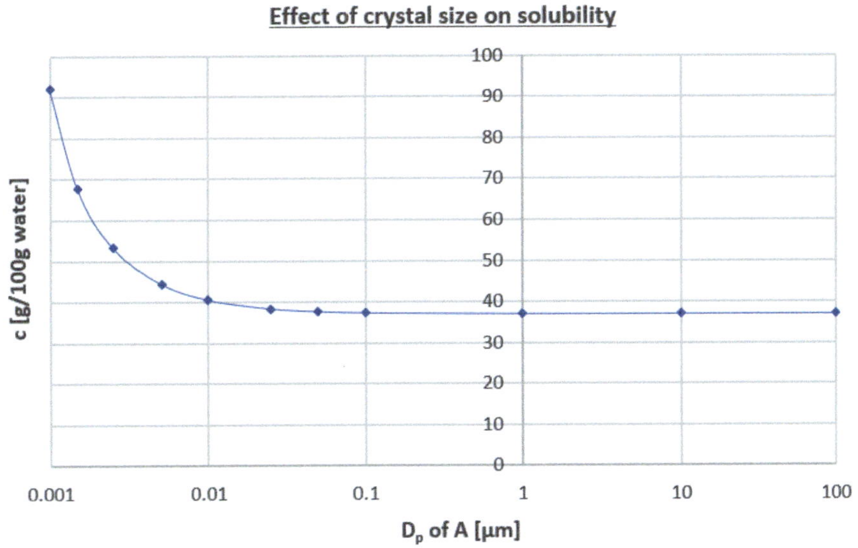

Understanding the impact of crystal size on solubility helps better design crystallization conditions. For example, for small crystals where solubility is sufficiently high, supersaturation can occur without agitation, especially if the cooling rate is slow.

## Problem 51

**The Mixed Suspension, Mixed Product Removal (MSMPR) model is a useful mathematical model developed to study a particular type of crystallizer known as the draft-tube baffled (DTB) crystallizer. It is useful for characterizing size distributions of crystals as experimental data can be easily obtained in the lab to determine useful parameters such as crystal nucleation rate and growth rate. Such parameters can then be applied in the design of industrial-scale crystallizer units.**

(a) **Explain what is meant by a particle size distribution and how it relates to the distribution function of the population. Plot the cumulative distribution for an arbitrary population to show how its distribution function relates to particle size and concentration of particles.**
(b) **One of the useful results from the MSMPR model is the population density function given as follows. Explain what the symbols in this equation mean and show how it is derived. Include in your derivation any key assumptions made.**

$$n = n_0 \exp\left(-\frac{t_L}{\tau}\right)$$

(c) **Nucleation rate $B_0$ is sometimes expressed as follows, whereby G denotes crystal growth rate. Show how this expression can be derived from the population density function $n$.**

$$B_0 = G n_0$$

## Solution 51

*Worked Solution*

(a) A distribution function helps us determine properties of interest for a population of particulates. An example of such properties is the particle concentration (i.e., number of particles of defined size or size range per unit volume).

The distribution function given in this problem is also referred to as "number density" or "population density" and can be denoted as $n(L)$. It represents the distribution of population about a size $L$ (assuming a linear length scale for size) and can be expressed mathematically as the number of crystal particles ($N$) per unit size of crystal ($L$) per unit volume of mother liquor ($V_{ML}$). [Note: Mother liquor refers to the solids-free liquid in a solid–liquid crystallization system.]

$$n(L) = \frac{d(N/V_{ML})}{dL} = \frac{1}{V_{ML}}\frac{dN}{dL}$$

It follows that the number of particles of size $L$ to $L + dL$ in volume $V_{ML}$ of liquid (solids-free) is therefore

$$dN = V_{ML}n(L)dL$$

We can determine cumulative distribution properties such as the total concentration of particles over a defined size range as follows, by integrating the population density function.

For the concentration (number per unit volume of mother liquor) of particles of size from $L_1$ to $L_2$, we have:

$$\int_{L_1}^{L_2} n(L)dL = \frac{N(L_1, L_2)}{V_{ML}}$$

We can also find the cumulative concentration, representing the number of particles per unit volume of mother liquor below a size $L$

$$\int_{0}^{L} n(L)dL = \frac{N(0, L)}{V_{ML}}$$

The total concentration of particles in our system can be determined by integrating over the limits of $L$, from 0 to the largest crystal in the population of size, e.g., $L_{\max}$:

$$\int_{0}^{L_{\max}} n(L)dL = \frac{N_T}{V_{ML}}$$

Graphically, we can see the significance of population density $n$ as the gradient of the graph of the cumulative number of particles per unit volume of mother liquor $N/V_{ML}$ against crystal size $L$.

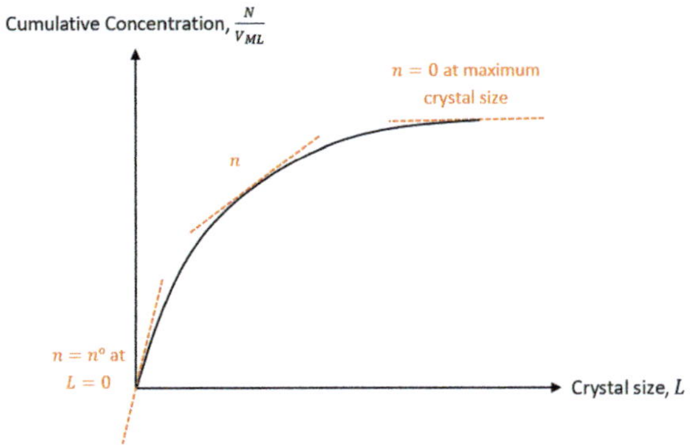

(b) One key assumption in the MSMPR model is that the crystal growth rate is the same for crystals of all sizes. Therefore, the growth rate $G$ can be simplified as a constant value.

$$G = \frac{dL}{dt} = \frac{\Delta L}{\Delta t} \tag{1}$$

Another important assumption for the MSMPR model is perfect mixing conditions such that the mixtures inside the crystallizer and in the exiting fluid are identical, uniform, and homogeneous in composition at all times. As such, the concentration of particles in the crystallizer can be expressed as equivalent to the concentration of particles in the product stream that is withdrawn as follows.

$$\frac{N_{\text{crystallizer}}}{V_{ML,\text{crystallizer}}} = \frac{N_{\text{withdrawn}}}{V_{ML,\text{withdrawn}}}$$

$$\frac{V_{ML,\text{withdrawn}}}{V_{ML,\text{crystallizer}}} = \frac{N_{\text{withdrawn}}}{N_{\text{crystallizer}}}$$

We assume a continuous, steady-state operations; hence, with a constant volumetric flow rate $Q_{ML}$ for the withdrawn product stream, we have

$$\frac{dV_{ML,\text{withdrawn}}}{dt} = Q_{ML}$$

$$V_{ML,\text{withdrawn}} = \int_{t_1}^{t_2} Q_{ML}dt = Q_{ML}\Delta t$$

Substituting back into the expression earlier, and noting the negative sign due to a decrease in the number of crystals left in the crystallizer upon withdrawal of crystals over time

$$\frac{Q_{ML}\Delta t}{V_{ML,\text{crystallizer}}} = \left.\frac{N_{\text{withdrawn}}}{N_{\text{crystallizer}}}\right|_{t_2} - \left.\frac{N_{\text{withdrawn}}}{N_{\text{crystallizer}}}\right|_{t_1} = \frac{-\Delta n}{n}$$

Substituting expression (1) to replace $\Delta t$, we have

$$\frac{Q_{ML}\Delta L}{GV_{ML,\text{crystallizer}}} = \frac{-\Delta n}{n}$$

$$\frac{\Delta n}{\Delta L} = -\frac{Q_{ML}n}{GV_{ML,\text{crystallizer}}} \xrightarrow{\text{limit } of\ \Delta L \to 0} \frac{dn}{dL} = -\frac{Q_{ML}n}{GV_{ML,\text{crystallizer}}}$$

The residence time $\tau$ is the time taken to completely exchange the contents of the crystallizer and is defined as follows. It is also referred to as turnover time or flushing time.

$$\tau = \frac{V_{ML,\text{crystallizer}}}{Q_{ML}}$$

$$\frac{dn}{dL} = -\frac{n}{G\tau}$$

Integrating the above using the lower limit of $n = n_0$ when $L = 0$, we obtain a function of $n$ with respect to $L$.

$$\int_{n=n_0}^{n} \frac{dn}{n} = \int_{L=0}^{L} -\frac{1}{G\tau} dL$$

$$\ln\left(\frac{n}{n_0}\right) = -\frac{L}{G\tau}$$

$$n(L) = n_0 \exp\left(-\frac{L}{G\tau}\right)$$

The time taken for crystals to grow to size $L$ can be further defined and denoted by $t_L$. $t_L$ may also be referred to as a "residence time," although this residence time is different from $\tau$. Instead it refers to "age" of particles in the crystallizer of size $L$; in other words, it represents the time spent by crystals of size $L$ in the crystallizer. Assuming a constant growth rate $G$ as shown in Eq. (1), we can also express number density in terms of $t_L$.

$$t_L = \frac{L}{G}$$

$$n = n_0 \exp\left(-\frac{t_L}{\tau}\right)$$

(c) $n_0$ refers to the concentration (number per volume of mother liquor per size/ length of crystal) of crystals of zero length (i.e., the nuclei) and it is also called the "zero size population density" or "nuclei population density". Therefore, the rate of formation of nuclei per unit volume of mother liquor $B_0$ can be determined by taking the limit of crystal size tends to zero.

$$B_0 = \frac{1}{V_{ML}} \lim_{L \to 0} \frac{dN}{dt} = \frac{1}{V_{ML}} \lim_{L \to 0} \left(\frac{dN}{dL}\frac{dL}{dt}\right)$$

We know that growth rate can be expressed as follows, and assuming that it is a constant under the MSMPR model, it can be taken out of the differential.

$$G = \frac{dL}{dt}$$

Therefore, we re-express nucleation rate as such,

$$B_0 = \frac{\mathrm{G}}{V_{ML}} \lim_{L \to 0} \left( \frac{dN}{dL} \right)$$

We know from earlier that by definition, we have

$$n(L) = \frac{1}{V_{ML}} \frac{dN}{dL}$$

We now set the condition when $L = 0$, $n = n_0$

$$n(L = 0) = n_0 = \frac{1}{V_{ML}} \frac{dN}{dL}$$

$$\lim_{L \to 0} \left( \frac{dN}{dL} \right) = V_{ML} n_0$$

Substituting this result back into the earlier expression for $B_0$, we obtain the given expression.

$$B_0 = \frac{\mathrm{G}}{V_{ML}} (V_{ML} n_0) = G n_0$$

## Problem 52

**In a DTB crystallizer, crystals of a particular monohydrate salt are to be produced at an operating temperature of 30 °C. The MSMPR model was used to study this process, whereby the population density function was given as follows:**

$$\boldsymbol{n(L) = n_0 \exp\left( -\frac{L}{G\tau} \right)}$$

(a) **Using a suitable dimensionless quantity for crystal size, derive an expression representing the fraction of crystals of size $L$ and below out of the total number of crystals in the crystallizer. Using your result, plot the cumulative crystal population distribution and the differential population distribution functions against the dimensionless size.**
(b) **The numbers fraction found in part a is also referred to as the zeroth moment for the population density function $n$. The form of this fraction**

**can be further generalized for a $k^{th}$ moment as shown in the expression below, where $z$ refers to dimensionless size.**

$$x_k = \frac{\int_0^z nz^k dz}{\int_0^\infty nz^k dz}$$

**Using the expression given above, determine the corresponding cumulative function $x_1$ and differential function $dx_1/dz$ for the first moment (i.e., $k = 1$).**

(c) **Given that the second and third moment equations for $n$ are as follows, plot the first, second, and third moment functions in the same diagram as the plots in part a and comment on any key features.**

$$x_2 = 1 - \left(1+z+\frac{z^2}{2}\right)e^{-z}$$

$$x_3 = 1 - \left(1+z+\frac{z^2}{2}+\frac{z^3}{6}\right)e^{-z}$$

(d) **The predominant (or modal) crystal size $L_d$ is a commonly used quantity that can be calculated using the expression shown below. This quantity is particularly useful in the design of crystallizers. Show how this expression may be derived from the mass distribution function.**

$$L_d = 3G\tau$$

## Solution 52

*Worked Solution*

(a) The population distribution function is given as follows:

$$n(L) = n_0 \exp\left(-\frac{L}{G\tau}\right)$$

To derive the said fraction, $x$, we have in the numerator an integral (summation over) from $L = 0$ to $L$ which represents the number of crystals of size $L$ and smaller, per unit volume of mother liquor. Similarly, by setting limits of integration from $L = 0$ to $\infty$, we have the denominator representing the total number of crystals for all sizes per unit volume of mother liquor.

$$x = \frac{\int_0^L n(L)dL}{\int_0^\infty n(L)dL} = \frac{\int_0^L n_0 \exp\left(-\frac{L}{G\tau}\right)dL}{\int_0^\infty n_0 \exp\left(-\frac{L}{G\tau}\right)dL}$$

We can also apply some useful results of integrals of exponential functions as shown below as Eqs. (1) and (2):

$$\int e^{ax}dx = \frac{1}{a}e^{ax} \tag{1}$$

$$\int_0^\infty e^{-ax}dx = \frac{1}{a} \tag{2}$$

Using result (1) for the numerator, where $a = -\frac{1}{G\tau}$ we have,

$$\int_0^L n_0 \exp\left(-\frac{L}{G\tau}\right)dL = n_0\left[(-G\tau)\exp\left(-\frac{L}{G\tau}\right)\right]_0^L$$

$$= n_0\left[(-G\tau)\exp\left(-\frac{L}{G\tau}\right) + G\tau\right] = n_0 G\tau\left(1 - \exp\left(-\frac{L}{G\tau}\right)\right)$$

Using result (2) for the denominator, where $a = \frac{1}{G\tau}$ we have,

$$\int_0^\infty n_0 \exp\left(-\frac{L}{G\tau}\right)dL = n_0 G\tau$$

Therefore, we have the fraction as follows

$$x = 1 - \exp\left(-\frac{L}{G\tau}\right)$$

We can non-dimensionalize crystal size, $L$ by dividing it over a characteristic length, $L^* = G\tau$ to obtain a dimensionless crystal size $z$ as follows

$$z = \frac{L}{L^*} = \frac{L}{G\tau}$$

$$n(z) = n_0 e^{-z}$$

We arrive at the following cumulative number distribution and differential distribution.

$$x = 1 - e^{-z}$$

$$\frac{dx}{dz} = e^{-z}$$

By plotting $x$ against $z$, we have the cumulative numbers distribution shown below.

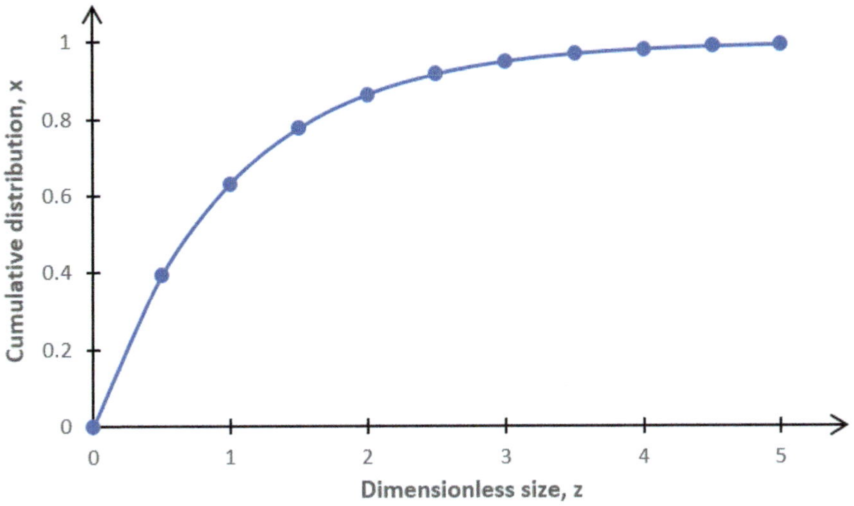


By plotting $\frac{dx}{dz}$ against $z$, we have the differential numbers distribution shown below.

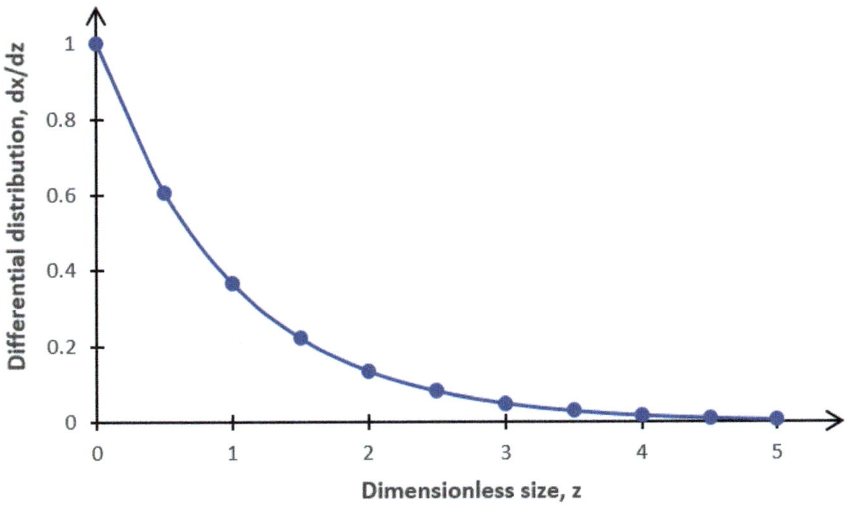

(b) Let us determine the first moment equation $x_1$ knowing that the population density function expressed in terms of dimensionless size is $n(z) = n_0 e^{-z}$

$$x_1 = \frac{\int_0^z nz\,dz}{\int_0^\infty nz\,dz} = \frac{\int_0^z (n_0 e^{-z}) z dz}{\int_0^\infty (n_0 e^{-z}) z dz} = \frac{\int_0^z z e^{-z}\,dz}{\int_0^\infty z e^{-z}\,dz}$$

We can use integration by parts to simplify the numerator. Recall that

$$\int u dv = uv - \int v du$$

We set $u = z$ and $dv = e^{-z} dz$, then applying the earlier results for integrals of exponential functions, we have

$$\int_0^z z e^{-z} dz = [-z e^{-z}]_0^z - \int_0^z (-e^{-z}) dz$$

$$= -z e^{-z} - [e^{-z}]_0^z = -z e^{-z} - [e^{-z} - 1] = 1 - (1+z) e^{-z}$$

For the denominator of the fraction $x_1$, we follow the same steps above but adjust to the new limits of integration

$$\int_0^\infty z e^{-z} dz = [-z e^{-z}]_0^\infty - \int_0^\infty (-e^{-z}) dz = 0 - \int_0^\infty (-e^{-z}) dz = \int_0^\infty e^{-z} dz$$

Applying result (2) from part a, we have

$$\int_0^\infty e^{-z} dz = 1$$

Therefore we derive the first moment equations (cumulative and differential) as follows.

$$x_1 = 1 - (1+z) e^{-z}$$

$$\frac{dx_1}{dz} = -(1+z)(-e^{-z}) + e^{-z}(-1) = z e^{-z}$$

(c) We are given the following moment equations. Upon examining the four moment equations, i.e., for $k = 0, 1, 2, 3$, we notice that these equations represent crystal size distributions in increasing dimension.

The numbers distribution function is obtained when $k = 0$.

$$x_0 = x_{\text{number}} = 1 - e^{-z}$$

$$\frac{dx_0}{dz} = e^{-z}$$

When $k = 1$, we obtain the size distribution in the linear form, i.e., length distribution.

$$x_1 = x_{\text{length}} = 1 - (1 + z)e^{-z}$$

$$\frac{dx_1}{dz} = ze^{-z}$$

By extension, it follows that the given expressions for the second and third moments represent the area (size in quadratic terms) distribution function and volume/mass (size in cubic terms) distribution function.

$$x_2 = x_{\text{area}} = 1 - \left(1 + z + \frac{z^2}{2}\right)e^{-z}$$

$$x_3 = x_{\text{volume/mass}} = 1 - \left(1 + z + \frac{z^2}{2} + \frac{z^3}{6}\right)e^{-z}$$

We can find the differential forms for $x_2$ and $x_3$ as follows

$$\frac{dx_2}{dz} = -\left(1 + z + \frac{z^2}{2}\right)(-e^{-z}) + e^{-z}(-1 - z) = \frac{z^2}{2}e^{-z}$$

$$\frac{dx_3}{dz} = -\left(1 + z + \frac{z^2}{2} + \frac{z^3}{6}\right)(-e^{-z}) + e^{-z}\left(-1 - z - \frac{z^2}{2}\right) = \frac{z^3}{6}e^{-z}$$

The cumulative and differential functions for the first, second, and third moments are added to the plot in part a (zeroth moment) as follows.

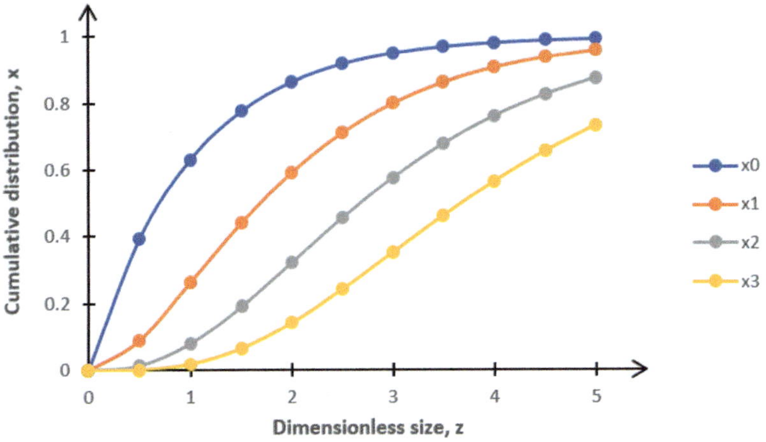

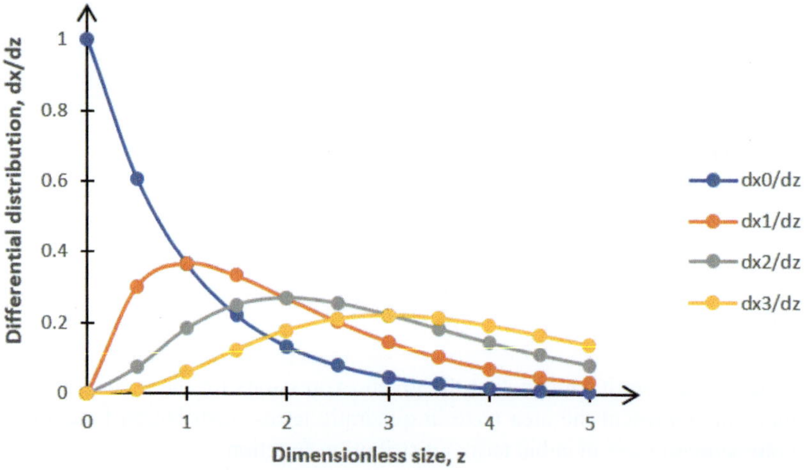


We notice that for the differential plots, the length, area, and volume/mass distributions experience a peak value. The peak value for the volume/mass distribution can give us insights on the predominant crystal size, and mathematically determined by setting the condition for the gradient of the differential plot to be equal to zero. This is also referred to as the predominant crystal size.

(d) The third moment for the population density function $n$ represents a volume or mass distribution. The predominant crystal size (or modal) crystal size $L_d$ occurs at the maximum point of the mass distribution (differential) function. To find this maximum point, we set the condition of gradient (i.e., derivative) equal to zero.

$$\frac{d\left(\frac{dx_3}{dz}\right)}{dz} = 0$$

$$\frac{d\left(\frac{z^3}{6}e^{-z}\right)}{dz} = -\frac{z^3}{6}e^{-z} + e^{-z}\left(\frac{z^2}{2}\right) = 0$$

$$z = 3$$

We know from part a that the dimensionless length $z$ can be expressed in terms of growth rate $G$ and residence (or turnover) time $\tau$; therefore, we derive the given expression for predominant crystal size.

$$z = \frac{L}{L^*} = \frac{L}{G\tau}$$

$$\frac{L_d}{G\tau} = 3$$

$$L_d = 3G\tau$$

# Reactor Kinetics

## Problem 1

**Using suitable examples, explain the rate of reaction *r*, rate law, and order of reaction.**

## Solution 1

***Worked Solution***

First we need to understand what this notation $r$ for rate of reaction means. Take for example the following reaction where 1 mole of reactant $A$ is converted to 1 mole of product $B$.

$$A \rightarrow B$$

Rate of reaction $r = r_B$, where $r > 0$, which refers to the rate of formation of product species $B$ and $r$ has units of mol $m^{-3}$ $s^{-1}$.

It follows that $r_B = -r_A$ since rate of production of species $B$ is equal to the rate of disappearance of species $A$. Note that the stoichiometric ratios may not always be 1:1 between $A$ and $B$; therefore, if we have a reaction that is described by the expression shown below, it should be noted that the rate of reaction is given by the expression, $r = r_B = -\frac{1}{2} r_A$.

$$2A \rightarrow B$$

The rate of reaction $r$ can also be expressed in terms of "per unit volume" such as in a CSTR and PFR, or in terms of "per unit mass of catalyst" such as in PBR, or "per unit area of reaction surface" such as for catalytic membranes. The specific choice depends on which parameter contributes to the rate of reaction. In a PBR, solid catalysts are packed in a tubular reactor and catalyze the reactions within. Hence, the



X. W. Ng, *Engineering Problems for Undergraduate Students*,
https://doi.org/10.1007/978-3-030-13856-1_4

mass of catalyst directly affects the rate of reaction, and it is logical that the rate of reaction is expressed per unit mass catalyst. In a CSTR, the volume of reactor contains the liquid reagents; hence, the volume that the CSTR can hold directly affects the rate of reaction and $r$ is therefore expressed per unit volume.

The rate law is a rate equation whereby $r$ is expressed as a product of reaction rate constant ($k$) and concentration of species ($C_i$). The specific correlation of species concentration to $r$ may be determined experimentally, and this correlation can be zeroth order, first order, second order, and up to $n$th order.

In a zeroth-order reaction, the rate of reaction is independent of species concentration, and is therefore simply the reaction rate constant. $k$ has units of mol m$^{-3}$ s$^{-1}$.

$$r = k$$

In a first-order reaction, the rate law for the example reaction below is as shown, and $k$ has units of s$^{-1}$. Note that there can exist reactants other than $A$ under this rate law; this is the case when these other reactants are zero order with respect to rate of reaction.

$$A \rightarrow B$$

$$r = kC_A$$

In a second-order reaction, the overall rate law equation will have concentration of species raised to power of 2. This may be from the second-order relationship with a single reactant species, or first-order relationship with respect to each of the two reactant species, as shown for the reaction below.

$$A + B \rightarrow C$$

$$r = kC_A C_B \text{ or } kC_A{}^2$$

**Problem 2**

**There are three main types of continuous flow reactors, the Continuous Stirred Tank Reactor (CSTR), the Plug Flow Reactor (PFR), and the Packed Bed Reactor (PBR).**

(a) **Derive the design equation for each type of reactor.**
(b) **Assuming we have a PFR containing a first-order reaction where species $A$ reacts to form species $B$. What is the reactor volume required to achieve a conversion of 90%, if the volumetric flow rate is constant at a value of $\dot{Q}$=0.3 m$^3$/s and the reaction rate constant $k = 0.05$ s$^{-1}$.**

## Solution 2

***Worked Solution***

(a) Design equations help us size the reactors for the reactions that we desire. Let us illustrate a simple CSTR.

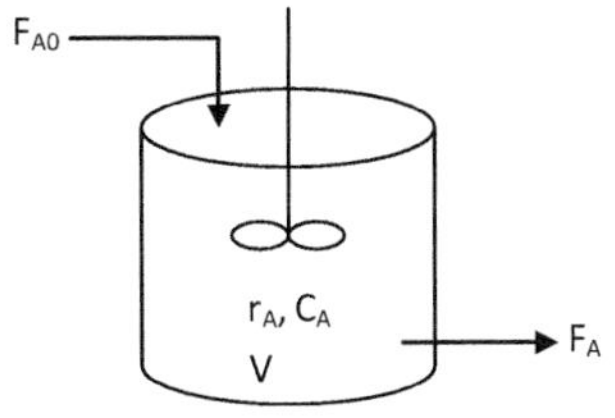


$F_{A0}$ denotes flow rate at the reactor inlet, $F_A$ denotes flow rate at the outlet, and $r = -r_A$ is the rate of reaction. It is assumed that the concentration of species $A$ in the tank is the same as that in the outlet stream; hence, it follows that $F_A = C_A\dot{Q}$ [mol s$^{-1}$]. We assume no spatial variations of $r$ within the volume; hence, a constant $V$ is multiplied to $r$.

The mass balance for species $A$ is as shown, note that $r_A < 0$:

$$F_{A0} - F_A + r_A V = \frac{dN_A}{dt}$$

$$F_{A0} - F_A + r_A V = 0 \text{ (steady state)}$$

Therefore the design equation for CSTR is as follows, where $X$ is conversion. [Note that for batch reactors, there are no inflows or outflows, therefore $r_A V = \frac{dN_A}{dt}$]

$$\boxed{V = \frac{F_{A0} - F_A}{-r_A} = \frac{F_{A0}X}{-r_A}}$$

Let us proceed to tubular reactors. A simple PFR (below left) and PBR (below right) are shown here.

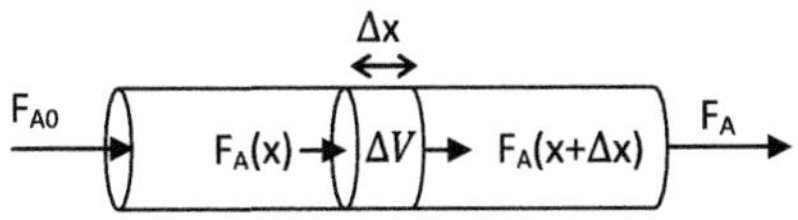


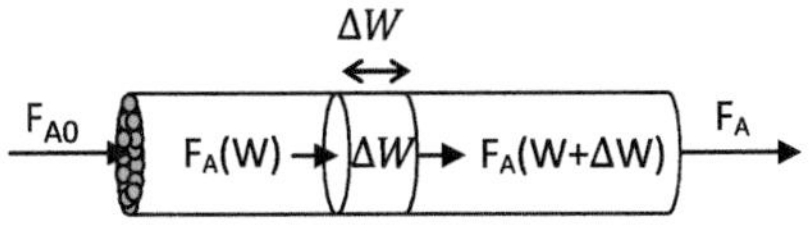


Tubular reactors are typically cylindrical with turbulent flow contained within, and we assume no spatial variation (in radial direction) of species concentrations under plug flow. However, species concentrations change in the axial direction as reactions proceed as the reagents flow along the axial direction.

For the PFR:

The mass balance for species $A$ in a differential volume element is below, note that $r_A < 0$:

$$F_A(x) - F_A(x + \Delta x) + r_A V = \frac{dN_A}{dt}$$

$$F_A(x) - F_A(x + \Delta x) + r_A \Delta V = 0 \text{ (steady state)}$$

$$r_A A \Delta x = F_A(x + \Delta x) - F_A(x)$$

$$r_A A = \frac{dF_A}{dx}$$

Therefore the design equation for the PFR is as follows,

$$\boxed{r_A = \frac{dF_A}{dV}}$$

As for a PBR, there is a solid catalyst, so rate of reaction $r'_A$ is defined per unit mass of catalyst instead of per unit volume.

Therefore the design equation for the PBR is as follows,

$$\boxed{r'_A = \frac{dF_A}{dW}}$$

(b) Given that the reaction $A \rightarrow B$ is first order, we can express the rate law equation as follows

$$-r_A = kC_A$$

Then, we apply the design equation for PFR which is derived from the mass balance for species $A$:

$$r_A = \frac{dF_A}{dV} = \frac{d(C_A \dot{Q})}{dV}$$

$$-\dot{Q}\frac{dC_A}{dV} = kC_A$$

$$-\frac{\dot{Q}}{k}\frac{dC_A}{C_A} = dV$$

We can integrate this differential equation using boundary conditions. We know that species $A$ is progressively converted to $B$ as it flows through the PFR (i.e., with increasing $V$); therefore, at the inlet when $V = 0$, $C_A = C_{A0}$.

$$-\frac{\dot{Q}}{k}\int_{C_{A0}}^{C_A}\frac{dC_A}{C_A} = \int_0^V dV$$

$$V = \frac{\dot{Q}}{k}\ln\frac{C_{A0}}{C_A} = \frac{\dot{Q}}{k}\ln\frac{1}{1-X} = \frac{0.3}{0.05}\ln\frac{1}{0.1} = 13.8$$

The reactor volume required is 13.8 $m^3$.

## Problem 3

**Consider a gas phase reaction $A \rightarrow B + C$ which occurs in a batch reactor of constant volume $V = 3$ m$^3$ and temperature. An initial amount of $A = 30$ moles was added.**

(a) **Assuming the reaction is first order with respect to $A$, find the time taken for 99% conversion to be reached. Assume that the rate constant $k = 0.67$ s$^{-1}$.**
(b) **Assuming the reaction is now second order with respect to $A$, find the time taken to deplete 27 moles of $A$, given that $k = 0.00035$ m$^3$ mol$^{-1}$ s$^{-1}$.**
(c) **Determine the final pressure when reaction is completed, if initial temperature is 105 °C.**

## Solution 3

***Worked Solution***

(a) Let us first state the rate law equation: $-r_A = kC_A$. We can apply the design equation for a batch reactor: $\frac{dN_A}{dt} = r_A V$. Combining both we have: $\frac{d\left(\frac{N_A}{V}\right)}{dt} = \frac{dC_A}{dt} = -kC_A$.

$$\frac{dC_A}{C_A} = -kdt$$

$$t = -\frac{1}{k}\ln\frac{C_A}{C_{A0}} = -\frac{1}{k}\ln(1-X)$$

$$t = -\frac{1}{0.67}\ln(1-0.99) = 6.9$$

The time taken for 99% conversion to be achieved is 6.9 s.

(b) Let us again state the rate law equation: $-r_A = kC_A{}^2$. Again, we can apply the design equation for a batch reactor: $\frac{dN_A}{dt} = r_A V$. Combining both we have,

$$\frac{d\left(\frac{N_A}{V}\right)}{dt} = \frac{dC_A}{dt} = -kC_A{}^2$$

$$\frac{dC_A}{C_A{}^2} = -kdt$$

$$-\frac{1}{C_A{}^2} + \frac{1}{C_{A0}{}^2} = -kt$$

$$C_{A0} = \frac{30}{3} = 10 \text{ mol m}^{-3}$$

$$C_A = \frac{27}{3} = 9 \text{ mol m}^{-3}$$

Therefore the time taken to deplete 27 moles of $A$ can be found as follows

$$t = -\frac{1}{k}\left(-\frac{1}{C_A{}^2} + \frac{1}{C_{A0}{}^2}\right) = -\frac{1}{0.00035}\left(-\frac{1}{9^2} + \frac{1}{10^2}\right)$$

$$t = 6.7 \text{ s}$$

(c) When reaction goes to completion, we will have 30 moles each of $B$ and $C$. Therefore using the ideal gas equation,

$$P = \frac{60 \times 8.314 \times (105 + 273)}{3} = 0.63 \times 10^5 \text{Pa}$$

**Problem 4**

**Consider a reversible gas phase reaction that is carried out in a CSTR. The equilibrium constant for the reaction $K = 120$. At 298 K, the $\Delta h = -50$ kJ mol$^{-1}$ and the $\Delta c_p = 7$ J mol$^{-1}$ K$^{-1}$ (assume $\Delta c_p$ is not temperature dependent). The forward and backward reactions follow first-order kinetics.**

$$P \leftrightarrow Q$$

(a) **Determine the maximum conversion that can be achieved at a temperature of 298 K and 500 K if we assume equilibrium is reached.**

(b) **Show that the below expression is true if we assume the rate constant obeys the Arrhenius equation of $k_{\text{forward}} = A \exp\left(-\frac{E_a}{RT}\right)$ where $A = 150$ s$^{-1}$ and $E_a = 18$ kJ mol$^{-1}$:**

$$-r_P = k_{\text{forward}} C_{P0}\left[1 - X_P\left(1 + \frac{1}{K}\right)\right]$$

(c) **The inlet volumetric flow rate into the CSTR of volume $V = 8\ \mathrm{m}^3$ is $1.5\ \mathrm{m}^3\ \mathrm{s}^{-1}$. Determine the conversion of $P$ that would be practically achieved, at 298 K and 500 K under isothermal conditions.**

## Solution 4

### *Worked Solution*

(a) We can find conversion $X$ at 298 K directly at equilibrium,

$$K = \frac{C_{Q,\ \mathrm{eq}}}{C_{p,\ \mathrm{eq}}} = \frac{X}{1-X} = 120$$

$$X = \frac{1}{1+1/K} = 0.99$$

Note that $\Delta h$ is temperature dependent. We can express $\Delta h$ as a function of $T$ to determine its value for $T = 500$ K.

$$\Delta h(T) = \Delta h(298\ \mathrm{K}) + \Delta c_p(T - 298)$$

$$= -50{,}000 + 7(T - 298) = -52{,}086 + 7T\ \mathrm{J\ mol}^{-1}$$

We recall a useful result from thermodynamics,

$$\Delta g_{rxn} = -RT \ln K$$

$$\frac{d \ln K}{dT} = -\frac{d\left(\frac{\Delta g_{\mathrm{rxn}}}{RT}\right)}{dT}$$

We can differentiate the right-hand side using chain rule

$$\frac{d \ln K}{dT} = \frac{\Delta g_{\mathrm{rxn}}}{RT^2} - \frac{1}{RT}\frac{d\Delta g_{\mathrm{rxn}}}{dT} = \frac{\Delta g_{\mathrm{rxn}}}{RT^2} + \frac{\Delta s_{\mathrm{rxn}}}{RT}$$

Finally we apply the relationship of $\Delta g_{\mathrm{rxn}} = \Delta h_{\mathrm{rxn}} - T\Delta s_{\mathrm{rxn}}$ using the values at 298 K as our lower limit of integration.

$$\frac{d \ln K}{dT} = \frac{\Delta h}{RT^2}$$

$$\int d \ln K = \frac{1}{R}\int \frac{-52{,}086 + 7T}{T^2} dT = \frac{1}{R}\int \frac{-52{,}086}{T^2} + \frac{7}{T} dT$$

$$\ln\left(\frac{K}{120}\right) = \frac{1}{R}\left[52{,}086\left(\frac{1}{T} - \frac{1}{298}\right) + 7\ln\frac{T}{298}\right]$$

At $T = 500$ K,

$$K = 120\ \exp\left\{\frac{1}{8.314}\left[52,086\left(\frac{1}{500} - \frac{1}{298}\right) + 7\ln\frac{500}{298}\right]\right\} = 0.04$$

$$X = 0.04$$

(b)

$$P \leftrightarrow Q$$

We can define the rate equation of this reversible reaction as follows

$$r = -r_P = k_{\text{forward}}\ C_P - k_{\text{backward}} C_Q$$
$$= k_{\text{forward}}\ C_{P0}(1 - X) - k_{\text{backward}} C_{P0} X$$
$$-r_P = k_{\text{forward}} C_{P0}\left[(1 - X) - \frac{k_{\text{backward}}}{k_{\text{forward}}} X\right]$$

At equilibrium, forward reaction rate is equal to backward reaction rate.

$$k_{\text{forward}} C_{P,\text{eq}} = k_{\text{backward}} C_{Q,\text{eq}}$$
$$K = \frac{C_{Q,\text{eq}}}{C_{p,\text{eq}}} = \frac{k_{\text{forward}}}{k_{\text{backward}}}$$

Therefore we can substitute back into the reaction rate equation

$$-r_P = k_{\text{forward}} C_{P0}\left[(1 - X) - \frac{1}{K} X\right]$$
$$= k_{\text{forward}} C_{P0}\left[1 - X_P\left(1 + \frac{1}{K}\right)\right]$$

(c) Based on the Arrhenius equation, we have

$$k_{\text{forward}} = 150\ \exp\left(-\frac{18,000}{RT}\right)$$

We know that the design equation for CSTR is as follows

$$-r_P = \frac{F_{P0} X_P}{V}$$

Substituting our results from part b,

$$k_{\text{forward}}C_{P0}\left[1 - X_P\left(1 + \frac{1}{K}\right)\right] = \frac{F_{P0}X_P}{V}$$

$$k_{\text{forward}}\left[1 - X_P\left(1 + \frac{1}{K}\right)\right] = \frac{(F_{P0}/C_{P0})X_P}{V} = \frac{(\dot{Q}_{\text{in}})X_P}{V}$$

Substituting parameters of the Arrhenius equation into $k_{\text{forward}}$ and using known values of $\dot{Q}_{\text{in}}$ and $V$,

$$150\ \exp\left(-\frac{18{,}000}{8.314 \times T}\right)\left[1 - X_P\left(1 + \frac{1}{K}\right)\right] = \frac{(1.5)X_P}{8}$$

$$X_P = \frac{1}{\left(\frac{8}{1.5 \times 150}\right)\exp\left(\frac{18{,}000}{8.314 \times T}\right) + \left(1 + \frac{1}{K}\right)}$$

At $T = 298$ K, $K = 120$ and at $T = 500$ K, $K = 0.04$ the conversions are $X_P(T = 298\text{ K}) = 0.02$ and $X_P(T = 500\text{ K}) = 0.03$.

## Problem 5

**Residence time distributions (RTDs) are useful in estimating the conversion in reactors that have non-ideal flow. The mean residence time $\tau$ is the time required to process one reactor volume of fluid based on entrance conditions.**

**Using Laplace transforms, and assuming 2 reactor vessels connected in series,**

(a) **Show that the overall RTD is independent of the order of the vessels. Define the transfer function for vessel 2.**

(b) **We can define the $n$th moment for a RTD as $\mu_n = \int_0^\infty t^n E(t)dt$. The zeroth moment is defined as follows:**

$$\int_0^\infty E(t)dt = \mu_0 = 1$$

**and the first moment which is also the mean residence time is defined as follows:**

$$\tau = \mu_1 = \int_0^\infty tE(t)dt$$

**Show that the mean residence time $\tau$ for the vessels in series is additive.**

(c) **Given that variance is defined as** $\sigma^2 = \int_0^\infty (t-\tau)^2 E(t)dt$, **show that** $\sigma^2 = \mu_2 - \mu_1^2$. **Show further that the variance** $\sigma^2$ **for the vessels in series is additive.**

## Solution 5

***Worked Solution***

(a) Before we begin, let us recall the concept of a residence time distribution.

---

***Background Concepts***

Residence Time Distribution (RTD) is a distribution of times spent by fluid elements taking different routes through the reactor and hence taking different times to exit the vessel. The distribution of these times for a specific stream of fluid (e.g., a delta pulse) is called the exit age distribution or RTD, or $E$ with units of $s^{-1}$. $E$ is defined as $\int_0^\infty E(t)dt = 1$ and it is normalized.

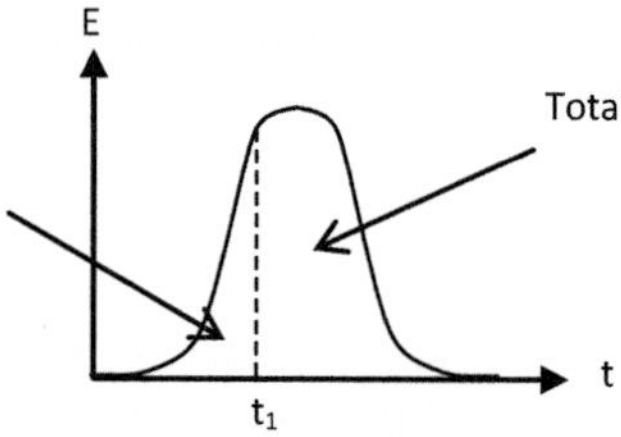


Separately it is useful to also recall the definition of a Laplace transform, which states that for a function $E(t)$, its transform $\bar{E}(s)$ is defined as follows

$$\bar{E}(s) = \mathcal{L}\{E(t)\} = \int_0^\infty e^{-st} E(t)dt$$

---

Now let us consider a system of vessels connected in series as shown below, and assume that we introduce a $\delta$-pulse of tracer into vessel 1 at $t = 0$. At a certain time $t = t_1$, a fraction of tracer leaves vessel 1 and enters vessel 2, and this fraction is denoted by $E_1(t_1)$. At time t, the fraction of tracer in exit stream of vessel 2 is denoted by $E_2(t-t_1)$.

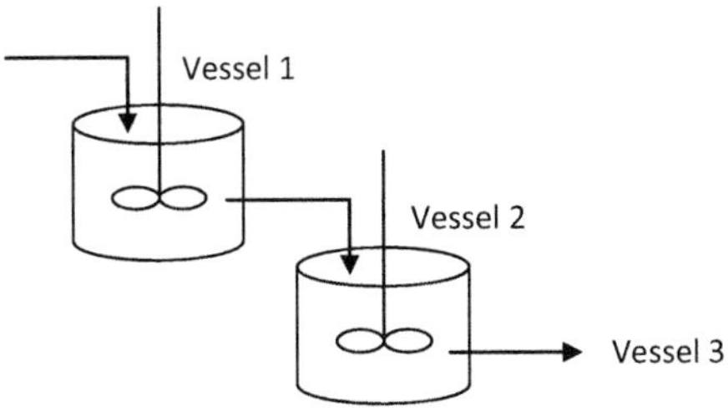


Since the fraction of vessel 2 is calculated using the exit fraction from vessel 1 as a basis, therefore the fractions multiply in the integral. This is mathematically shown as

$$E_3(t) = \int_0^t E_1(t_1)E_2(t - t_1)dt_1 = \int_0^\infty E_1(t_1)E_2(t - t_1)dt_1$$

Since there is fixed amount of tracer injected in a $\delta$-pulse, extending the upper limit of time integral makes no difference to the value.

After performing Laplace transform on $E_3(t)$, and substituting the integral for $E_3(t)$

$$\bar{E}_3(s) = \int_0^\infty e^{-st}E_3(t)dt = \int_0^\infty e^{-st}dt \int_0^\infty E_1(t_1)E_2(t - t_1)dt_1$$
$$= \int_0^\infty e^{-st}E_2(t - t_1)dt \int_0^\infty E_1(t_1)dt_1$$

Let us define a new variable $t' = t - t_1$

$$\bar{E}_3(s) = \int_{-t_1}^\infty e^{-s(t'+t_1)}E_2(t')dt' \int_0^\infty E_1(t_1)dt_1$$

Note that $\int_{-t_1}^\infty e^{-s(t'+t_1)}E_2(t')dt' = \int_0^\infty e^{-s(t'+t_1)}E_2(t')dt'$ because $E_2(t') = 0$ for $t' < 0$.

$$\bar{E}_3(s) = \int_0^\infty e^{-st'}E_2(t')dt' \int_0^\infty E_1(t_1)e^{-st_1}dt_1 = \bar{E}_1(s)\bar{E}_2(s)$$

The overall RTD of two vessels in series is the convolution of the individual RTDs. Therefore, the order of vessels does not affect the overall RTD. The transfer function of vessel 2 is defined as $G(s)$ such that

$$G(s) = \frac{\bar{E}_3(s)}{\bar{E}_1(s)} = \bar{E}_2(s)$$

(b) From the definition of a Laplace transform, we can take $\frac{d}{ds}$ of $\bar{E}(s)$ as follows

$$\bar{E}(s) = \int_0^\infty e^{-st} E(t)dt$$

$$\frac{d}{ds}\bar{E}(s) = -t\int_0^\infty e^{-st} E(t)dt$$

For a derivative of $k$th order with respect to $s$, we have the general form

$$\frac{d^k}{ds}\bar{E}(s) = (-1)^k \int_0^\infty t^k e^{-st} E(t)dt$$

When we take the limit of $s \to 0$, we obtain the $k$th moment $\mu_k$

$$\lim_{s\to 0}\frac{d^k}{ds}\bar{E}(s) = (-1)^k \int_0^\infty t^k E(t)dt$$

$$(-1)^k \lim_{s\to 0}\frac{d^k}{ds}\bar{E}(s) = \int_0^\infty t^k E(t)dt = \mu_k$$

In our problem, $\bar{E}_3(s) = \bar{E}_1(s) \times \bar{E}_2(s)$, and taking first moment (since it is equal to $\tau$)

$$\mu_{1,vessel\ 3} = (-1)\lim_{s\to 0}\frac{d}{ds}\bar{E}_3(s) = (-1)\lim_{s\to 0}\frac{d}{ds}[\bar{E}_1(s)\bar{E}_2(s)]$$

Using product rule to evaluate the differential,

$$\mu_{1,vessel\ 3} = (-1)\lim_{s\to 0}\left[\bar{E}_1(s)\frac{d\bar{E}_2(s)}{ds} + \bar{E}_2(s)\frac{d\bar{E}_1(s)}{ds}\right]$$

$$= \lim_{s\to 0}\left[-\bar{E}_1(s)\frac{d\bar{E}_2(s)}{ds} - \bar{E}_2(s)\frac{d\bar{E}_1(s)}{ds}\right]$$

Since we know that $\lim_{s\to 0}\bar{E}(s) = \lim_{s\to 0}\int_0^\infty e^{-st}E(t)dt = \int_0^\infty E(t)dt = 1$

$$\mu_{1,vessel\ 3} = \lim_{s\to 0}\left[-\frac{d\bar{E}_2(s)}{ds} - \frac{d\bar{E}_1(s)}{ds}\right]$$

$$= (-1)\lim_{s\to 0}\frac{d}{ds}\bar{E}_2(s) + (-1)\lim_{s\to 0}\frac{d}{ds}\bar{E}_1(s)$$

$$\mu_{1,vessel\ 3} = \mu_{1,vessel\ 2} + \mu_{1,vessel\ 1}$$

$$\tau_{vessel\ 3} = \tau_{vessel\ 2} + \tau_{vessel\ 1}$$

(c)

$$\sigma^2 = \int_0^\infty (t-\tau)^2 E(t)dt$$

$$= \int_0^\infty t^2 E(t)dt + \int_0^\infty \tau^2 E(t)dt - 2\tau \int_0^\infty tE(t)dt$$

$$= \mu_2 + \tau^2 \int_0^\infty E(t)dt - 2\tau(\mu_1)$$

$$= \mu_2 + \tau^2 - 2\tau^2 = \mu_2 - \tau^2$$

$$\sigma^2 = \mu_2 - {\mu_1}^2$$

Now we proceed to show that $\sigma^2$ is additive. In our problem, $\bar{E}_3(s) = \bar{E}_1(s) \times \bar{E}_2(s)$, and taking second moment this time round

$$\mu_{2,vessel\ 3} = (-1)^2 \lim_{s\to 0} \frac{d^2}{ds^2} \bar{E}_3(s) = \lim_{s\to 0} \frac{d^2}{ds^2} [\bar{E}_1(s)\bar{E}_2(s)]$$

Using the product rule to evaluate the derivative,

$$\mu_{2,vessel\ 3} = \lim_{s\to 0} \frac{d}{ds}\left[\bar{E}_1(s)\frac{d\bar{E}_2(s)}{ds} + \bar{E}_2(s)\frac{d\bar{E}_1(s)}{ds}\right]$$

$$= \lim_{s\to 0}\left[\bar{E}_1(s)\frac{d^2\bar{E}_2(s)}{ds^2} + \frac{d\bar{E}_1(s)}{ds}\frac{d\bar{E}_2(s)}{ds} + \bar{E}_2(s)\frac{d^2\bar{E}_1(s)}{ds^2} + \frac{d\bar{E}_1(s)}{ds}\frac{d\bar{E}_2(s)}{ds}\right]$$

$$= \lim_{s\to 0}\left[\bar{E}_1(s)\frac{d^2\bar{E}_2(s)}{ds^2} + 2\frac{d\bar{E}_1(s)}{ds}\frac{d\bar{E}_2(s)}{ds} + \bar{E}_2(s)\frac{d^2\bar{E}_1(s)}{ds^2}\right]$$

We know that $\lim_{s\to 0} \bar{E}(s) = \lim_{s\to 0} \int_0^\infty \mathrm{e}^{-st}E(t)dt = \int_0^\infty E(t)dt = 1$

$$\mu_{2,vessel\ 3} = \lim_{s\to 0}\left[(1)\frac{d^2\bar{E}_2(s)}{ds^2} + 2\left(-\frac{d\bar{E}_1(s)}{ds}\right)\left(-\frac{d\bar{E}_2(s)}{ds}\right) + (1)\frac{d^2\bar{E}_1(s)}{ds^2}\right]$$

$$= \mu_{2,vessel\ 2} + 2(\mu_{1,vessel\ 1})(\mu_{1,vessel\ 2}) + \mu_{2,vessel\ 1}$$

$$\mu_{2,vessel\ 3} - {\mu_{1,vessel\ 3}}^2 = \mu_{2,vessel\ 2} + 2(\mu_{1,vessel\ 1})(\mu_{1,vessel\ 2}) + \mu_{2,vessel\ 1} - {\mu_{1,vessel\ 3}}^2$$

We know from the earlier part that $\mu_{1,\text{vessel }3} = \mu_{1,\text{vessel }1} + \mu_{1,\text{vessel }2}$ and we have also shown earlier that $\sigma^2 = \mu_2 - {\mu_1}^2$; therefore

$$\sigma_{vessel\ 3}^{\ 2} = \mu_{2,vessel\ 2} + 2(\mu_{1,vessel\ 1})(\mu_{1,vessel\ 2}) + \mu_{2,vessel\ 1} - (\mu_{1,vessel\ 1} + \mu_{1,vessel\ 2})^2$$
$$= \mu_{2,vessel\ 2} + \mu_{2,vessel\ 1} - \mu_{1,vessel\ 1}^{\ 2} - \mu_{1,vessel\ 2}^{\ 2}$$
$$= \sigma_{vessel\ 1}^{\ 2} + \sigma_{vessel\ 2}^{\ 2}$$

## Problem 6

**Given a CSTR and a PFR, there are two possible ways to arrange them in series. Option 1 was to have the CSTR placed before the PFR, and Option 2 was to have the PFR placed before the CSTR. The reaction to be carried out is second order. We are provided with the mean residence time for the CSTR which is 2 min, while that of the PFR is 1.5 min. The reaction rate constant $k = 3 \times 10^{-4}$ m$^3$ mol$^{-1}$ s$^{-1}$ and the initial concentration of reactant is 1200 mol m$^{-3}$.**

(a) **Assuming all reactors are well mixed, explain which option should be chosen for maximum conversion.**
(b) **If the reaction was first order, how would your choice in part a change?**

## Solution 6

*Worked Solution*

(a) Let the reactant that is first order to reaction rate be $A$. Then we have the same rate law for both reactors as follows: $-r_A = kC_A^{\ 2}$.

For CSTR, $-r_A V = F_{A0} - F_A$, we have the following equation, knowing that $\tau = \frac{V}{\dot{Q}}$:

$$kC_A^{\ 2}V = F_{A0} - F_A = \dot{Q}\,(C_{A0} - C_A)$$
$$kC_A^{\ 2}\tau = C_{A0} - C_A$$

For PFR, $r_A dV = dF_A$, similarly, we have the equation:

$$-kC_A^{\ 2}dV = dF_A = \dot{Q}\,dC_A$$
$$\int_0^V \frac{k}{\dot{Q}}dV = \int_{C_{A0}}^{C_A} -\frac{1}{C_A^{\ 2}}dC_A = \frac{1}{C_A} - \frac{1}{C_{A0}}$$
$$k\tau = \frac{1}{C_A} - \frac{1}{C_{A0}}$$

*Option 1: CSTR Followed by PFR*

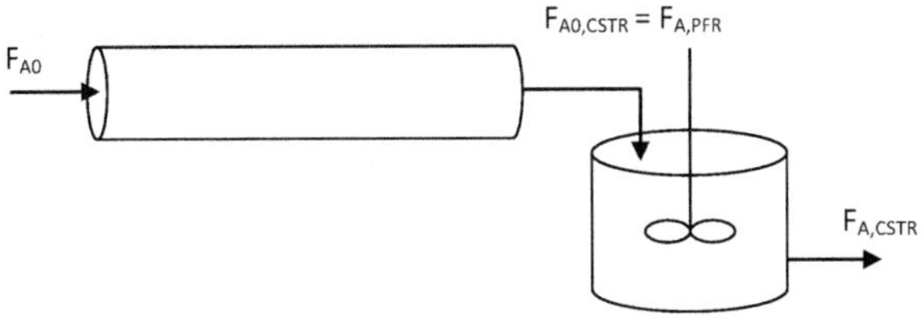


In the CSTR, we substitute the given values, $C_{A0} = 1200 \text{ mol m}^{-3}$, $k = 3 \times 10^{-4} \text{ m}^3 \text{ mol}^{-1} \text{ s}^{-1}$, $\tau = 120$ s.

$$kC_A{}^2\tau = (C_{A0} - C_A)$$

$$\left(3 \times 10^{-4}\right)C_A{}^2(120) = (1200 - C_A)$$

$$0.036C_A{}^2 + C_A - 1200 = 0$$

$$C_A = \frac{-1 \pm \sqrt{1 - 4(0.036)(-1200)}}{2(0.036)} = 169 \text{ mol m}^{-3}$$

In the PFR, at $V = 0$, $C_{A0} = 169 \text{ mol m}^{-3}$. $\tau = 90$ s and $k = 3 \times 10^{-4} \text{ m}^3 \text{ mol}^{-1} \text{ s}^{-1}$.

$$k\tau = \frac{1}{C_A} - \frac{1}{C_{A0}}$$

$$3 \times 10^{-4} \times 90 = \frac{1}{C_A} - \frac{1}{169}$$

$$C_A = 30.4 \text{ mol m}^{-3}$$

$$X_{\text{option 1}} = \frac{C_{A0} - C_A}{C_{A0}} = \frac{1200 - 30.4}{1200} = 0.975$$

*Option 2: PFR Followed by CSTR*

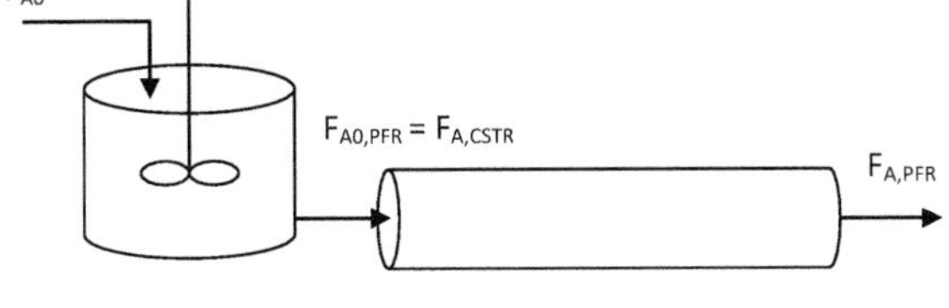

In the PFR, at $V = 0$, $C_{A0} = 1200 \text{ mol m}^{-3}$. $\tau = \frac{V}{Q} = 90$ s and $k = 3 \times 10^{-4} \text{ m}^3 \text{ mol}^{-1} \text{ s}^{-1}$.

$$\frac{kV}{\dot{Q}} = \frac{1}{C_A} - \frac{1}{C_{A0}}$$

$$3 \times 10^{-4} \times 90 = \frac{1}{C_A} - \frac{1}{1200}$$

$$C_A = 35.9 \text{ mol m}^{-3}$$

In the CSTR, we substitute the given values, $C_{A0} = 35.9 \text{ mol m}^{-3}$, $k = 3 \times 10^{-4} \text{ m}^3 \text{ mol}^{-1} \text{ s}^{-1}$, $\tau = \frac{V}{\dot{Q}} = 120$ s.

$$kC_A{}^2\left(\frac{V}{\dot{Q}}\right) = (C_{A0} - C_A)$$

$$\left(3 \times 10^{-4}\right)C_A{}^2(120) = (35.9 - C_A)$$

$$0.036C_A{}^2 + C_A - 35.9 = 0$$

$$C_A = \frac{-1 \pm \sqrt{1 - 4(0.036)(-35.9)}}{2(0.036)} = 20.6 \text{ mol m}^{-3}$$

$$X_{\text{option 2}} = \frac{C_{A0} - C_A}{C_{A0}} = \frac{1200 - 20.6}{1200} = 0.983$$

In this case, there is higher conversion when the CSTR is placed after the PFR. Note that the order of the vessels affect the overall conversion but the order does not affect the overall residence time distribution since the overall RTD of two vessels in series is the convolution of the individual RTDs and $\bar{E}_{\text{overall}}(s) = \bar{E}_1(s) \times \bar{E}_2(s) = \bar{E}_2(s) \times \bar{E}_1(s)$.

(b) If reaction was first order, then $-r_A = kC_A$.

In the CSTR,

$$kC_A\tau = C_{A0} - C_A$$

$$k\tau = \frac{C_{A0}}{C_A} - 1 = \frac{1}{1 - X} - 1$$

In the PFR,

$$-kC_A dV = \dot{Q}\, dC_A$$

$$\int_0^V -\frac{k}{\dot{Q}} dV = \int_{C_{A0}}^{C_A} \frac{1}{C_A} dC_A = \ln\frac{C_A}{C_{A0}}$$

$$-k\tau = \ln(1 - X)$$

In both cases, the conversion becomes only a function of $\tau$. Since overall RTD is the same regardless of order of vessel, the same conversion will be achieved regardless of the option chosen.

## Problem 7

**Describe methods for measuring residence time distributions and provide some examples of reactor RTDs.**

## Solution 7

***Worked Solution***

Residence time distributions can be measured using an inert tracer in a pulse experiment or a step function experiment.

*Pulse Input:*

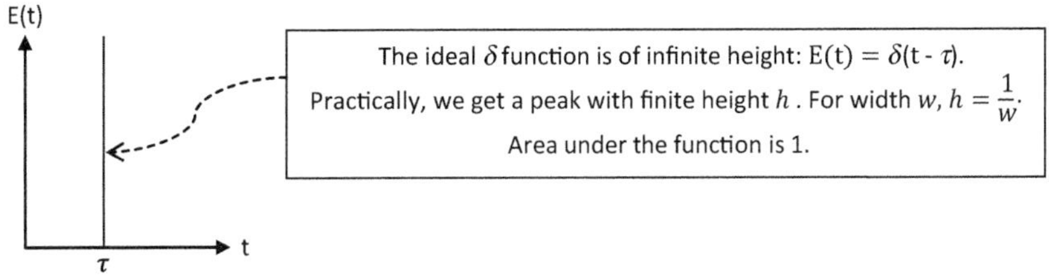
E(t)
The ideal δ function is of infinite height: E(t) = δ(t - τ).
Practically, we get a peak with finite height h. For width w, h = 1/w.
Area under the function is 1.
t
τ

A typical setup comprises a syringe that instantaneously injects a $\delta$-pulse of tracer into a reactor. The concentration of tracer is plotted against time elapsed. The tracer concentration profile is the same as residence time distribution E, where $\int_0^\infty E(t)dt = 1 = \int_0^\infty C(t)dt$. In this setup we assume closed vessel, i.e., no variations introduced at the boundaries of the system (i.e., vessel inlet/outlet) for the length of time concerned. Note that the concentration response at the reactor exit is normalized, and that the Laplace transform of a $\delta$ function is 1.

*Unit Step Function Input:*

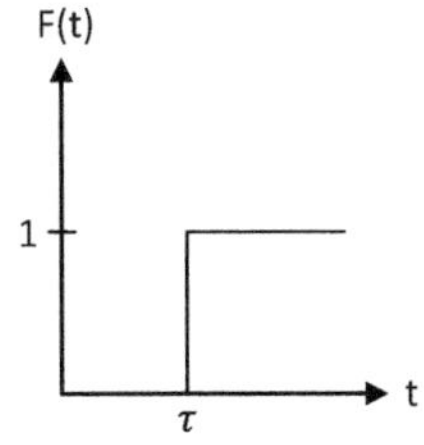


A step function of unit size is introduced at $t = 0$, and the fraction of tracer in the outlet stream at time $t$ can be measured as concentration, and this will provide the fraction of fluid that has an age $\leq t$. This fraction is equivalent to the cumulative distribution of this tracer up to time $t$, $F(t) = \int_0^t E(t)dt$. To derive the residence time $E$ from this fraction, we can use $E(t) = \frac{dF}{dt}$. Note that $E(t)dt$ represents the fraction of fluid elements leaving between $t$ and $dt$. Note that the Laplace transform of a unit step function is 1/s.

_Typical Reactor RTDs_

In ideal reactors, we can formulate simple expressions for the mean residence time $\tau$ from the RTD profiles, where $\tau = V/\dot{Q}$. In an ideal PFR, $E(t) = \delta(t - \tau)$, while in an ideal CSTR, $E(t) = \frac{1}{\tau}e^{-t/\tau}$. For a CSTR, the $E(t)$ and $F(t)$ graphs are exponential.

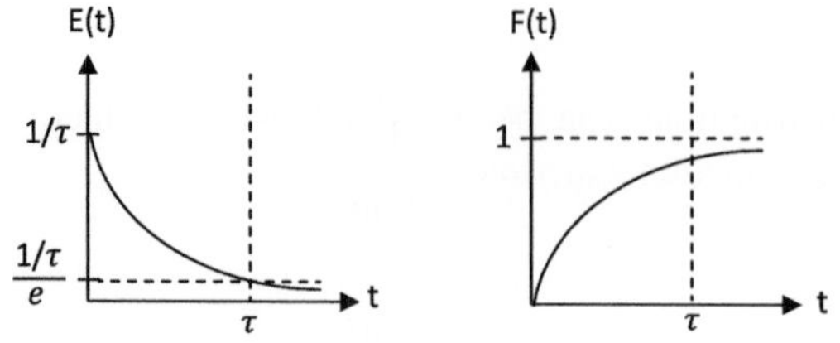


**Problem 8**

**The "mixed cup" method is sometimes used in measuring RTD to remove the flow variations near the boundaries of the reactor. Two identical CSTRs are placed upstream and downstream of the test reactor each, as shown in the configuration below. A $\delta$-function pulse is introduced into the first CSTR at $t = 0$ and the RTD functions are recorded at points marked with $X$. The plotted data points are fit into a best fit curve, and it was found that after the first CSTR, the distribution correlated to $e^{-t}$, while at the end of the series of reactors, it was $\frac{1}{3}\left[\frac{4}{3}\left(e^{-\frac{t}{4}} - e^{-t}\right)\right] - te^{-t}$. Using Laplace transform, determine the second and third moments of the RTD of the test reactor.**

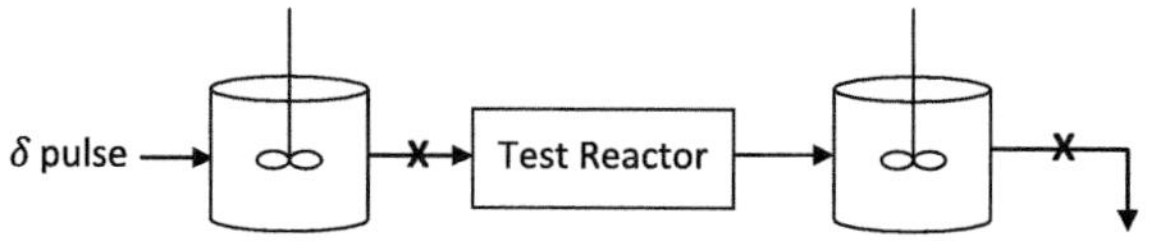


**Solution 8**

***Worked Solution***

Before we begin the solution, let us revisit some key related concepts.

- - - - - - - - - - - - - - - - - - - - - - - - - - - - - - - - - - - - - - - -

***Background Concepts***

A $\delta$ pulse and step function pulse are in reality, difficult to administer because they require an instantaneous shot of tracer. However, there will be time lag and velocity gradients at the reactor inlet; this problem repeats itself at the outlet of the reactor, and is more apparent with a $\delta$ pulse function. To get over this, CSTRs are placed before and after the test reactor, and the analysis of the test reactor can be done by deconvolution of the series. The "imperfections" present in the CSTRs before and after the test reactor become well defined, and hence acceptable as they can be measured and eliminated from subsequent consideration.

---

Let us label the vessels as shown below.

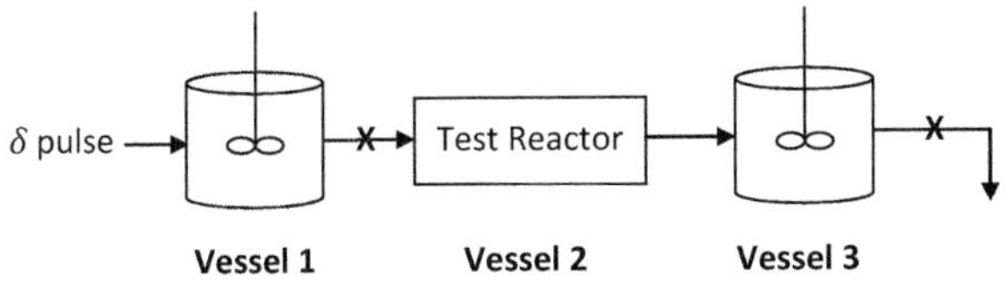


We know that for this system of vessels in series, individual RTDs convolute. For the first point of measurement,

$$E_1(t) = E_3(t) = e^{-t}$$

For the second point of measurement,

$$E_1(t) * E_2(t) * E_3(t) = E_1(t) * E_2(t) * E_1(t) = \frac{1}{3}\left[\frac{4}{3}\left(e^{-\frac{t}{4}} - e^{-t}\right)\right] - te^{-t}$$

Let us convert the RTDs into Laplace forms. By definition, Laplace transform is defined as

$$\bar{E}(s) = \mathcal{L}\{E(t)\} = \int_0^{\infty} e^{-st}E(t)dt$$

$$\bar{E}_1(s) = \bar{E}_3(s) = \int_0^{\infty} e^{-st}e^{-t}dt = \int_0^{\infty} e^{-(s+1)t}dt = \left[\frac{e^{-(s+1)t}}{-(s+1)}\right]_0^{\infty} = \frac{1}{1+s}$$

$$\bar{E}_1(s) \times \bar{E}_3(s) = \frac{1}{(1+s)^2}$$

$$E_1(t) * E_2(t) * E_1(t) = \frac{1}{3}\left[\frac{4}{3}\left(e^{-\frac{t}{4}} - e^{-t}\right)\right] - te^{-t}$$

$$\bar{E}_1(s) \times \bar{E}_2(s) \times \bar{E}_1(s) = \int_0^\infty e^{-st}\left\{\frac{1}{3}\left[\frac{4}{3}\left(e^{-\frac{t}{4}} - e^{-t}\right)\right] - te^{-t}\right\}dt$$

$$= \frac{1}{3}\int_0^\infty \left\{\left[\frac{4}{3}\left(e^{-\left(s+\frac{1}{4}\right)t} - e^{-(s+1)t}\right)\right] - te^{-(s+1)t}\right\}dt$$

$$= \frac{1}{3}\int_0^\infty \left[\frac{4}{3}\left(e^{-\left(s+\frac{1}{4}\right)t} - e^{-(s+1)t}\right)\right]dt + \left[\frac{t}{(s+1)}e^{-(s+1)t}\right]_0^\infty - \frac{1}{(s+1)}\int_0^\infty e^{-(s+1)t}dt$$

$$= \frac{1}{3}\left[\frac{4}{3}\left(-\frac{1}{s+\frac{1}{4}}e^{-\left(s+\frac{1}{4}\right)t} + \frac{1}{(s+1)}e^{-(s+1)t}\right)\right]_0^\infty + \frac{1}{(s+1)}\left[\frac{1}{s+1}e^{-(s+1)t}\right]_0^\infty$$

$$= \frac{1}{3}\left[\frac{4}{3}\left(\frac{1}{s+\frac{1}{4}} - \frac{1}{(s+1)}\right)\right] - \frac{1}{(s+1)^2} = \frac{1}{3}\left[\frac{4}{3}\left(\frac{\frac{3}{4}}{\left(s+\frac{1}{4}\right)(s+1)}\right)\right] - \frac{1}{(s+1)^2}$$

$$= \frac{1}{3}\left[\frac{\frac{3}{4}}{\left(s+\frac{1}{4}\right)(s+1)^2}\right] = \frac{1}{4\left(s+\frac{1}{4}\right)(s+1)^2}$$

The transfer function is also the RTD of the test reactor and is defined as

$$G = \bar{E}_2(s) = \frac{\bar{E}_1(s)\bar{E}_2(s)\bar{E}_3(s)}{\bar{E}_1(s)\bar{E}_3(s)} = \frac{\frac{1}{4\left(s+\frac{1}{4}\right)(s+1)^2}}{\frac{1}{(s+1)^2}} = \frac{1}{4\left(s+\frac{1}{4}\right)}$$

We need to find the second and third moments of the RTD for the test reactor, we will first define $k$th moment of an RTD, $\mu_k$ as follows

$$\mu_k = (-1)^k \lim_{s\to 0}\frac{d^k}{ds^k}\bar{E}(s)$$

First moment is also the mean residence time, $\tau$

$$\mu_1 = (-1)\lim_{s\to 0}\frac{d}{ds}\bar{E}_2(s) = (-1)\lim_{s\to 0}\left[-\frac{1}{4\left(s+\frac{1}{4}\right)^2}\right] = 4\text{ s}$$

The second moment can be derived as follows

$$\mu_2 = (-1)^2\lim_{s\to 0}\frac{d^2}{ds^2}\bar{E}_2(s) = \lim_{s\to 0}\left[\frac{2}{4\left(s+\frac{1}{4}\right)^3}\right] = 32\text{ s}$$

And finally the third moment is derived as follows

$$\mu_3 = (-1)^3 \lim_{s\to 0} \frac{d^3}{ds^3} \bar{E}_2(s) = (-1) \lim_{s\to 0} \left[ -\frac{6}{4\left(s+\frac{1}{4}\right)^4} \right] = 384 \text{ s}$$

## Problem 9

**Consider a tubular reactor packed with non-reactive solid particles, and operating at constant temperature. The packed bed reactor is 16.5 cm in length, and particles have radius of 7.5 mm. A liquid phase reaction of first order occurs in the reactor, and it was advised that this packed bed reactor be modeled using a cascade of CSTRs model. The mean residence time of the packed bed reactor is $\tau = 120$ s and the reaction rate constant $k = 0.04\ s^{-1}$.**

(a) **Describe the cascade of N CSTRs model and using Laplace transform, derive the transfer function of the cascade.**
(b) **Determine the conversion achieved in the packed bed reactor.**
(c) **Comment on how your answer in part a would change if it was a PFR.**
(d) **Derive the first and second moments of the RTD for the cascade of CSTRs model.**
(e) **Tracer experiments produced data showing that the variance of $E(t)$ for the packed bed reactor is 3600 $s^2$. How does the conversion based on the experimental data compare with the value found in part b.**

## Solution 9

*Worked Solution*

(a) In the modeling of non-ideal flow in real reactors, we seldom see the exact behavior as in a CSTR or PFR which are ideal models. Actual reactors operate between limits defined in the ideal situation. Therefore, we can model real reactors as combinations of ideal reactors. One model is assuming a cascade of N CSTRs. In this model, N identical CSTRs are connected in series.

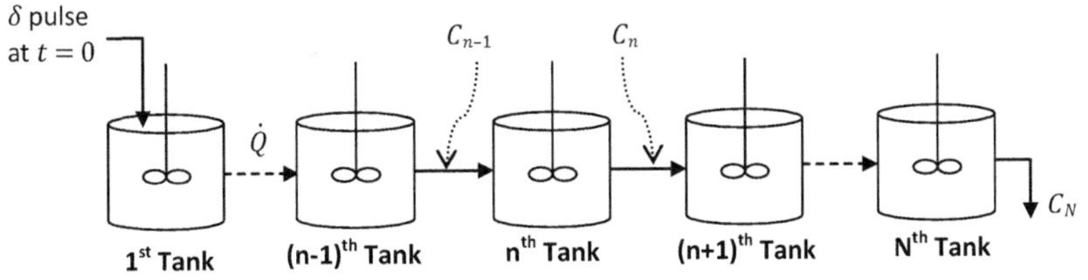


Let $V$ denote the total volume of the $N$ tanks, and $\dot{Q}$ denote the volumetric flow rate. Therefore the mean residence time for the entire cascade, $\tau = V/\dot{Q}$, and the mean residence time per tank is $\frac{\tau}{N}$.

Assuming that we inject a $\delta$ pulse of tracer at $t = 0$, let us analyze its behavior by taking a mass balance around the nth tank.

$$\dot{Q}C_{n-1} - \dot{Q}C_n = \frac{V}{N}\frac{dC_n}{dt}$$

$$C_{n-1} - C_n = \frac{\tau}{N}\frac{dC_n}{dt}$$

The boundary condition for this differential equation is such that at $t \leq 0$, $C_n = 0$. The solution to this equation is as follows, where it is normalized.

$$C_n(t) = \frac{N^n}{(n-1)!\tau^n}t^{n-1}\exp\left(-\frac{Nt}{\tau}\right)$$

Expanding to the another cascade, we can obtain the residence time *E*(*t*) of the cascade,

$$E(t) = C_N(t) = \frac{N^N}{(N-1)!\tau^N}t^{N-1}\exp\left(-\frac{Nt}{\tau}\right)$$

We can take Laplace transform of the differential equation in $t$ to find the transfer function of one tank, and of the whole cascade.

$$\bar{C}_{n-1}(s) - \bar{C}_n(s) = \frac{\tau}{N}\left[s\bar{C}_n(s) - C_n(t=0)\right]$$

$$\frac{\tau}{N}s\bar{C}_n(s) = \bar{C}_{n-1}(s) - \bar{C}_n(s)$$

$$\bar{C}_n(s)\left[\frac{s\tau}{N} + 1\right] = \bar{C}_{n-1}(s)$$

The transfer function across one tank is found as follows:

$$\frac{\bar{C}_n(s)}{\bar{C}_{n-1}(s)} = \left(\frac{s\tau}{N} + 1\right)^{-1}$$

The transfer function across $N$ tanks is therefore:

$$\frac{\bar{C}_N(s)}{\bar{C}_0(s)} = \frac{\bar{C}_1(s)}{\bar{C}_0(s)} \times \frac{\bar{C}_2(s)}{\bar{C}_1(s)} \cdots = \left(\frac{s\tau}{N} + 1\right)^{-N}$$

As $N \to \infty$, the CSTR cascade will be closer to the PFR, whereby at each point along the length of the PFR is an instantaneous CSTR (perfect mixing across the circular cross section)

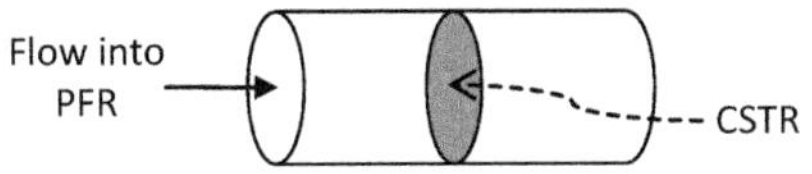


(b) In a PBR, we can assume that the void space between each layer of solid particles is a CSTR. We can also assume that the size of the gap between particles is approximately equivalent to the characteristic diameter of the solid particle. This hypothetical CSTR is assumed to be well mixed and this can be ensured by maintaining a high superficial velocity (high Re) in the PBR to allow good mixing in the gaps.

Therefore, the number of CSTRs required for the cascade model in this problem is:

$$N = \frac{16.5}{(0.75 \times 2)} = 11$$

The mass balance around the *n*th CSTR in the cascade gives for species *A*, where *V* is the total volume of the *N* reactors in the cascade model and $\tau = V/N$ is the mean residence time for the whole cascade:

$$\dot{Q}\,C_{A,n-1} - \dot{Q}\,C_{A,n} + r_A\left(\frac{V}{N}\right) = \frac{V}{N}\frac{dC_{A,n}}{dt} = 0 \text{ (steady state)}$$

The rate law for a first-order reaction is $-r_A = kC_{A,n}$; therefore

$$\dot{Q}\,C_{A,n-1} - \dot{Q}\,C_{A,n} + (-kC_{A,n})\left(\frac{V}{N}\right) = 0$$

$$\dot{Q}\,C_{A,n-1} - \dot{Q}\,C_{A,n} = k\left(\frac{V}{N}\right)C_{A,n}$$

$$C_{A,n-1} = C_{A,n}\left(\frac{\tau k}{N} + 1\right)$$

$$\frac{C_{A,n}}{C_{A,n-1}} = \frac{1}{\frac{\tau k}{N} + 1}$$

Extrapolating this result to the whole cascade, we have

$$\frac{C_{A,N}}{C_{A,0}} = \frac{1}{\left(\frac{\tau k}{N} + 1\right)^N}$$

Conversion achieved by the whole cascade (i.e., the packed bed reactor) can be expressed as

$$X = 1 - \frac{C_{A,N}}{C_{A,0}} = 1 - \frac{1}{\left(\frac{\tau k}{N} + 1\right)^N} = 1 - \frac{1}{\left(\frac{120\times0.04}{11} + 1\right)^{11}} = 0.981$$

(c) If the reactor was a PFR, we can use the design equation for a PFR for the analysis.

$$r_A = \frac{dF_A}{dV} = \frac{d(C_A\dot{Q})}{dV}$$

$$-kC_A = \dot{Q}\,\frac{dC_A}{dV}$$

$$V = -\frac{\dot{Q}}{k}\ln\left(\frac{C_A}{C_{A0}}\right)$$

$$-\tau k = \ln(1 - X)$$

$$X = 1 - \mathrm{e}^{-\tau k} = 1 - \mathrm{e}^{-120(0.04)} = 0.992$$

(d) To find the first and second moments, we can first assume that we introduce a $\delta$ function tracer into the cascade at $t = 0$. A mass balance around the $n$th CSTR gave us this result from part a,

$$\bar{E}_{\text{cascade}}(s) = \frac{\bar{C}_N(s)}{\bar{C}_0(s)} = \left(\frac{s\tau}{N} + 1\right)^{-N}$$

First moment is defined as follows

$$\mu_1 = (-1)\lim_{s\to0}\frac{d}{ds}\bar{E}_{\text{cascade}}(s) = (-1)\lim_{s\to0}\frac{d}{ds}\left[\left(\frac{s\tau}{N}+1\right)^{-N}\right] = (-1)\lim_{s\to0}\frac{d}{ds}\left[-\tau\left(\frac{s\tau}{N}+1\right)^{-N-1}\right] = \tau$$

Second moment can be derived as follows

$$\mu_2 = (-1)^2\lim_{s\to0}\frac{d^2}{ds^2}\bar{E}_{\text{cascade}}(s) = \lim_{s\to0}\left[\frac{-\tau^2(-N-1)}{N}\left(\frac{s\tau}{N}+1\right)^{-N-2}\right] = \frac{\tau^2(N+1)}{N}$$

(e) Variance $\sigma^2$ is defined as $\mu_2 - {\mu_1}^2$. From the experimental data, we have a variance of 3600 $s^2$; therefore

$$\sigma^2 = \frac{\tau^2(N+1)}{N} - \tau^2 = \frac{\tau^2}{N} = 3600 \rightarrow N = \frac{120^2}{3600} = 4$$

This means that the number of particles corresponding to one hypothetical CSTR in the cascade of CSTRs is

$$\text{Number of particles per CSTR} = \frac{1}{4}\frac{16.5}{(0.75 \times 2)} = 2.75$$

Typically, for liquids at low Re, the liquid phase is not turbulent to ensure ideal mixing in the inter-particle spaces if only one particle diameter space is available. It requires approximately 4 cavities per CSTR for optimum mixing, in this case it is about 3, which is close to the ideal 4. Note that the conversion achieved with a cascade of 2.75 CSTRs, instead of the earlier 11, is lower as shown below.

$$X = 1 - \frac{C_{A,N}}{C_{A,0}} = 1 - \frac{1}{\left(\frac{\tau k}{N} + 1\right)^N}$$

$$X = 1 - \frac{1}{\left(\frac{120 \times 0.04}{2.75} + 1\right)^{2.75}} = 0.938$$

## Problem 10

**The residence time distribution of a series of $N$ CSTRs in series is given as follows where $\tau$ denotes the residence time for the entire cascade and $t$ is time:**

$$E(t) = \frac{N^N}{(N-1)!} t^{N-1} \exp\left(-\frac{Nt}{\tau}\right)$$

**In the form of Laplace transform, it is given that**

$$\bar{E}(s) = \left[\frac{s\tau}{N} + 1\right]^{-N}$$

$$\lim_{N\to\infty} \left[\frac{s\tau}{N} + 1\right]^{-N} = \exp(-s\tau)$$

(a) **Identify the extreme types of reactors described by this distribution, and explain how this equation can be applied to model a real reactor.**

(b) **For a given reactor of volume 4 m$^3$, given that the overall residence time for the cascade is 2 min and that the inlet flow rate is 0.025 m$^3$/s, propose a model comprising of a combination of reactors which would provide the RTD as shown below. Comment on the presence/absence of any dead volumes.**

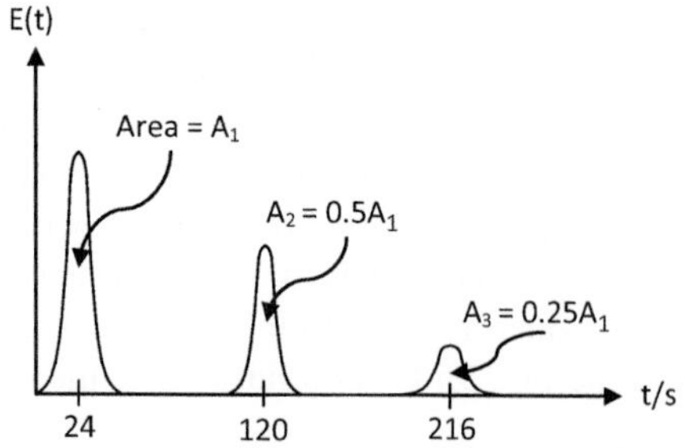


(c) **Given that the variance of the first peak is 8 × 10$^{-7}$s$^{2}$, find the number of CSTRs that should be used to model each of the reactors in part b.**

## Solution 10

### *Worked Solution*

(a) When $N = 1$, we have the residence time distribution of a CSTR.

$$E(t) = \exp\left(-\frac{t}{\tau}\right)$$

When $N \to \infty$, we have a Plug Flow Reactor (PFR), we observe from the Laplace expression given above that:

$$\lim_{N\to\infty}\left[\frac{s\tau}{N} + 1\right]^{-N} = \exp(-s\tau)$$

The RTD of a PFR is a delta function which is of value 0 everywhere except at $t = \tau$.

$$E(t) = \delta(t - \tau)$$

For an arbitrary function, $y(t)$, its Laplace transform $\bar{y}(s)$ is defined as follows.

$$\bar{y}(s) = \int_0^\infty y(t)\exp(-st)dt$$

Therefore the Laplace expression for a PFR is:

$$\int_0^\infty \delta(t - \tau)\exp(-st)dt = \exp(-s\tau) = \bar{E}(s)|_{N\to\infty}$$

CSTRs and PFRs are extreme cases of real reactors. The CSTR is assumed perfectly mixed, while the PFR is assumed to have perfect radial mixing but no axial mixing. Therefore the PFR is in fact a cascade of infinite number of CSTRs in series.

(b) We can represent the given reactor using a configuration of 2 PFRs with a recycle loop. Each peak in the graph (also called an RTD "fingerprint") is observed to be sharp and this corresponds to the feature of a plug flow. Let $\dot{Q}$ denote volumetric flowrate and $R$ denote recycle ratio which is the ratio of recycled fluid to fluid removed from the reactor.

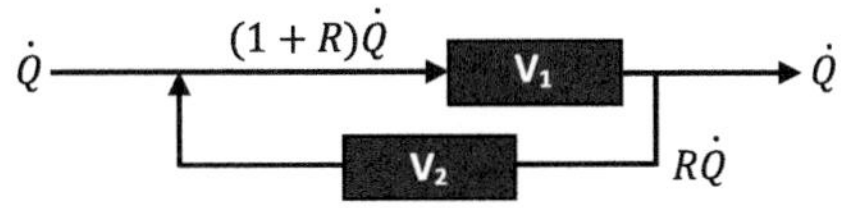


$$\frac{A_2}{A_1} = \frac{1}{2} = \frac{R}{R+1} \rightarrow R = 1$$

$$\frac{A_3}{A_1} = \left(\frac{R}{R+1}\right)^2 = \frac{1}{4}$$

$$A_1 = \frac{1}{R+1} = \frac{1}{2}$$

Let $t_n$ denote the time at the $n$th peak,

$$\Delta t = t_2 - t_1 = t_3 - t_2 = 96 \text{ s}$$

$$\frac{V_1}{(1+R)\dot{Q}} = t_1 = 24 \text{ s}$$

$$\frac{V_1}{(1+1)0.025} = 24 \rightarrow V_1 = 1.2 \text{ m}^3$$

$$\Delta t = \frac{V_1}{(1+R)\dot{Q}} + \frac{V_2}{R\dot{Q}} = 96 \text{ s}$$

$$24 + \frac{V_2}{1(0.025)} = 96 \rightarrow V_2 = 1.8 \text{ m}^3$$

$$\text{Active volume} = V_1 + V_2 = 1.2 + 1.8 = 3 \text{ m}^3 < \text{Total volume of } 4 \text{ m}^3$$

Therefore there is dead volume of 1 $\text{m}^3$.

(c) For $N$ CSTR tanks in series, variance is related to mean residence time as follows:

$$\sigma^2 = \frac{\tau^2}{N}$$

Variance of the first peak, $\sigma_1^2$, is given as $8 \times 10^{-7} s^2$. The mean residence time for the first peak, $\tau_1$, can be found as follows.

$$\tau_1 = \frac{V_1}{(1+R)\dot{Q}} = 24 \text{ s}$$

Combining the results, we can find $N_1$, the number of CSTRs that should be used to model the first reactor of volume $V_1$ in part b.

$$\sigma_1^2 = \frac{\tau_1^2}{N_1}$$

$$N_1 = \frac{24^2}{\left(8 \times 10^{-7}\right)^2} = 9 \times 10^{14}$$

$$N_2 = \frac{V_2}{V_1} N_1 = \frac{1.8}{1.2}\left(9 \times 10^{14}\right) = 1.4 \times 10^{15}$$

It follows that the number of CSTRs to model the second reactor of volume $V_2$ is $1.4 \times 10^{15}$.

## Problem 11

**Consider a plug flow reactor as shown in the diagram below. The PFR has volume *V* and recycle feed with a recycle ratio *R*.**

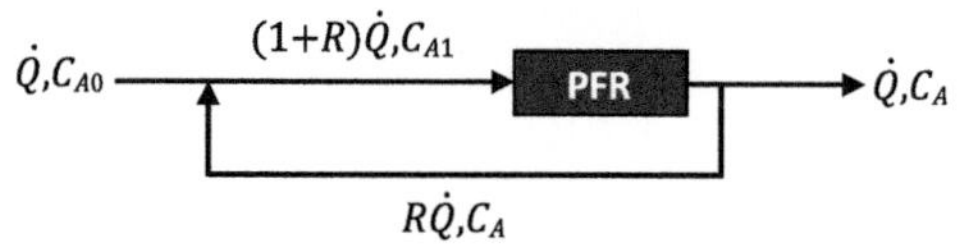


(a) **Derive the design equation for this PFR with recycle such that it correlates the residence time in the PFR section to conversion.**
(b) **Sketch the residence time distributions for this reactor system at various values of recycle ratio, $R = 0,1,5$ and $\infty$.**

## Solution 11

***Worked Solution***

(a) Consider the mass balance for a typical PFR without recycle feed, except that the volumetric flow rate instead of the typical $\dot{Q}$ is replaced by $\dot{Q}_{new} = (1+R)\dot{Q}$.

So the mass balance around this PFR is as follows, where $F_A$ denotes the molar flow rate and $x$ is in the axial direction of flow.

$$F_A(x) - F_A(x + \Delta x) + r_A V = \frac{dN_A}{dt}$$

$$F_A(x) - F_A(x + \Delta x) + r_A \Delta V = 0 \text{ (steady state)}$$

$$r_A \Delta V = F_A(x + \Delta x) - F_A(x)$$

For small values of $\Delta x$ and $\Delta V$

$$r_A = \frac{dF_A}{dV} = \frac{d(C_A \dot{Q}_{\text{new}})}{dV}$$

$$C_A = C_{A0}(1 - X) \rightarrow dC_A = -C_{A0} dX$$

$$\dot{Q}_{\text{new}} C_{A0} dX = -r_A dV$$

$$(1 + R)\dot{Q} C_{A0} dX = -r_A dV$$

We can integrate this differential equation using boundary conditions. Let the conversion achieved be denoted by $X_1$, and $\tau$ represents the residence time.

$$(1 + R)\dot{Q} C_{A0} dX = -r_A V$$

$$\frac{V}{\dot{Q}} = \tau = \frac{C_{A0}(1 + R)}{-r_A} \int_{X_1}^{X} dX$$

By conservation of mass for species $A$, we deduce that

$$C_{A1} = \frac{C_{A0} + RC_A}{1 + R}$$

$$X_1 = 1 - \frac{C_{A1}}{C_{A0}} = 1 - \frac{\frac{C_{A0} + RC_A}{1 + R}}{C_{A0}}$$

$$X_1 = \frac{R\left(1 - \frac{C_A}{C_{A0}}\right)}{1 + R}$$

Therefore, putting the results together,

$$\tau = \frac{C_{A0}(1 + R)}{-r_A} \int_{X_1}^{X} dX, \text{ where } X_1 = \frac{R\left(1 - \frac{C_A}{C_{A0}}\right)}{1 + R}$$

(b) At $R = 0$, the residence time distribution is that of a perfect PFR, which is an infinite delta pulse at $t = \tau$. As recycle ratio $R$ increases, the number of peaks increase until the other extreme at $R \to \infty$, the reactor's residence time distribution becomes equivalent to that of a perfect CSTR.

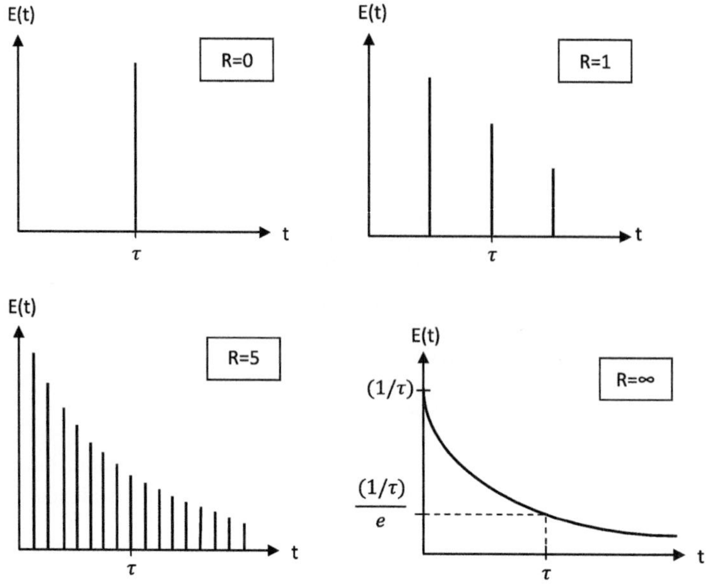


## Problem 12

**Experiments were conducted to determine the rate of reaction for a first-order reaction occurring in catalytic particles shaped as rectangular slabs. The half-width $L$ of the catalytic particles were varied and the corresponding reaction rates per unit area are shown below. Both faces of the slab particles are exposed to surrounding reactant fluid of concentration 12000 mol $m^{-3}$, which is equivalent to the concentration of reactant $A$ at the surface of the slabs (i.e., external mass transfer effects may be ignored). Determine the reaction rate constant per unit volume $k_v$ and effective diffusivity $\mathcal{D}$.**

| $L$[m] | 0.0002 | 0.0008 | 0.0012 | 0.015 | 0.05 | 0.15 | 0.25 | 0.35 |
|---|---|---|---|---|---|---|---|---|
| $-r_A''$[mol $m^{-2}$ $s^{-1}$] | 1.3 | 2.1 | 3.0 | 40.1 | 56.0 | 61.5 | 64.8 | 64.9 |

## Solution 12

### *Worked Solution*

Observe from the data that as $L$ tends to a large size, the reaction rate tapers off to a constant value of around $-r_A'' = 64.9$ mol $m^{-2}$ $s^{-1}$. This extreme corresponds to the case when internal mass transfer effect is limiting. This also means that reaction rate is much faster than internal mass transfer rate and the overall rate of reaction is determined largely by the rate of internal mass transfer of reactant within the catalyst particle.

We can also consider the other extreme, which is at the smallest particle size of $L = 0.0002$ m, whereby we may assume that the rate-limiting process is reaction rate. This means that the internal mass transfer rate is much faster than chemical reaction rate at the active sites of the catalytic particles, and therefore overall rate of reaction is largely a reflection of chemical reaction rate. This is also denoted by the ideal rate which is the maximum possible rate achievable without mass transfer effects.

$$\text{Ideal rate} = -r_A V = k_v C_{A,s}(2LA)$$

$$-r_A V = -r_A'' A = 1.3A$$

$$k_v C_{A,s}(2LA) = 1.3A$$

We know that $L = 0.0002$ m, $C_{A,s} = 12000$ mol m$^{-3}$; therefore,

$$k_v = \frac{1.3}{2(12000)(0.0002)} = 0.27\ \text{s}^{-1}$$

We can relate the observed rate under mass transfer limitation to the ideal rate using effectiveness factor $\eta$ at large particle sizes when $\eta \sim 1/\text{Ø}$.

$$\eta = \frac{1}{\text{Ø}} = \frac{\text{Observed rate}|_{L=0.35\ \text{m}}}{\text{Ideal rate}|_{L=0.35\ \text{m}}}$$

$$\text{Ideal rate}|_{L=0.35\ \text{m}} = \frac{\text{Ideal rate}|_{L=0.0002\ \text{m}}}{0.0002} \times 0.35 = 1750 \times 1.3A$$

$$\text{Observed rate}|_{L=0.35\ \text{m}} = -r_A'' A = 64.9A$$

$$\frac{1}{\text{Ø}} = \frac{64.9A}{1750 \times 1.3A} \rightarrow \text{Ø} = 35$$

By definition, we know that for a slab, $\text{Ø} = L\sqrt{\frac{k_v}{\mathcal{D}}}$

$$35 = 0.35\sqrt{\frac{0.27}{\mathcal{D}}}$$

$$\mathcal{D} = 2.7 \times 10^{-5}\ \text{m}^2\text{s}^{-1}$$

## Problem 13

**We have a CSTR of volume $V = 25\ \text{m}^3$ that contains a liquid phase reaction that is experimentally shown to obey the following rate law equation, where $k_f$ refers to the reaction rate constant of the forward reaction and $K$ is the equilibrium constant. Given that $K = 0.012\ \text{m}^3\ \text{mol}^{-1}$, $\dot{Q} = 0.2\ \text{m}^3$, $k_f = 0.89\ \text{s}^{-1}$, and $C_{A0} = 1800\ \text{mol m}^{-3}$**

$$A \leftrightarrow B$$

$$-r_A = \frac{k_f C_A}{(1 + K C_A)^2}$$

(a) **Derive the solutions for $C_A$ at steady-state conditions.**
(b) **Using suitable plots, determine the value of $C_A$ that gives the most stable steady state.**

## Solution 13

*Worked Solution*

(a) Let us start with the mass balance equation using the rate law information provided,

$$\dot{Q}C_{A0} - \dot{Q}C_A + r_A V = V\frac{dC_A}{dt} = 0 \text{ (steady state)}$$

$$\dot{Q}C_{A0} - \dot{Q}C_A = \frac{V k_f C_A}{(1 + K C_A)^2}$$

$$\frac{\dot{Q}}{V k_f}(C_{A0} - C_A) = \frac{C_A}{(1 + K C_A)^2}$$

$$\frac{0.2}{25(0.89)}(1800 - C_A) = \frac{C_A}{(1 + 0.012 C_A)^2}$$

$$0.009(1800 - C_A) = \frac{C_A}{(1 + 0.012 C_A)^2}$$

An easy way to solve this equation is to plot a straight line (for left-hand side of equation) and the curve (right-hand side of equation) using a series of hypothetical $C_A$ values starting from the $C_{A0}$ value, and observing the intersection points which represent the solutions to the equation. A plot is shown below and the solutions (i.e., steady state) occur at $C_A = 180, 380, 1250\ \text{mol m}^{-3}$, as marked by black crosses.

(b) The value of $C_A = 180$ mol m$^{-3}$ gives the most stable steady state, because when $C_A$ is reduced from this value, $\left|\frac{C_A}{(1+0.012C_A)^2}\right| < |0.009(1800 - C_A)|$, this means that the rate of consumption of $A$ via the reaction will be less than the net inflow rate of species $A$ into the tank, this serves to replenish the decrease in $A$, and bring up the $C_A$ value, in a direction back to its steady-state position of $C_A = 180$ mol m$^{-3}$. The other value of $C_A = 1250$ mol m$^{-3}$ behaves similarly but is not as stable as compared to the point at $C_A = 180$ mol m$^{-3}$.

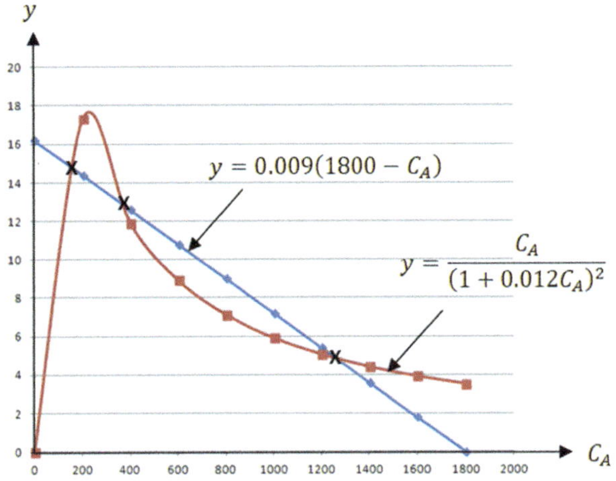


## Problem 14

**We have an exothermic reaction $A \rightarrow B$ with the enthalpy of reaction $\Delta h_{rxn}$ given as $-39560$ J mol$^{-1}$. The reactants are in the liquid phase and the reaction obeys first-order kinetics. The reaction rate constant $k$ is represented by the Arrhenius equation where the pre-exponential factor is $10^7$ and the activation energy is 58200 J mol$^{-1}$. You may assume that the rate of backward reaction is negligible. If the mean residence time in the reactor is 100 s,**

(a) **Express conversion $X$ as a function of temperature $T$.**
(b) **To ensure isothermal conditions at 298 K, a coolant is introduced whereby the overall heat transfer coefficient between the inside of reactor and the coolant $U = 2500$ J s$^{-1}$ m$^{-2}$ K$^{-1}$ and the heat transfer area $A = 68$ m$^2$. The inlet flow of pure reactant into the reactor is at 298 K and at a flow rate of 450 mol s$^{-1}$. The reaction mixture has an average $C_p$ of 70 J mol$^{-1}$ K$^{-1}$. Using an energy balance, find a correlation between conversion $X$ and temperature.**
(c) **Based on the results in earlier parts and a suitable plot, explain why the reactor can operate at about 310 K and 377 K but not at 325 K.**

## Solution 14

***Worked Solution***

(a) Let us first perform mass balance on reactant $A$, and combine it with the rate law equation.

$$F_{A0} - F_A + r_A V = 0 \text{ (steady state)}$$
$$F_{A0} - F_A + (-kC_A)V = 0$$
$$F_{A0}X = kC_A V = kVC_{A0}(1 - X)$$

From the Arrhenius equation, we derive an expression for reaction rate constant,

$$k = A\exp\left(\frac{-E_a}{RT}\right) = 10^7 \exp\left(\frac{-58,200}{RT}\right)$$
$$\tau = \frac{V}{\dot{Q}} = \frac{C_{A0}V}{F_{A0}} = 100 \text{ s}$$

Combining the expressions we have,

$$F_{A0}X = kVC_{A0}(1 - X)$$
$$X = k\tau(1 - X)$$
$$\frac{X}{1 - X} = 10^7 \tau \exp\left(\frac{-58,200}{RT}\right)$$
$$\frac{1}{\frac{1}{X} - 1} = 10^7(100)\exp\left(\frac{-58,200}{RT}\right)$$
$$X = \frac{1}{10^{-9}\exp\left(\frac{58,200}{RT}\right) + 1}$$

(b) We can construct a heat balance equation in [J s$^{-1}$] as follows

$$\text{Rate of heat generated from reaction} = (-\Delta h_{\text{rxn}})F_{A0}X = 39560(450)X = 17,802,000X$$

$$\text{Rate of heat removal by coolant} = UA\Delta T = 2500(68)(T - 298) = 170,000(T - 298)$$

Rate of net heat inflow = Heat inflow − Heat outflow*.

*Note that because this is an irreversible reaction of $A \rightarrow B$ with a 1:1 stoichiometric conversion, due to conservation of mass, the outlet stream is also at the same molar flow rate of 450 mol s$^{-1}$ = $F_{A0}$.

$$\text{Rate of net heat inflow} = F_{A0}c_p(298) - F_{A0}c_p(T) = 450(70)(298 - T)$$
$$= 31500(298 - T)$$

Combining the expressions, we have the overall heat balance equation as shown below:

$$17,802,000X + 31,500(298 - T) - 170,000(T - 298) = 0$$

$$X = \frac{170,000(T - 298) + 31,500(T - 298)}{17,802,000} = 0.0113(T - 298)$$

(c) We can construct two simple plots, one for heat generation and the other for heat removal. For the function representing heat generation, we have the following, note that gas constant $R = 8.314$ J mol$^{-1}$ K$^{-1}$.

$$\text{Heat generated per unit time} = 17,802,000X = \frac{17,802,000}{10^{-9}\exp\left(\frac{58,200}{8.314T}\right) + 1}$$

$$\text{Heat removed per unit time} = 31,500(T - 298) + 170,000(T - 298)$$
$$= 201,500(T - 298)$$

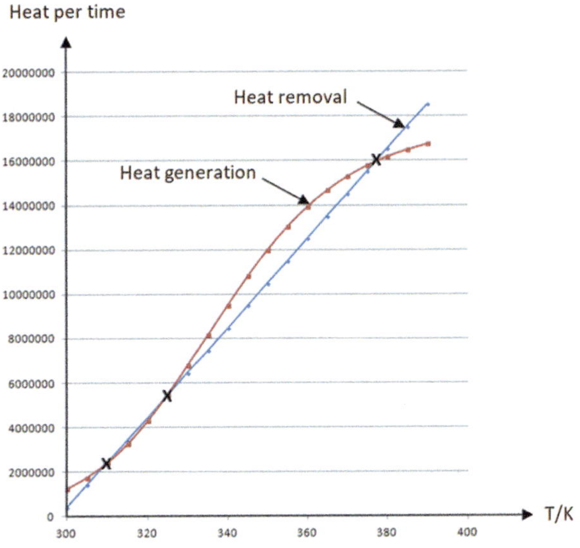

The reactor achieves heat balance at the three intersection points of the heat generation and heat removal plots. However the point at $T = 325$ K is unstable since any slight deviation from the point will cause a further deviation. For example, a small temperature increase from 325 K will move the point to the right where heat generation is greater than removal and the temperature increases further (hence unstable). The heat balances at 310 K and 377 K are more stable since any slight deviation will be corrected. For example, any slight increase in temperature above 377 K will cause heat removal to be faster than heat generation, and this will bring the temperature back down to 377 K.

## Problem 15

**In reaction kinetics, we may sometimes model heterogeneous non-catalytic reactions occurring in spherical particles using the Shrinking Core Model (SCM). State the assumptions of the SCM.**

## Solution 15

***Worked Solution***

For heterogeneous non-catalytic reactions, we can adopt the SCM assuming the following:

- The spherical particle itself (e.g., solid $B$) is one of the reactants of the reaction, with the other reactant (e.g., gas $A$) occurring in the surrounding bulk fluid.
- The spherical particle $B$ is non-porous (impervious) to the reactant $A$ in its surrounding.
- Reaction proceeds only at the surface boundary of the unreacted core of $B$, and this boundary is also termed the "reaction front".
- The reaction front moves radially inwards with time as the unreacted core shrinks in size as reaction proceeds to consume (and hence "shrink") the unreacted core.
- The overall size of the particle may or may not reduce. It may not change in size if the product of reaction is a solid that collects at the surface of the particle, restoring the shrinkage of the unreacted core. Conversely, the overall size of the particle will reduce if the solid product falls off easily, or the product of reaction is gaseous.
- Reaction is assumed to go to completion at the front, i.e., complete conversion.

## Problem 16

**In heterogeneous reactor systems, we have different phases present which may consist of a fluid (liquid or gas) and a solid. In non-catalytic cases, the solid particle does not change in size throughout the reaction; however, the composition of the solid particle changes as the unreacted core of the solid reduces in size as reaction progresses. The overall size of the particle remains unchanged as the volume previously occupied by the unreacted core is replaced by solid products (e.g., ash) of reaction.**

(a) **With the help of a diagram, describe the sequence of events that occur as reaction progresses for a shrinking core model, whereby the size of particle remains unchanged. You may assume that a solid spherical particle of radius $R$ reacts with a gas $A$ in the reaction below and $R_c$ is the radius of the unreacted core of $B$.**

$$A(g)+bB\ (s)\rightarrow cC(g)+dD(s)$$

(b) **Under steady-state conditions, if diffusion through the gas film surrounding the solid particle is the rate-controlling step, derive an expression that shows how the radius of the unreacted core changes with time. Show further that the fractional time to complete conversion is equivalent to the fractional conversion of solid reactant $B$. Let $\tau$ denote the time taken for complete conversion, and assume that the reaction is almost irreversible.**

(c) **Repeat part b assuming diffusion through the ash layer is the rate-controlling step.**

(d) **Repeat part b assuming chemical reaction at $r = R_c$ is the rate-controlling step. You may assume that the reaction is first order with respect to $A$, with a reaction rate constant per area denoted as $k_r''$.**

## Solution 16

### *Worked Solution*

(a) We have the following reaction whereby gas $A$ reacts with a solid $B$, at the interface area between the solid particle and surrounding gas. The reaction produces a gas $C$ and a solid ash by-product $D$.

$$A(g) + bB\ (s) \rightarrow cC(g) + dD(s)$$

This process can be described in a series of steps as summarized below, corresponding to the diagram below:

1. Bulk gas $A$ at concentration $C_{A,1}$ diffuses to a region near to the solid particle surface ($B$).
2. At a small distance δ from the solid particle surface, i.e., between $r = R$ and $r = R + \delta$, a gas film forms whereby reactant gas $A$ diffuses through (this distance δ) to reach the surface of the solid. The concentration of $A$ decreases from $C_{A,1}$ to $C_{A,2}$ at the solid surface. The rate of diffusion through the gas film is dependent on the concentration gradient across δ, which is proportional to $C_{A,1} - C_{A,2}$.
3. At the surface of the solid particle, i.e., $r = R$, there is by-product solid ash ($D$). Reactant gas $A$ has to diffuse through this ash layer in order to react with $B$ which is in the inner core of the particle. The diffusion of gas $A$ through this ash layer is driven by a concentration gradient proportional to $C_{A,2} - C_{A,3}$, between $C_{A,2}$ at $r = R$ and $C_{A,3}$ at the surface of the unreacted core $r = R_c$.

4. Gas $A$ passes through the ash layer and reaches the interface with the unreacted core. Gas $A$ reacts with solid $B$, as per the reaction equation, to produce gas $C$ and solid ash $D$. Ash $D$ collects on the surface of the unreacted core to keep the particle size constant by replacing the lost volume of solid $B$ that is consumed by the reaction.
5. The gaseous product $C$ is produced at the surface of the unreacted core, $r = R_c$ and starts to diffuse outwards through the ash layer. This diffusion is driven by a concentration gradient proportional to $C_{C,3} - C_{C,2}$, between $C_{C,3}$ at $r = R_c$ at the surface of the unreacted core and $C_{C,2}$ at the surface of the solid particle $r = R$.
6. After diffusion through the ash layer, gas product $C$ reaches the surface of the solid particle where it starts to diffuse through the same gas film layer (of thickness δ) as encountered at step 2, into the bulk gas region. The diffusion through this gas film is driven by a concentration gradient proportional to $C_{C,2} - C_{C,1}$, between $C_{C,2}$ at $r = R$ at the surface of the solid particle and $C_{C,1}$ at the outer boundary of the gas film at $r = R + \delta$.

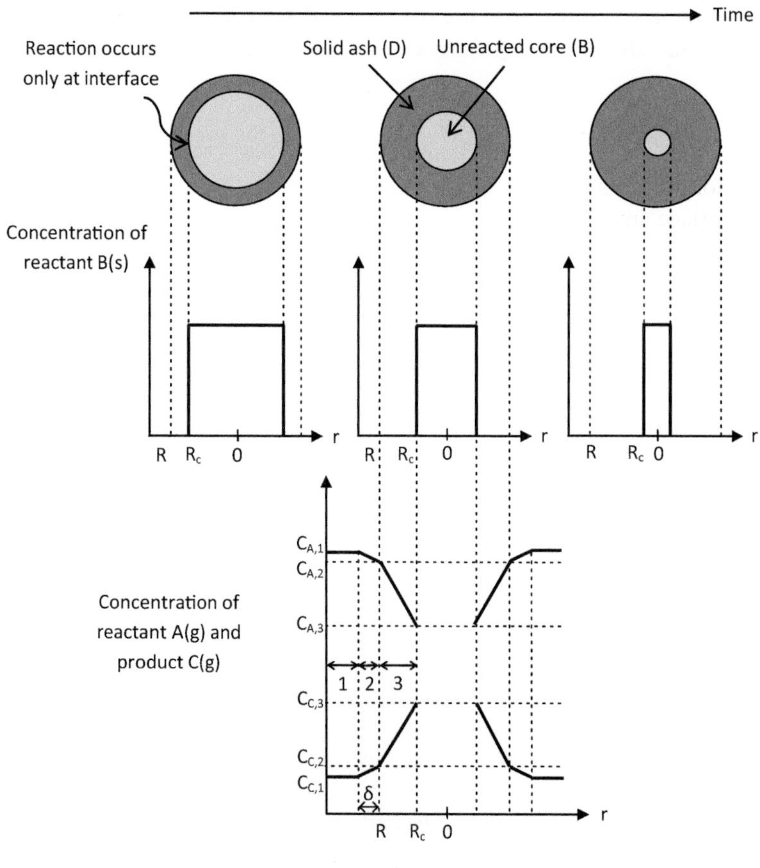

(b) Note that in a shrinking core model, due to the processes that occur in series, either one of them can be the rate-limiting or rate-controlling step. Recall that we have the following reaction.

$$A(g) + bB\,(s) \rightarrow cC(g) + dD(s)$$

Consider the case whereby diffusion through the gas film is the rate-limiting step, as shown in yellow highlighted. Hence all of the concentration gradient occurs in the gas film layer, and with an irreversible reaction, $C_{A,2} = 0$.

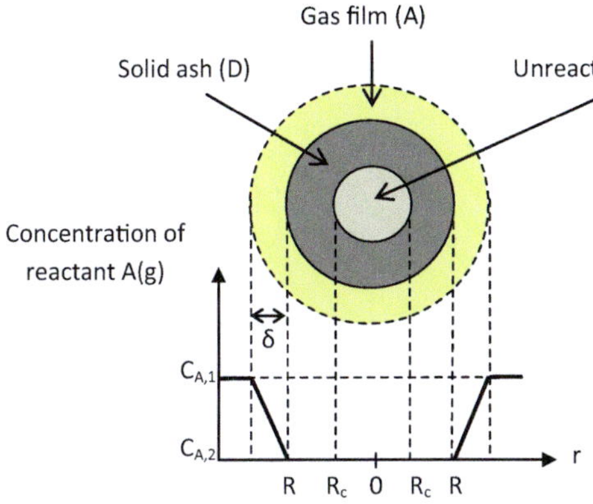


It follows from a mass balance on species $B$ that the rate of loss of $B(s)$ per unit surface area of the particle is

$$\text{Rate of loss of } B \text{ per area} = -\frac{1}{A_s}\frac{dN_B}{dt} = -\frac{1}{4\pi R^2}\frac{dN_B}{dt}$$

A mass balance on species $A$ gives the following, where $k$ is the mass transfer coefficient for the diffusion through the gas film. $C_{A,1}$ is the concentration of $A$ $(g)$ in the bulk gas, and $C_{A,2}$ is the concentration of $A(g)$ at the solid particle surface ($r = R$). When diffusion through the gas film is rate limiting, there is no accumulation of $A(g)$ at the surface of the particle at steady state; therefore $C_{A,2} = 0$.

The rate of loss of $A(g)$ per unit surface area of the particle is

$$\text{Rate of loss of } A \text{ per area} = -\frac{1}{4\pi R^2}\frac{dN_A}{dt} = k(C_{A,1} - C_{A,2}) = kC_{A,1}$$

By reaction stoichiometry, $dN_B = bdN_A$. We can therefore convert the rate of loss of $B(s)$ to rate of loss of $A(g)$ as follows

$$-\frac{1}{4\pi R^2}\frac{dN_B}{dt} = -\frac{b}{4\pi R^2}\frac{dN_A}{dt} = bkC_{A,1}$$

The amount of $B$ present in the particle is given by $N_B = \rho_B\left(\frac{4}{3}\pi {R_C}^3\right)$. The loss of $B(s)$, $-dN_B$, can be further expressed as:

$$-dN_B = -\rho_B dV = -\rho_B d\left(\frac{4}{3}\pi {R_C}^3\right) = -4\pi\rho_B {R_C}^2 dR_C$$

Substituting back into the mass balance equation, we have

$$-\frac{1}{4\pi R^2}\frac{dN_B}{dt} = -\frac{4\pi\rho_B {R_C}^2}{4\pi R^2}\frac{dR_C}{dt} = -\frac{\rho_B {R_C}^2}{R^2}\frac{dR_C}{dt} = bkC_{A,1}$$

$$-\frac{\rho_B}{R^2}\int_R^{R_C} {R_C}^2 dR_C = bkC_{A,1}\int_0^t dt$$

$$-\frac{\rho_B}{R^2}\left(\frac{{R_C}^3 - R^3}{3}\right) = \frac{\rho_B R}{3}\left[1 - \left(\frac{R_C}{R}\right)^3\right] = bkC_{A,1}t$$

$$t = \frac{\rho_B R}{3bkC_{A,1}}\left[1 - \left(\frac{R_C}{R}\right)^3\right]$$

This gives us the profile of $R_C$ with respect to time. $\tau$ is the time taken for complete conversion, which means that at $t = \tau$, $R_C = 0$.

$$\tau = \frac{\rho_B R}{3bkC_{A,1}}$$

And the function $R_C(t)$ can be expressed in terms of the fractional time for complete conversion:

$$\frac{t}{\tau} = 1 - \left(\frac{R_C}{R}\right)^3$$

The fractional conversion $X$ for the solid reactant $B$ (which forms the unreacted core at time $t$) is such that

$$1 - X = \frac{\frac{4}{3}\pi {R_C}^3}{\frac{4}{3}\pi R^3} = \left(\frac{R_C}{R}\right)^3$$

$$X = \frac{t}{\tau}$$

Therefore it is shown that the fractional time to complete conversion is the same as the fractional conversion of solid reactant $B$.

(c)

$$A(g) + bB\,(s) \rightarrow cC(g) + dD(s)$$

Consider the case whereby diffusion through the ash layer is now the rate-limiting step, as shown in yellow highlighted. Hence all of the concentration gradient occurs in the ash layer, and with an irreversible reaction, $C_{A,3} = 0$. Note that there is actually a small concentration gradient for the diffusion across the gas film ($A$) from $r = R$ to $r = R + \delta$; however, as diffusion through the gas film is not rate controlling , it is assumed to be fast and hence the concentration gradient is negligible; therefore, $C_{A,2} \cong C_{A,1}$.

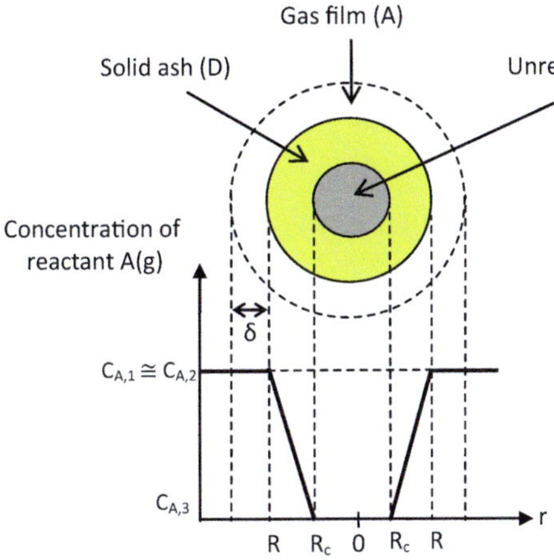


Let $q''(r)$ denote the molar flux of gas $A$ in the radial direction towards the core of the particle at a certain $r$ whereby $R \leq r \leq R_C$. By mass conservation, it follows that

$$-\frac{dN_A}{dt} = \text{constant at any } r$$

$$-\frac{dN_A}{dt} = j''(r = R)A_s\big|_{r=R} = j''(r)A_s\big|_{r=r} = j''(R_C)A_s\big|_{r=R_C}$$

$$-\frac{dN_A}{dt} = j''(r = R)4\pi R^2 = j''(r)4\pi r^2 = j''(R_C)4\pi {R_C}^2$$

By Fick's Law, we can relate mass flux $j''(r)$ to concentration gradient via diffusivity $\mathcal{D}$

$$j''(r) = \mathcal{D}\frac{dC_A}{dr}$$

Therefore the rate of loss of $A(g)$ can be expressed as

$$-\frac{dN_A}{dt} = 4\pi r^2 \mathcal{D}\frac{dC_A}{dr} = \text{constant}$$

$$-\frac{dN_A}{dt}\int_R^{R_C}\frac{1}{r^2}dr = 4\pi\mathcal{D}\int_{C_{A,1}=C_{A,2}}^{C_{A,3}=0} dC_A$$

$$\frac{dN_A}{dt}\left(-\frac{1}{R_C}+\frac{1}{R}\right) = 4\pi\mathcal{D}C_{A,1}$$

To arrive at the time-dependent profile of $R_C$, we will try to express $dN_A$ in terms of $R_C$,

$$bdN_A = dN_B = \rho_B dV = \rho_B d\left(\frac{4}{3}\pi R_C{}^3\right) = 4\pi\rho_B R_C{}^2 dR_C$$

Therefore, we have the following equation only in terms of $R_C$ and time,

$$\frac{\left(\frac{4\pi\rho_B R_C{}^2}{b}\right)dR_C}{dt}\left(-\frac{1}{R_C}+\frac{1}{R}\right) = 4\pi\mathcal{D}C_{A,1}$$

$$\int_R^{R_C}\left(-\frac{1}{R_C}+\frac{1}{R}\right)R_C{}^2 dR_C = \int_0^t \frac{b\mathcal{D}C_{A,1}}{\rho_B}dt$$

$$\left[-\frac{R_C{}^2}{2}+\frac{R_C{}^3}{3R}\right]_R^{R_C} = -\frac{R_C{}^2}{2}+\frac{R_C{}^3}{3R}+\frac{R^2}{2}-\frac{R^2}{3} = \frac{b\mathcal{D}C_{A,1}}{\rho_B}t$$

$$\frac{R^2}{6}\left[-3\left(\frac{R_C}{R}\right)^2+2\left(\frac{R_C}{R}\right)^3+1\right] = \frac{b\mathcal{D}C_{A,1}}{\rho_B}t$$

$$t = \frac{\rho_B R^2}{6b\mathcal{D}C_{A,1}}\left[1-3\left(\frac{R_C}{R}\right)^2+2\left(\frac{R_C}{R}\right)^3\right]$$

$\tau$ is the time taken for complete conversion, which means that at $t=\tau$, $R_C=0$.

$$\tau = \frac{\rho_B R^2}{6b\mathcal{D}C_{A,1}}$$

And the function $R_C(t)$ can be expressed in terms of the fractional time for complete conversion:

$$\frac{t}{\tau} = 1-3\left(\frac{R_C}{R}\right)^2+2\left(\frac{R_C}{R}\right)^3$$

The fractional conversion for the solid reactant $B$ (which forms the unreacted core at time $t$), $X$ is such that

$$1 - X = \frac{\frac{4}{3}\pi {R_C}^3}{\frac{4}{3}\pi R^3} = \left(\frac{R_C}{R}\right)^3$$

$$\frac{t}{\tau} = 1 - 3(1 - X)^{2/3} + 2(1 - X)$$

Therefore we have derived the fractional time to complete conversion in terms of the fractional conversion of solid reactant $B$.

(d)

$$A(g) + bB\,(s) \rightarrow cC(g) + dD(s)$$

Consider the case whereby reaction at the surface of the unreacted core is now the rate-limiting step, as shown by the yellow line. Assuming an irreversible reaction, concentration of $A$ at the reaction surface, i.e., $C_{A,3} = 0$. Note that there is actually a small concentration gradient for the diffusion across the gas film from $r = R$ to $r = R + \delta$, and through the ash layer from $r = R$ to $r = R_C$, however as diffusion through the gas film and ash layer are not rate controlling, they are assumed to be fast and hence the concentration gradients are negligible all the way to the reaction surface where it assumed to instantaneously deplete to zero.

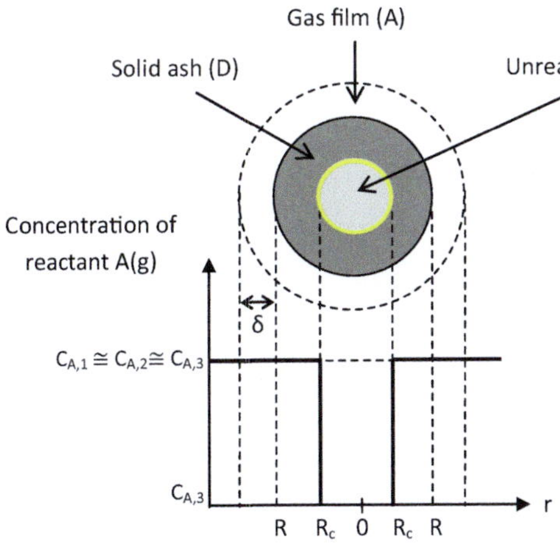


The rate law for this reaction is as follows, noting that reaction occurs at $r = R_c$

$$-{r_A}'' = {k_r}'' C_{A,3} = {k_r}'' C_{A,1}$$

$$-\frac{1}{4\pi {R_C}^2}\frac{dN_A}{dt} = {k_r}'' C_{A,1} = -\frac{1}{4\pi {R_C}^2}\left(\frac{1}{b}\right)\frac{dN_B}{dt}$$

$$-\frac{1}{4\pi R_C{}^2}\frac{\rho_B d\left(\frac{4}{3}\pi R_C{}^3\right)}{dt} = -\frac{1}{4\pi R_C{}^2}\frac{4\pi\rho_B R_C{}^2 dR_C}{dt} = bk_r{}'' C_{A,1}$$

$$-\rho_B\int_R^{R_C} dR_C = bk_r{}'' C_{A,1}\int_0^t dt$$

$$\rho_B(R - R_C) = bk_r{}'' C_{A,1} t$$

$$t = \frac{\rho_B(R - R_C)}{bk_r{}'' C_{A,1}}$$

$\tau$ is the time taken for complete conversion, which means that at $t = \tau$, $R_C = 0$.

$$\tau = \frac{\rho_B R}{bk_r{}'' C_{A,1}}$$

And the function $R_C(t)$ can be expressed in terms of the fractional time for complete conversion:

$$t = \frac{\rho_B R}{bk_r{}'' C_{A,1}}\left(1 - \frac{R_C}{R}\right)$$

$$\frac{t}{\tau} = 1 - \frac{R_C}{R}$$

The fractional conversion for the solid reactant $B$ (which forms the unreacted core at time $t$), $X$ is such that

$$1 - X = \frac{\frac{4}{3}\pi R_C{}^3}{\frac{4}{3}\pi R^3} = \left(\frac{R_C}{R}\right)^3$$

$$\frac{t}{\tau} = 1 - (1 - X)^{1/3}$$

Therefore we have derived the fractional time to complete conversion in terms of the fractional conversion of solid reactant $B$.

## Problem 17

**Consider the following experimental data obtained for a heterogeneous non-catalytic reaction occurring in spherical particles, where $\tau = 500$ s and represents the fractional time to complete conversion, $X$ represents conversion, and $R_C$ represents the radius of unreacted core. The particles do not change in size as reaction proceeds due to the deposition of solid product ash at the surface, and the particles remain at a size of radius $R = 1.2$ mm. However, it**

**is expected that the size of unreacted core shrinks with progress of reaction and hence the shrinking core analysis may be used to model this system. What is the rate-controlling step (e.g., diffusion through gas film, or reaction or diffusion through solid ash layer) for this reaction and explain.**

| *t*/s | 50 | 150 | 250 | 350 | 450 |
|---|---|---|---|---|---|
| $R_C$/m | 0.00108 | 0.00084 | 0.00060 | 0.00036 | 0.00012 |

## Solution 17

### *Worked Solution*

Based on the shrinking core model, we know that the overall reaction rate is limited by a particular reaction step if they display the characteristic correlations between fractional time to conversion $t/\tau$ and conversion $X$.

- Diffusion across gas film: $\frac{t}{\tau} = X$
- Chemical reaction: $\frac{t}{\tau} = 1 - (1 - X)^{1/3}$
- Diffusion across ash layer: $\frac{t}{\tau} = 1 - 3(1 - X)^{2/3} + 2(1 - X)$

We can use the experimental data provided to calculate values of conversion $X$, knowing that $R = 0.0012$ m and $\tau = 500$ s. Regardless of which is the rate-limiting step, the following is true:

$$1 - X = \frac{\frac{4}{3}\pi {R_C}^3}{\frac{4}{3}\pi R^3} = \left(\frac{R_C}{R}\right)^3$$

We can therefore obtain the following table of values, where we note that the experimental data fits with the correlation that shows that the overall reaction is limited by chemical reaction as shown below.

$$\text{Chemical reaction limited:} \ \frac{t}{\tau} = 1 - (1 - X)^{1/3}$$

| *t*[s] | 50 | 150 | 250 | 350 | 450 |
|---|---|---|---|---|---|
| $t/\tau$ | 0.1 | 0.3 | 0.5 | 0.7 | 0.9 |
| $R_C$[m] | 0.00108 | 0.00084 | 0.00060 | 0.00036 | 0.00012 |
| $1 - X$ | 0.729 | 0.343 | 0.125 | 0.027 | 0.001 |
| $1 - (1 - X)^{1/3}$ | 0.1 | 0.3 | 0.5 | 0.7 | 0.9 |

## Problem 18

**Consider a reaction as follows, whereby the reactant solid particle ($B$) shrinks as reaction progresses.**

$$A(g) + bB\ (s) \rightarrow cC(g) + dD(s)$$

**The particle of $B$ is also in a free fall whereby Re < 1 and inertial effects are negligible. The mass transfer coefficient for the particle is described by $k = \frac{\mathcal{D}}{R}$, where $\mathcal{D}$ is diffusivity and $R$ is the radius of the particle at time $t$. Assuming diffusion across the gas film surrounding the solid particle is the rate-limiting step,**

(a) **Derive an expression showing how particle radius $R$ changes with time. The radius of the particle is $R_1$ at time $t = 0$. The molar density of solid particle $B$ is $\rho_B$ and the bulk concentration of reactant gas $A$ is $C_{A,\text{bulk}}$.**
(b) **Express the fractional time for complete conversion as a function of conversion $X$, if $\tau$ is defined as the time taken for complete conversion.**

## Solution 18

### *Worked Solution*

(a) It follows from a mass balance on species $B$ that the rate of disappearance of solid reactant $B$ per unit surface area of particle is

$$-\frac{1}{A_s}\frac{dN_B}{dt} = -\frac{1}{4\pi R^2}\frac{dN_B}{dt}$$

Since diffusion through the gas film is rate limiting, therefore we assume that $C_A|_{r=R} = 0$, as all of the concentration gradient occurs within the gas film.

$$-\left(\frac{1}{b}\right)\frac{1}{4\pi R^2}\frac{dN_B}{dt} = k\left(C_{A,\text{bulk}} - C_A|_{r=R}\right) = kC_{A,\text{bulk}}$$

We can express $dN_B$ in terms of $dR$,

$$dN_B = \rho_B dV = \rho_B d\left(\frac{4}{3}\pi R^3\right) = 4\pi\rho_B R^2 dR$$

Substituting this result back into the earlier mass balance equation and integrating, we have

$$-\left(\frac{1}{b}\right)\frac{1}{4\pi R^2}\frac{4\pi\rho_B R^2 dR}{dt} = kC_{A,\text{bulk}} = \frac{\mathcal{D}}{R}C_{A,\text{bulk}}$$

$$-\int_{R_1}^{R} R dR = \int_0^t \frac{b\mathcal{D}C_{A,\text{bulk}}}{\rho_B} dt$$

$$\frac{{R_1}^2}{2} - \frac{R^2}{2} = \frac{b\mathcal{D}C_{A,\text{bulk}}}{\rho_B} t$$

$$t = \frac{\rho_B {R_1}^2}{2b\mathcal{D}C_{A,\text{bulk}}}\left[1 - \left(\frac{R}{R_1}\right)^2\right]$$

(b) $\tau$ is the time taken for complete conversion, which means that at $t = \tau$, $R = 0$.

$$\tau = \frac{\rho_B {R_1}^2}{2b\mathcal{D}C_{A,\text{bulk}}}$$

The fractional time for complete conversion is therefore:

$$\frac{t}{\tau} = 1 - \left(\frac{R}{R_1}\right)^2$$

The conversion $X$ is defined as follows,

$$1 - X = \frac{\frac{4}{3}\pi R^3}{\frac{4}{3}\pi {R_1}^3} = \left(\frac{R}{R_1}\right)^3$$

$$\frac{t}{\tau} = 1 - (1 - X)^{2/3}$$

## Problem 19

**In heterogeneous reactions, we may encounter adsorption systems, and one of the types is physical adsorption due to attractive Van der Waals interactions. It is common in physisorption that multilayer adsorption occurs.**

(a) **Explain how physical adsorption works using the example of nitrogen gas as adsorbate for an adsorbent surface of material $X$. State your assumptions and show how the following expression can be obtained for the $n$th layer where $k_n$ is the rate of adsorption to form the $n$th layer and $k_{-n}$ is the rate of desorption from the $n$th layer. $\theta_n$ is the fraction of total number of sites $N$ occupied by n adsorbed molecules and $P$ is the pressure of gas above the surface.**

$$\theta_n = P^n \left(\frac{k_n}{k_{-n}}\right)\left(\frac{k_{n-1}}{k_{-(n-1)}}\right)\left(\frac{k_{n-2}}{k_{-(n-2)}}\right)\cdots\left(\frac{k_1}{k_{-1}}\right)\theta_0$$

(b) **How can we determine the surface area of *X* from experimental data given the following BET model equation? *V* is the volume of nitrogen gas adsorbed at pressure *P*, $P^{sat}$ is the saturated vapor pressure of nitrogen gas, $V_m$ is the volume of a monolayer of gas molecules adsorbed onto the surface, *C* is a constant, and cross-sectional area of one nitrogen molecule $A_C$ is 0.16 nm$^2$.**

$$\frac{P}{V(P^{sat} - P)} = \frac{(C-1)P}{CV_m P^{sat}} + \frac{1}{CV_m}$$

## Solution 19

*Worked Solution*

(a) Since multilayer adsorption is possible, it means that more than one nitrogen molecule can adsorb to each site of the surface. It is worth noting the following assumptions for this analysis:

- Ideal gas behavior of nitrogen
- Monolayer adsorption is possible as each adsorbed molecule provides a site for adsorption of molecule in the layer above it
- All sites on the surface of adsorbent is identical
- There are no interactions between adsorbates (i.e., between nitrogen molecules)
- Adsorbed molecules are stationary.
- The layers of nitrogen molecules from the second layer and above are liquid like; therefore, adsorption energy of higher layers are determined by enthalpy of vaporization, $\Delta H_{vap}$.

We have below an illustration of the surface of material X, where a circle represents one nitrogen molecule. $\theta_n$ is the fraction of total number of sites $N$ ($N = 11$ in this example) occupied by $n$ adsorbed molecules. For our example, we have $\theta_0 = \frac{2}{11}, \theta_1 = \frac{2}{11}, \theta_2 = \frac{4}{11}, \theta_3 = \frac{2}{11}, \theta_4 = \frac{1}{11}$. $k_n$ denotes rate of adsorption to form the $n$th layer and $k_{-n}$ is the rate of desorption from the $n$th layer.

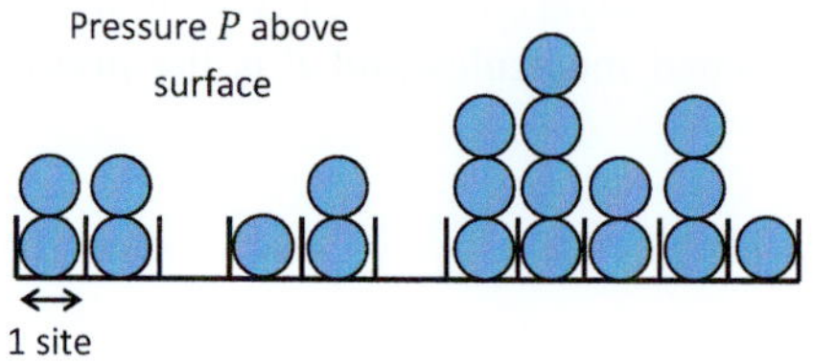

At equilibrium, $\frac{d\theta_n}{dt} = r_{\text{ads}} - r_{\text{des}} = 0$ for all $n$, which means rate of adsorption = rate of desorption.

Where $n = 0$, i.e., for surface sites with no adsorbed molecules

$$k_{-1}(\theta_1 N) - k_1(\theta_0 N)P = 0$$

$$\theta_1 = P\frac{k_1}{k_{-1}}\theta_0$$

Similarly for surface sites with 1 adsorbed molecule,

$$k_{-2}(\theta_2 N) + k_1(\theta_0 N)P - k_2(\theta_1 N)P - k_{-1}(\theta_1 N) = 0$$

$$k_{-2}(\theta_2 N) - k_2(\theta_1 N)P = 0$$

$$k_{-2}\theta_2 = k_2\theta_1 P$$

$$\theta_2 = P\frac{k_2}{k_{-2}}\theta_1 = P\frac{k_2}{k_{-2}}\left(P\frac{k_1}{k_{-1}}\theta_0\right) = P^2\left(\frac{k_2}{k_{-2}}\right)\left(\frac{k_1}{k_{-1}}\right)\theta_0$$

We can repeat the process for the further layers, and obtain the following general expression for the $n$th layer

$$\theta_n = P^n\left(\frac{k_n}{k_{-n}}\right)\left(\frac{k_{n-1}}{k_{-(n-1)}}\right)\left(\frac{k_{n-2}}{k_{-(n-2)}}\right)\cdots\left(\frac{k_1}{k_{-1}}\right)\theta_0$$

(b) We can experimentally determine values of $P$ and $V$ upon contacting the solid surface with a known volume of gas. $P$ is obtained by measuring the pressure of gas in equilibrium, and $V$ is obtained by measuring volume change on adsorption assuming ideal gas behavior. Hence, we can plot an adsorption isotherm using the BET model equation, using values of $\frac{P}{V(P^{\text{sat}} - P)}$ as the vertical axis and $\frac{P}{P^{\text{sat}}}$ as the horizontal axis.

$$\frac{P}{V(P^{\text{sat}} - P)} = \frac{C-1}{CV_m}\left(\frac{P}{P^{\text{sat}}}\right) + \frac{1}{CV_m}$$

The vertical axis intercept will give the value of $\frac{1}{CV_m}$ while the gradient will give $\frac{C-1}{CV_m}$. We can then obtain values of $V_m$ and $C$ by reading off the plot. If $N_T$ denotes the total number of nitrogen molecules forming the monolayer on the surface of $X$, and $V_{N_2}$ denotes the volume of a single nitrogen molecule, then

$$N_T = \frac{V_m}{V_{N_2}}$$

Surface area of material $X = N_T(A_C)$

## Problem 20

**Consider the heterogeneous reaction $P + Q \rightarrow R$ whereby all the molecules involved, i.e., $P$, $Q$, and $R$ adsorb onto the catalyst surface.**

(a) **Explain the key concepts and assumptions in the Langmuir-Hinshelwood model for adsorption with reaction.**
(b) **Show that the rate equation can be expressed as shown below, using the Langmuir-Hinshelwood model, where $k_r$ is the reaction rate constant and $K_i$ is the adsorption equilibrium constant for species $i$:**

$$-r_P = \frac{k_r K_P P_P K_Q P_Q}{\left(1 + K_P P_P + K_Q P_Q + K_R P_R\right)^2}$$

(c) **How would your result in part b change if $P$ is strongly adsorbed but $Q$ and $R$ are weakly adsorbed?**

## Solution 20

*Worked Solution*

(a) The Langmuir-Hinshelwood model is a further development from the Langmuir model. In chemisorptions, a chemical bond is formed between the adsorbate and adsorbent surface; therefore, the enthalpy of adsorption for chemisorption is typically higher than that for physisorption, as the latter involves weaker Van der Waals interactions between the adsorbate and adsorbent.

The Langmuir model has the following key assumptions:

- One molecule is specifically adsorbed to one site; hence, the maximum adsorption is a monolayer of molecules. [The Langmuir model is unsuitable if molecules vary significantly in size, as there may be co-adsorption of molecules of different sizes at single site.]
- The probability of adsorption is not site specific, the surface energy is uniform. [Note that in reality, most surfaces will not achieve perfectly uniform surface energy. This assumption is particularly unsuitable for larger molecules as the adsorption of a large molecule at one site will more significantly affect the adsorption probability of its adjacent sites.]
- Molecules do not interact with each other, i.e., negligible attractive or repulsive inter-molecular forces [The Langmuir model is unsuitable if molecules are charged, due to the significant intermolecular ionic forces.]

Following from the Langmuir model, the Langmuir-Hinshelwood model applies when there is a reaction occurring between adsorbed species. When multiple species are present, they compete for adsorption sites. If there is a reaction occurring between adsorbed species (e.g., $P$ and $Q$), then the adsorbed species must be sufficiently near to each other, i.e., adsorbed on adjacent sites, in order for a reaction to take place and it follows that the rate of reaction is proportional to $\theta_P \theta_Q$, whereby $\theta_P$ represents the fraction of total number of sites occupied by $P$.

(b) We have the following reaction: $P + Q \rightarrow R$

Let $\theta_P$, $\theta_Q$, and $\theta_R$ denote the fractions of total number sites occupied by $P$, $Q$, and $R$, respectively. The total number of occupied sites $\theta_T$ is such that $\theta_T = \theta_P + \theta_Q + \theta_R$

Then the reaction rate is given as shown, since species $A$ and $B$ must be adsorbed on neighboring sites in order for reaction to occur.

$$-r_P = k_r \theta_P \theta_Q$$

Separately, by definition, the adsorption equilibrium constant for $P$ is defined as:

$$K_P = \frac{k_{\text{ads}}}{k_{\text{des}}}$$

The rate of adsorption for $P$ is given by $k_{\text{ads}} P_P(1 - \theta_T)$, while the rate of desorption of $P$ is $k_{\text{des}}\theta_P$. Therefore, at equilibrium when rate of adsorption is equivalent to the rate of desorption,

$$k_{\text{ads}} P_P(1 - \theta_T) = k_{\text{des}} \theta_P$$

$$\theta_P = \frac{k_{\text{ads}}}{k_{\text{des}}} P_P(1 - \theta_T) = K_P P_P(1 - \theta_T)$$

We can repeat the same for species $Q$ and $R$ to obtain:

$$\theta_Q = K_Q P_Q(1 - \theta_T)$$

$$\theta_R = K_R P_R(1 - \theta_T)$$

$$\theta_T = \theta_P + \theta_Q + \theta_R = (K_P P_P + K_Q P_Q + K_R P_R)(1 - \theta_T)$$

$$\frac{1}{\frac{1}{\theta_T} - 1} = K_P P_P + K_Q P_Q + K_R P_R$$

$$1 - \theta_T = \frac{1}{1 + K_P P_P + K_Q P_Q + K_R P_R}$$

We can now substitute this result into the earlier rate law expression to arrive at the given expression.

$$-r_P = k_r \theta_P \theta_Q$$

$$-r_P = k_r [K_P P_P(1 - \theta_T)][K_Q P_Q(1 - \theta_T)]$$

$$-r_P = \frac{k_r K_P P_P K_Q P_Q}{(1 + K_P P_P + K_Q P_Q + K_R P_R)^2}$$

(c) The definition of the adsorption equilibrium constant $K$ is such that $K = \frac{k_{ads}}{k_{des}}$. It follows that for strong adsorption, $K$ is large and vice versa. If $P$ is strongly adsorbed, and $Q$ and $R$ are weakly adsorbed, then $K_P$ is large, while $K_Q$ and $K_R$ are small.

$$-r_P = \frac{k_r K_P P_P K_Q P_Q}{(1 + K_P P_P + K_Q P_Q + K_R P_R)^2}$$

$$-r_P \cong \frac{k_r K_P P_P K_Q P_Q}{(K_P P_P)^2} = \frac{k_r K_Q P_Q}{K_P P_P}$$

## Problem 21

**Consider a bimolecular reaction occurring at a catalyst surface between reactants $A$ and $B$, producing a desired product $C$ and a by-product $D$.**

$$A + B \rightarrow C + D$$

(a) **Discuss how we can obtain equilibrium constant for adsorption $K$, and enthalpy of adsorption $\Delta h_{ads}$ from experimental data.**
(b) **Show that the rate of reaction is given by:**

$$-r_A = \frac{k_r K_A P_A K_B P_B}{(1 + K_A P_A + K_B P_B + K_C P_C + K_D P_D)^2}$$

(c) **Based on experimental results, we find that the observed activation energy of this reaction differs from the actual activation energy for this reaction. We are also told that $D$ is more strongly adsorbed as compared to the other species. Assuming the observed activation energy is $E_{A,app}$, show that $E_{A,app}$ can be expressed in terms of enthalpies of adsorption and the true activation energy for the reaction, $E_A$ as shown below.**

$$-E_{A,\,app} = -E_A - \Delta h_{ads,\,A} - \Delta h_{ads,\,B} + 2\Delta h_{ads,\,D}$$

## Solution 21

***Worked Solution***

(a) Let us first define the equilibrium constant $K_A$ for adsorption of a single species $A$ onto a surface. Let $\theta_A$ represent the fraction of surface sites occupied by $A$, and $N_A$ represents the number of molecules of $A$. The rate of adsorption of $A$ depends on the fraction of empty sites, i.e., $(1 - \theta_A)$, as well as the partial pressure of $A$, $P_A$. It follows that the rate of adsorption is as follows:

$$\left.\frac{dN_A}{dt}\right|_{\text{ads}} = k_{\text{ads}}(1 - \theta_A)P_A$$

The rate of desorption depends on the fraction of occupied sites and therefore this rate is as follows:

$$\left.\frac{dN_A}{dt}\right|_{\text{des}} = k_{\text{des}}\theta_A$$

At steady state, we have adsorption rate equivalent to desorption rate, and therefore we have the following whereby the equilibrium constant for adsorption is defined as $K_A = \frac{k_{\text{ads}}}{k_{\text{des}}}$.

$$k_{\text{ads}}(1 - \theta_A)P_A = k_{\text{des}}\theta_A$$

$$\theta_A = \frac{k_{\text{ads}}P_A}{k_{\text{ads}}P_A + k_{\text{des}}}$$

$$= \frac{K_A P_A}{K_A P_A + 1}$$

$$\theta_A \cong K_A P_A \text{ (at low values of } P_A\text{)}$$

We are able to measure and plot values of $\theta_A$ against $P_A$ and find the value of $K_A$ from the initial gradient of the graph near to the limit where $P_A$ is small (purple line).

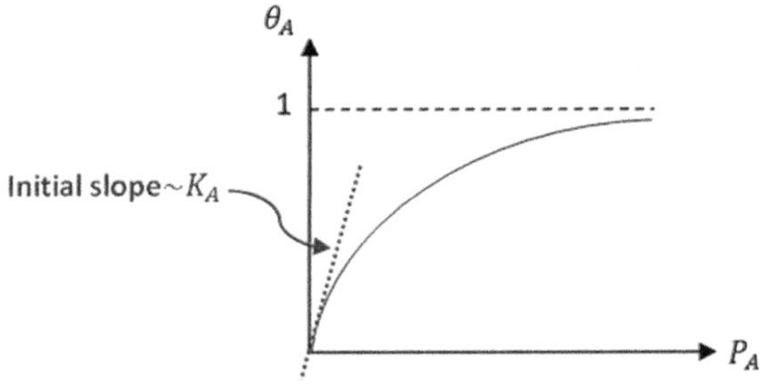


In general, we can relate equilibrium constant $K$ to thermodynamic properties, since we know that:

$$\Delta g_{\text{rxn}} = -RT\ln K$$

$$-\frac{d(\Delta g_{\text{rxn}}/RT)}{dT} = \frac{d\ln K}{dT}$$

Applying chain rule, we can expand the left-hand side, and simplify further knowing that this is true $\left(\frac{\partial \Delta g_{\text{rxn}}}{\partial T}\right)_P = -\Delta s_{\text{rxn}}$.

$$-\frac{1}{RT}\frac{d\Delta g_{\text{rxn}}}{dT}+\frac{\Delta g_{\text{rxn}}}{RT^2}=\frac{d\ln K}{dT}$$

$$\frac{\Delta s_{\text{rxn}}}{RT}+\frac{\Delta g_{\text{rxn}}}{RT^2}=\frac{d\ln K}{dT}$$

We know further that $\Delta g_{\text{rxn}} = \Delta h_{\text{rxn}} - T\Delta s_{\text{rxn}}$; therefore

$$\frac{d\ln K}{dT}=\frac{\Delta h_{\text{rxn}}}{RT^2}$$

For an adsorption process for species $A$, it follows the temperature dependence of $K_A$ can be derived. $C$ is a constant of integration.

$$\frac{d\ln K_A}{dT}=\frac{\Delta h_{\text{ads},\,A}}{RT^2}$$

$$K_A = C\exp(-\Delta h_{\text{ads},\,A}/RT)$$

From this correlation, we can then derive values of $K_A$ at different temperatures, to obtain $\Delta h_{\text{ads},A}$. The gradient of a straight line plot of $\ln K_A$ against $\frac{1}{T}$ is $-\frac{\Delta h_{\text{ads},\,A}}{R}$. $\Delta h_{\text{ads},\,A}$ can be subsequently calculated since gas constant $R$ is of known value.

$$\ln K_A = -\frac{\Delta h_{\text{ads},\,A}}{R}\left[\frac{1}{T}\right]+\ln C$$

(b) We have the following reaction: $A + B \rightarrow C + D$

Let $\theta_A$, $\theta_B$, $\theta_C$, and $\theta_D$ denote the fractions of total number sites occupied by $A$, $B$, $C$, and $D$, respectively. The total number of occupied sites $\theta_T$ is such that $\theta_T = \theta_A + \theta_B + \theta_C + \theta_D$

Then the reaction rate is given as shown, since species $A$ and $B$ must be adsorbed on neighboring sites in order for reaction to occur.

$$-r_A = k_r\theta_A\theta_B$$

And by definition, the adsorption equilibrium constant is defined as:

$$K=\frac{k_{\text{ads}}}{k_{\text{des}}}$$

The rate of adsorption for $A$ is given by $k_{\text{ads}}P_A(1 - \theta_T)$, while the rate of desorption of $A$ is $k_{\text{des}}\theta_A$. Therefore, at equilibrium when rate of adsorption is equivalent to the rate of desorption,

$$k_{\text{ads}} P_A (1 - \theta_T) = k_{\text{des}} \theta_A$$

$$\theta_A = \frac{k_{\text{ads}}}{k_{\text{des}}} P_A (1 - \theta_T) = K_A P_A (1 - \theta_T)$$

We can repeat the same for species *B*, *C*, and *D* to obtain:

$$\theta_B = K_B P_B (1 - \theta_T)$$

$$\theta_C = K_C P_C (1 - \theta_T)$$

$$\theta_D = K_D P_D (1 - \theta_T)$$

$$\theta_T = \theta_A + \theta_B + \theta_C + \theta_D = (K_A P_A + K_B P_B + K_C P_C + K_D P_D)(1 - \theta_T)$$

$$\frac{1}{\frac{1}{\theta_T} - 1} = K_A P_A + K_B P_B + K_C P_C + K_D P_D$$

$$1 - \theta_T = \frac{1}{1 + K_A P_A + K_B P_B + K_C P_C + K_D P_D}$$

We can now substitute this result into the earlier rate law expression to arrive at the given expression.

$$-r_A = k_r \theta_A \theta_B$$

$$= k_r [K_A P_A (1 - \theta_T)][K_B P_B (1 - \theta_T)]$$

$$= \frac{k_r K_A P_A K_B P_B}{(1 + K_A P_A + K_B P_B + K_C P_C + K_D P_D)^2}$$

(c) From part a, we have shown that in general for a species *i*, we have the following relationship,

$$K_i = C_i \exp(-\Delta h_{\text{ads},\, i}/RT)$$

Therefore, we can write down the equilibrium constants for the species in this problem as follows:

$$K_A = C_A \exp(-\Delta h_{\text{ads},\, A}/RT)$$

$$K_B = C_B \exp(-\Delta h_{\text{ads},\, B}/RT)$$

$$K_C = C_C \exp(-\Delta h_{\text{ads},\, C}/RT)$$

$$K_D = C_D \exp(-\Delta h_{\text{ads},\, D}/RT)$$

The reaction rate constant $k_r$ for the reaction is also temperature dependent according to the Arrhenius equation, where $E_A$ is the activation energy:

$$k_r = A\exp(-E_A/RT)$$

We are told that $D$ is strongly adsorbed but $A$, $B$, and $C$ are only weakly adsorbed. Therefore $K_D$ and $\theta_D$ are large, since $\theta_i$ is related to $K_i$ as follows.

$$\theta_i = \frac{K_i P_i}{K_i P_i + 1}$$

It follows that the rate of reaction found in part b can be approximated to

$$-r_A = \frac{k_r K_A P_A K_B P_B}{(1 + K_A P_A + K_B P_B + K_C P_C + K_D P_D)^2} \cong \frac{k_r K_A P_A K_B P_B}{K_D {P_D}^2}$$

$$A\exp(-E_{A,\text{app}}/RT) \sim \frac{A C_A C_B \exp[(-E_A - \Delta h_{\text{ads},\,A} - \Delta h_{\text{ads},\,B})/RT]}{{C_D}^2 \exp(-2\Delta h_{\text{ads},\,D}/RT)}$$

$$-E_{A,\text{app}} = -E_A - \Delta h_{\text{ads},\,A} - \Delta h_{\text{ads},\,B} + 2\Delta h_{\text{ads},\,D}$$

## Problem 22

(a) **Outline the main principles of the Langmuir-Hinshelwood mechanism which is used to model adsorption of reactant species onto a catalytic surface in a heterogeneous reaction.**

(b) **Explain how the Langmuir-Hinshelwood model differs from the Eley–Rideal model.**

## Solution 22

### *Worked Solution*

(a) The Langmuir-Hinshelwood mechanism is a progression from the Langmuir model; hence, it follows from the assumptions and principles underlying the Langmuir model.

The Langmuir model describes adsorption whereby the adsorbed species are held to specific active sites on the surface of a catalytic solid. The surface sites are equivalent, due to uniform surface energy across all sites and hence an equal chance of adsorption between them. The maximum number of adsorbed species occurs when a monolayer of molecules is formed, with one molecule taking up one site (no sharing of sites, or single molecule occupying more than one site). It is also assumed in the Langmuir model that adsorbed species do not interact with each other, i.e., negligible attractive or repulsive forces.

The Langmuir-Hinshelwood model further stipulates that for two or more types of species, they will compete for adsorption at a fixed number of sites. Also, chemical reaction rate is rate controlling, i.e., adsorption and desorption are relatively much faster than reaction rate.

Some reactions that fit well with the Langmuir-Hinshelwood model are listed below, note that the species involved are typically small and uncharged molecules. Processes such as the catalytic cracking of long chain hydrocarbons, or other reactions involving long chain alkanes and/or alkenes, are unlikely to be well represented by the Langmuir-Hinshelwood model as the hydrocarbon chain molecules involved are much bigger in size:

- $N_2 + 3H_2 \leftrightarrow 2NH_3$
- $CO + 2H_2 \leftrightarrow CH_3OH$

Let us consider the following illustration for an example reaction, $A + B \rightarrow$ product $P$, where the reaction rate is first order with respect to $A$ and first order with respect to $B$ (overall order 2). The adsorption process is fast, and each species occupies one site. Similarly, desorption is assumed fast. Note that the activated complex formed in the intermediate stage is a transition state complex that cannot be isolated.

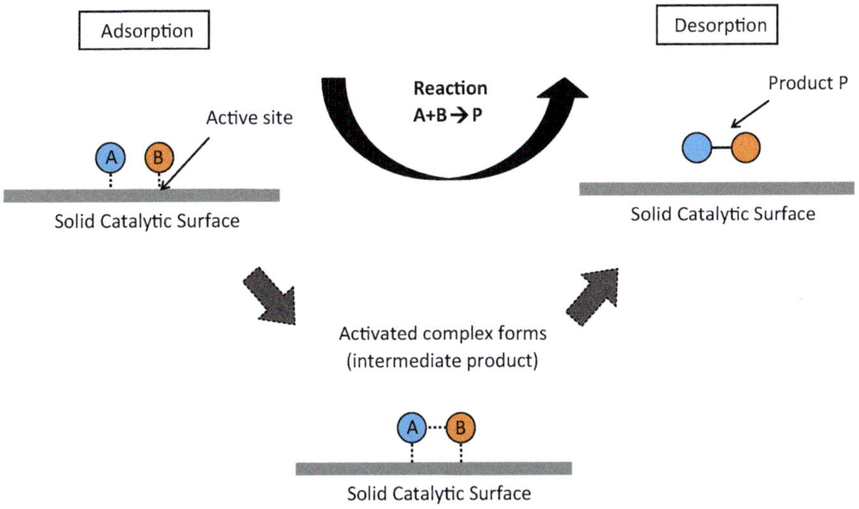


Let the fraction of total sites which is occupied by species $A$ be defined as $\theta_A$. Then the overall rate of reaction, which is also the chemical reaction rate (rate-limiting step) is expressed as follows where $k$ is the reaction rate constant:

$$r = k\theta_A\theta_B$$

(b) In the Eley–Rideal model, one of the reactants does not adsorb prior to reaction (e.g., $A$). Instead it reacts directly in its gas phase with the adsorbed reactant (e.g., $B$). Alternatively, it can adsorb but it becomes inactive and does not participate in the reaction. As such the reaction rate is as shown below:

$$r = k\theta_B P_A$$

Species $B$ is adsorbed, while species $A$ reacts directly with $B$ in its gas phase.

## Problem 23

**Consider the reaction $A + B \rightarrow P$ whereby all the species $A$, $B$, and $P$ are adsorbed on a catalytic surface. Derive expressions for the rate of reaction for the cases below, assuming the reaction is first order with respect to $A$ and $B$:**

(a) **All three species compete for the same sites.**
(b) **$A$ and $B$ compete for different sites, while $P$ adsorbs onto both types of sites.**

## Solution 23

***Worked Solution***

(a) For both cases, the rate law is the same as follows where $k$ is the reaction rate constant.

$$r = k\theta_A\theta_B$$

What differs when the adsorbed sites are different is the expression for fraction of occupied sites. The occupied fractions of each species and fraction of empty sites always sum to one, for scenario in (a) whereby all $A$, $B$, and $P$ compete for the same sites, we have the following:

$$1 = \theta_A + \theta_B + \theta_P + \theta_{\text{empty}}$$

$$\theta_{\text{empty}} = 1 - \sum_i \theta_i$$

We assume that adsorption and desorption are fast, i.e., rate-limiting step is the chemical reaction. Therefore adsorption equilibrium is reached for each adsorbed species and is expressed as follows for species $A$ (same equation applies for other adsorbed species $B$ and $P$), whereby $k_{\text{ads}}$ and $k_{\text{des}}$ are the adsorption and desorption rate constants, respectively, $P_A$ is the partial pressure of $A$, and $K_A$ is the adsorption equilibrium constant for $A$:

$$k_{\text{ads}}\theta_{\text{empty}}P_A = k_{\text{des}}\theta_A$$

$$K_A = \frac{k_{\text{ads}}}{k_{\text{des}}}$$

Substituting $K_A$ into the equation, we have

$$K_A k_{\text{des}} \theta_{\text{empty}} P_A = k_{\text{des}} \theta_A$$

$$\theta_A = K_A P_A \left(1 - \sum_i \theta_i\right)$$

Similarly equation applies for species $B$ and $P$:

$$\theta_B = K_B P_B \left(1 - \sum_i \theta_i\right)$$

$$\theta_P = K_P P_P \left(1 - \sum_i \theta_i\right)$$

We can express fraction of occupied sites as:

$$\sum_i \theta_i = \theta_A + \theta_B + \theta_P = \left(1 - \sum_i \theta_i\right)(K_A P_A + K_B P_B + K_P P_P)$$

$$\sum_i \theta_i = (K_A P_A + K_B P_B + K_P P_P) - \left(\sum_i \theta_i\right)(K_A P_A + K_B P_B + K_P P_P)$$

$$\left(\sum_i \theta_i\right)[1 + (K_A P_A + K_B P_B + K_P P_P)] = K_A P_A + K_B P_B + K_P P_P$$

$$\sum_i \theta_i = \frac{K_A P_A + K_B P_B + K_P P_P}{1 + K_A P_A + K_B P_B + K_P P_P}$$

$$\left(1 - \sum_i \theta_i\right) = \frac{1}{1 + K_A P_A + K_B P_B + K_P P_P}$$

The rate equation can be re-expressed as follows:

$$r = k\theta_A \theta_B = kK_A P_A \left(1 - \sum_i \theta_i\right) K_B P_B \left(1 - \sum_i \theta_i\right)$$

$$= \frac{kK_A P_A K_B P_B}{(1 + K_A P_A + K_B P_B + K_P P_P)^2}$$

(b) For this second scenario whereby $A$ and $B$ compete for different sites, while $P$ adsorbs onto both types of sites, we have two types of sites (defined here as type 1 and 2) and the following two equations:

$$1 = \theta_A + \theta_P + \theta_{\text{empty,type 1}}$$

$$1 = \theta_B + \theta_P + \theta_{\text{empty,type 2}}$$

$$\text{For type 1 site:}\ \left(1 - \sum_i \theta_i\right) = \frac{1}{1 + K_A P_A + K_P P_P}$$

$$\text{For type 2 site:}\ \left(1 - \sum_i \theta_i\right) = \frac{1}{1 + K_B P_B + K_P P_P}$$

The rate equation can be expressed as follows:

$$r = k\theta_A\theta_B = kK_A P_A \left(1 - \sum_i \theta_i\right)_{\text{type 1}} K_B P_B \left(1 - \sum_i \theta_i\right)_{\text{type 2}}$$

$$= \frac{kK_A P_A K_B P_B}{(1 + K_A P_A + K_P P_P)(1 + K_B P_B + K_P P_P)}$$

## Problem 24

**Consider the reaction $A + B \rightarrow P$ whereby all the species $A$, $B$, and $P$ are adsorbed on a catalytic surface and compete for the same sites. The general rate equation for this scenario is provided as follows, where $k$ is the reaction rate constant, $K_i$ is the adsorption equilibrium constant for species $i$, and $P_i$ is the partial pressure of species $i$:**

$$r = \frac{kK_A P_A K_B P_B}{(1 + K_A P_A + K_B P_B + K_P P_P)^2}$$

**Two different types of catalysts were tested using this reaction, and experimental data showed that for catalyst 1, the reaction was first order with respect to $B$, while for catalyst 2, the reaction was first order with respect to $A$ and negative order with respect to $B$. Explain how this is possible and the relative adsorption strength of species $B$ to the two types of catalysts.**

## Solution 24

### *Worked Solution*

For catalyst 1, the reaction rate equation can be expressed as follows based on observed kinetics:

$$r = k' P_B$$

This can be called a pseudo-first-order reaction whereby the reaction rate can be approximated to depend only on concentration (or partial pressure for gaseous species) of $B$, if $P_A$ and $P_P$ remain relatively constant compared to $P_B$, for example, if $A$ and $P$ are present in large excess and $P_B$ is small. Therefore the rest of the terms in the general rate equation provided in the question are absorbed into the pseudo-first-order rate constant, $k'$ as shown here.

For $K_B P_B \ll 1$,

$$r = \frac{kK_A P_A K_B P_B}{(1 + K_A P_A + K_B P_B + K_P P_P)^2} \approx \frac{kK_A P_A K_B P_B}{(1 + K_A P_A + K_P P_P)^2}$$

$$r = k' P_B, \text{where } k' = \frac{kK_A P_A K_B}{(1 + K_A P_A + K_P P_P)^2}$$

As for catalyst 2, the reaction rate can be expressed as follows based on the observed kinetics:

$$r = \frac{k' P_A}{P_B}$$

In this case, we adjust the terms in the general rate equation that should be absorbed into the pseudo-first-order rate constant, $k'$, in order to fit with the observed data.

For $K_A P_A$, $K_P P_P \ll 1$, and $K_B P_B \gg 1$,

$$r = \frac{kK_A P_A K_B P_B}{(1 + K_A P_A + K_B P_B + K_P P_P)^2} \approx \frac{kK_A P_A K_B P_B}{(K_B P_B)^2} = \frac{kK_A P_A}{K_B P_B}$$

$$r = \frac{k' P_A}{P_B}, \text{where } k' = \frac{kK_A}{K_B}$$

We can deduce that the reactant $B$ is much more strongly adsorbed onto catalyst 2 than onto catalyst 1.

## Problem 25

(a) **For non-porous solids with reactions occurring at the solid particle surface, reaction rate may be affected by the rate of external mass transport of the reactant from bulk fluid to the particle surface. Discuss the meaning of effectiveness factor, $\eta$, a dimensionless quantity that measures the deviation of observed reaction kinetics from ideal kinetics. Show that $\eta$ can be expressed as shown below, whereby $C_A{}^0$ is the concentration of reactant $A$ in bulk fluid, $k_{r,s}$ is the reaction rate constant at the surface of the particle, and $k$ is the external mass transfer coefficient. You may assume first-order reaction kinetics.**

$$\eta = \frac{1}{1 + \frac{k_{r,s} C_A{}^0}{k C_A{}^0}}$$

(b) **Explain the meaning of the Damköhler number $Da$ in terms of the implications of its value, and how it is related to effectiveness factor, $\eta$.**

## Solution 25

### *Worked Solution*

(a) For non-porous solids with surface reactions, the observed rate of reaction, $-r_A$, depends on rate of mass transport of $A$ from bulk fluid to the particle surface, $r_{\mathrm{mt}}$, and rate of reaction, $r_{\mathrm{rxn}}$, at the particle surface. Under steady-state conditions, where $C_{A,s}$ is the concentration of $A$ at the particle surface, we have the following

$$-r_A = r_{\mathrm{mt}} = r_{\mathrm{rxn}}$$

$$-r_A = k\left(C_A{}^0 - C_{A,s}\right) = k_{r,s} C_{A,s}$$

$$C_{A,s} = \frac{1}{1 + \frac{k_{r,s}}{k}} C_A{}^0 \qquad (1)$$

Consider the extreme (ideal case) where mass transfer rate was not at all limiting, then $C_{A,s} = C_A{}^0$, and the observed rate of reaction will be the maximum possible in the ideal case,

$$-r_{A,\mathrm{ideal}} = k_{r,s} C_A{}^0$$

However, when mass transfer effects are felt, then $C_{A,s} \neq C_A{}^0$, and the rate of reaction will be $k_{r,s} C_{A,s}$, whereby $C_{A,s} < C_A{}^0$ with the exact correlation as found earlier in (1)

The effectiveness factor $\eta$ is a dimensionless quantity whereby value of 1 means reaction rate is ideal (highest possible) at the extreme when mass transfer has no effect. $\eta$ measures the ratio between the actual observed rate of reaction to the ideal rate of reaction. As such, this ratio can be expressed as follows:

$$\eta = \frac{k_{r,s}C_{A,s}}{k_{r,s}{C_A}^0} = \frac{1}{1 + \frac{k_{r,s}}{k}} = \frac{1}{1 + \frac{k_{r,s}{C_A}^0}{k{C_A}^0}}$$

(b) Following from part a, we note that since the mass transport of reactant from bulk to surface, and the reaction that occurs at the particle surface are processes that happen "in series", the slower of the two processes will be the rate-determining (or rate-limiting or rate-controlling) step.

We introduce another dimensionless quantity known as the Damköhler number or *Da* which measures the ratio of the ideal rate of reaction where mass transfer effects are absent to the maximum rate of mass transfer to the particle surface. This ratio is therefore a direct comparison between the influence of reaction and mass transfer.

Da is defined as follows in general for a reaction of order n with respect to reactant $A$:

$$\text{Da} = \frac{k_{r,s}{C_A}^{0n}}{k{C_A}^0}$$

For first-order kinetics, where $n = 1$,

$$\text{Da} = \frac{k_{r,s}{C_A}^0}{k{C_A}^0} = \frac{k_{r,s}}{k}$$

$$\eta = \frac{1}{1 + \frac{k_{r,s}}{k}} = \frac{1}{1 + \text{Da}}$$

For cases where Da is small, i.e., $\ll 1$, then $k \gg k_{r,s}$, the rate of mass transfer is much larger than the rate of reaction; therefore, mass transfer has little effect on the overall observed rate of reaction, and the overall rate will be largely influenced by the slower rate of reaction (i.e., controlled by intrinsic chemical reaction rate). As $Da \to 0$, $\eta \to 1$.

Conversely, when *Da* is large, mass transfer is dominating, hence the overall rate of reaction is controlled by the rate of transport of reactant from bulk fluid to the particle surface. When $Da \gg 1$, then $k \ll k_{r,s}$ and $\eta \to \frac{1}{Da}$. The rate of mass transfer is much smaller than the rate of reaction; therefore, rate of reaction has little effect on the overall observed rate of reaction, and the overall rate will be largely influenced by the slower mass transfer rate. The observed rate of reaction will approximate to $-r_A = k{C_A}^0$.

## Problem 26

**Consider a slab of porous material whereby a reaction occurs under isothermal conditions within the entire porous structure of the slab. The reaction $A \rightarrow B$ is first order with respect to $A$. You may assume that the active reaction sites are distributed uniformly throughout the interior of the slab and the slab has a thickness of $2H$.**

(a) **Explain the effects of intra-particle mass transfer on the overall reaction rate, and explain how the effectiveness factor $\eta$ can be determined, where**

$$\eta = \frac{\text{Observed rate}}{\text{Ideal rate}}$$

(b) **Show the derivation of the following differential equation which can be used to find the concentration profile of reactant $A$ within the slab. $\mathcal{D}$ is the effective diffusivity which is assumed constant, $C_A$ is the concentration of reactant $A$, $k_v$ is the reaction rate constant per unit volume, and $x$ is the distance measured from the axial center line of the slab.**

$$\mathcal{D}\frac{d^2C_A}{dx^2} = k_v C_A$$

(c) **Using an appropriate substitution of variables, show that the differential equation can be simplified to the following whereby $C_s$ is the concentration at the surface of the slab and $\varnothing$ is the Thiele modulus where $\varnothing = H\sqrt{\frac{k_v}{\mathcal{D}}}$.**

$$C_A = C_{A,s}\frac{\cosh\left(\varnothing \frac{x}{H}\right)}{\cosh\varnothing}$$

(d) **Find the effectiveness factor $\eta$ for this solid slab.**

## Solution 26

*Worked Solution*

(a) When reaction occurs within the entire internal structure of a solid particle, there must exist a concentration gradient within the particle material which serves as the driving force for reactant to diffuse from the particle surface inwards, into the interior of the particle. Fick's law tells us that mass flux $j''(x)$ is correlated to concentration gradient via diffusivity, $\mathcal{D}$ as follows

$$j''(x) = \mathcal{D}\frac{dC_A}{dx}$$

Due to the concentration gradient within the particle, $C_A$ must therefore be lower within the particle, than at the surface. Assuming first-order kinetics, this leads to the fact that the rate of reaction will be lower in the interior than at the surface. The effectiveness factor $\eta$ is defined as follows

$$\eta = \frac{\text{Observed rate}}{\text{Ideal rate}}$$

To find $\eta$, we need to first determine the total rate of reaction (i.e., the observed rate) within the particle. Since reaction rate varies with position within the particle, we will need to then first derive an expression for the concentration profile and hence reaction rate profile relative to position in the particle. This expression can be integrated over the entire particle to obtain the total rate of reaction as required.

As for the ideal rate of reaction, it simply assumes that throughout the particle, the concentration is constant and equivalent to the concentration at the surface, which means the extreme case where internal mass transfer effects are absent.

(b) Let us first illustrate the system,

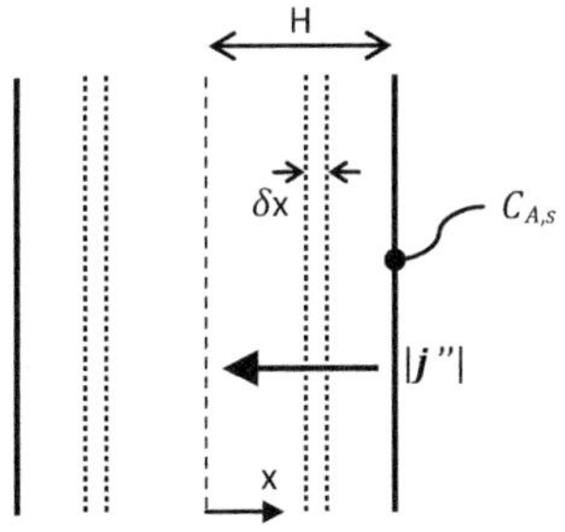


The inward diffusive flux of reactant $A$ at positions $x$ and $x + \delta x$ is given below, where $A$ is the surface area of the slab,

$$\text{Flux into slice } (-x \text{ direction}) = -\left[A\left(-\mathcal{D}\frac{dC_A}{dx}\right)_{x+\delta x}\right]$$

$$\text{Flux out of slice } (-x \text{ direction}) = -\left[A\left(-\mathcal{D}\frac{dC_A}{dx}\right)_{x}\right]$$

Other than diffusive flux, there is also reaction occurring which serves to reduce $N_A$, assuming first-order kinetics, we have the following expression for reaction rate

$$\frac{dN_A}{dt} = r_A V = -k_v C_A V = -k_v C_A A \delta x$$

By mass balance over the differential element of width $\delta x$ at steady state, we have

$$-\left[A\left(-\mathcal{D}\frac{dC_A}{dx}\right)_{x+\delta x}\right]-\left\{-\left[A\left(-\mathcal{D}\frac{dC_A}{dx}\right)_x\right]\right\}-k_v C_A A\delta x=0$$

$$\frac{\left(\frac{dC_A}{dx}\right)_{x+\delta x}-\left(\frac{dC_A}{dx}\right)_x}{\delta x}-\frac{k_v}{\mathcal{D}}C_A=0$$

For $\delta x \rightarrow 0$,

$$\frac{d^2C_A}{dx^2}-\frac{k_v}{\mathcal{D}}C_A=0$$

(c) Let us consider the boundary conditions for this second-order differential equation. When $x=0$ at the centerline of the slab, $\frac{dC_A}{dx}=0$, while at $x=H$, $C_A=C_{A,s}$.
A useful way to simplify the equation and its boundary conditions is to introduce dimensionless variables. Let $C_A{}^*=C_A/C_{A,s}$ and $x^*=x/H$. Hence we can re-express the earlier differential equation in terms of the new variables.

$$\frac{d^2C_A}{dx^2}-\frac{k_v}{\mathcal{D}}C_A=0$$

$$dx^2=H^2dx^{*2};\ d^2C_A=C_{A,s}d^2C_A{}^*$$

$$\frac{\mathcal{D}C_{A,s}}{H^2}\frac{d^2C_A{}^*}{dx^{*2}}-k_vC_{A,s}\,C_A{}^*=0$$

$$\frac{d^2C_A{}^*}{dx^{*2}}-H^2\frac{k_v}{\mathcal{D}}C_A{}^*=0$$

It is given that $\emptyset=H\sqrt{\frac{k_v}{\mathcal{D}}}$; therefore,

$$\frac{d^2C_A{}^*}{dx^{*2}}-\emptyset^2\,C_A{}^*=0$$

We can try a trial solution of exponential form based on our knowledge that derivatives of the exponential function can give the same exponential function, multiplied by a constant. Therefore, we can try a solution of the form $C_A{}^*=c_1\exp(\emptyset x^*)+c_2\exp(-\emptyset x^*)$, where $c_1$ and $c_2$ are constants to be determined by applying boundary conditions. We know that when $x^*=0$, $\frac{dC_A{}^*}{dx^*}=0$, and when $x^*=1$, $C_A{}^*=1$.

[Note that another possible trial solution is $C_A{}^* = c_1 \cosh\ \text{Ø}x^* + c_2 \sinh\ \text{Ø}x^*$.]

$$\frac{dC_A{}^*}{dx^*} = c_1\text{Ø}\exp(\text{Ø}x^*) - c_2\text{Ø}\exp(-\text{Ø}x^*)$$

$$\left.\frac{dC_A{}^*}{dx^*}\right|_{x^*=0} = c_1\text{Ø} - c_2\text{Ø} = 0 \rightarrow c_1 = c_2$$

$$1 = c_1 \exp\text{Ø} + c_2 \exp(-\text{Ø}) = c_1\left[\mathrm{e}^{\text{Ø}} + \mathrm{e}^{-\text{Ø}}\right]$$

$$c_1 = \frac{1}{\mathrm{e}^{\text{Ø}} + \mathrm{e}^{-\text{Ø}}}$$

Therefore, we obtain the expression,

$$C_A{}^* = \frac{1}{\mathrm{e}^{\text{Ø}} + \mathrm{e}^{-\text{Ø}}}\mathrm{e}^{\text{Ø}x^*} + \frac{1}{\mathrm{e}^{\text{Ø}} + \mathrm{e}^{-\text{Ø}}}\mathrm{e}^{-\text{Ø}x^*}$$

$$C_A{}^* = \frac{1}{2\cosh\text{Ø}}\left[\mathrm{e}^{\text{Ø}x^*} + \mathrm{e}^{-\text{Ø}x^*}\right] = \frac{1}{2\cosh\text{Ø}}[2\cosh(\text{Ø}x^*)]$$

$$C_A{}^* = \frac{\cosh(\text{Ø}x^*)}{\cosh\text{Ø}}$$

$$C_A = C_{A,s}\frac{\cosh\left(\text{Ø}\frac{x}{H}\right)}{\cosh\text{Ø}}$$

(d) To find the effectiveness factor $\eta$, we start with its definition.

$$\eta = \frac{\text{Observed rate}}{\text{Ideal rate}}$$

The observed rate of reaction at steady state will be equivalent to the rate at which $A$ diffuses into the slab at both of its faces (i.e., total area of $2A$ comprising $A$ at $x = H$ and $A$ at $x = -H$). Therefore,

$$\text{Observed rate} = 2A\left(-\mathcal{D}\left.\frac{dC_A}{dx}\right|_{x=H}\right) = -2A\mathcal{D}\left[\frac{C_{A,\,s}\text{Ø}}{H\cosh\text{Ø}}\sinh\left(\text{Ø}\frac{x}{H}\right)\right]_{x=H}$$

$$= -\frac{2A\mathcal{D}\text{Ø}C_{A,\,s}}{H}\tanh\text{Ø}$$

We know that the ideal (or maximum possible) rate is when mass transfer is perfect, and therefore reaction rate operates only at the maximum concentration of reactant throughout the slab, and this maximum concentration is the concentration at the slab surface, $C_{A,s}$

$$\text{Ideal rate} = -k_v V C_{A,s} = -k_v 2AHC_{A,s}$$

Therefore substituting the expressions, and knowing $\varnothing = H\sqrt{\frac{k_v}{\mathcal{D}}}$, we obtain the effectiveness factor as follows:

$$\eta = \frac{-\frac{2A\mathcal{D}\varnothing C_{A,s}}{H}\tanh\varnothing}{-k_v 2AHC_{A,s}}$$

$$\eta = \frac{\tanh\varnothing}{\varnothing}$$

## Problem 27

**Consider a reaction that occurs within a porous rectangular slab of thickness $x = 2L$. The active reaction sites are uniformly distributed within the slab, and the reaction $A \rightarrow B$ is zero order. Assuming a Thiele modulus of the form as shown below, where $\mathcal{D}$ is the effective diffusivity, $C_{A,s}$ is the concentration of $A$ at the surface of the slab, and $k_v$ is the reaction rate constant per unit volume, find the internal effectiveness factor $\eta$ and comment on the result.**

$$\varnothing^2 = \frac{k_v L^2}{\mathcal{D} C_{A,s}}$$

## Solution 27

### *Worked Solution*

We have the following system:

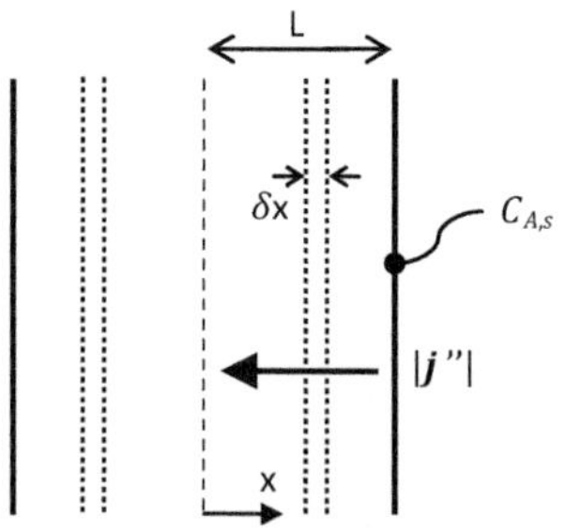


The boundary conditions are when $x = 0$ at the centerline of the slab, $\frac{dC_A}{dx} = 0$, while at $x = L$, $C_A = C_{A,s}$.

By mass balance over the differential element of width $\delta x$ at steady state, we have:

$$\text{Flux into shell } (-x \text{ direction}) = -\left[A\left(-\mathcal{D}\frac{dC_A}{dx}\right)_{x+\delta x}\right]$$

$$\text{Flux out of shell } (-x \text{ direction}) = -\left[A\left(-\mathcal{D}\frac{dC_A}{dx}\right)_{x}\right]$$

$$\left\{-\left[A\left(-\mathcal{D}\frac{dC_A}{dx}\right)_{x+\delta x}\right]\right\} - \left\{-\left[A\left(-\mathcal{D}\frac{dC_A}{dx}\right)_{x}\right]\right\} - k_v A\delta x = 0$$

$$\frac{\left(\mathcal{D}\frac{dC_A}{dx}\right)_{x+\delta x} - \left(\mathcal{D}\frac{dC_A}{dx}\right)_{x}}{\delta x} - k_v = 0$$

Where $\delta x \to 0$,

$$\frac{d}{dx}\left[\mathcal{D}\frac{dC_A}{dx}\right] - k_v = 0$$

$$\frac{d^2 C_A}{dx^2} - \frac{k_v}{\mathcal{D}} = 0$$

Let $C_A{}^* = C_A/C_{A,s}$ and $x^* = x/L$. And knowing that $\text{Ø}^2 = \frac{k_v L^2}{\mathcal{D}C_{A,s}}$, we can re-express the differential equation in terms of the new dimensionless variables and Ø.

$$\frac{C_{A,s}}{L^2}\frac{d^2 C_A{}^*}{dx^{*2}} = \frac{k_v}{\mathcal{D}}$$

$$\frac{d^2 C_A{}^*}{dx^{*2}} = \text{Ø}^2$$

The boundary conditions in terms of the new variables are when $x^* = 0, \frac{dC_A{}^*}{dx^*} = 0$, and when $x^* = 1$, $C_A{}^* = 1$. We can integrate the differential equation and determine the constants of integration.

$$\frac{dC_A{}^*}{dx^*} = \text{Ø}^2 x^* + c_1 \to c_1 = 0$$

$$C_A{}^* = \frac{\text{Ø}^2 x^{*2}}{2} + c_2 \to c_2 = 1 - \frac{\text{Ø}^2}{2}$$

$$C_A{}^* = \frac{\text{Ø}^2 x^{*2}}{2} + 1 - \frac{\text{Ø}^2}{2} = 1 + \frac{\text{Ø}^2}{2}\left[x^{*2} - 1\right]$$

$$\frac{C_A}{C_{A,s}} = 1 + \frac{\text{Ø}^2}{2}\left[\frac{x^2}{L^2} - 1\right]$$

Note that this solution is only valid where $\frac{\text{Ø}^2}{2} \leq 1$ since $C_A{}^*$ cannot be negative. For a zero-order reaction, reaction rate is a constant.

Effectiveness factor $\eta$, is defined as

$$\eta = \frac{\text{Observed rate}}{\text{Ideal rate}}$$

The observed rate of reaction at steady state will be equivalent to the rate at which $A$ diffuses into the slab at both of its faces (i.e., total area of $2A$ comprising $A$ at $x = L$ and $A$ at $x = -L$). Therefore,

$$\text{Observed rate} = 2A\left(-\mathcal{D}\frac{dC_A}{dx}\bigg|_{x=L}\right) = -2A\mathcal{D}\frac{C_{A,s}\text{Ø}^2}{L}$$

We know that for a zero-order reaction, the reaction rate is a constant equivalent to the reaction rate constant, and is therefore independent of concentration of $A$, hence

$$\text{Ideal rate} = -k_v V = -k2AL$$

Therefore substituting the expressions, and knowing $\text{Ø}^2 = \frac{kL^2}{\mathcal{D}C_{A,s}}$, we obtain the effectiveness factor as follows:

$$\eta = \frac{-2A\mathcal{D}\frac{C_{A,s}\text{Ø}^2}{L}}{-k_v 2AL} = 1$$

This means that everywhere in the slab material, there is a uniform rate of reaction at the highest possible, which is the same as the rate of reaction at the surface of the slab where $C_A = C_{A,s}$.

## Problem 28

**The graph below is a plot of effectiveness factor $\eta$ against the Thiele modulus Ø using the log scale, for a first-order reaction within a porous catalytic slab with internal mass transfer effects.**

**For a solid slab, the internal effectiveness factor is given by**

$$\eta = \frac{\tanh\text{Ø}}{\text{Ø}}$$

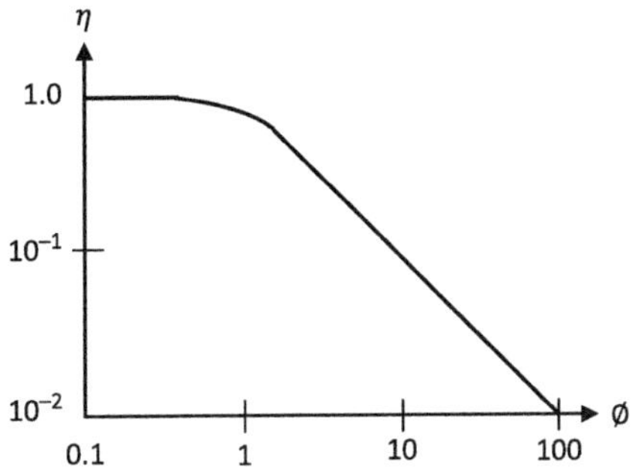


(a) **Explain the behavior of the plot, at extreme values of $\emptyset$.**

(b) **Explain the first generalization of the Thiele modulus in terms of how it relates to different particle geometries. Does the above $\eta$-$\emptyset$ plot that is drawn for a slab of half width $L$ (measured from centerline to edge) also apply for different particle geometries?**

(c) **For a general reaction of order $n$, where $n \geq 0$, it can be shown that by including a factor $(n+1)/2$ to the Thiele modulus $\emptyset$, we obtain the same trend as shown in the plot above at large values of $\emptyset$. Show that for an $n$th order reaction, $\emptyset^2 = L^2\left[\frac{(n+1)k_v C_{A,s}{}^{n-1}}{2\mathcal{D}}\right]$. $k_v$ is the reaction rate constant per unit volume, $\mathcal{D}$ is the effective diffusivity, and $C_{A,s}$ is the concentration of reactant $A$ at the surface of the slab.**

(d) **Using information provided in the plot, explain the general method to obtain the reduction factor for the mean particle size if we want to increase the effectiveness of the catalyst from a value of $\eta = 0.25$ to 0.95? [Actual calculations not required]**

## Solution 28

### *Worked Solution*

(a) From the equation provided for effectiveness factor, we can observe that at large values of Ø, tanh Ø ~1 and $\eta \sim \frac{1}{\emptyset}$. Therefore, the plot becomes a straight line with constant gradient of value −1, and this region is dominated by mass transfer effects. While at small values of Ø, it is reaction limited, and $\eta$~1.

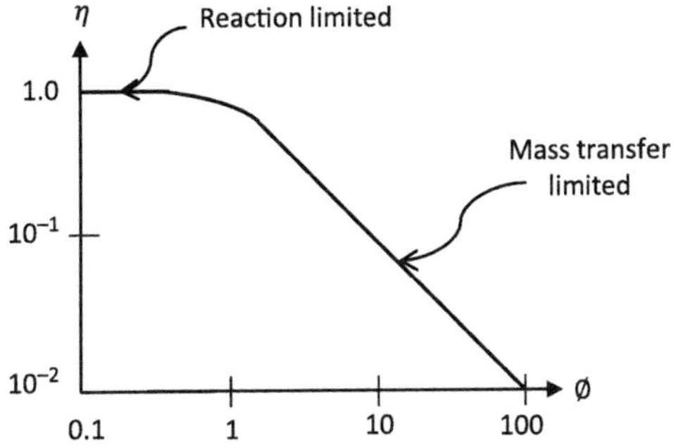

(b) The first generalization of the Thiele modulus $\varnothing_1$ is defined such that $\varnothing_1 = L_{\mathrm{char}}\sqrt{\frac{k_v}{\mathcal{D}}}$, where $L_{\mathrm{char}}$ is the characteristic length for the particle geometry. In general, $L_{\mathrm{char}} = \frac{V}{A}$, where $V$ is the volume of the particle and $A$ is the external surface area of the particle (i.e., the area exposed to mass transfer effects).

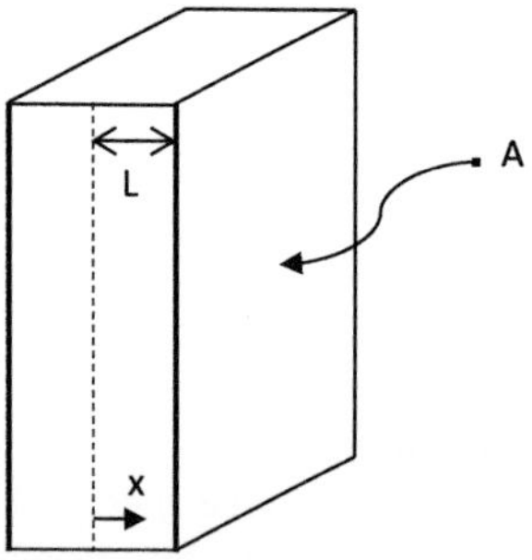


In the case of a rectangular slab:

$$L_{char} = \frac{2LA}{2A} = L$$

$$\varnothing_1 = \varnothing_{slab} = L\sqrt{\frac{k_v}{\mathcal{D}}}$$

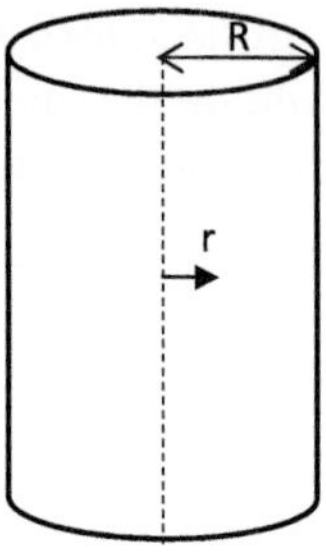


In the case of a cylindrical particle:

$$L_{char} = \frac{\pi R^2 L}{2\pi R L} = \frac{R}{2}$$

$$\varnothing_1 = \frac{R}{2}\sqrt{\frac{k_v}{\mathcal{D}}}, whereby\ \varnothing_{cyl} = R\sqrt{\frac{k_v}{\mathcal{D}}}$$

$$\varnothing_1 = \frac{\varnothing_{cyl}}{2}$$

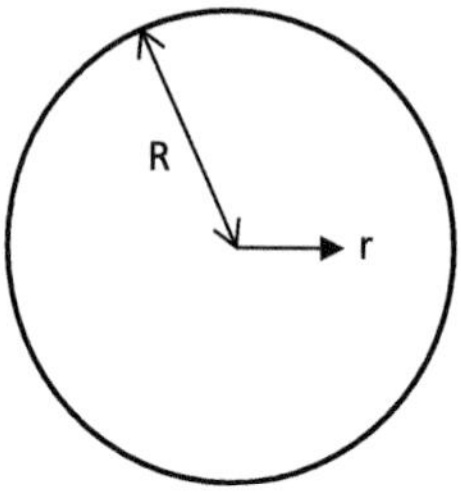


In the case of a spherical particle:

$$L_{char} = \frac{\frac{4}{3}\pi R^3}{4\pi R^2} = \frac{R}{3}$$

$$\varnothing_1 = \frac{R}{3}\sqrt{\frac{k_v}{\mathcal{D}}}, whereby\ \varnothing_{sph} = R\sqrt{\frac{k_v}{\mathcal{D}}}$$

$$\varnothing_1 = \frac{\varnothing_{sph}}{3}$$

The definition of the first Thiele modulus applies for any geometry, and the $\eta$-$\varnothing$ plot does not vary significantly with particle geometry. Regardless of geometry, the plot in part a follows the same general behavior that at small values of $\varnothing_1$, the effectiveness factor $\eta$ tends to 1. And at large values of $\varnothing_1$, $\eta$ tends to $\frac{1}{\varnothing_1}$, and for different geometries, we have $\eta \rightarrow \frac{1}{\varnothing_{\mathrm{slab}}}$ or $\eta \rightarrow \frac{2}{\varnothing_{\mathrm{cyl}}}$ or $\eta \rightarrow \frac{3}{\varnothing_{\mathrm{sph}}}$ depending on which shape we have.

(c) For a reaction of order $n$, the mass balance equation will result in the following differential equation.

$$\mathcal{D}\frac{d^2C_A}{dx^2} = k_vC_A{}^n$$

We can non-dimensionalize the variables by substituting $C_A$ and $x$ using $C_A{}^* = C_A/C_{A,s}$ and $x^* = x/L$, where $C_{A,s}$ is the concentration at the slab surface at $x = L$.

$$\frac{C_{A,s}}{L^2}\frac{d^2C_A{}^*}{dx^{*2}} = \frac{k_v}{\mathcal{D}}C_A{}^{*n}C_{A,s}{}^n$$

$$\frac{d^2C_A{}^*}{dx^{*2}} = \left[\frac{L^2k_vC_{A,s}{}^{n-1}}{\mathcal{D}}\right]C_A{}^{*n}$$

By analogy with the derivation for a first-order reaction, the dimensionless group $\text{Ø}^2$ will be proportional to the group of parameters, $\text{Ø}^2 \sim \frac{L^2k_vC_{A,s}{}^{n-1}}{\mathcal{D}}$. We find that in order to achieve the same trend for Ø for any $n$th order of reaction, whereby at large values of Ø, tanh Ø ~1 and $\eta \sim \frac{1}{\text{Ø}}$ and at small values of Ø, $\eta$~1, we can add a factor of $\frac{n+1}{2}$ to the Thiele modulus; therefore, we arrive at the following second generalization of the Thiele modulus:

$$\text{Ø}^2 = \left(\frac{n+1}{2}\right)\frac{L^2k_vC_{A,s}{}^{n-1}}{\mathcal{D}}, n \geq 0$$

Except for the case where reaction order $n = 0$, the effectiveness factor $\eta$ behaves similarly at small and large values of phi for all reaction orders ($n > 0$), and their plots on the same graph coincide.

(d) From the plot provided, we can first obtain values of Ø for $\eta = 0.25$ and $\eta = 0.95$. Say for example, we obtain values of Ø = 3.5 for $\eta = 0.25$ and Ø = 0.55 for $\eta = 0.95$.

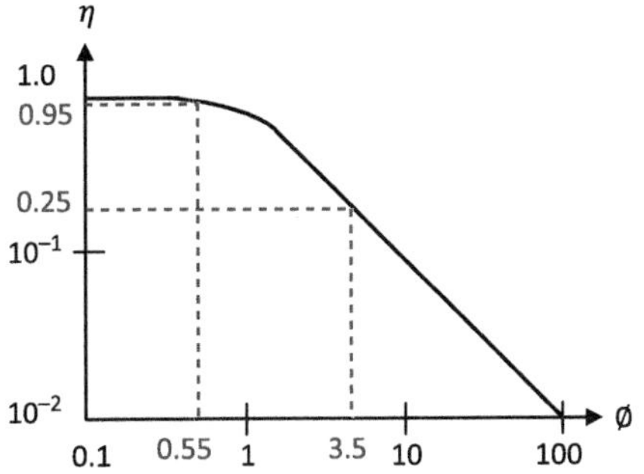

Then we can find the reduction in particle size required to achieve the increase in $\eta$:

$$\varnothing|_{\eta=0.25} = 3.5 = L|_{\eta=0.25}\sqrt{\frac{k_v}{\mathcal{D}}}$$

$$\varnothing|_{\eta=0.95} = 0.55 = L|_{\eta=0.95}\sqrt{\frac{k_v}{\mathcal{D}}}$$

Hence reduction factor in mean particle size is given by:

$$\frac{L|_{\eta=0.95}}{L|_{\eta=0.25}} = \frac{0.55}{3.5} = 0.16$$

## Problem 29

**Consider a spherical catalytic particle whereby a reaction occurs under isothermal conditions within the entire porous structure of the particle. You may assume that the active reaction sites are distributed uniformly throughout the interior of the particle and the particle has a radius $R$.**

(a) **Show the following expression for a reaction of order $n$, where $\mathcal{D}$ is the effective diffusivity assumed constant, $C_A$ is the concentration of reactant $A$, $k_v$ is the reaction rate constant per unit volume, and $r$ is the radial distance measured from the center of the sphere.**

$$\frac{d^2C_A}{dr^2} + \frac{2}{r}\frac{dC_A}{dr} - \frac{k_v}{\mathcal{D}}C_A{}^n = 0$$

(b) **Using an appropriate substitution of variables, show that the differential equation can be simplified to the following dimensionless form, the Thiele modulus for a reaction of order $n$ is defined $\varnothing_n = R\sqrt{\frac{k_v}{\mathcal{D}}C_{A,s}{}^{n-1}}$. Explain the physical significance of $\varnothing_n$.**

$$\frac{d^2C_A{}^*}{dr^{*2}} + \frac{2}{r^*}\frac{dC_A{}^*}{dr^*} - \varnothing_n{}^2 C_A{}^{*n} = 0$$

(c) **Assuming first-order kinetics, solve the differential equation in part b, using the substitution, $\varepsilon = C_A{}^* r^*$, and show that the solution is as follows. Sketch a plot of $\frac{C_A}{C_{A,s}}$ against $r/R$ showing the effect of increasing $\varnothing_1$.**

$$\frac{C_A}{C_{A,s}} = \frac{\sinh[\varnothing_1(r/R)]}{(r/R)\sinh\varnothing_1}$$

(d) **Using the expression found in part c, show that the effectiveness factor for a spherical particle is given by the following.**

$$\eta_{\mathbf{sph}} = \frac{\mathbf{3}}{\boldsymbol{\emptyset}_1}\left(\frac{\mathbf{1}}{\tanh\boldsymbol{\emptyset}_1} - \frac{\mathbf{1}}{\boldsymbol{\emptyset}_1}\right)$$

**Solution 29**

***Worked Solution***

(a) When reaction occurs within the entire internal structure of a particle, there must exist a concentration gradient within the particle material which serves as the driving force for reactant to diffuse from the particle surface inwards, into the interior of the particle. Fick's law tells us that mass flux $j''(r)$ is correlated to concentration gradient via diffusivity, $\mathcal{D}$ as follows

$$j''(r) = -\mathcal{D}\frac{dC_A}{dr}$$

Let us illustrate the system, and perform a mass balance for a differential element in the radial direction.

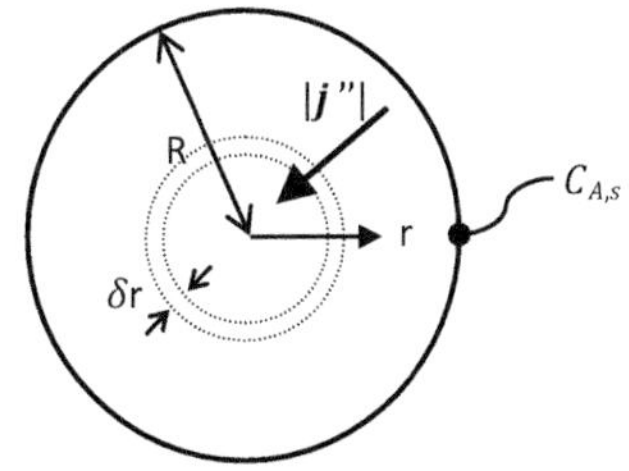


The inward molar flux of reactant $A$ at positions $r$ and $r + \delta r$ is given below, where $A = 4\pi r^2$ is the surface area of the spherical shell at $r$,

$$\text{Flux into shell } (-r \text{ direction}) = -\left[4\pi(r+\delta r)^2\left(-\mathcal{D}\frac{dC_A}{dr}\right)_{r+\delta r}\right]$$

$$\text{Flux out of shell } (-r \text{ direction}) = -\left[4\pi r^2\left(-\mathcal{D}\frac{dC_A}{dr}\right)_{r}\right]$$

Other than diffusive flux, there is also reaction of order $n$ occurring which serves to reduce $N_A$. Since $\delta r$ is small, we may assume that the differential volume in the spherical shell is $4\pi r^2\delta r$, and the reaction rate expression is as follows,

$$\frac{dN_A}{dt} = r_A V = -k_v {C_A}^n V = -k_v {C_A}^n 4\pi r^2 \delta r$$

By mass balance over the differential element of width $\delta r$ at steady state, we have

$$-\left\{4\pi(r+\delta r)^2\left(-\mathcal{D}\frac{dC_A}{dr}\right)_{r+\delta r}\right\} - \left\{-\left[4\pi r^2\left(-\mathcal{D}\frac{dC_A}{dr}\right)_r\right]\right\} - k_v {C_A}^n 4\pi r^2 \delta r = 0$$

$$4\pi(r+\delta r)^2\left(\mathcal{D}\frac{dC_A}{dr}\right)_{r+\delta r} - 4\pi r^2\left(\mathcal{D}\frac{dC_A}{dr}\right)_r - k_v {C_A}^n 4\pi r^2 \delta r = 0$$

$$\frac{(r+\delta r)^2\left(\mathcal{D}\frac{dC_A}{dr}\right)_{r+\delta r} - r^2\left(\mathcal{D}\frac{dC_A}{dr}\right)_r}{\delta r} - k_v {C_A}^n r^2 = 0$$

Where $\delta r \to 0$,

$$\frac{d}{dr}\left[\left(\mathcal{D}\frac{dC_A}{dr}\right)r^2\right] - k_v {C_A}^n r^2 = 0$$

$$2\mathcal{D}r\frac{dC_A}{dr} + \mathcal{D}r^2\frac{d^2C_A}{dr^2} - k_v {C_A}^n r^2 = 0$$

$$\frac{d^2C_A}{dr^2} + \frac{2}{r}\frac{dC_A}{dr} - \frac{k_v}{\mathcal{D}}{C_A}^n = 0$$

(b) Let us consider the boundary conditions for this second-order differential equation. When $r = 0$ at the center of the sphere, $\frac{dC_A}{dr} = 0$, while at $x = R$, $C_A = C_{A,s}$. A useful way to simplify the equation and its boundary conditions is to introduce dimensionless variables. Let ${C_A}^* = C_A/C_{A,s}$ and $r^* = r/R$. Hence we can re-express the earlier differential equation in terms of the new variables.

$$\frac{d^2C_A}{dr^2} + \frac{2}{r}\frac{dC_A}{dr} - \frac{k_v}{\mathcal{D}}{C_A}^n = 0$$

$$dr = Rdr^*; dC_A = C_{A,s}d{C_A}^*; dr^2 = R^2 dr^{*2}; d^2C_A = C_{A,s}d^2{C_A}^*$$

$$\frac{C_{A,s}}{R^2}\frac{d^2{C_A}^*}{dr^{*2}} + \frac{2}{r}\frac{C_{A,s}}{R}\frac{d{C_A}^*}{dr^*} - \frac{k_v}{\mathcal{D}}{C_{A,s}}^n {C_A}^{*n} = 0$$

$$\frac{d^2{C_A}^*}{dr^{*2}} + \frac{2}{r^*}\frac{d{C_A}^*}{dr^*} - \left[\frac{k_v R^2}{\mathcal{D}}{C_{A,s}}^{n-1}\right]{C_A}^{*n} = 0$$

It is given that ${\varnothing_n}^2 = \frac{k_v R^2}{\mathcal{D}}{C_{A,s}}^{n-1}$; therefore,

$$\frac{d^2C_A{}^*}{dr^{*2}} + \frac{2}{r^*}\frac{dC_A{}^*}{dr^*} - \text{Ø}_n{}^2 C_A{}^{*n} = 0$$

There is physical significance in the Thiele modulus, $\text{Ø}_n{}^2$ as a ratio, of the surface reaction rate to the rate of internal mass transfer of reactant into the center of the particle. Therefore when $\text{Ø}_n$ is large, the rate-limiting (or slower) process is mass transfer, and for small values of $\text{Ø}_n$, surface reaction rate is limiting.

$$\text{Ø}_n{}^2 = \frac{k_v C_{A,s}{}^n R}{\mathcal{D}\left(\frac{C_{A,s}-0}{R}\right)}$$

(c) Using the substitution $\varepsilon = C_A{}^* r^*$ or $C_A{}^* = \frac{\varepsilon}{r^*}$

$$\frac{dC_A{}^*}{dr^*} = \frac{1}{r^*}\frac{d\varepsilon}{dr^*} - \frac{\varepsilon}{r^{*2}}$$

$$\frac{d^2C_A{}^*}{dr^{*2}} = \frac{1}{r^*}\frac{d^2\varepsilon}{dr^{*2}} - \frac{1}{r^{*2}}\frac{d\varepsilon}{dr^*} - \frac{1}{r^{*2}}\frac{d\varepsilon}{dr^*} + 2\varepsilon\frac{1}{r^{*3}} = \frac{1}{r^*}\frac{d^2\varepsilon}{dr^{*2}} - \frac{2}{r^{*2}}\frac{d\varepsilon}{dr^*} + \frac{2\varepsilon}{r^{*3}}$$

Combining the above with the differential equation, we have

$$\frac{d^2C_A{}^*}{dr^{*2}} + \frac{2}{r^*}\frac{dC_A{}^*}{dr^*} - \text{Ø}_n{}^2 C_A{}^{*n} = 0$$

$$\frac{1}{r^*}\frac{d^2\varepsilon}{dr^{*2}} - \frac{2}{r^{*2}}\frac{d\varepsilon}{dr^*} + \frac{2\varepsilon}{r^{*3}} + \frac{2}{r^*}\left(\frac{1}{r^*}\frac{d\varepsilon}{dr^*} - \frac{\varepsilon}{r^{*2}}\right) - \text{Ø}_n{}^2 C_A{}^{*n} = 0$$

$$\frac{1}{r^*}\frac{d^2\varepsilon}{dr^{*2}} - \text{Ø}_n{}^2 C_A{}^{*n} = 0$$

For first-order kinetics, $n = 1$; therefore

$$\text{Ø}_1{}^2 = \frac{k_v R^2}{\mathcal{D}}$$

$$\frac{1}{r^*}\frac{d^2\varepsilon}{dr^{*2}} - \frac{k_v R^2}{\mathcal{D}} C_A{}^* = 0$$

$$\frac{d^2\varepsilon}{dr^{*2}} - \text{Ø}_1{}^2 \varepsilon = 0$$

We can try a trial solution of the form below, where $c_1$ and $c_2$ are constants to be determined using boundary conditions. When $r^* = 0$, $\varepsilon$ is finite; therefore $c_1 = 0$. When $r^* = 1$, $\varepsilon = 1$.

$$\varepsilon = c_1 \cosh Ø_1 r^* + c_2 \sinh Ø_1 r^* = c_2 \sinh Ø_1 r^*$$

$$c_2 = \frac{1}{\sinh Ø_1}$$

$$\varepsilon = \frac{\sinh Ø_1 r^*}{\sinh Ø_1}$$

Therefore converting dimensionless variables back to the original variables, the solution for a spherical particle is:

$$(C_A/C_{A,s})(r/R) = \frac{\sinh[Ø_1(r/R)]}{\sinh Ø_1}$$

$$\frac{C_A}{C_{A,s}} = \frac{\sinh[Ø_1(r/R)]}{(r/R)\sinh Ø_1}$$

A sketch of $\frac{C_A}{C_{A,s}}$ against $\frac{r}{R}$ is such that at $\frac{r}{R} = 1$, $\frac{C_A}{C_{A,s}} = 1$. The effect of increasing $Ø_1$ is to go from surface reaction limited to becoming more diffusion-limited. Therefore the plot will be as follows:

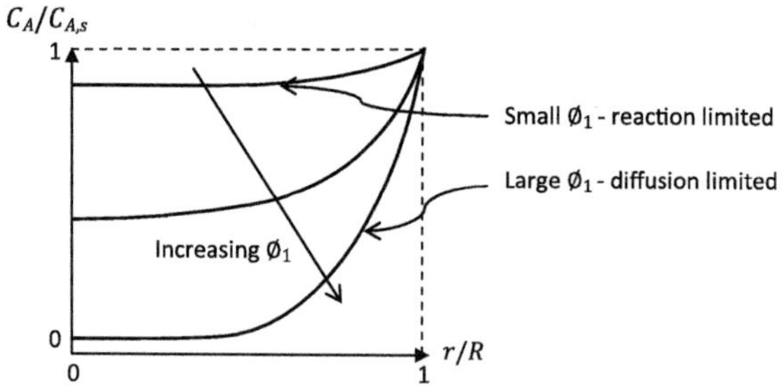


(d) To find the effectiveness factor $\eta$, we start with its definition.

$$\eta = \frac{\text{Observed rate}}{\text{Ideal rate}}$$

The observed rate of reaction at steady state will be equivalent to the rate at which $A$ diffuses into the sphere at its surface. Therefore,

$$\text{Observed rate} = 4\pi R^2 \left( -\mathcal{D} \left. \frac{dC_A}{dr} \right|_{r=R} \right) = -\frac{4\pi R^2 \mathcal{D} C_{A,s}}{\sinh Ø_1} \frac{d}{dr} \left[ \frac{\sinh[Ø_1(r/R)]}{(r/R)} \right]_{r=R}$$

$$= -\frac{4\pi R^2 \mathcal{D} C_{A,s}}{\sinh\varnothing_1}\left[\frac{\left(\frac{r\varnothing_1}{R^2}\right)\cosh\left[\varnothing_1\left(\frac{r}{R}\right)\right] - \frac{1}{R}\sinh\left[\varnothing_1\left(\frac{r}{R}\right)\right]}{\left(\frac{r}{R}\right)^2}\right]_{r=R}$$

$$= -4\pi R\mathcal{D}\varnothing_1 C_{A,s}\left(\frac{1}{\tanh\varnothing_1} - \frac{1}{\varnothing_1}\right)$$

We know that the ideal (or maximum possible) rate is when mass transfer is perfect, and therefore reaction rate operates only at the maximum concentration of reactant throughout the sphere, and this maximum concentration is the concentration at the surface, $C_{A,s}$

$$\text{Ideal rate} = -k_v V C_{A,s} = -\frac{4}{3}\pi R^3 k_v C_{A,s}$$

Therefore substituting the expressions, and knowing $\varnothing_1 = R\sqrt{\frac{k_v}{\mathcal{D}}}$, we obtain the effectiveness factor for a spherical particle as follows:

$$\eta = \frac{-4\pi R\mathcal{D}\varnothing_1 C_{A,s}\left(\frac{1}{\tanh\varnothing_1} - \frac{1}{\varnothing_1}\right)}{-\frac{4}{3}\pi R^3 k_v C_{A,s}} = \frac{3}{\varnothing_1}\left(\frac{1}{\tanh\varnothing_1} - \frac{1}{\varnothing_1}\right)$$

## Problem 30

**Consider a heterogeneous reaction occurring within a spherical catalyst particle of radius $R = 5 \times 10^{-5}$ m. The reaction $A \rightarrow B$ is first order and irreversible. Experimental data showed that the concentration of $A$ at a distance of $0.6R$ from the center of the particle is $0.1C_{A,s}$, where $C_{A,s}$ is the reactant concentration at the particle surface. The bulk concentration of $A$ is $2 \times 10^{-9}$ mol m$^{-3}$, and the diffusion coefficient $\mathcal{D} = 1.5 \times 10^{-4}$ m$^2$ s$^{-1}$. External mass transfer effects may be ignored.**

(a) **Determine the concentration of $A$ at $r = 1 \times 10^{-5}$ m. You may use without derivation, the following concentration profile for a spherical particle with first-order reaction, where $\varnothing_1 = R\sqrt{\frac{k_v}{\mathcal{D}}}$.**

$$\frac{C_A}{C_{A,s}} = \frac{\sinh[\varnothing_1(r/R)]}{(r/R)\sinh\varnothing_1}$$

(b) **What should the catalyst particle size be reduced to, in order to achieve an effectiveness factor $\eta$ of 0.9? You are to use without derivation the following expression for the effectiveness factor of a spherical particle.**

$$\boldsymbol{\eta_{sph} = \frac{3}{\emptyset_1{}^2}(\emptyset_1 \coth\emptyset_1 - 1)}$$

## Solution 30

### *Worked Solution*

(a) The concentration profile of a spherical catalytic particle with internal mass transfer effects and reaction of order 1 is as follows whereby the Thiele modulus is defined as $\emptyset_1 = R\sqrt{\frac{k_v}{\mathcal{D}}}$.

$$\frac{C_A}{C_{A,s}} = \frac{\sinh[\emptyset_1(r/R)]}{(r/R)\sinh\emptyset_1}$$

The non-dimensional form of this concentration profile is as follows, where $r^* = r/R$ and $C_A{}^* = \frac{C_A}{C_{A,s}}$

$$C_A{}^* = \frac{1}{r^*}\frac{\sinh\emptyset_1 r^*}{\sinh\emptyset_1}$$

Given that at $r^* = 0.6$, $C_A{}^* = 0.1$, we substitute into the equation to obtain $\emptyset_1$.

$$0.1 = \frac{1}{0.6}\frac{\sinh(0.6\emptyset_1)}{\sinh\emptyset_1}$$

$$\emptyset_1 = 7$$

Given that $R = 5 \times 10^{-5}$ m, $\mathcal{D} = 1.5 \times 10^{-4}$ m$^2$ s$^{-1}$, and $C_{A,s} = 2 \times 10^{-9}$ mol m$^{-3}$, we are required to determine $C_A$ at $r = 1 \times 10^{-5}$ m.

$$r^* = \frac{10^{-5}}{5 \times 10^{-5}} = 0.2$$

$$C_A{}^* = \frac{1}{0.2}\frac{\sinh(7 \times 0.2)}{\sinh 7} = 0.017$$

$$C_A = 2 \times 10^{-9} \times 0.017 = 3.5 \times 10^{-11} \text{ mol m}^{-3}$$

(b) For a spherical particle, effectiveness factor is given as follows.

$$\eta_{sph} = \frac{3}{\emptyset_1{}^2}(\emptyset_1 \coth\emptyset_1 - 1)$$

$$\eta_{\text{sph}} = \frac{3}{7^2}(7\coth 7 - 1) = 0.37$$

The original effectiveness factor is 0.37. We can increase this to 0.9 by reducing particle size.

$$0.9 = \frac{3}{\text{Ø}_1^{\,2}}(\text{Ø}_1\coth\text{Ø}_1 - 1) \rightarrow \text{Ø}_1 = 1.32$$

$$\frac{\text{Ø}_1|_{\eta=0.37}}{\text{Ø}_1|_{\eta=0.9}} = \frac{R|_{\eta=0.37}}{R|_{\eta=0.9}}$$

$$\frac{7}{1.32} = \frac{5 \times 10^{-5}}{R|_{\eta=0.9}} \rightarrow R|_{\eta=0.9} = 9.4 \times 10^{-6}\text{ m}$$

The reduced particle radius should be $9.4 \times 10^{-6}$ m to achieve an effectiveness factor of 0.9.

## Problem 31

**Consider a cylindrical pore of total length $2L = 3 \times 10^{-5}$ m which is coated with platinum at its surface walls along the length of the pore. The coating provides active sites for chemical reaction $A \rightarrow B$ to take place. The reaction is first order with respect to $A$. Due to diffusive flux of $A$ into the pore, a concentration gradient develops along the axial length of the pore. The concentration of $A$ at the exposed surfaces of the pore $C_{A,s}$ is $2 \times 10^{-6}$ mol m$^{-3}$.**

(a) **Determine the concentration profile for reactant $A$ with respect to axial length, $z$. Then show that the effectiveness factor $\eta$ for this cylindrical catalytic pore is given by:**

$$\eta = \frac{\tanh\text{Ø}}{\text{Ø}}$$

(b) **At a distance $z = L$, the concentration of reactant $A$ drops to 20% of $C_{A,s}$. Find the concentration of $A$ at a distance of $z = \frac{L}{3}$.**

(c) **A reduction in pore length is required to improve the effectiveness of this cylindrical catalyst. Determine the new pore length that would achieve an effectiveness factor of 0.9.**

## Solution 31

### *Worked Solution*

(a) Let us illustrate the system.

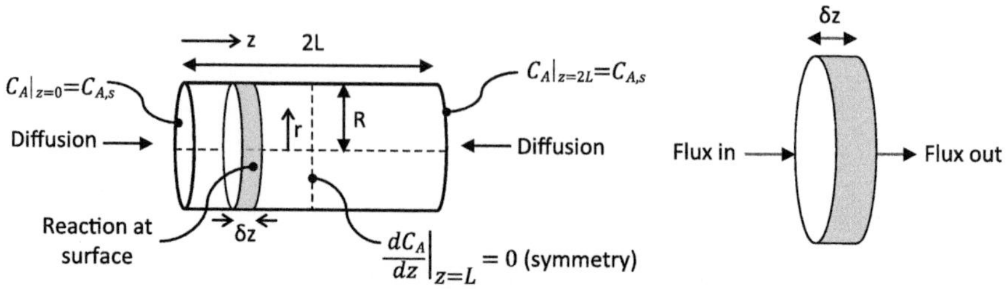


The inward molar flux of reactant $A$ at positions $z$ and $z + \delta z$ is given below, where $A = \pi R^2$ is the cross-sectional area of a differential slice,

$$\text{Flux into slice} = \pi R^2 \left(-\mathcal{D}\frac{dC_A}{dz}\right)_z$$

$$\text{Flux out of slice} = \pi R^2 \left(-\mathcal{D}\frac{dC_A}{dz}\right)_{z+\delta z}$$

Other than diffusive flux, there is also reaction of order 1 occurring at the coating on the cylindrical surface area; this reaction acts to reduce $N_A$. Note that $r_A''$ has units of mol per area catalyst per time, and $A$ is the surface area of the cylindrical pore at which the reactions occur.

$$-\frac{1}{A}\frac{dN_A}{dt} = -r_A'' = k''C_A$$

By mass balance over the differential slice of width $\delta z$ at steady state, we have

$$\pi R^2 \left(-\mathcal{D}\frac{dC_A}{dz}\right)_z - \pi R^2 \left(-\mathcal{D}\frac{dC_A}{dz}\right)_{z+\delta z} - k''C_A(2\pi R\delta z) = 0$$

$$\frac{\left(\frac{dC_A}{dz}\right)_{z+\delta z} - \left(\frac{dC_A}{dz}\right)_z}{\delta z} - \frac{2k''C_A}{\mathcal{D}R} = 0$$

Where $\delta z \to 0$,

$$\frac{d^2C_A}{dz^2} - \frac{2k''C_A}{\mathcal{D}R} = 0$$

We can convert $k''$ into a reaction rate constant per unit volume,

Note that $k_v V = k'' A$

$$k'' = \frac{k_v\left(\pi R^2 L \times 2\right)}{\left(2\pi RL \times 2\right)} = \frac{k_v R}{2}$$

Substituting the result, we have

$$\frac{d^2 C_A}{dz^2} - \frac{k_v}{\mathcal{D}} C_A = 0$$

We can non-dimensionalize this equation using the substitutions, $z^* = z/L$, ${C_A}^* = C_A/C_{A,s}$, and ${\emptyset_1}^2 = L^2 \frac{k_v}{\mathcal{D}}$.

$$\frac{C_{A,s}}{L^2}\frac{d^2 {C_A}^*}{dz^{*2}} - \frac{k_v C_{A,s}}{\mathcal{D}} {C_A}^* = 0$$

$$\frac{d^2 {C_A}^*}{dz^{*2}} - \frac{k_v L^2}{\mathcal{D}} {C_A}^* = 0$$

$$\frac{d^2 {C_A}^*}{dz^{*2}} - {\emptyset_1}^2 {C_A}^* = 0$$

We can use the trial solution of the form below, where $c_1$ and $c_2$ are constants to be determined by applying boundary conditions.

$${C_A}^* = c_1 \cosh\emptyset_1 z^* + c_2 \sinh\emptyset_1 z^*$$

When $z^* = 0$, ${C_A}^* = 1$; therefore

$$1 = c_1$$

When $z^* = 1$, $\frac{d{C_A}^*}{dz^*} = 0$ due to symmetry; therefore

$${C_A}^* = \cosh\emptyset_1 z^* + c_2 \sinh\emptyset_1 z^*$$

$$\frac{d{C_A}^*}{dz^*} = \emptyset_1 \sinh\emptyset_1 z^* + c_2 \emptyset_1 \cosh\emptyset_1 z^*$$

$$0 = \emptyset_1 \sinh\emptyset_1 + c_2 \emptyset_1 \cosh\emptyset_1$$

$$c_2 = -\frac{\sinh\emptyset_1}{\cosh\emptyset_1}$$

We can now determine the concentration profile in the cylindrical pore,

$$C_A{}^* = \cosh\varnothing_1 z^* - \left(\frac{\sinh\varnothing_1}{\cosh\varnothing_1}\right)\sinh\varnothing_1 z^*$$

$$= \frac{\cosh\varnothing_1 \cosh\varnothing_1 z^* - \sinh\varnothing_1 \sinh\varnothing_1 z^*}{\cosh\varnothing_1}$$

$$= \frac{\cosh(\varnothing_1 - \varnothing_1 z^*)}{\cosh\varnothing_1}$$

$$\frac{C_A}{C_{A,s}} = \frac{\cosh\left[\varnothing_1\left(1 - \frac{z}{L}\right)\right]}{\cosh\varnothing_1}$$

To find the effectiveness factor $\eta$, we start with its definition as shown.

$$\eta = \frac{\text{Observed rate}}{\text{Ideal rate}}$$

The observed rate of reaction at steady state will be equivalent to the rate at which $A$ diffuses into the pore through its cylindrical surface,

$$\text{Observed rate} = \pi R^2\left(-\mathcal{D}\frac{dC_A}{dz}\bigg|_{z=0}\right) = \pi R^2\left[\frac{-\mathcal{D}C_{A,s}\varnothing_1}{L\cosh\varnothing_1}\sinh\left[\varnothing_1\left(1 - \frac{z}{L}\right)\right]\right]_{z=0}$$

$$= \frac{-\pi R^2 \mathcal{D}C_{A,s}\varnothing_1}{L}\tanh\varnothing_1$$

We know that the ideal (or maximum possible) rate is when mass transfer is perfect, and therefore it is the reaction rate at the maximum concentration of reactant at the surface, $C_{A,s}$. Note that the volume to be used here is over half length, i.e., $\pi R^2 L$ to match the basis of the preceding analysis.

$$\text{Ideal rate} = -k_v V C_{A,s} = -k_v\left(\pi R^2 L\right)C_{A,s}$$

Therefore substituting the two reaction rates and knowing that $\varnothing_1 = L\sqrt{\frac{k_v}{\mathcal{D}}}$, we obtain the effectiveness factor as follows:

$$\eta = \frac{\frac{-\pi R^2 \mathcal{D}C_{A,s}\varnothing_1}{L}\tanh\varnothing_1}{-k_v\left(\pi R^2 L\right)C_{A,s}}$$

$$= \frac{\tanh\varnothing}{\varnothing}$$

(b) When $z = L$, $C_A = 0.2C_{A,s}$; therefore, we can find $\varnothing_1$

$$0.2 = \frac{1}{\cosh Ø_1}$$

$$Ø_1 = 2.3$$

At $z = L/3$, the concentration $C_A$ can be found as follows, given that $C_{A,s} = 2 \times 10^{-6}$ mol m$^{-3}$

$$\frac{C_A}{2 \times 10^{-6}} = \frac{\cosh\left[2.3\left(1 - \frac{1}{3}\right)\right]}{\cosh 2.3}$$

$$C_A = 9.6 \times 10^{-7} \text{mol m}^{-3}$$

(c) Note that effectiveness factor was found earlier to be given by:

$$\eta = \frac{\tanh Ø}{Ø}$$

$$\eta_{\text{old}} = \frac{\tanh Ø_{\text{old}}}{Ø_{\text{old}}} = \frac{\tanh 2.3}{2.3} = 0.43$$

$$\eta_{\text{new}} = 0.9 = \frac{\tanh Ø_{\text{new}}}{Ø_{\text{new}}} \rightarrow Ø_{\text{new}} = 0.58$$

Since $Ø_1 = L\sqrt{\frac{k_v}{\mathcal{D}}}$, therefore,

$$\frac{Ø_{\text{old}}}{Ø_{\text{new}}} = \frac{2L_{\text{old}}}{2L_{\text{new}}}$$

$$\frac{2.3}{0.58} = \frac{3 \times 10^{-5}}{2L_{\text{new}}} \rightarrow 2L_{\text{new}} = 7.6 \times 10^{-6} \text{ m}$$

Therefore the pore length should be reduced from $3 \times 10^{-5}$ m to $7.6 \times 10^{-6}$ m for the effectiveness factor to be 0.9.

## Problem 32

**Consider a first-order reaction $A \rightarrow B$ that occurs within the porous material of a catalytic slab. The reaction sites are distributed uniformly within the slab. The slab is of thickness 2 H in the $z$ direction and of length $L$ in the axial $x$ direction such that $L \gg H$. There is a bulk fluid flowing along the axial direction of the slab on its exterior and the concentration of reactant $A$ in the bulk fluid is $C_{A,0}$. The reactant concentration at the surface of the slab is $C_{A,s}$ where $C_{A,0} \neq C_{A,s}$**

**due to the presence of external mass transfer effects. The reaction rate constant per unit volume for the first-order reaction is $k_1$ and $\mathcal{D}$ is effective diffusivity.**

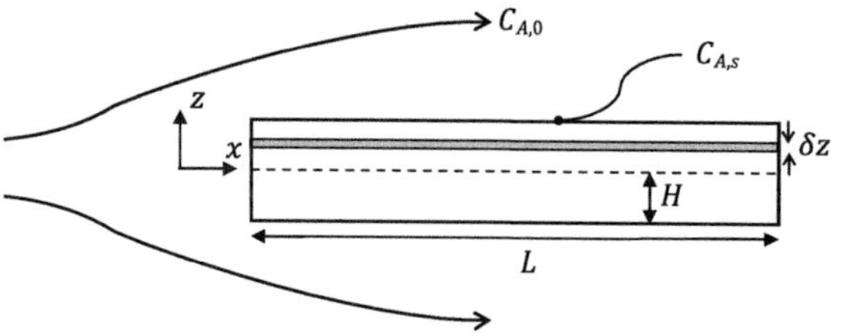


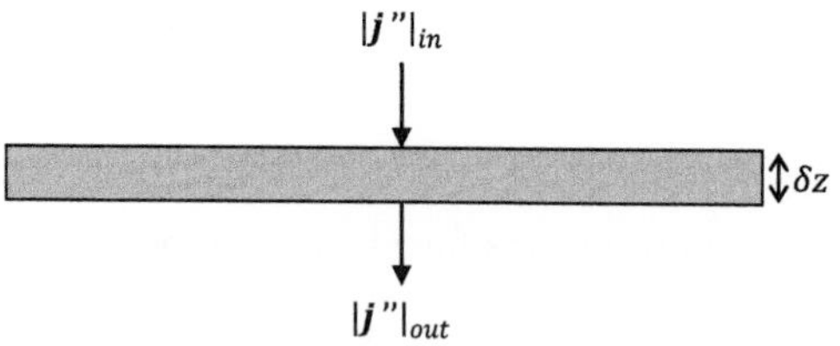


(a) **Derive an expression for the internal effectiveness factor $\eta$ of the catalytic slab.**
(b) **Show that the expression for the external effectiveness factor $\Omega$ for the catalytic slab is given by:**

$$\boldsymbol{\Omega = \frac{\frac{\tanh\emptyset}{\emptyset}}{1 + \frac{\mathcal{D}\emptyset}{k_c H}\tanh\emptyset}, \text{ where } \emptyset_1 = H\sqrt{\frac{k_1}{\mathcal{D}}}}$$

**Solution 32**

***Worked Solution***

(a) We may assume that since $L \gg H$, variations in the $x$ direction are negligible, and only variations in the $z$ direction need to be considered.

The inward diffusive flux of reactant $A$ at positions $z$ and $z + \delta z$ is given below, where $A$ is the surface area of the slab,

$$\text{Flux into slice } (-z \text{ direction}) = -\left[A\left(-\mathcal{D}\frac{dC_A}{dz}\right)_{z+\delta z}\right]$$

$$\text{Flux out of slice } (-z \text{ direction}) = -\left[A\left(-\mathcal{D}\frac{dC_A}{dz}\right)_{z}\right]$$

Other than diffusive flux, there is also reaction occurring which serves to reduce $N_A$, assuming first-order kinetics, we have the following expression for reaction rate

$$\frac{dN_A}{dt} = r_A V = -k_1 C_A V = -k_1 C_A A \delta z$$

By mass balance over the differential element of width $\delta z$ at steady state, we have

$$-\left[A\left(-\mathcal{D}\frac{dC_A}{dz}\right)_{z+\delta z}\right] - \left\{-\left[A\left(-\mathcal{D}\frac{dC_A}{dz}\right)_z\right]\right\} - k_1 C_A A \delta z = 0$$

$$\frac{\left(\frac{dC_A}{dz}\right)_{z+\delta z} - \left(\frac{dC_A}{dz}\right)_z}{\delta z} - \frac{k_1}{\mathcal{D}} C_A = 0$$

For $\delta z \to 0$,

$$\frac{d^2 C_A}{dz^2} - \frac{k_1}{\mathcal{D}} C_A = 0$$

Let us consider the boundary conditions for this second-order differential equation. When $z = 0$ at the centerline of the slab, $\frac{dC_A}{dz} = 0$, while at $z = H$, $C_A = C_{A,s}$.

A useful way to simplify the equation and its boundary conditions is to introduce dimensionless variables. Let $C_A{}^* = C_A/C_{A,s}$ and $z^* = z/H$. Hence we can re-express the earlier differential equation in terms of the new variables.

$$\frac{d^2 C_A}{dz^2} - \frac{k_1}{\mathcal{D}} C_A = 0$$

$$dz^2 = H^2 dz^{*2}; d^2 C_A = C_{A,s} d^2 C_A{}^*$$

$$\frac{\mathcal{D} C_{A,s}}{H^2} \frac{d^2 C_A{}^*}{dz^{*2}} - k_1 C_{A,s}\, C_A{}^* = 0$$

$$\frac{d^2 C_A{}^*}{dz^{*2}} - H^2 \frac{k_1}{\mathcal{D}} C_A{}^* = 0$$

It is given that $\varnothing_1 = H\sqrt{\frac{k_1}{\mathcal{D}}}$; therefore,

$$\frac{d^2 C_A{}^*}{dx^{*2}} - \varnothing_1{}^2\, C_A{}^* = 0$$

We can try a trial solution of exponential form based on our knowledge that derivatives of the exponential function can give the same exponential function, multiplied by a constant. Therefore, we can try a solution of the form $C_A^* = c_1 \exp(\text{Ø}_1 z^*) + c_2 \exp(-\text{Ø}_1 z^*)$, where $c_1$ and $c_2$ are constants to be determined by applying boundary conditions. We know that when $z^* = 0$, $\frac{dC_A^*}{dx^*} = 0$, and when $z^* = 1$, $C_A^* = 1$.

[Note that another possible trial solution is $C_A^* = c_1 \cosh \text{Ø}_1 z^* + c_2 \sinh \text{Ø}_1 z^*$.]

$$\frac{dC_A^*}{dz^*} = c_1 \text{Ø} \exp(\text{Ø}_1 z^*) - c_2 \text{Ø} \exp(-\text{Ø}_1 z^*)$$

$$\left.\frac{dC_A^*}{dz^*}\right|_{z^*=0} = c_1 \text{Ø}_1 - c_2 \text{Ø}_1 = 0 \rightarrow c_1 = c_2$$

$$1 = c_1 \exp \text{Ø}_1 + c_2 \exp(-\text{Ø}_1) = c_1 \left[e^{\text{Ø}_1} + e^{-\text{Ø}_1}\right]$$

$$c_1 = \frac{1}{e^{\text{Ø}_1} + e^{-\text{Ø}_1}}$$

Therefore, we obtain the expression,

$$C_A^* = \frac{1}{e^{\text{Ø}_1} + e^{-\text{Ø}_1}} e^{\text{Ø}_1 z^*} + \frac{1}{e^{\text{Ø}_1} + e^{-\text{Ø}_1}} e^{-\text{Ø}_1 z^*}$$

$$C_A^* = \frac{1}{2\cosh \text{Ø}_1}\left[e^{\text{Ø}_1 z^*} + e^{-\text{Ø}_1 z^*}\right] = \frac{1}{2\cosh \text{Ø}_1}[2\cosh(\text{Ø}_1 z^*)]$$

$$C_A^* = \frac{\cosh(\text{Ø}_1 z^*)}{\cosh \text{Ø}_1}$$

$$C_A = C_{A,s} \frac{\cosh\left(\text{Ø}_1 \frac{z}{H}\right)}{\cosh \text{Ø}_1}$$

To find the internal effectiveness factor $\eta$, we start with its definition.

$$\eta = \frac{\text{Observed rate}}{\text{Ideal rate}}$$

The observed rate of reaction at steady state will be equivalent to the rate at which $A$ diffuses into the slab at both of its faces (i.e., total area of $2A$ comprising $A$ at $z = H$ and $A$ at $z = -H$). Therefore,

$$\text{Observed rate} = 2A\left(-\mathcal{D}\left.\frac{dC_A}{dz}\right|_{z=H}\right) = -2A\mathcal{D}\left[\frac{C_{A,s}\text{Ø}_1}{H\cosh \text{Ø}_1} \sinh\left(\text{Ø}_1 \frac{z}{H}\right)\right]_{z=H}$$

$$= -\frac{2A\mathcal{D}\varnothing_1 C_{A,s}}{H}\tanh\varnothing_1$$

We know that the ideal (or maximum possible) rate is when mass transfer is perfect, and therefore reaction rate operates only at the maximum concentration of reactant throughout the slab, and this maximum concentration is the concentration at the slab surface, $C_{A,s}$

$$\text{Ideal rate} = -k_1 V C_{A,s} = -k_1 2AHC_{A,s}$$

Therefore substituting the expressions, and knowing $\varnothing_1 = H\sqrt{\frac{k_1}{\mathcal{D}}}$, we obtain the effectiveness factor as follows:

$$\eta = \frac{-\frac{2A\mathcal{D}\varnothing_1 C_{A,s}}{H}\tanh\varnothing_1}{-k_1 2AHC_{A,s}} = \frac{\tanh\varnothing_1}{\varnothing_1}$$

(b)

$$\eta = \frac{2A\left(-\mathcal{D}\frac{dC_A}{dz}\Big|_{z=H}\right)}{-k_1 C_{A,s} 2AH}$$

The ideal rate for $\Omega$ is the rate of reaction when no external mass transfer effects and hence the reaction operates at $C_{A,s} = C_{A,0}$

$$\Omega = \frac{2A\left(-\mathcal{D}\frac{dC_A}{dz}\Big|_{z=H}\right)}{-k_1 C_{A,0} 2AH}$$

Observed rate remains the same whether expressed in $\Omega$ or $\eta$

$$\text{Observed rate} = 2A\left(-\mathcal{D}\frac{dC_A}{dz}\Big|_{z=H}\right)$$

$$\Omega(-k_1 C_{A,0} 2AH) = \eta(-k_1 C_{A,s} 2AH)$$

$$\frac{\Omega}{\eta} = \frac{C_{A,s}}{C_{A,0}}$$

We can also equate the mass flux due to mass conservation, using $k_c$ as the external mass transfer coefficient,

$$-\mathcal{D}\frac{dC_A}{dz}\Big|_{z=H} = -k_c(C_{A,0} - C_{A,s})$$

We have earlier obtained the expression for left-hand side by differentiating the concentration profile, using the result, we have

$$\frac{\mathcal{D}\text{Ø}_1 C_{A,s}}{H}\tanh\text{Ø}_1 = k_c(C_{A,0} - C_{A,s})$$

$$C_{A,s}\left[1 + \frac{\mathcal{D}\text{Ø}_1}{k_c H}\tanh\text{Ø}_1\right] = C_{A,0}$$

$$\frac{C_{A,s}}{C_{A,0}} = \frac{\Omega}{\eta} = \frac{1}{1 + \frac{\mathcal{D}\text{Ø}_1}{k_c H}\tanh\text{Ø}_1}$$

$$\Omega = \frac{\frac{\tanh\text{Ø}_1}{\text{Ø}_1}}{1 + \frac{\mathcal{D}\text{Ø}_1}{k_c H}\tanh\text{Ø}_1}, \text{ where } \text{Ø}_1 = H\sqrt{\frac{k_1}{\mathcal{D}}}$$

**Problem 33**

**Consider a zero-order irreversible reaction $A \rightarrow B$ that occurs within a porous rectangular catalytic slab. The slab is of thickness $2L$ in the $y$ direction and of length $L$ in the axial $x$ direction. There is a bulk fluid around the slab with a concentration of reactant $A$ denoted by $C_{A,0}$. The reactant concentration at the surface of the slab is $C_{A,s}$. The reaction rate constant for the reaction is $k_0$ with units of $s^{-1}$ and $\mathcal{D}$ is effective diffusivity.**

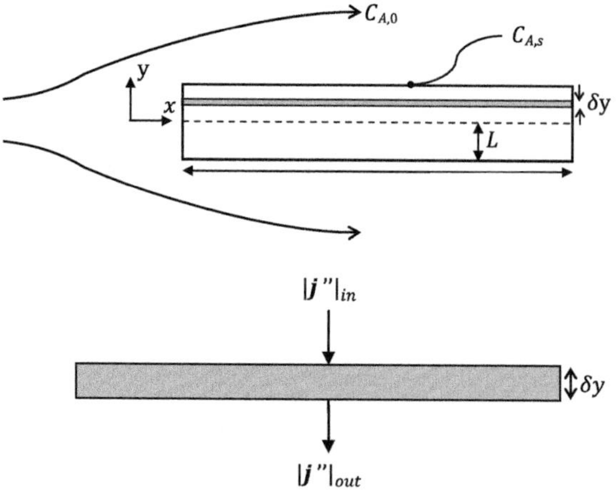


(a) **Determine the concentration profile of $A$ where $\text{Ø}_0^2 = \frac{k_0 L^2}{\mathcal{D} C_{A,s}}$.**
(b) **Determine the positions in the slab when $C_A = 0$ for $\text{Ø}_0^2 = 2$ and $\text{Ø}_0^2 = 9$.**
(c) **Given that the effectiveness factor is defined as follows where $A$ is the area of the face of the slab normal to mass transport of $A$ into the slab along the $y$-axis. Determine expressions for the effectiveness factor for different values of $\text{Ø}_0^2$. Plot a graph of $\eta$ against $\text{Ø}_0$.**

$$\eta = \frac{2\left(\int_0^L -r_A A dy\right)}{2(-r_{A,s} AL)}$$

(d) **Sketch the concentration profiles for values of $\emptyset_0^2 < 2$, $\emptyset_0^2 = 2$ and $\emptyset_0^2 = 9$.**

## Solution 33

### *Worked Solution*

(a) If the reaction was zero order, then the reaction term in the mass balance of $A$ becomes,

$$\frac{dN_A}{dt} = r_A V = -k_0 V = -k_0 A\delta y$$

Following through the same process of derivation via mass balance of $A$, we will arrive at

$$-\left[A\left(-\mathcal{D}\frac{dC_A}{dy}\right)_{y+\delta y}\right] - \left\{-\left[A\left(-\mathcal{D}\frac{dC_A}{dy}\right)_y\right]\right\} - k_0 A\delta y = 0$$

$$\frac{\left(\frac{dC_A}{dy}\right)_{y+\delta y} - \left(\frac{dC_A}{dy}\right)_y}{\delta y} - \frac{k_0}{\mathcal{D}} = 0$$

For $\delta y \to 0$,

$$\frac{d^2 C_A}{dy^2} - \frac{k_0}{\mathcal{D}} = 0$$

We can apply boundary conditions to solve this equation. When $y = 0$ at the centerline of the slab, $\frac{dC_A}{dy} = 0$, while at $z = L$, $C_A = C_{A,s}$. This differential equation is simple to integrate, hence dimensionless variables are not necessary.

$$\frac{dC_A}{dy} = \frac{k_0}{\mathcal{D}} y + c_1 \to c_1 = 0$$

$$C_A = \frac{k_0}{2\mathcal{D}} y^2 + c_2 \to c_2 = C_{A,s} - \frac{k_0}{2\mathcal{D}} L^2$$

$$C_A = C_{A,s} + \frac{k_0}{2\mathcal{D}}y^2 - \frac{k_0}{2\mathcal{D}}L^2 = C_{A,s}\left[1 + \frac{\text{Ø}_0{}^2}{2}\left(\frac{y}{L}\right)^2 - \frac{\text{Ø}_0{}^2}{2}\right], \text{ where } \text{Ø}_0{}^2 = \frac{k_0 L^2}{\mathcal{D}C_{A,s}}$$

$$\frac{C_A}{C_{A,s}} = 1 + \frac{\text{Ø}_0{}^2}{2}\left[\left(\frac{y}{L}\right)^2 - 1\right]$$

(b) Substitute $\text{Ø}_0{}^2 = 2$ and $C_A = 0$ into the concentration profile obtained in part a to find the position $y$.

$$\frac{0}{C_{A,s}} = 1 + \frac{2}{2}\left[\left(\frac{y}{L}\right)^2 - 1\right]$$

$$y = 0 \text{ (centerline of slab)}$$

Next we can find the same for a different value of $\text{Ø}_0{}^2 = 9$ where $C_A = 0$,

$$\frac{0}{C_{A,s}} = 1 + \frac{9}{2}\left[\left(\frac{y}{L}\right)^2 - 1\right]$$

$$y = \sqrt{\frac{2}{9}\left(\frac{9}{2} - 1\right)}L = 0.88L$$

(c) We are given the expression for internal effectiveness factor as follows.

$$\eta = \frac{2\left(\int_0^L -r_A A dy\right)}{2(-r_{A,s}AL)}$$

For a zero-order reaction, $-r_A = k_0$ for $C_A > 0$, and $-r_A = 0$ for $C_A = 0$.

For $\text{Ø}_0{}^2 < 2$, $\eta = 1$ since $C_A > 0$ throughout the slab and therefore $-r_A = -r_{A,s} = k_0$.

For $\text{Ø}_0{}^2 > 2$, $\eta < 1$ since $C_A = 0$ (i.e., is completely depleted) in the inner region of the slab and $-r_A = 0$.

$$\eta = \frac{2\left(\int_0^L -r_A A dy\right)}{2(-r_{A,s}AL)} = \frac{\int_0^{y^*} -r_A dy + \int_{y^*}^L -r_A dy}{-r_{A,s}L}$$

$$= \frac{\int_0^{y^*} 0 dy + \int_{y^*}^L k_0 dy}{k_0 L}$$

$$= \frac{k_0(L - y^*)}{k_0 L} = 1 - \frac{y^*}{L}$$

In general, the expression for $y^*$ can be found by setting $C_A = 0$ in the concentration profile, assuming the region $\text{Ø}_0{}^2 > 2$.

$$\frac{C_A}{C_{A,s}} = 1 + \frac{\text{Ø}_0{}^2}{2}\left[\left(\frac{y}{L}\right)^2 - 1\right]$$

$$0 = 1 + \frac{\text{Ø}_0{}^2}{2}\left[\left(\frac{y^*}{L}\right)^2 - 1\right]$$

$$y^* = L\sqrt{1 - \frac{2}{\text{Ø}_0{}^2}}$$

Therefore the profile of $\eta$ is:

$$\eta = 1 \text{ where } \text{Ø}_0{}^2 < 2$$

$$= 1 - \sqrt{1 - \frac{2}{\text{Ø}_0{}^2}} \text{ where } \text{Ø}_0{}^2 > 2$$

A plot of $\eta$ against the Thiele modulus for a zero-order reaction is as shown below.

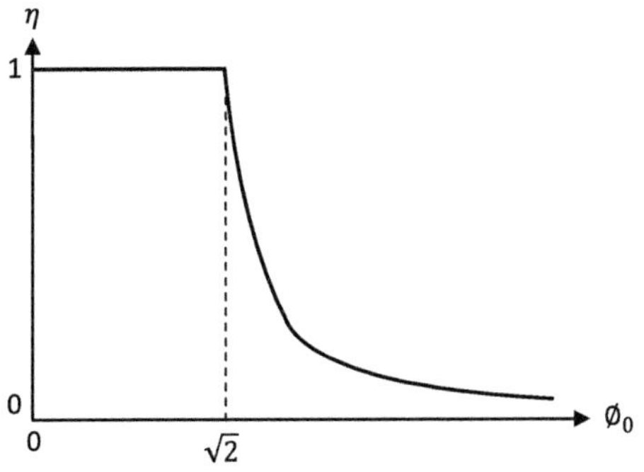


(d) The concentration profiles for values of $\text{Ø}_0{}^2 < 2$, $\text{Ø}_0{}^2 = 2$, and $\text{Ø}_0{}^2 = 9$ are shown below.

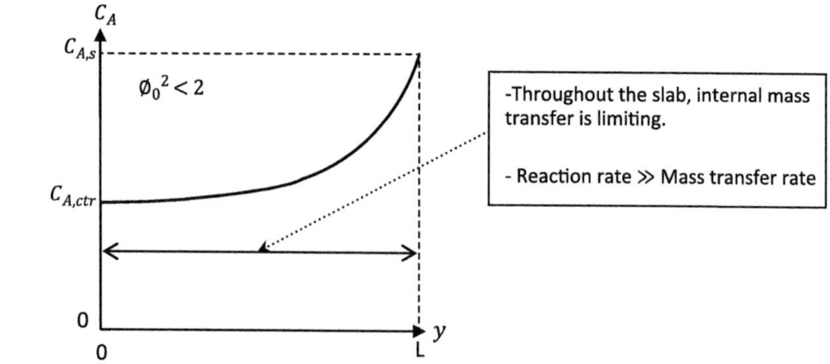


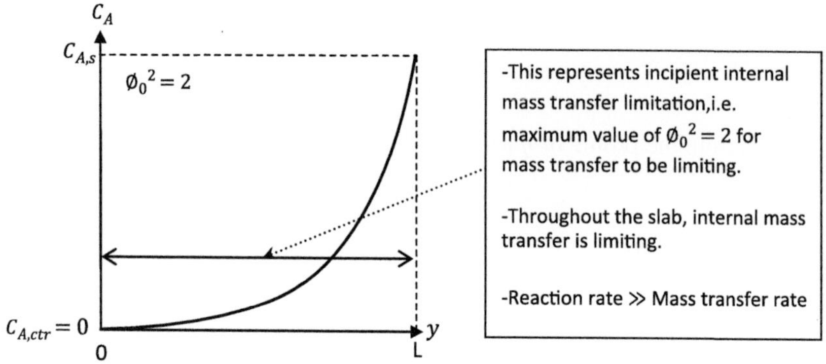
$C_A$
$C_{A,s}$
$\emptyset_0^2 = 2$
$C_{A,ctr} = 0$
0
L
$y$
-This represents incipient internal mass transfer limitation,i.e. maximum value of $\emptyset_0^2 = 2$ for mass transfer to be limiting.
-Throughout the slab, internal mass transfer is limiting.
-Reaction rate ≫ Mass transfer rate

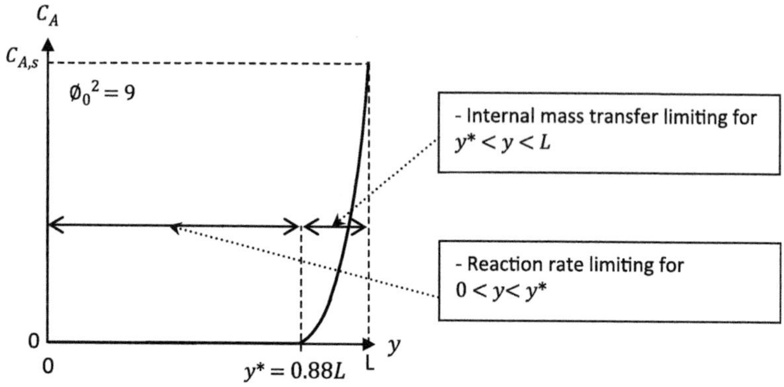


## Problem 34

**Consider a CSTR that has a volume of $V = 3\ \text{m}^3$. The inlet stream contains species $A$ of concentration $C_0$ while the outlet stream contains species $A$ at concentration $C_1$. The inlet and outlet flow rates are maintained at $\dot{Q} = 0.06\ \text{m}^3/\text{s}$.**

(a) **Derive a first-order differential equation that relates the outlet concentration of $A$ to the inlet concentration.**
(b) **Assuming that at $t = 0$, $C_0 = C_1 = 0$, determine an expression for $C_1(t)$ in terms of Laplace transform variables.**
(c) **Prior to $t = 0$, the concentration of $A$ in the inlet pipe is 80 units (measured in normalized dimensionless units). In the initial duration of 0.5 s, i.e., from $t = 0$ to 0.5 s, the inlet concentration is a constant value of 80 units after which it is maintained at 15 units during the operation of the CSTR. The concentration profile of $A$ is as shown below.**

i. **Derive the Laplace expression for the outlet concentration of $A$, $\bar{C}_1(s)$.**
ii. **Find the outlet concentration profile of $A$, $C_1(t)$ and show it in a plot.**

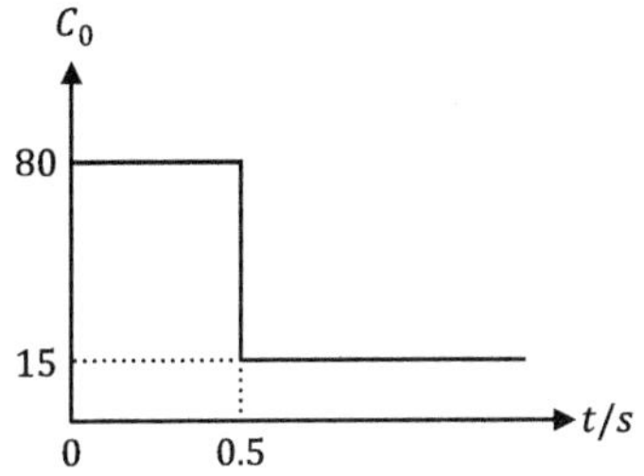


**Solution 34**

*Worked Solution*

(a) The CSTR is assumed well mixed, whereby the outlet concentration $C_1$ is equivalent to the concentration in the tank at time $t$. We are told that the inlet flow rate is equivalent to outlet flow rate, i.e., $\dot{Q}$ is constant. Therefore we can perform a mass balance for species $A$ as follows, where $\tau$ is the mean residence time in the CSTR:

$$\text{Inflow} - \text{Outflow} = \text{Rate of accumulation}$$

$$\dot{Q}C_0 - \dot{Q}C_1 = \frac{d(C_1 V)}{dt}$$

$$C_0 - C_1 = \frac{V}{\dot{Q}}\frac{dC_1}{dt} = \tau\frac{dC_1}{dt}$$

(b) Laplace transform converts the terms as functions of time $t$ to functions of Laplace variable $s$. The Laplace transform of the first-order differential equation in part a is:

$$C_0(t) - C_1(t) = \tau \frac{dC_1}{dt}(t)$$

$$\bar{C}_0(s) - \bar{C}_1(s) = \tau[s\bar{C}_1(s) - C_1(0)]$$

We know that $C_1(0) = 0$; therefore,

$$\frac{1}{\tau s + 1}\bar{C}_0(s) = \bar{C}_1(s), \text{ where by } \tau = \frac{V}{\dot{Q}} = \frac{3}{0.06} = 50 \text{ s}$$

(c) (i) Based on the description of the problem, we can come up with an expression for a step-like function that depicts the changes in inlet concentration of *A* from $t = 0$ s, where *H*(*t*) denotes a Heaviside function.

$$C_0(t) = 80H(t) - 65H(t - 0.5)$$

Since $C_0(0) = 0$, the Laplace transform of the inlet concentration expression is as follows:

$$\bar{C}_0(s) = \frac{80}{s} - \frac{65}{s}\text{e}^{-0.5 \text{ s}}$$

The outlet concentration is related to the inlet concentration via our result in part b, hence we can find out the Laplace expression for outlet concentration, $\bar{C}_1(s)$.

$$\bar{C}_1(s) = \frac{1}{\tau s + 1}\bar{C}_0(s)$$

$$= \frac{1}{\tau s + 1}\left(\frac{80}{s} - \frac{65}{s}\text{e}^{-0.5 \text{ s}}\right)$$

$$= \frac{80}{s(\tau s + 1)} - \frac{65}{s(\tau s + 1)}\text{e}^{-0.5 \text{ s}}$$

(ii) We can mathematically convert our result from (i) into partial fractions, to help us with our inverse Laplace transform back to variable *t*.

$$\bar{C}_1(s) = 80\left(\frac{1}{s} - \frac{\tau}{\tau s + 1}\right) - 65\left(\frac{1}{s} - \frac{\tau}{\tau s + 1}\right)\text{e}^{-0.5 \text{ s}}$$

$$= 80\left(\frac{1}{s} - \frac{1}{s + \frac{1}{\tau}}\right) - 65\left(\frac{1}{s}\text{e}^{-0.5 \text{ s}} - \frac{1}{s + \frac{1}{\tau}}\text{e}^{-0.5 \text{ s}}\right)$$

We can break down the inverse Laplace transform into parts for clarity.

$$\mathcal{L}^{-1}\left(\frac{1}{s}\right) = 1, \quad \mathcal{L}^{-1}\left(\frac{1}{s+\frac{1}{\tau}}\right) = \mathrm{e}^{-t/\tau}$$

$$\mathcal{L}^{-1}\left(\frac{1}{s}\mathrm{e}^{-0.5\ s}\right) = H(t-0.5)$$

$$\mathcal{L}^{-1}\left(\frac{1}{s+\frac{1}{\tau}}\mathrm{e}^{-0.5\ s}\right) = H(t-0.5)\ \mathrm{e}^{-(t-0.5)/\tau}$$

Combining our results, we have the following outlet concentration profile.

$$C_1(t) = 80\left(1-\mathrm{e}^{-t/\tau}\right) - 65H(t-0.5)\left(1-\mathrm{e}^{-(t-0.5)/\tau}\right)$$

We can now plot this profile, noting that there are two key time-periods, $t \le 0.5$ and $t \ge 0.5$. When $t \le 0.5$ s,

$$C_1(t) = 80\left(1-\mathrm{e}^{-t/\tau}\right)$$

At $t = 0.5$ s, $C_1(t) = 0.8$. When $t \ge 0.5$ s,

$$C_1(t) = 80\left(1-\mathrm{e}^{-t/\tau}\right) - 65H(t-0.5)\left(1-\mathrm{e}^{-(t-0.5)/\tau}\right)$$

$$= 80 - 80\mathrm{e}^{-\frac{t}{\tau}} - 65 - 65\mathrm{e}^{\frac{0.5}{\tau}}\mathrm{e}^{-\frac{t}{\tau}} = 15 - \mathrm{e}^{-\frac{t}{\tau}}\left(80 - 65\mathrm{e}^{\frac{0.5}{\tau}}\right), \text{where } \tau = \frac{V}{\dot{Q}} = \frac{3}{0.06} = 50\,\mathrm{s}$$

$$= 15 - \mathrm{e}^{-\frac{t}{50}}\left(80 - 65\mathrm{e}^{\frac{0.5}{50}}\right) = 15 - 14.3\mathrm{e}^{-\frac{t}{50}}$$

As $t \to \infty$, $C_1(t) \to 15$.

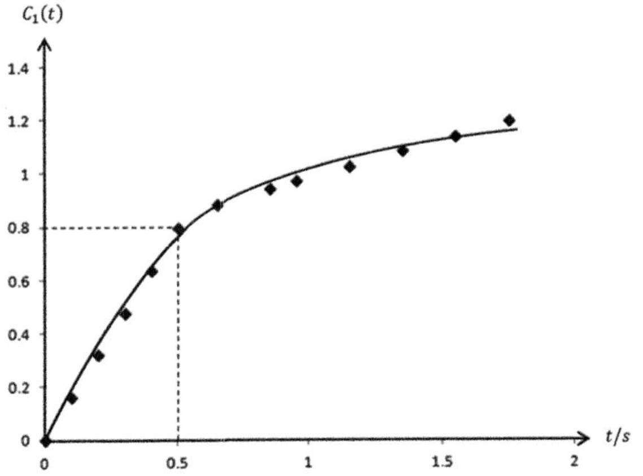

## Problem 35

**Consider the reaction $A \rightarrow B + 2C$ whereby the overall rate of reaction is limited by external diffusion of bulk reactant $A$ to the particle surface. The reaction occurs in a packed bed reactor that contains spherical catalytic particles of diameter $D_p = 0.007$ m. Gaseous reactant $A$ flows through the reactor in the axial direction at a superficial molar average velocity $V_s = 5$ m/s. The bed porosity $\emptyset = 0.5$ and it may be assumed that the surface area of catalyst per unit volume reactor, $a = \frac{6(1-\emptyset)}{D_p}$.**

(a) **By performing a mass balance, derive a differential equation for concentration of $A$.**
(b) **Determine the profile of reaction rate in the axial direction. You may assume that convective flux dominates axial diffusive flux.**
(c) **Determine the length of PBR required to achieve a conversion of 85%. The mass transfer coefficient, $k_c = 1.2$ m/s and $= 40$ m$^2$ cat m$^{-3}$ reactor.**

## Solution 35

### *Worked Solution*

(a) Let us start with a mole balance by referring to the PBR as shown below, where bulk flow is in the axial direction ($z$ direction).

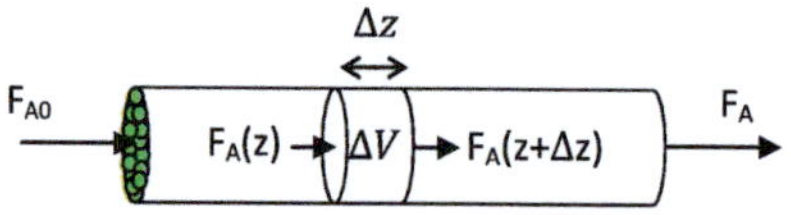


A mole balance for species $A$ in a differential element follows, note that $r_A < 0$:

$$F_A(z) - F_A(z + \Delta z) + r_A V = \frac{dN_A}{dt}$$

$$F_A(z) - F_A(z + \Delta z) + r_A V = 0 \text{ (steady state)}$$

Let $A$ refer to the cross-sectional area of the packed bed reactor. $r_A''$ refers to the reaction rate per unit surface area of catalyst particle with units of mol s$^{-1}$ m$^{-2}$ cat, and $a$ refers to the surface area of catalyst particle per unit volume of reactor with units of m$^2$ cat m$^{-3}$ reactor.

For packed beds, $a$ is related to bed porosity Ø and diameter of particle $D_p$ by the equation:

$$a = \frac{6(1-\emptyset)}{D_p} = \frac{6(1-0.5)}{0.007} = 430$$

$$F_A(z) - F_A(z + \Delta z) + (r_A{}'' \, a)(A\Delta z) = 0$$

Dividing throughout by $A\Delta z$ and taking the limit where $\Delta z \to 0$,

$$-\left(\frac{1}{A}\right)\frac{dF_A}{dz} + r_A{}'' \, a = 0$$

In the general form, we know that the molar flow rate of $A$ in the axial direction $z$ is contributed by diffusion (or dispersion) in the axial direction denoted here by $j_{Az}{}''$ and convection (or bulk flow) in the axial direction denoted here by $b_{Az}{}''$.

$$\frac{F_A}{A} = j_{Az}{}'' + b_{Az}{}''$$

We often assume that $b_{Az}{}'' \gg j_{Az}{}''$ for packed bed reactors, as they are often operated in turbulent regimes where bulk flow transport dominates.

$$\frac{F_A}{A} = b_{Az}{}'' = V_s C_A$$

Substituting this back into the differential equation, we have the following, assuming $V_s$ is held constant.

$$-V_s \frac{dC_A}{dz} + r_A{}'' \, a = 0$$

We can equate $r_A{}''$ [mol s$^{-1}$ m$^{-2}$ cat] which is the rate of disappearance of $A$ at the particle surface to the flux of $A$ from the bulk fluid to particle surface, under steady state. For the purpose of this, we define $k_c$ as the mass transfer coefficient whereby $k_c[\text{m s}^{-1}] = \frac{\mathcal{D}}{\delta'}$ and $\mathcal{D}$ represents diffusivity and $\delta$ represents the diffusion boundary layer around the particle surface.

$$-r_A{}'' = k_c(C_{A,\text{bulk}} - C_{A,s})$$

Therefore we substitute this back into the differential equation, note that $C_A$ is the same as $C_{A,\text{bulk}}$ above.

$$-V_s \frac{dC_A}{dz} - k_c(C_A - C_{A,s}) \, a = 0$$

(b) In most cases where mass transfer is limiting, reaction rate is assumed much faster so $C_{A,s} = 0$. Or in other words, $C_{A,\text{bulk}} \gg C_{A,s}$.

$$-V_s\frac{dC_A}{dz} - ak_cC_A = 0$$

We can now integrate with the boundary condition that at $z = 0$, $C_A = C_{A0}$.

$$\frac{dC_A}{dz} = -\frac{ak_c}{V_s}C_A$$

$$C_A = C_{A0}\exp\left(-\frac{ak_c}{V_s}z\right)$$

Therefore the rate of reaction varies along the $z$ direction as follows

$$-r_A{''} = k_cC_{A0}\exp\left(-\frac{ak_c}{V_s}z\right)$$

(c) To find out the length of bed $L$ required to achieve a specific conversion $X = 0.85$, we can express conversion in terms of concentration

$$X = 1 - \frac{C_{AL}}{C_{A0}} = 0.85$$

$$C_{AL} = C_{A0}\exp\left(-\frac{ak_c}{V_s}L\right)$$

$$\ln\left(\frac{C_{AL}}{C_{A0}}\right) = \ln(1 - X) = -\frac{ak_c}{V_s}L$$

$$\ln(1 - 0.85) = -\frac{40(1.2)}{5}L$$

$$L = 0.198 \text{ m}$$

**Problem 36**

**Consider a reactor of volume 6 m$^3$ whereby gas is fed and bubbled through it from the bottom inlet and exits from the top. At the same time, liquid flows through at a flow rate of 0.003 m$^3$/s. A delta pulse tracer of total mass 120 g was injected at the liquid inlet to study the residence time distribution and flow pattern in the tank. The following plot was drawn from the measured outlet concentration of the tracer.**

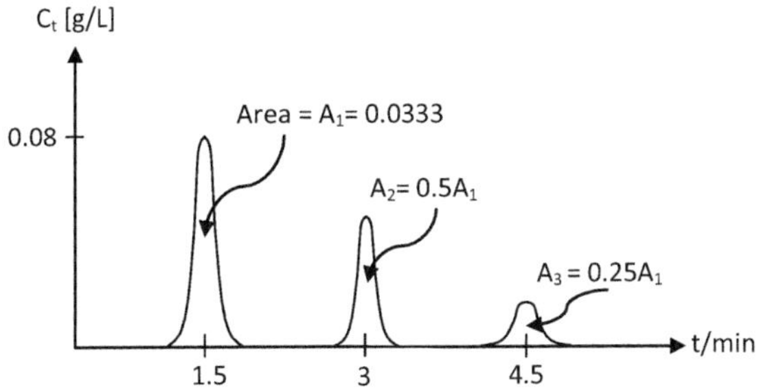


(a) **Verify the soundness of the experimental results obtained by applying mass balance on the mass of tracer injected.**
(b) **Determine the vapor fraction in the reactor.**

## Solution 36

***Worked Solution***

(a) We are told that the total mass of tracer injected is 120 g at 3 L/s. Based on this, we should expect the total area under the curve of all the tracer peaks to add up to

$$\text{Area}\left[\text{g min L}^{-1}\right] = \frac{120}{3(60)} = 0.0666$$

From the peaks, we observe the pattern as follows, where $A_n$ refers to area under the $n$th peak:

$$A_1 = 0.0333$$

$$A_2 = \left(\frac{1}{2}\right)A_1 \; ; A_3 = \left(\frac{1}{2}\right)^2 A_1$$

$$A_n = \left(\frac{1}{2}\right)^{n-1} A_1$$

We note that the areas of the peaks converge to a specific value:

$$\frac{1}{2} + \left(\frac{1}{2}\right)^2 + \left(\frac{1}{2}\right)^3 \ldots = 2 \text{ g min L}^{-1}$$

$$\text{Total area} = A_1 + A_2 + A_3 \ldots = A_1\left(1 + \frac{1}{2} + \left(\frac{1}{2}\right)^2 + \left(\frac{1}{2}\right)^3 + \ldots\right) = 0.0333(2)$$

$$\text{Total area} = 0.0666 \text{ (verified)}$$

We verify that the experimental results are sound since total area adds up to 0.0666.

(b)

$$\bar{t}_t = \frac{\int tC_t dt}{\int C_t dt}$$

$$= \frac{1}{0.0666}\left[1.5(0.0333) + 3\left(\frac{1}{2}\right)(0.0333) + 4.5\left(\frac{1}{2}\right)^2(0.0333)\ldots\right]$$

$$= \frac{0.2}{0.0666} = 3 \text{ min}$$

The liquid flow rate is 3 L/s or 1800 L/min. Therefore, the total liquid volume in one minute of flow is

$$V_l = 3(1800) = 5400 \text{ L}$$

The volume of gas and gas fraction are therefore

$$V_g = 6000 - 5400 = 600 \text{ L}$$

$$\text{Vapor fraction} = \frac{600}{6000} = 0.1 \text{ or } 10\%$$

## Problem 37

(a) **Explain what it means by homogeneous and heterogeneous reactions and list some of the advantages and disadvantages of each.**

(b) **Describe the structure of a zeolite, and explain their applications for commercial use.**

## Solution 37

***Worked Solution***

(a)
- Homogeneous reactions are reactions whereby the catalyst and reactants are in the same phase, for example, both gases or both liquids. An example of a homogeneous reaction is the industrial liquid phase process for producing isobutyl aldehyde using a cobalt complex.
- Conversely, heterogeneous reactions are reactions whereby catalyst and reactant are in different phases, it is commonly the case that the catalyst is a solid, while the reactant and products are in gas or liquid state. An example of a heterogeneous reaction is the hydrogenation of vegetable oils into saturated fats over nickel metal mounted on an inert solid support. The

metal is dispersed on the support to maximize its exposed surface area to optimize contact with the reactants. Other heterogeneous reactions that occur in catalytic industrial processes include dehydrogenation of alkanes into alkenes, steam reforming, oxidation of carbon monoxide and hydrocarbons and zeolitic hydrocracking.

- The advantages of homogeneous reactions include high selectivity which is favorable for achieving optimal conversion in batch reactions. However homogeneous reactions are expensive and its process conditions may be difficult to maintain. Separation and purification processes after the reaction is complete may be tedious due to the need to separate substances in the same phase.
- The advantages of heterogeneous reactions include being inexpensive and process conditions are relatively robust and hence easier to manage. Post-reaction purification techniques required are more straightforward. However, these reactions tend to have lower selectivity as compared to homogeneous reactions since the latter allows better mixing and contact between reactants and catalyst.

(b)
- Zeolites are a specific type of crystalline aluminosilicate catalyst, with stacked alumina and silica tetrahedra. Structurally, they are open, three-dimensional honeycomb shaped with a negative charge within their pores. This negative charge is neutralized by positive cations (e.g., $Na^+$ or $H^+$ ions which are interchangeable via ion exchange).
- The specific shape of its pores allows selectivity at the molecular level, and its charge provides it with acid activity.
- The shape selectivity allows differentiation even between isomers. Hence, the choice of zeolites can reduce feed purification steps as its specific shape filters out "unsuitably sized" substances in the feed. Furthermore, product separation steps can also be minimized as they can help inhibit the production of certain undesirable intermediates in the reaction that they catalyze.

## Problem 38

**Pulse tracer experiments are useful in studying the flow patterns within reactors and may help us diagnose and resolve undesirable flow characteristics. Describe some examples of undesirable flow patterns for the PFR and CSTR, via simple plots of their tracer concentration profiles.**

## Solution 38

***Worked Solution***

Pulse tracer experiments can provide us with insights on the residence time distribution of reactors, and are sometimes termed "RTD fingerprints".

Here are some examples of the pulse tracer response curves of an ideal PFR and "misbehaving" PFRs. Let $\tau = \frac{V}{Q}$.

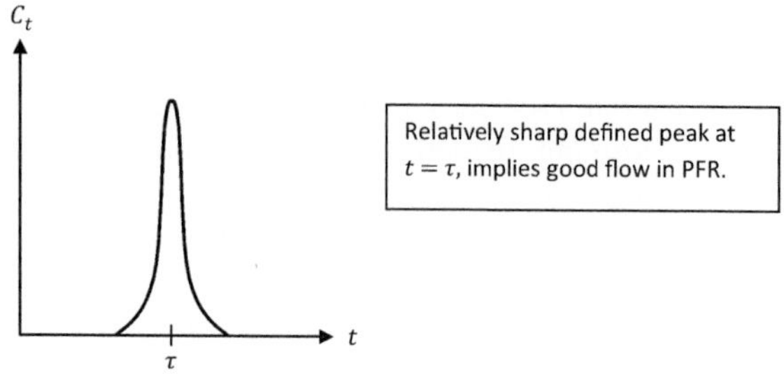


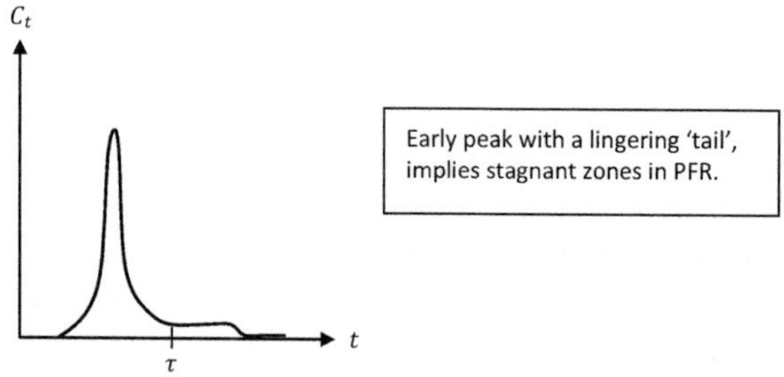
$C_t$
$t$
$\tau$
Early peak with a lingering 'tail', implies stagnant zones in PFR.

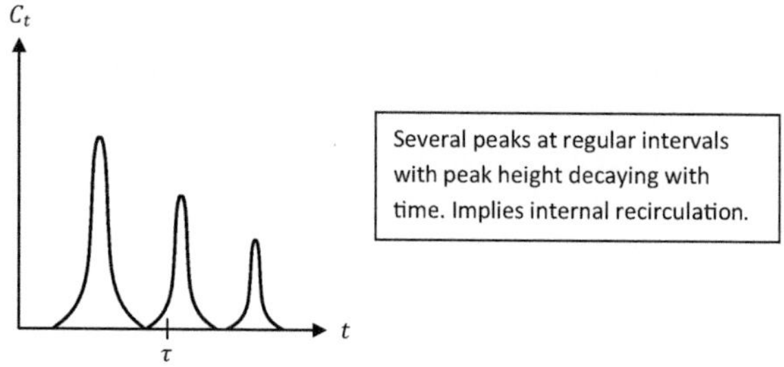
$C_t$
$t$
$\tau$
Several peaks at regular intervals with peak height decaying with time. Implies internal recirculation.

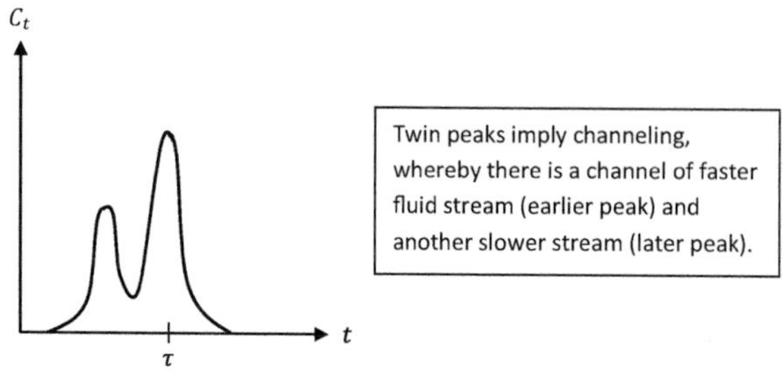
$C_t$
$t$
$\tau$
Twin peaks imply channeling, whereby there is a channel of faster fluid stream (earlier peak) and another slower stream (later peak).

Here are some examples of the pulse tracer response curves of an ideal CSTR and "misbehaving" CSTRs. Let $\tau = \frac{V}{\dot{Q}}$.

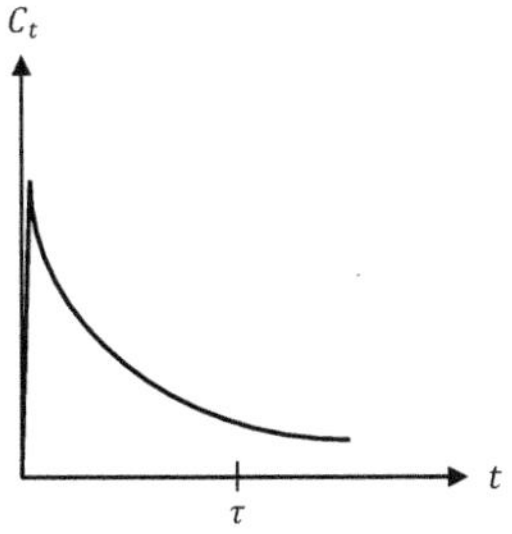


Exponential decay with $t = \tau$ occurring at where the decay is about 1/e. Implies good flow in CSTR.

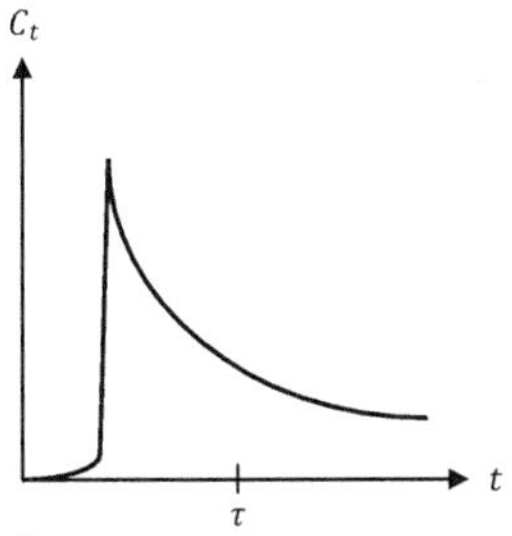


Time lag before tracer response implies that plug flow in series prior to mixed flow region. Could imply presence of a long pipe upstream.

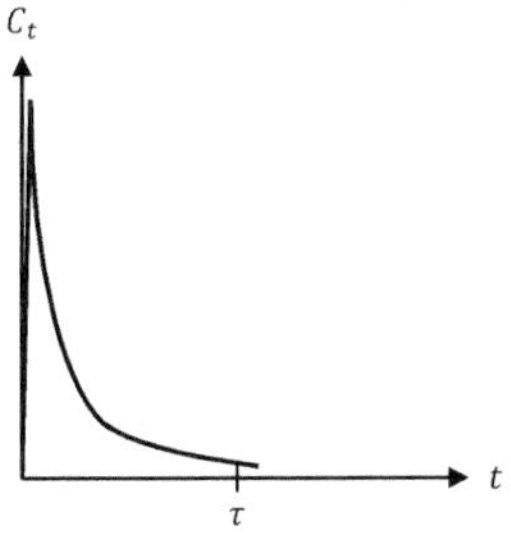


Exponential decay but decay occurs prematurely, implies stagnant zone present. If peak is very sharp and very early, implies short-circuiting from inlet to outlet.

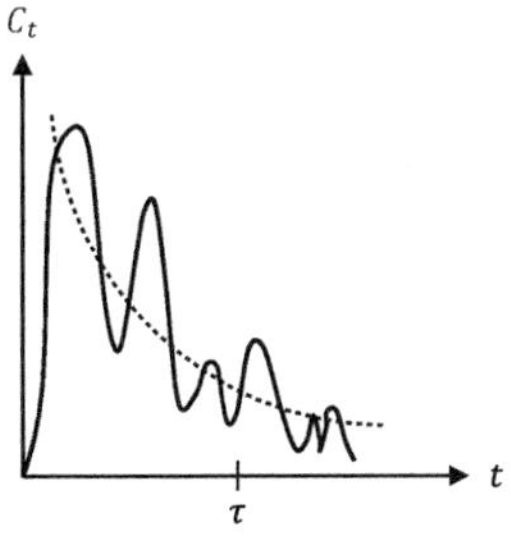


General exponential decay but comprise of several peaks. Implies zones of poor, non-uniform mixing.

## Problem 39

**Consider a packed bed tubular vessel with a voidage of 0.4, with a circular cross section of radius $R = 0.7$ m, and length of bed $L = 15$ m. Gas is bubbled through the reactor from the bottom inlet, at a volumetric flow rate of 0.35 m$^3$/s. Liquid enters from the top and flows through at a flow rate of 0.05 m$^3$/s. A pulse of inert tracer was injected, and the tracer response curves are shown below for the liquid and gas, respectively. Determine the vapor and liquid fractions in the tubular vessel, and comment on whether there are "dead" volumes.**

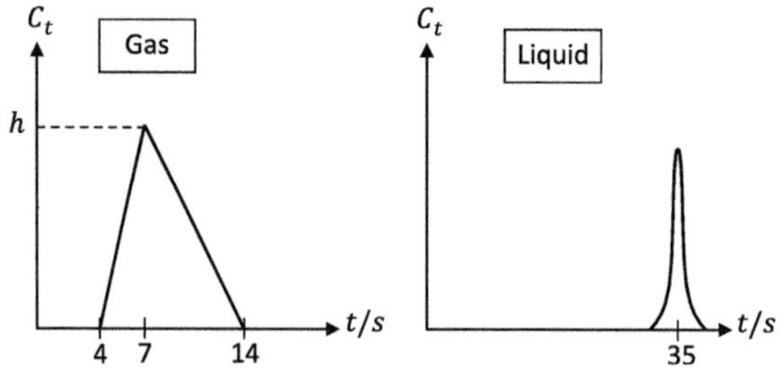


**Solution 39**

***Worked Solution***

The mean residence time for the gas is calculated as follows

$$\bar{t}_g = \frac{\sum tC}{\sum C} = \frac{0.5\,\mathrm{h}(7-4)(6) + 0.5\,\mathrm{h}(14-7)(9.3)}{(14-4)0.5\,\mathrm{h}} = 8.3\ \mathrm{s}$$

The mean residence time for the liquid is obtained directly from the plot

$$\bar{t}_l = 35\ \mathrm{s}$$

Therefore, we can find the volumes of gas phase and liquid phase

$$V_g = \bar{t}_g \dot{Q}_g = 8.3(0.35) = 2.9\ \mathrm{m}^3$$

$$V_l = \bar{t}_l \dot{Q}_l = 35(0.05) = 1.75\ \mathrm{m}^3$$

We can calculate the total volume and void volume of the packed bed vessel.

$$\text{Volume} = \pi R^2 L = \pi\left(0.7^2\right)(15) = 23\ \mathrm{m}^3$$

$$\text{Void volume} = 0.4(23) = 9.2\ \mathrm{m}^3$$

$$\text{Gas fraction} = \frac{2.9}{9.2} = 0.32 \text{ or } 32\%$$

$$\text{Liquid fraction} = \frac{1.75}{9.2} = 0.19 \text{ or } 19\%$$

$$\text{Dead volume (stagnant) fraction} = 0.49 \text{ or } 49\%$$

There is significant dead volume, hence this setup is poorly designed as there are large stagnant zones which represents poor mixing and contact between gas and liquid. One of the ways to reduce the fraction of stagnant volume is to increase the gas and liquid flow rates.

**Problem 40**

**Consider a homogeneous reaction in the gas phase $A \rightarrow 3B$ that occurs at 490 K and 3 bar in a PFR. The feed consists of 50% $A$ and 50% of inert.**

(a) **Show that for this reaction of varying volume, the concentration of $A$ can be expressed as follows, where $C_{A0}$ is the inlet feed concentration, $X$ is the conversion, and $\varepsilon$ is the fractional change in volume of the system between $X = 0$ and $X = 1$.**

$$C_A = C_{A0}\left(\frac{1-X}{1+\varepsilon X}\right)$$

(b) **Find the space time $\tau$ needed for $X = 0.90$ if $C_{A0} = 5.4 \times 10^{-5}$ mol m$^{-3}$. The rate equation is given as follows in units of mol m$^{-3}$ s$^{-1}$.**

$$-r_A = 0.015{C_A}^{0.5}$$

**Solution 40**

*Worked Solution*

(a) For varying volume batch reactor systems, the volume as reaction progresses changes with conversion. We can express this as follows, where $V_0$ is the initial volume.

$$V = V_0(1 + \varepsilon X)$$

Therefore, we can express concentration of $A$ as follows,

$$C_A = \frac{N_A}{V} = \frac{N_{A0}(1-X)}{V_0(1+\varepsilon X)} = C_{A0}\left(\frac{1-X}{1+\varepsilon X}\right)$$

(b) $\varepsilon$ represents the fractional change in volume of the system between $X = 0$ and $X = 1$. For our reaction, the volume at $X = 0$ consists of one part inert and one part $A$, hence a total of 2 parts of gas in the total volume. Separately, the volume at $X = 1$ consists of 1 part inert and 3 parts $B$, hence a total of 4 parts of gas in the total volume. Putting together these results into the equation, we have $\varepsilon = 1$.

$$\varepsilon = \frac{4-2}{2} = 1$$

Space time $\tau$ is defined as the time required to process one reactor volume of feed, and is also called mean residence time. $\tau = \frac{V}{Q}$, and it is usually measured using the inlet conditions. For a PFR, we can derive an expression for space time $\tau$ starting from a mass balance on $A$.

$$F_{A,\text{in}} - F_{A,\text{out}} + r_A dV = 0$$
$$-dF_A + r_A dV = 0$$
$$-d[F_{A0}(1-X)] + r_A dV = 0$$
$$F_{A0} dX + r_A dV = 0$$
$$-r_A \int_0^V dV = C_{A0}\dot{Q} \int_0^X dX$$
$$\frac{V}{\dot{Q}} = \tau = C_{A0} \int_0^X \frac{dX}{-r_A}$$

We know the rate law for this reaction which we can substitute into the equation above.

$$\tau = C_{A0} \int_0^X \frac{dX}{0.015 {C_A}^{0.5}}$$

$$\tau = C_{A0} \int_0^X \frac{dX}{0.015\left[C_{A0}\left(\frac{1-X}{1+\varepsilon X}\right)\right]^{0.5}} = C_{A0} \int_0^X \frac{dX}{0.015 {C_{A0}}^{0.5}\left(\frac{1-X}{1+X}\right)^{0.5}}$$

We are told that $C_{A0} = 5.4 \times 10^{-5}$ mol m$^{-3}$ and we require $X = 0.90$; therefore,

$$\tau = \frac{\left(5.4 \times 10^{-5}\right)^{0.5}}{0.015} \int_0^{0.9} \left(\frac{1+X}{1-X}\right)^{0.5} dX$$

To solve this integral, we can apply Simpson's rule, for an even number, $n$ of equally spaced intervals. We can create 4 equally spaced intervals of 0.225 each, i.e., $n = 4$

| $X$ | $\left(\frac{1+X}{1-X}\right)^{0.5}$ |
|---|---|
| 0 | 1 |
| 0.225 | 1.257 |
| 0.45 | 1.624 |
| 0.675 | 2.270 |
| 0.90 | 4.359 |

Then, using Simpson's rule, we can evaluate the integral as follows.

$$\int_0^{0.9} \left(\frac{1+X}{1-X}\right)^{0.5} dX \approx \frac{0.9-0}{3(4)}[1(1)+4(1.257)+2(1.624)+4(2.270)+1(4.359)] = 1.7$$

$$\tau = \frac{\left(5.4 \times 10^{-5}\right)^{0.5}}{0.015}(1.7)$$

$$\tau = 0.83 \text{ s}$$

Hence, the space time for the required conversion is 0.83 s.

## Problem 41

**In a batch culture of bacterial cells, fresh nutrient medium is inoculated with an initial population of cells and incubated over a period of time. A characteristic growth pattern can then be observed by taking periodic measurements of cell count and plotting this data to obtain the bacterial growth curve.**

(a) **Plot a typical bacteria growth curve assuming a constant growth rate and describe key aspects of each phase. Explain what is meant by the term "doubling time".**
(b) **Briefly explain the term "growth yield coefficient" and how it can be derived theoretically and experimentally.**
(c) **Describe the Monod growth model, highlighting key characteristics.**
(d) **A strain of bacteria was batch-cultured in a growth medium at a concentration of 30 g growth nutrient/L. The initial inoculum had a cell density of 0.2 g/L and the final required cell density was 12 g/L. Given that the doubling time for this strain of bacteria is 30 min and the growth yield coefficient is 0.4 g/g, determine the time taken to reach the final cell density if the bacterial growth was modeled as:**

   i. **Exponential growth**
   ii. **Monod growth with a Monod constant equivalent to 2 g/L**

(e) **Plot the bacterial growth profiles for the two models indicated in part d, and comment on their shapes.**

## Solution 41

*Worked Solution*

(a) Bacterial growth typically follows exponential growth in a closed system (e.g., batch culture in test tube). Cellular growth can be described using the 4-stage growth curve as shown below where the change in the number of viable cells is recorded over time. [Note that a logarithmic scale is used for the vertical axis, hence the straight lines.]

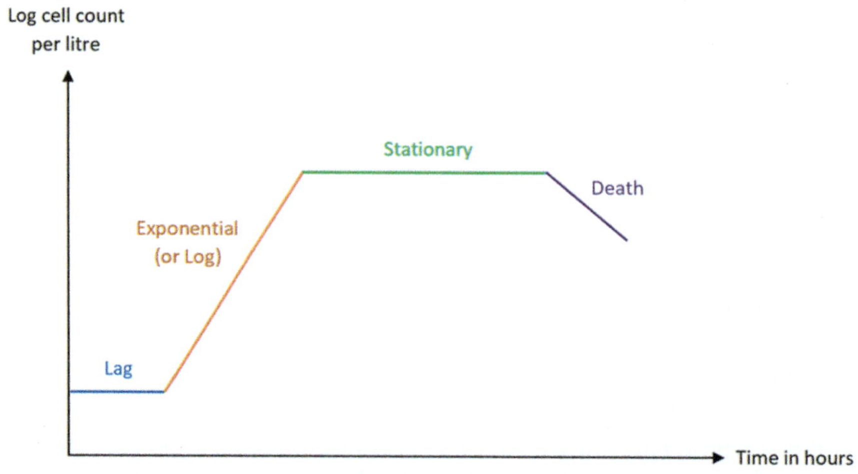


*Lag Phase*

In the initial lag phase, cells are just adjusting to new growth conditions such as the nutrient medium and so have not begun multiplying.

Although there may not be cell division occurring as observed from no increase in cell count, the existing cells in the inoculum are using this time to ramp up their metabolic activities and grow in size as they synthesize new cell matter such as proteins, enzymes, and ribonucleic acids. The length of the lag phase depends on factors including the size of inoculum, the time needed for the cells to recover from any shock in being transferred to a new culture medium, and the time required to metabolize nutrients in the media substrate, and the time required to synthesize essential enzymes necessary for cell division in the subsequent log phase.

As there is no increase in cell numbers in the lag phase, it is typically desired to optimize cell culture productivity by minimizing the duration of this period. Some methods to achieve this include using relatively active cells undergoing the exponential phase as the inoculum, pre-acclimatizing the inoculum to growth media, and ensuring a sufficiently high cell density in the inoculum (e.g., approximately 5–10% by volume).

*Exponential/Log Phase*

After the lag phase, we enter into rapid and balanced growth in the exponential or log phase whereby cells divide at a constant rate by binary fission and cell number (or mass) increases exponentially. As cells multiply by binary fission, the increase in cell count follows a geometric progression. This means that a single cell splits into two in the first generation, and two cells split into four cells in the second generation, etc. Therefore, we can describe cellular multiplication as follows whereby $N_0$ and $N$ refer to the initial and final numbers of cells, and $n$ refers to the number of times the cell population doubles (i.e., the number of generations).

$$N = N_0(2^n)$$

Taking natural log (you can choose any logarithmic base) on both sides, we have

$$\ln N = \ln[N_0(2^n)] = \ln N_0 + \ln 2^n$$

$$\ln N - \ln N_0 = \ln \frac{N}{N_0} = n \ln 2$$

The *"doubling (or generation) time"* is the time needed for cells to double (or divide) and we can denote it as $t_d$. The generation time for the common Escherichia Coli bacteria under optimal conditions using glucose as substrate is approximately 17 mins. Mathematically, the total time $t$ taken to reach a final number of $N$ cells can be written as such

$$nt_d = t$$

Substituting this result into our earlier equation, we obtain the growth rate (units of time$^{-1}$) which we denote here as $\mu$.

$$\ln \frac{N}{N_0} = \frac{t}{t_d} \ln 2$$

$$\mu = \frac{\ln \frac{N}{N_0}}{t} = \frac{\ln 2}{t_d}$$

We can observe that this can be written in another form as shown below which shows more clearly the exponential pattern of growth.

$$N = N_0 e^{\mu t}$$

$$\ln \frac{N}{N_0} = \mu t$$

In the exponential phase, nutrient (or substrate) concentrations are in excess and growth rate is independent of nutrient (or substrate) concentrations and is at a maximum value where $\mu = \mu_{\max}$. [Note that $\mu \neq \mu_{\max}$ if other (i.e., not exponential) growth patterns are relevant, e.g., Monod growth model]

$$\frac{dN}{dt} = \mu N = \mu_{\max} N$$

*Stationary Phase*

Between the exponential and stationary phases there is usually a short transition period when cell growth decelerates (deceleration phase not shown in diagram above). This occurs because in a closed system, growth conditions change with time. Cells get "overcrowded," nutrients become depleted, and toxic by-products of growth may accumulate. Cell growth decelerates as metabolism shifts into survival mode and growth becomes unbalanced and limited.

The slowing down of growth eventually leads to the stationary phase whereby cells stop dividing and there is no net increase in cell numbers. Cell growth rate is equivalent to cell death rate. The number of cells in existence are maintained through endogenous metabolism of energy stores.

*Death Phase*

Eventually cells start to die as their energy stores become depleted and the closed system does not provide additional resources to sustain cellular activities required for continued survival. Cells may undergo lysis and this phase may be modeled as a first-order reaction similar to that of the exponential phase except that a "death rate" denoted below as $k_d$ is used instead of growth rate $\mu$ and the negative sign accounts for a reduction in cell count (or mass).

$$\frac{dN}{dt} = -k_d t$$

$$\ln \frac{N}{N_0} = -k_d t$$

(b) Cell mass of microorganisms such as bacteria can be expressed quantitatively by "*yield coefficients*" in terms of mass of cells per unit mass of substrate consumed. $Y_{x/s}$ is a typical notation to represent the growth coefficient whereby subscripts $x$ and $s$ refer to cells and substrate (or growth-limiting nutrient), respectively. $Y_{x/s}$ is useful as it can help in the analysis of material balances involving cells, substrate, and product.

Consider the following general example of a growth equation below whereby S refers to substrate, N refers to a nitrogen source, X refers to cell mass, and P refers to product. The alphabets *a*, *b*, *c*, *d*, and e are stoichiometric coefficients.

The theoretical yield coefficient can be determined from the stoichiometry of the growth equation below, if the chemical formula for S, N, X, and P are specified. A common substrate for cellular growth is glucose, $C_6H_{12}O_6$ and an example of product is ethanol, $C_2H_5OH$. The stoichiometric coefficients should be such that the growth equation is balanced for all elements on both sides of the arrow.

$$a\mathrm{S} + b\mathrm{N} + c\mathrm{O}_2 \rightarrow \mathrm{X} + d\mathrm{P} + e\mathrm{H}_2\mathrm{O} + f\mathrm{CO}_2$$

$$Y_{\mathrm{x/s}} = \frac{MW_{\mathrm{x}}}{aMW_{\mathrm{s}}}$$

The value of $Y_{x/s}$ can also be determined experimentally (more common), by making measurements of cell mass and substrate consumed. If bacterial growth occurs in a batch reactor of fixed volume, then the growth rate of cells can be expressed as follows whereby $\mu$ represents specific growth rate.

$$\frac{dx}{dt} = \mu x$$

The rate of substrate (growth-limiting) consumption can then be expressed as follows. We indicate "growth-limiting" in this case for the simplified assumption that all substrate consumed contributes to cell growth. In reality, substrate consumed by bacterial cells can also be used for to make extracellular products and produce energy to sustain cellular activities, in addition to being assimilated as cellular biomass.

$$-\frac{ds}{dt} = \frac{1}{Y_{x/s}} \frac{dx}{dt} \quad (1)$$

$$Y_{x/s} = \frac{\Delta x}{\Delta s} \quad (2)$$

From (1), we observe that the substrate consumption rate is dependent on cell growth rate and this interdependence can also be understood from the earlier growth equation. From (2), we also notice that $Y_{x/s}$ may not be a constant value, as it depends on the consumption rate of substrate which changes with growth condition.

(c) The Monod growth model was developed when it was observed that in certain continuous cultures of bacteria, growth rate $\mu$ was not constant (i.e., $\mu \neq \mu_{\max}$) and it varied in a non-linear manner with respect to the growth-limiting substrate as shown in the diagram below.

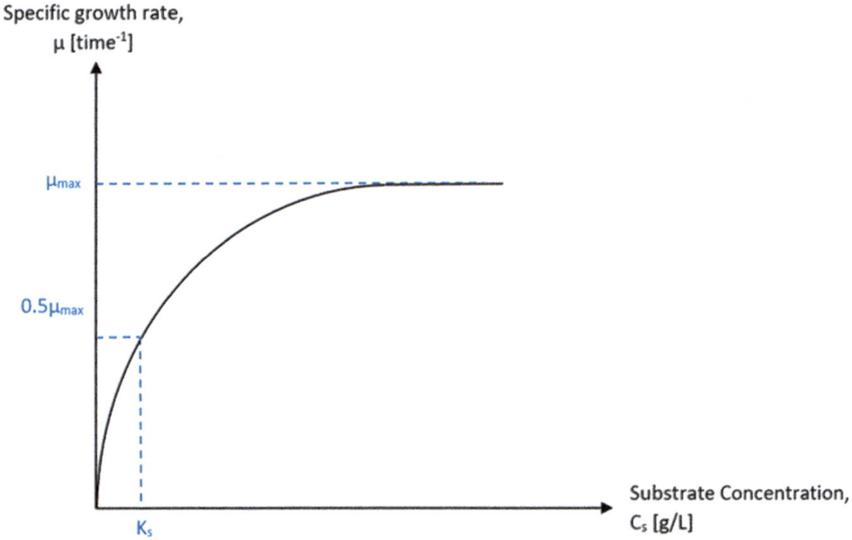

The above diagram illustrates the change in $\mu$ with substrate (growth-limiting) concentration, $C_s$. It is worth noting that the Monod model expressed in the equation below assumes a single culture and a single limiting substrate. $K_s$ is the Monod constant or half-velocity constant and is equivalent to substrate concentration when $\mu = \frac{\mu_{\max}}{2}$.

$$\mu = \mu_{\max} \frac{C_s}{C_s + K_s}$$

Under the Monod growth model, we can have zero-order or first-order kinetics depending on the concentration of substrate. At very high and low substrate concentrations, we tend towards limits for the analysis of enzymes and substrates in an enzyme-catalyzed reaction.

At low substrate concentration, we have the analytical region for the analysis of substrates where we observe first-order kinetics. Since $C_s \ll K_s$, the specific growth rate can be approximated as follows and growth rate becomes linearly proportional to the rate-limiting substrate concentration:

$$\mu \cong \mu_{\max} \frac{C_s}{K_s}$$

At high substrate concentration, we have the analytical region for the analysis of enzymes, where we observe zero-order kinetics. Growth rate is a constant at its maximum value, and this is the same conditions as in part a of this problem, where an exponential growth pattern was assumed. Since $C_s \gg K_s$, the specific growth rate is approximated as a constant value.

$$\mu \cong \mu_{\max}$$

(d) (i) If simple exponential growth was assumed, then the specific growth rate $\mu$ is constant as follows.

$$\mu = \mu_{\max}$$

We know from earlier in part a that $\mu$ can also be expressed in terms of doubling time $t_d$ as shown

$$\mu = \frac{\ln 2}{t_d}$$

Combining both expressions and substituting known values, we have

$$\mu_{\max} = \frac{\ln 2}{t_d} = \frac{\ln 2}{\left(\frac{30}{60}\right)} = 1.4\ \mathrm{h}^{-1}$$

To find the total time $t$ taken to reach a final cell concentration, $C_{x,2} = 12$ g/L, we construct a differential equation that describes the change in cell concentration (denoted as $C_x$ here) over time.

$$\frac{dC_x}{dt} = \mu_{\max} C_x$$

$$\int_{C_{x,1}}^{C_{x,2}} \frac{dC_x}{C_x} = \mu_{\max} \int_0^t dt$$

$$\int_{0.2}^{12} \ln C_x = \ln \frac{12}{0.2} = \mu_{\max} t = 1.4t$$

$$t = \frac{4.1}{1.4} = 2.9 \text{ h}$$

(ii) The cell concentration created is given as

$$12 - 0.2 = 11.8 \text{ g/L}$$

The amount of nutrient used up to create these new cells can be found using the growth yield coefficient $Y_{x/s} = 0.4$ g/g.

$$\Delta C_s = -\frac{11.8}{0.4} = -29.5 \text{ g/L}$$

Given that the initial nutrient concentration was $C_{s,1} = 30$ g/L, the nutrient remaining is therefore

$$C_{s,2} = C_{s,1} - \frac{C_{x,2} - C_{x,1}}{Y_{x/s}} = 30 - 29.5 = 0.5 \text{ g/L}$$

Assuming a Monod growth model, specific growth rate is not a constant and follows the expression below.

$$\mu = \mu_{\max} \frac{C_s}{C_s + K_s}$$

We know the initial and final nutrient (or growth-limiting substrate) concentrations to be 30 g/L and 0.5 g/L. This gives us a hint that working with a differential equation in terms of nutrient concentration, $C_s$ with respect to time would be ideal in determining time taken since we have known limits for nutrient concentrations for the said time interval. We also know the initial cell concentration given as 0.2 g/L. With these in mind, we go about writing out all equations that describe our system as follows:

$$\frac{dC_x}{dt} = \mu C_x = \mu_{\max} \frac{C_s C_x}{C_s + K_s}$$

We can use the growth yield coefficient $Y_{x/s}$ to convert the above differential equation which is a derivative of $C_x$ with respect to time, into one that is a derivative of $C_s$ with respect to time. By definition, $Y_{x/s}$ relates the two derivatives as follows:

$$\frac{dC_x}{dt} = -Y_{x/s} \frac{dC_s}{dt} \qquad (1)$$

Therefore, we have

$$\mu_{\max} \frac{C_s C_x}{C_s + K_s} = -Y_{x/s} \frac{dC_s}{dt}$$

We notice that the variable $C_s$ appears twice (once in numerator and once in denominator) in the fraction on the left-hand side of the above equation. A common "trick" in working with differential equations with many variables is to create a new variable that helps simplify and reduce the appearance of variables. Let us define a new variable $Z$ as follows.

$$Z = \frac{C_s}{K_s}$$

$$K_s dZ = dC_s$$

Substituting this into our earlier equation, we simplify the fraction such that the variable $Z$ appears only once. In doing so, we just need to ensure the equation still balances.

$$\mu_{\max} \frac{C_s C_x}{C_s + K_s} = -Y_{x/s} \frac{dC_s}{dt}$$

$$Y_{x/s} \frac{dC_s}{dt} = -\mu_{\max} \left( \frac{C_x}{1 + \frac{K_s}{C_s}} \right) = -\mu_{\max} \left( \frac{C_x}{1 + \frac{1}{Z}} \right)$$

Now we need to try to convert the other variable $C_x$ into an expression in terms of $C_s$ so that the differential equation is consistent in having only one variable $C_s$. We can do so using the growth yield coefficient. Following from equation (1), we can integrate from initial values $C_{x,1}$ and $C_{s,1}$ at $t = 0$ to an arbitrary time $t$

$$\frac{dC_x}{dt} = -Y_{x/s} \frac{dC_s}{dt}$$

$$C_x - C_{x,1} = -Y_{x/s}(C_s - C_{s,1})$$

$$C_x = C_{x,1} + Y_{x/s}C_{s,1} - Y_{x/s}C_s$$

We can now substitute this expression into our earlier differential equation

$$Y_{x/s}\frac{dC_s}{dt} = -\mu_{\max}\frac{C_{x,1} + Y_{x/s}C_{s,1} - Y_{x/s}C_s}{1 + \frac{1}{Z}}$$

$$\left(-\frac{Y_{x/s}}{\mu_{\max}K_s}\right)\frac{dC_s}{dt} = \frac{\left(\frac{C_{x,1}+Y_{x/s}C_{s,1}}{K_s}\right) - Y_{x/s}Z}{1 + \frac{1}{Z}}$$

We can also substitute $K_s dZ = dC_s$ to obtain a differential equation in terms of our new variable $Z$

$$\left(-\frac{Y_{x/s}}{\mu_{\max}}\right)\frac{dZ}{dt} = \frac{\left(\frac{C_{x,1}+Y_{x/s}C_{s,1}}{K_s}\right) - Y_{x/s}Z}{1 + \frac{1}{Z}}$$

When equations look too complicated to solve, we can group together constant values into a single new constant so that the equation appears more manageable to work through. So let us define a constant $H$ whereby

$$H = C_{x,1} + Y_{x/s}C_{s,1}$$

Therefore, we obtain the following,

$$\left(-\frac{Y_{x/s}}{\mu_{\max}}\right)\frac{dZ}{dt} = \frac{\left(\frac{H}{K_s}\right) - Y_{x/s}Z}{1 + \frac{1}{Z}}$$

$$\left(-\frac{Y_{x/s}}{\mu_{\max}}\right)\frac{dZ}{dt} = \frac{\left[\left(\frac{H}{K_s}\right) - Y_{x/s}Z\right]Z}{Z + 1}$$

$$\frac{Z + 1}{\left[\left(\frac{H}{K_s}\right) - Y_{x/s}Z\right]Z}dZ = \left(-\frac{\mu_{\max}}{Y_{x/s}}\right)dt$$

$$\frac{Z + 1}{Z\left[Z - \left(\frac{H}{Y_{x/s}K_s}\right)\right]}dZ = \mu_{\max}dt$$

In order to integrate the above, we can use partial fractions to simplify the fraction on the left-hand side of the equation into two separate fractions. Note that this technique only works if the degree of the numerator is no more than the degree of the denominator. In this example, the numerator is linear in $Z$ (degree of 1) and the denominator is quadratic in $Z$ (degree 2), hence we can use this method.

$$\frac{Z+1}{Z\left[Z-\left(\frac{H}{Y_{x/s}K_s}\right)\right]}=\frac{A}{Z}+\frac{B}{Z-\left(\frac{H}{Y_{x/s}K_s}\right)}$$

The fraction above can be reduced to two fractions, each with a constant as the numerator and denoted here as $A$ and $B$. To find the constants, we can use the "cover up rule".

To find $A$, we let $Z = 0$ and covering up $Z$ in the fraction, we have

$$A=\frac{Z+1}{Z-\left(\frac{H}{Y_{x/s}K_s}\right)}=\frac{0+1}{0-\left(\frac{H}{Y_{x/s}K_s}\right)}=\frac{1}{-\frac{H}{Y_{x/s}K_s}}=-\frac{Y_{x/s}K_s}{H}$$

Similarly to find $B$, we make the second factor in the denominator equal to zero.

$$Z-\left(\frac{H}{Y_{x/s}K_s}\right)=0\rightarrow Z=\frac{H}{Y_{x/s}K_s}$$

Covering up $\left[Z-\left(\frac{H}{Y_{x/s}K_s}\right)\right]$, we can find $B$

$$B=\frac{Z+1}{Z}=\frac{\frac{H}{Y_{x/s}K_s}+1}{\frac{H}{Y_{x/s}K_s}}=1+\frac{Y_{x/s}K_s}{H}$$

Therefore our fraction can be expressed as follows:

$$\frac{Z+1}{Z\left[Z-\left(\frac{H}{Y_{x/s}K_s}\right)\right]}=\left(-\frac{Y_{x/s}K_s}{H}\right)\frac{1}{Z}+\left(1+\frac{Y_{x/s}K_s}{H}\right)\left(\frac{1}{Z-\left(\frac{H}{Y_{x/s}K_s}\right)}\right)$$

Returning to our earlier differential equation, we integrate from $t = 0$ to final time $t$ in order to subsequently find the final cell concentration, $C_{s,2}$

$$\int_{Z_1}^{Z_2}\left(-\frac{Y_{x/s}K_s}{H}\right)\frac{1}{Z}+\left(1+\frac{Y_{x/s}K_s}{H}\right)\left(\frac{1}{Z-\left(\frac{H}{Y_{x/s}K_s}\right)}\right)dZ=\mu_{\max}t$$

$$\left[-\frac{Y_{x/s}K_s}{H}\ln Z\right]_{Z_1}^{Z_2}+\left[\left(1+\frac{Y_{x/s}K_s}{H}\right)\ln\left(Z-\left(\frac{H}{Y_{x/s}K_s}\right)\right)\right]_{Z_1}^{Z_2}=\mu_{\max}t$$

$$-\frac{Y_{x/s}K_s}{H}\ln\left(\frac{Z_2}{Z_1}\right)+\left(1+\frac{Y_{x/s}K_s}{H}\right)\ln\left(\frac{Z_2-\left(\frac{H}{Y_{x/s}K_s}\right)}{Z_1-\left(\frac{H}{Y_{x/s}K_s}\right)}\right)=\mu_{\max}t$$

We can now substitute back the original notations

$$Z_1=\frac{C_{s,1}}{K_s}$$

$$Z_2=\frac{C_{s,2}}{K_s}$$

$$-\frac{Y_{x/s}K_s}{H}\ln\left(\frac{C_{s,2}}{C_{s,1}}\right)+\left(1+\frac{Y_{x/s}K_s}{H}\right)\ln\left(\frac{\frac{C_{s,2}}{K_s}-\left(\frac{H}{Y_{x/s}K_s}\right)}{\frac{C_{s,1}}{K_s}-\left(\frac{H}{Y_{x/s}K_s}\right)}\right)=\mu_{\max}t$$

$$K_s\ln\left(\frac{C_{s,1}}{C_{s,2}}\right)+\left(\frac{H}{Y_{x/s}}+K_s\right)\ln\left(\frac{\frac{C_{s,2}}{K_s}-\left(\frac{H}{Y_{x/s}K_s}\right)}{\frac{C_{s,1}}{K_s}-\left(\frac{H}{Y_{x/s}K_s}\right)}\right)=\left(\frac{H\mu_{\max}}{Y_{x/s}}\right)t$$

$$H=C_{x,1}+Y_{x/s}C_{s,1}$$

$$t=\frac{Y_{x/s}}{\mu_{\max}\left(C_{x,1}+Y_{x/s}C_{s,1}\right)}\left[K_s\ln\left(\frac{C_{s,1}}{C_{s,2}}\right)+\left(\frac{C_{x,1}+Y_{x/s}C_{s,1}}{Y_{x/s}}+K_s\right)\ln\left(\frac{\frac{C_{s,2}}{K_s}-\left(\frac{C_{x,1}+Y_{x/s}C_{s,1}}{Y_{x/s}K_s}\right)}{\frac{C_{s,1}}{K_s}-\left(\frac{C_{x,1}+Y_{x/s}C_{s,1}}{Y_{x/s}K_s}\right)}\right)\right]$$

$$t=\frac{Y_{x/s}}{\mu_{\max}\left(C_{x,1}+Y_{x/s}C_{s,1}\right)}\left[K_s\ln\left(\frac{C_{s,1}}{C_{s,2}}\right)+\left(\frac{C_{x,1}+Y_{x/s}(C_{s,1}+K_s)}{Y_{x/s}}\right)\ln\left(\frac{C_{x,1}+Y_{x/s}(C_{s,1}-C_{s,2})}{C_{x,1}}\right)\right]$$

Substituting in known values, we can find the value of $t$

$$t=\frac{0.4}{1.4(0.2+0.4(30))}\left[2\ln\left(\frac{30}{0.5}\right)+\left(\frac{0.2+0.4(30+2)}{0.4}\right)\ln\left(\frac{0.2+0.4(30-0.5)}{0.2}\right)\right]$$

$$t=0.023(8.2+32.5\ln 60)=3.2\text{ h}$$

The time taken using the Monod growth model is longer than that calculated using the exponential growth model. This makes sense since the Monod model does not assume growth rate to be maximum at all times.

(e) We can plot graphs to better illustrate the difference in growth patterns between the two models, by using the general equations for both models. Taking a range of values for time $t$, we can calculate the corresponding values of $C_s$ and hence $C_x$ for the plots.

*Monod Model (Orange Line):*

$$t = \frac{Y_{x/s}}{\mu_{\max}\left(C_{x,1} + Y_{x/s}C_{s,1}\right)}\left[K_s \ln\left(\frac{C_{s,1}}{C_s}\right) + \left(\frac{C_{x,1} + Y_{x/s}(C_{s,1} + K_s)}{Y_{x/s}}\right)\ln\left(\frac{C_{x,1} + Y_{x/s}(C_{s,1} - C_s)}{C_{x,1}}\right)\right]$$

$$C_x = C_{x,1} + Y_{x/s}C_{s,1} - Y_{x/s}C_s$$

*Exponential Model (Blue Line):*

$$\ln\frac{C_{x,2}}{C_{x,1}} = \mu_{\max}t$$

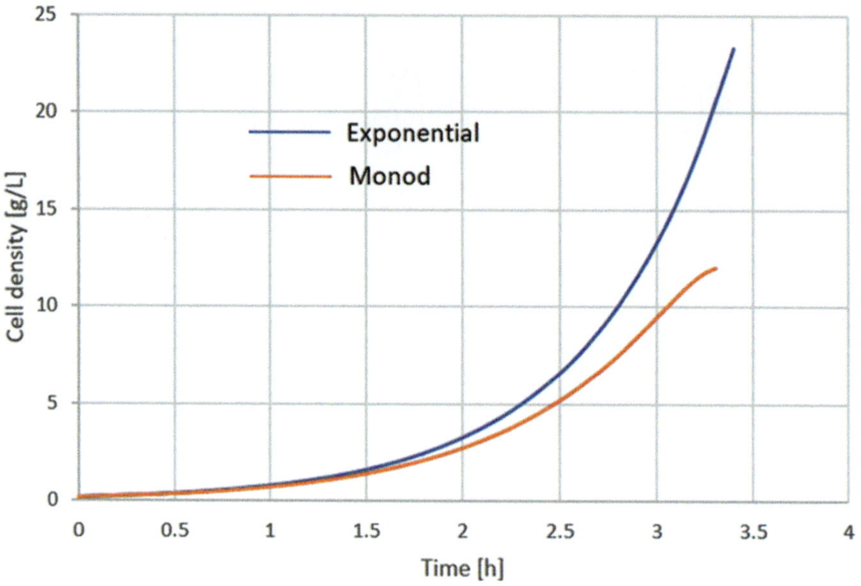


We observe from the plots that it takes a longer time to achieve final cell density under the Monod growth model. This is due to the differentiating assumption between the exponential and Monod growth models, whereby the former assumes that growth rate $\mu$ is always at its maximum value, while the latter follows a growth rate that varies with substrate concentration, according to the equation below. Under the Monod model, growth rate $\mu$ tends to maximum only at higher substrate concentrations, and is less than its maximum value at other (lower) substrate concentrations.

$$\mu = \mu_{\max}\frac{C_s}{C_s + K_s}$$

## Problem 42

**One of the most widely used models to describe enzymatic reactions is the Michaelis–Menten equation as shown below whereby $v_0$ denotes initial rate of enzymatic reaction, $v_{\max}$ denotes maximum reaction rate, $K_M$ denotes the Michaelis constant, and $C_s$ represents substrate concentration.**

$$v_0 = \frac{v_{\max} C_s}{C_s + K_M}$$

(a) **Show how the above equation is derived and comment on any assumptions and key characteristics.**

(b) **Soluble enzymes are sometimes used in industrial applications such as the making of cheese. These enzymes are typically added to substrate in a batch reactor that operates similarly to a CSTR (Continuous Stirred Tank Reactor). The mixture is assumed well-mixed and of a homogeneous phase. Over time, substrate will be converted to product and the process is stopped when the desired level of conversion is reached. Using the expression in part a, derive an expression of the form below for time taken to achieve conversion of substrate, *X*. Explain what *R* means.**

$$t = \frac{C_{S0} X - K_M \ln(1 - X)}{R}$$

(c) **In reality, conversion may be close to, but rarely reaches 100%. Discuss possible reasons for this.**

(d) **Briefly discuss any disadvantages of homogeneous phase enzymatic reactions and suggest alternative methods to overcome these problems.**

## Solution 42

*Worked Solution*

(a) Biochemical reactions involving a single substrate are often assumed to follow Michaelis–Menten (MM) kinetics. Under the MM model, we have a free enzyme *E* that binds reversibly to a substrate *S* to form an Enzyme–Substrate complex denoted as *ES*. The reaction rate constants for the forward and backward directions of this reversible reaction are denoted by $k_1$ and $k_{-1}$, respectively. This *ES* complex then breaks down to yield the desired product *P* as well as the free enzyme *E* which is recovered and returned to the reaction mixture for further reactions. The reaction rate constant for this second step is denoted by $k_2$.

$$E + S \overset{k_1, k_{-1}}{\longleftrightarrow} ES \overset{k_2}{\rightarrow} P + E \tag{1}$$

Since we are deriving *initial* rate of reaction, we assume that there is relatively little product ($C_P \ll C_S$) and hence ignore any backward reaction for the second step. Another key assumption is that enzyme concentration is much less than substrate concentration ($C_E$, $C_{ES} \ll C_S$). Therefore we can approximate that the substrate concentration remains relatively constant and equivalent to the initial substrate concentration $C_{S0}$ throughout the reaction since it is in large excess.

$$C_{S0} = C_S + C_{ES} + C_P \cong C_S$$

The reaction rate for the second step can thus be written as follows, where $C_{ES}$ refers to concentration of the *ES* complex:

$$v_0 = \frac{dC_P}{dt} = k_2 C_{ES} \tag{2}$$

It is experimentally difficult to determine $C_{ES}$ since it forms briefly during the two-step reaction. Therefore, we need to think of a way to replace this term with another expression of known/measurable values.

To do so, we first assume steady-state conditions such that $C_{ES}$ is a low value that remains almost constant since its rate of formation (left-hand side of equation below) is equivalent to its rate of consumption (right-hand side of equation below).

$$k_1 C_E C_S = k_{-1} C_{ES} + k_2 C_{ES}$$

Since it is also challenging to measure $C_E$, the concentration of free enzymes alone (since some enzymes are bound and some are free), we perform a simple mass balance on the enzyme which will help us express the above equation in terms of total enzyme concentration $C_{TE}$ (also equivalent to the initial enzyme concentration $C_{E0}$) instead of $C_E$.

$$C_E + C_{ES} = C_{TE}$$

$$C_E = C_{TE} - C_{ES}$$

Substituting the above, we have the following which gives us an expression for $C_{ES}$. We can group rate constants together to simplify the expression, which gives rise to the familiar Michaelis constant $K_M$ which is in fact a grouping of rate constants. The Michaelis constant $K_M$ is also equivalent to the substrate concentration at which the reaction rate is half of the maximum, $v_{\max}$. The reciprocal of this constant (i.e., $1/K_M$) is useful in analyzing the degree of affinity of an enzyme to its substrate, whereby the greater the affinity, the higher the value of $1/K_M$. The value of $K_M$ does not depend on enzyme concentration or purity, but does vary with operating conditions such as temperature and pH, as well as the specific enzyme–substrate pair used.

$$k_1(C_{TE} - C_{ES})C_S = k_{-1}C_{ES} + k_2C_{ES}$$

$$k_1C_{TE}C_S = C_{ES}(k_1C_S + k_{-1} + k_2)$$

$$C_{ES} = \frac{k_1C_{TE}C_S}{k_1C_S + k_{-1} + k_2} = \frac{C_{TE}C_S}{C_S + \frac{k_{-1}+k_2}{k_1}} = \frac{C_{TE}C_S}{C_S + K_M}$$

$$K_M = \frac{k_{-1} + k_2}{k_1}$$

Substituting the above expression for $C_{ES}$ into equation (2), we have the following.

$$v_0 = k_2\left(\frac{C_{TE}C_S}{C_S + K_M}\right)$$

We may refer to $k_2$ as the catalytic constant $k_{cat}$ as it serves as the rate constant in this rate equation. $k_{cat}$ may also be referred to as the turnover number defined as the maximum number of substrate molecules converted to product per enzyme molecule per second. Hence, the turnover number is also representative of the maximum number of conversions of substrate molecules per second that a single catalytic site can execute for a given enzyme concentration $C_{TE}$ (or $C_{E0}$).

$$v_0 = k_{cat}\left(\frac{C_{TE}C_S}{C_S + K_M}\right)$$

The above equation is now more useful since it is expressed in terms of values $C_S$ and $C_{TE}$ which can be measured experimentally. The maximum rate of reaction $v_{max}$ occurs when all the enzyme molecules are used for reactions, i.e., bound to substrate, which means

$$C_{TE} = C_{ES}$$

$$v_{max} = k_{cat}C_{TE}$$

Therefore, we can express initial rate of reaction in terms of $v_{max}$ and derive the given expression for the MM equation as shown below.

$$v_0 = \frac{v_{max}C_S}{C_S + K_M}$$

(b) To find time taken, we need to have an equation that is related to time, and one such equation is the differential equation as follows for consumption rate of substrate. Note that sign convention is such that $-r_s$ is a positive value.

$$-r_s = -\frac{dC_s}{dt}$$

Assuming MM kinetics, we can also express rate of reaction as follows where $v_{\max} = k_2 C_{TE}$ where $k_2$ is as defined in equation (1) of part a. Note that total enzyme concentration $C_{TE}$ is also equivalent to the initial concentration of enzyme, $C_{E0}$.

$$-r_s = \frac{v_{\max} C_S}{C_S + K_M}$$

Equating both expressions for $-r_s$, we have

$$-\frac{dC_s}{dt} = \frac{v_{\max} C_S}{C_S + K_M}$$

$$\left(1 + \frac{K_M}{C_S}\right) dC_s = -v_{\max} dt$$

Integrating from initial conditions to an arbitrary time $t$, we have

$$\int_{C_{S0}}^{C_S} \left(1 + \frac{K_M}{C_S}\right) dC_s = \int_0^t -v_{\max} dt$$

$$[C_S]_{C_{S0}}^{C_S} + [K_M \ln C_s]_{C_{S0}}^{C_S} = -v_{\max} t$$

$$(C_S - C_{S0}) + K_M \ln\left(\frac{C_S}{C_{S0}}\right) = -v_{\max} t$$

$$t = \frac{(C_{S0} - C_S) + K_M \ln\left(\frac{C_{S0}}{C_S}\right)}{v_{\max}}$$

The fractional conversion of substrate to product is defined as

$$X = \frac{C_{S0} - C_S}{C_{S0}} = 1 - \frac{C_S}{C_{S0}}$$

We therefore obtain the given expression, which shows how the fractional conversion of substrate to product varies with time, where $v_{\max} = R$. It follows that $R$ refers to the maximum reaction rate when enzyme is saturated with substrate.

$$t = \frac{C_{S0} X + K_M \ln\left(\frac{1}{1-X}\right)}{v_{\max}}$$

$$t = \frac{C_{S0} X - K_M \ln(1 - X)}{R}$$

(c) Incomplete reactions may occur when the enzymes are denatured during the reaction, as such conversion of substrate to product may be slowed down or stopped. Certain products may also inhibit the enzyme's catalytic activities; therefore, as more product is formed over time, enzymatic reactions slow down. The reaction mixture may also have reached equilibrium for the reversible enzymatic reaction, in which case it is not thermodynamically possible to achieve complete conversion.

(d) The key disadvantage of homogeneous reactions lies in the fact that the desired product is mixed with enzymes and unused reactants. The product is therefore not separated out once formed and this could lead to contamination of product. This is especially undesirable if the product is a biopharmaceutical or food product that has to comply with stringent purity and hygiene standards.

In a batch process, if the enzyme is not separated out after use, it is also wasteful that it has to be discarded after just one batch cycle and cannot be reused. The cost of producing new enzyme has to be balanced with the cost of separating out the enzyme from the products before subsequent reuse. It is often difficult and expensive to isolate enzymes from the reaction mixture.

In order to overcome these constraints, the use of immobilized enzymes has become popular and this approach uses a heterogenous phase system. The main method to immobilize enzymes involves binding enzyme molecules to an inert and insoluble support material such that the enzyme is exposed to the reaction mixture and is therefore still able to catalyze reactions at the surface of the support. These enzymes may be chemically bonded or physically adsorbed onto the surface of the support. Support materials can vary from polymer matrices to bead particles and they should be mechanically stable and easy to recover between batches.

In some cases, highly porous solids may be used as supports since they offer a large surface area for enzyme immobilization which allows a high amount of enzyme to be loaded onto the support. However, caution has to be taken to ensure pore sizes are not too small such that they block substrates from accessing the enzymes or limit the diffusion of substrates and products to and from the enzymes.

Some examples of support materials used industrially for enzyme immobilization include organic materials such as cellulose, agarose, and dextrans, as well as inorganic materials such as porous glass and silicas.

## Problem 43

**The Monod model can be used to study growth of microbial cells in a continuous stirred tank fermenter. Given the expression below which describes Monod kinetics, whereby $\mu$ refers to the specific growth rate, $\mu_{max}$ refers to the maximum value for $\mu$, $C_s$ refers to substrate concentration, and $K_s$ refers to the Monod constant.**

$$\mu = \frac{\mu_{\max} C_s}{C_s + K_s}$$

**Assuming that the following lab equipment and reagents are provided, and that the growth yield coefficient $Y_{X/S}$ for the strain of cells to be cultured in the fermenter is 0.5 g/g,**

- **400 mL continuous stirred tank fermenter**
- **Growth-limiting substrate of concentration 1.2 g/L**

**Explain how experiments can be designed to determine $K_s$ and $\mu_{\max}$.**

## Solution 43

### *Worked Solution*

We can design lab experiments to determine the key parameters of the Monod model, i.e., $K_s$ and $\mu_{\max}$. First, we set up a continuous process for our cell growth in the continuous stirred tank fermenter as shown below.

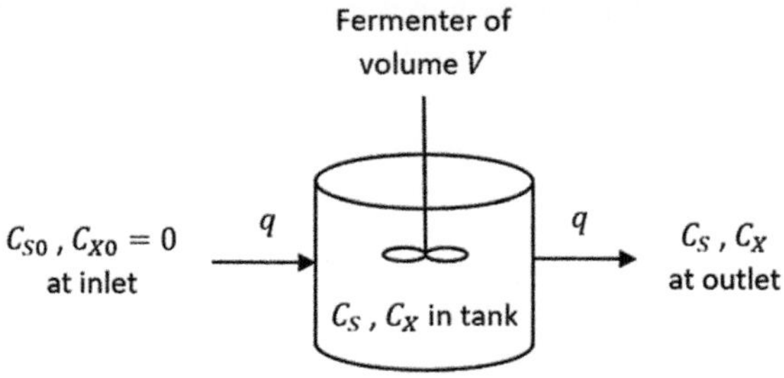


We know that volume of the fermenter $V = 4 \times 10^{-4}$ m$^3$ and $C_{SO} = 1.2$ g/L. There are no cells in the inlet, i.e., $C_{XO} = 0$ which means all cells in the outlet stream come from growth in the fermenter. The inlet and outlet volumetric flow rates are equivalent for a CSTR (continuous stirred tank reactor) type vessel at steady state, and this is denoted as $q$. Also, assuming perfect mixing in a CSTR, the concentrations of cells and substrate ($C_X$, $C_S$) in the tank will be equal to that in the outlet stream.

To see how we can make use of experimental measurements to determine the Monod parameters, we can start with a mass balance for cells in the CSTR under steady state as shown below.

$$\text{Inlet} - \text{Outlet} + \text{Grown} = 0$$

We know that the number of cells grown per unit time can be expressed in terms of specific growth rate $\mu$.

$$\frac{dC_X}{dt} = \mu C_X$$

Therefore, we use the above expression in our mass balance for the "grown" term. There are no cells entering the tank from the inlet stream. Hence our mass balance as follows:

$$0 - qC_X + \mu C_X V = 0$$

$$\mu = \frac{q}{V} = \frac{q}{4 \times 10^{-4}} \tag{1}$$

We now apply Monod kinetics expressed as shown below.

$$\mu = \frac{\mu_{\max} C_s}{C_s + K_s}$$

Taking reciprocal for both sides of the equation, we obtain a useful form of the equation which can be fit to a straight line equation $y = mx + c$ whereby $x$ and $y$ are the horizontal and vertical axes variables, respectively, $m$ is the gradient, and $c$ is the vertical axis intercept.

$$\frac{1}{\mu} = \frac{C_s + K_s}{\mu_{\max} C_s}$$

$$\frac{1}{\mu} = \frac{1}{\mu_{\max}} + \frac{K_s}{\mu_{\max} C_s} \tag{2}$$

To determine $K_s$ and $\mu_{\max}$, we can plot $1/\mu$ against $1/C_s$. The vertical axis intercept value will be $1/\mu_{\max}$, from which we can easily derive the value of $\mu_{\max}$. Separately, the gradient of the line will be $K_s/\mu_{\max}$. Using the value of $\mu_{\max}$ found earlier from the vertical axis intercept, we can then determine the value of $K_s$. To create such a plot, we need to conduct experiments to gather data points for $\mu$ and $C_s$, which we then fit to a straight line and determine its equation.

To obtain data points for $\mu$, we can vary flow rate $q$ over a range by using equation (1). As for data points for $C_s$, some further treatment is useful as it is difficult to determine $C_s$ directly experimentally.

We can use the definition of growth yield coefficient $Y_{X/S}$ to express substrate concentration, $C_S$ in terms of cell concentration, $C_X$. This is useful since it is experimentally easier to measure cell concentration in the outlet stream than substrate concentration. Substrate is typically in large excess, hence its concentration will be high and remain relatively constant, thereby making any small changes in its concentration more difficult to measure.

$$Y_{X/S}(C_{S0} - C_S) = C_X - C_{X0} = C_X$$

$$C_S = C_{S0} - \frac{C_X}{Y_{X/S}} = 1.2 - \frac{C_X}{0.5}$$

Substituting equation (1) and the above expression for $C_S$ into equation (2), we obtain

$$\frac{1}{\mu} = \frac{1}{\mu_{\max}} + \frac{K_s}{\mu_{\max} C_S}$$

$$\frac{4 \times 10^{-4}}{q} = \frac{1}{\mu_{\max}} + \frac{K_s}{\mu_{\max}\left(1.2 - \frac{C_X}{0.5}\right)}$$

Below are some arbitrary experimental data for $q$ and $C_X$ which we use to demonstrate how the Monod parameters can be determined with actual numbers. Assuming that a range of feed flow rates are experimentally set, we can measure the corresponding cell concentrations in the outlet stream as follows.

| Feed flow rate $q$ [m$^3$/s] | Cell concentration at outlet $C_X$ [g/L] |
|---|---|
| 4.0E-08 | 0.51 |
| 4.8E-08 | 0.48 |
| 5.6E-08 | 0.45 |
| 6.6E-08 | 0.38 |
| 7.1E-08 | 0.34 |

From the values above, we can derive the necessary quantities for the straight line plot.

| Feed flow rate $q$ [m$^3$/s] | Vertical axis values, $1/\mu$ [s] | Cell concentration at outlet $C_X$ [g/L] | Horizontal axis values, $1/C_s$ [L/g] |
|---|---|---|---|
| 4.0E-08 | $\frac{4\times10^{-4}}{q} = 1.0 \times 10^4$ | 0.51 | $\frac{1}{1.2-\frac{C_X}{0.5}} = 5.6$ |
| 4.8E-08 | 8.3E+03 | 0.48 | 4.2 |
| 5.6E-08 | 7.1E+03 | 0.45 | 3.3 |
| 6.6E-08 | 6.1E+03 | 0.38 | 2.3 |
| 7.1E-08 | 5.6E+03 | 0.34 | 1.9 |

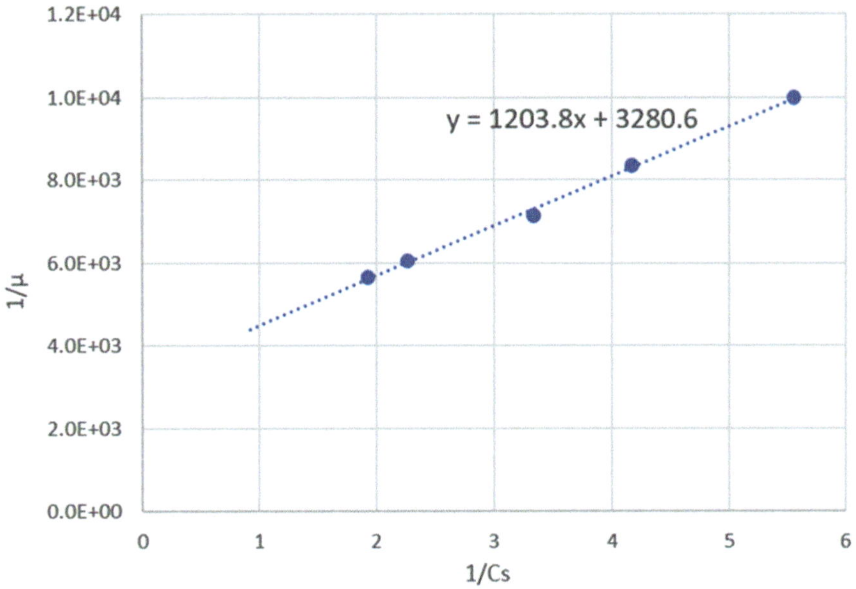


The plot is shown below, and we can determine the Monod parameters from the equation of the line of best fit.

$$\frac{1}{\mu_{\max}} = 3280.6 \rightarrow \mu_{\max} = 3.0 \times 10^{-4}\ \mathrm{s}^{-1}$$

$$\frac{K_s}{\mu_{\max}} = 1203.8 \rightarrow K_s = 1203.8\left(3.0 \times 10^{-4}\right) = 0.4\ \mathrm{g/L}$$

## Problem 44

**The Michaelis–Menten model is used widely in studying enzyme-catalyzed biochemical reactions. It is expressed as follows where $r$ denotes initial reaction rate, and $R$ denotes maximum reaction. $C_S$ and $K_M$ represent substrate concentration and the Michaelis constant, respectively.**

$$r = \frac{RC_S}{C_S + K_M}$$

(a) **List down some examples of biochemical reactions that may be modeled using the Michaelis–Menten model.**

(b) **The production of nicotinamide-adenine dinucleotide ($P$) is catalyzed by an enzyme known as nicotinamide mononucleotide adenylyltransferase. Experimental measurements were made for this reaction over the initial 3-minute reaction duration. The following table shows recorded data for**

**substrate (nicotinamide mononucleotide) concentrations and amount of product $P$ formed during the reaction duration.**

| Amount of $P$ formed [μmol] | Substrate concentration [mM] |
|---|---|
| 0.231 | 0.129 |
| 0.308 | 0.210 |
| 0.361 | 0.281 |
| 0.515 | 0.549 |
| 0.633 | 0.752 |
| 0.800 | 1.448 |

i. **Explain what a Lineweaver–Burke plot is and apply it to the data above to show that the above reaction obeys Michaelis–Menten kinetics.**
ii. **Determine the Michaelis constant $K_M$ [mM] and maximum reaction velocity $R$ [μmol min$^{-1}$] for this enzyme.**

(c) **Discuss the pros and cons of the Lineweaver–Burke method and suggest any alternative methods to study enzyme kinetics.**

## Solution 44

### *Worked Solution*

(a) Biochemical processes that may apply the Michaelis–Menten (MM) model in the analysis of enzyme kinetics include enzyme–substrate interactions, antigen–antibody binding, DNA–DNA hybridization, and protein–protein interactions. Common enzymes that have been studied and characterized include digestive enzymes like amylase and chymotrypsin for the breakdown of starch and protein, respectively. Another example is ribonuclease which is found in the pancreas and aids the breakdown of ribonucleic acid (RNA) that cells no longer require, as well as serves an immune function against RNA viruses through degradation of their RNA. The parameters of the MM model such as the catalytic constant (or enzyme turnover number) $k_{cat}$ and the Michaelis constant $K_M$ can be experimentally determined and these values vary widely between different enzymes.

(b) (i) The Lineweaver–Burke plot is also known as the double reciprocal plot and is especially used for its ability to graphically represent enzyme kinetics using straight line plots. The Y-intercept and gradient of this straight line can then be used to determine Michaelis–Menten (MM) parameters such as the Michaelis constant $K_M$ and the catalytic constant $k_{cat}$. The MM equation can be expressed in the form of a straight line equation as shown below, and the inverse of reaction rate $1/r$ can be plotted against the inverse of substrate concentration $1/C_S$. The gradient will be $K_M/R$ while the Y-intercept will be $1/R$.

$$r = \frac{RC_S}{C_S + K_M} \rightarrow \frac{1}{r} = \frac{1}{R} + \frac{K_M}{R}\left(\frac{1}{C_S}\right)$$

$R$ is the maximum reaction velocity (or reaction rate) and is defined as follows where $k_{cat}$ is the catalytic constant (or enzyme turnover number) and $C_{E0}$ is the initial enzyme concentration.

$$R = k_{cat} C_{E0}$$

Note here that the amount of $P$ formed during the initial reaction duration of 3 minutes is directly proportional to reaction rate.

$$r\left[\mu\text{mol min}^{-1}\right] = \frac{dP}{dt} = \frac{\Delta P}{3} \rightarrow \frac{1}{r} = \frac{3}{\Delta P}$$

| Amount of $P$ formed, $\Delta P$ [μmol] | $1/r$ [min μmol$^{-1}$] | Substrate concentration [mM] | $1/C_S$ [mM$^{-1}$] |
|---|---|---|---|
| 0.231 | $\frac{3}{0.231} = 13.0$ | 0.129 | $\frac{1}{0.129} = 7.75$ |
| 0.308 | 9.74 | 0.210 | 4.76 |
| 0.361 | 8.31 | 0.281 | 3.56 |
| 0.515 | 5.83 | 0.549 | 1.82 |
| 0.633 | 4.74 | 0.752 | 1.33 |
| 0.800 | 3.75 | 1.448 | 0.69 |

The plot is shown below with the equation of the line of best fit indicated. Since the data points fit well to a straight line, this reaction obeys Michaelis–Menten kinetics.

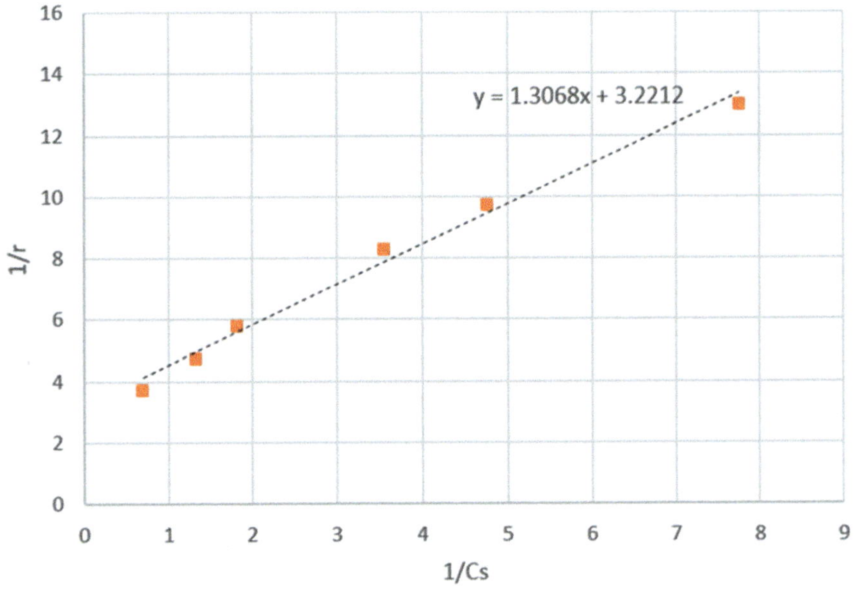

(ii) The Michaelis constant $K_M$ is 0.4 mM and the maximum reaction velocity is 0.31 μmol $min^{-1}$.

$$\frac{1}{R} = 3.22 \rightarrow R = 0.31\ \mu\text{mol}\ \text{min}^{-1}$$

$$\frac{K_M}{R} = 1.3 \rightarrow K_M = 1.3(0.31) = 0.4\ \text{mM}$$

(c) The Lineweaver–Burke plot was more useful during earlier days when there was a lack of sophisticated computing software that were able to fit experimental data easily using regression methods. For back-of-envelope estimations, it is useful as the method is relatively simple, and key parameters for enzyme kinetics can be determined graphically from a simple straight-line plot. However for more rigorous analysis, one should note some of the limitations of the Lineweaver–Burke method:

- As the axes values are reciprocal values, any small measurement errors will be magnified after taking reciprocal of an experimentally measured data value.
- Most of the data points will tend to congregate towards the right end, away from the *y*-axis, due to the practical limitation of having large values of substrate concentration $C_S$ as solubility reaches its limit for the particular solute (substrate). Therefore, there is a need for back extrapolation in drawing the line of best fit in order to reach the *y*-axis to determine the *y*-intercept value.

To overcome some of these limitations, computational methods that make use of non-linear regression software to make precise calculations can be used.

Graphically, some simple treatment of the MM equation can help address some of the downsides mentioned above.

$$r = \frac{RC_S}{C_S + K_M}$$

$$\frac{R}{r} = \frac{R(C_S + K_M)}{RC_S} = \frac{C_S + K_M}{C_S} = 1 + \frac{K_M}{C_S}$$

$$R = r + \frac{rK_M}{C_S}$$

$$r = -K_M\frac{r}{C_S} + R$$

In the above equation, we can obtain a straight line as well by plotting $r$ against $r/C_S$. The gradient will then be $-K_M$ and the y-intercept value will give maximum reaction rate $R$. Unlike the Lineweaver–Burke equation, this plot does not use reciprocal values and hence more equally distributes weightage to the impact of errors from data points in any range of substrate concentration or reaction rate. However note that for this equation, there is a new downside in terms of having both axes values dependent on reaction rate $r$ (experimentally determined), thus any measurement errors will propagate in both axes.

Other than determining model parameters (e.g., Michaelis constant), the Lineweaver–Burke plot is also useful in identifying the type of enzyme inhibition:

| Type of enzyme inhibition | $y$-intercept value (compared to uninhibited) | Slope of line (compared to uninhibited) | $x$-intercept value (compared to uninhibited) |
|---|---|---|---|
| Competitive | Same | Different | Different |
| Non-competitive | Different | Different | Same |
| Uncompetitive | Different | Same | Different |

## Problem 45

**Enzymes play a critical role in numerous biological processes and strict control of its activity is often necessary. In some cases, the catalytic function of enzymes may be inhibited by molecules that bind reversibly and specifically to the enzyme via non-covalent interactions. As a result, catalytic activity is slowed or may even stop, as the substrate-binding sites become obstructed.**

(a) **These inhibitor molecules generally fall under three types of mechanisms, competitive, uncompetitive, and substrate inhibition/non-competitive. Briefly explain these mechanisms.**

(b) **An enzyme was found to be inhibited competitively and reversibly by an isomeric form of its intended substrate. This inhibitor has an apparent Michaelis constant $K'_M$ that is expressed as follows whereby $C_i$ denotes the concentration of inhibitor and $K_i$ denotes the inhibition constant. The Michaelis constant $K_M$ for the enzyme is 0.18 × $10^{-3}$ kmol/m$^3$, while $K_i$ has a value of 9.8 × $10^{-6}$ kmol/m$^3$.**

$$K'_M = K_M\left(1 + \frac{C_i}{K_i}\right)$$

**1.2 g of enzyme was added to 2 kg of substrate (contaminated with 1.8% weight of inhibitor) in a reaction vessel that is 12 m$^3$ in volume. Given that this enzyme has a molecular weight of 68,000 daltons and a turnover number of 250 s$^{-1}$, and the substrate and inhibitor have molecular weights of 350 and 360, respectively, determine the time taken to achieve 97% conversion of substrate to product. You may assume that this enzyme obeys Michaelis–Menten kinetics.**

## Solution 45

### *Worked Solution*

(a) There are three common mechanisms under which enzyme inhibitors operate—competitive, non-competitive, and uncompetitive. In all three cases, it is assumed that an enzyme–substrate complex *ES* is temporarily formed after substrate *S* binds to an enzyme's catalytic (or active) site *E*. *ES* then converts to product *P* and the unbound enzyme *E* is made available again for further reactions.

$$E + S \leftrightarrow ES \rightarrow E + P$$

<u>*Competitive Inhibition*</u>

Competitive inhibitors are molecularly similar to the actual substrate, often resembling the substrate in terms of chemical structure, shape, and polarity characteristics. Hence they compete with the substrate for the same substrate-binding site of the enzyme. When the inhibitor binds to enzyme instead of the substrate, an enzyme–inhibitor complex *EI* is formed and the enzyme becomes inactive, causing a slowing of reaction rate. This inhibitory mechanism can be described using a dissociation constant $K_i$ whereby

$$EI \overset{K_i}{\leftrightarrow} E + I$$

$$K_i = \frac{[E][I]}{[EI]}$$

Under competitive inhibition, two equilibria coexist, as both substrate and inhibitor bind reversibly to the enzyme active sites.

$$E + S \leftrightarrow ES \rightarrow E + P \tag{1}$$

$$E + I \leftrightarrow EI \tag{2}$$

The effect of competitive inhibition for enzymes that obey Michaelis–Menten kinetics can be observed graphically from the straight line plot of reciprocal of reaction rate $1/v$ against reciprocal of substrate concentration $1/C_S$.

$$\frac{1}{v} = \frac{1}{v_m} + \frac{K_M}{v_m}\left(\frac{1}{C_S}\right)$$

In the presence of a competitive inhibitor, the line is steeper than if uninhibited because $K_M$ is increased and thus gradient ($K_M/v_m$) is increased. $K_M$ represents the substrate concentration needed to achieve half of the maximum reaction rate. It makes sense that this value is higher since more substrate is required to achieve the same reaction rate (i.e., having more substrate moves equilibrium position of equation (1) to the right) to make up for some active sites being occupied by inhibitor molecules.

However, the y-intercept of the plot remains the same for both inhibited and uninhibited cases since $v_m$ remains the same. This follows from the definition of maximum reaction rate $v_m$ which occurs at infinitely high substrate concentrations. Under this limiting scenario, it is thermodynamically favored for equation (1) to proceed.

*Uncompetitive Inhibition*

Uncompetitive inhibitors bind only to the enzyme–substrate complex, and not to the free enzyme.

$$E + S \leftrightarrow ES \rightarrow E + P \tag{1}$$

$$ES + I \leftrightarrow ESI \tag{2}$$

The dissociation constant $K_i$ for this mechanism is defined below.

$$ESI \overset{K_i}{\leftrightarrow} ES + I$$

$$K_i = \frac{[ES][I]}{[ESI]}$$

In the double reciprocal straight line plot for this case, the uncompetitive inhibitor leads to a lower $v_m$ and thus a higher value of y-intercept ($1/v_m$). This makes sense since the amount of *ES* that goes on to form product is reduced due to some of the *ES* being bound to inhibitor molecules. Although binding of *ES* to *I* is reversible, increasing substrate concentration does not help reverse uncompetitive inhibition (unlike in competitive inhibition) because inhibitors bind to *ES* and not to *E*. The maximum amount of product that can be formed even under infinitely high substrate concentrations (for $v_m$) will be lower in the inhibited case.

The value of $K_M$ also decreases, and to the same extent as $v_m$ is decreased, therefore the gradient of the straight line ($K_M/v_m$) remains constant and does not vary between the inhibited and uninhibited cases.

*Noncompetitive Inhibition*

This type of inhibition is sometimes referred to as “mixed inhibition”. It is a combination of competitive and uncompetitive mechanisms with the addition of one more reaction equation (4) as shown below.

$$E + S \leftrightarrow ES \rightarrow E + P \tag{1}$$

$$E + I \leftrightarrow EI \tag{2}$$

$$ES + I \leftrightarrow ESI \tag{3}$$

$$EI + S \leftrightarrow ESI \tag{4}$$

The inhibitor $I$ can bind to the free enzyme $E$, as well as to the enzyme–substrate $ES$ complex. The substrate can also bind to free enzyme $E$ and the enzyme–inhibitor complex $EI$. The inhibitor and substrate do not affect each other's binding to the enzyme. However one should note that if inhibitor is bound (either in $EI$ or $ESI$), the enzyme cannot catalyze product formation, hence the result of this type of inhibition is in reducing the amount of functional enzyme molecules that can carry out reactions to form product. As a result, the reaction can never reach its normal $v_m$ (uninhibited) regardless of how much substrate is added. A portion of enzyme molecules will always be made non-functional by the inhibitors present and the effective concentration of the enzyme is reduced. Graphically, a lower $v_m$ means a higher y-intercept value ($1/v_m$).

As for $K_M$, its value remains unchanged between the inhibited and uninhibited cases, since the binding of inhibitor to enzyme does not affect the binding of substrate to enzyme. Therefore, the substrate concentration to reach half of the maximum reaction rate $v_m$ (i.e., $K_M$) also remains unchanged. This type of inhibition only acts to decrease the amount of usable enzyme. Graphically, the gradient ($K_M/v_m$) will be higher under inhibition since $K_M$ is unchanged while $v_m$ is reduced.

(b) Under competitive inhibition in a batch system, we have the following reaction mechanism whereby $k_{cat}$ is the enzyme turnover number:

$$E + S \leftrightarrow ES \xrightarrow{k_{cat}} E + P \quad (1)$$

$$E + I \leftrightarrow EI \quad (2)$$

Also for a batch system, we can write down rate equations as follows:

$$r = r_p = \frac{dC_P}{dt} = -r_s = -\frac{dC_S}{dt}$$

Given that this enzyme obeys Michaelis–Menten kinetics, and knowing that this reaction is inhibited, we can express rate of reaction $r$ as follows whereby $r'_{max}$ is the maximum rate under inhibition, $C_{E0}$ is the initial enzyme concentration, and $K'_M$ is the apparent Michaelis constant under inhibition:

$$r = \frac{r'_{max} C_S}{C_S + K'_M}$$

$$K'_M = K_M\left(1 + \frac{C_i}{K_i}\right)$$

$$r'_{max} = k_{cat} C_{E0}$$

Combining both the rate equation and MM kinetics, we have the following differential equation which we can integrate from initial conditions to an arbitrary time $t$.

$$-\frac{dC_S}{dt} = \frac{r'_{\max} C_S}{C_S + K'_M}$$

$$\int_0^t dt = -\frac{1}{r'_{\max}} \int_{C_{S0}}^{C_S} \left(\frac{C_S + K'_M}{C_S}\right) dC_S = -\frac{1}{r'_{\max}} \int_{C_{S0}}^{C_S} \left(1 + \frac{K'_M}{C_S}\right) dC_S$$

$$t = -\frac{1}{r'_{\max}} \left( [C_S]_{C_{S0}}^{C_S} + \left[K'_M \ln C_S\right]_{C_{S0}}^{C_S} \right)$$

$$= \frac{1}{r'_{\max}} \left( (C_{S0} - C_S) + K'_M \ln\left(\frac{C_{S0}}{C_S}\right) \right)$$

We can compute the values of maximum reaction rate $r'_{\max}$, inhibitor concentrations $C_i$, and initial substrate concentration $C_{S0}$ using the given values.

$$r'_{\max} = k_{\text{cat}} C_{E0} = 250\left(\frac{1.2 \times 10^{-3}}{68,000(12)}\right) = 3.7 \times 10^{-7}\ \text{kmol m}^{-3}\ \text{s}^{-1}$$

$$C_i = \frac{0.018(2)}{360(12)} = 8.3 \times 10^{-6}\ \text{kmol m}^{-3}$$

$$C_{S0} = \frac{(1 - 0.018)(2)}{350(12)} = 4.7 \times 10^{-4}\ \text{kmol m}^{-3}$$

$$K'_M = K_M\left(1 + \frac{C_i}{K_i}\right) = 0.18 \times 10^{-3}\left(1 + \frac{8.3 \times 10^{-6}}{9.8 \times 10^{-6}}\right) = 3.3 \times 10^{-4}\ \text{kmol m}^{-3}$$

The end point of the reaction is defined by a conversion $X$ equivalent to 97%; therefore, we can find substrate concentration when this is reached.

$$X = \frac{C_{S0} - C_S}{C_{S0}} = 1 - \frac{C_S}{C_{S0}}$$

$$C_S = C_{S0}(1 - X) = 4.7 \times 10^{-4}(1 - 0.97) = 1.4 \times 10^{-5}\ \text{kmol m}^{-3}$$

Substituting all values found above into our earlier expression for $t$, we find that the time taken to reach desired conversion of 97% is approximately 4370 s or 1.2 h.

$$t = \frac{1}{3.7 \times 10^{-7}}\left( \left( (4.7 \times 10^{-4}) - (1.4 \times 10^{-5}) \right) + 3.3 \times 10^{-4} \ln\left(\frac{4.7 \times 10^{-4}}{1.4 \times 10^{-5}}\right) \right) = 4370\ \text{s}$$

## Problem 46

**Due to the difficulty in recovering soluble enzymes from homogeneous mixtures post-reaction, methods of enzyme immobilization have been widely considered for industrial applications. When an enzyme is attached to a support material, some of its properties may differ from those of the original soluble enzyme due to mass transfer effects.**

(a) **Outline the key steps of heterogenous catalysis in a continuous flow reactor operating at steady state and with an enzyme immobilized on a non-porous support. Explain briefly how mass transfer affects reaction kinetics, assuming Michaelis–Menten kinetics.**

(b) **Explain the terms effectiveness factor $\eta$ and Damköhler number $Da$ in the context of mass transfer effects on immobilized enzyme systems.**

(c) **Consider a packed bed reactor of total volume 1.3 m$^3$ filled with spherical support particles (diameter $D_p$) immobilized with catalytic enzymes.**

i. **Show that the total number of moles of enzymes per unit total volume of the packed bed can be expressed as follows, whereby $\varepsilon$ denotes void fraction of the packed bed, and $N''$ denotes the total number of moles of enzyme immobilized per unit surface area of the support.**

$$q_0 = \frac{6N''(1-\varepsilon)}{D_p}$$

ii. **Given the following data for the immobilized enzyme, determine whether the catalytic reaction is mass-transfer limited, or chemical reaction limited in the packed bed for a required conversion to product of 0.7.**

| | |
|---|---|
| Substrate concentration at inlet [M] | 0.03 |
| Residence time of liquid feed in reactor [h] | 0.43 |
| Voidage of packed bed | 0.5 |
| Diameter of spherical solid supports with immobilized enzyme [μm] | 263 |
| Mass transfer coefficient [m s$^{-1}$] | $1.3 \times 10^{-5}$ |

## Solution 46

*Worked Solution*

(a) One of the main differences in the functionality of soluble and immobilized enzymes originates from the significance of mass transfer effects in the latter. When an enzyme is immobilized on a support material, the following steps in an overall reaction should be considered:

- Mass transport of substrate from bulk solution to the support surface.

- (Note: If the support material was porous, an extra step of diffusion occurs here as the substrate moves along the pores in order to reach the enzyme surface.)
- Enzyme is in contact with substrate and reaction occurs to produce product.
- (Note: If the support material was porous, an extra step of diffusion occurs here as product moves along the pores to the other surface of the support material.)
- Mass transport of product back to the surrounding bulk solution.

It is observed from the above steps that mass transfer plays a significant role in immobilized enzyme kinetics. It is worth noting that a consequence of this effect is that the concentration of substrate at the enzyme surface $C_S$ can no longer be assumed to be equivalent to bulk substrate concentration $C_B$. In fact, $C_S < C_B$ and the difference in values depends on the rate of mass transfer relative to reaction rate at the enzyme surface.

In the original Michaelis–Menten equation for soluble enzymes ($C_S \cong C_B$), we had the following equation whereby $r$ is the reaction rate.

$$r = \frac{r_{\max} C_S}{K_M + C_S} = \frac{r_{\max} C_B}{K_M + C_B}$$

For immobilized enzymes, the above equation still holds true (i.e., $C_S \cong C_B$) provided that the reaction rate at the enzyme surface is much slower than mass transfer rate for bringing substrate to the enzyme, then in this case, the slower of the two rates (i.e., reaction) will be the limiting factor, and the overall rate of reaction is said to be dominated by chemical reaction.

However, if the assumption that $C_S \cong C_B$ (in fact $C_S < C_B$) is not true, we may consider the following equation under steady state when the rate of mass transfer of substrate from bulk solution to enzyme surface is equal to the chemical reaction rate at the enzyme surface. $k_c$ refers to the mass transfer coefficient across the stagnant liquid film surrounding the enzyme particle which poses resistance to mass transfer. $r'_{\max}$ denotes maximum reaction rate per unit surface area of the support material.

$$k_c(C_B - C_S) = \frac{r'_{\max} C_S}{K_M + C_S}$$

We should note that the Michaelis–Menten (MM) model implicitly assumes that substrate concentration is in excess of enzyme concentration ($C_S \gg C_{E0}$). Since we are told that the immobilized enzyme still obeys MM kinetics, we may assume here that even though mass transfer effects are present and significant (effect of reducing value of $C_S$ as compared to soluble free enzyme), the reduced value of $C_S$ is still not so severe that MM assumption breaks down and we can still assume that the local concentration of substrate is still sufficiently higher than enzyme concentration ($C_S \gg C_{E0}$ still valid). As such we can still express the reaction rate at enzyme surface as the MM expression as shown on the right-hand side of the above equation. [However, note that there will no longer be a MM type dependence on $C_B$ as $C_B \neq C_S$.]

(b) The effectiveness factor $\eta$ is a useful quantity that provides a measure of how much reaction rate is affected (reduced) by mass transfer effects. It is defined as the ratio of the observed (i.e., when experimentally measured) reaction rate to the rate if no mass transfer resistance was present (i.e., no mass transfer effects considered).

$$\eta = \frac{\frac{r'_{\max} C_S}{K_M + C_S}}{\frac{r'_{\max} C_B}{K_M + C_B}} = \frac{C_S}{C_B}\left(\frac{K_M + C_B}{K_M + C_S}\right)$$

We may observe that the range of values that $\eta$ can take is as follows.

$$0 \leq \frac{C_S}{C_B} \leq 1 \rightarrow \eta \leq 1$$

When chemical reaction rate is very fast, then $C_S \cong 0$ since substrate is consumed/used up by reaction almost immediately upon reaching the enzyme's catalytic surface, and therefore $\eta \cong 0$.

The Damköhler number Da is a dimensionless quantity defined as the ratio of the maximum chemical reaction rate to the maximum mass transfer rate. The maximum mass transfer rate occurs when there is very fast chemical reaction at the enzyme surface ($C_S \cong 0$), because in order to still achieve steady state, the mass transfer rate has to be equivalently fast ($k_c(C_B - C_S) \cong k_c C_B$).

$$\text{Da} = \frac{r'_{\max}}{k_c C_B}$$

There are two useful regimes to consider with the use of *Da*.

When Da $\ll$ 1, we are in the reaction rate-limited regime whereby reaction rate is much slower than mass transfer rate.

$$\eta = 1$$

$$C_S \cong C_B$$

$$\text{Rate of substrate consumption} = -r_s = \frac{r'_{\max} C_B}{K_M + C_B}$$

When Da $\gg$ 1, we are in the diffusion-limited regime whereby the reaction rate is very fast and the rate of mass transfer is also equivalently fast under steady state.

$$\text{Da} \rightarrow \infty$$

$$C_S \cong 0$$

$$\text{Rate of substrate consumption} = -r_s = k_c C_B$$

It is worth noting that under this regime, the observed reaction rate does not depend on the properties of the enzyme related to $K_M$ and $r'_{\max}$.

(c) (i) Assuming we have the following reaction mechanism,

$$\text{Enzyme} + \text{Substrate} \overset{k_1,\ k_{-1}}{\longleftrightarrow} ES\ \text{Complex} \overset{k_{\text{cat}}}{\rightarrow} \text{Product} + \text{Enzyme}$$

Then maximum reaction rate $r_{\max}$ can be expressed as follows in terms of the catalytic constant (or enzyme turnover number) $k_{\text{cat}}$ and initial enzyme concentration $C_{E0}$.

$$r_{\max} = k_{\text{cat}} C_{E0}$$

In a continuous flow stirred tank reactor with soluble enzymes, we can express $C_{E0}$ in terms of total number of moles of enzymes, $N$, and volume of liquid in the reactor, $V_{\text{tank}}$, as shown

$$C_{E0} = \frac{N}{V_{\text{tank}}}$$

The analogous form of the above expression can be obtained for a packed bed reactor whereby $q_0$ is the equivalent quantity for $C_{E0}$ with units of total number of moles of enzymes per unit total volume of packed bed. The total volume in this case comprises both voidage (liquid can flow through), and packed portions (solid support particles).

First, we can find the area-to-volume ratio for spherical support particles:

$$\frac{\text{Volume of sphere}}{\text{Surface area of sphere}} = \frac{\frac{4}{3}\pi\left(\frac{D_p}{2}\right)^3}{4\pi\left(\frac{D_p}{2}\right)^2} = \frac{D_p}{6}$$

We know that the volume of solid spherical particles can be converted to the total volume of packed bed by using the void fraction, $\varepsilon$.

$$\varepsilon = \frac{\text{Void volume}}{\text{Total packed bed volume}} = \frac{\text{Void volume}}{\text{Solid volume} + \text{void volume}}$$

$$1 - \varepsilon = \frac{\text{Solid volume}}{\text{Solid volume} + \text{void volume}}$$

$$\frac{1}{1 - \varepsilon} = \frac{\text{Solid volume} + \text{void volume}}{\text{Solid volume}}$$

We can use the factor of $1/(1-\varepsilon)$ to re-express our earlier area-to-volume ratio in terms of total packed bed volume.

$$\frac{D_p}{6}\left(\frac{1}{1-\varepsilon}\right)=\frac{\text{Volume of spheres}}{\text{Surface area of spheres}}\left(\frac{\text{Volume of spheres}+\text{void volume}}{\text{Volume of spheres}}\right)$$

Taking reciprocal, we have

$$\frac{6(1-\varepsilon)}{D_p}=\frac{\text{Surface area of solids}}{\text{Total packed bed volume}}$$

We know further that $N''$ denotes the moles of enzyme immobilized per unit surface area of the solid spherical support particles; therefore,

$$\frac{6(1-\varepsilon)}{D_p}(N'')=\frac{\text{Surface area of solids}}{\text{Total packed bed volume}}\left(\frac{\text{Mole enzyme}}{\text{Surface area of solids}}\right)$$

Therefore, we have shown that the total number of moles of enzyme per unit total packed bed reactor volume, $q_0$, is given as

$$q_0=\frac{6N''(1-\varepsilon)}{D_p}$$

(ii) We can first find out the surface area of the solid supports in the packed bed (denoted as $A$) since we know the total packed bed volume is 1.3 $m^3$ and using our earlier result from part c(i)

$$\frac{6(1-\varepsilon)}{D_p}=\frac{\text{Surface area of solids}}{\text{Total packed bed volume}}$$

$$\text{Surface area of solids, } A=\frac{6(1-\varepsilon)(1.3)}{D_p}=\frac{6(1-0.5)(1.3)}{263\times10^{-6}}=14,800\ \text{m}^2$$

We are given the liquid feed residence time, $\tau=0.43$ h. Knowing also that the void volume is where liquid flows through, therefore the volumetric flow rate for the liquid, $q$ can be found as follows

$$\text{Void volume}=\varepsilon(\text{total packed bed volume})=0.5(1.3)=0.65\ \text{m}^3$$

$$q=\frac{1}{0.43}(0.65)=1.5\text{m}^3/\text{h}=4.199\times10^{-4}\text{m}^3/\text{s}$$

Note that the substrate concentration has units of moles of substrate per unit volume of liquid. Knowing that we need to achieve a conversion of $X=0.7$, we can use the value of $q$ to compute the number of moles of substrate converted to product as follows

$$C_{s,\text{out}} = (1 - X)C_{s,\text{in}}$$

$$\text{Moles of substrate converted} = q(C_{s,\text{in}} - C_{s,\text{out}}) = qC_{s,\text{in}}X$$

$$= (4.199 \times 10^{-4})(0.03)(0.7) = 8.82 \times 10^{-6}\ \text{kmol/s}$$

We can determine the maximum mass transfer rate by setting the condition that there is a maximum (very fast) chemical reaction rate such that substrate concentration at enzyme surface is almost zero. Under steady-state condition both rates are equal and of a high value in this case.

$$\text{Maximum mass transfer rate} = k_cA(C_B - C_S) \approx k_cAC_B$$

We can use an estimate for $C_B$ that is the average between the inlet and outlet substrate concentrations to give some consideration to the varying concentration profile along the packed bed reactor. Therefore,

$$C_B \approx \frac{C_{s,\text{in}} + C_{s,\text{out}}}{2} = \frac{0.03 + (1 - 0.7)0.03}{2} = 0.0195$$

$$\text{Maximum mass transfer rate} = k_cAC_B = 1.3 \times 10^{-5}(14,800)(0.0195)$$

$$= 3.75 \times 10^{-3}\ \text{kmol/s}$$

In order to find out if the process is mass transfer or reaction rate limited, we can calculate the Damköhler number, Da

$$\text{Da} = \frac{8.82 \times 10^{-6}}{3.75 \times 10^{-3}} \ll 1$$

We conclude that since Da $\ll$ 1, this process is chemical reaction rate limited, and not mass transfer limited. In other words, the mass transfer rate is much faster than the reaction rate at the enzyme surface.

## Problem 47

**The steady-state approximation is useful in analyzing sequential reactions involving an intermediate product as it simplifies the solution of differential equations by removing time dependence.**

(a) **Consider the reaction scheme shown below:**

$$X \xrightarrow{k_1} Y \qquad (1)$$

$$Y \xrightarrow{k_2} Z \quad (2)$$

i. **Show how the steady-state hypothesis can achieve the abovementioned simplification, assuming that $k_2 \gg k_1$. Comment on possible scenarios whereby the steady-state hypothesis may be invalid.**
ii. **Using the reaction scheme given above, sketch the concentration profiles for $X$, $Y$, and $Z$ for the cases whereby $k_1 \gg k_2$ and $k_2 \gg k_1$. Comment on the shapes of the graphs.**

(b) **An ester P is hydrolyzed under acidic conditions to form product $X$ as shown below. It is said that three different rate laws are possible for this reaction scheme, depending on the relative rates of the reactions involved.**

$$\mathrm{P} + \mathrm{H_3O^+} \xrightarrow{k_1} \mathrm{PH^+} + \mathrm{H_2O} \quad (1)$$

$$\mathrm{PH^+} + \mathrm{H_2O} \xrightarrow{k_2} \mathrm{P} + \mathrm{H_3O^+} \quad (2)$$

$$\mathrm{PH^+} + \mathrm{H_2O} \xrightarrow{k_3} \text{Product } X \quad (3)$$

i. **Briefly discuss the purpose of rate law equations.**
ii. **Given that reaction (1) is much slower than reactions (2) and (3), show how the three different rate laws may be derived, highlighting any assumptions.**

## Solution 47

### *Worked Solution*

(a) (i) We are given the reaction sequence as follows and know that the rate of reaction (2) is much greater than that of reaction (1) ($k_2 \gg k_1$). The implication of this is that the intermediate product $Y$ will be reacted once it is formed and therefore the concentration of $Y$ will always be low.

$$X \xrightarrow{k_1} Y \quad (1)$$

$$Y \xrightarrow{k_2} Z \quad (2)$$

The steady-state approximation can be made in this case and it states that $Y$ is in steady state. Therefore the concentration of $Y$ denoted as $[Y]$ is assumed to remain constant with time.

$$\left.\frac{d[Y]}{dt}\right|_{ss} = 0$$

This is a useful expression because it removes time dependence in the solution of differential equations, as explained further below.

We can write the rate of change of $[Y]$ in terms of rate constants as follows:

$$\frac{d[Y]}{dt} = k_1[X] - k_2[Y]$$

Substituting the steady-state expression for $[Y]$, we find an expression for $[Y]_{ss}$

$$\left.\frac{d[Y]}{dt}\right|_{ss} = 0 = k_1[X] - k_2[Y]_{ss}$$

$$[Y]_{ss} = \frac{k_1[X]}{k_2}$$

The rate of formation of product can also be expressed in terms of rate constant as follows,

$$\left.\frac{d[Z]}{dt}\right|_{ss} = k_2[Y]_{ss}$$

Substituting our earlier expression for $[Y]_{ss}$, we obtain

$$\left.\frac{d[Z]}{dt}\right|_{ss} = k_2\left(\frac{k_1[X]}{k_2}\right) = k_1[X]$$

This result is consistent with the assumption that $k_2 \gg k_1$ which means that reaction 1 is the rate-determining step in the reaction scheme. In steady-state analysis, we first identify the intermediate product, and for an expression for its concentration under steady-state conditions. We can then re-express the rate of product formation in terms of the rate constant that dominates in the reaction scheme.

It should be noted that the steady-state hypothesis may not be suitable to be applied to all species in the reaction mixture, for example, reactants or products that accumulate during the reaction. It is most appropriate for intermediate products, these are species that are relatively slow to form but once formed they are highly reactive and are immediately reacted away by subsequent reactions. It is also worth noting that rate laws obtained from the steady-state approximation are only valid once the reaction is at steady state. The expressions are not appropriate to describe the initial phase of the reaction or when the reaction is close to completion.

(ii) We note that reaction (1) shows the formation of $Y$ while reaction (2) describes the consumption of $Y$.

For the first case whereby $k_1 \gg k_2$, we note that $X$ is converted quickly to $Y$, but $Y$ is subsequently slow to be consumed by reaction (2). This would lead to an accumulation of $Y$. It can also be expected that the formation of $Z$ will mirror the drop in $Y$ by nature of reaction (2). Since the overall rate of reaction is dependent on reaction (2), we call it the rate-determining step. The concentration profiles are shown as follows.

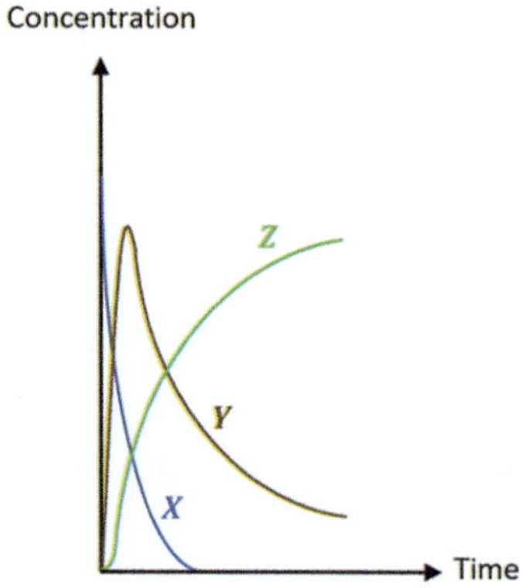


For the second case whereby $k_2 \gg k_1$, we note that $X$ is converted slowly to $Y$, but $Y$ is subsequently quickly consumed by reaction (2) once formed. Hence in this case, there is no accumulation of $Y$, in fact there is a small amount of $Y$ at any time. It is expected here that the formation of $Z$ will mirror the drop in $X$. And in this case, the rate-determining step is reaction (1). The concentration profiles are shown as follows.

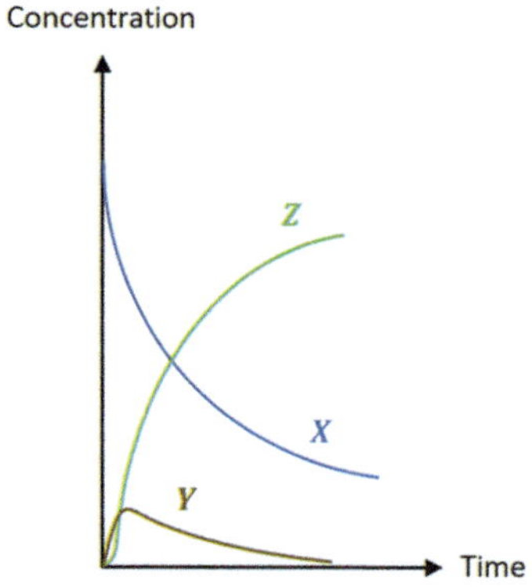


In general, for sequential reactions, the reaction step that has the smallest rate constant (slowest reaction) is typically the rate-determining step, such that the overall rate of reaction for the entire reaction scheme is most dependent on this step.

(b) (i) The study of chemical kinetics and rate laws helps us understand the factors that influence the rate of reactions so that we may better predict reaction outcomes and design appropriate reaction schemes and process conditions to achieve desired products. It is important to consider kinetics in conjunction with thermodynamics because a reaction may be thermodynamically favorable to occur but may still be slow to form product if the process is not kinetically favored.

We can experimentally determine rates of reactions by measuring the concentrations of substances involved in the reaction scheme. A rate law equation is essentially a mathematical expression used to relate the rate of reaction to the measured concentrations. It is also possible to theoretically deduce rate law equations.

In the following example of a rate law, reaction rate is denoted by $r$, concentrations of reactants $A$ and $B$ are denoted by $[A]$ and $[B]$, respectively, and $k$ is the rate constant. The reaction is said to be first order with respect to $A$ and second order with respect to $B$. The overall order of the reaction is the sum of both orders (i.e., 3). Note that the units for $r$ is concentration per unit time. Therefore, the units for the rate constant $k$ differ for reactions of different orders, in order for the equation to be consistent in units.

$$r = k[A][B]^2$$

From the form of the rate law equation, we can see that it is useful in helping us predict the reaction rate for a set of experimental conditions (e.g., concentrations).

(ii) We are given the reaction scheme below, and told that reaction (1) is slower than reactions (2) and (3). This means that $PH^+$ is removed quickly by reactions (2) and (3) the moment it is formed by reaction (1). We therefore identify our intermediate to be $PH^+$ and apply our steady-state approximation.

$$\mathrm{P} + \mathrm{H_3O^+} \xrightarrow{k_1} \mathrm{PH^+} + \mathrm{H_2O} \qquad (1)$$

$$\mathrm{PH^+} + \mathrm{H_2O} \xrightarrow{k_2} \mathrm{P} + \mathrm{H_3O^+} \qquad (2)$$

$$\mathrm{PH^+} + \mathrm{H_2O} \xrightarrow{k_3} \text{Product } X \qquad (3)$$

We can express the concentration of the intermediate under steady-state conditions,

$$\frac{d[\mathrm{PH^+}]}{dt} = k_1[\mathrm{P}][\mathrm{H_3O^+}] - k_3[\mathrm{PH^+}][\mathrm{H_2O}] - k_2[\mathrm{PH^+}][\mathrm{H_2O}] = 0$$

$$[\mathrm{PH^+}]_{ss} = \frac{k_1[\mathrm{P}][\mathrm{H_3O^+}]}{(k_2 + k_3)[\mathrm{H_2O}]}$$

The rate of formation of product $X$ under steady state can be expressed as follows:

$$r_p = k_3[\mathrm{PH}^+]_{ss}[\mathrm{H_2O}] = \frac{k_3 k_1[\mathrm{P}][\mathrm{H_3O}^+]}{(k_2 + k_3)}$$

From here we may consider different scenarios for the relative values of the rate constants $k_2$ and $k_3$.

*Scenario 1:*

If reaction (3) is much faster than reaction (2), then,

$$k_3[\mathrm{PH}^+][\mathrm{H_2O}] \gg k_2[\mathrm{PH}^+][\mathrm{H_2O}]$$

$$k_3 \gg k_2$$

$$r_p \cong \frac{k_3 k_1[\mathrm{P}][\mathrm{H_3O}^+]}{k_3} = k_1[\mathrm{P}][\mathrm{H_3O}^+]$$

The above expression means that the overall reaction rate only depends on $k_1$ and in this case then reaction (1) is the rate-determining step.

*Scenario 2:*

It is also possible that reaction (2) is much faster than reaction (3), then,

$$k_2[\mathrm{PH}^+][\mathrm{H_2O}] \gg k_3[\mathrm{PH}^+][\mathrm{H_2O}]$$

$$k_2 \gg k_3$$

$$r_p \cong \frac{k_3 k_1[\mathrm{P}][\mathrm{H_3O}^+]}{k_2} = \left(\frac{k_3 k_1}{k_2}\right)[\mathrm{P}][\mathrm{H_3O}^+]$$

We can further express the ratio $k_1/k_2$ as an equilibrium constant $K_{eq}$ since we observe that reactions (1) and (2) are simply the reverse of each other and hence describe the forward and backward reactions of a reversible reaction.

The overall rate equation is therefore expressed as follows where we see that it is second order and reaction (3) is now the rate-determining step.

$$r_p = K_{eq} k_3[\mathrm{P}][\mathrm{H_3O}^+]$$

*Scenario 3:*

Finally, it is possible that reactions (2) and (3) have comparable rates, in this case, no simplifications can be made and the rate law would remain as follows.

$$r_p = \frac{k_3 k_1[\mathrm{P}][\mathrm{H_3O}^+]}{(k_2 + k_3)}$$

## Problem 48

**An enzyme $E$ catalyzes the reversible reaction shown below, where compounds $P$ and $Q$ are isomers of each other and $ES$ is the enzyme–substrate complex. The rate constants for the forward and backward reactions are denoted by $k_1$ (and $k_2$) and $k_{-1}$ (and $k_{-2}$), respectively.**

$$E + P \overset{k_1,\,k_{-1}}{\longleftrightarrow} ES \overset{k_2,\,k_{-2}}{\longleftrightarrow} E + Q$$

**Given that the enzyme has a size of 55,000 Daltons, and the rate of reaction in a batch system is as shown below, whereby $K_P = \frac{k_{-1}+k_2}{k_1}$, $K_Q = \frac{k_{-1}+k_2}{k_{-2}}$, $R_p = k_2 C_{E0}$, $R_Q = k_{-1} C_{E0}$ and $C_{E0}$ refers to the total enzyme concentration.**

$$r = -\frac{dC_P}{dt} = \frac{\frac{R_p C_P}{K_P} - \frac{R_Q C_Q}{K_Q}}{1 + \frac{C_P}{K_P} + \frac{C_Q}{K_Q}}$$

(a) **Derive the following expression for the equilibrium constant for the reaction.**

$$K_{\text{eq}} = \frac{R_P K_Q}{R_Q K_P}$$

(b) **It is required to isomerize 12 moles of pure compound $P$ in a batch reactor of volume 0.6 m$^3$ using 12 g of enzyme. Given that $K_P = 14$ mol/m$^3$, $K_Q = 29$ mol/m$^3$, $k_2 = 1400$ min$^{-1}$, and $k_{-1} = 950$ min$^{-1}$, show that the time taken (in minutes) for the reaction to reach 97% of the equilibrium position can be expressed as follows:**

$$t|_{97\%} \approx \int_{5.2}^{20} \left( \frac{1.7 + 0.037 C_P}{0.048 C_P - 0.23} \right) dC_P$$

**Briefly outline the method for computing the above integral using graphical methods. [Actual solution not necessary.]**

## Solution 48

*Worked Solution*

(a) We are given the expression for rate of reaction which we can equate to zero by imposing the equilibrium condition, in order to derive the expression for the equilibrium constant, $K_{\text{eq}}$

$$r = -\frac{dC_P}{dt} = \frac{\frac{R_pC_P}{K_P} - \frac{R_QC_Q}{K_Q}}{1 + \frac{C_P}{K_P} + \frac{C_Q}{K_Q}} = 0$$

$$\frac{R_pC_P}{K_P} - \frac{R_QC_Q}{K_Q} = 0$$

$$K_{\text{eq}} = \frac{C_Q}{C_P} = \frac{R_PK_Q}{R_QK_P}$$

(b) In order to find time taken to achieve a certain defined end state, we need to work with a differential equation with respect to time, so that we may integrate this equation to compute time taken based on defined limits. With this in mind, we calculate the following quantities to try to specify our initial and final states.

The initial enzyme concentration (also the total enzyme concentration) can be found as follows where molar (M) is equivalent to moles per liter.

$$C_{E0} = \frac{12}{55,000(0.6 \times 1000)} = 3.6 \times 10^{-7} \text{ M}$$

$$R_P = k_2C_{E0} = 1400 \times 3.6 \times 10^{-7} = 5.1 \times 10^{-4} \text{ M/min}$$

$$R_Q = k_{-1}C_{E0} = 950 \times 3.6 \times 10^{-7} = 3.4 \times 10^{-4} \text{ M/min}$$

Substituting into our result in part a, we can compute $K_{\text{eq}}$.

$$K_{\text{eq}} = \frac{R_PK_Q}{R_QK_P} = \frac{5.1(29)}{3.4(14)} = 3.1$$

When compound $P$ isomerizes to compound $Q$ until equilibrium position is reached, we can denote the mole fraction of $Q$ as $x_Q|_{\text{eq}} = a$. It follows that the mole fraction of $P$ is $x_P|_{\text{eq}} = 1 - a$ since mole fractions of all species present should sum to 1.

By definition at equilibrium,

$$K_{\text{eq}} = \frac{C_Q}{C_P} = \frac{a}{1-a} = 3.1$$

$$a = 0.76$$

Therefore to reach 97% of the equilibrium position, the mole fractions of $Q$ and $P$ are

$$x_Q|_{97\%} = 0.97(0.76) = 0.74$$

$$x_P|_{97\%} = 1 - 0.74 = 0.26$$

The initial concentration of reactant $P$ can be found from the initial mass of $P$ and volume of batch reactor,

$$C_{P0} = \frac{12}{0.6 \times 1000} = 0.02 \text{ M}$$

At 97% of the equilibrium position, the concentration of $P$ can be expressed as follows, knowing that one molecule of $P$ isomerizes to one molecule of $Q$.

$$C_P|_{97\%} = C_{P0}\left(x_P|_{97\%}\right) = 0.02(0.26) = 5.2 \times 10^{-3}\text{M}$$

We have now computed the values that define limits of integration with respect to time as shown below. Starting with the differential equation, we have

$$-\frac{dC_P}{dt} = \frac{\frac{R_pC_P}{K_P} - \frac{R_QC_Q}{K_Q}}{1 + \frac{C_P}{K_P} + \frac{C_Q}{K_Q}}$$

$$-\left(\frac{1 + \frac{C_P}{K_P} + \frac{C_Q}{K_Q}}{\frac{R_pC_P}{K_P} - \frac{R_QC_Q}{K_Q}}\right) dC_P = dt$$

$$-\int_{C_{P0}}^{C_P|_{97\%}} \left(\frac{1 + \frac{C_P}{K_P} + \frac{C_Q}{K_Q}}{\frac{R_pC_P}{K_P} - \frac{R_QC_Q}{K_Q}}\right) dC_P = \int_0^{t|_{97\%}} dt$$

We know the following correlation should hold between species $P$ and $Q$,

$$C_Q = C_{P0} - C_P$$

Substituting the above expression into the integral equation, we re-express it in terms of only one variable $C_P$ in order to integrate with respect to $C_P$.

$$-\int_{C_{P0}}^{C_P|_{97\%}} \left(\frac{1 + \frac{C_P}{K_P} + \frac{C_{P0}-C_P}{K_Q}}{\frac{R_pC_P}{K_P} - \frac{R_Q(C_{P0}-C_P)}{K_Q}}\right) dC_P = t|_{97\%}$$

Substituting the values computed earlier, we can compute the integral to find time taken. Take note of units, as 1 mole per $m^3$ is equal to $10^{-3}$ mol/L (or M).

$$t|_{97\%} = -\int_{0.02}^{0.0052} \left( \frac{1 + \frac{C_P}{14 \times 10^{-3}} + \frac{0.02 - C_P}{29 \times 10^{-3}}}{\frac{(5.1 \times 10^{-4})C_P}{14 \times 10^{-3}} - \frac{(3.4 \times 10^{-4})(0.02 - C_P)}{29 \times 10^{-3}}} \right) dC_P$$

Converting the units for concentration from molar (M) to millimolar (mM) for ease of handling of the numbers involved, we derive the following expression for time taken in minutes.

$$t|_{97\%} = \int_{5.2}^{20} \left( \frac{1 + \frac{C_P}{14} + \frac{20 - C_P}{29}}{\frac{0.51C_P}{14} - \frac{(0.34)(20 - C_P)}{29}} \right) dC_P \approx \int_{5.2}^{20} \left( \frac{1.7 + 0.037C_P}{0.048C_P - 0.23} \right) dC_P$$

We can solve the above integral by graphical or numerical methods. For the graphical method, we can tabulate a series of data points by defining a function $f(C_P)$ as shown below and compute a range of values for $f(C_P)$ from the lower to upper limits of integration, i.e., from $C_P = 5.2$ to $C_P = 20$.

$$f(C_P) = \frac{1.7 + 0.037C_P}{0.048C_P - 0.23}$$

$$t|_{97\%} \approx \int_{5.2}^{20} f(C_P) dC_P$$

| $C_P$ | $f(C_P)$ |
|---|---|
| 5.2 | 96.6 |
| 5.4 | 65.1 |
| 5.7 | 43.8 |
| 6.4 | 25.1 |
| 7.5 | 15.2 |
| 10 | 8.28 |
| 12.5 | 5.84 |
| 15 | 4.60 |
| 17.5 | 3.85 |
| 20 | 3.34 |

The time taken is equivalent to the value of the integral. By definition, the value of the integral for a plot of $f(C_P)$ against $C_P$ is equivalent to the area under the graph between specified limits (i.e., area under the blue line, between the blue line and the horizontal axis, for the range defined by the black dotted lines).

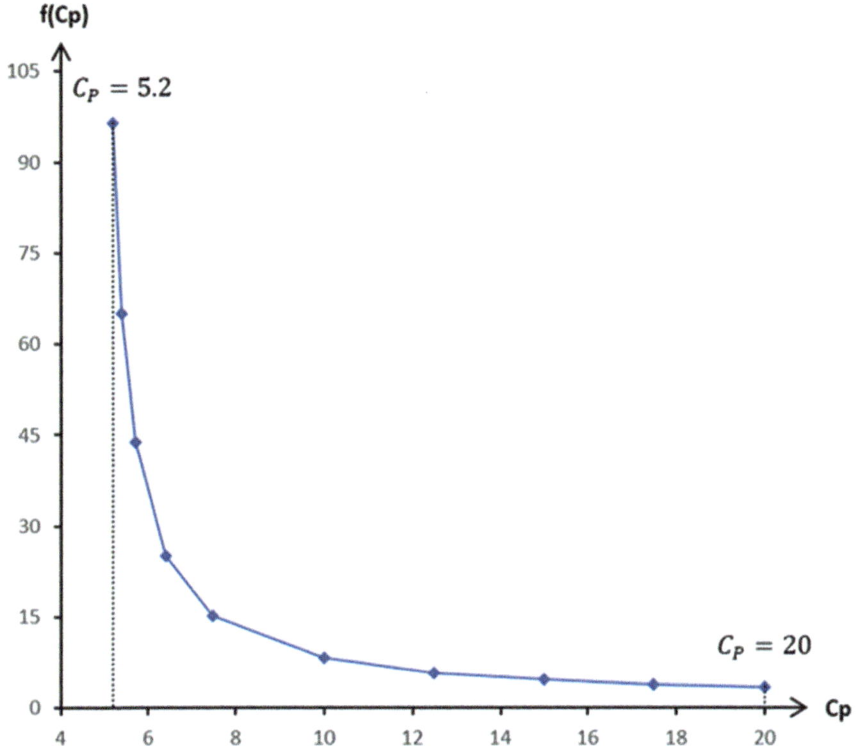


**Problem 49**

**The continuous culture of yeast cells can be done in a chemostat, a type of bioreactor that functions like a Continuous Stirred Tank Reactor (CSTR). In a chemostat, fresh medium is continuously introduced at a constant rate while culture liquid (containing medium, cells and any metabolic products) is continuously removed at the same rate, hence keeping a constant culture volume. You may assume perfect mixing in the chemostat and that glucose is the single nutrient (limiting substrate) controlling cell growth rate.**

(a) **Explain briefly how a chemostat works, making reference to dilution rate, *D*, and its implications on cell growth.**
(b) **In a particular yeast cell culture under aerobic conditions, glucose was used as the growth-limiting substrate, while ammonia ($NH_3$) provided the nitrogen source. The chemical formula of the yeast strain was given as $CH_{1.7}O_{0.4}N_{0.1}$. When dilution rate exceeds a critical value, ethanol is formed as a metabolic by-product. Given that the cell growth process can be represented by the equation below**

$$aCH_2O + bNH_3 + cO_2 \rightarrow CH_{1.7}O_{0.4}N_{0.1} + dCH_3O_{0.5} + eH_2O + fCO_2$$

i. **Derive expressions for the following:**

- **Growth yield coefficient, $Y_{X/S}$**
- **Ethanol yield coefficient, $Y_{P/S}$**
- **Oxygen uptake rate (molar units), $r_{O_2}$**
- **Respiratory quotient (molar units), $K$**

ii. **When ethanol is produced, derive an expression that relates $Y_{X/S}$ to the oxygen uptake rate $r_{O_2}$ and respiratory quotient $K$.**

iii. **Assuming that measured data for flow rates and compositions of the species involved in the cell culture are available, briefly describe an experimental approach to determine the critical dilution rate.**

iv. **Find an expression for $Y_{X/S}$ when dilution rate is less than its critical value.**

v. **Given that $Y_{X/S} = 0.22$ and $Y_{P/S} = 0.42$ at a particular dilution rate, calculate the oxygen uptake rate.**

## Solution 49

### *Worked Solution*

(a) Dilution rate, $D$, has units of time$^{-1}$ and is defined as the flowrate of medium into the chemostat, $v$, divided by the volume of culture in the chemostat, $V$, as shown below. Dilution rate therefore measures the rate of nutrient exchange in the bioreactor and is useful as it is an experimentally variable parameter.

$$D = \frac{v}{V}$$

Under steady-state conditions in a chemostat, the biomass concentration and limiting substrate concentration in the culture remain constant as shown in the plot below. This can be expressed as follows whereby $Y_{X/S}$ is the growth yield coefficient, $C_X$ and $C_S$ are the cell and substrate concentrations in the vessel, and $C_{S,\text{in}}$ is the substrate concentration at the inlet.

$$C_X = Y_{X/S}(C_{S,\text{in}} - C_S)$$

In a chemostat, we typically assume a single limiting substrate (e.g., glucose), which means that almost all the glucose is used by the cells to maintain cell concentration at a constant value under steady state. It follows that this value of cell concentration will also be proportional to the glucose concentration in the inlet stream containing fresh medium. We can therefore control the specific growth rate of cells in the vessel by controlling the concentration of this single limiting nutrient at the inlet.

At steady state, the specific growth rate of the cells, $\mu$, is equal to dilution rate. By changing the rate at which fresh medium is added (i.e., dilution rate), we can control the specific growth rate of cells to the desired level (can be varied from just above zero to just below $\mu_{\max}$). It is important to have control of dilution rate as cell growth rate also affects products of cell metabolism. For example, yeast cells start to form ethanol as a metabolic by-product when dilution rate exceeds a certain value.

Note that prior to steady-state continuous culture, there is a short transition period when batch culture occurs (as shown in plot below). During this period, cell count increases rapidly to reach a desirably sizeable concentration (i.e., the starting culture (or inoculum) for the subsequent continuous culture phase), and this concentration is maintained under steady-state conditions during the continuous culture phase. Also during this transition period, operating conditions in the chemostat take time to ramp up to steady-state conditions.

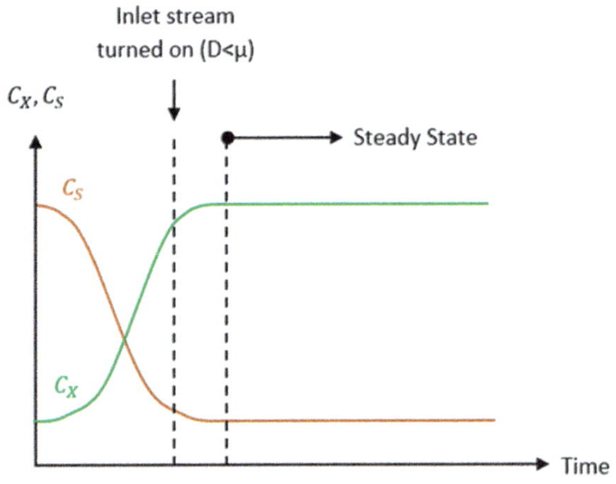


It is common to encounter a term known as the critical dilution rate, $D_{\text{crit}}$. When dilution rate exceeds this value, steady state can no longer be maintained as some cells are washed out of the vessel and this loss cannot be replaced by further cell growth. Consequently, unused substrate starts to accumulate in the vessel causing substrate concentration to rise. The value of $D_{\text{crit}}$ is equivalent to the maximum specific growth rate of the cells, $\mu_{\max}$. It is therefore crucial to ensure that the dilution rate in a chemostat does not go close to or exceed $D_{\text{crit}}$ (or $\mu_{\max}$) in order to maintain steady-state culture. Dilution rate is also related to the mean residence time $\tau$ of cells in the vessel as follows.

$$\tau = \frac{1}{D}$$

(b) (i) Given the following equation, we can make use of stoichiometry to derive the required expressions.

$$aCH_2O + bNH_3 + cO_2 \rightarrow CH_{1.7}O_{0.4}N_{0.1} + dCH_3O_{0.5} + eH_2O + fCO_2$$

We know that the single limiting substrate is glucose ($CH_2O$) and we know that the cell biomass is represented by $CH_{1.7}O_{0.4}N_{0.1}$. Therefore, growth yield coefficient is expressed as follows:

$$Y_{X/S} = \frac{1}{a}$$

We know that the metabolic product in this case is ethanol, given by $CH_3O_{0.5}$, therefore, product yield coefficient can be expressed as follows:

$$Y_{P/S} = \frac{d}{a}$$

Oxygen uptake rate is defined as the number of moles of oxygen gas taken in by the cells per mole of substrate (growth-limiting). Therefore this rate may be expressed as follows.

$$r_{O_2} = \frac{c}{a}$$

Finally respiratory quotient is defined as the number of moles of carbon dioxide per mole of oxygen taken in. This follows from the fact that in aerobic respiration, cells take in oxygen and glucose to produce carbon dioxide, water, and energy.

$$K = \frac{f}{c}$$

(ii) Note that $Y_{X/S}$ is a function of $a$, $r_{O_2}$ is a function of $a$ and $c$, and $K$ is a function of $c$ and $f$. We know the relevant oxidation numbers for the elements involved in the chemical formulae for glucose substrate ($CH_2O$) and cells ($CH_{1.7}O_{0.4}N_{0.1}$).

$$C = +4$$

$$H = +1$$

$$O = -2$$

$$N = -3$$

The sum of the oxidation numbers for the species involved can be computed as shown:

$$CH_2O = +4 + 2(1) + (-2) = +4$$

$$CH_{1.7}O_{0.4}N_{0.1} = +4 + 1.7(1) + 0.4(-2) + 0.1(-3) = +4.6$$

$$O_2 = 2(-2) = -4$$

$$CH_3O_{0.5} = +4 + 3(1) + 0.5(-2) = +6$$

The oxidation numbers for ammonia, water, and carbon dioxide are zero (this is the case for pure elements or neutral compounds).

We can now formulate an equation for the degree of reduction for the cell growth equation as follows.

$$a(+4) + b(0) + c(-4) = +4.6 + d(+6) + e(0) + f(0)$$

$$4a - 4c = 4.6 + 6d \qquad (1)$$

We need to come up with more equations so that we can solve all unknown variables (number of equations equal to number of unknowns). Another useful equation that correlates the stoichiometric coefficients for the cell growth equation is a mass balance on carbon atoms as shown below. We can further substitute our earlier derived expression for $K$.

$$a = 1 + d + f = 1 + d + Kc$$

$$d = a - 1 - Kc$$

Substituting the above expression for $d$ back into equation (1), we have

$$4a - 4c = 4.6 + 6(a - 1 - Kc)$$

$$a = 0.7 - c(2 - 3K)$$

We can also make use of our earlier expression for oxygen uptake rate to substitute $c$ as follows

$$a = 0.7 - a(r_{O_2})(2 - 3K)$$

$$a[1 + r_{O_2}(2 - 3K)] = 0.7$$

Going back to the expression for growth yield coefficient, we derive the required expression as follows

$$Y_{X/S} = \frac{1}{a}$$

$$Y_{X/S} = \frac{1}{0.7}[1 + r_{O_2}(2 - 3K)]$$

(iii) At low dilution rates (and correspondingly low specific growth rates), glucose metabolism is full aerobic and respiratory. However, when dilution rate increases to reach a certain critical value, respiro-fermentative metabolism sets in whereby ethanol starts being produced. This ethanol production increases as dilution rate increases. This is in fact the typical behavior of the commercial baker's yeast strain known as Saccharomyces cerevisiae. Under high substrate concentrations and high specific growth rates, alcoholic fermentation is triggered even under fully aerobic conditions. This process is usually undesired in cell cultures as it reduces the amount of cell biomass obtained from the substrate feedstock. Therefore, yeast cell production is usually done under aerobic and substrate-limited conditions.

It follows that when $d \to 0$, most of the substrate is used for cell growth, and the growth yield coefficient will also tend to a maximum value. This point can be reached experimentally by reducing the dilution rate by reducing feed flow rate gradually until ethanol production just reaches zero. At this point, we have reached the critical dilution rate.

(iv) We need to find an expression for growth yield coefficient when $D < D_{\text{crit}}$, in other words, when ethanol is not produced.

When $D = D_{\text{crit}}$, $d = 0$ and equation (1) becomes

$$4a - 4c = 4.6$$

$$1 = \frac{1.15}{a} + \frac{c}{a}$$

Substituting $r_{O_2}$ and $Y_{X/S}$ into the equation, we obtain

$$1 = 1.15 Y_{X/S} + r_{O_2}$$

$$r_{O_2} = 1 - 1.15 Y_{X/S} \qquad (2)$$

We can do the same for the other carbon balance equation, i.e., impose the condition when $D = D_{\text{crit}}$ and $d = 0$.

$$a = 1 + Kc$$

$$1 = \frac{1}{a} + \frac{Kc}{a} = Y_{X/S} + K r_{O_2}$$

Substituting our earlier expression (2) into the above, we have

$$1 = Y_{X/S} + K\left(1 - 1.15 Y_{X/S}\right) = Y_{X/S}(1 - 1.15K) + K$$

$$Y_{X/S} = \frac{1 - K}{1 - 1.15K}$$

(v) Given that $Y_{X/S} = 0.22$ and $Y_{P/S} = 0.42$, we can substitute into equation (1) since we know that $Y_{X/S} = 1/a$ and $Y_{P/S} = d/a$.

$$4a - 4c = 4.6 + 6d$$

$$1 - \frac{c}{a} = \frac{1.15}{a} + \frac{1.5d}{a}$$

$$r_{O_2} = \frac{c}{a} = 1 - 1.15Y_{X/S} - 1.5Y_{P/S}$$

$$r_{O_2} = 1 - 1.15(0.22) - 1.5(0.42) \approx 0.12$$

# Fluid Mechanics

## Problem 1

**We have a hydropower system shown below whereby water flows from reservoir A to reservoir B through a circular pipe to the turbine located near reservoir B. The flow rate is 10 $m^3$/s and the electrical power generated from the turbine supplies the power grid. The difference in height between the reservoirs is 15 m. The reservoirs are large enough to assume that this difference in height remains relatively constant. A distance of 20 km separates the reservoirs and the circular pipe may be assumed to be of similar length. Assume that the density of water is 1000 kg/$m^3$, and the kinematic viscosity is $10^{-6}$ $m^2$/s.**

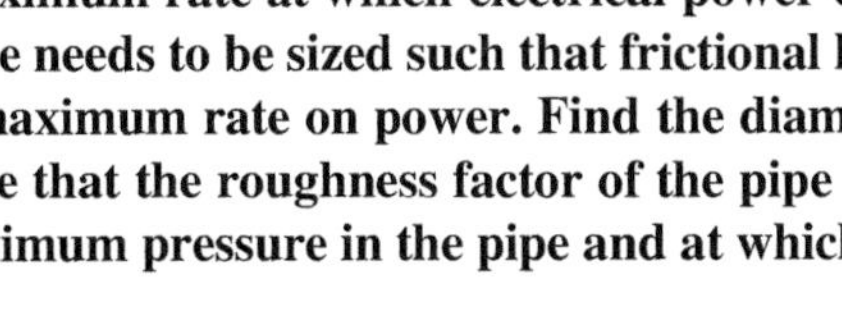


(a) **Find out the maximum rate at which electrical power can be produced.**
(b) **The circular pipe needs to be sized such that frictional losses in the pipe are 0.005% of the maximum rate on power. Find the diameter of pipe needed. You may assume that the roughness factor of the pipe is 0.0002.**
(c) **What is the maximum pressure in the pipe and at which point in the pipe is that located?**



X. W. Ng, *Engineering Problems for Undergraduate Students*,
https://doi.org/10.1007/978-3-030-13856-1_5

## Solution 1

### *Worked Solution*

(a) When we tackle such a problem, where we have information such as power generated, height difference, density of the fluid involved, and frictional losses, we may begin to think about using the Bernoulli equation.

$$\frac{V_a{}^2}{2} + \frac{P_a}{\rho} + gh_a - l_v + \delta W_s = \frac{V_b{}^2}{2} + \frac{P_b}{\rho} + gh_b$$

We know that in this case, $P_a = P_b = P_{\text{atm}}$, $V_a = V_b = 0$ since the reservoirs contain static bodies of water, and for the maximum amount of power generated we assume that frictional losses $l_v = 0$. Note that $\delta W_s$ is defined as work input; hence, a negative value denotes work produced.

$$\delta W_{s,\max} = g(h_b - h_a) = 9.81 \times (-15) = -150 \text{ Jkg}^{-1}$$

The maximum rate of power generation can be found by using the data of flowrate $Q = 10 \text{ m}^3\text{s}^{-1}$ and density $\rho = 1000 \text{ kgm}^{-3}$. Note that $Js^{-1} = W$.

$$\text{Max power generated} = |\delta W_{s,\max}| \times \rho \times Q = 150 \times 1000 \times 10 = 1.5 \text{ MW}$$

(b) The viscous losses take up 0.005% of the max power limit; therefore

$$l_v = 0.00005 \times 1500000 = 75 \text{ W}$$

---

### *Background Concepts*

Before we figure out the sizing of pipe required, we may first recall how to express pipeline frictional (or viscous) losses in terms of known variables. For a horizontal pipe and assuming the fluid is incompressible, we consider two points along the pipe, points 1 and 2, and the Bernoulli equation simplifies to

$$V_1 = V_2$$

$$h_1 = h_2$$

$$\frac{P_1 - P_2}{\rho} = l_v$$

The pressure difference may be further expressed in terms of a friction factor *f*, whereby

$$f = \frac{P_1 - P_2}{2\rho V^2}\frac{D}{L}$$

- - - - - - - - - - - - - - - - - - - - - - - - - - - - - - - - - - - - - - - -

So we can express viscous losses in the pipe in terms of friction factor,

$$l_v = \frac{2LfV^2}{D}$$

$$V = \frac{Q}{\pi D^2/4}$$

$$l_v = \frac{2Lf}{D}\left[\frac{Q}{\pi D^2/4}\right]^2$$

Roughness factor *k*/*D* is related to Reynolds number and friction factor, and hence can be looked up from data booklets. We are told further that $\frac{k}{D} = 0.0002$. We may assume large flowrates (Re > $10^7$) whereby the friction factor tends to a constant value. From the data booklet, we will be able to find that $f = 0.0035$. Note that if Re is low, there are other correlations that are appropriate for finding $f$ whereby $f = f$ (Re).

$$l_v = \frac{2Lf}{D}\left[\frac{Q}{\pi D^2/4}\right]^2 = \frac{2(20000)(0.0035)}{D}\left[\frac{10}{\pi D^2/4}\right]^2 = 2.3 \times 10^4 \times D^{-5}$$

$$75 = \frac{2.3 \times 10^4}{D^5}$$

$$D = 3\text{ m}$$

Now we can check if our earlier assumption on Reynolds number was valid. The result below reaffirms the assumption.

$$\text{Re} = \frac{VD}{\nu} = \frac{10}{\pi D\nu/4} = \frac{40}{\pi(3)10^{-6}} = 0.4 \times 10^7 > 10^7$$

(c) Maximum pressure occurs at the point just before the turbine. As such we may pick this point (indicated as point C in the diagram below) as well as a second point at reservoir A, to be used in our Bernoulli equation.

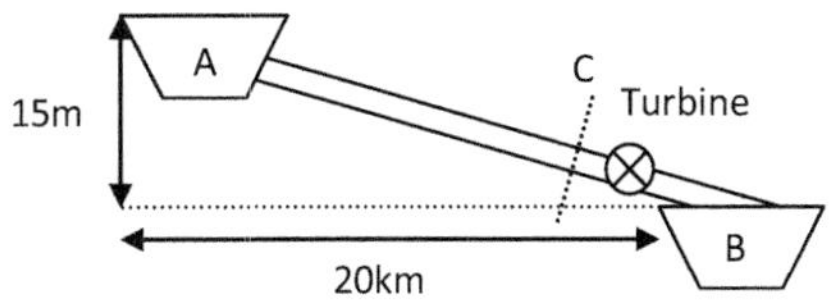

$$\frac{V_a^{\,2}}{2} + \frac{P_a}{\rho} + gh_a - l_v = \frac{V_c^{\,2}}{2} + \frac{P_c}{\rho} + gh_c$$

$$V_a = 0, P_a = P_{\text{atm}},\; l_v = 75$$

$$V_c = \frac{Q}{\pi D^2/4} = \frac{10(4)}{\pi(3^2)} = 1.4\ \text{m/s}$$

$$\frac{(P_c - P_{\text{atm}})}{\rho} = g(h_a - h_c) - \frac{1.4^2}{2} - 75$$

$$\frac{(P_c - 10^5)}{1000} = 9.81(15) - \frac{1.4^2}{2} - 75$$

$$P_c = 1.7\ \text{bar}$$

## Problem 2

**The diagram below shows a viscous polymer that is injected into an injection mould (disc-shaped).**

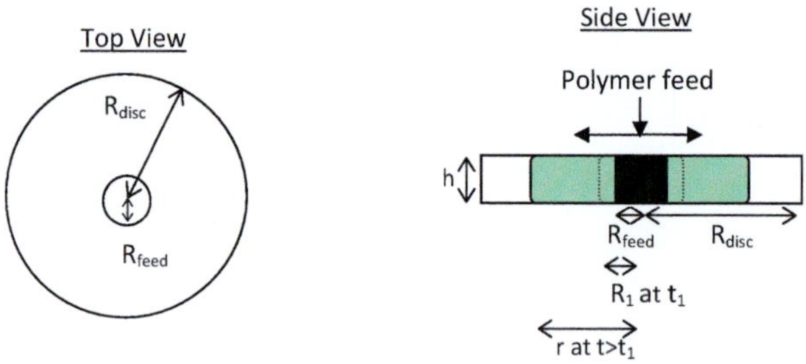


**The polymer melt is injected axially through the central filling point which is of radius $R_{\text{feed}}$. From the top view of the mold, we observe that the outer radius of the mould is $R_{\text{disc}}$. The polymer melt flows radially out from the central point slowly, filling the mould. The polymer melt has density $\rho$ and viscosity $\eta$. The thickness of the disc is $h$ as observed from the side view. After time $t_1$, which is shortly after the filling process is started, the polymer has filled the mould up to a radius $R_1$. The mould is vented at the outer edge so that pressure downstream of the polymer–air interface remains constant at $\mathcal{P}_{\text{atm}}$ At this time and all later times, the pressure in the polymer at the central point is equal to feed pressure $\mathcal{P}_{\text{feed}}$.**

(a) **Sketch the position of the polymer front as a function of time ($r$ vs. $t$), starting from $R_1$ at $t_1$.**
(b) **Determine the velocity profile of the polymer melt. State any assumptions.**

(c) **Show how we can find out the time needed to completely fill the mould, i.e., $(t - t_1)$. You may leave your answer in integral form.**

(d) **Show by order of magnitude analysis, to explain the conditions that must exist to justify our assumptions in simplifying the Navier–Stokes equation.**

## Solution 2

### *Worked Solution*

(a)

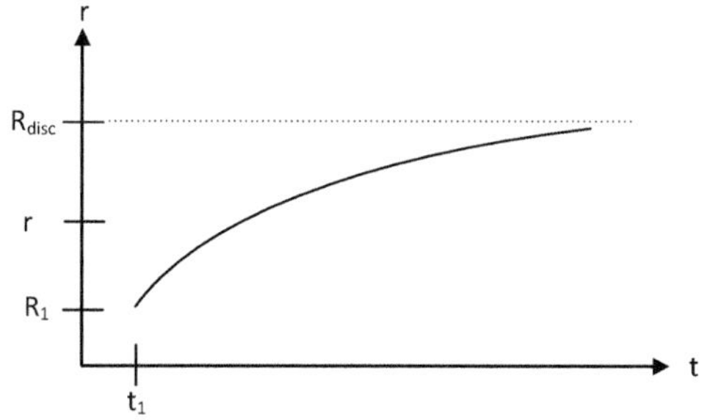


$\frac{dr}{dt}$ decreases with time; hence, there is a decreasing gradient as $t$ increases. The amount of polymer melt required to move a small radial distance $dr$ can be represented by differential volume $dV = (2\pi r)(dr)(h)$ which increases as $r$ increases. As such, the rate at which $r$ increases will reduce with $t$, given a constant pressure differential $\Delta\mathcal{P} = \mathcal{P}_{\text{feed}} - \mathcal{P}_{\text{atm}}$. The $\Delta\mathcal{P}$ acts to overcome the shear force (=shear stress × area) opposing the movement of the polymer melt. The area over which the shear force acts is $A_c = \pi(r^2 - R_1{}^2)$. As this area increases as $r$ increases, shear force increases. Since pressure difference $\Delta\mathcal{P}$ is constant, volumetric flow rate $\dot{Q}$ will decrease (and hence $\frac{dr}{dt}$ will decrease) as $r$ increases (or as $t$ increases).

(b) First, we identify the dominant velocity direction in this problem which is radial $v_r$. Cylindrical coordinates are suitable to analyze this problem. We can refer to the data booklet for mass continuity equations, which we can assume the following for this problem:

- $v_\theta = v_z = 0$
- Incompressible fluid ($\rho$ is constant)
- Steady state ($\frac{\partial}{\partial t} = 0$)

So we have the following mass continuity equation:

$$\frac{1}{r}\frac{\partial}{\partial r}(rv_r) = 0$$

$$rv_r \neq f(r)$$

We can now refer to the Navier–Stokes (NS) equations, which are differential forms of the energy balance. The macroscopic version of this would be the Bernoulli equation. We have the following assumptions:

- $v_\theta = v_z = 0$
- Constant viscosity for a Newtonian fluid
  Recap that Newtonian fluids are fluids that exhibit a straight line (passing through origin) correlation between shear stress and shear rate. Viscosity is the ratio of shear stress to shear rate. Newtonian properties may be assumed for gases and liquids with low molecular weight.
- Assume inertial terms are insignificant as compared to the viscous and pressure terms. This is valid due to the small gap $h \ll R$. Inertial terms are the terms on the left-hand side of the Navier–Stokes equations, for example, the inertial terms for the $r$ component: $\rho\left(v_r \frac{\partial v_r}{\partial r} + v_z \frac{\partial v_r}{\partial z}\right)$
- Axisymmetry which means $\frac{\partial}{\partial \theta} = 0$
- Assume that the main pressure gradient exists in the radial direction; hence only $\frac{\partial \mathcal{P}}{\partial r} \neq 0$. $\frac{\partial \mathcal{P}}{\partial z} = 0$ and $\frac{\partial \mathcal{P}}{\partial \theta} = 0$ due to the small gap and axisymmetry, respectively.

So we have the following simplified NS equation:
$r$ component:

$$0 = -\frac{\partial \mathcal{P}}{\partial r} + \eta \frac{\partial^2 v_r}{\partial z^2}$$

$$\frac{\partial \mathcal{P}}{\partial r} = \eta \frac{\partial^2 v_r}{\partial z^2} = g(r)$$

$$\frac{\partial^2 v_r}{\partial z^2} = \frac{g(r)}{\eta}$$

$$\frac{\partial v_r}{\partial z} = \frac{g(r)}{\eta} z + C_1$$

$$v_r = \frac{g(r)}{2\eta} z^2 + C_1 z + C_2$$

To solve for the integration constants, we need to establish boundary conditions. When $z = 0$ and $z = h$, $v_r = 0$. This is called the no-slip condition which means that the fluid sticks to the solid surface and follows the velocity of the surface. If the surface is stationary, then the fluid velocity at this interface is also zero.

$$C_2 = 0$$

$$0 = \frac{g(r)}{2\eta}h^2 + C_1 h$$

$$C_1 = -\frac{g(r)}{2\eta}h$$

$$v_r = \frac{g(r)}{2\eta}z^2 - \frac{g(r)}{2\eta}hz = \frac{g(r)}{2\eta}z(z-h)$$

We recall the result from mass continuity and can define another function $g' = rg(r)$

$$rv_r \neq f(r)$$

$$rv_r = \frac{g(r)}{2\eta}z(z-h)r = \frac{g'}{2\eta}z(z-h)$$

We observe that $g' \neq f(r) = K$, so $\frac{K}{r} = g$. Also we may assume that the pressure $\mathcal{P} = \mathcal{P}_{\text{feed}}$ at both $R_{\text{feed}}$ and $R_1$ as the distance apart is minimal.

$$\frac{\partial \mathcal{P}}{\partial r} = \frac{K}{r}$$

$$\int_{\mathcal{P}_{\text{feed}}}^{\mathcal{P}_{\text{atm}}} d\mathcal{P} = K \ln\left(\frac{r}{R_1}\right)$$

$$\mathcal{P}_{\text{atm}} - \mathcal{P}_{\text{feed}} = K \ln\left(\frac{r}{R_1}\right)$$

$$K = \frac{(\mathcal{P}_{\text{atm}} - \mathcal{P}_{\text{feed}})}{\ln\left(\frac{r}{R_1}\right)}$$

Note that $K$ is a negative value. So we can express the velocity profile as follows

$$v_r = \frac{K}{2\eta r}z(z-h) = \frac{(\mathcal{P}_{\text{atm}} - \mathcal{P}_{\text{feed}})}{2\eta r \ln\left(\frac{r}{R_1}\right)}z(z-h)$$

(c) We can deduce the expression $v_r dt = dr$. So we let $\alpha = \frac{(\mathcal{P}_{\text{atm}} - \mathcal{P}_{\text{feed}})}{2\eta}$ to simplify this expression to

$$dr = \frac{\alpha}{r \ln\left(\frac{r}{R_1}\right)} z(z-h)dt$$

We will need boundary conditions to integrate. We know that when $t = t_1$, $r = R_1$

$$r \ln\left(\frac{r}{R_1}\right) dr = \alpha z(z-h)dt$$

We assume that when we completely fill the mould, the midpoint of the polymer front ($z = h/2$) just reaches $r = R_{\text{disc}}$

$$\int_{R_1}^{R_{\text{feed}}} r \ln\left(\frac{r}{R_1}\right) dr = \int_{t_1}^{t} -\alpha\left(h^2/4\right)dt$$

Solve the integration for $t$ and the time needed to fill the mould can be found to be $(t - t_1)$.

(d) In simplifying the NS equations, we assumed that inertial terms are insignificant when compared to the viscous terms. The main inertial term is $\rho v_r \frac{\partial v_r}{\partial r}$ while the main viscous term is $\eta \frac{\partial^2 v_r}{\partial z^2}$. Let $v_r$~U, $r$~$R$, $z$~$h$

$$\rho v_r \frac{\partial v_r}{\partial r} \sim \rho U \frac{U}{R}$$

$$\eta \frac{\partial^2 v_r}{\partial z^2} \sim \eta \frac{U}{h^2}$$

$$\frac{\text{Inertial}}{\text{Viscous}} = \frac{\rho U \frac{U}{R}}{\eta \frac{U}{h^2}} = \frac{\rho U h}{\eta}\left(\frac{h}{R}\right)$$

Since we know that $\frac{h}{R} \ll 1$, hence the assumption is valid.

## Problem 3

**Coating of a solid sheet takes place by sliding it in a coating liquid at a steady velocity $U$, through a slot in a die. The die is fabricated such that the liquid gap (between solid sheet and die surface) is of thickness $H$ on both sides of the sheet. Further downstream from the die, the coating becomes a thickness of $H_\infty$. The pressure in the upstream liquid is $\mathcal{P}_{\text{atm}}$. The die length is $L$, and the system width is $W$. The liquid is Newtonian and Re is such that the flow of the coating liquid is laminar.**

(a) **Find $v_x(y)$ for the liquid in the upper gap. State any assumptions to facilitate the analysis.**

(b) **Determine the force required to pull the sheet through the die. You may assume that the shear force on the sheet in the upstream liquid where $x < 0$ is negligible.**
(c) **Express $H_\infty$ in terms of $H$.**

## Solution 3

### *Worked Solution*

(a) Let us first illustrate the problem to better visualize it.

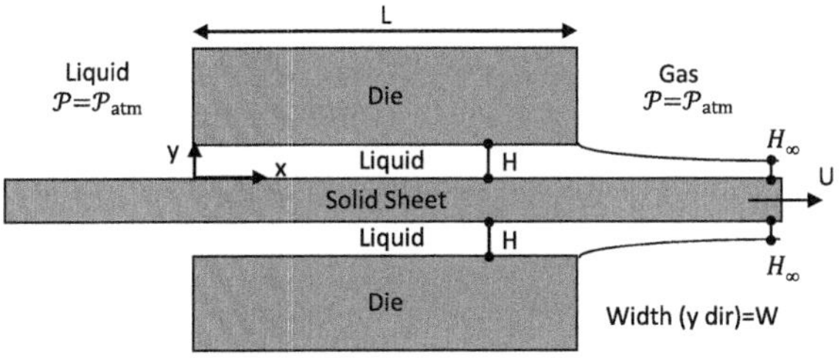


These are some of the assumptions we can make in this problem:

- We may assume fully developed flow. Fully developed flow means that we may ignore entrance or exit effects, this is valid as the flow development regions at these terminal portions are negligible as compared to the total distance of flow, this is usually reasonable if $H \ll L$. The implication of this is that flow field at all $x$ positions are the same, $\frac{\partial v_x}{\partial x} = \frac{\partial v_y}{\partial x} = \frac{\partial v_z}{\partial x} = 0$.
- The dominant velocity direction is the $x$ direction; hence $v_y = v_z = 0$. On a related note, we assume a 2D flow where $z$ direction may be ignored.
- The flow is not turbulent as we are told that it is laminar.
- The flow is at steady state ($\frac{\partial}{\partial t} = 0$) since velocity is at a constant U.
- The liquid is incompressible $\rho$ is constant.
- Constant viscosity for a Newtonian fluid.
- Gravitational and surface tension effects are ignored.

So we have the following simplified NS equation.
$x$ component:

$$0 = -\frac{\partial \mathcal{P}}{\partial x} + \eta \frac{\partial^2 v_x}{\partial y^2}$$

We know the boundary conditions for pressure, $\mathcal{P}(x = 0\,\&\, x = L) = \mathcal{P}_{\text{atm}}$, we may then also deduce that $\frac{\partial \mathcal{P}}{\partial x} = 0$ since there is no pressure change with the change in $x$ from 0 to $L$.

$$0 = \frac{\partial^2 v_x}{\partial y^2}$$

$$v_x = ay + b$$

We can apply boundary conditions for velocity which would be the no-slip condition at the surface of the die and solid sheet, $y = 0$, $v_x = U$ and $y = H$, $v_x = 0$.

$$b = U$$

$$0 = aH + U \rightarrow a = -\frac{U}{H}$$

$$v_x = -\left(\frac{U}{H}\right)y + U = U\left[1 - \frac{y}{H}\right]$$

(b) The force required to pull the sheet can be denoted by $F_x$ and it needs to overcome the shear stress caused by friction against the solid surfaces that the liquid polymer interfaces with on both sides. The force balance equation is as follows.

$$F_x = (-\tau_{yx})(A_s) = (-\tau_{yx})(2LW)$$

At this point, recall the convention for tensor notation which we have used for the shear stress term. $\tau_{yx}$ refers to shear force direction in the $x$ direction, in the $y$-plane. The $y$-plane is the plane where the normal to the plane is the $y$-dir axis. The shear stress acts in the negative $x$ direction, hence $\tau_{yx}$ is a negative value, and we accounted for this by adding a negative sign at the right-hand side of the equation so that $F_x$ is a positive value when it acts in the positive $x$ direction which it rightfully does. By definition, $-\tau_{yx} = -\eta \frac{dv_x}{dy} = \eta \frac{U}{H}$.

$$F_x = \left(\eta \frac{U}{H}\right)(2LW) = \frac{2\eta ULW}{H}$$

(c) To find $H_\infty$, we need to start with a differential equation that involves the $y$ direction, i.e., $dy$. We then think about whether we are able to identify points along the flow that help us link to $H_\infty$ and are able to correlate to each other with commonalities. We note that we have yet to apply the mass continuity equation which proves useful here, as mass is conserved both at a point within the die slot (where $y = H$) and at a further downstream portion out of the slot (where $y = H_\infty$).

$$\dot{Q}_{\text{slot}} = \dot{Q}_{\text{downstream}}$$

$$\int_0^H v_x dy = \int_0^{H_\infty} v_x dy$$

There is an easy way to figure out the right-hand side of the equation, because we can assume that there is negligible shear stress between the liquid and gas interface, hence the entire liquid film is moving at $U$ across all values of $y$ at the downstream portion.

$$\int_0^{H_\infty} v_x dy = UH_\infty$$

There is a neat way to evaluate the integral on the left-hand side, by letting $Y = y/H$. When $Y = 0, y = 0$ and when $Y = 1, y = H$ and $dy = HdY$. This method allows us to circumvent having second-order terms after integration, because we have converted the integration limit to value 1 which is indifferent to being raised to second order (i.e., $1^2 = 1$)

$$v_x = U\left[1 - \frac{y}{H}\right] = U[1 - Y]$$

$$\int_0^H v_x dy = \int_0^1 UH[1 - Y]dY = UH\left[Y - \frac{Y^2}{2}\right]_0^1 = \frac{UH}{2}$$

Finally we can equate both sides of the equation and find an expression for $H_\infty$.

$$UH_\infty = \frac{UH}{2}$$

$$H_\infty = \frac{H}{2}$$

## Problem 4

**We have a vertical cylinder (with two open ends on each side) of radius $R$ and length $L$ that is filled with a Bingham fluid. Initially the cylinder rests on a solid and impermeable horizontal plane that blocks any flow, and the top end is exposed to the atmosphere. When the bottom plane is removed, the bottom end of the cylinder is exposed and contents of the cylinder may begin to empty.**

(a) **If no flow occurs when the bottom plane is removed, find $\tau_{rz}$. State any assumptions.**

(b) **What is the minimum cylinder radius $R_{\min}$ needed for flow to occur when the bottom plane is removed?**

(c) **For fully developed downward flow, the following equation holds when $\tau_{rz}$ is greater or equal to the yield stress, $\tau^*$. Determine $v_z$ for $R = 2R_{\min}$. Sketch and explain the velocity profile for a fluid of column height $L$. You may assume that $\eta^*$ is a constant.**

$$\tau_{rz} = \tau^* + \eta^* \frac{dv_z}{dr}$$

**Solution 4**

***Worked Solution***

(a) Recall that a Bingham fluid is a special type of fluid that is able to endure a certain amount of shear stress without flowing. If the shear stress stays below a threshold value which is commonly known as yield stress ($\tau^*$), this material would behave like a solid. When $\tau^*$ is exceeded, the material starts to deform continuously like a liquid.

Let us illustrate the problem to better visualize it.

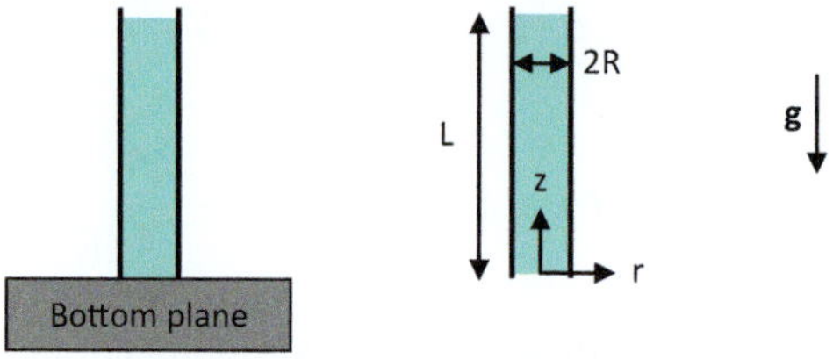


When no flow occurs when the bottom plane is removed, it means that forces balance. The pressures at the exposed top and bottom ends of the cylinder are ambient at $P_{atm}$. We may analyze this case with the following assumptions:

- Surface tension is negligible, only viscous and gravitational forces are dominant in the force balance.
- There is stress symmetry, which means $\tau_{zr} = \tau_{rz}$. The subscript notation is defined in such a way that the first subscript denotes the direction of the stress, while the second subscript denotes the plane.

The force balance equation for a static fluid at a certain value of $r$ for a fluid column $L$ is as shown:

$$\text{Upward shear stress} = \text{Downward weight}$$

$$\tau_{zr}(r) \cdot 2\pi r L = \rho\left(\pi r^2 L\right)g$$

$$\tau_{zr}(r) = \tau_{rz}(r) = \frac{\rho g r}{2}$$

(b) The minimum radius for flow to occur is when $\tau_{rz}(r)$ first reaches the yield stress $\tau^*$. This means that

$$\tau_{rz}(r = R_{\min}) = \tau_{\text{wall}} = \tau^*$$

$$\tau^* = \frac{\rho g R_{\min}}{2}$$

$$R_{\min} = \frac{2\tau^*}{\rho g}$$

(c) When $r \geq R_{\min} = \frac{R}{2}$, the Bingham fluid begins to deform like a liquid. Force balance still holds in this case, in terms of the balance of shear stress and gravitational force.

$$\tau_{rz} = \tau^* + \eta^* \frac{dv_z}{dr} = \frac{\rho g r}{2}$$

$$\frac{dv_z}{dr} = \frac{\rho g r}{2\eta^*} - \frac{\tau^*}{\eta^*}$$

$$v_z = \frac{\rho g r^2}{4\eta^*} - \frac{\tau^*}{\eta^*} r + c_1$$

When $r = R$, we hit the wall of the cylinder and assuming no-slip condition, we have $v_z = 0$.

$$0 = \frac{\rho g R^2}{4\eta^*} - \frac{\tau^*}{\eta^*} R + c_1$$

$$c_1 = \frac{\tau^*}{\eta^*} R - \frac{\rho g R^2}{4\eta^*}$$

Hence the velocity profile is as follows in the liquid region where $\frac{R}{2} \leq r \leq R$.

$$v_z = \frac{\rho g r^2}{4\eta^*} - \frac{\tau^*}{\eta^*} r + \frac{\tau^*}{\eta^*} R - \frac{\rho g R^2}{4\eta^*} = -\frac{\rho g R^2}{4\eta^*}\left[1 - \left(\frac{r}{R}\right)^2\right] + \frac{\tau^* R}{\eta^*}\left[1 - \frac{r}{R}\right]$$

For the velocity profile to be complete, we will need to figure out the profile for the solid region. This occurs when $0 \leq r \leq \frac{R}{2}$. At the point $r = \frac{R}{2}$, the velocities between both regions must be continuous. Hence the velocity at this point will also satisfy the equation for the liquid region.

$$v_z\left(r = \frac{R}{2}\right) = -\frac{\rho g R^2}{4\eta^*}\left[1 - \left(\frac{1}{2}\right)^2\right] + \frac{\tau^* R}{\eta^*}\left[1 - \frac{1}{2}\right] = -\frac{3\rho g R^2}{16\eta^*} + \frac{\tau^* R}{2\eta^*}$$

Even though $v_z\left(r = \frac{R}{2}\right)$ is mathematically the velocity at the point when the liquid transitions from liquid to solid region, we know that it makes sense that this value remains constant throughout the entire solid region as we move to the center of the cylinder at $r = 0$. This is because throughout the solid region, forces balance and the fluid will not experience any acceleration (i.e., change in velocity) due to an absence of any net force. Hence, velocity has to stay constant. Therefore we can deduce that for the solid region where $0 \le r \le \frac{R}{2}$.

$$v_{\text{solid}} = -\frac{3\rho g R^2}{16\eta^*} + \frac{\tau^* R}{2\eta^*}$$

A sketch of this velocity profile is as shown below. The direction convention for $v_z$ is such that $v_z \le 0$ for downward flow in the negative $z$ direction.

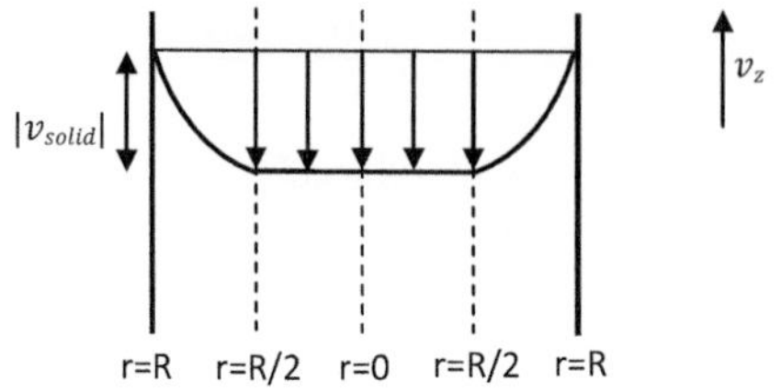


**Problem 5**

**Consider steady flow of a Newtonian liquid past a spherical vapor bubble of radius $R$. It may be assumed that this flow occurs at low Re, and for a frame of reference whereby the bubble is stationary, the liquid far away from the bubble has a constant velocity $U$ in the $z$ direction. The flow is axisymmetric and there is no flow about the axis of symmetry.**

(a) **Given that the $\theta$ component of the velocity in the liquid around the bubble is as shown, determine the radial velocity.**

$$v_\theta = -U \sin\theta \left(1 - \frac{R}{2r}\right)$$

(b) **Also given that the dynamic pressure in the surrounding liquid is as follows where $\eta$ is the liquid viscosity, find the form drag on the bubble.**

$$\mathcal{P} = -\frac{\eta U R}{r^2} \cos\theta$$

(c) **Given that the velocity profile is as found in a, find the viscous component of the drag force exerted by the liquid on the bubble. How does this drag force compare with that on a solid sphere?**

(d) **Derive the terminal velocity of a rising bubble. If we have an air bubble in water at atmospheric conditions, how small must the bubble be such that Re < 1?**

## Solution 5

### *Worked Solution*

(a) First, we establish a suitable coordinate system to solve this problem. Due to the spherical nature of the bubble, we will use spherical coordinates in terms of $r$, $\theta$, and Ø. Steady flow with constant velocity $U$ in the surrounding fluid is illustrated below:

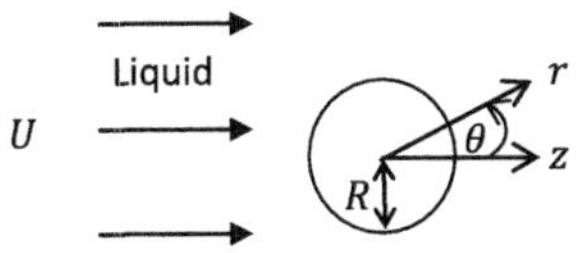


First we may consider the mass continuity equation, where the following assumptions are useful in simplifying the continuity and NS equations:

- The liquid is incompressible $\rho$ is constant.
- The flow around the bubble is at steady state ($\frac{\partial}{\partial t} = 0$).
- We are told that the flow is axisymmetric, this means that $\frac{\partial}{\partial Ø} = 0$.
- We also note from the problem statement that $v_{Ø} = 0$.
- For low Re, creeping flow may be assumed, which means inertial terms may be ignored.
- Constant viscosity for a Newtonian fluid.

The continuity equation is simplified to the following:

$$0 = \frac{1}{r^2}\frac{\partial}{\partial r}\left(r^2 v_r\right) + \frac{1}{r\sin\theta}\frac{\partial}{\partial\theta}\left(v_\theta \sin\theta\right)$$

$$\frac{\partial}{\partial r}\left(r^2 v_r\right) = -\frac{r^2}{r\sin\theta}\frac{\partial}{\partial\theta}\left(v_\theta \sin\theta\right)$$

We are given the following expression, which allows us to simplify the right-hand side of the above equation

$$v_\theta = -U\sin\theta\left(1 - \frac{R}{2r}\right)$$

$$\frac{\partial}{\partial\theta}\left(v_\theta \sin\theta\right) = \frac{\partial}{\partial\theta}\left[-U\sin^2\theta\left(1 - \frac{R}{2r}\right)\right] = -2U\sin\theta\cos\theta\left(1 - \frac{R}{2r}\right)$$

Substitute back into the continuity equation,

$$\frac{\partial}{\partial r}\left(r^2 v_r\right) = \frac{r^2}{r \sin\theta}\left[2U \sin\theta \cos\theta\left(1 - \frac{R}{2r}\right)\right] = 2U\cos\theta\left(r - \frac{R}{2}\right)$$

$$r^2 v_r = 2U\cos\theta\left(\frac{r^2}{2} - \frac{Rr}{2}\right) + f(\theta) = U\cos\theta\left(r^2 - Rr\right) + f(\theta)$$

In order to solve for the integration constant $f(\theta)$, we need boundary conditions. We know that when $r = R$, $v_r = 0$ as there is no flow that is normal to the bubble surface.

$$R^2(0) = U\cos\theta\left(R^2 - R^2\right) + f(\theta)$$

$$f(\theta) = 0$$

Hence the radial velocity profile is as follows:

$$v_r = U\cos\theta\left(1 - \frac{R}{r}\right)$$

(b) We need to calculate form drag, so let us recall some relevant concepts at this point.

---

***Background Concepts***

There are two common types of drag force, one is form drag, and the other is skin drag (or friction drag). Form drag acts on the surface and arises from a pressure difference caused by the flow. Form drag is the dynamic pressure force in the direction of approach flow ($U$ in this case) and acting normal to the surface. Skin drag on the other hand, arises from shear stress acting at the surface and it acts in the direction along the interface line at the surface.

In this problem, we are introduced to the term dynamic pressure $\mathcal{P}$. This is different from total pressure, whereby total pressure is a sum of dynamic pressure and static pressure. A useful result is in terms of pressure gradients whereby $\nabla P = \nabla\mathcal{P} + \rho\mathbf{g}$. Typically, static pressure would be the gravitational term. In the Navier–Stokes equations, we may observe two forms where the difference lies in whether we see a gravitational term in the likes of $\rho\mathbf{g}$. If the gravitational term is apparent, then the pressure term is referring only to dynamic pressure, and conversely if the gravitational term was absent, then one should be mindful that it is already subsumed under the differential pressure term $\nabla P$ which would refer to the total pressure.

---

So back to the problem, we are required to find form drag. We apply the definition of form drag in the dot product as shown. The approach flow is in the +$z$ direction, so we use $\boldsymbol{e}_z$ in the outer dot product as this is the direction that the form drag acts. $\boldsymbol{e}_z$ is the unit vector in the $z$ direction, which has unit magnitude. $\boldsymbol{e}_z$ is commonly also denoted as $\boldsymbol{\delta}_z$.

The $\boldsymbol{n}$ vector refers to the normal pointing into surrounding fluid so a negative sign is added to correct this to find the direction on the bubble which leads to the force on the bubble which the question asks for. The inner dot product helps to extract all components of the pressure term that act normal to the bubble surface, after that this set of relevant pressure components go through the second dot product which serves to further extract the components only in the +z direction which is the direction of approach flow.

$$F_{D,\text{form}} = \mathbf{e}_z \cdot \left[ - \int \boldsymbol{n}\mathcal{P}dS \right] = - \int (\mathbf{e}_z \cdot \boldsymbol{n})\mathcal{P}dS$$

For a sphere,

$$\boldsymbol{n} = \mathbf{e}_r = \sin\theta\cos\text{Ø}\mathbf{e}_x + \sin\theta\sin\text{Ø}\mathbf{e}_y + \cos\theta\mathbf{e}_z$$

$$\mathbf{e}_z \cdot \boldsymbol{n} = \cos\theta$$

$$dS = R^2 \sin\theta d\theta d\text{Ø}$$

We can now evaluate the integral at the bubble surface

$$F_{D,\text{form}} = - \int_{\text{Ø}=0}^{\text{Ø}=2\pi} \int_{\theta=0}^{\theta=\pi} \mathcal{P}\cos\theta R^2 \sin\theta d\theta d\text{Ø} = -2\pi R^2 \int_{\theta=0}^{\theta=\pi} \mathcal{P}\cos\theta\sin\theta d\theta$$

We can determine pressure as follows, and substitute back into the expression for form drag.

$$\mathcal{P}(r = R) = -\frac{\eta UR}{R^2}\cos\theta = -\frac{\eta U}{R}\cos\theta$$

$$F_{D,\text{form}} = 2\pi R^2 \int_{\theta=0}^{\theta=\pi} \frac{\eta U}{R}\cos^2\theta\sin\theta d\theta = 2\pi\eta UR \int_{\theta=0}^{\theta=\pi} \cos^2\theta\sin\theta d\theta$$

Evaluating the integral separately and substituting the result back into the expression for form drag.

$$\int_{\theta=0}^{\theta=\pi} \cos^2\theta\sin\theta d\theta = \left[ -\frac{\cos^3\theta}{3} \right]_0^\pi = \frac{2}{3}$$

$$F_{D,\text{form}} = 2\pi\eta UR\left(\frac{2}{3}\right) = \frac{4}{3}\pi\eta UR$$

(c) We wish to find out now the viscous contribution to drag force, in other words skin drag or frictional drag.

By definition, skin drag can be obtained using the following dot product. Similar to the derivation of form drag, as the approach flow is in the +$z$ direction, we apply $\boldsymbol{e}_z$ in the dot product in the outer bracket as this is also the direction in which we want to find out the skin drag. Inside the integral, the $\boldsymbol{n}$ vector refers to the normal pointing into surrounding fluid so it is equivalent to $\boldsymbol{e}_r$. The dot product of the stress tensor $\boldsymbol{T}$ with $\boldsymbol{n}$ helps to extract all components of shear stress that are in the $r$ plane ($r$ plane is defined as the plane with a normal that is in the $\boldsymbol{e}_\text{r}$ direction, which refers to tangential planes to the surface of the bubble). With the integration over the surface, we essentially sum all these shear stress components over the entire surface, after which the second dot product with $\boldsymbol{e}_z$ will extract the components that are in the direction of approach flow.

$$F_{D,\text{viscous}} = \mathbf{e}_z \cdot \int \boldsymbol{n} \cdot \boldsymbol{T} dS$$

$$dS = R^2 \sin\theta d\theta d\text{Ø}$$

$$\boldsymbol{n} = \mathbf{e}_r$$

$$\mathbf{e}_r \cdot \boldsymbol{T} = \sigma_{rr}\mathbf{e}_r + \tau_{r\theta}\mathbf{e}_\theta + \tau_{r\text{Ø}}\mathbf{e}_\text{Ø} = (\tau_{rr} - p)\mathbf{e}_r + \tau_{r\theta}\mathbf{e}_\theta + \tau_{r\text{Ø}}\mathbf{e}_\text{Ø}$$

$$\mathbf{e}_r \cdot \boldsymbol{T} = \tau_{rr}\mathbf{e}_r + \tau_{r\theta}\mathbf{e}_\theta$$

We know that $\tau_{\text{rØ}}\boldsymbol{e}_\text{Ø} = 0$ because it was established earlier that $v_\text{Ø} = 0$, so there cannot exist a shear stress component that is in the r plane that is acting in the Ø direction. We have already considered dynamic pressure effects ($p$) in form drag derivation earlier, so we will ignore its effects here to avoid double counting and consider $\tau_{\text{rr}}$ instead of $\sigma_{\text{rr}}$. From the data booklet, we can obtain the expression for $\tau_{\text{r}\theta}$:

$$\tau_{r\theta} = \eta\left[r\frac{\partial}{\partial r}\left(\frac{v_\theta}{r}\right) + \frac{1}{r}\frac{\partial v_r}{\partial\theta}\right]$$

We know from earlier that

$$v_\theta = -U\sin\theta\left(1 - \frac{R}{2r}\right)$$

$$\frac{v_\theta}{r} = -U\sin\theta\left(\frac{1}{r} - \frac{R}{2r^2}\right)$$

$$\frac{\partial}{\partial r}\left(\frac{v_\theta}{r}\right) = -U\sin\theta\left(-\frac{1}{r^2} + \frac{R}{r^3}\right) = 0 \text{ at } r = R$$

We found from part a that

$$v_r = U\cos\theta\left(1 - \frac{R}{r}\right)$$

$$\frac{\partial v_r}{\partial \theta} = -U\sin\theta\left(1 - \frac{R}{r}\right) = 0 \text{ at } r = R$$

$$\tau_{r\theta}(r = R) = 0$$

So we are left with this expression

$$\mathbf{e}_r \cdot \boldsymbol{T} = \tau_{rr}\mathbf{e}_r$$

From the data booklet, we can obtain the expression for $\tau_{\mathrm{rr}}$, whereby $\tau_{\mathrm{rr}} = \sigma_{\mathrm{rr}} + p$:

$$\tau_{rr} = \eta\left[2\frac{\partial v_r}{\partial r} - \frac{2}{3}(\nabla\cdot\boldsymbol{v})\right]$$

We know further from the continuity equation that $\nabla \cdot \boldsymbol{v} = 0$; therefore

$$\tau_{rr} = \eta\left[2\frac{\partial v_r}{\partial r}\right]$$

$$\frac{\partial v_r}{\partial r} = U\cos\theta\left(\frac{R}{r^2}\right) = \frac{U\cos\theta}{R} \text{ at } r = R$$

$$\tau_{rr}(r = R) = \frac{2\eta U\cos\theta}{R}$$

Finally we can combine our results of shear stress components at $r = R$,

$$\mathbf{e}_r \cdot \boldsymbol{T} = \tau_{rr}\mathbf{e}_r = \frac{2\eta U\cos\theta}{R}\mathbf{e}_r$$

$$\mathbf{e}_z \cdot \mathbf{e}_r = \cos\theta$$

$$F_{D,\text{viscous}} = \int_{\varnothing=0}^{\varnothing=2\pi}\int_{\theta=0}^{\theta=\pi} \frac{2\eta U\cos^2\theta}{R} R^2 \sin\theta d\theta d\varnothing = 4\pi\eta UR \int_{\theta=0}^{\theta=\pi} \cos^2\theta \sin\theta d\theta$$

$$= 4\pi\eta UR\left[-\frac{\cos^3\theta}{3}\right]_0^\pi = \frac{8\pi\eta UR}{3}$$

For a solid sphere at low Re, the total drag force is as follows

$$F_D = F_{D,\text{viscous}} + F_{D,\text{form}} = 4\pi\eta UR + 2\pi\eta UR$$

We can observe that the viscous drag for an air bubble is less than that for a solid sphere

$$\frac{F_{D,\text{viscous, bubble}}}{F_{D,\text{viscous, solid sphere}}} = \frac{2}{3}$$

(d) To find terminal velocity of a rising bubble, it requires a force balance on the bubble since terminal velocity is constant velocity at zero acceleration and hence zero net force, which means forces balance.

The volume of a spherical bubble is $\frac{4}{3}\pi R^3$. The three forces acting on the bubble are buoyancy, weight and drag force. We assume that the positive direction is upwards in the $+z$ direction. Note that drag force acts in the direction opposite to motion (rising bubble); hence it acts downwards.

$$-\frac{4}{3}\pi R^3 g\rho_{\text{bubble}} + \frac{4}{3}\pi R^3 g\rho_L - F_D = 0$$

$$\frac{4}{3}\pi R^3 g(\rho_L - \rho_{\text{bubble}}) = \frac{8\pi\eta_L U_{\text{terminal}} R}{3} + \frac{4}{3}\pi\eta_L U_{\text{terminal}} R = 4\pi\eta_L U_{\text{terminal}} R$$

$$U_{\text{terminal}} = \frac{R^2 g(\rho_L - \rho_{\text{bubble}})}{3\eta_L} \approx \frac{R^2 g\rho_L}{3\eta_L} = \frac{R^2 g}{3\nu_L}$$

It is commonly the case that $\rho_{\text{bubble}} \ll \rho_L$, and we know that kinematic viscosity $\nu_L = \frac{\eta_L}{\rho_L}$. Finally we are asked to find out what is the maximum size of bubble such that Re < 1. Let us start with the definition of Re,

$$\text{Re}_{\text{bubble}} = \frac{2UR}{\nu_{\text{bubble}}}$$

$$\text{Re}_L = \frac{2UR}{\nu_L}$$

We can find out kinematic viscosities from the data booklet as follows.

$$\nu_{\text{bubble}} = \nu_{\text{air}} = 2 \times 10^{-5}\ \text{m}^2/\text{s}$$

$$\nu_L = \nu_{\text{water}} = 1 \times 10^{-6}\ \text{m}^2/\text{s}$$

So $\text{Re}_{\text{bubble}} < \text{Re}_L$. For this case, we need both the internal and external Re to be small. The limiting Re to ensure Re is small is $\text{Re}_L$, i.e., the external flow is limiting.

The maximum velocity that the bubble can reach is terminal velocity.

$$\text{Re}_L = 1 = \frac{2U_{\text{terminal}}R_{\text{max}}}{\nu_L} = \frac{2{R_{\text{max}}}^3 g}{3{\nu_L}^2}$$

$$R_{\text{max}} = \left(\frac{3{\nu_L}^2}{2g}\right)^{1/3} = \left(\frac{3\left(10^{-6}\right)^2}{2(9.81)}\right)^{1/3} = 5.4 \times 10^{-5}\ \text{m}$$

## Problem 6

**A fluid is contained in the annular region between two concentric cylinders of radii *αR* and *R*. The inner cylinder is rotating at an angular velocity of ω. Find the velocity profile for the fluid in the annulus, and the force required to keep the outer cylinder stationary. Fully developed flow may be assumed for the length of the cylinder.**

## Solution 6

***Worked Solution***

Let us illustrate the problem to visualize it better.

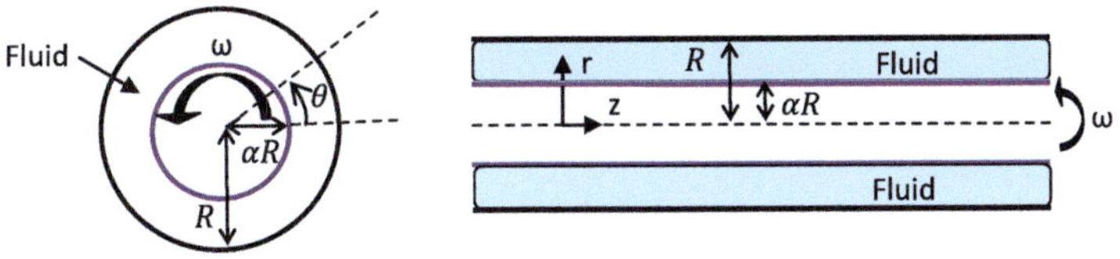


Let us consider some assumptions:

- The assumption of a fully developed flow means that we may ignore entrance/end effects. In this case, we may ignore the end effects even though in reality, there will be such effects due to the finite length of the cylinders. $v_\theta = v_\theta(r)$ only.
- We may assume axisymmetry such that $\frac{\partial}{\partial \theta} = 0$.
- The liquid is incompressible $\rho$ is constant.
- The flow is at steady state ($\frac{\partial}{\partial t} = 0$).
- Constant viscosity for a Newtonian fluid.
- The dominant velocity of the fluid is driven by the rotating inner cylinder; hence, we may ignore $v_r = v_z = 0$, and focus on $v_\theta$.

- This is a wall-driven flow without significant pressure differences. Hence inertial terms dominate and pressure terms may be neglected, e.g., $\frac{\partial P}{\partial \theta}$.

Looking at the $\theta$ component of the Navier–Stokes equation in cylindrical coordinates, we have the following simplified expression.

$$0 = \eta \left[ \frac{\partial}{\partial r} \left( \frac{1}{r} \frac{\partial}{\partial r} (r v_\theta) \right) \right]$$

$$\frac{1}{r} \frac{\partial}{\partial r} (r v_\theta) = c_1$$

$$r v_\theta = \frac{c_1 r^2}{2} + c_2$$

$$v_\theta = \frac{c_1 r}{2} + \frac{c_2}{r}$$

To solve for the integration constants, we need to apply boundary conditions. When $r = \alpha R$, $v = \alpha R \omega$. When $r = R$, $v = 0$.

$$\alpha R \omega = \frac{c_1 \alpha R}{2} + \frac{c_2}{\alpha R}$$

$$0 = \frac{c_1 R}{2} + \frac{c_2}{R} \rightarrow c_2 = -\frac{c_1 R^2}{2}$$

$$\alpha R \omega = \frac{c_1 \alpha R}{2} - \frac{c_1 R}{2\alpha}$$

$$c_1 = \frac{\alpha R \omega}{\left( \frac{\alpha R}{2} - \frac{R}{2\alpha} \right)} = -\frac{2\omega \alpha^2}{1 - \alpha^2}$$

$$c_2 = -\frac{c_1 R^2}{2} = \frac{\omega \alpha^2 R^2}{1 - \alpha^2}$$

Combining the above results, we obtain the velocity profile below.

$$v_\theta = -\frac{\omega \alpha^2}{1 - \alpha^2} (r) + \frac{\omega \alpha^2 R^2}{1 - \alpha^2} \left( \frac{1}{r} \right) = \frac{\omega \alpha^2 R}{1 - \alpha^2} \left[ \frac{R}{r} - \frac{r}{R} \right]$$

Now we know that the force required to keep the outer cylinder stationary can be expressed in terms of a torque, denoted as $G$ here.

$$G = F \times \text{Perpendicular distance}$$

$$G = (|\tau_{r\theta}(r = R)| \cdot 2\pi R L) \times R = 2\pi R^2 L |\tau_{r\theta}(r = R)|$$

From the data booklet, we can obtain the expression for $\tau_{r\theta}$:

$$\tau_{r\theta} = \eta\left[r\frac{\partial}{\partial r}\left(\frac{v_\theta}{r}\right)\right]$$

$$\frac{v_\theta}{r} = \frac{\omega\alpha^2 R}{1-\alpha^2}\left[\frac{R}{r^2} - \frac{1}{R}\right]$$

$$\frac{\partial}{\partial r}\left(\frac{v_\theta}{r}\right) = -\frac{2\omega\alpha^2 R^2}{(1-\alpha^2)r^3}$$

$$\tau_{r\theta} = -\frac{2\eta\omega\alpha^2 R^2}{(1-\alpha^2)r^2}$$

$$\tau_{r\theta}(r = R) = -\frac{2\eta\omega\alpha^2}{1-\alpha^2}$$

Finally we can derive an expression for torque, $G$.

$$G = \left(\frac{2\eta\omega\alpha^2}{1-\alpha^2}\right)\left(2\pi R^2 L\right) = \frac{4\pi R^2 L\eta\omega\alpha^2}{1-\alpha^2}$$

## Problem 7

**A polymer melt is extruded through a gap in a die onto a sheet that is moving at speed $U$. The pressure when the melt leaves the die is $P_1$. The gap size between the sheet and the die surface is $h$. There is flow in the positive $x$ direction in the region of length $L_1$ of the sheet. As for the region of length $L_2$, there is a tendency of backflow of the polymer melt, and this is balanced by the movement of the sheet in the opposite direction. We need to find out the backflow length $L_2$, the final coating thickness $h_\infty$, and the shearing force that the liquid exerts on the sheet, $F$. You may assume that the flow is fully developed and the effects of surface tension and gravity may be neglected.**

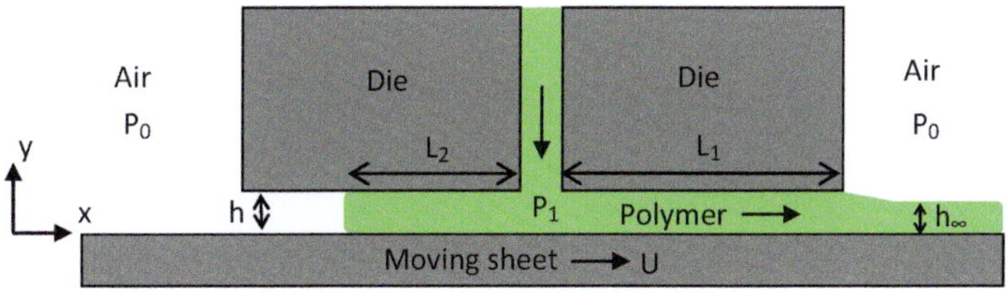


(a) **Using rectangular coordinates, determine $v_x$ in the forward flow region in terms of the parameters $P_1$ and $U$.**
(b) **Find $v_x$ in the backflow region and find $L_2$.**
(c) **Determine $h_\infty$.**
(d) **Find the force $F$ for a sheet of width $W$. Is the net force in the $+x$ or $-x$ direction? Outside the die, where the liquid is exposed to ambient conditions, is this force negligible and why?**

## Solution 7

***Worked Solution***

(a) We can refer to the data booklet for the Navier–Stokes equation in rectangular Cartesian coordinates. Let us examine possible simplifying assumptions:

- Constant viscosity for a Newtonian fluid.
- Fully developed flow where entrance/end effects are ignored. $v_x = v_x(y)$ only.
- The liquid is incompressible $\rho$ is constant.
- The flow is at steady state ($\frac{\partial}{\partial t} = 0$).
- The dominant velocity of the fluid is driven by the moving sheet; hence we may assume $v_y = v_z = 0$, and focus on $v_x$.

$x$-component NS equation can be simplified as follows

$$0 = -\frac{\partial P}{\partial x} + \eta\left(\frac{\partial^2 v_x}{\partial y^2}\right)$$

$$\frac{\partial^2 v_x}{\partial y^2} = \frac{1}{\eta}\frac{\partial P}{\partial x}$$

$$v_x = \frac{1}{2\eta}\frac{\partial P}{\partial x}y^2 + c_1 y + c_2$$

To solve for integration constants, we need to establish boundary conditions. We know that at $y = 0$, $v_x = U$ and at $y = h$, $v_x = 0$.

$$c_2 = U$$

$$0 = \frac{1}{2\eta}\frac{\partial P}{\partial x}h^2 + c_1 h + U$$

$$c_1 = \frac{-\frac{1}{2\eta}\frac{\partial P}{\partial x}h^2 - U}{h}$$

$$v_x = \frac{1}{2\eta}\frac{\partial P}{\partial x}y^2 + \left[\frac{-\frac{1}{2\eta}\frac{\partial P}{\partial x}h^2 - U}{h}\right]y + U$$

$$\frac{\partial P}{\partial x} = \frac{P_0 - P_1}{L_1}$$

We now substitute $\frac{\partial P}{\partial x}$ using the expression in terms of $L_1$

$$v_x = \frac{1}{2\eta}\left(\frac{P_0 - P_1}{L_1}\right)y^2 + \left[\frac{-\frac{1}{2\eta}\left(\frac{P_0 - P_1}{L_1}\right)h^2 - U}{h}\right]y + U$$

$$v_x = h^2\left(\frac{P_0 - P_1}{2\eta L_1}\right)\left[\left(\frac{y}{h}\right)^2 - \left(\frac{y}{h}\right)\right] - \frac{Uy}{h} + U$$

$$v_x = U\left[1 - \left(\frac{y}{h}\right)\right] + h^2\left(\frac{P_1 - P_0}{2\eta L_1}\right)\left[\left(\frac{y}{h}\right) - \left(\frac{y}{h}\right)^2\right]$$

Note that in this case $P = \mathcal{P}$, i.e., total pressure is the same as dynamic pressure because static pressure (i.e., gravitational effects) is negligible in a horizontal flow.

(b) To find $v_x$ in the backflow region, let us revisit the general expression for $v_x$ in terms of $\frac{\partial P}{\partial x}$ and the variable $y$.

$$v_x = U\left[1 - \left(\frac{y}{h}\right)\right] - \frac{h^2}{2\eta}\left(\frac{\partial P}{\partial x}\right)\left[\left(\frac{y}{h}\right) - \left(\frac{y}{h}\right)^2\right]$$

Let $\varepsilon = y/h$ to simplify the expression, then $\varepsilon = 0$ when $y = 0$, $\varepsilon = 1$ when $y = h$, and $hd\varepsilon = dy$

$$v_x = U[1 - \varepsilon] - \frac{h^2}{2\eta}\left(\frac{\partial P}{\partial x}\right)\left[\varepsilon - \varepsilon^2\right]$$

We can express the mean velocity $\langle v_x \rangle$ in the backflow region and this is zero, since we are told that forces balance in this region.

$$\langle v_x \rangle = \frac{1}{h}\int_0^h v_x dy = \frac{1}{h}\int_0^1 \left(U[1 - \varepsilon] - \frac{h^2}{2\eta}\left(\frac{\partial P}{\partial x}\right)\left[\varepsilon - \varepsilon^2\right]\right) hd\varepsilon = 0$$

$$0 = \int_0^1 \left(U[1 - \varepsilon] - \frac{h^2}{2\eta}\left(\frac{\partial P}{\partial x}\right)\left[\varepsilon - \varepsilon^2\right]\right) d\varepsilon = U\left[\varepsilon - \frac{\varepsilon^2}{2}\right]_0^1 - \frac{h^2}{2\eta}\left(\frac{\partial P}{\partial x}\right)\left[\frac{\varepsilon^2}{2} - \frac{\varepsilon^3}{3}\right]_0^1$$

$$0 = \frac{U}{2} - \frac{h^2}{12\eta}\left(\frac{\partial P}{\partial x}\right)$$

$$\frac{\partial P}{\partial x} = \frac{6\eta U}{h^2}$$

Recall that we have earlier expressed pressure gradient as shown below, inherently we have set $x = 0$ at the die slot where the polymer first enters and hits the sheet below. This is how we obtain $L_1$ when integrated over distance of the pressure difference, $P_0 - P_1$

$$\frac{\partial P}{\partial x} = \frac{P_0 - P_1}{L_1}$$

We can re-express the pressure gradient in terms of $L_2$, we need to correct for the direction in the reverse $x$ direction by adding a negative sign.

$$\frac{\partial P}{\partial x} = \frac{P_0 - P_1}{-L_2} = \frac{P_1 - P_0}{L_2}$$

Now we combine both expressions for pressure gradient,

$$\frac{\partial P}{\partial x} = \frac{6\eta U}{h^2} = \frac{P_1 - P_0}{L_2}$$

$$L_2 = \frac{h^2(P_1 - P_0)}{6\eta U}$$

$$v_x = U\left[1 - \left(\frac{y}{h}\right)\right] - \frac{h^2}{2\eta}\left(\frac{P_1 - P_0}{\frac{h^2(P_1 - P_0)}{6\eta U}}\right)\left[\left(\frac{y}{h}\right) - \left(\frac{y}{h}\right)^2\right]$$

$$v_x = U\left[1 - \left(\frac{y}{h}\right)\right] - 3U\left[\left(\frac{y}{h}\right) - \left(\frac{y}{h}\right)^2\right] = U\left[1 - 4\left(\frac{y}{h}\right) + 3\left(\frac{y}{h}\right)^2\right]$$

(c) To find $h_\infty$, we can apply mass continuity

$$\dot{Q}_{\text{in die}} = \dot{Q}_{\text{outside die}}$$

$$\int_0^h v_x dy = \int_0^{h_\infty} v_x dy$$

There is an easy way to figure out the right-hand side of the equation, because we can assume that there is negligible shear stress between the liquid and gas interface, hence the entire liquid film is moving at $U$ across all values of $y$ at the downstream portion.

$$\int_0^{h_\infty} v_x dy = Uh_\infty$$

The left-hand side of the equation, $\int_0^h v_x dy$ was evaluated earlier,

$$\int_0^h v_x dy = h\left[\frac{U}{2} - \frac{h^2}{12\eta}\left(\frac{\partial P}{\partial x}\right)\right]$$

So substituting both results in the continuity equation,

$$h\left[\frac{U}{2}-\frac{h^2}{12\eta}\left(\frac{\partial P}{\partial x}\right)\right]=Uh_\infty$$

$$\frac{h_\infty}{h}=\frac{1}{2}-\frac{h^2}{12\eta U}\left(\frac{P_0-P_1}{L_1}\right)=\frac{1}{2}+\frac{h^2(P_1-P_0)}{12\eta UL_1}$$

(d) To find the force required to be exerted on the sheet, we need to know the shear force that this applied force has to overcome. We have set $x=0$ at the die slot where the polymer vertically enters the sheet, where $P=P_1$. The total shear force is the sum of the relevant shear component over the surface area that is in contact between polymer and sheet.

$$F=W\int_{-L_2}^{L_1}\tau_{yx}dx$$

$$F=WL_2\tau_{yx}\big|_{\text{backflow}}(y=0)+WL_1\tau_{yx}\big|_{\text{forward}}(y=0)$$

By definition, we know that $\tau_{yx}=\eta\frac{dv_x}{dy}$ and so for the backflow region,

$$\tau_{yx}\big|_{\text{backflow}}(y=0)=\eta\left[\frac{d}{dy}\left(U\left[1-4\left(\frac{y}{h}\right)+3\left(\frac{y}{h}\right)^2\right]\right)\right]_{y=0}=-\frac{4\eta U}{h}$$

In the forward flow region,

$$\tau_{yx}\big|_{\text{forward}}(y=0)=v_x=\eta\left[\frac{d}{dy}\left(U\left[1-\left(\frac{y}{h}\right)\right]+h^2\left(\frac{P_1-P_0}{2\eta L_1}\right)\left[\left(\frac{y}{h}\right)-\left(\frac{y}{h}\right)^2\right]\right)\right]_{y=0}$$

$$\tau_{yx}\big|_{\text{forward}}(y=0)=-\frac{\eta U}{h}+h\left(\frac{P_1-P_0}{2L_1}\right)$$

Total shear force in the $x$ direction,

$$\frac{F}{W}=-\frac{4\eta UL_2}{h}-\frac{\eta UL_1}{h}+h\left(\frac{P_1-P_0}{2}\right)=h\left(\frac{P_1-P_0}{2}\right)-\frac{\eta U}{h}(L_1+4L_2)$$

We know from our earlier result in part b that

$$L_2=\frac{h^2(P_1-P_0)}{6\eta U}$$

Hence the total shear force is as shown below, and it is a negative value hence it acts in the negative x direction. The force needed to be applied to the sheet has to be in the positive x direction to counteract this shear force.

$$\text{Total shear force per unit width} = \frac{F}{W} = -\left[\frac{\eta U L_1}{h} + h\left(\frac{P_1 - P_0}{6}\right)\right]$$

We may assume that $\tau_{yx} \approx 0$ outside the die as it is exposed to air, and hence the shear force is negligible.

## Problem 8

**We have a distribution system that can help to deliver a viscous fluid with a uniform exit velocity as the fluid leaves a thin gap between two large parallel planes. The fluid enters and flows in the pipe in the +*z* direction, the flow leaves the pipe through the gap and this is in the +*x* direction. Both the pipe and parallel planes lie on the *y*-plane (i.e., no gravitational effects). The other end of the circular pipe is closed.**

(a) **Show that if $Q_{\text{in}}$ is the volumetric flow rate of the fluid entering the circular tube, then the volumetric flow rate across any *z* plane is as follows**

$$Q(z) = Q_{\text{in}}\left(1 - \frac{z}{\alpha}\right)$$

(b) **You may assume Poiseuille's law for flow through the circular tube. Derive the following differential equation.**

$$Q_{\text{in}}\left(1 - \frac{z}{\alpha}\right) = \frac{\pi R^4}{8\eta}\left(-\frac{dP}{dz}\right)$$

(c) **Show further the following expression where $\varepsilon = \frac{z}{\alpha}$**

$$P - P_{\text{atm}} = \frac{4\eta\alpha}{\pi R^4} Q_{\text{in}}(1 - \varepsilon)^2$$

(d) **Derive the expression for the exit velocity *V* if you are given that the average velocity for pressure driven flow in the *x* direction through a slit of width *H* is given by the following where *L* is the length of the slit.**

$$\langle v_x \rangle = \frac{H^2}{12\eta}(-\Delta P/L)$$

(e) **Find an equation for the function $x(\varepsilon)$ which will be required to maintain the uniform exit velocity *V*.**

## Solution 8

### *Worked Solution*

(a) We can illustrate this system to visualize the problem in appropriate coordinates.

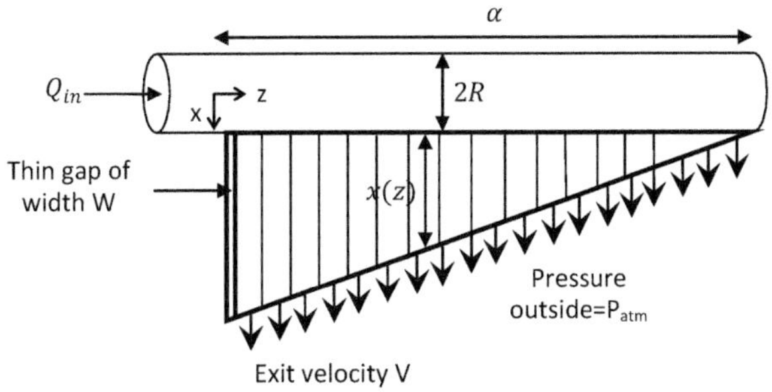


Let's state our assumptions to be clear about the problem.

- Constant viscosity for a Newtonian fluid.
- Fully developed Poiseuille flow in the circular tube, where entrance/end effects are ignored. $v_z = v_z(x)$ only. We may recall properties of Poiseuille flow. It describes a laminar flow (Re < 2000), and pressure terms and viscous terms are significant.
- The dominant flow direction is $v_z$. $v_r = v_\theta = 0$ if we consider cylindrical coordinates for the pipe.
- The liquid is incompressible $\rho$ is constant.
- The flow is at steady state ($\frac{\partial}{\partial t} = 0$).
- The flow is axisymmetric ($\frac{\partial}{\partial \theta} = 0$), where $\theta$ is the angle around the circumferential direction of the circular cross section of the pipe.

Before we try to find out the volumetric flow rate at any $z$ plane, let us consider a differential control volume of the circular pipe, by taking a small section at an arbitrary $z$ value. This differential volume will have a differential length $dz$. We will define positive as volume into the element.

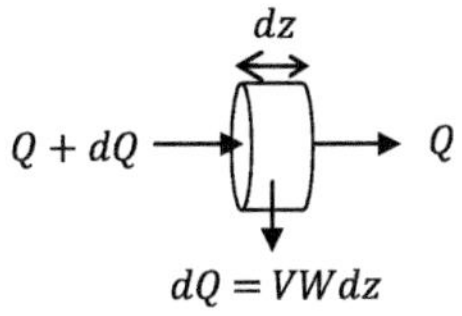


$$dQ = -VWdz$$

$$\int_{Q_{\text{in}}}^{Q} dQ = \int_{0}^{z} -VWdz$$

$$Q(z) - Q_{\text{in}} = -VWz$$

Let us establish some boundary conditions. At $z = \alpha$, $Q = 0$ since all the fluid has left the tube through the gap.

$$Q_{\text{in}} = VW\alpha \rightarrow \alpha = \frac{Q_{\text{in}}}{VW}$$

$$Q(z) = Q_{in} - VWz = Q_{in}\left(1 - \frac{z}{\alpha}\right)$$

(b) Now we need to derive the following differential equation: $Q_{\text{in}}\left(1 - \frac{z}{\alpha}\right) = \frac{\pi R^4}{8\eta}\left(-\frac{dP}{dz}\right)$. Let's start with analyzing Poiseuille flow. We will consider the relevant Navier–Stokes equation in cylindrical coordinates to analyze flow through a circular pipe.
$r$ component:

$$0 = -\frac{dP}{dr}$$

$\theta$ component:

$$0 = -\frac{1}{r}\frac{dP}{d\theta}$$

$z$ component:

$$0 = -\frac{dP}{dz} + \eta\frac{1}{r}\frac{d}{dr}\left(r\frac{dv_z}{dr}\right)$$

From the $r$ and $\theta$ components, we can deduce that $P = P(z)$. We can express $dP/dz$ for a finite length of pipe of length $\alpha$, $\frac{dP}{dz} = \frac{\Delta P}{\alpha} = [P(z = \alpha) - P(z = 0)]/\alpha$. Note that $\frac{dP}{dz}$ is a negative value, pressure reduces along the direction of flow as it is used to overcome viscous forces along the way.

$$\frac{dP}{dz} = \eta\frac{1}{r}\frac{d}{dr}\left(r\frac{dv_z}{dr}\right) = \frac{\Delta P}{\alpha}$$

$$\frac{d}{dr}\left(r\frac{dv_z}{dr}\right) = \frac{r\Delta P}{\eta\alpha}$$

$$r\frac{dv_z}{dr} = \frac{r^2\Delta P}{2\eta\alpha} + c_1$$

$$\frac{dv_z}{dr} = \frac{r\Delta P}{2\eta\alpha} + \frac{c_1}{r}$$

$$v_z = \frac{r^2\Delta P}{4\eta\alpha} + c_1 \ln r + c_2$$

As $r$ goes to zero at the center of the pipe, $\ln r$ becomes undefined which is not possible. Hence we may deduce that $c_1 = 0$.

$$v_z = \frac{r^2\Delta P}{4\eta\alpha} + c_2$$

At $r = R$, $v_z = 0$ due to the no-slip condition at the pipe wall

$$c_2 = -\frac{R^2\Delta P}{4\eta\alpha}$$

$$v_z = \frac{r^2\Delta P}{4\eta\alpha} - \frac{R^2\Delta P}{4\eta\alpha} = \frac{R^2}{4\eta}\left(-\frac{\Delta P}{\alpha}\right)\left[1 - \left(\frac{r}{R}\right)^2\right]$$

Volumetric flow rate in a circular pipe can be derived as follows

$$Q = \int \frac{R^2}{4\eta}\left(-\frac{\Delta P}{\alpha}\right)\left[1 - \left(\frac{r}{R}\right)^2\right] dA = \int_0^R \frac{R^2}{4\eta}\left(-\frac{\Delta P}{\alpha}\right)\left[1 - \left(\frac{r}{R}\right)^2\right] 2\pi r dr$$

$$Q = \frac{\pi R^2}{2\eta}\left(-\frac{\Delta P}{\alpha}\right)\left[\frac{r^2}{2} - \frac{r^4}{4R^2}\right]_0^R$$

$$Q = \frac{\pi R^4}{8\eta}\left(-\frac{\Delta P}{\alpha}\right)$$

Hence we have shown the equation where $\frac{dP}{dz} = \frac{\Delta P}{\alpha}$

$$Q = Q_{\text{in}}\left(1 - \frac{z}{\alpha}\right) = \frac{\pi R^4}{8\eta}\left(-\frac{dP}{dz}\right)$$

(c) We notice that the left-hand side of the equation that we need to prove has a pressure difference. This is a hint to integrate our differential equation in pressure. When $P = P_{\text{atm}}$, $z = \alpha$. And we are told that $\varepsilon = z/\alpha$.

$$dP = -\frac{8\eta}{\pi R^4} Q_{\text{in}} \left(1 - \frac{z}{\alpha}\right) dz$$

$$P_{\text{atm}} - P = -\frac{8\eta}{\pi R^4} Q_{\text{in}} \left[z - \frac{z^2}{2\alpha}\right]_z^{\alpha} = -\frac{8\eta}{\pi R^4} Q_{\text{in}} \left[\frac{\alpha}{2} - z + \frac{z^2}{2\alpha}\right]$$

$$P - P_{\text{atm}} = \frac{4\eta\alpha}{\pi R^4} Q_{\text{in}} \left[1 - \frac{2z}{\alpha} + \frac{z^2}{\alpha^2}\right]$$

$$P - P_{\text{atm}} = \frac{4\eta\alpha}{\pi R^4} Q_{\text{in}} (1 - \varepsilon)^2$$

(d) Now we wish to derive the exit velocity $V$. We are given the expression below

$$\langle v_x \rangle = \frac{H^2}{12\eta} (-\Delta P/L)$$

In our system, the following are analogous:

- $H$ can be replaced by $W$.
- $-\frac{\Delta P}{L}$ can be replaced by $-\frac{(P_{\text{atm}} - P)}{x(z)}$ or $\frac{P - P_{\text{atm}}}{x(z)}$.
- $\langle v_x \rangle$ can be replaced by $V$.

$$V = \frac{W^2}{12\eta} \left(\frac{P - P_{\text{atm}}}{x(z)}\right)$$

(e) Note that since $\varepsilon$ is a function of $z$, $x(z)$ and $x(\varepsilon)$ are interchangeable. We know from part c that pressure difference can be expressed in terms of $\varepsilon$. We can substitute this result into the expression for $V$ as found in part $d$.

$$P - P_{\text{atm}} = \frac{4\eta\alpha}{\pi R^4} Q_{\text{in}} (1 - \varepsilon)^2$$

$$V = \frac{W^2}{12\eta} \left(\frac{P - P_{\text{atm}}}{x(z)}\right) = \frac{W^2}{12\eta} \left(\frac{\frac{4\eta\alpha}{\pi R^4} Q_{\text{in}} (1 - \varepsilon)^2}{x(z)}\right) = \frac{\alpha W^2}{3\pi R^4} Q_{\text{in}} \frac{(1 - \varepsilon)^2}{x(\varepsilon)}$$

We can make $x(\varepsilon)$ the subject of the equation by rearranging, and then re-expressing $V$ in terms of known parameters.

$$x(\varepsilon) = \frac{\alpha W^2 (1 - \varepsilon)^2}{3\pi R^4 V} Q_{\text{in}}$$

We know that since the terminal end of the circular pipe is closed, the length of pipe is such that it just empties all of the inlet volume into the pipe at $z = 0$.

Therefore mass balance tells us that the exit velocity through the gap, $V\alpha W = Q_{\text{in}}$.

$$x(\varepsilon) = \frac{\alpha W^2(1-\varepsilon)^2}{3\pi R^4 V} Q_{\text{in}} = \frac{\alpha W^2(1-\varepsilon)^2}{3\pi R^4 (Q_{\text{in}}/\alpha W)} Q_{\text{in}} = \frac{\alpha^2 W^3 (1-\varepsilon)^2}{3\pi R^4}$$

$$x(z) = \frac{\alpha^2 W^3}{3\pi R^4}\left(1 - \frac{z}{\alpha}\right)^2$$

## Problem 9

**We have a circular tube of radius $R$ and length $L$ which contains liquid in a steady pressure-driven flow. The tube material is porous, allowing fluid to permeate through at the wall at a velocity $V_w$ that remains constant throughout the length of the tube. The driving force for this flow to occur is a pressure difference maintained between the inside and outside of the tube. As more fluid exits the tube as it flows along the tube, its axial velocity drops as we travel downstream along the tube. Assume that $V_w$ and mean inlet velocity $U_{\text{in}}$ are known values.**

(a) **Find out the mean (averaged over the circular cross-sectional area) axial velocity $\langle v_z \rangle$ as a function of $z$. What is the maximum tube length for our assumptions in this analysis to remain valid.**
(b) **If $V_w/U_{\text{in}} \ll 1$ and assuming $\text{Re}_w = V_w R/\nu \ll 1$, comment on assumptions that can be made. Then find the axial velocity profile $v_z$.**
(c) **Determine the radial velocity profile $v_r$.**

## Solution 9

***Worked Solution***

(a) Let us illustrate this problem to visualize it better.

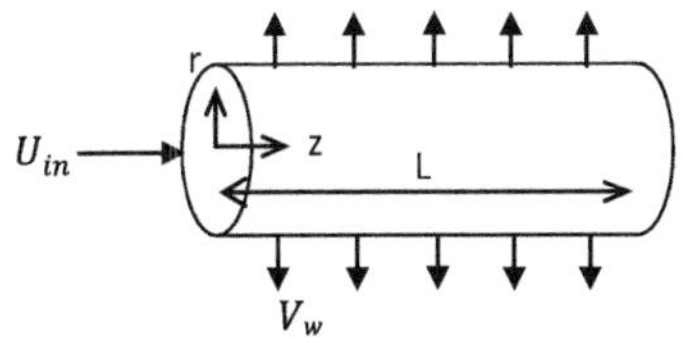


In order to find out the mean axial velocity profile $\langle v_z \rangle$ as a function of $z$, we can first examine a differential element of differential width $dz$,

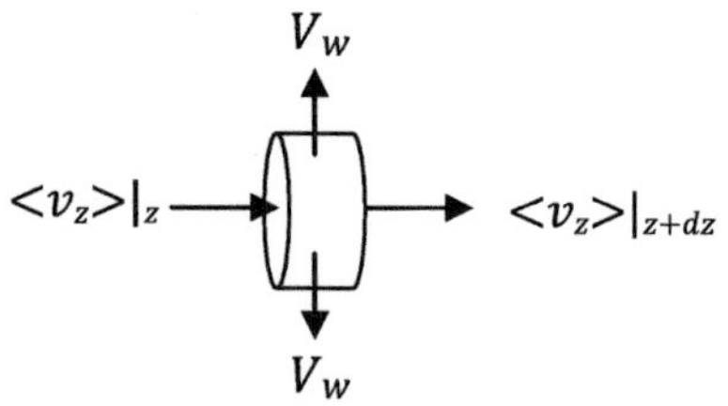


Now we construct a mass balance equation for this system at steady state

$$0 = (\rho\pi R^2)[v_z(z) - v_z(z + dz)] - (2\rho\pi R dz)V_w$$

$$\frac{-2V_w}{R} = \frac{v_z(z + dz) - v_z(z)}{dz} = \frac{dv_z}{dz}$$

To integrate this differential equation, we need to establish boundary conditions. When $z = 0$, $\langle v_z \rangle = U_{\text{in}}$. When $z = L$, $\langle v_z \rangle = 0$.

$$\langle v_z \rangle = \frac{-2V_w}{R} z + c_1$$

$$c_1 = U_{\text{in}}$$

Therefore we have found the mean axial velocity as a function of $z$

$$\langle v_z \rangle = \frac{-2V_w}{R} z + U_{\text{in}}$$

For the maximum tube length $L_{\text{max}}$, it is when the fluid just empties fully at this end of the tube of this length

$$0 = \frac{-2V_w}{R} L_{\text{max}} + U_{\text{in}}$$

$$L_{\text{max}} = \frac{RU_{\text{in}}}{2V_w}$$

(b) Now we are asked to find axial velocity profile again, i.e., an expression for $v_z$ similar as in part a; however, this time, we are given conditions to be satisfied.

- $V_w/U_{\text{in}} \ll 1$
- $\text{Re}_w = V_w R/\nu \ll 1$

These conditions make it possible for the Lubrication Approximation.

---

***Background Concepts***

Let us recall some key concepts regarding the Lubrication Approximation. It is an approximation made for converging or diverging flows whereby the angle of converge/diverge is small. Small angle approximations may also be used when solving such problems. This approximation assumes that the planes within which the flow exists are "nearly parallel".

- - - - - - - - - - - - - - - - - - - - - - - - - - - - - - - - - - - - - - - - - - - - - - - - - - -

So in this problem, we will look at the Navier–Stokes equations in cylindrical coordinates and apply the Lubrication Approximation. The assumptions that we can make are as follows:

- Constant viscosity for a Newtonian fluid.
- Fully developed pressure-driven flow in the circular tube, where entrance/end effects are ignored. $v_z \neq v_z(z)$.
- This is a laminar flow (Re < 2000). Pressure and viscous effects are significant.
- The dominant flow direction is $v_z$. $v_r = v_\theta = 0$ if we consider cylindrical coordinates for the pipe.
- The liquid is incompressible $\rho$ is constant.
- The flow is at steady state ($\frac{\partial}{\partial t} = 0$).
- The flow is axisymmetric ($\frac{\partial}{\partial \theta} = 0$), where $\theta$ is the angle around the circumferential direction of the circular cross section of the pipe.

The $z$ component of the NS equation simplifies to

$$0 = -\frac{dP}{dz} + \eta\frac{1}{r}\frac{d}{dr}\left(r\frac{dv_z}{dr}\right)$$

We can now apply boundary conditions to integrate this equation. When $r = R$, $v_z = 0$. Note that this is the case because even though there is exit velocity, we are only looking at $z$ component velocity here. Also, when $r = 0$, $\frac{dv_z}{dr} = 0$. Note that the flow pattern is symmetrical about the centerline; hence the maxima/minima velocity will occur at the centerline where the gradient $\frac{dv_z}{dr} = 0$.

$$\frac{r}{2\eta}\frac{dP}{dz} + c_1 = \frac{dv_z}{dr} \rightarrow c_1 = 0$$

$$\frac{r^2}{4\eta}\frac{dP}{dz} + c_2 = v_z \rightarrow c_2 = -\frac{R^2}{4\eta}\frac{dP}{dz}$$

By substituting the boundary conditions, we can evaluate the integration constants. After which we can find that the velocity profile is

$$v_z = \frac{r^2}{4\eta}\frac{dP}{dz} - \frac{R^2}{4\eta}\frac{dP}{dz} = \frac{1}{4\eta}\frac{dP}{dz}\left(r^2 - R^2\right)$$

This expression requires further work so that we can express the $\frac{dP}{dz}$ in terms of known values/measurable parameters. To do this, we recall that in part a, we have found the mean axial velocity which can be used to relate to volumetric flow rate.

$$\langle v_z \rangle \pi R^2 = \int_0^R v_z(2\pi r dr)$$

$$\langle v_z \rangle = \frac{2}{R^2}\frac{dP}{dz}\int_0^R \frac{1}{4\eta}\left(r^3 - R^2 r\right)dr$$

$$\langle v_z \rangle = \frac{1}{R^2}\frac{dP}{dz}\frac{1}{2\eta}\left[-\frac{R^4}{4}\right] = -\frac{dP}{dz}\frac{R^2}{8\eta}$$

$$\frac{dP}{dz} = \frac{-8\eta\langle v_z \rangle}{R^2}$$

Substitute this result back into the velocity profile earlier

$$v_z = \frac{1}{4\eta}\frac{dP}{dz}\left[r^2 - R^2\right] = \frac{1}{4\eta}\left(\frac{-8\eta\langle v_z \rangle}{R^2}\right)\left[r^2 - R^2\right]$$

$$v_z = 2\langle v_z \rangle\left[1 - \left(\frac{r}{R}\right)^2\right] = 2\left[\frac{-2V_w}{R}z + U_{\text{in}}\right]\left[1 - \left(\frac{r}{R}\right)^2\right]$$

(c) We need to find now the radial velocity profile. We have earlier derived the axial velocity profile. A useful "trick" to relate velocity profiles of different coordinate directions is to use the mass continuity equation.

In cylindrical coordinates,

$$0 = \frac{1}{r}\frac{\partial}{\partial r}(r v_r) + \frac{\partial v_z}{\partial z}$$

$$\frac{1}{r}\frac{\partial}{\partial r}(r v_r) = \left(\frac{4V_w}{R}\right)\left[1 - \left(\frac{r}{R}\right)^2\right]$$

$$\frac{\partial}{\partial r}(r v_r) = \left(\frac{4V_w}{R}\right)\left[r - \frac{r^3}{R^2}\right]$$

$$r v_r = \left(\frac{4V_w}{R}\right)\left[\frac{r^2}{2} - \frac{r^4}{4R^2}\right] + c_1$$

We know that at $r = 0$, $v_r = 0$ due to the symmetry about the centerpoint

$$v_r = \left(\frac{4V_w}{R}\right)\left[\frac{r}{2} - \frac{r^3}{4R^2}\right] + \frac{c_1}{r} \rightarrow c_1 = 0$$

$$v_r = 2V_w\left[\frac{r}{R} - \frac{1}{2}\left(\frac{r}{R}\right)^3\right]$$

## Problem 10

**Rollers are often used to spread liquid polymer melts uniformly onto a solid sheet. A motor drives the rotation of the rollers at a linear velocity of $V$, which then leads the adhering liquid melt to flow out in the +$z$ direction at velocity $V$. We have an illustration below of the roller-film setup whereby the gap size between the rollers is fixed at $H_0$ as measured from the centerline. The local gap size between the rollers is $h(z)$ as measured from the centerline. The rollers have radii $R$ each. The final thickness of the film exiting the rollers is $H_f$. The location at which the film leaves the rolls is $z_{exit}$.**

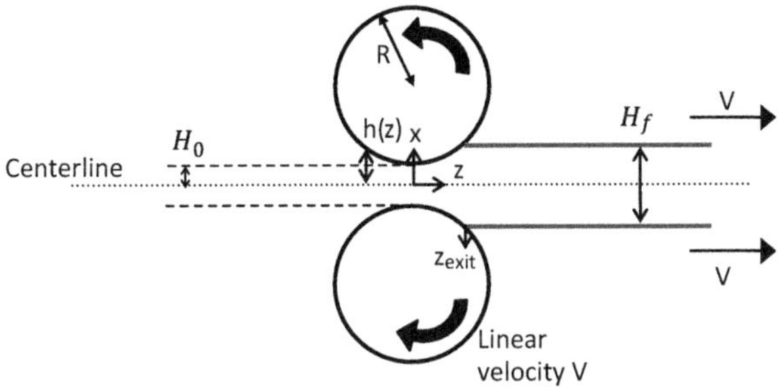


(a) **Show that near to the region where $z = 0$, the gap between the rollers is**

$$h(z) = H_0\left[1 + \frac{1}{2}\frac{z^2}{RH_0}\right]$$

(b) **In the region near to $z = 0$, we can assume the lubrication approximation to determine the velocity profile. Derive an expression for $v_z(x)$ at any $z$, in terms of the local gap size $h(x)$ and local pressure gradient $\beta = \frac{\partial P}{\partial z}$.**

(c) **At $z_{exit}$ where the film separates from the rollers, what is the condition that $\beta$ must have? Assuming no net pressure drop during the passage of the film through the rollers, develop an equation that can be used to solve for $z_{exit}$ but you need not solve it.**

(d) **Sketch the velocity profile at a location slightly upstream of $z = 0$, at $z = 0$ and far downstream.**

(e) **Sketch the pressure profile from far upstream to the region where $z = z_{exit}$.**

(f) **Find out the force by the fluid on the top roller in the +$x$ direction.**

## Solution 10

### *Worked Solution*

(a) We can first zoom into the region near $z = 0$ to study the relevant geometries

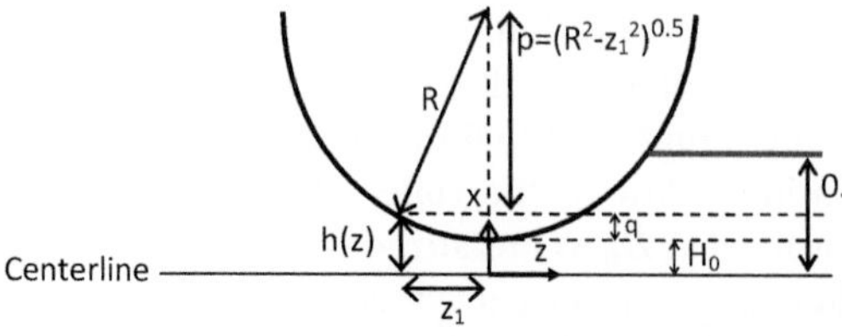


We can define new variables $z_1$, $p$, and $q$ to help the analysis. Using Pythagoras theorem, we can express $p$ in terms of $R$ and $z_1$.

$$q = R - p$$

$$q = R - \left(R^2 - {z_1}^2\right)^{0.5}$$

$$h(z) = H_0 + q$$

$$h(z) = H_0 + R - \left(R^2 - {z_1}^2\right)^{0.5}$$

When we have a term raised to a power index that may be difficult to evaluate, we can think about using Taylor's series expansion to simplify. Over here we assume $\frac{z_1}{R}$ is small to neglect higher order terms in the series expansion.

$$h(z) = H_0 + R - R\left(1 - \left(\frac{z_1}{R}\right)^2\right)^{0.5}$$

$$h(z) = H_0 + R - R\left(1 - 0.5\left(\frac{z_1}{R}\right)^2 + \ldots\right)$$

$$h(z) = H_0\left(1 + \frac{1}{2}\left(\frac{z^2}{RH_0}\right)\right)$$

We have replaced $z_1$ with $z$ as our expression was derived assuming a specific $z = z_1$ but remains valid for a more generic $z$ local to the region.

(b) Near to the region where $z = 0$, the curvature of the rollers become less apparent relative to the horizontal; hence we can adopt the lubrication approximation for the flow in this region. We will look at the Navier–Stokes equation, with the following assumptions in mind:

- Constant viscosity for a Newtonian fluid.
- The liquid is incompressible, $\rho$ is constant.
- The flow is at steady state ($\frac{\partial}{\partial t} = 0$).

- The flow in the $z$ direction is fully developed and hence $v_z \neq v_z(z)$.
- The dominant flow is $v_z$; hence $v_x = v_y = 0$.

Looking at the $z$ component,

$$0 = -\frac{dP}{dz} + \eta\left(\frac{d^2 v_z}{dx}\right) = -\beta + \eta\left(\frac{d^2 v_z}{dx}\right)$$

We may consider boundary conditions in integrating this equation. When $x = 0, \frac{dv_z}{dx} = 0$ due to symmetry about the centerline. When $x = h(z)$ or $h$, $v_z = V$.

$$\frac{d^2 v_z}{dx} = \frac{\beta}{\eta}$$

$$\frac{dv_z}{dx} = \frac{\beta}{\eta}x + c_1 \rightarrow c_1 = 0$$

$$v_z = \frac{\beta}{2\eta}x^2 + c_2$$

$$V = \frac{\beta}{2\eta}h^2 + c_2 \rightarrow c_2 = V - \frac{\beta}{2\eta}h^2$$

$$v_z = \frac{\beta}{2\eta}x^2 + V - \frac{\beta}{2\eta}h^2$$

$$v_z = \frac{\beta}{2\eta}\left[x^2 - h^2\right] + V$$

(c) Where the film separates from the rollers, i.e., where $z = z_{\text{exit}}$, the fluid is exposed to ambient conditions and the shear stress at the air–liquid interface is assumed negligible, while the interface between the fluid and solid moving sheet adopts the no-slip boundary condition (i.e., fluid moves at same velocity $V$). There is no shear stress (or viscous forces); hence the counterbalancing shear force in the Navier–Stokes equation must also be zero. Ignoring gravitational effects (since pressure is a sum of dynamic pressure and static gravitational pressure), the pressure gradient $\beta$ in the $z$ direction has to be zero for a steady-state flow.

Now we have to find $z_{\text{exit}}$ which is a specified point $z$. We recall the mass continuity equation which is can be used to relate the constant mass (or volume) flow rates for a steady-state flow at various cross-section locations of the flow. We attempt to express volumetric flow rate in this case. Let $W$ denote the width of the sheet in the $y$ direction:

$$\dot{Q} = 2W\int_0^{h(z)} v_z dx$$

$$\dot{Q} = 2W\int_0^{h(z)}\left[\frac{\beta}{2\eta}(x^2 - h^2) + V\right]dx$$

$$\dot{Q} = 2W\left[\frac{\beta}{2\eta}\left(\frac{x^3}{3} - h^2x\right) + Vx\right]_0^{h(z)} = 2W\left[\frac{\beta}{2\eta}\left(\frac{h^3}{3} - h^3\right) + Vh\right]$$

$$\dot{Q} = 2WVh\left[1 - \frac{\beta}{\eta V}\left(\frac{h^2}{3}\right)\right]$$

$$\beta = \frac{dP}{dz} = \frac{3\eta V}{h^2}\left(1 - \frac{\dot{Q}}{2WVh}\right) = \frac{3\eta V}{h^2} - \frac{3\eta\dot{Q}}{2Wh^3}$$

To solve the differential equation in pressure, we establish boundary conditions for $z = z_{\text{exit}}$, $P = P_{\text{atm}}$

$$P - P_{\text{atm}} = \int_{z_{\text{exit}}}^{z}\left(\frac{3\eta V}{h^2} - \frac{3\eta\dot{Q}}{2Wh^3}\right)dz$$

We note that h is a function of $z$; hence we need to either express $h$ in terms of $z$ or express $dz$ in terms of $dh$ in order to do the integration. The latter is easier; hence from the earlier expression obtained, we find $dh$.

$$h(z) = H_0\left(1 + \frac{1}{2}\left(\frac{z^2}{RH_0}\right)\right)$$

$$\frac{dh}{dz} = \frac{z}{R}$$

$$z = \sqrt{2R(h - H_0)}$$

$$dz = \frac{R}{z}dh = \frac{R}{\sqrt{2R(h - H_0)}}dh$$

$$P - P_{\text{atm}} = \int_{H_f/2}^{h}\left(\frac{3\eta V}{h^2} - \frac{3\eta\dot{Q}}{2Wh^3}\right)\frac{R}{\sqrt{2R(h - H_0)}}dh$$

Let us introduce a new term $z_{\text{in}}$ that defines the $z$ value at which the fluid enters the system. At this point, $P_{\text{in}} = P_{\text{atm}}$ and the value of $h(z = z_{\text{in}})$ is also fixed based on the structure of the system setup, and we denote it $H_{\text{in}}$. We assume that the small gap approximation is valid for the region upstream of $z = 0$ all the way to $z_{\text{in}}$.

$$z_{\text{in}} = \sqrt{2R(H_{\text{in}} - H_0)}$$

$$P_{\text{in}} - P_{\text{atm}} = 0 = \int_{z_{\text{exit}}}^{z_{\text{in}}} \left( \frac{3\eta V}{h^2} - \frac{3\eta \dot{Q}}{2Wh^3} \right) dz$$

$$0 = \int_{z_{\text{exit}}}^{\sqrt{2R(H_{\text{in}} - H_0)}} \left( \frac{3\eta V}{h^2} - \frac{3\eta \dot{Q}}{2Wh^3} \right) \frac{R}{\sqrt{2R(h - H_0)}} dh$$

This differential equation may be solved to obtain $z_{\text{exit}}$.

(d) The velocity profiles at the three locations, (1) locally upstream of $z = 0$, (2) at $z = 0$, and (3) downstream of $z = 0$, can be illustrated as shown. Here are some guiding points for the sketch.

- At the centerline, the gradient of the velocity profile should be zero, i.e., symmetry about the centerline.
- The magnitude of the velocities should be such that at various values of $z$, mass is conserved; hence the area under the curves (in red) for the velocity sketches should be the same. As we move away from the roller surface, the velocity will increase as viscous forces reduce away from the roller surface, until a velocity maxima is reached at the centerline.
- From the result obtained in part a as shown below, we note that the velocity profile is parabolic in shape with respect to $x$ (second-order $x$ term).
- The $z$ component of fluid velocity $v_z$ at the location $z$ upstream of $z = 0$ where $h = h(z)$ will be smaller than $V$. This is because $V$ is the linear velocity tangential to the curved roller surface; hence, the horizontal $z$ component will be a fraction of $V$.

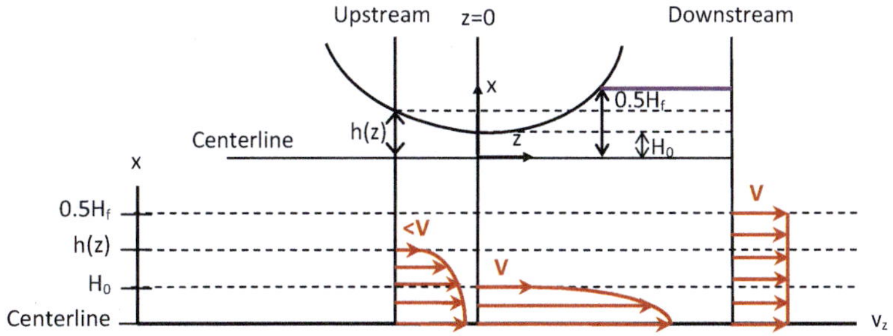


(e) In order to sketch the pressure profile $P(z)$ from upstream to zexit, we can start by making logical deductions from our knowledge of the flow pattern, to gather key information about the pressures and/or pressure gradients to help us with this sketch.

- Pressure far upstream of $z$ where the fluid first enters our roller system is $P = P_{\text{atm}}$. $z_{\text{in}}$ was earlier defined as the $z$ value for this entry point, and at this point, $P_{\text{in}} = P_{\text{atm}}$. For the small angle assumption to remain valid for the region upstream of $z = 0$ all the way to $z_{\text{in}}$, the distance between $z = 0$ and $z_{\text{in}}$ should be small.

- Upon exit of the film where $z \geq z_{exit}$, $P = P_{atm}$. Since shear stress at the air/liquid interface is assumed negligible, pressure gradient $\frac{dP}{dz}$ is zero in order to satisfy the Navier–Stokes equation where pressure terms counterbalance viscous terms in this problem.
- Since we know that shear stresses balance pressure gradient in our system, we can deduce that pressure gradient is highest at the point where shear stress is the highest, i.e., at $z = 0$ where the gap is the narrowest gap.
- As our system is symmetrical about $z = 0$, so will the pressure gradient. Hence we deduce that the pressure gradient profile mirrors itself from $z = 0$ to $z = -z_{exit}$, in the same way when we go from $z = 0$ to $z = z_{exit}$.

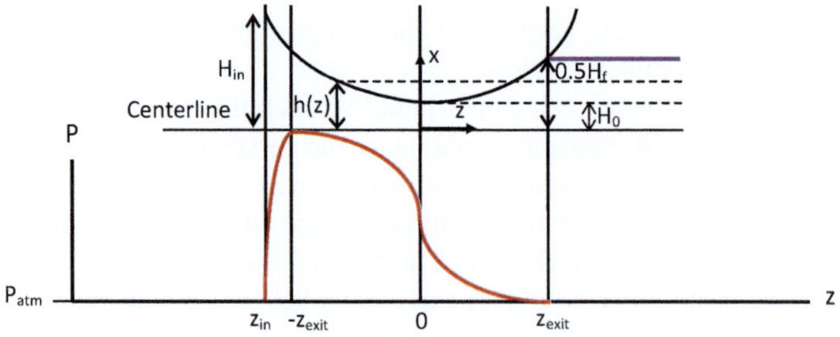


(f) Let us illustrate a close-up of our system at an arbitrary point on the curved surface of the top roller. Note that $\boldsymbol{e}_z$ refers to the unit vector (i.e., of unit magnitude) in the $z$ direction. $\boldsymbol{e}_z$ is commonly also denoted as $\boldsymbol{\delta}_z$. We perform a change in coordinate system from $x$-$z$-$y$ to $r$-$\theta$-$y$ to help with our analysis.

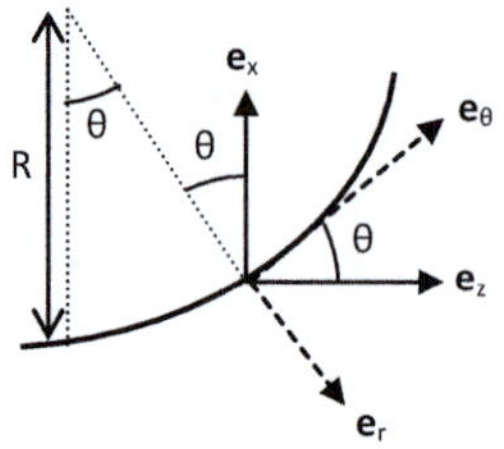


From the above geometry, we can re-express $\boldsymbol{e}_x$ and $\boldsymbol{e}_z$ in terms of $\boldsymbol{e}_r$ and $\boldsymbol{e}_\theta$.

$$\mathbf{e}_r = -\cos\theta\mathbf{e}_x + \sin\theta\mathbf{e}_z$$

$$\mathbf{e}_\theta = \sin\theta\mathbf{e}_x + \cos\theta\mathbf{e}_z$$

To derive an expression for the force exerted by the fluid on the top roller in the $x$ direction, we first make the assumption of small angles so that the lubrication approximation holds.

- - - - - - - - - - - - - - - - - - - - - - - - - - - - - - - - - - - - - - - - - - - - - -

***Background Concepts***

The force exerted by the fluid on the roller is used to overcome the shear (or viscous) force due to friction between the roller surface and the fluid. The directions of stress tensor components that are relevant are all the stress components acting tangential to the roller surface (i.e., in $\boldsymbol{e}_r$ direction). The dot product of the stress tensor $\boldsymbol{T}$ with $\boldsymbol{e}_r$ helps us extract all these components of shear stress. These shear stress components have a first subscript of r as they act in the $r$ plane (normal to plane in the $\boldsymbol{e}_r$ direction). The $r$ plane is the plane tangential to the surface. After this dot product, we perform a second dot product with $\boldsymbol{e}_x$ on our earlier result, to extract only the component of force in the $x$ direction as indicated in our problem statement. Note that $F$ is the force on the roller by the fluid.

$$F = \mathbf{e}_x \cdot \int \mathbf{e}_r \cdot \boldsymbol{T} dS$$

$$dS = WRd\theta$$

$$\mathbf{e}_r \cdot \boldsymbol{T} = \sigma_{rr}\mathbf{e}_r + \tau_{r\theta}\mathbf{e}_\theta + \tau_{ry}\mathbf{e}_y$$

We know that $\tau_{ry} = 0$ because we have no movement in the $y$ direction. Therefore the following

$$\mathbf{e}_r \cdot \boldsymbol{T} = \sigma_{rr}\mathbf{e}_r + \tau_{r\theta}\mathbf{e}_\theta$$

At this point let us recall some relevant concepts. Recall that by definition, $\tau_{rr} = \sigma_{rr} + p$ where $p$ refers to the total pressure gradient term (i.e., $-\frac{dP}{dx}$ assuming $x$ component) in a typical Navier–Stokes equation. The sum of this total pressure gradient (negative sign), together with the gravitational term (e.g., $\rho g_x$) gives the dynamic pressure gradient. In tensor notation, $\sigma$ denotes stress components normal to the face and physically means a tension or compression, while $\tau$ is used to denote stress components parallel to the face and represent shear.

$$\mathbf{e}_r \cdot \boldsymbol{T} = \sigma_{rr}\mathbf{e}_r + \tau_{r\theta}\mathbf{e}_\theta = (\tau_{rr} - p)\mathbf{e}_r + \tau_{r\theta}\mathbf{e}_\theta$$

The above dot product is valid; however it is tricky to solve. So we are going to convert the cylindrical coordinates into Cartesian coordinates for simplification of this dot product. Note that by definition $\mathbf{e}_i \cdot \mathbf{e}_j = \mathbf{e}_{ij}$

$$\mathbf{e}_x \cdot [\mathbf{e}_r \cdot \boldsymbol{T}] = [\mathbf{e}_x \cdot (-\cos\theta\mathbf{e}_x + \sin\theta\mathbf{e}_z)] \cdot \boldsymbol{T}$$

$$= -\cos\theta\boldsymbol{T}_{xx} + \sin\theta\boldsymbol{T}_{xz} = -\cos\theta(\sigma_{xx}) + \sin\theta(\tau_{xz})$$

$$= -\cos\theta(\tau_{xx} - p) + \sin\theta(\tau_{xz})$$

Note that by definition, and also available in data booklets as one of the stress constitutive equations, the following expressions for the stress terms are valid for an incompressible fluid of constant ρ that obeys the mass continuity equation. The below expressions also assume $v_z$ is the dominant flow direction ($v_x = 0$).

$$\tau_{xx} = 2\eta \frac{\partial v_x}{\partial x} = 0$$

$$\tau_{xz} = \eta \left[ \frac{\partial v_x}{\partial z} + \frac{\partial v_z}{\partial x} \right] = \eta \frac{\partial v_z}{\partial x} = \eta \left( \frac{\beta}{\eta} x \right) = \frac{\partial P}{\partial z} x$$

$$\mathbf{e}_x \cdot [\mathbf{e}_r \cdot \boldsymbol{T}] = -\cos\theta(0 - p) + \sin\theta\left(\frac{\partial P}{\partial z} x\right) = p\cos\theta + \sin\theta\left(\frac{\partial P}{\partial z} x\right)$$

---

Now let us return to our problem and determine the integral below to find force $F$

$$F = \mathbf{e}_x \cdot \int_{\theta_{\text{in}}}^{\theta_{\text{exit}}} \mathbf{e}_r \cdot \boldsymbol{T}(WRd\theta)$$

$$F = WR \int_{\theta_{\text{in}}}^{\theta_{\text{exit}}} \left[ p\cos\theta + \sin\theta\left(\frac{\partial P}{\partial z} x\right) \right] d\theta$$

We will try to convert the integration limits to parameters that we know, i.e., from $\theta$ to $z$ using some geometric analysis. Also we know that $x$ is a function of $z$, i.e., $x = h(z)$

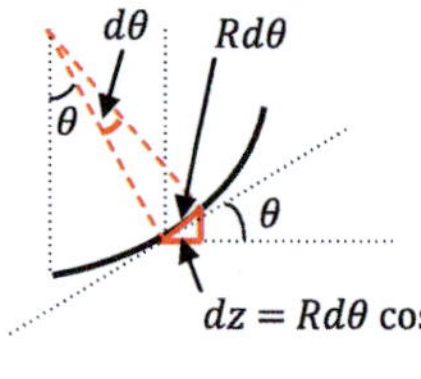


$$F = W \int_{z_{\text{in}}}^{z_{\text{exit}}} \left[ p + \tan\theta\left(\frac{\partial P}{\partial z} h(z)\right) \right] dz$$

We can express tan$\theta$ in terms of variable $z$ using $\tan\theta = \frac{z}{\sqrt{R^2 - z^2}}$. The geometry is shown below, where Pythagora's theorem is used. Now we have an equation that is only in terms of variable $z$ that can be solved using the integration limits.

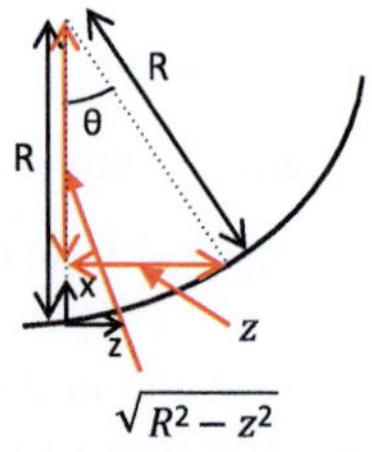

The integration may be further simplified if we assume small angle $\theta$ and that the lubrication approximation holds. Then $\tan\theta \approx \theta$ which is negligible, and we obtain the following expression for $F$.

$$F = W \int_{z_{\text{in}}}^{z_{\text{exit}}} p dz$$

## Problem 11

**A factory manufactures a chemical at an upstream location $x = 0$ and this chemical is distributed to customers at values of $x > 0$ downstream in the $+x$ direction. The price of this chemical, $\beta$, depends on the time elapsed after production completion (i.e., $t$), as well as the location that the chemical is sold (i.e., $x$). Because the chemical has a short shelf life, its price decreases with time $t$. We are given the following correlation, where $\alpha$ is a positive value in \$/hour.**

$$\frac{\partial p}{\partial t} = -\alpha$$

**We are also informed that the price increases with distance of the point of sale from the factory according to the following correlation, where $\beta$ is a positive value in \$/meter.**

$$\frac{\partial p}{\partial x} = \beta$$

**If the chemical producer wants to sell the product at the same price at any location along the distribution route, find out the speed of distribution that is necessary. Comment on how the above problem relates to the concept of convective derivative used in Fluid Mechanics analysis.**

## Solution 11

***Worked Solution***

- - - - - - - - - - - - - - - - - - - - - - - - - - - - - - - - - - - - - - - - - - - - - - - -

***Background Concepts***

Recall what convective derivative means. It refers to a derivative taken with respect to a moving coordinate system, whereby in fluid mechanics applications, we "follow" the fluid particle as it moves. The convective derivative is also termed substantive derivative or total derivative and denoted as follows, where $\mathbf{\nabla}$ is a gradient operator and $\mathbf{v}$ is the velocity of the fluid. This convective derivative is often applied to velocity (i.e., $\frac{D\mathbf{v}}{Dt}$) in fluid mechanics, and is usually at the left-hand side of the Navier–Stokes equation representing the convective/inertial terms.

$$\frac{D}{Dt} = \frac{\partial}{\partial t} + \boldsymbol{v} \cdot \boldsymbol{\nabla}$$

For example, we may take the convective derivative of $v_x$ under Cartesian coordinates for the $x$ component of the velocity vector.

$$\frac{Dv_x}{Dt} = \frac{\partial v_x}{\partial t} + \boldsymbol{v} \cdot \boldsymbol{\nabla} v_x$$

$$\boldsymbol{\nabla} v_x = \frac{\partial v_x}{\partial x}\boldsymbol{i} + \frac{\partial v_x}{\partial y}\boldsymbol{j} + \frac{\partial v_x}{\partial z}\boldsymbol{k}$$

$$\boldsymbol{v} \cdot \boldsymbol{\nabla} v_x = v_x \frac{\partial v_x}{\partial x} + v_y \frac{\partial v_x}{\partial y} + v_z \frac{\partial v_x}{\partial z}$$

$$\frac{Dv_x}{Dt} = \frac{\partial v_x}{\partial t} + v_x \frac{\partial v_x}{\partial x} + v_y \frac{\partial v_x}{\partial y} + v_z \frac{\partial v_x}{\partial z}$$

This convective derivative is useful as it allows us to determine the rate of change of parameters associated with the fluid particle even as it moves about. We can find out the positional information at specified times. We can also find out velocity at a given position and time.

---

In this problem, we may apply the concept of convective derivative by "following the chemical" as it moves along the distribution route.

$$\frac{Dp}{Dt} = \frac{\partial p}{\partial t} + v_x \frac{\partial p}{\partial x} + v_y \frac{\partial p}{\partial y} + v_z \frac{\partial p}{\partial z}$$

To sell the product at the same price at any location along the distribution route, the condition is that the price is steady, i.e., the convective derivative of the price is zero or $\frac{Dp}{Dt} = 0$. We may assume that $v_y = v_z = 0$ for a distribution route that is only in the $x$ direction.

$$0 = \frac{\partial p}{\partial t} + v_x \frac{\partial p}{\partial x} + v_y \frac{\partial p}{\partial y} + v_z \frac{\partial p}{\partial z}$$

$$\frac{\partial p}{\partial t} = -\alpha$$

$$\frac{\partial p}{\partial x} = \beta$$

$$0 = -\alpha + \beta v_x \rightarrow v_x = \frac{\alpha}{\beta} \text{m/h}$$

Therefore the speed at which the chemical producer needs to travel is $\frac{\alpha}{\beta}$ m/h.

## Problem 12

**We have laminar flow in a circular pipe that is held horizontal.**

(a) **Comment on the streamlines within the pipe.**
(b) **Assuming Cartesian coordinates, is it possible to use the below equation to model the $z$ component of velocity in terms of whether (1) the boundary conditions are satisfied and (2) the continuity equation is satisfied? It is given that $R$ refers to the radius of the pipe cross section, and $x^2 + y^2 = R^2$.**

$$v_z = K\left(R^2 - x^2 - y^2\right)$$

(c) **Given the velocity distribution in part b, find out the correlation between the pressure gradient along the pipe and the volumetric flow rate through the tube.**

## Solution 12

*Worked Solution*

(a) We have a circular pipe as shown below. This is an example of a pressure driven flow in the laminar region. The streamlines in the pipe are all straight and parallel. Note that in turbulent flow, the velocity changes rapidly at each point in the flow field about a mean value and it is the mean velocity vector that may be referred to under turbulent regimes. However for a steady-state turbulent flow, the mean velocity vector does not change with time.

Recall that streamlines are a family of lines to which the velocity vectors are tangent to (i.e., same gradient). Velocity vectors may be drawn at any point within the flow field. At steady state, a fluid particle will move along a streamline.

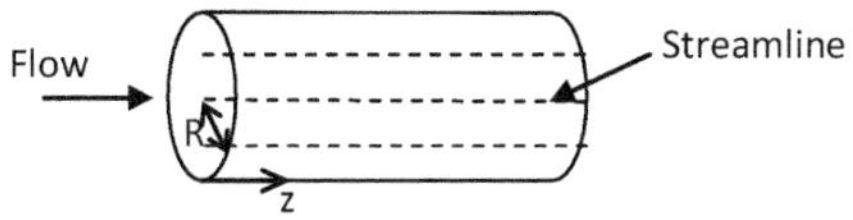


(b) We can apply the $z$ component of the Navier–Stokes equation. Note that we have used a form of the NS equation below which consists of the gravitational term $\rho g_z$; hence the $P$ in $\frac{\partial P}{\partial z}$ refers to total pressure (as opposed to dynamic pressure). We have also assumed:

- Constant viscosity for a Newtonian fluid.
- The liquid is incompressible, $\rho$ is constant.
- The flow is at steady state ($\frac{\partial}{\partial t} = 0$).
- The dominant flow direction is in the $z$ direction; hence we assume $v_x = v_y = 0$.
- The flow is fully developed in the $z$ direction; hence $v_z \neq v_z(z)$.
- Horizontal pipe hence gravitational effects are negligible.

$$\rho\left(\frac{\partial v_z}{\partial t}+v_x\frac{\partial v_z}{\partial x}+v_y\frac{\partial v_z}{\partial y}+v_z\frac{\partial v_z}{\partial z}\right)=-\frac{\partial P}{\partial z}+\eta\left(\frac{\partial^2 v_z}{\partial x^2}+\frac{\partial^2 v_z}{\partial y^2}+\frac{\partial^2 v_z}{\partial z^2}\right)+\rho g_z$$

$$0=-\frac{\partial P}{\partial z}+\eta\left(\frac{\partial^2 v_z}{\partial x^2}+\frac{\partial^2 v_z}{\partial y^2}\right)$$

We are given a suggested form of the velocity profile which we can test here

$$v_z=K\left(R^2-x^2-y^2\right)=K\left(R^2-r^2\right)$$

$$\frac{\partial v_z}{\partial r}=-2Kr$$

At the pipe wall, $x^2+y^2=r^2=R^2$, $v_z=0$. This is satisfied by the velocity profile provided. Also at the centerline of the pipe $x=y=0$ $(r=0)$, the velocity gradient $\frac{dv_z}{dr}$ is zero due to axisymmetry, and since a maxima in velocity occurs along the centerline. This is also satisfied by the velocity profile provided.

We will now test the validity of the continuity equation

$$\nabla\cdot\boldsymbol{v}=0$$

$$\frac{\partial v_x}{\partial x}+\frac{\partial v_y}{\partial y}+\frac{\partial v_z}{\partial z}=0$$

$$\frac{\partial v_z}{\partial z}=0$$

This condition is satisfied with the velocity profile $v_z=K(R^2-x^2-y^2)$ provided.

(c) To find an expression for the pressure gradient, let us return to the Navier–Stokes equation

$$0=-\frac{\partial P}{\partial z}+\eta\left(\frac{\partial^2 v_z}{\partial x^2}+\frac{\partial^2 v_z}{\partial y^2}\right)$$

$$\frac{\partial^2 v_z}{\partial x^2}=-2K$$

$$\frac{\partial^2 v_z}{\partial y^2}=-2K$$

$$0=-\frac{\partial P}{\partial z}+\eta(-4K)\rightarrow\frac{\partial P}{\partial z}=-4K\eta$$

To find the volumetric flow rate, we need to express a differential area element for the integral. This can be more easily done in cylindrical coordinates; hence we may analyze this problem in cylindrical coordinates.

After applying the same assumptions as shown earlier in part b, the $z$ component of the NS equation in cylindrical coordinates is

$$0 = -\frac{\partial P}{\partial z} + \eta \frac{1}{r}\frac{d}{dr}\left(r\frac{dv_z}{dr}\right)$$

$$r\frac{dv_z}{dr} = \frac{r^2}{2\eta}\frac{\partial P}{\partial z} + c_1$$

$$\frac{dv_z}{dr} = \frac{r}{2\eta}\frac{\partial P}{\partial z} + \frac{c_1}{r} \rightarrow c_1 = 0\left(r = 0, \frac{dv_z}{dr} = 0\right)$$

$$v_z = \frac{r^2}{4\eta}\frac{\partial P}{\partial z} + c_2 \rightarrow c_2 = -\frac{R^2}{4\eta}\frac{\partial P}{\partial z}\ (r = R, v_z = 0)$$

$$v_z = \frac{r^2}{4\eta}\frac{\partial P}{\partial z} - \frac{R^2}{4\eta}\frac{\partial P}{\partial z} = \frac{1}{4\eta}\frac{\partial P}{\partial z}\left[r^2 - R^2\right]$$

$$v_z = \frac{1}{4\eta}(-4K\eta)\left(r^2 - R^2\right) = -K\left(r^2 - R^2\right)$$

Now we can relate volumetric flow rate to velocity as shown below.

$$\dot{Q} = \int_0^R v_z 2\pi r dr$$

$$\dot{Q} = \int_0^R K\left(R^2 - r^2\right)2\pi r dr = 2K\pi\left[\frac{R^2 r^2}{2} - \frac{r^4}{4}\right]_0^R$$

$$\dot{Q} = \frac{K\pi R^4}{2} \rightarrow K = \frac{2\dot{Q}}{\pi R^4}$$

The pressure gradient can be expressed in terms of volumetric flow rate.

$$\frac{\partial P}{\partial z} = -4K\eta = -4\frac{2\dot{Q}}{\pi R^4}\eta = -\frac{8\eta\dot{Q}}{\pi R^4}$$

## Problem 13

**We have a fluid that flows between two large flat plates. There is steady flow in the +*x* direction and the separation distance between the plates is *H*. The top plate is moving at a velocity of *U*. The bottom plate is stationary.**

(a) **Find the velocity profile.**
(b) **Comment on the effect of the pressure gradient and the implications if this was absent.**
(c) **Comment on the flow behavior if the top plate was still.**

## Solution 13

### *Worked Solution*

(a) Let us visualize the flow in a simple diagram as shown below.

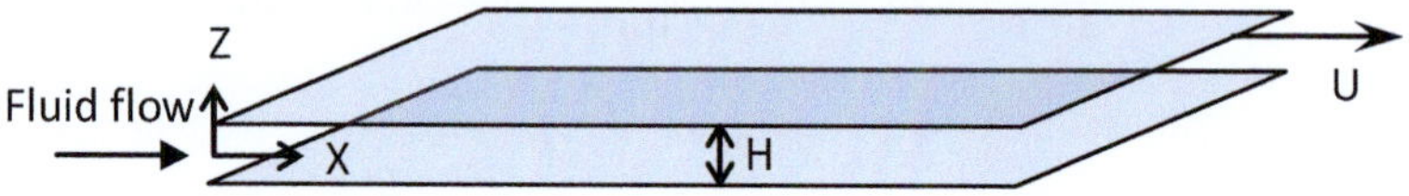


To find the velocity profile, let's examine the boundary conditions and the relevant Navier–Stokes equation.

The assumptions are as follows:

- Constant viscosity for a Newtonian fluid.
- The liquid is incompressible, $\rho$ is constant.
- The flow is at steady state ($\frac{\partial}{\partial t} = 0$).
- The dominant flow direction is in the $x$ direction; hence $v_y = v_z = 0$.
- The flow is fully developed in the $x$ direction; hence $v_x \neq v_x(x)$.
- Horizontal plates, hence gravitational effects are zero.

$$0 = -\frac{\partial P}{\partial x} + \eta\left(\frac{\partial^2 v_x}{\partial z^2}\right)$$

$$\frac{\partial^2 v_x}{\partial z^2} = \frac{1}{\eta}\left(\frac{\partial P}{\partial x}\right)$$

$$\frac{\partial v_x}{\partial z} = \frac{1}{\eta}\left(\frac{\partial P}{\partial x}\right)z + c_1$$

$$v_x = \frac{\left(\frac{\partial P}{\partial x}\right)z^2}{2\eta} + c_1 z + c_2$$

We know that when $z = -H/2$, $v_x = 0$ and when $z = H/2$, $v_x = U$

$$0 = \left(\frac{\partial P}{\partial x}\right)\frac{H^2}{8\eta} + \frac{c_1 H}{2} + c_2$$

$$V = \left(\frac{\partial P}{\partial x}\right)\frac{H^2}{8\eta} - \frac{c_1 H}{2} + c_2$$

$$-U = c_1 H \rightarrow c_1 = -\frac{U}{H}$$

$$c_2 = -\left(\frac{\partial P}{\partial x}\right)\frac{H^2}{8\eta} - \frac{H}{2}\left(-\frac{U}{H}\right) = -\left(\frac{\partial P}{\partial x}\right)\frac{H^2}{8\eta} + \frac{HU}{2H}$$

$$v_x = \frac{\left(\frac{\partial P}{\partial x}\right)z^2}{2\eta} - \left(\frac{U}{H}\right)z - \left(\frac{\partial P}{\partial x}\right)\frac{H^2}{8\eta} + \frac{HU}{2H}$$

$$v_x = \frac{1}{2\eta}\left(\frac{\partial P}{\partial x}\right)\left(z^2 - \frac{H^2}{4}\right) + U\left(\frac{1}{2} - \frac{z}{H}\right)$$

(b) We can observe from the simplified Navier–Stokes equation that the pressure gradient balances the viscous/shear stresses. This force balance explains the steady state of flow. If pressure gradient was absent, the velocity profile will change to the following. From observation we notice that the initial profile was the sum of a parabolic part and a linear part. If pressure gradient was zero, the profile changes to a linear profile.

$$v_x = U\left(\frac{1}{2} - \frac{z}{H}\right)$$

(c) If the upper plate was stationary, then $U = 0$. Then the velocity field will change to the following parabolic profile.

$$v_x = \frac{1}{2\eta}\left(\frac{\partial P}{\partial x}\right)\left(z^2 - \frac{H^2}{4}\right)$$

## Problem 14

**We have a long cylindrical tube within which an incompressible fluid flows steadily in laminar flow. The velocity field is provided as follows:**

$$v_z = V\left[1 - \varepsilon^2\right]$$

**Given that $V$ is the velocity along the centerline, $\varepsilon = r/R$ where $R$ is the radius of the cross section, and the dominant flow direction is in the $z$ direction only.**

**The tube gets heated with passing fluid due to viscous dissipation. The rate of this heat generation per unit volume fluid, *H* is as follows**

$$H_{\text{visc}} = \eta \left( \frac{\partial v_z}{\partial r} \right)^2$$

**Determine the temperature profile *T*(*r*) if**

(a) **Heating due to viscous dissipation is negligible.**
(b) **Heating due to viscous dissipation was significant.**

## Solution 14

*Worked Solution*

(a) Before we begin, let us think about the approach to solving this problem. We need to determine temperature profile in this problem. Therefore instead of the typical approach of working with differential equations in velocity (i.e., the typical Navier–Stokes momentum equations), we need equivalent differential equations in temperature. The typical NS equations come about by applying conservation of momentum on a differential volume element of fluid. In the same way we can apply conservation of energy (or heat flux) on a differential volume element of fluid.

The general form of the energy equation is as follows.

$$\rho c_p \frac{\partial T}{dt} + \rho c_p (\boldsymbol{v} \cdot \nabla T) = k \nabla^2 T + H_{\text{visc}}$$

$$\frac{\partial T}{dt} + \boldsymbol{v} \cdot \nabla T = \frac{k}{\rho c_p} \nabla^2 T + \frac{H_{\text{visc}}}{\rho c_p}$$

Some points to note about the above equation are as follows:

- The rate of heat energy accumulation is given by $\rho c_p \frac{\partial T}{dt}$.
- The convective derivative is represented by $\rho c_p(\boldsymbol{v} \cdot \nabla T)$ which denotes heat energy changes due to positional changes of the fluid particle caused by fluid flow.
- The diffusion/conduction effect is denoted by $k\nabla^2 T$, whereby $k$ is the thermal conductivity. This term takes care of the heat flux that enters or leaves the differential element of fluid. Another useful property is thermal diffusivity which is $\frac{k}{\rho c_p}$.
- $H_{\text{visc}}$ represents any heat source or sinks, and in this case, we have been given an expression for a heat source/generation due to viscous dissipation and this is used for this source/sink term.

In the cylindrical coordinates, this is our energy equation.

$$\frac{\partial T}{dt} + \boldsymbol{v} \cdot \nabla T = \frac{k}{\rho c_p} \nabla^2 T + \frac{H_{\text{visc}}}{\rho c_p}$$

For part a, we assume $H_{\text{visc}} = 0$

$$\frac{\partial T}{dt} + \boldsymbol{v} \cdot \nabla T = \frac{k}{\rho c_p} \nabla^2 T$$

$$\frac{\partial T}{dt} + (v_r, v_\theta, v_z) \cdot \left( \frac{\partial T}{dr}, \frac{1}{r} \frac{\partial T}{d\theta}, \frac{\partial T}{dz} \right) = \frac{k}{\rho c_p} \left( \frac{1}{r} \frac{\partial}{dr} \left( r \frac{\partial T}{dr} \right) + \frac{1}{r^2} \frac{\partial^2 T}{d\theta^2} + \frac{\partial^2 T}{dz^2} \right)$$

We can simplify the above expression using the assumptions below:

- We have a steady-state flow, so the temperatures are also constant. $\frac{\partial T}{dt} = 0$
- The dominant flow direction is in the $z$ direction, $v_r = v_\theta = 0$
- We are told that the temperature profile is a function of $r$ only, $T(r)$; hence $\frac{\partial T}{d\theta} = \frac{\partial T}{dz} = 0$

The energy equation is therefore simplified to the following

$$0 = \frac{k}{\rho c_p} \left( \frac{1}{r} \frac{\partial}{dr} \left( r \frac{\partial T}{dr} \right) \right)$$

$$\frac{\partial}{dr} \left( r \frac{\partial T}{dr} \right) = 0$$

$$\frac{\partial T}{dr} = \frac{c_1}{r}$$

$$T = c_1 \ln r + c_2$$

We can apply boundary conditions to evaluate the integration constants. At $r = R$, $T = T_w$. At $r = 0$, $\frac{\partial T}{dr} = 0$ due to axisymmetry.

$$c_1 = 0$$

$$T = c_2 = T_w$$

In this case, the other boundary condition at $r = R$ is not necessary as temperature is constant at $T_w$ regardless of $r$.

(b) Assuming viscous dissipation is present,

$$0 = \frac{k}{\rho c_p}\left(\frac{1}{r}\frac{\partial}{dr}\left(r\frac{\partial T}{dr}\right)\right) + \frac{H_{\text{visc}}}{\rho c_p}$$

$$0 = \frac{k}{\rho c_p}\left(\frac{1}{r}\frac{\partial}{dr}\left(r\frac{\partial T}{dr}\right)\right) + \eta\frac{\left(\frac{\partial v_z}{\partial r}\right)^2}{\rho c_p}$$

We can find an expression for velocity gradient, $\frac{\partial v_z}{\partial r}$ in terms of $V$ and $R$.

$$v_z = V\left[1 - \left(\frac{r}{R}\right)^2\right]$$

$$\frac{\partial v_z}{\partial r} = -\frac{2Vr}{R^2}$$

Substituting this expression back into the earlier energy equation, we have

$$0 = \frac{k}{\rho c_p}\left(\frac{1}{r}\frac{\partial}{dr}\left(r\frac{\partial T}{dr}\right)\right) + \eta\frac{\frac{4V^2r^2}{R^4}}{\rho c_p}$$

$$0 = k\left(\frac{1}{r}\frac{\partial}{dr}\left(r\frac{\partial T}{dr}\right)\right) + \frac{4\eta V^2 r^2}{R^4}$$

$$\frac{\partial}{dr}\left(r\frac{\partial T}{dr}\right) = -\frac{4\eta V^2 r^3}{kR^4}$$

Integrating with boundary conditions, we obtain the temperature profile.

$$\frac{\partial T}{dr} = -\frac{\eta V^2 r^3}{kR^4} + \frac{c_1}{r} \rightarrow c_1 = 0\ \left(r = 0, \frac{\partial T}{dr} = 0\right)$$

$$T = -\frac{\eta V^2 r^4}{4kR^4} + c_2 \rightarrow c_2 = T_w + \frac{\eta V^2}{4k}$$

$$T = T_w + \frac{\eta V^2}{4k}\left[1 - \varepsilon^4\right]$$

## Problem 15

**Consider the vertical flow of a thin film of viscous fluid up a conveyor belt. The belt has a large width *W* and is moving at a constant upward velocity of *V*. The fluid's viscosity allows it to adhere to the belt surface with a thickness of *h*. Gravity causes the fluid to flow back down into a container below. You may**

**assume that the fluid is Newtonian, and the flow is laminar, fully developed and at steady state.**

(a) **Show that the pressure in the liquid film is constant at a specified height.**
(b) **Derive an expression for the velocity field in the liquid film and sketch this profile for a fluid with high viscosity and a fluid with low viscosity. Comment on the effect of viscosity.**
(c) **Find the volumetric flow rate of the liquid film. And explain how belt speed relates to viscosity of fluid for this system to work.**

## Solution 15

### *Worked Solution*

(a) We have a simple illustration of the problem below.

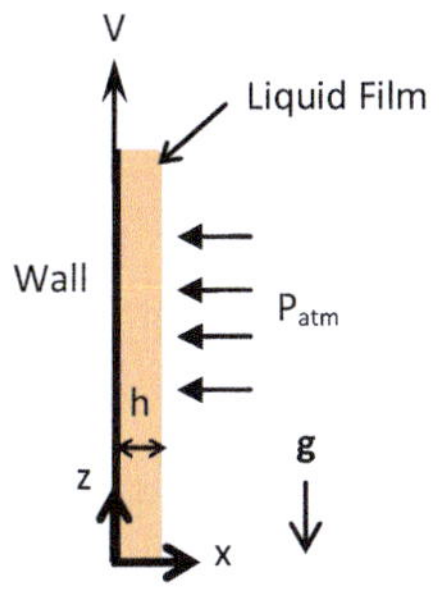


We are required to show that pressure is constant at a specified height. Implicitly it means that pressure is not a function of $x$, and only a function of $z$ (assuming no flow in $y$ direction).

Let us examine our assumptions before looking at the Navier–Stokes equation

- The dominant flow direction is in the $z$ direction; hence $v_x = v_y = 0$.
- We know that $g_z = -9.81$, and $g_x = g_y = 0$.
- Constant viscosity for a Newtonian fluid.
- The liquid is incompressible, $\rho$ is constant.
- The flow is at steady state ($\frac{\partial}{\partial t} = 0$).
- The belt was described as wide, implying that entrance/end effects would be negligible and hence the flow may be assumed fully developed $v_z \neq v_z(z)$.
- The fluid is described as viscous; hence viscous terms dominate, and inertial terms may be assumed negligible.

With the above assumptions, the $x$ component of the Navier–Stokes equation can be simplified to

$$\frac{dP}{dx} = 0$$

We have thus shown that $P$ is not a function of $x$, and hence pressure in the fluid is constant at a given height ($z$).

(b) Knowing that the direction of flow is in the $z$ direction, we can similarly simplify the $z$ component of the Navier–Stokes equation. We also know from part a that $\frac{dP}{dx} = 0$.

$$0 = \rho g_z + \eta\left(\frac{d^2 v_z}{dx^2}\right)$$

$$\frac{\rho(-g_z)}{\eta}x + c_1 = \frac{dv_z}{dx}$$

We can now apply the boundary condition whereby when $x = h, \frac{dv_z}{dx} = 0$ as we assume negligible shear stress at the liquid–air interface.

$$\frac{\rho(-g_z)}{\eta}h + c_1 = 0 \rightarrow c_1 = \frac{\rho g_z}{\eta}h$$

$$\frac{\rho(-g_z)}{\eta}x + \frac{\rho g_z}{\eta}h = \frac{dv_z}{dx}$$

$$v_z = \frac{\rho(-g_z)}{\eta}\frac{x^2}{2} + \frac{\rho g_z}{\eta}hx + c_2$$

We can apply the other boundary condition of no-slip at the belt surface; hence $v_z = V$ when $x = 0$.

$$V = c_2$$

Let us denote $g_z = -9.81 = -g$. Therefore the velocity profile in the liquid film is

$$v_z = \frac{\rho g}{\eta}\left(\frac{x^2}{2} - hx\right) + V$$

$$v_z = \frac{\rho g h^2}{\eta}\left(\frac{1}{2}\left(\frac{x}{h}\right)^2 - \left(\frac{x}{h}\right)\right) + V$$

Let us now sketch this profile for a fluid with high viscosity and a fluid with low viscosity. We can observe from the velocity profile equation some key features below

- At $x = h$, $v_z = V - \frac{\rho g h^2}{2\eta}$. So the greater the viscosity $\eta$ of the fluid, the greater the value of $v_z$ at $x = h$. This makes sense since the more viscous the fluid, the closer it will follow the speed of the moving belt, and less easily "affected" by the downward gravity pull, hence it will maintain a higher upward velocity.
- Regardless of viscosity, the fluid velocity is fixed at $V$ when $x = 0$.

- The shape of the velocity is second order (parabolic) with respect to $x$.
- At the air–fluid interface the gradient $\frac{dv_z}{dx}$ is zero as we assume shear stress is zero at that interface.

Therefore we can develop the sketch below.

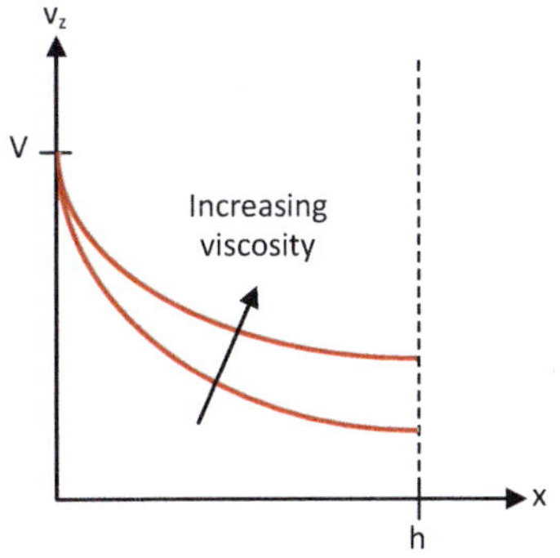


(c) To find the volumetric flow rate of the liquid film $\dot{Q}$, we can integrate across the entire belt from a differential element.

$$\dot{Q} = W\int_0^h v_z dx$$

$$\frac{\dot{Q}}{W} = \int_0^h \left[\frac{\rho g h^2}{\eta}\left(\frac{1}{2}\left(\frac{x}{h}\right)^2 - \left(\frac{x}{h}\right)\right) + V\right] dx$$

$$\frac{\dot{Q}}{W} = \left[\frac{\rho g h^2}{\eta}\left(\frac{1}{6}\left(\frac{x^3}{h^2}\right) - \left(\frac{x^2}{2h}\right)\right) + Vx\right]_0^h = Vh - \frac{\rho g h^3}{3\eta}$$

We can observe from the above expression that in order to lift the liquid film upwards, the term $Vh > \frac{\rho g h^3}{3\eta}$. The lower the viscosity of the fluid, the greater is the required belt speed to still satisfy upward fluid film movement ($\dot{Q} > 0$, where $\dot{Q}$ is positive in the $+z$ direction upwards).

## Problem 16

**Starting from first principles, show how the continuity equation in cylindrical coordinates as shown below is derived.**

$$\frac{1}{r}\frac{\partial(rv_r)}{\partial r} + \frac{1}{r}\frac{\partial v_\theta}{\partial \theta} + \frac{\partial v_z}{\partial z} = 0$$

## Solution 16

### *Worked Solution*

The first step of this derivation is to accurately visualize the system. It helps to understand how the Cartesian coordinates $(x, y, z)$ and cylindrical coordinates $(r, \theta, z)$ relate to each other. This is shown on the bottom right diagram. Then we take an infinitesimal volume element of fluid, and develop the mass (or volume) balance equation for this control volume. The bottom left diagram is a zoom-in on this small volume element. The angle of curvature for the volume element may be ignored if the element is small enough, i.e., $d\theta$ is sufficiently small such that the length of $rd\theta$ may be assumed straight. Then this volume element can be treated as a rectangular block as a simplified geometry.

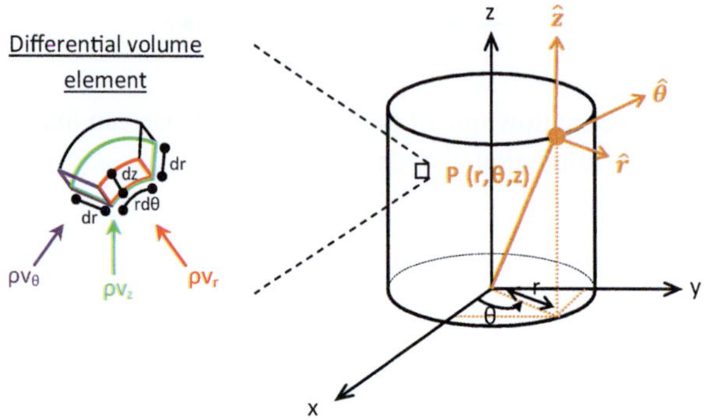


Volume of small element follows that of a rectangular block (length × breadth × height). Note that arc length $S$ for a curvature over an angle measured by $\theta$, with a radius of curve at $R$, is defined mathematically as $S = R\theta$. Therefore in our problem, for an arbitrary $r$, $dS = rd\theta$

$$dV = (rd\theta)drdz$$

Mass flowrate into the volume element can be found by looking at each face. The product of the area of the face and the velocity into the element, will give the volumetric flowrate, which is then converted to mass flowrate using density.

$$\rho v_r rd\theta dz + \rho v_z rd\theta dr + \rho v_\theta drdz$$

Mass flowrate out of the volume element is expressed as follows

$$\left(\rho v_r + \frac{\partial(\rho v_r)}{\partial r}dr\right)(r + dr)d\theta dz + \left(\rho v_z + \frac{\partial(\rho v_z)}{\partial z}dz\right)rd\theta dr + \left(\rho v_\theta + \frac{\partial(\rho v_\theta)}{\partial \theta}d\theta\right)drdz$$

Conceptually, mass balance is governed by this rationale

Rate of mass accumulation = Mass inflow rate – Mass outflow rate

Rate of mass accumulation is expressed as follows where $V$ is the volume

$$\frac{d(\rho V)}{dt} = \frac{d(\rho(rd\theta)drdz)}{dt}$$

Therefore the mass balance equation becomes

$$\begin{aligned}\frac{d(\rho V)}{dt} &= \rho v_r rd\theta dz - \left(\rho v_r + \frac{\partial(\rho v_r)}{\partial r}dr\right)(r+dr)d\theta dz + \rho v_z rd\theta dr \\ &\quad - \left(\rho v_z + \frac{\partial(\rho v_z)}{\partial z}dz\right)rd\theta dr + \rho v_\theta drdz - \left(\rho v_\theta + \frac{\partial(\rho v_\theta)}{\partial\theta}d\theta\right)drdz\end{aligned}$$

$$\begin{aligned}\frac{d(\rho(rd\theta)drdz)}{dt} &= -\rho v_r drd\theta dz - \left(\frac{\partial(\rho v_r)}{\partial r}dr\right)(r+dr)d\theta dz - \left(\frac{\partial(\rho v_z)}{\partial z}dz\right)rd\theta dr \\ &\quad - \left(\frac{\partial(\rho v_\theta)}{\partial\theta}d\theta\right)drdz\end{aligned}$$

$$\frac{d\rho}{dt} = -\frac{\rho v_r}{r} - \left(\frac{\partial(\rho v_r)}{\partial r}\right)\frac{(r+dr)}{r} - \left(\frac{\partial(\rho v_z)}{\partial z}\right) - \left(\frac{\partial(\rho v_\theta)}{\partial\theta}\right)\left(\frac{1}{r}\right)$$

For an incompressible fluid, $\rho$ is constant; therefore, we simplify the equation to

$$0 = \frac{v_r}{r} + \left(\frac{\partial v_r}{\partial r}\right)\left(1+\frac{dr}{r}\right) + \left(\frac{\partial v_z}{\partial z}\right) + \left(\frac{\partial v_\theta}{\partial\theta}\right)\left(\frac{1}{r}\right)$$

We may assume that $\frac{dr}{r} \ll 1$ and $\left(1+\frac{dr}{r}\right) \approx 1$. We also know that $\frac{v_r}{r} + \left(\frac{\partial v_r}{\partial r}\right) = \frac{1}{r}\left(r\frac{\partial v_r}{\partial r} + v_r\right)$; therefore, we have the following expression

$$0 = \frac{1}{r}\left(r\frac{\partial v_r}{\partial r} + v_r\right) + \left(\frac{\partial v_z}{\partial z}\right) + \left(\frac{\partial v_\theta}{\partial\theta}\right)\left(\frac{1}{r}\right)$$

$$\frac{1}{r}\left(\frac{\partial(rv_r)}{\partial r}\right) + \frac{1}{r}\left(\frac{\partial v_\theta}{\partial\theta}\right) + \left(\frac{\partial v_z}{\partial z}\right) = 0$$

## Problem 17

**An incompressible Newtonian fluid flows through a tube with permeable walls. The tube is of radius $R$ and fluid enters the pipe and flows through it. In addition, more fluid enters the tube through the porous walls at a constant velocity of $U$. The axial velocity $v_z$ can be expressed as follows where $L$ is the length of the tube and $\beta$ is a constant.**

$$\frac{v_z}{v_z(z=0)} = \left(1 - \left(\frac{r}{R}\right)^2\right)\left(1 + (\beta - 1)\frac{z}{L}\right)$$

**Determine the radial velocity profile and express it in terms of $\beta$ and $U$.**

## Solution 17

***Worked Solution***

Let us illustrate this system

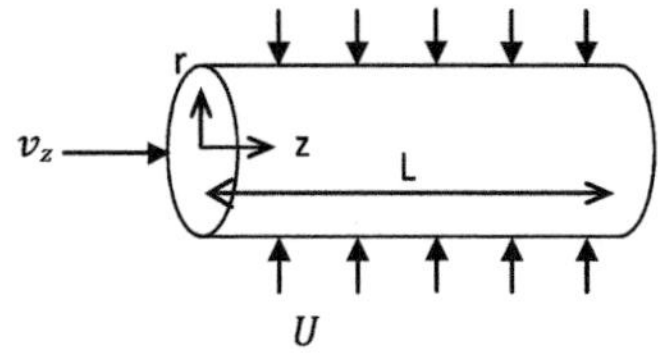


Let us apply the mass continuity equation in cylindrical coordinates

$$\frac{1}{r}\left(\frac{\partial(rv_r)}{\partial r}\right) + \frac{1}{r}\left(\frac{\partial v_\theta}{\partial \theta}\right) + \left(\frac{\partial v_z}{\partial z}\right) = 0$$

We have the following assumptions

- The flow is axisymmetric; therefore $\frac{\partial}{\partial \theta} = 0$
- The dominant flow directions are in the $r$ and $z$ directions; hence we assume that $v_\theta = 0$
- Constant viscosity for a Newtonian fluid.
- The liquid is incompressible, $\rho$ is constant.

$$\frac{1}{r}\left(\frac{\partial(rv_r)}{\partial r}\right) + \frac{\partial v_z}{\partial z} = 0$$

We are given the axial velocity profile below, let's denote $v_z(z = 0) = v_{z,0}$

$$\frac{v_z}{v_{z,0}} = \left(1 - \left(\frac{r}{R}\right)^2\right)\left(1 + (\beta - 1)\frac{z}{L}\right)$$

$$\frac{\partial v_z}{\partial z} = v_{z,0}\left(1 - \left(\frac{r}{R}\right)^2\right)\frac{(\beta - 1)}{L}$$

Substituting back into the continuity equation

$$\frac{1}{r}\left(\frac{\partial(rv_r)}{\partial r}\right) + v_{z,0}\left(1 - \left(\frac{r}{R}\right)^2\right)\frac{(\beta - 1)}{L} = 0$$

$$\frac{\partial(rv_r)}{\partial r} = v_{z,0}\left(\frac{r^3}{R^2} - r\right)\frac{(\beta - 1)}{L}$$

Note that velocity is a function of two variables, $r$ and $z$; hence the "integration constant" is a function.

$$v_r = v_{z,0}\left(\frac{r^3}{4R^2} - \frac{r}{2}\right)\frac{(\beta - 1)}{L} + \frac{f(z)}{r}$$

Now we apply boundary conditions to figure out an expression for $f(z)$. When $r = 0$, $v_r$ is finite. Hence $f(z)$ has to be zero.

$$v_r = v_{z,0}\left(\frac{r^3}{4R^2} - \frac{r}{2}\right)\frac{(\beta - 1)}{L}$$

We have now obtained the radial velocity as shown

$$v_r = -rv_{z,0}\left(2 - \left(\frac{r}{R}\right)^2\right)\frac{(\beta - 1)}{4L}$$

When $r = R$, $v_r = -U$ (inward direction is negative of the $r$ coordinate direction).

$$-U = v_{z,0}\left(-\frac{R}{4}\right)\frac{(\beta - 1)}{L}$$

$$\beta = \frac{4LU}{Rv_{z,0}} + 1$$

## Problem 18

**We have a system that comprises of two concentric cylinders which are both rotating. The inner cylinder has a radius of $R_a$ and rotates at an angular frequency of $\omega_a$. The outer cylinder has a radius of $R_b$ and rotates at an angular frequency of $\omega_b$. There is a very viscous fluid contained in the annular region between the two cylinders, which is Newtonian.**

(a) **Show that the linear velocity of this fluid is as shown below:**

$$v_\theta = \left(\frac{R_b{}^2\omega_b - R_a{}^2\omega_a}{R_b{}^2 - R_a{}^2}\right)\left(r - \frac{R_b{}^2}{r}\right) + \frac{R_b{}^2\omega_b}{r}$$

(b) **Information about velocity profile is useful in many practical applications as it helps us relate to useful properties such as shear stress and force, which can then be used to find out fluid viscosities through measured quantities. We will use the velocity profile for $v_\theta$ obtained in a, to show how a viscometer can be used to determine fluid viscosity. You are told that the measured torque of the inner cylinder is 0.003 Nm, whereby the inner cylinder rotates at four revolutions per second and has a diameter of 0.05 m and the outer cylinder is stationary with a diameter of 0.07 m. The fluid in the annular region is up to a depth of 0.04 m.**

## Solution 18

*Worked Solution*

(a)

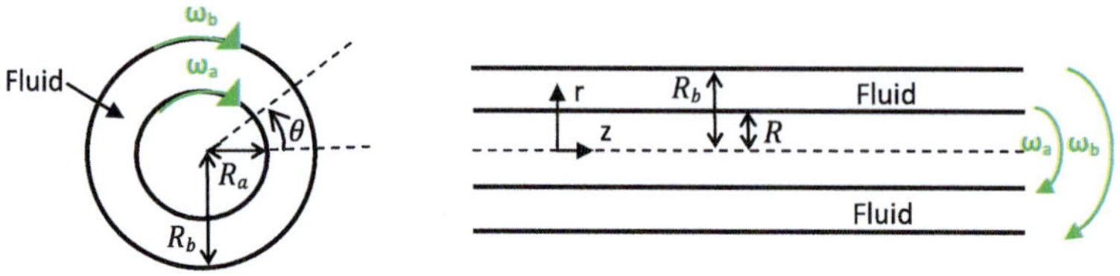


To find the velocity profile, let us examine the continuity equation and Navier–Stokes equation. To simplify the equations we may assume the following:

- The flow is axisymmetric; therefore $\frac{\partial}{\partial\theta} = 0$.
- The dominant flow direction is in the $\theta$ direction; hence we assume that $v_r = v_z = 0$.
- Constant viscosity for a Newtonian fluid.
- The liquid is incompressible, $\rho$ is constant.
- The flow is at steady state ($\frac{\partial}{\partial t} = 0$).
- The fluid is described as very viscous; hence viscous terms dominate, and inertial terms may be assumed negligible.

So we end up with the following simplified continuity equation

$$\frac{1}{r}\frac{\partial(rv_r)}{\partial r}+\frac{1}{r}\frac{\partial v_\theta}{\partial \theta}+\frac{\partial v_z}{\partial z}=0$$

$$\frac{\partial v_\theta}{\partial \theta}=0$$

This result makes sense as it means the fluid flow is axisymmetric. It also means that that flow is fully developed in the $\theta$ direction. We can therefore deduce that

$$v_\theta = v_\theta(r,z)$$

Let us now examine the $\theta$-component Navier–Stokes equation

$$0=\frac{\partial}{\partial r}\left(\frac{1}{r}\frac{\partial}{\partial r}(rv_\theta)\right)$$

The $r$-component equation is as follows

$$-\frac{\rho {v_\theta}^2}{r}=-\frac{\partial P}{\partial r}$$

The $z$-component equation is as follows

$$0=-\frac{\partial P}{\partial z}-\rho g$$

We will now integrate the $\theta$-component differential equation as we want to find out velocity field for $v_\theta$.

$$0=\frac{\partial}{\partial r}\left(\frac{1}{r}\frac{\partial}{\partial r}(rv_\theta)\right)$$

$$\frac{1}{r}\frac{\partial}{\partial r}(rv_\theta)=c_1$$

$$v_\theta=\frac{c_1 r}{2}+\frac{c_2}{r}$$

We can now apply boundary conditions to solve for the velocity in. When $r=R_a$, $v_\theta=R_a\omega_a$ and when $r=R_b$, $v_\theta=R_b\omega_b$.

$$R_a\omega_a = \frac{c_1 R_a}{2} + \frac{c_2}{R_a} \rightarrow {R_a}^2\omega_a = \frac{c_1 {R_a}^2}{2} + c_2$$

$$R_b\omega_b = \frac{c_1 R_b}{2} + \frac{c_2}{R_b} \rightarrow {R_b}^2\omega_b = \frac{c_1 {R_b}^2}{2} + c_2$$

$${R_b}^2\omega_b - {R_a}^2\omega_a = \frac{c_1 {R_b}^2}{2} - \frac{c_1 {R_a}^2}{2}$$

$$c_1 = \frac{2\left({R_b}^2\omega_b - {R_a}^2\omega_a\right)}{{R_b}^2 - {R_a}^2}$$

$$c_2 = {R_b}^2\omega_b - \frac{{R_b}^2\left({R_b}^2\omega_b - {R_a}^2\omega_a\right)}{{R_b}^2 - {R_a}^2}$$

Substituting the constants back into the expression, we obtain the velocity profile.

$$v_\theta = \frac{{R_b}^2\omega_b - {R_a}^2\omega_a}{{R_b}^2 - {R_a}^2} r + \frac{{R_b}^2\omega_b}{r} - \frac{{R_b}^2\left({R_b}^2\omega_b - {R_a}^2\omega_a\right)}{\left({R_b}^2 - {R_a}^2\right)r}$$

$$v_\theta = \left(\frac{{R_b}^2\omega_b - {R_a}^2\omega_a}{{R_b}^2 - {R_a}^2}\right)\left(r - \frac{{R_b}^2}{r}\right) + \frac{{R_b}^2\omega_b}{r}$$

(b) From the data booklet, we can find expressions for different shear stress components. Torque in this case is used to overcome the shear stress contributed by the $\tau_{r\theta}$ component at $r = R_a$.

$$\tau_{r\theta} = \eta\left[r\frac{\partial}{\partial r}\left(\frac{v_\theta}{r}\right) + \frac{1}{r}\frac{\partial v_r}{\partial \theta}\right]$$

Due to axisymmetry which was earlier established and still holds, $\frac{\partial v_r}{\partial \theta} = 0$.

$$\tau_{r\theta} = \eta\left[r\frac{\partial}{\partial r}\left(\left(\frac{{R_b}^2\omega_b - {R_a}^2\omega_a}{{R_b}^2 - {R_a}^2}\right)\left(1 - \frac{{R_b}^2}{r^2}\right) + \frac{{R_b}^2\omega_b}{r^2}\right)\right]$$

$$\tau_{r\theta} = \eta r\left[\left(2\left(\frac{{R_b}^2\omega_b - {R_a}^2\omega_a}{{R_b}^2 - {R_a}^2}\right)\frac{{R_b}^2}{r^3}\right) - \frac{2{R_b}^2\omega_b}{r^3}\right]$$

$$\tau_{r\theta} = \eta\left[\frac{2{R_b}^2{R_a}^2(\omega_b - \omega_a)}{\left({R_b}^2 - {R_a}^2\right)r^2}\right]$$

At $r = R_a$ we can find the relevant shear stress component. Note that $\omega_b$ in this case is zero since the outer cylinder is stationary.

$$\tau_{r\theta}\big|_{r=R_a} = \eta\left[\frac{-2{R_b}^2\omega_a}{\left({R_b}^2 - {R_a}^2\right)}\right]$$

Shear force $F$ is related to shear stress and torque $T$ is related to the shear force which it has to overcome. Note that the force expressed below is the force exerted by the inner cylinder on the fluid as is apparent from the direction of $\tau_{r\theta}$.

$$dF_{\text{cylinder on fluid}} = \left(\tau_{r\theta}\big|_{r=R_a}\right)dA = \left(\tau_{r\theta}\big|_{r=R_a}\right)(H(R_a d\theta))$$

$$dT_{\text{cylinder on fluid}} = R_a dF = \left(\tau_{r\theta}\big|_{r=R_a}\right)\left(H\left({R_a}^2 d\theta\right)\right)$$

$$dF_{\text{fluid on cylinder}} = -dF_{\text{cylinder on fluid}}$$

$$dT_{\text{fluid on cylinder}} = \eta\left[\frac{2{R_b}^2\omega_a}{\left({R_b}^2 - {R_a}^2\right)}\right]H\left({R_a}^2 d\theta\right)$$

To find the total torque exerted by the fluid on the cylinder, we need to integrate over the entire cylinder surface

$$T_{\text{fluid on cylinder}} = \int_0^{2\pi}\eta\left[\frac{2{R_b}^2\omega_a}{\left({R_b}^2 - {R_a}^2\right)}\right]H{R_a}^2 d\theta = \frac{4\pi\eta H{R_a}^2{R_b}^2\omega_a}{{R_b}^2 - {R_a}^2}$$

Now it is straightforward to solve for $T_{\text{fluid on cylinder}}$ by substituting the values of the variables that have been given in the problem.

$$\left|T_{\text{fluid on cylinder}}\right| = 0.003 \text{ Nm}$$

$$\omega_a = 4 \text{ rev/s} = 8\pi \text{ rad/s}$$

$$\omega_b = 0$$

$$R_a = \frac{0.05}{2} = 0.025 \text{ m}$$

$$R_b = \frac{0.07}{2} = 0.035 \text{ m}$$

$$H = 0.04 \text{ m}$$

## Problem 19

**Consider diverging flow of a very viscous fluid through a channel as shown in the diagram below. This channel is the narrow gap between two large parallel plates of width *W*. The top plate leads on to a downstream portion that has a**

**wedge shape that diverges the flow at a small angle of $\theta$. You may assume that the fluid is Newtonian and at steady state.**

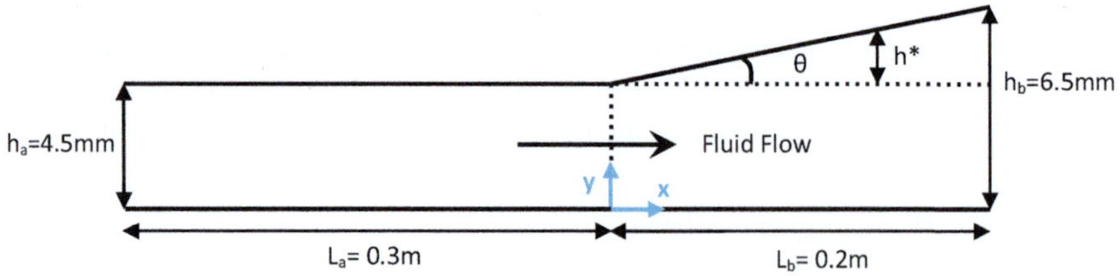


(a) **Show that the pressure difference across the diverging region $L_b$ is as shown below.**

$$\Delta P_b = \frac{6\eta\dot{Q}}{W\theta}\left(\frac{1}{{h_a}^2} - \frac{1}{{h_b}^2}\right)$$

(b) **Calculate the total pressure difference across both $L_a$ and $L_b$ if the flow rate per unit width of plate is 0.0003 m$^2$/s and the liquid has a viscosity of 25 Pa.s.**

**Solution 19**

*Worked Solution*

(a) We can apply the mass continuity and Navier–Stokes momentum equations here. Let us first list down our assumptions for this flow.

- Constant viscosity for a Newtonian fluid.
- The liquid is incompressible, $\rho$ is constant.
- The flow is at steady state ($\frac{\partial}{\partial t} = 0$).
- The angle of divergence at region $L_b$ is small. Hence in this region, $v_y$ also assumed negligible. Also small angle approximation for trigonometric functions may be used.
- The dominant flow direction is in the $x$ direction; hence we assume that $v_y = v_z = 0$
- Given that the plates are large, we may assume the flow is fully developed in the $x$ direction and any entrance/end effects may be ignored. Hence $v_x \neq v_x(x)$.
- Gravitational effects are ignored.
- The fluid is described as very viscous; hence viscous terms dominate, and inertial terms may be assumed negligible.

Now we may obtain the following simplified continuity equation

$$\frac{\partial v_x}{\partial x} = 0$$

This result is consistent with the fully developed flow assumption, and we deduce that $v_x = v_x(y)$. The $x$-component Navier–Stokes equation is simplified to the following:

$$0 = -\frac{\partial P}{\partial x} + \eta\left(\frac{\partial^2 v_x}{\partial y}\right)$$

$$v_x = \frac{1}{2\eta}\frac{\partial P}{\partial x}y^2 + c_1 y + c_2$$

Let us consider boundary conditions to evaluate the integration constants. When $y = 0$, $v_x = 0$. When $y = h^*$, $v_x = 0$.

$$h^* = h_a + x\tan\theta \approx h_a + x\theta$$

$$c_2 = 0$$

$$0 = \frac{1}{2\eta}\frac{\partial P}{\partial x}(h_a + x\theta)^2 + c_1(h_a + x\theta) \rightarrow c_1 = -\frac{1}{2\eta}\frac{\partial P}{\partial x}(h_a + x\theta)$$

$$v_x = \frac{1}{2\eta}\frac{\partial P}{\partial x}y^2 - \frac{1}{2\eta}\frac{\partial P}{\partial x}(h_a + x\theta)y = \frac{1}{2\eta}\frac{\partial P}{\partial x}\left[y^2 - (h_a + x\theta)y\right]$$

Now that we have found the velocity profile in the $x$ direction, we can find the volumetric flow rate $\dot{Q}$

$$\frac{\dot{Q}}{W} = \int_0^{(h_a+x\theta)} v_x dy = \frac{1}{2\eta}\frac{\partial P}{\partial x}\int_0^{(h_a+x\theta)}\left[y^2 - (h_a + x\theta)y\right]dy$$

$$\frac{\dot{Q}}{W} = \frac{1}{2\eta}\frac{\partial P}{\partial x}\left[\frac{y^3}{3} - (h_a + x\theta)\frac{y^2}{2}\right]_0^{(h_a+x\theta)} = -\frac{1}{12\eta}\frac{\partial P}{\partial x}(h_a + x\theta)^3$$

$$\frac{\partial P}{\partial x} = -\frac{12\eta\dot{Q}}{W(h_a + x\theta)^3}$$

Let us now consider the boundary conditions for pressure to evaluate this differential equation. The limits of integration for region $L_b$ are $x = 0$ and $x = L_b$. However we note that the expression we need to prove is in terms of $h_a$ and $h_b$; hence we may rewrite $L_b$ in terms of $h_a$ and $h_b$. Note also that $\frac{\partial P}{\partial x}$ is a negative value; hence we need to account for the sign change to express the magnitude (positive value) of pressure drop $\Delta P_b$.

$$L_b = \frac{h_b - h_a}{\tan\theta} \approx \frac{h_b - h_a}{\theta}$$

$$\Delta P_b = -\frac{\partial P}{\partial x} = \frac{12\eta\dot{Q}}{W}\int_0^{\frac{h_b-h_a}{\theta}} (h_a + x\theta)^{-3}dx$$

$$\Delta P_b = \frac{6\eta\dot{Q}}{W}\left[\frac{-1}{\theta(h_a + x\theta)^2}\right]_0^{\frac{h_b-h_a}{\theta}} = \frac{6\eta\dot{Q}}{W}\left[-\frac{1}{\theta {h_b}^2} + \frac{1}{\theta {h_a}^2}\right]$$

$$\Delta P_b = \frac{6\eta\dot{Q}}{W\theta}\left(\frac{1}{{h_a}^2} - \frac{1}{{h_b}^2}\right)$$

(b) We are given that $\frac{\dot{Q}}{W} = 0.0003\ \mathrm{m^2/s}$. $\eta = 25$ Pas. $L_a = 0.3$ m, $L_b = \frac{h_b - h_a}{\theta} = 0.2$ m. $h_a = 4.5$ mm and $h_b = 6.5$ mm. Note that due to mass continuity, the mass flow rate is the same for sections $L_a$ and $L_b$. This translates into the same volumetric flow rate due to the same constant density $\rho$ for the same fluid flowing through both sections.

For the region $L_a$, $\theta = 0$; therefore, we can find an expression for $\frac{\dot{Q}}{W}$ by integrating the velocity profile $v_x$ for this region

$$v_x|_{L_a} = \frac{1}{2\eta}\frac{\partial P}{\partial x}\left[y^2 - (h_a + x\theta)y\right] = \frac{1}{2\eta}\frac{\partial P}{\partial x}\left[y^2 - h_a y\right]$$

$$\left.\frac{\dot{Q}}{W}\right|_{L_a} = \frac{1}{2\eta}\frac{\partial P}{\partial x}\int_0^{h_a}(y^2 - h_a y)dy = \frac{1}{2\eta}\frac{\partial P}{\partial x}\left[\frac{y^3}{3} - \frac{h_a y^2}{2}\right]_0^{h_a}$$

$$\left.\frac{\dot{Q}}{W}\right|_{L_a} = -\frac{1}{12\eta}\frac{\partial P}{\partial x}{h_a}^3$$

$$-\left.\frac{\partial P}{\partial x}\right|_{L_a} = \frac{\Delta P_a}{L_a} = \frac{12\eta}{{h_a}^3}\left(\left.\frac{\dot{Q}}{W}\right|_{L_a}\right)$$

$$\Delta P_a = \frac{0.3 \times 12 \times 25 \times 0.0003}{0.0045^3} = 3 \times 10^5\ \mathrm{Pa}$$

For the region $L_b$, we found from earlier

$$\Delta P_b = \frac{6\eta\dot{Q}}{W\theta}\left(\frac{1}{{h_a}^2} - \frac{1}{{h_b}^2}\right) = \frac{6 \times 25 \times 0.0003}{\left(\frac{0.0065-0.0045}{0.2}\right)}\left(\frac{1}{0.0045^2} - \frac{1}{0.0065^2}\right)$$

$$\Delta P_b = 1.2 \times 10^5\ \mathrm{Pa}$$

Therefore, combining both results, we have the total pressure drop

$$\Delta P_{\text{total}} = \Delta P_a + \Delta P_b = 4.2 \times 10^5 \text{ Pa}$$

## Problem 20

**Consider a pipe bend as shown below. The pipe may be assumed smooth and axially horizontal. A fluid flows through the pipe at ambient conditions. The pipe diameter is 8 cm, its total length is 2 m and the mean fluid velocity is 2 m/s. The internal (fluid) pressures at the entry and exit points of the pipe bend are $P_{\text{in,internal fluid}} = P_{\text{atm}}$ + 450 Pa and $P_{\text{out, internal fluid}} = P_{\text{atm}}$. The fluid density is 1050 kg/m$^3$, its kinematic viscosity is 0.8 $\times$ 10$^{-6}$ m$^2$/s and $P_{\text{atm}} = 10^5$ Pa.**

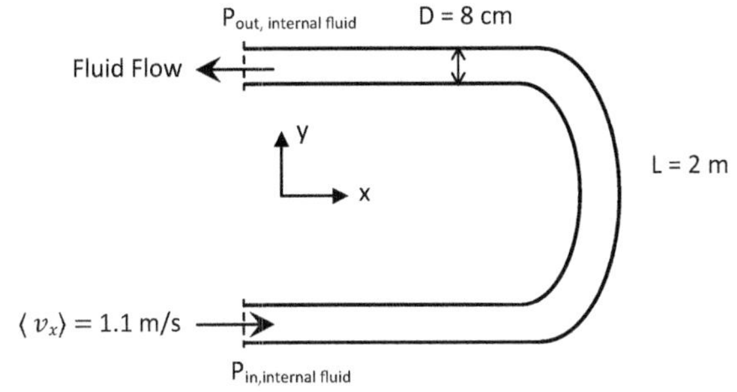


(a) **Find the $x$ component of the force exerted by the fluid on the pipe. Then find the total pressure on the pipe contributed by both the force by the fluid and by the surrounding air.**
(b) **Determine the viscous loss in the pipe, and determine the $K_f$ value for the bend.**

## Solution 20

***Worked Solution***

(a) Before we tackle this problem, let us recall key concepts about pipeline losses.

---

***Background Concepts***

The energy losses in a pipeline network can be contributed by straight sections and other sections (typically called fittings) such as bends, expansions/contractions, etc. Mathematically, this may be expressed as shown below, whereby subscript $i$ refers to the lengths of straight pipe sections and fittings which are summed over. In this case, we are assuming no gravitational effects, i.e., pipe is on a horizontal plane. Also, we assume that no shaft work is done and the fluid is incompressible.

$$\text{Viscous loss } l_v = \sum_{\text{straight lengths}} \frac{2\langle v\rangle_i^2 L_i f_i}{D_i} + \sum_{\text{fittings}} \frac{1}{2}\langle v\rangle_i^2 K_{fi}$$

The derivation of the above expression is done via a macroscopic force balance using the Bernoulli equation.

$$\frac{p_1 - p_2}{\rho} = l_v$$

Separately we know that friction factor $f$ is a ratio between shear stress at the pipe wall and the bulk kinetic energy, $f = \frac{\tau_w}{\frac{1}{2}\rho\langle v\rangle^2}$ or $\tau_w = \frac{1}{2}\rho\langle v\rangle^2 f$.

$$(p_1 - p_2)(\pi D^2/4) = \tau_w(\pi DL) = \left(\frac{1}{2}\rho\langle v\rangle^2 f\right)(\pi DL)$$

$$\frac{p_1 - p_2}{\rho} = \frac{2fL\langle v\rangle^2}{D}$$

Pressure drop is used to overcome shear forces caused by friction between the fluid and the pipe wall. We can therefore express pressure drop in terms of friction factor, in order to relate frictional/viscous losses with friction factor $f$ and other pipeline-related parameters (e.g., length and diameter) that are either known or measurable.

$$l_v = \frac{2fL\langle v\rangle^2}{D}$$

This is how we arrived at the earlier expression, for the viscous loss component due to straight lengths of pipe, $\sum_{\text{straight lengths}} \frac{2\langle v\rangle_i^2 L_i f_i}{D_i}$.

As for the viscous loss term for pipe fittings, we have simply further "grouped" the pipeline-related parameters (e.g., $f$, $L$, $D$) into a new parameter denoted by $K_f = 4Lf/D$ which can be evaluated from pipe specifications. $K_f$ is defined as the number of "velocity heads" lost by flow through a fitting, and we can usually look up known values of $K_f$ from data booklets, for specified fitting types.

$$l_{v,\text{fitting}} = \frac{2fL\langle v\rangle^2}{D} = \frac{1}{2}\langle v\rangle^2\left[\frac{4Lf}{D}\right] = \frac{1}{2}\langle v\rangle^2 K_f$$

To find the force by the fluid on the pipe, we can start with understanding the conservation of linear momentum for a control volume. In defining the control volume for a $2D$ flow, it is useful to define two reference points represented by taking the surface normal to the flow direction. For entry and exit points, we then get two surfaces of reference which we may denote 1 and 2 here. Note that linear momentum is a vector and linear momentum per unit mass is the velocity vector. Hence the conservation equation is such that:

$$\begin{array}{c}\text{Rate of change}\\\text{of linear}\\\text{momentum in}\\\text{control volume}\end{array} = \begin{array}{c}\text{Linear momentum}\\\text{inflow rate}\end{array} - \begin{array}{c}\text{Linear momentum}\\\text{outflow rate}\end{array} + \begin{array}{c}\text{All forces acting on}\\\text{system (e.g. pressure,}\\\text{gravitational etc.)}\end{array}$$

In mathematical form, the general form of the equation is as follows

$$\frac{d}{dt}\int_1^2 \rho\langle \boldsymbol{v}\rangle A dS = \rho_1\langle \boldsymbol{v}V\rangle_1 A_1 - \rho_2\langle \boldsymbol{v}V\rangle_2 A_2 + p_1\boldsymbol{A}_1 - p_2\boldsymbol{A}_2 - \boldsymbol{F} + \left(\int_1^2 \rho A dz\right)\boldsymbol{g}$$

It is useful to note the following key points about this equation:

- All terms on right-hand side of equation has a positive sign for direction into the control volume.
- The inflow term has two different notations for velocity, where $V$ is the magnitude of velocity vector $\boldsymbol{v}$
- $F$ is the net force by the fluid on the solid surface (e.g., pipe wall). Hence a negative sign is added to the term to change it in terms of direction into control volume; hence under this frame of reference it will be a negative value represented by "$-\boldsymbol{F}$".
- $p_1\boldsymbol{A}_1$ is the force by the ambient fluid (e.g., air) outside, acting on fluid inside the control volume at position 1. Therefore it is a positive sign acting into the control volume. Conversely, $p_2\boldsymbol{A}_2$ acts out of the control volume; hence a negative sign is added to give a negative value for the term "$-p_2\boldsymbol{A}_2$".

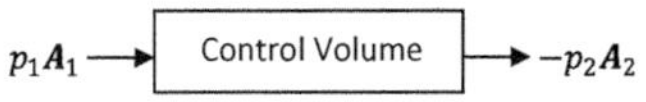


- $\boldsymbol{A}$ is defined as a vector with magnitude equal to total surface area over which the force acts and has a direction normal to the planar surface in the direction of mean flow.
- $\left(\int_1^2 \rho A dz\right)$ represents the total mass of the fluid in the control volume and $\boldsymbol{g}$ refers to the gravitational acceleration

Let us now make three common simplifications, so that we arrive at the form of the equation that is more useful.

1. When the velocity vector $\boldsymbol{v}$ is normal to the cross section over the entire section from entrance to exit of fluid. We can then substitute $\boldsymbol{v}$ with $V$.

$$\frac{d}{dt}\int_1^2 \rho\langle \boldsymbol{v}\rangle A dS = \rho_1\langle V^2\rangle_1 \boldsymbol{A}_1 - \rho_2\langle V^2\rangle_2 \boldsymbol{A}_2 + p_1\boldsymbol{A}_1 - p_2\boldsymbol{A}_2 - \boldsymbol{F} + \left(\int_1^2 \rho A dz\right)\boldsymbol{g}$$

2. We define a ratio $\beta = \langle V^2\rangle/\langle V\rangle^2$ which is useful in differentiating flow regimes, where $\beta \approx 1$ when flow is turbulent and $\beta = 4/3$ when flow is laminar (pipeline). We can therefore express the equation in terms of $\beta$.

$$\frac{d}{dt}\int_1^2 \rho\langle \boldsymbol{v}\rangle A dS = \beta\rho_1\langle V\rangle^2{}_1 \boldsymbol{A}_1 - \beta\rho_2\langle V\rangle^2{}_2 \boldsymbol{A}_2 + p_1\boldsymbol{A}_1 - p_2\boldsymbol{A}_2 - \boldsymbol{F} + \left(\int_1^2 \rho A dz\right)\boldsymbol{g}$$

3. For most simplified problems, we seldom deal with $\beta$ (i.e., assume turbulent regime after checking Re) and assume the same fluid flows from entrance to exit. If we assume further that there is steady state and gravitational effects may be ignored, then we have

$$0 = \rho\langle V\rangle^2{}_1 \boldsymbol{A}_1 - \rho\langle V\rangle^2{}_2 \boldsymbol{A}_2 + p_1\boldsymbol{A}_1 - p_2\boldsymbol{A}_2 - \boldsymbol{F}$$

---

For our problem, we can first check if our assumption of turbulent flow is correct by calculating Re

$$\mathrm{Re} = \frac{\langle v_x\rangle D}{\nu} = \frac{1.1(0.08)}{0.8\times 10^{-6}} = 1\times 10^5 \rightarrow \beta = 1$$

Now we return to the equation which has applied $\beta = 1$

$$\rho\langle V\rangle^2{}_1 \boldsymbol{A}_1 - \rho\langle V\rangle^2{}_2 \boldsymbol{A}_2 + p_1\boldsymbol{A}_1 - p_2\boldsymbol{A}_2 = \boldsymbol{F}$$

Since our pipe is of uniform size and contains the same fluid throughout, $\langle V\rangle_1 = \langle V\rangle_2 = \langle V\rangle$ and $A_1 = A_2 = A$. Also, we only require the $x$-component force; therefore, we need to correct $\boldsymbol{A}$ at the exit point for the terms $\rho\langle V\rangle^2{}_2\boldsymbol{A}_2$ and $p_2\boldsymbol{A}$ as $\boldsymbol{A}$ has an opposite direction in the $-x$ direction at the exit of the pipe bend.

$$\rho\langle V\rangle^2 A - \rho\langle V\rangle^2(-A) + p_1A - p_2(-A) = F_x$$

$$F_x = A\left[2\rho\langle V\rangle^2 + p_1 + p_2\right]$$

Note that in our problem, we need to also calculate the pressure by the ambient air (another fluid) on the pipe. For this, we need to note that control volume has changed to the body of air. Hence the contribution at the entrance is $-p_{atm}A$ since it acts in a direction going out of the control volume of air (on the pipe), and at the exit it is the same, $-p_{atm}A$. For this case, $\boldsymbol{A}$ is already in the $+x$ direction (i.e., normal to the cross-sectional area) in the direction of flow (air "flow direction" acts to push against the pipe towards the right in the $+x$ direction).

Therefore, we have the total $x$ component of the force on pipe as follows:

$$F_{\text{on pipe}} = A\left[2\rho\langle V\rangle^2 + p_1 + p_2\right] - 2p_{atm}A$$

where $A = \frac{\pi D^2}{4}$

$$F_{\text{on pipe}} = \frac{\pi(0.08)^2}{4}\left[2(1050)\left(1.1^2\right) + 450\right] = 15\ \text{N}$$

(b) To find viscous loss, we can first look at the general Bernoulli equation

$$\Delta\left(\frac{\alpha}{2}\langle V\rangle^2 + gh + \frac{p}{\rho}\right) = \delta W_s - l_v$$

where $\alpha = \langle V^3\rangle/\langle V\rangle^3$ which is useful in differentiating flow regimes, where $\alpha \approx 1$ when there is turbulent pipe flow and $\alpha = 2$ when flow is laminar in a round pipe.

For our problem, we can simplify the equation to the following as the pipe size (hence fluid velocity via mass continuity) is the same from entrance to exit, and the pipe is horizontal throughout the fluid flow.

$$-\Delta\left(\frac{p}{\rho}\right) = l_v = \frac{p_{\text{in, internal fluid}} - p_{\text{out, internal fluid}}}{\rho}$$

$$l_v = \frac{450}{1050} = 0.4\ \text{m}^2\text{s}^{-2}$$

This value of viscous loss consists of losses in the straight sections and losses in the bend.

$$0.4 = l_{v,\text{straight}} + l_{v,\text{bend}} = \frac{2\langle v\rangle^2 Lf}{D} + \frac{1}{2}\langle v\rangle^2 K_f$$

For our flow problem, the friction factor for the straight sections of pipe can be found by using the appropriate correlation between $f$ and Re. For our flow parameters, $\text{Re} = 1 \times 10^5$; hence we can use the Blasius equation that is valid for $4000 \le \text{Re} \le 1 \times 10^5$.

$$f = 0.079\text{Re}^{-1/4} = 4.4 \times 10^{-3}$$

$$l_{v,\text{straight}} = \frac{2 \times 1.1^2 \times 2 \times 4.4 \times 10^{-3}}{0.08} = 0.27$$

$$l_{v,\text{bend}} = 0.4 - 0.27 = 0.13 = \frac{1}{2}\langle v \rangle^2 K_f$$

$$K_f = \frac{2 \times 0.13}{1.1^2} = 0.2$$

## Problem 21

**Consider two large circular discs of radius $R$ each and axially aligned. Both are held horizontally with a small gap of height $H$ separating their surfaces. A viscous Newtonian fluid fills this gap between the discs. The top disc is rotated at an angular velocity $\omega$ while the lower disc is stationary.**

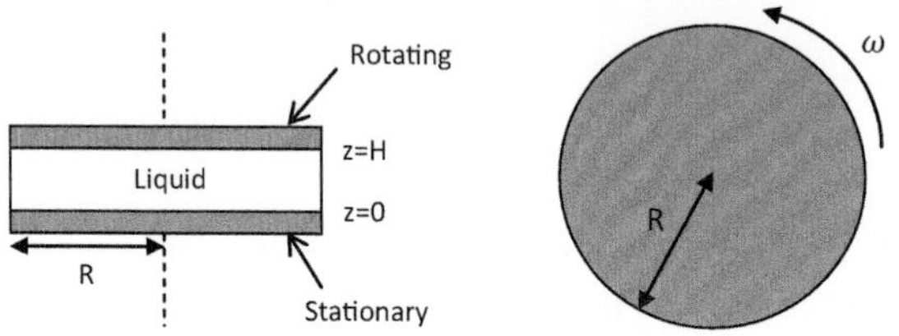


(a) **Assuming creeping flow, show that $v_\theta(r,z) = \omega r F(z)$ satisfies the Stoke's equation if the function $F(z)$ is appropriately defined. Find the differential equation and boundary conditions that constrain $F(z)$ and solve $F(z)$.**
(b) **Find the torque that is required to be applied to the upper disk to ensure steady rotation.**

## Solution 21

### *Worked Solution*

(a) Before we begin solving this problem, let us recall basic concepts about a particular flow regime called Creeping Flow.

---

### *Background Concepts*

In fluid mechanics, we may refer to a particular flow regime called Stokes flow whereby Re is small, typically <1. In this region inertia effects are assumed negligible compared to viscous forces. [In fact, Reynolds number, Re, represents a dimensionless group that may be interpreted as the ratio of inertial to viscous forces]

This phenomenon is also referred to as creeping flow, and appropriate approximations can be made to simplify computation.

As for Newtonian fluids, we have the following Navier–Stokes equation when inertial effects are assumed negligible and it is also called Stoke's equation (dominated by viscosity) used for creeping flow situations.

$$0 = -\nabla P + \eta \nabla^2 \boldsymbol{v}$$

---

Now we can return to this problem. To find *F*(*z*), let us examine the $\theta$ component Navier–Stokes equation in cylindrical coordinates.

Our assumptions are as follows:

- Constant viscosity for a Newtonian fluid.
- The liquid is incompressible, $\rho$ is constant.
- The system is axisymmetric; hence $\frac{\partial}{\partial \theta} = 0$.
- The flow is at steady state ($\frac{\partial}{\partial t} = 0$).
- The dominant flow direction is in the $\theta$ direction; hence we assume that $v_r = v_z = 0$.
- Given that the discs are large, we may assume the flow is fully developed in the $\theta$ direction and any entrance/end effects may be ignored. Hence $v_\theta \neq v_\theta(\theta)$.
- Gravitational effects are ignored.
- The fluid is described as viscous; hence viscous terms dominate, and inertial terms may be assumed negligible, i.e., creeping flow.

After assuming the above, the Navier–Stokes equation becomes

$$0 = \frac{\partial}{\partial r}\left(\frac{1}{r}\frac{\partial}{\partial r}(r v_\theta) + \frac{\partial^2 v_\theta}{\partial z^2}\right)$$

Given that $v_\theta(r, z) = \omega r F(z)$,

$$0 = \frac{\partial}{\partial r}\left(\frac{1}{r}\frac{\partial}{\partial r}(\omega r^2 F(z)) + \omega r \frac{\partial^2 F}{\partial z^2}\right)$$

$$0 = \frac{\partial}{\partial r}\left(2\omega F(z) + \omega r \frac{\partial^2 F}{\partial z^2}\right) = \frac{\partial}{\partial r}\left(2F(z) + r\frac{\partial^2 F}{\partial z^2}\right)$$

$$\frac{\partial^2 F}{\partial z^2} = 0$$

$$F = c_1 z + c_2$$

To solve this differential equation, we require boundary conditions. We know that when $z = 0$ at any $r$, $v_\theta = 0$, $F = c_2 = \frac{v_\theta}{\omega r} = 0$. When $z = H$ at any $r$, $v_\theta = r\omega$, $F = c_1 H = 1 \rightarrow c_1 = \frac{1}{H}$

$$F(z) = \frac{z}{H}$$

$$v_\theta(r, z) = r\omega\left(\frac{z}{H}\right)$$

(b) The torque $T$ required to maintain steady rotation is to overcome shear stress $\tau$

$$dT = r\left(\tau_{z\theta}\big|_{z=H}\right)(2\pi r dr)$$

$$T = 2\pi\int_0^R r^2\left(\tau_{z\theta}\big|_{z=H}\right)dr$$

We can find the expression for $\tau_{z\theta}|_{z\,=H}$ from data booklets

$$T = 2\pi\int_0^R r^2\left(\eta\left(\frac{\partial v_\theta}{\partial z}\right)\right)dr = \frac{2\pi\eta\omega}{H}\left(\frac{R^4}{4}\right)$$

$$T = \frac{\pi\eta\omega R^4}{2H}$$

## Problem 22

**Consider two large flat parallel planes containing a liquid within the separation space between the planes of distance $h$ apart. The upper plate is rotating about the $z$-axis with a uniform angular velocity $\omega$. You may assume that the velocity $v_\theta$is the dominant flow direction and flow is axisymmetric. Gravitational effects are assumed negligible.**

(a) **Show that the pressure gradient in $z$ direction is zero.**
(b) **Referring to the $r$-component Navier–Stokes equation, comment on the velocity profile $v_\theta$.**
(c) **Using the $\theta$-component Navier–Stokes equation, and assuming creeping flow, determine the velocity profile in the $\theta$ direction assuming the form $v_\theta = rF(z)$, where $F(z)$ is a function of $z$.**
(d) **If both planes are in fact circular discs of radius $R$ each, find the torque required on each disc.**

## Solution 22

*Worked Solution*

(a) We have a system that can be visualized in the diagrams below:

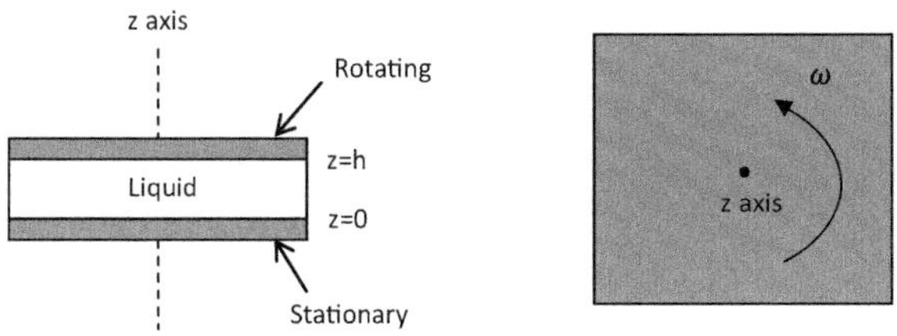


We have the following assumptions for this flow:

- Constant viscosity for a Newtonian fluid.
- The liquid is incompressible, $\rho$ is constant.
- The system is axisymmetric; hence $\frac{\partial}{\partial\theta}=0$.
- The flow is at steady state ($\frac{\partial}{\partial t}=0$).
- The dominant flow direction is in the $\theta$ direction; hence we assume that $v_r=v_z=0$
- Given that the planes are large, we may assume the flow is fully developed in the $\theta$ direction and any entrance/end effects may be ignored. Hence $v_\theta \neq v_\theta(\theta)$.
- Gravitational effects are ignored.

Consider the $z$-component Navier–Stokes equation simplified according to the assumptions above. Hence we note that pressure is independent of $z$.

$$0=-\frac{\partial P}{\partial z}$$

(b) Now we refer to the $r$-component Navier–Stokes equation, which can be simplified to the following.

$$\rho\frac{{v_\theta}^2}{r}=\frac{\partial P}{\partial r}$$

Note that if the conclusion from part a holds, i.e., pressure is independent of $z$, then it follows that $\frac{\partial P}{\partial r}$ is also not a function of $z$, and by the above equation, $v_\theta$ should also be independent of $z$. However, this is incorrect as $v_\theta$ varies with $z$ as we go from $z=0$ where $v_\theta=0$ to $z=h$, where $v_\theta\neq 0$.

(c) Creeping flow approximation means inertial effects are negligible as compared to viscous effects. From the $\theta$-component Navier–Stokes equation, this translates to

$$0=\frac{\partial}{\partial r}\left(\frac{1}{r}\frac{\partial}{\partial r}(rv_\theta)+\frac{\partial^2 v_\theta}{\partial z^2}\right)$$

Using the trial solution of the form $v_\theta=rF(z)$, we can find the differentials,

$$\frac{\partial}{\partial r}\left(\frac{\partial^2 v_\theta}{\partial z^2}\right) = F''$$

$$\begin{aligned}\frac{\partial}{\partial r}\left(\frac{1}{r}\frac{\partial}{\partial r}(rv_\theta)\right) &= \frac{\partial}{\partial r}(rv_\theta)\frac{\partial}{\partial r}\left(\frac{1}{r}\right) + \frac{1}{r}\left(\frac{\partial^2(rv_\theta)}{\partial r^2}\right) \\ &= \left[\frac{\partial}{\partial r}(r^2F)\right]\left[-\frac{1}{r^2}\right] + \frac{1}{r}\left(\frac{\partial^2(r^2F)}{\partial r^2}\right) = -\frac{2rF}{r^2} + \frac{2F}{r} \\ &= -\frac{2F}{r} + \frac{2F}{r}\end{aligned}$$

$$0 = -\frac{2F}{r} + \frac{2F}{r} + F'' = F''$$

$$F(z) = c_1 z + c_2$$

Applying boundary conditions, at any $r$ when $z = 0$, $v_\theta = rF = 0$, $F = 0$, and at any $r$ when $z = h$, $v_\theta = rF = r\omega$, so $F = \omega$. Therefore, $c_1 = \frac{\omega}{h}$ and $c_2 = 0$.

$$F(z) = \frac{\omega}{h}z \rightarrow v_\theta = \frac{r\omega}{h}z$$

(d) To find torque, we may first determine the shear stress it has to overcome. Note that in this case $\frac{\partial v_z}{\partial \theta} = 0$.

$$\tau_{z\theta} = \eta e_{z\theta} = \eta\left(\frac{\partial v_\theta}{\partial z} + \frac{1}{r}\frac{\partial v_z}{\partial \theta}\right) = \eta\left(\frac{\partial v_\theta}{\partial z}\right) = \eta\left(\frac{r\omega}{h}\right)$$

$$\text{Torque} = \int_0^R \eta\left(\frac{r\omega}{h}\right)r(2\pi r dr) = \frac{2\pi\eta\omega}{h}\left[\frac{r^4}{4}\right]_0^R = \frac{\pi\eta\omega R^4}{2h}$$

## Problem 23

**A team of engineers sought to find out the turbulent boundary layer for a fluid on a flat plate using particle-image velocimetry (PIV), whereby tiny particles seeded in the flow were used to study the speed and direction of the flow. The instantaneous velocity profile was made and the formation of a boundary layer was observed. The boundary layer was initiated at the point $x = 0$, with transition occurring at the lowest possible value of $Re_x$. The PIV measurement was done at a position $x_{PIV} = 1.2$ m downstream.**

**The flow of this fluid was at a steady rate of constant velocity $V_\infty = 0.5$ m/s and at constant pressure in a large channel that was 4 m by 0.4 m by 0.1 m. The fluid properties are $T = 290$ K, $\rho = 1000$ kg/m$^3$, viscosity $\eta = 1 \times 10^{-3}$ Pas and $P_\infty = 1 \times 10^5$ Pa.**

(a) **Show that the boundary layer at $x_{PIV}$ is turbulent.**

**(b) At $x_{\text{PIV}}$, find the estimated boundary layer thickness $\delta$ m, average friction factor $\bar{f}$ and average wall shear stress $\overline{\tau_w}$ Pa.**

**(c) The actual measurement of $\tau_w$ was 0.45 Pa. Comment on how this value compares with that obtained in (b). The friction velocity for the measured $\tau_w$ found from literature is $u_\tau = 0.021$ m/s. Is this value of $u_\tau$ consistent with the measured $\tau_w$?**

## Solution 23

### *Worked Solution*

(a) This problem helps us to understand some key concepts about the turbulent boundary layer.

---

### *Background Concepts*

A boundary layer (typically denoted as $\delta$) develops when a flow gradually adjusts from its no-slip condition at a solid surface to the flow in the free stream.

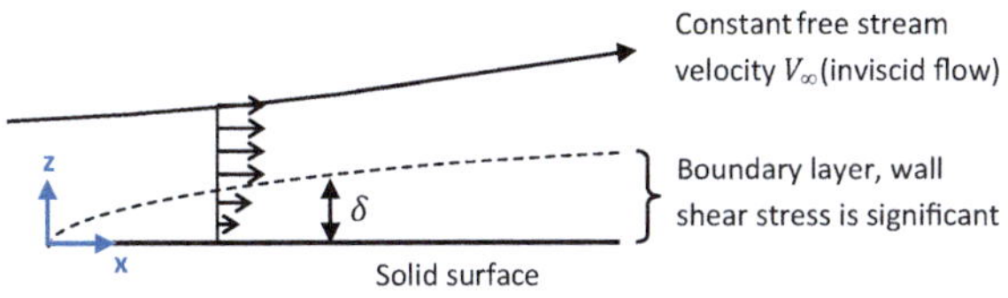


It has been noted that the use of constant-property relations (in laminar boundary layers) is in good agreement for turbulent boundary layers as well. To simplify calculations, the relationship between average friction factor $\bar{f}$ (or average friction coefficient $C_f$) can be established in the form of Blasius equations for turbulent boundary layer over a flat plate.

The solutions to the Navier–Stokes equations are mainly developed for laminar flow regimes. However, turbulent flows are erratic and characterized by random variations. When we measure a macroscopic property such as friction factor, we are in fact measuring a mean value, as the variations occur too rapidly for physical limits of our measurement methods. We can still apply NS equations to turbulent flows but we must be conscious that our solutions are only approximate solutions that are averaged over a sufficiently long time scale (larger that the characteristic time of fluctuations about the mean).

---

We are required to prove that the boundary layer at $x_{\text{PIV}}$ is turbulent. So we begin by calculating $\text{Re}_{\text{PIV}}$.

$$\mathrm{Re_{PIV}} = \frac{x_{\mathrm{PIV}} V_\infty}{\nu} = \frac{1.2 \times 0.5}{(\eta/\rho)} = 6 \times 10^5$$

For flat plates, the transition from laminar to turbulent occurs between $3 \times 10^5 < \mathrm{Re}_{tr} < 3 \times 10^6$. In our case, transition was triggered at the lowest possible Re; therefore since $\mathrm{Re_{PIV}} > \mathrm{Re}_{tr}$, the boundary layer will be turbulent at the PIV location.

Transition first occurs when Re reaches $3 \times 10^5$, this gives a downstream distance of $x_{tr}$

$$x_{tr} = \frac{(3 \times 10^5)\nu}{V_\infty} = 0.6\mathrm{m}$$

We know that $x_{\mathrm{PIV}} = 1.2$ m. Therefore a laminar boundary layer will occur between $x = 0$ and $x = 0.6$ m, and a turbulent boundary layer will occur from 0.6 m till the end of the surface plate. The PIV location occurs within the latter region.

(b) Turbulent boundary layer thickness is given by:

$$\frac{\delta}{L} = 0.37\mathrm{Re}^{-0.2}$$

Therefore we can find the thickness of our boundary layer

$$\delta = 0.37{\mathrm{Re_{PIV}}}^{-0.2} x_{\mathrm{PIV}} = 0.03 \text{ m}$$

We know further that average friction factor follows the correlation below, where $A = 1050$ for $\mathrm{Re}_{tr} = 3 \times 10^5$:

$$\bar{f} = \frac{0.455}{(\log\mathrm{Re})^{2.58}} - \frac{A}{\mathrm{Re}}$$

We can now calculate average friction factor, $\bar{f}$ and relate it to average wall shear stress

$$\bar{f} = \frac{0.455}{(\log\mathrm{Re_{PIV}})^{2.58}} - \frac{1050}{\mathrm{Re_{PIV}}} = 0.0032$$

$$\bar{f} = \frac{\overline{\tau_w}}{\frac{1}{2}\rho V_\infty^2}$$

$$\overline{\tau_w} = 0.0032\left[\frac{1000 \times 0.5^2}{2}\right] = 0.4 \text{ Pa}$$

(c) The actual measurement was 0.45 Pa while we obtained 0.4 Pa in part b. Note that 0.4 Pa refers to a mean value over the entire boundary layer length from $x = 0$ to $x = x_{\text{PIV}}$. However, the value of 0.45 Pa refers to a local value at the PIV location. As the boundary layer grows with downstream distance, the difference in $\tau_w$ values is expected.

Moreover, upstream of the PIV location, the turbulent boundary layer is still small. Hence when $\tau_w$ is averaged over the entire boundary layer length, its value is reduced by the inclusion of laminar boundary layer (more significant at the upstream portion) in our mean value. This is the reason for our calculated mean value in part b to be slightly lower than the actual local value.

By definition, friction velocity $u_\tau$ is given by

$$u_\tau = \sqrt{\frac{\tau_w}{\rho}} = \sqrt{\frac{0.45}{1000}} = 0.021 \text{ m/s}$$

Therefore we verify that the value of $u_\tau$ obtained from literature for this fluid is consistent with the measured $\tau_w$.

## Problem 24

**We have the following experimental data obtained from the pipe flow characteristics of two fluids, A and B. We are also given a chart that plots on a graph of $1/\sqrt{f}$ against $\log\left(Re\sqrt{f}\right)$ based on the friction factor (*f*) data obtained.**

| | Fluid A | Fluid B |
|---|---|---|
| Pipe diameter | 0.038 m | 0.038 m |
| Bulk velocity | 0.25 m/s | 0.55 m/s |
| Density | 1000 kg/m$^3$ | 1000 kg/m$^3$ |
| Kinematic viscosity | $1 \times 10^{-6}$ m$^2$/s | $2 \times 10^{-6}$ m$^2$/s |

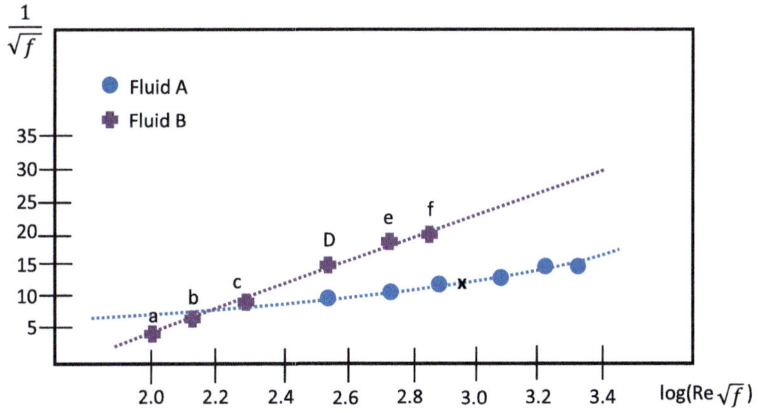

**(a) Two measuring stations are located at a distance of $x = 10$ m and $x = 20$ m downstream of the flow from the pipe entrance respectively. Determine if the flow is fully developed when the first station is reached.**

**(b) Find out the pressure drop $\Delta P$/Pa across the two measuring stations and determine the wall shear stress $\tau_w$ and friction velocity $u_\tau$.**

**(c) Using the data provided, verify if the data point marked by the black cross in the chart corresponds to that of fluid A.**

**(d) Identify which data point for fluid B corresponds to the data provided. Also find the percentage of drag reduction (%DR) for the identified point, where $f_A$ and $f_B$ refer to the friction factors for the two fluids.**

$$\%\mathbf{DR} = \left(1 - \frac{f_B}{f_A}\right)\Bigg|_{\mathbf{Re}}$$

**Solution 24**

*Worked Solution*

(a) Let us revisit some key concepts related to this problem before we begin solving.

---

*Background Concepts*

Note that the graph that is provided is a plot of Prandtl–Karman coordinates of $\frac{1}{\sqrt{f}}$ against $\log(\mathrm{Re}\sqrt{f})$. This plot is useful as it gives us straight line correlations between friction factor $f$ and Re especially in the turbulent region where this plot would give straight lines. [Recall that for turbulent flow when Re > 4000: $\frac{1}{\sqrt{f}} = 4.0\log(\mathrm{Re}\sqrt{f}) - 0.4$ as represented by the Karman–Nikuradse equation.]

Turbulent flow is often preferred for industrial processes due to its intense mixing. The key is to grasp the transition point from laminar to turbulent flow and experiments may be used to determine this as it also varies from apparatus to apparatus. The plot of $f$–Re is not as well defined experimentally; hence the Prandtl–Karman coordinates are often used for experimental plots instead. When Re < 2100, we have laminar flow. Between Re = 2100 and 4000, the transition point occurs. And when Re > 4000, we usually have fully developed turbulent flow.

In order to determine if the flow is developed upon reaching the first measuring station, we should identify the flow regime (i.e., laminar or turbulent) so that the appropriate correlation for fully developed verification can be used.

---

For fluid A, we have

$$\text{Re} = \frac{VD}{\nu} = \frac{0.038 \times 0.25}{10^{-6}} = 9500 > 4000$$

Entry length is a term used to describe a portion upstream near the entrance of the pipe (assuming for pipe flow) where the flow is still adjusting. Beyond the entry length, the flow is fully developed and the velocity profile steadies itself. Similarly this phenomenon can happen near the end of pipe, in which case it is called end effects. Entry and end effects are less significant when the Re number is high. Pressure drops caused by entry and end effects should be considered when very viscous liquids are used or during slow flow situation, and corrections should be made accordingly.

In our problem, as Re > 4000, the flow is turbulent. Therefore we may use the following correlation for finding entry length for our turbulent regime.

$$L_{\text{entry}} = 40D = 30 \times 0.038 = 1.52$$

Since the entry length of 1.52 m $\ll$ 10 m, the flow is fully developed at the location of the first measuring station.

(b) To find the pressure drop, we may recall that pressure drop is related to wall shear stress. This can be easily established from a force balance between pressure force and shear force which have to balance for steady flow.

$$\text{Pressure Force} = \text{Shear Force}$$

$$\Delta P\left(\frac{\pi D^2}{4}\right) = \tau_w(\pi DL)$$

$$\Delta P = 4\tau_w\left(\frac{L}{D}\right)$$

Separately, we know that shear stress is related to friction factor $f$ since $f$ is a ratio between shear stress at the pipe wall and the bulk kinetic energy; therefore

$$f = \frac{\tau_w}{\frac{1}{2}\rho\langle v\rangle^2}$$

$$\tau_w = \frac{1}{2}\rho\langle v\rangle^2 f$$

Therefore to find pressure drop, we need a relationship between Re (which we know) and $f$. After we determine $f$, we can find $\Delta P$ via $\tau_w$.

In our case, for fluid A, Re = 9500. The relevant correlation to find $f$ is the Blasius equation as follows, that is valid for $4000 \ll \text{Re} \leq 10^5$.

$$f = 0.079\text{Re}^{-\frac{1}{4}} = 0.079(9500)^{-\frac{1}{4}} = 0.008$$

$$\tau_w = \frac{1}{2}(1000)\left(0.25^2\right)(0.008) = 0.25\ \text{Pa}$$

$$\Delta P = 4\tau_w\left(\frac{L}{D}\right) = 4(0.25)\frac{(20-10)}{0.038} = 260\ \text{Pa}$$

$$u_\tau = \sqrt{\tau_w/\rho} = \sqrt{0.25/1000} = 0.016\ \text{m/s}$$

(c) Based on the calculations for fluid A, we should have the following coordinate values:

$$\frac{1}{\sqrt{f}} = \frac{1}{\sqrt{0.008}} = 11$$

$$\log\left(\text{Re}\sqrt{f}\right) = \log\left(9500 \times \sqrt{0.008}\right) = 2.9$$

The point at the black cross has coordinate values similar to the above as observed from the chart; hence the data point corresponds to fluid A.

(d) Let us calculate Re for fluid B.

$$\text{Re}_B = \frac{0.55 \times 0.038}{2 \times 10^{-6}} = 10450 > 9500$$

To find out which of the data points on the chart corresponds to our case for fluid B, we may first deduce that since $\text{Re}_B > \text{Re}_A$, then $f_B < f_A \left(or \frac{1}{\sqrt{f_B}} > \frac{1}{\sqrt{f_A}}\right)$. We can observe from the chart that points a, b and c are not possible.

We can find from the chart provided, the coordinates of points d, e and f.

For point d,

$$\frac{1}{\sqrt{f_B}} = 15$$

From this coordinate value, $f_B = \left(\frac{1}{15}\right)^2 = 0.0044$. This gives $\log(\text{Re}_B\sqrt{f_B}) = \log\left(10450\sqrt{0.0044}\right) = 2.84$.

$$\log\left(\text{Re}_B\sqrt{f_B}\right) = 2.55$$

The value of 2.84 does not correspond to the coordinate value of 2.55 found. Hence this point is incorrect.

For point e,

$$\frac{1}{\sqrt{f_B}} = 19$$

From this coordinate value, $f_B = \left(\frac{1}{19}\right)^2 = 0.0028$. This gives $\log(\text{Re}_B\sqrt{f_B}) = \log\left(10450\sqrt{0.0028}\right) = 2.74$.

$$\log\left(\text{Re}_B\sqrt{f_B}\right) = 2.75$$

The value of 2.74 is close to the coordinate value of 2.75 found. Hence this point is correct.

For point f,

$$\frac{1}{\sqrt{f_B}} = 20$$

From this coordinate value, $f_B = \left(\frac{1}{20}\right)^2 = 0.0025$. This gives $\log(\text{Re}_B\sqrt{f_B}) = \log\left(10450\sqrt{0.0025}\right) = 2.72$.

$$\log\left(\text{Re}_B\sqrt{f_B}\right) = 2.85$$

The value of 2.72 does not correspond to the coordinate value of 2.85 found. Hence this point is incorrect.

Therefore we can conclude that the data point that matches fluid B best is point e.

Finally to find the percentage of drag reduction (%DR) for point e (i.e., fluid B) relative to fluid A, we can calculate $f_B = 0.0028$ while $f_A = 0.008$ from earlier.

$$\%DR = \left(1 - \frac{0.0028}{0.008}\right)\Bigg|_{\text{Re}} = 65\%$$

We have assumed that Re does not differ significantly between the two fluid flows, for the above approximate calculation.

## Problem 25

(a) **Express the energy equation that is appropriate to describe instantaneous turbulent flow. You may assume there are no energy sources/sinks, and no viscous losses.**

(b) **Derive the corresponding energy equation that is averaged over time and express heat transfer fluxes for the turbulent flow.**

## Solution 25

### *Worked Solution*

(a) The general form of an energy equation is as follows assuming no source/sink terms and no viscous contributions, where $\frac{k}{\rho c_p}$ is also known as the thermal diffusivity.

$$\frac{\partial T}{\partial t} + \boldsymbol{v} \cdot \nabla T = \frac{k}{\rho c_p} \nabla^2 T$$

Turbulent flow is typically characterized by its rapid fluctuations. As such instantaneous flow properties (e.g., velocity, shear stress, pressure etc.) may be expressed as a sum between a time-averaged component (e.g., $\bar{T}$) and a fluctuating component (e.g., $T'$). The properties $T$ and $\boldsymbol{v}$ are further defined as follows to describe instantaneous turbulent flow.

$$T = \bar{T} + T' \rightarrow \frac{\partial T}{\partial t} = \frac{\partial(\bar{T} + T')}{\partial t}$$

$$\boldsymbol{v}(t) = \bar{\boldsymbol{v}} + \boldsymbol{v}'$$

$$\frac{\partial(\bar{T} + T')}{\partial t} + (\bar{\boldsymbol{v}} + \boldsymbol{v}') \cdot \nabla(\bar{T} + T') = \frac{k}{\rho c_p} \nabla^2(\bar{T} + T')$$

(b) We can obtain a time-averaged energy equation starting with the time-average for the temperature differential

$$\overline{\frac{\partial(\bar{T} + T')}{\partial t}} = \overline{\frac{\partial(\bar{T})}{\partial t}} = \frac{\partial \bar{T}}{\partial t} \tag{1}$$

Next we find out the time-average for the dot product $\boldsymbol{v} \cdot \nabla T$

$$\boldsymbol{v} \cdot \nabla T = (v_x, v_y, v_z) \cdot \left(\frac{\partial T}{\partial x}, \frac{\partial T}{\partial y}, \frac{\partial T}{\partial z}\right) = v_x \frac{\partial T}{\partial x} + v_y \frac{\partial T}{\partial y} + v_z \frac{\partial T}{\partial z}$$

$$\overline{v_x \frac{\partial T}{\partial x}} = \overline{(\overline{v_x} + v_x') \frac{\partial(\bar{T} + T')}{\partial x}}$$

$$= \overline{\overline{v_x} \frac{\partial \bar{T}}{\partial x}} + \overline{v_x' \frac{\partial \bar{T}}{\partial x}} + \overline{\overline{v_x} \frac{\partial T'}{\partial x}} + \overline{v_x' \frac{\partial T'}{\partial x}}$$

$$= \overline{v_x} \frac{\partial \bar{T}}{\partial x} + 0 + 0 + \overline{v_x' \frac{\partial T'}{\partial x}} = \overline{v_x} \frac{\partial \bar{T}}{\partial x} + \overline{v_x' \frac{\partial T'}{\partial x}}$$

The same can be done for the $y$ and $z$ components of velocity; hence we obtain the following expressions

$$\overline{v_x \frac{\partial T}{\partial x}} = \overline{v_x} \frac{\partial \bar{T}}{\partial x} + \overline{v_x{}' \frac{\partial T'}{\partial x}}$$

$$\overline{v_y \frac{\partial T}{\partial y}} = \overline{v_y} \frac{\partial \bar{T}}{\partial y} + \overline{v_y{}' \frac{\partial T'}{\partial y}}$$

$$\overline{v_z \frac{\partial T}{\partial z}} = \overline{v_z} \frac{\partial \bar{T}}{\partial z} + \overline{v_z{}' \frac{\partial T'}{\partial z}}$$

We can now put together the three expressions to obtain the time-average of the dot product

$$\overline{\boldsymbol{v} \cdot \nabla T} = \overline{v_x} \frac{\partial \bar{T}}{\partial x} + \overline{v_x{}' \frac{\partial T'}{\partial x}} + \overline{v_y} \frac{\partial \bar{T}}{\partial y} + \overline{v_y{}' \frac{\partial T'}{\partial y}} + \overline{v_z} \frac{\partial \bar{T}}{\partial z} + \overline{v_z{}' \frac{\partial T'}{\partial z}}$$

$$\overline{\boldsymbol{v} \cdot \nabla T} = \bar{\boldsymbol{v}} \cdot \nabla \bar{T} + \overline{\boldsymbol{v}' \cdot \nabla T'} \tag{2}$$

Next we look at the term at the right-hand side of the energy equation

$$\nabla^2 T = \nabla^2 (\bar{T} + T') = \frac{\partial^2 (\bar{T} + T')}{\partial x^2} + \frac{\partial^2 (\bar{T} + T')}{\partial y^2} + \frac{\partial^2 (\bar{T} + T')}{\partial z^2}$$

$$\overline{\nabla^2 T} = \nabla^2 \bar{T} \tag{3}$$

Finally, we combine the expressions (1–3) to obtain the time averaged energy equation as shown

$$\frac{\partial \bar{T}}{\partial t} + \bar{\boldsymbol{v}} \cdot \nabla \bar{T} + \overline{\boldsymbol{v}' \cdot \nabla T'} = \frac{k}{\rho c_p} \nabla^2 \bar{T}$$

The turbulent heat transfer flux $\boldsymbol{q}$ is contributed by the fluctuating variable, and we know that temperature is related to heat flux via heat capacity and velocity; therefore $\boldsymbol{q}$ is expressed as shown

$$\boldsymbol{q} = (q_x, q_y, q_z) = \rho c_p \left(\overline{\boldsymbol{v}' \cdot \nabla T'}\right) = \rho c_p \left(\overline{v_x{}' \cdot T'} + \overline{v_y{}' \cdot T'} + \overline{v_z{}' \cdot T'}\right)$$

## Problem 26

**We are provided with the velocity profile of the turbulent boundary layer for a pipe flow as follows:**

$$u^+ = y^+ \text{ for } y^+ < 5$$

$$u^+ = 5\ln y^+ - 3.05 \text{ for } 5 < y^+ < 30$$

$$u^+ = 2.5\ln y^+ + 5.5 \text{ for } y^+ > 30$$

(a) **Show that the mean velocity in the pipe is**

$$\langle u \rangle^+ = 2.5\ln R^+ + 1.75$$

**where $R$ is the pipe radius. You may assume that mass flow rates in the viscous and buffer layers are negligible relative to that in the turbulent core.**

(b) **A certain fluid with $\eta = 9 \times 10^{-4}$ Ns/m$^2$ and $\rho = 1050$ kg/m$^3$ has a mean velocity of 2.8 m/s in a pipe of radius 0.025 m. Determine the friction velocity.**

(c) **Find the friction factor.**

(d) **Find the thickness of the viscous sublayer and buffer layer.**

(e) **Find the velocity at a distance of 0.75 mm from the wall.**

(f) **Assuming there is a heat flux of 6.5 kW/m$^2$ transmitted through the wall, determine the temperature gradient at the wall, and at a distance 0.75 mm from the wall. You are given the thermal conductivity $k = 0.61$ W/mK and heat capacity $C_p = 4.2$ kJ/kgK.**

## Solution 26

### *Worked Solution*

(a)

---

### *Background Concepts*

Before going into details about the characteristics of turbulent flow velocity profile, it helps to first understand how laminar flow and turbulent flow profiles differ. Fully developed laminar and turbulent flows are shown below, with a zoom-in on the turbulent boundary layer near the wall.

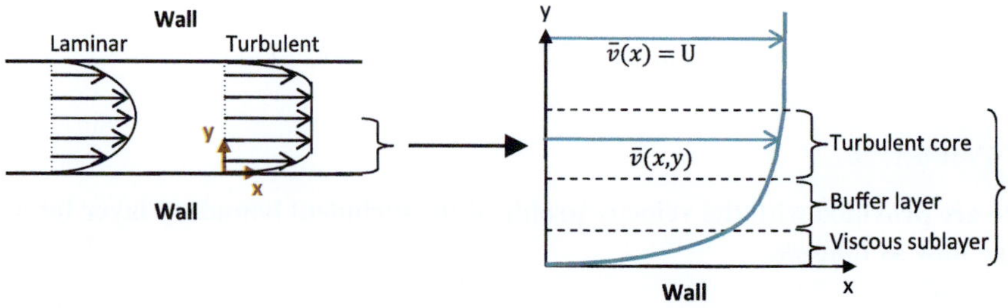


In the above, note that the laminar profile has a parabolic shape. The turbulent profile, however, has a much steeper velocity gradient near the walls, and a flatter portion near the core. The entire turbulent boundary layer has thickness, $\delta$. Also note that the turbulent velocity profile plots time-averaged velocities or mean velocities, $\bar{v}$. Useful parameters for turbulent flow to note:

- Friction velocity, $u^* = \sqrt{\frac{\tau_w}{\rho}}$, is a quantity which has units of velocity (m/s) but is not an actual flow velocity. $u^*$ is used to non-dimensionalize turbulent mean velocity (time-averaged) to obtain a dimensionless velocity $u^+ = \frac{u}{u^*}$, assuming smooth walls. $\tau_w$ refers to wall shear stress and $\rho$ is fluid density.
- Another useful dimensionless parameter is the dimensionless distance from wall, $y^+ = \frac{y}{y^*}$, also called viscous length. Similar to friction velocity, $y^*$ is used to non-dimensionalize distance and $y^* = \frac{\nu}{u^*} = \frac{\eta}{\sqrt{\tau_w \rho}}$, where $\nu$ is kinematic viscosity and $\eta$ is the fluid viscosity (or shear viscosity).
- The thickness of the viscous sublayer is approximately $y^+ = 5$.

The velocity profiles of the three sublayers of the turbulent boundary layer are as follows.

***Viscous sublayer***

Viscous effects almost completely dominate in this sublayer near the walls. The steep velocity gradient means a high shear stress $\tau$, since $\tau = \eta \frac{dv(x)}{dy}$. This wall shear stress is much greater in turbulent flow than laminar flow. The viscous sublayer may be thin, but its viscous effects are significant within this region. This layer can be modeled using a linear profile:

$$u^+ = y^+$$

where $y^+ < 5$. Note that the no-slip condition applies at the walls, i.e., $y^+ = 0$, $u^+ = 0$.

***Buffer layer***

In this layer, there is a mix of viscous stress and turbulent stresses. It connects the two adjacent layers (i.e., inner viscous sublayer and outer turbulent core) to give a continuous velocity profile. Turbulent energy is dissipated here, where the fluid near the wall and in the turbulent core are rapidly exchanged. The buffer layer can be modeled as follows:

$$u^+ = 5 \ln y^+ - 3.05$$

where $5 < y^+ < 30$.

***Turbulent core***

Turbulent shear stress dominates. Velocity almost reaches a constant value that does not vary much with distance from the wall. Beyond this layer, flow is dominated by inertia and velocity reaches a constant value, $U$. This turbulent core can be modeled as follows:

$$u^+ = 2.5\ln y^+ + 5.5$$

where $y^+ > 30$.

---

Now let's return to the solving the problem.

To obtain the mean velocity, we need to consider the individual velocity profiles for each sublayer of the turbulent boundary layer and average over the entire cross section of the pipe.

The differential area element for cross section of a cylindrical pipe in the general form is $2\pi rdr$. In terms of the variables provided in this problem, we have

$$r = R - y \rightarrow dr = -dy$$
$$dA = 2\pi(R - y)(-dy)$$

Note that the limits of integration are converted from $r$ coordinates to $y$ coordinates. When $r = 0$, $y = R$ and when $r = R$, $y = 0$.
The mean velocity can be found as follows

$$\pi R^2\langle u\rangle = \int_{r=0}^{r=R} u2\pi rdr = \int_{y=R}^{y=0} u2\pi(R - y)(-dy)$$

Dividing throughout by $u^* y^{*2}$,

$$\pi R^{+2}\langle u\rangle^+ = \int_{R^+}^{0} u^+ 2\pi(R^+ - y^+)(-dy^+) = 2\pi\int_0^{R^+} u^+(R^+ - y^+)(dy^+)$$

$$R^{+2}\langle u\rangle^+ = 2\int_0^5 y^+(R^+ - y^+)(dy^+) + 2\int_5^{30}(5\ln y^+ - 3.05)(R^+ - y^+)(dy^+)$$
$$+ 2\int_{30}^{R^+}(2.5\ln y^+ + 5.5)(R^+ - y^+)(dy^+)$$

$$R^{+2}\langle u\rangle^+ = 2\left[\frac{R^+ y^{+2}}{2} - \frac{y^{+3}}{3}\right]_0^5 + 2\int_5^{30}(5\ln y^+ - 3.05)(R^+ - y^+)dy^+$$
$$+ 2\int_{30}^{R^+}(2.5\ln y^+ + 5.5)(R^+ - y^+)dy^+$$

In order to integrate the logarithmic expressions for the buffer layer and the turbulent core, we can use integration by parts $\int udv = uv - \int vdu$

For the second term's integral,

$$\int_5^{30} (5\ln y^+ - 3.05)(R^+ - y^+)dy^+$$

$$= \left[(5\ln y^+ - 3.05)\left(R^+y^+ - \frac{y^{+2}}{2}\right)\right]_5^{30} - \int_5^{30}\left(R^+y^+ - \frac{y^{+2}}{2}\right)\left(\frac{5}{y^+}\right)dy^+$$

$$= [13.96(30R^+ - 450) - 5(5R^+ - 12.5)] - \left[5R^+y^+ - \frac{5y^{+2}}{4}\right]_5^{30}$$

$$= 394R^+ - 6220 - [(150R^+ - 1125) - (25R^+ - 31.25)]$$

$$= 394R^+ - 6220 - [125R^+ - 1094]$$

$$= 269R^+ - 5126$$

For the third term's integral,

$$\int_{30}^{R^+} (2.5\ln y^+ + 5.5)(R^+ - y^+)dy^+ = \left[(2.5\ln y^+ + 5.5)\left(R^+y^+ - \frac{y^{+2}}{2}\right)\right]_{30}^{R^+} - \int_{30}^{R^+}\left(R^+y^+ - \frac{y^{+2}}{2}\right)\left(\frac{2.5}{y^+}\right)dy^+$$

$$= \left[(2.5\ln R^+ + 5.5)\left(\frac{R^{+2}}{2}\right) - 14(30R^+ - 450)\right] - \left[2.5R^+y^+ - \frac{2.5y^{+2}}{4}\right]_{30}^{R^+}$$

$$= 1.25R^{+2}\ln R^+ + 2.75R^{+2} - 420R^+ + 6300 - 1.875R^{+2} + 75R^+ - 562.5$$

$$= 1.25R^{+2}\ln R^+ + 0.875R^{+2} - 345R^+ + 5737.5$$

Combining our results, we have

$$R^{+2}\langle u\rangle^+ = 25R^+ - 83 + 538R^+ - 10252 + 2.5R^{+2}\ln R^+ + 1.75R^{+2} - 690R^+ + 11475$$

$$\langle u\rangle^+ = \frac{-127}{R^+} + \frac{1140}{R^{+2}} + 2.5\ln R^+ + 1.75$$

The first two terms which are reciprocals of $R^+$ (i.e., $\frac{-127}{R^+} + \frac{1140}{R^{+2}}$) are negligible compared to the last two terms, hence

$$\langle u\rangle^+ \cong 2.5\ln R^+ + 1.75$$

(b) We are given the radius is $R = 0.0025$ m and $\langle u \rangle = 2.8$ m/s; therefore, friction velocity $u^*$ can be found

$$u^* = \frac{\langle u \rangle}{\langle u \rangle^+} = \frac{2.8}{2.5 \ln R^+ + 1.75}$$

$$R^+ = \frac{R}{y^*}$$

$$y^* = \frac{\nu}{u^*} = \frac{\eta}{\rho u^*} \rightarrow R^+ = \frac{R\rho u^*}{\eta} = \frac{0.025 \times 1050 \times u^*}{9 \times 10^{-4}}$$

$$u^* = \frac{2.8}{2.5 \ln (29167u^*) + 1.75} = \frac{2.8}{27.45 + 2.5 \ln u^*}$$

$$u^* = 0.126 \, \mathrm{m/s}$$

(c) To find friction factor, we need to understand the relationship between friction velocity and wall shear stress, which then relates to friction factor.

$$u^* = \sqrt{\frac{\tau_w}{\rho}} \rightarrow \tau_w = \rho u^{*2}$$

$$\langle u \rangle^+ = \frac{\langle u \rangle}{u^*} = \frac{2.8}{0.126} = 22$$

$$f = \frac{\tau_w}{\frac{1}{2}\rho \langle u \rangle^2} = \frac{2\tau_w}{\rho u^{*2} \langle u \rangle^{+2}} = \frac{2}{\langle u \rangle^{+2}} = 0.00413$$

(d) The thickness of the viscous sublayer is found by finding the distance from the wall at the upper limit of the viscous sublayer region, i.e., when $y^+ = 5$.

$$5 = \frac{y}{y^*} = \frac{y\rho u^*}{\eta} \rightarrow y = \frac{5 \times 9 \times 10^{-4}}{1050 \times 0.126} = 0.034 \, \mathrm{mm}$$

As for the thickness of the buffer layer,

$$y^+ = 30 - 5 = 25$$

$$y = \frac{25 \times 9 \times 10^{-4}}{1050 \times 0.126} = 0.17 \, \mathrm{mm}$$

(e) To find the velocity $\langle u \rangle$ at a distance $y = 0.00075$ m, we can first find out which sublayer the flow belongs to so as to apply the right equation.

$$y^{+} = \frac{0.00075 \times \rho u^{*}}{\eta} = 110 > 30 \text{ (turbulent core)}$$

$$\langle u \rangle^{+} = 2.5 \ln 110 + 5.5 = 17.25$$

$$\langle u \rangle = u^{*} \langle u \rangle^{+} = 0.126 \times 17.25 = 2.2\,\text{m/s}$$

(f) To relate to turbulence heat flux near the wall, let us revisit some relevant concepts. The energy equation for turbulent flow is shown below, where $\frac{k}{\rho c_p}$ is also known as the thermal diffusivity. A key characteristic about turbulent flow is that its flow properties such as velocity may be expressed as a sum of a mean value and a fluctuating component. Noted that in the energy equation, the turbulent heat transfer flux $\boldsymbol{q}_{\mathbf{turb}}$ is from the fluctuating component $\rho c_p \left( \overline{\boldsymbol{v}' \cdot \nabla \mathrm{T}'} \right)$.

$$\rho c_p \left[ \frac{\partial \bar{T}}{\partial t} + \bar{\boldsymbol{v}} \cdot \nabla \bar{T} + \overline{\boldsymbol{v}' \cdot \nabla T'} \right] = k \nabla^2 \bar{T}$$

Let us express total heat flux (starting with a breakdown of the shear stress contributions) as a sum of two components to try to solve this problem. The "turbulent version" of kinematic viscosity is called eddy viscosity $\epsilon$.

Total wall shear stress = Viscous stress + Turbulent stress

$$\tau_w = \rho \nu \frac{dv_x}{dy} + \rho \epsilon \frac{dv_x}{dy}$$

$$\frac{\tau_w}{\rho} = (\nu + \epsilon) \frac{dv_x}{dy}$$

We know that the friction velocity $u^{*}$ is the turbulent flow velocity scale and it is related to $\tau_w$ via $\frac{\tau_w}{\rho} = u^{*2}$. Also let $\varepsilon^{+} = \frac{\epsilon}{\nu'} \frac{v_x}{u^{*}} = \frac{\langle u \rangle}{u^{*}} = \langle u \rangle^{+}$ and $y^{+} = \frac{y}{y^{*}}$.

$$(\nu + \epsilon) \frac{dv_x}{dy} = u^{*2}$$

$$\left( 1 + \frac{\epsilon}{\nu} \right) \frac{dv_x}{dy} = \frac{u^{*2}}{\nu}$$

$$\frac{dv_x}{u^{*}} = \frac{1}{(1 + \varepsilon^{+})} dy \frac{u^{*}}{\nu} = \frac{1}{(1 + \varepsilon^{+})} \frac{dy}{y^{*}}$$

$$d \langle u \rangle^{+} = \frac{1}{(1 + \varepsilon^{+})} dy^{+} \rightarrow \varepsilon^{+} = \frac{1}{\frac{d \langle u \rangle^{+}}{dy^{+}}} - 1$$

Total heat flux = Heat flux from viscous stress + Heat flux from turbulent stress

Note that the turbulent heat flux in the energy equation is the term $\rho c_p \left(\overline{\boldsymbol{v}' \cdot \nabla T'}\right)$, which is the equivalent of the term $-\rho c_p \epsilon \frac{d\bar{T}}{dy}$ below.

$$|\boldsymbol{q}| = -k\frac{d\bar{T}}{dy} - \rho c_p \epsilon \frac{d\bar{T}}{dy} = -\rho c_p \left[\frac{k}{\rho c_p} + \varepsilon^+ \nu\right] \frac{d\bar{T}}{dy}$$

When distance from wall is 0.75 mm, the governing equation is $\langle u \rangle^+ = 2.5 \ln y^+ + 5.5$. Therefore, $\frac{d\langle u \rangle^+}{dy^+} = \frac{2.5}{y^+}$. We also know that at this distance from wall, $y^+ = 110$, $\langle u \rangle^+ = 17.25$ from part e.

$$\varepsilon^+ = \frac{1}{\frac{d\langle u \rangle^+}{dy^+}} - 1 = \frac{110}{2.5} - 1 = 43$$

$$\nu = \frac{\eta}{\rho} = \frac{9 \times 10^{-4}}{1050} = 9.6 \times 10^{-7}$$

$$|\boldsymbol{q}| = 6500 = \rho c_p \left[\frac{k}{\rho c_p} + \varepsilon^+ \nu\right] \frac{d\bar{T}}{dy}$$

$$= 1050 \times 4200 \left[\frac{0.61}{1050 \times 4200} + 43\left(9.6 \times 10^{-7}\right)\right] \frac{d\bar{T}}{dy}$$

$$\left.\frac{d\bar{T}}{dy}\right|_{y=0.75\text{mm}} = 35.6 \text{ K/m}$$

The temperature gradient at the wall occurs in the viscous sublayer region where $\langle u \rangle^+ = y^+$ and $\frac{d\langle u \rangle^+}{dy^+} = 1$. Hence $\varepsilon^+ = 0$. This makes sense as there is no turbulent stress component (or turbulent heat flux contribution) right at the wall surface.

$$|\boldsymbol{q}| = 6500 = k\frac{d\bar{T}}{dy} = 0.61\frac{d\bar{T}}{dy}$$

$$\left.\frac{d\bar{T}}{dy}\right|_{y=0} = 1.1 \times 10^4 \text{ K/m}$$

## Problem 27

**A viscous liquid flows steadily down a long and gentle slope of angle $\theta$ to the horizontal. It forms a film of uniform thickness $h$. Find the velocity profile of the flowing liquid in terms of angle $\theta$ and distance $y$ which is the perpendicular distance from the slope surface. State any assumptions made.**

## Solution 27

***Worked Solution***

We can first sketch a diagram to better visualize this problem. The key in solving this is to be familiar with converting between our usual $x$, $y$ Cartesian coordinate system to one that is at an angle to the horizontal.

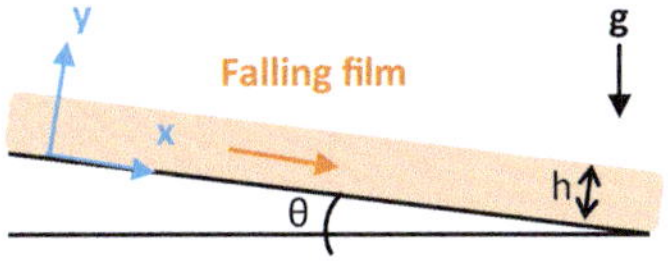


Let us start by stating our assumptions for this flow

- Constant viscosity for a Newtonian fluid.
- The liquid is incompressible, $\rho$ is constant.
- The flow is at steady state ($\frac{\partial}{\partial t} = 0$).
- The dominant flow direction is in the $x$ direction (as defined in the diagram above); hence we assume that $v_y = v_z = 0$
- Given that the slope is long, we may assume the flow is fully developed in the $x$ direction and any entrance/end effects may be ignored. Hence $v_x \neq v_x(x)$.
- Note that gravity is driving the flow; hence its effects are significant. The $g$ vector consists of both non-zero $x$ and $y$ components. In the $+x$ direction, we have $\rho g \sin\theta$ and in the $+y$ direction, we have $-\rho g \cos\theta$.

We can now look at the $x$ and $y$ components of the Navier–Stokes equation

$$0 = -\frac{\partial P}{\partial x} + \rho g \sin\theta + \eta\left(\frac{\partial^2 v_x}{\partial y^2}\right)$$

$$0 = -\frac{\partial P}{\partial y} + (-\rho g \cos\theta)$$

Now we may establish known boundary conditions. For pressure, at $y = h$, $P = P_{\text{atm}}$ and for velocity, at $y = 0$, $v_x = 0$ and at $y = h$, $\frac{\partial v_x}{\partial y} = 0$ since the air–liquid interface can be assumed to have zero shear stress if the viscosity of liquid is much greater than that of air, which is a valid assumption.

$$\frac{\partial P}{\partial y} = -\rho g \cos\theta$$

$$P = -\rho g(\cos\theta)\mathrm{y} + \mathrm{c}_1 \rightarrow \mathrm{c}_1 = P_{\text{atm}} + \rho g(\cos\theta)\mathrm{h}$$

$$P = -\rho g(\cos\theta)\mathrm{y} + P_{\text{atm}} + \rho g(\cos\theta)\mathrm{h} = P_{\text{atm}} + \rho g(\cos\theta)(h - y)$$

We can observe that $\frac{\partial P}{\partial x} = 0$ since $P$ is not a function of $x$. Therefore going back to the $x$-component Navier–Stokes equation, we have

$$0 = \rho g \sin \theta + \eta \left( \frac{\partial^2 v_x}{\partial y^2} \right)$$

$$\frac{-\rho g \sin \theta}{\eta} = \frac{\partial^2 v_x}{\partial y^2}$$

$$\frac{-\rho g \sin \theta}{\eta} y + c_1 = \frac{\partial v_x}{\partial y} \rightarrow c_1 = \frac{\rho g \sin \theta}{\eta} h$$

$$v_x = \frac{-\rho g \sin \theta}{2\eta} y^2 + \frac{\rho g \sin \theta}{\eta} hy + c_2 \rightarrow c_2 = 0$$

Substituting the integration constants back into the velocity profile, we obtain the following.

$$v_x = \frac{-\rho g \sin \theta}{2\eta} y^2 + \frac{\rho g \sin \theta}{\eta} hy$$

$$v_x = \frac{h^2 \rho g \sin \theta}{2\eta} \left[ 2\left(\frac{y}{h}\right) - \left(\frac{y}{h}\right)^2 \right]$$

## Problem 28

**Consider a fluid above a flat plate that is moving at a velocity of $V \cos \omega t$ in the $x$ direction. You may assume the no-slip boundary condition at the interface between the fluid and plate surface where $z = 0$. The velocity of the fluid can be described as dominated by the $x$-component velocity which is a function of the perpendicular distance from the plate, $z$ and time, $t$.**

(a) **Comment on the pressure gradient for this flow, assuming there is no externally applied pressure gradient.**
(b) **Derive a solution of the form $v_x =$ Re $[G(z)e^{i\omega t}]$ whereby Re refers to the real part of the complex function $[G(z)e^{i\omega t}]$ and $i^2 = 1$. Show the following expression and find $m$.**

$$v_x(z, t) = Ve^{-mz} \cos(\omega t - mz)$$

(c) **Using your results in b, find the magnitude of temperature fluctuation at a distance 3.5 m below ground level if the ground temperature is assumed to vary sinusoidally with an amplitude of 18 K and a period of 6 months. The density of the ground material is 2700 kg/m$^3$, the specific heat capacity is 0.9 kJ/kgK and thermal conductivity is 0.7 W/mK.**

## Solution 28

***Worked Solution***

(a) We have below an illustration of the problem.

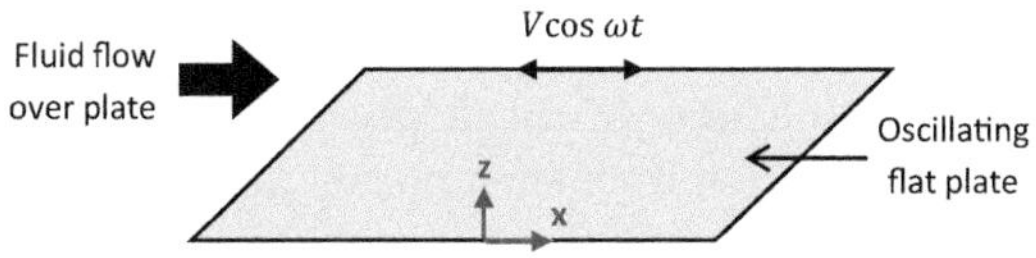


We note from the problem statement that there is no applied pressure gradient. Let us look at the $x$ component Navier–Stokes equation. Note that in this case, there is no steady state (i.e., no change in velocity with time $t$) as the flat plate's movement varies with time; hence the fluid velocity is also a function of time. Our assumptions are as follows:

- Constant viscosity for a Newtonian fluid.
- The liquid is incompressible, $\rho$ is constant.
- The dominant flow direction is in the $x$ direction; hence $v_z = v_y = 0$.
- The flow is fully developed where entry/end effects are assumed negligible; hence $v_x \neq v_x(x)$
- Gravity effects are absent.

The $x$-component Navier–Stokes equation is simplified to the following

$$\rho \frac{\partial v_x}{\partial t} = -\frac{\partial P}{\partial x} + \eta \left( \frac{\partial^2 v_x}{\partial z^2} \right)$$

Since we know that $v_x \neq v_x(x)$, it follows that

$$\frac{\partial}{\partial x}\left[\rho \frac{\partial v_x}{\partial t}\right] = 0$$

We can substitute this result into the Navier–Stokes equation

$$\frac{\partial}{\partial x}\left[-\frac{\partial P}{\partial x} + \eta \left( \frac{\partial^2 v_x}{\partial z^2} \right)\right] = 0 \rightarrow \frac{\partial^2 P}{\partial x^2} = 0$$

$$\frac{\partial P}{\partial x} = c_1$$

Pressure gradient for this flow is uniform. Since there is no applied pressure gradient, we know that $c_1 = 0$.

$$P = c_1 x + c_2 = c_2$$

$$\frac{\partial v_x}{\partial t} = \frac{\eta}{\rho}\left(\frac{\partial^2 v_x}{\partial z^2}\right) = \nu\left(\frac{\partial^2 v_x}{\partial z^2}\right)$$

(b) To derive the complex solution, we can adopt a trial and error approach since we know the form of the solution from the problem statement. Let us try a solution of the form $v_x = G(z)\mathrm{e}^{i\omega t}$

$$\frac{\partial v_x}{\partial t} = i\omega G(z)\mathrm{e}^{i\omega t}$$

$$\frac{\partial^2 v_x}{\partial z^2} = G''\mathrm{e}^{i\omega t}$$

$$i\omega G(z)\mathrm{e}^{i\omega t} = \nu G''\mathrm{e}^{i\omega t} \rightarrow G'' = \frac{i\omega}{\nu}G$$

$$G = c_1 \mathrm{e}^{\sqrt{\frac{i\omega}{\nu}}z} + c_2 \mathrm{e}^{-\sqrt{\frac{i\omega}{\nu}}z}$$

$$G = c_2 \mathrm{e}^{-\sqrt{\frac{i\omega}{\nu}}z} = c_2 \mathrm{e}^{-\sqrt{i}\sqrt{\frac{\omega}{\nu}}z}$$

We can deduce that $c_1 = 0$ as $G$ should be diminishing with increasing distance $z$ from the oscillating plate. Also we know that $i^2 = -1$.

We recall a useful property mathematical "trick" whereby

$$(1+i)^2 = 1 + 2i + i^2 = 1 + 2i - 1 = 2i$$

$$1 + i = \sqrt{2}\sqrt{i} \rightarrow \sqrt{i} = \frac{1+i}{\sqrt{2}}$$

Therefore we can substitute for $\sqrt{i}$

$$G = c_2 \mathrm{e}^{-\left(\frac{1+i}{\sqrt{2}}\right)\sqrt{\frac{\omega}{\nu}}z} = c_2 \mathrm{e}^{-(1+i)\sqrt{\frac{\omega}{2\nu}}z}$$

Let $m = \sqrt{\frac{\omega}{2\nu}}$, and taking the real part of the complex function for the solution

$$v_x = \mathrm{Re}\left[G(z)\mathrm{e}^{i\omega t}\right] = \mathrm{Re}\left[c_2 \mathrm{e}^{-(1+i)mz}\mathrm{e}^{i\omega t}\right] = \mathrm{Re}\left[c_2 \mathrm{e}^{-mz}\mathrm{e}^{i(\omega t - mz)}\right]$$

We know that when $z = 0$, $v_x = V \sin \omega t$

$$V \cos \omega t = \mathrm{Re}\left[c_2 \mathrm{e}^{i\omega t}\right] = \mathrm{Re}[c_2 \cos \omega t + (c_2 \sin \omega t)i] = c_2 \cos \omega t$$

$$c_2 = V$$

Therefore the velocity profile is as follows where $m = \sqrt{\frac{\omega}{2\nu}}$.

$$v_x = \mathrm{Re}\left[V\mathrm{e}^{-mz}\mathrm{e}^{i(\omega t - mz)}\right]$$

$$v_x = V\mathrm{e}^{-mz}\cos(\omega t - mz)$$

(c) We know that the amplitude of this sinusoidal function is 18 K, with a period of 6 months or 0.5 years.

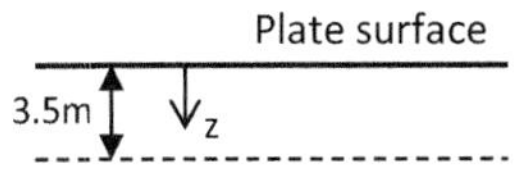


We may now adapt the concept of the oscillating plate which contributes to a sinusoidal velocity in surrounding fluid, to a plate that is a heat source contributing temperature changes in the surrounding. In part b, velocity decreases with increasing $z$. In this case, we defined a similar system whereby temperature decreases with increasing $z$.

The analogous form of the earlier Navier–Stokes equation in terms of velocity for temperature is as shown below, where we assume that for temperature changes, the key term is the diffusion (thermal energy) contribution and kinematic viscosity $\nu = \frac{\eta}{\rho}$ is replaced by thermal diffusivity $\frac{k}{\rho c_p}$ where $k$ is the thermal conductivity.

$$\frac{\partial v_x}{\partial t} = \nu\left(\frac{\partial^2 v_x}{\partial z^2}\right) \leftrightarrow \frac{\partial T}{\partial t} = \frac{k}{\rho c_p}\left(\frac{\partial^2 T}{\partial z^2}\right) = \alpha\left(\frac{\partial^2 T}{\partial z^2}\right)$$

The equivalent of kinematic viscosity for velocity profile is thermal diffusivity α in this case. Substituting values of material properties, we have the following

$$\frac{\partial T}{\partial t} = \frac{0.7}{2700(900)}\left(\frac{\partial^2 T}{\partial z^2}\right) = 2.9 \times 10^{-7}\left(\frac{\partial^2 T}{\partial z^2}\right)$$

$$\alpha = 2.9 \times 10^{-7}$$

We also know that the period of the sinusoidal variation of temperature is 0.5 years. Therefore we can find $\omega$.

$$\omega = \frac{2\pi}{0.5 \times 365 \times 24 \times 60 \times 60} = 4 \times 10^{-7}\ \mathrm{rad/s}$$

$$m = \sqrt{\frac{\omega}{2\nu}} \leftrightarrow \sqrt{\frac{\omega}{2\alpha}} = \sqrt{\frac{4 \times 10^{-7}}{2(2.9 \times 10^{-7})}} = 0.7$$

The amplitude of temperature was given as 18 K at $z = 0$. Adapting the solution in part b, $v_x = Ve^{-mz} \cos(\omega t - mz)$ where $m = \sqrt{\frac{\omega}{2\nu}}$, magnitude of temperature fluctuation (i.e., amplitude) is given by $Ve^{-mz}$. Therefore,

$$Ve^{-mz}|_{z=0} = 18 \rightarrow V = 18$$

At a distance of $z = 3.5$ m, the magnitude of temperature fluctuation is

$$Ve^{-mz}|_{z=3.5} = 18e^{-0.7(3.5)} = 1.55\,\text{K}$$

## Problem 29

**We typically encounter problems with circular pipes for fluid flow within. Let us now consider a case where the cross section of the pipe is an isosceles triangle.**

(a) **Find the velocity profile of a laminar flow in this horizontal pipe with a triangular cross section as illustrated below, using the solution for $v_y$ in the form shown below. You may assume $\alpha$ is a constant, the angle $\theta = 30^\circ$ and the base of the triangle has length $m$, while the two other lengths are of the same length $p$.**

$$v_y = \alpha\left(x - \frac{m\sqrt{3}}{2}\right)(9z^2 - 3x^2)$$

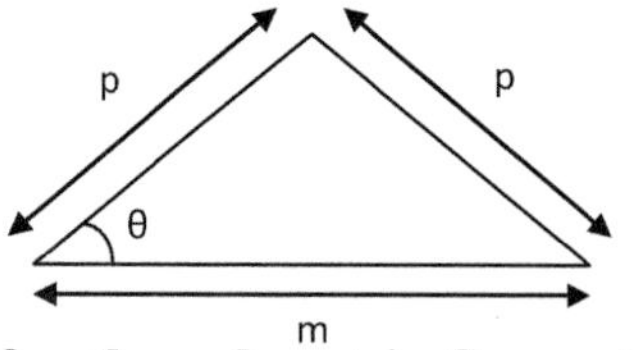


(b) **Find an expression for the volumetric flow rate per unit length (in $y$ direction) that relates to pressure gradient in $y$.**

## Solution 29

*Worked Solution*

(a) We can project the triangle onto a suitable coordinate system, i.e., Cartesian. Based on this coordinate system, the dominant velocity of the fluid in this triangular pipe will be in the $y$ direction.

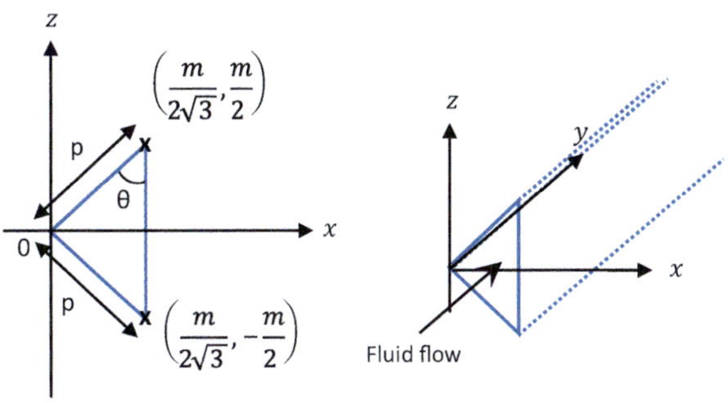


Let us state some assumptions before we apply the Navier–Stokes equation to describe the flow.

- Constant viscosity for a Newtonian fluid.
- The liquid is incompressible, $\rho$ is constant.
- The flow is at steady state ($\frac{\partial}{\partial t} = 0$).
- The dominant flow direction is in the $y$ direction; hence $v_x = v_z = 0$.
- The flow is fully developed where entry/end effects are assumed negligible; hence $v_y \neq v_y(y)$
- Gravity effects are absent as the pipe is held horizontal.

Looking at the $y$-component equation

$$0 = -\frac{\partial P}{\partial y} + \eta \left[ \frac{\partial^2 v_y}{\partial x^2} + \frac{\partial^2 v_y}{\partial z^2} \right]$$

We are given the solution; hence we can evaluate the differentials

$$v_y = \alpha \left( x - \frac{m\sqrt{3}}{2} \right) (9z^2 - 3x^2)$$

$$\frac{\partial v_y}{\partial x} = \alpha \left( x - \frac{m\sqrt{3}}{2} \right) (-6x) + \alpha (9z^2 - 3x^2) = \alpha \left( -6x^2 + 3\sqrt{3}mx \right) + \alpha (9z^2 - 3x^2)$$

$$\frac{\partial^2 v_y}{\partial x^2} = \alpha \left( -12x + 3\sqrt{3}m \right) - 6\alpha x = \alpha \left( -18x + 3\sqrt{3}m \right)$$

$$\frac{\partial v_y}{\partial z} = 18\alpha \left( x - \frac{m\sqrt{3}}{2} \right) z$$

$$\frac{\partial^2 v_y}{\partial z^2} = 18\alpha\left(x - \frac{m\sqrt{3}}{2}\right)$$

Substituting back into the flow equation

$$0 = -\frac{\partial P}{\partial y} + \eta\left[\alpha\left(-18x + 3\sqrt{3}m\right) + 18\alpha\left(x - \frac{m\sqrt{3}}{2}\right)\right]$$

$$\alpha = \left(-\frac{\partial P}{\partial y}\right)\left(\frac{1}{6\sqrt{3}\eta m}\right)$$

Therefore the velocity profile is shown in the required form as follows.

$$v_y = \left(-\frac{\partial P}{\partial y}\right)\left(\frac{1}{6\sqrt{3}\eta m}\right)\left(x - \frac{m\sqrt{3}}{2}\right)(9z^2 - 3x^2)$$

(b) The volumetric flow rate per unit length in $y$, $L$ can be found by integrating the velocity profile over the entire cross-sectional area in the $y$ plane.

$$\frac{\dot{Q}}{L} = 2\int_{x=0}^{x=\frac{m}{2\sqrt{3}}}\int_{z=0}^{z=\sqrt{3}x}\left(-\frac{\partial P}{\partial y}\right)\left(\frac{1}{6\sqrt{3}\eta m}\right)\left(x - \frac{m\sqrt{3}}{2}\right)(9z^2 - 3x^2)dzdx$$

$$= \left(-\frac{\partial P}{\partial y}\right)\left(\frac{1}{6\sqrt{3}\eta m}\right)\int_{x=0}^{x=\frac{m}{2\sqrt{3}}}\int_{z=0}^{z=\sqrt{3}x}\left(9xz^2 - 3x^3 - \frac{9\sqrt{3}}{2}mz^2 + \frac{3\sqrt{3}mx^2}{2}\right)dzdx$$

$$= \left(-\frac{\partial P}{\partial y}\right)\left(\frac{1}{6\sqrt{3}\eta m}\right)\int_{x=0}^{x=\frac{m}{2\sqrt{3}}}\left[3xz^3 - 3x^3z - \frac{3\sqrt{3}mz^3}{2} + \frac{3\sqrt{3}mx^2z}{2}\right]_{z=0}^{z=\sqrt{3}x}dx$$

$$= \left(-\frac{\partial P}{\partial y}\right)\left(\frac{1}{6\sqrt{3}\eta m}\right)\int_{x=0}^{x=\frac{m}{2\sqrt{3}}}\left[9\sqrt{3}x^4 - 3\sqrt{3}x^4 - \frac{27mx^3}{2} + \frac{9mx^3}{2}\right]dx$$

$$= \left(-\frac{\partial P}{\partial y}\right)\left(\frac{1}{6\sqrt{3}\eta m}\right)\left[\frac{9\sqrt{3}x^5}{5} - \frac{3\sqrt{3}x^5}{5} - \frac{27mx^4}{8} + \frac{9mx^4}{8}\right]_{x=0}^{x=\frac{m}{2\sqrt{3}}}$$

$$= \left(-\frac{\partial P}{\partial y}\right)\left(\frac{1}{6\sqrt{3}\eta m}\right)\left(\frac{m^5}{32} - \frac{3m^5}{1440} - \frac{27m^5}{1152} + \frac{9m^5}{1152}\right)$$

$$= \left(-\frac{\partial P}{\partial y}\right)\left(\frac{m^5}{6\sqrt{3}\eta m}\right)\left(\frac{13}{960}\right) = \frac{13m^4}{5760\sqrt{3}\eta}\left(-\frac{\partial P}{\partial y}\right)$$

## Problem 30

**Consider a viscous fluid in a cylindrical pipe of radius $R$ that is in steady, laminar and fully developed flow. Streamlines are parallel to the $x$-axis. Assume gravity effects are negligible.**

(a) **Starting from derivation of energy dissipation per unit volume, show that the rate of mechanical energy dissipation per unit length of pipe (in the $z$ direction) is as follows:**

$$\frac{\pi R^4}{8\eta}\left(\frac{\partial P}{\partial z}\right)^2$$

(b) **Determine the net rate of PV work done by upstream fluid on downstream fluid per unit length of pipe, and comment on your result.**

## Solution 30

*Worked Solution*

(a) Before we begin solving the problem, let us recall some relevant concepts on mechanical energy dissipation.

---

*Background Concepts*

When a fluid with non-zero viscosity flows, its viscosity contributes to energy dissipation. Therefore it requires more energy to make a highly viscous fluid flow, so as to overcome the energy that will be dissipated.

To better understand energy dissipation, let us consider the shear deformation of a small element of fluid of unit depth $\delta z = 1$ over an infinitesimal amount of time $\delta t$. Assume $v_x = v_x(y)$ and $v_y = v_z = 0$. $\tau$ denotes shear stress and $P$ refers to pressure.

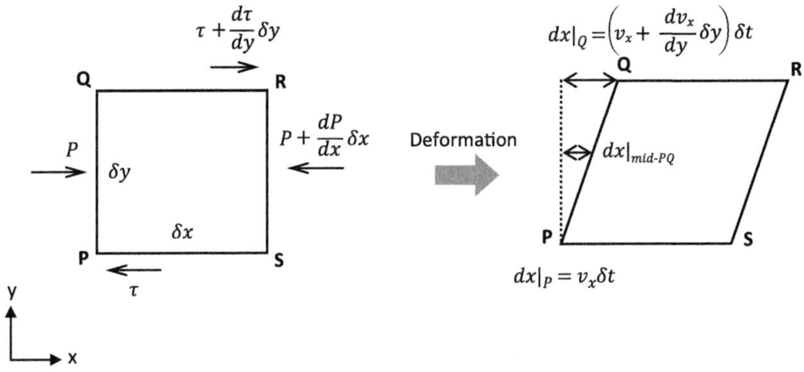


*PQ face* ☐

To express the distance moved by PQ over time $\delta t$, we take the average distance moved at the midpoint of PQ (i.e., $dx$). $dx$ is the average of the distance moved by lengths PS and QR. Since $v_x$ is assumed a linear function in $y$ for this small element, it has different values at PS and QR.

$$dx|_{\text{mid-PQ}} = \frac{dx|_{\text{P}} + dx|_{\text{Q}}}{2} = \frac{(v_x\delta t) + \left(v_x + \frac{dv_x}{dy}\delta y\right)\delta t}{2} = \left(v_x + \frac{1}{2}\frac{dv_x}{dy}\delta y\right)\delta t$$

The force (per unit depth $z$) acting on PQ is pressure force. This force acts to deform the fluid element in the clockwise direction.

$$dF|_{\text{PQ}} = P\delta y$$

Work done (per unit $z$) on the fluid element along PQ over time $\delta t$ is the product of force and distance moved by this force. We previously found the average distance at the midpoint of PQ, which when multiplied by $\delta y$ (which is the area (per unit $z$) over which the force (i.e., pressure) acts) gives the total work done on the PQ face.

$$dW|_{\text{PQ}} = P\delta y\left(v_x + \frac{1}{2}\frac{dv_x}{dy}\delta y\right)\delta t$$

<u>*RS face*</u> □

In a similar manner, the work done along RS by pressure force at RS can be determined. The average distance moved by the RS face is as follows.

$$dx|_{\text{mid-RS}} = dx|_{\text{Pmid-PQ}} = \left(v_x + \frac{1}{2}\frac{dv_x}{dy}\delta y\right)\delta t$$

The force (per unit $z$) acting on RS is also pressure force. This force acts to deform the fluid element in the anticlockwise direction (hence a negative sign).

$$dF|_{\text{RS}} = -\left(P + \frac{dP}{dx}\delta x\right)\delta y$$

Work done (per unit $z$) on the fluid element on the RS face over time $\delta t$ is as follows.

$$dW|_{\text{RS}} = -\left(P + \frac{dP}{dx}\delta x\right)\left(v_x + \frac{1}{2}\frac{dv_x}{dy}\delta y\right)\delta t\delta y$$

<u>*PS face*</u> □

The force (per unit $z$) acting on the PS face is shear stress. This force acts to deform the fluid element in the anticlockwise direction (hence a negative sign).

$$dF|_{\text{PS}} = -\tau\delta x$$

The distance moved by the PS face is as follows.

$$dx|_{\text{PS}} = v_x\delta t$$

Work done (per unit $z$) on the fluid element on the RS face over time $\delta t$ is as follows.

$$dW|_{\text{PS}} = -\tau\delta x v_x\delta t$$

<u>*QR face*</u> □

The force (per unit $z$) acting on the QR face is shear stress. This force acts to deform the fluid element in the clockwise direction.

$$dF|_{\text{QR}} = \left(\tau + \frac{d\tau}{dy}\delta y\right)\delta x$$

The distance moved by the QR face is as follows.

$$dx|_{\text{QR}} = \left(v_x + \frac{dv_x}{dy}\delta y\right)\delta t$$

Work done (per unit $z$) on the fluid element on the QR face over time $\delta t$ is as follows.

$$dW|_{\text{QR}} = \left(\tau + \frac{d\tau}{dy}\delta y\right)\delta x\left(v_x + \frac{dv_x}{dy}\delta y\right)\delta t$$

The total work done (per unit $z$) on the fluid element is the sum of all the components

$$\text{Total work done per unit } z = P\delta y\left(v_x + \frac{1}{2}\frac{dv_x}{dy}\delta y\right)\delta t - \left(P + \frac{dP}{dx}\delta x\right)$$
$$\left(v_x + \frac{1}{2}\frac{dv_x}{dy}\delta y\right)\delta t\delta y - \tau\delta x v_x\delta t + \left(\tau + \frac{d\tau}{dy}\delta y\right)\delta x\left(v_x + \frac{dv_x}{dy}\delta y\right)\delta t$$

$$= \delta x \delta y \delta t \left[ P \frac{\left( v_x + \frac{1}{2} \frac{dv_x}{dy} \delta y \right)}{\delta x} - \frac{\left( P + \frac{dP}{dx} \delta x \right) \left( v_x + \frac{1}{2} \frac{dv_x}{dy} \delta y \right)}{\delta x} - \frac{\tau v_x}{\delta y} + \frac{\left( \tau + \frac{d\tau}{dy} \delta y \right) \left( v_x + \frac{dv_x}{dy} \delta y \right)}{\delta y} \right]$$

$$= \delta x \delta y \delta t \left[ -\frac{dP}{dx} v_x - \frac{1}{2} \frac{dv_x}{dy} \delta y \frac{dP}{dx} + \tau \frac{dv_x}{dy} + \frac{d\tau}{dy} v_x + \frac{d\tau}{dy} \frac{dv_x}{dy} \delta y \right]$$

We substitute the relationship between shear stress and velocity gradient $\tau = \eta \frac{dv_x}{dy}$, and $\frac{d\tau}{dy} = \eta \frac{d^2 v_x}{dy^2}$,

Total work done per unit $z$

$$= \delta x \delta y \delta t \left[ -\frac{dP}{dx} v_x - \frac{1}{2} \frac{dv_x}{dy} \delta y \frac{dP}{dx} + \eta \frac{dv_x}{dy} \frac{dv_x}{dy} + \eta \frac{d^2 v_x}{dy^2} v_x + \eta \frac{d^2 v_x}{dy^2} \frac{dv_x}{dy} \delta y \right]$$

$$= \delta x \delta y \delta t \left[ v_x \left( -\frac{dP}{dx} + \eta \frac{d^2 v_x}{dy^2} \right) - \frac{1}{2} \frac{dv_x}{dy} \frac{dP}{dx} \delta y + \eta \left( \frac{dv_x}{dy} \right)^2 + \eta \frac{d^2 v_x}{dy^2} \frac{dv_x}{dy} \delta y \right]$$

Ignoring the second-order derivatives and above, we obtain

$$\text{Total work done per unit } z = \delta x \delta y \delta t \left[ v_x \left( -\frac{dP}{dx} + \eta \frac{d^2 v_x}{dy^2} \right) + \eta \left( \frac{dv_x}{dy} \right)^2 \right]$$

It follows from the Navier–Stokes equation that for a flow that is at steady state, i.e., no net force (or forces balance) on fluid element, then the following should be true. Note that we still have a non-zero work done when net force is zero, since the constant velocity still contributes to a constant kinetic energy of the element over time. This kinetic energy is how the element dissipates its energy (mechanical) and this "lost energy" is supplied back by the applied energy which also relates to the rate of doing work on the element.

$$\text{Net force} = 0 = -\frac{dP}{dx} + \eta \frac{d^2 v_x}{dy^2}$$

The above expression can also be arrived at by simplifying the Navier–Stokes equation using our assumptions of steady state, laminar, and fully developed flow in the dominant direction ($x$).

After substituting the Navier–Stokes result, the expression for work done becomes

$$\text{Total work done per unit } z = \delta x \delta y \delta t \left[\eta \left(\frac{dv_x}{dy}\right)^2\right]$$

$$\text{Rate of work done per unit } z = \delta x \delta y \left[\eta \left(\frac{dv_x}{dy}\right)^2\right]$$

$$\text{Rate of work done per unit volume} = \Phi = \eta \left(\frac{dv_x}{dy}\right)^2$$

Let us expand this result to the general form of the expression for $\Phi$. The overall rate of strain/deformation, $e_{ij}$ is related to shearing stress $\tau_{yx}$ by $\tau_{yx} = \eta e_{yx}$. In general, $e_{ij}$ is defined as a sum of two components of strain rates. For example, in $x$–$y$ coordinates,

$$e_{yx} = e_{xy} = \frac{dv_y}{dx} + \frac{dv_x}{dy}$$

$$\tau_{yx} = \tau_{xy} = \eta \left(\frac{dv_y}{dx} + \frac{dv_x}{dy}\right)$$

And in general, the rate of energy dissipation per volume is given by the following, where the sum is over nine terms where $i = 1$ to 3 and $j = 1$ to 3.

$$\Phi = \frac{1}{2}\eta \sum_{ij} e_{ij}{}^2$$

In our problem, $e_{yx} = e_{xy} = \frac{dv_x}{dy}$ because $\frac{dv_y}{dx} = 0$, and we have only the $x$–$y$ coordinates which are relevant. Therefore,

$$\Phi = \frac{1}{2}\eta\left(e_{xy}{}^2 + e_{yx}{}^2\right) = \frac{1}{2}\eta\left[\left(\frac{dv_x}{dy}\right)^2 + \left(\frac{dv_x}{dy}\right)^2\right] = \eta\left(\frac{dv_x}{dy}\right)^2$$

---

Let us now return to our problem, we have the following system

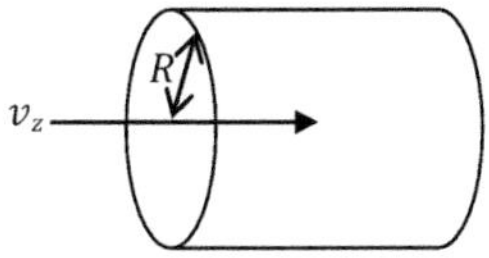


Let us state some assumptions before we apply the Navier–Stokes equation to describe the flow.

- Constant viscosity for a Newtonian fluid.
- The liquid is incompressible, $\rho$ is constant.
- The flow is at steady state ($\frac{\partial}{\partial t} = 0$).
- The dominant flow direction is in the $z$ direction; hence $v_r = v_\theta = 0$.
- The flow is fully developed where entry/end effects are assumed negligible; hence $v_z \neq v_z(z)$
- The system is axisymmetric; therefore $\frac{\partial}{\partial \theta} = 0$.
- Gravity effects are absent as the pipe is held horizontal.

The simplified $z$ component of the Navier–Stokes equation is

$$0 = -\frac{\partial P}{\partial z} + \eta\left[\frac{1}{r}\frac{\partial}{\partial r}\left(r\frac{\partial v_z}{\partial r}\right)\right]$$

$$\frac{r^2}{2\eta}\frac{\partial P}{\partial z} + c_1 = r\frac{\partial v_z}{\partial r}$$

$$v_z = \frac{r^2}{4\eta}\frac{\partial P}{\partial z} + c_1 \ln r + c_2$$

Let us establish boundary conditions, when $r = 0$, $v_z$ has to be defined. Therefore $c_1 = 0$. When $r = R$, $v_z = 0$; therefore $c_2 = -\frac{R^2}{4\eta}\frac{\partial P}{\partial z}$. Therefore the velocity profile can be written as follows.

$$v_z = \frac{r^2}{4\eta}\frac{\partial P}{\partial z} - \frac{R^2}{4\eta}\frac{\partial P}{\partial z} = \left(-\frac{\partial P}{\partial z}\right)\left(\frac{1}{4\eta}\right)\left[R^2 - r^2\right]$$

Now let us evaluate the relevant rate of strain $e_{rz}$ for the system.

$$e_{rz} = e_{zr} = \frac{dv_r}{dz} + \frac{dv_z}{dr} = 0 + \frac{dv_z}{dr} = \frac{dv_z}{dr} = \left(\frac{\partial P}{\partial z}\right)\left(\frac{r}{2\eta}\right)$$

The rate of energy dissipation per unit volume is

$$\Phi = \eta\left(\frac{dv_z}{dr}\right)^2 = \left(\frac{\partial P}{\partial z}\right)^2\left(\frac{r^2}{4\eta}\right)$$

The rate of mechanical energy dissipation per unit $z$ of pipe can be found by integrating over the cross-sectional area, considering the area element in cylindrical coordinates

$$\int_0^R \int_0^{2\pi} \Phi r d\theta dr = \int_0^R \int_0^{2\pi} \left(\frac{\partial P}{\partial z}\right)^2 \left(\frac{r^3}{4\eta}\right) d\theta dr = \int_0^R \left(\frac{\partial P}{\partial z}\right)^2 \left(\frac{r^3}{4\eta}\right) 2\pi dr$$

$$\int_0^R \int_0^{2\pi} \Phi r d\theta dr = \left[\left(\frac{\partial P}{\partial z}\right)^2 \left(\frac{r^4}{8\eta}\right)\pi\right]_0^R = \frac{\pi R^4}{8\eta}\left(\frac{\partial P}{\partial z}\right)^2$$

This energy that deforms the fluid element comes from pressure gradient. Pressure gradient contributes to a net rate of flow work or PV work done. The deformation work increases the internal energy of the fluid (or macroscopic rate of energy production). This energy is also equivalent to the local rate of energy dissipation.

(b) Note that the net rate of PV work done (or flow work) by upstream fluid on downstream fluid is caused by pressure gradient and is equivalent to the local rate of mechanical energy dissipation found in part a.

Rate of PV work is done by finding the product of pressure and distance moved per unit time; hence we integrate $\left[\left(-\frac{\partial P}{\partial z}\right)v_z\right]$ over the cross-sectional area element. $\frac{\partial P}{\partial z}$ is negative; hence we add a negative sign to find the magnitude of the difference in rate of work done.

Net rate of PV work done = Difference in rate of work done by upstream and downstream fluid

$$\int_0^R \int_0^{2\pi} \left(-\frac{\partial P}{\partial z}\right) v_z r d\theta dr = \int_0^R \int_0^{2\pi} \left(-\frac{\partial P}{\partial z}\right)\left(-\frac{\partial P}{\partial z}\right)\left(\frac{1}{4\eta}\right)\left[R^2 - r^2\right] r d\theta dr$$

$$= \int_0^R \left(-\frac{\partial P}{\partial z}\right)^2 \left(\frac{\pi}{2\eta}\right)\left[R^2 - r^2\right] r dr$$

$$= \left(-\frac{\partial P}{\partial z}\right)^2 \left(\frac{\pi}{2\eta}\right)\left[\frac{R^2 r^2}{2} - \frac{r^4}{4}\right]_0^R = \frac{\pi R^4}{8\eta}\left(\frac{\partial P}{\partial z}\right)^2$$

You may observe that it follows that this result is the same as in part a.

## Problem 31

**We have a viscous fluid contained within the annular region of two concentric cylinders, whereby the inner cylinder is held stationary while the other cylinder rotates at an angular velocity $\omega_2$. The radius of the inner and outer cylinders are $R_1$ and $R_2$ respectively. Assuming steady flow of the fluid,**

(a) **Determine the velocity profile in the $\theta$ direction.**
(b) **Find the rate of mechanical energy dissipation per unit axial length in $z$.**

(c) **Find the rate of work done required to rotate the outer cylinder. Comment on the result.**

## Solution 31

### *Worked Solution*

(a) Let us sketch a diagram for our problem.

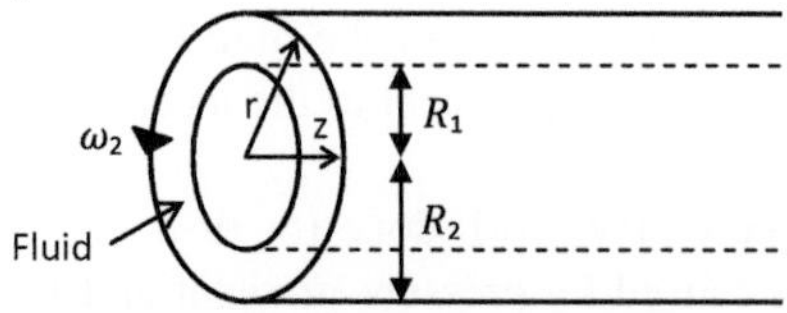


This system is best described by cylindrical coordinates. Let us examine the $r$-component and $\theta$-component Navier–Stokes equations. We have the following simplifying assumptions

- Constant viscosity for a Newtonian fluid.
- The liquid is incompressible, $\rho$ is constant.
- The flow is at steady state ($\frac{\partial}{\partial t} = 0$).
- The dominant flow direction is in the $\theta$ direction (as defined in the diagram above); hence we assume that $v_r = v_z = 0$.
- The flow is axisymmetric, therefore $\frac{\partial}{\partial \theta} = 0$, and we assume the flow is fully developed in the $\theta$ direction, so $\frac{\partial v_\theta}{\partial \theta} = 0$ and $\frac{\partial P}{\partial \theta} = 0$.
- Gravity effects are absent.

The $r$-component Navier–Stokes equation tells us some information about the pressure gradient. In this system, the pressure gradient balances centripetal force.

$$-\rho \frac{{v_\theta}^2}{r} = -\frac{\partial P}{\partial r}$$

The $\theta$-component Navier–Stokes equation tells us more about the velocity profile for $v_\theta$.

$$\frac{\partial}{\partial r}\left(\frac{1}{r}\frac{\partial r v_\theta}{\partial r}\right) = 0$$

$$v_\theta = \frac{c_1 r}{2} + \frac{c_2}{r} = {c_1}' r + \frac{c_2}{r}$$

To solve for the integration constants, we can look at boundary conditions. When $r = R_1$, $v_\theta = 0$ and when $r = R_2$, $v_\theta = R_2 \omega_2$

$$0 = c_1{}'R_1 + \frac{c_2}{R_1} \rightarrow c_1{}' = -\frac{c_2}{R_1{}^2}$$

$$R_2\omega_2 = c_1{}'R_2 + \frac{c_2}{R_2} = -\frac{c_2R_2}{R_1{}^2} + \frac{c_2}{R_2} \rightarrow c_2 = \frac{R_2\omega_2}{\frac{1}{R_2} - \frac{R_2}{R_1{}^2}}$$

$$c_1{}' = -\frac{c_2}{R_1{}^2} = -\frac{R_2\omega_2}{\frac{R_1{}^2}{R_2} - R_2}$$

Substituting the integration constants, we obtain the velocity profile.

$$v_\theta = -\frac{R_2\omega_2 r}{\frac{R_1{}^2}{R_2} - R_2} + \frac{R_2\omega_2}{r\left(\frac{1}{R_2} - \frac{R_2}{R_1{}^2}\right)} = \frac{R_2\omega_2\left(\frac{r}{R_1}\right)}{\frac{R_2}{R_1} - \frac{R_1}{R_2}} - \frac{R_2\omega_2\left(\frac{R_1}{r}\right)}{\left(\frac{R_2}{R_1} - \frac{R_1}{R_2}\right)}$$

$$v_\theta = R_2\omega_2\left[\frac{\left(\frac{r}{R_1}\right) - \left(\frac{R_1}{r}\right)}{\frac{R_2}{R_1} - \frac{R_1}{R_2}}\right]$$

(b) To find the rate of energy dissipation per unit $z$, we can integrate the expression for energy dissipation per unit volume, $\Phi$ over the cross-sectional area of the cylindrical pipe.

$$\Phi = \frac{1}{2}\eta\sum_{ij} \mathrm{e}_{ij}{}^2$$

In our problem, $\mathrm{e}_{r\theta} = \mathrm{e}_{\theta r} = \frac{dv_\theta}{dr}$ because $\frac{dv_r}{d\theta} = 0$, and we have only the $r$–$\theta$ coordinates which are relevant. Therefore,

$$\Phi = \frac{1}{2}\eta\left(\mathrm{e}_{r\theta}{}^2 + \mathrm{e}_{\theta r}{}^2\right)$$

Now let us evaluate the relevant rate of strain $\mathrm{e}_{r\theta}$ for the system. Note that the definition for cylindrical coordinates is different from that for Cartesian coordinates.

$$\mathrm{e}_{r\theta} = \mathrm{e}_{\theta r} \neq \frac{dv_\theta}{dr} + \frac{dv_r}{d\theta}$$

$$\mathrm{e}_{r\theta} = \mathrm{e}_{\theta r} = r\frac{\partial}{\partial r}\left(\frac{v_\theta}{r}\right) + \frac{1}{r}\frac{\partial v_r}{\partial\theta} = r\frac{\partial}{\partial r}\left(\frac{v_\theta}{r}\right) = \frac{2R_1R_2\omega_2}{\left(\frac{R_2}{R_1} - \frac{R_1}{R_2}\right)r^2}$$

The rate of energy dissipation per unit volume is

$$\Phi = \frac{\eta}{r^4}\left[\frac{2R_1R_2\omega_2}{\left(\frac{R_2}{R_1} - \frac{R_1}{R_2}\right)}\right]^2$$

The rate of mechanical energy dissipation per unit $z$ of pipe can be found by integrating over the cross-sectional area, considering the area element in cylindrical coordinates

$$\int_{R_1}^{R_2}\int_0^{2\pi} \Phi r d\theta dr = \int_{R_1}^{R_2}\int_0^{2\pi} \frac{\eta}{r^3}\left[\frac{2R_1R_2\omega_2}{\left(\frac{R_2}{R_1} - \frac{R_1}{R_2}\right)}\right]^2 d\theta dr = \int_{R_1}^{R_2} \frac{\eta}{r^3}\left[\frac{2R_1R_2\omega_2}{\left(\frac{R_2}{R_1} - \frac{R_1}{R_2}\right)}\right]^2 2\pi dr$$

$$= 2\pi\eta\left[\frac{2R_1R_2\omega_2}{\left(\frac{R_2}{R_1} - \frac{R_1}{R_2}\right)}\right]^2\left[-\frac{1}{2r^2}\right]_{R_1}^{R_2} = 2\pi\eta\left[\frac{2R_1R_2\omega_2}{\left(\frac{R_2}{R_1} - \frac{R_1}{R_2}\right)}\right]^2\left[\frac{1}{2R_1{}^2} - \frac{1}{2R_2{}^2}\right]$$

$$= \frac{4\pi\eta\omega_2{}^2R_1{}^2R_2{}^2}{R_2{}^2 - R_1{}^2}$$

(c) Work is required to be done to rotate the outer cylinder so as to maintain the steady-state velocity profile of the system, in order to offset the continuous energy dissipation due to viscosity of fluid.

To determine the rate of work done, let us first find out the force required to be applied, which depends on the shear stress to be overcome. The component of shear stress relevant to our problem is given by

$$\tau_{r\theta} = \eta e_{r\theta} = \eta\left[\frac{2R_1R_2\omega_2}{\left(\frac{R_2}{R_1} - \frac{R_1}{R_2}\right)r^2}\right]$$

The force is applied at the outer cylinder, which is at radial distance $r = R_2$. Force is equivalent to shear stress summed over the area which it acts ($2\pi R_2 L$ or $2\pi R_2$ per unit length).

$$\tau_{r\theta}\big|_{r=R_2} = \eta\left[\frac{2R_1R_2\omega_2}{\left(\frac{R_2}{R_1} - \frac{R_1}{R_2}\right)R_2{}^2}\right] = \frac{2\eta R_1{}^2\omega_2}{R_2{}^2 - R_1{}^2}$$

Rate of work done per unit length of cylinder is the rate of torque $T$ per length, and the rate of torque $T$ per unit length = Shear Force × Tangential distance per time (or tangential velocity $v = r\omega$).

$$\text{Rate of } T \text{ per unit length} = \left(\tau_{r\theta}\big|_{r=R_2}\right)(2\pi R_2)(R_2\omega_2)$$

$$= \left(\frac{2\eta {R_1}^2\omega_2}{{R_2}^2 - {R_1}^2}\right)(2\pi R_2)(R_2\omega_2)$$

$$= \frac{4\eta\pi {R_2}^2{R_1}^2{\omega_2}^2}{{R_2}^2 - {R_1}^2}$$

Hence we can observe that the rate of work done required to maintain the rotation of the outer cylinder (answer in part c) is equivalent to the rate of energy dissipation due to the viscosity of the fluid (answer in part b).

## Problem 32

**Consider a large square plate of length $L$ that is moving away from a solid plane, where the separation distance between the two is $h(t)$ such that $h(t) \ll L$. There is a fluid contained within the separation space.**

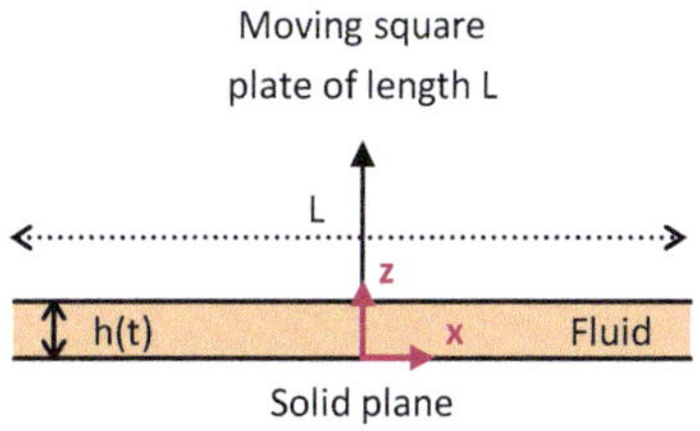


(a) **Assuming creeping flow, show that $0 = -\frac{\partial P}{\partial x} + \eta\left(\frac{\partial^2 v_x}{\partial z^2}\right)$, where $P$ is a function of time $t$ and distance $x$ only.**

(b) **Show that the velocity profile in the $x$ direction is given by**

$$v_x = \frac{1}{2\eta}\left(\frac{\partial P}{\partial x}\right)z[z - h(t)]$$

(c) **Derive the differential equation in $P$ given below, and find the force (per unit area of plate) required to move the square plate. Assume that the pressure $P$ at the edge of the square plate is equivalent to atmospheric pressure and at $x = 0$, $P = P_{ctr}$**

$$\frac{\partial^2 P}{\partial x^2} = \frac{12\eta}{h^3}\frac{dh}{dt}$$

## Solution 32

### *Worked Solution*

(a) Let us list down the assumptions for this flow

- Constant viscosity for a Newtonian fluid.
- The liquid is incompressible, $\rho$ is constant.
- Gravity effects are absent and inertial effects are ignored under the creeping flow approximation for the thin film of fluid, i.e., convective derivative portion of the Navier–Stokes equation is zero.

$$\text{Convective derivative} = \rho\left(\frac{\partial v_x}{\partial t} + v_x\frac{\partial v_x}{\partial x} + v_y\frac{\partial v_x}{\partial y} + v_z\frac{\partial v_x}{\partial z}\right) = 0$$

Looking at the $x$-component Navier–Stokes equation, we have

$$0 = -\frac{\partial P}{\partial x} + \eta\left(\frac{\partial^2 v_x}{\partial x^2} + \frac{\partial^2 v_x}{\partial z^2}\right)$$

For thin film, $\left|\frac{\partial^2 v_x}{\partial x^2}\right| \ll \left|\frac{\partial^2 v_x}{\partial z^2}\right|$, therefore

$$0 = -\frac{\partial P}{\partial x} + \eta\left(\frac{\partial^2 v_x}{\partial z^2}\right)$$

(b) To find the velocity profile of $v_x$, we integrate and impose boundary conditions where when $z = 0$, $v_x = 0$ and when $z = h(t)$, $v_x = 0$.

$$v_x = \frac{1}{\eta}\left(\frac{\partial P}{\partial x}\right)\frac{z^2}{2} + c_1 z + c_2$$

$$c_2 = 0 \text{ and } c_1 = -\frac{1}{\eta}\left(\frac{\partial P}{\partial x}\right)\frac{h(t)}{2}$$

$$v_x = \frac{1}{\eta}\left(\frac{\partial P}{\partial x}\right)\frac{z^2}{2} - \frac{1}{\eta}\left(\frac{\partial P}{\partial x}\right)\frac{h(t)}{2}z = \frac{1}{2\eta}\left(\frac{\partial P}{\partial x}\right)z[z - h(t)]$$

(c) Since we know $v_x$, we can find the velocity profile in the $z$ direction, using the continuity relationship.

$$\frac{\partial v_x}{\partial x} + \frac{\partial v_z}{\partial z} = 0$$

$$\frac{\partial v_z}{\partial z} = -\frac{\partial v_x}{\partial x} = -\frac{\partial}{\partial x}\left[\frac{1}{2\eta}\left(\frac{\partial P}{\partial x}\right)z(z-h(t))\right] = -\frac{1}{2\eta}\frac{\partial}{\partial x}\left[\left(\frac{\partial P}{\partial x}\right)(z^2 - zh)\right]$$

When $z = 0$, $v_z = 0$; therefore we can integrate to find $v_z$

$$v_z = -\frac{1}{2\eta}\frac{\partial}{\partial x}\left[\left(\frac{\partial P}{\partial x}\right)\left(\frac{z^3}{3} - \frac{z^2 h}{2}\right)\right] + H(x)$$

Applying the boundary condition at $z = 0$ gives us $H(x) = 0$.

$$v_z = \frac{dh}{dt} = -\frac{1}{2\eta}\frac{\partial}{\partial x}\left[\left(\frac{\partial P}{\partial x}\right)\left(\frac{z^3}{3} - \frac{z^2 h}{2}\right)\right]$$

At $z = h$, we have

$$\frac{dh}{dt} = -\frac{1}{2\eta}\frac{\partial}{\partial x}\left[\left(\frac{\partial P}{\partial x}\right)\left(\frac{z^3}{3} - \frac{z^2 h}{2}\right)\right] = \frac{1}{2\eta}\frac{\partial}{\partial x}\left[\left(\frac{\partial P}{\partial x}\right)\left(\frac{h^3}{6}\right)\right] = \frac{h^3}{12\eta}\frac{\partial^2 P}{\partial x^2}$$

$$\frac{\partial^2 P}{\partial x^2} = \frac{12\eta}{h^3}\frac{dh}{dt}$$

We note that the boundary condition for pressure is such that at $x = 0$, $P = P_{\mathrm{ctr}}$ and at $x = \frac{L}{2}$, $P = P_{\mathrm{atm}}$. We can integrate the differential equation with respect to $x$ as follows.

$$P = \frac{6\eta}{h^3}\frac{dh}{dt}x^2 + c_1 x + c_2$$

$$c_2 = P_{\mathrm{ctr}}$$

$$c_1 = \frac{(P_{\mathrm{atm}} - P_{\mathrm{ctr}}) - \frac{6\eta}{h^3}\frac{dh}{dt}\left(\frac{L}{2}\right)^2}{\frac{L}{2}} = 2\left(\frac{(P_{\mathrm{atm}} - P_{\mathrm{ctr}})}{L}\right) - \frac{3\eta}{h^3}\frac{dh}{dt}L$$

$$P = \frac{6\eta}{h^3}\frac{dh}{dt}x^2 + \left[2\left(\frac{(P_{\mathrm{atm}} - P_{\mathrm{ctr}})}{L}\right) - \frac{3\eta}{h^3}\frac{dh}{dt}L\right]x + P_{\mathrm{ctr}}$$

The force required to move the plate is used to balance the pressure differential in the fluid relative to the external atmospheric pressure (note: choice of control volume for force balance automatically excludes viscous force acting along the fluid/solid surface interfaces as they do not act at the boundary of the control volume. Only pressure force and applied force $F$ on the upper plate are relevant here).

$$\text{Force to move the plate, } F = \int_{-\frac{L}{2}}^{\frac{L}{2}} \int_0^L (P - P_{\text{atm}}) dy dx$$

$$\frac{F}{L} = \int_{-L/2}^{L/2} \left( \frac{6\eta}{h^3} \frac{dh}{dt} x^2 + \left[ 2\left( \frac{P_{\text{atm}} - P_{\text{ctr}}}{L} \right) - \frac{3\eta}{h^3} \frac{dh}{dt} L \right] x + P_{\text{ctr}} - P_{\text{atm}} \right) dx$$

$$\frac{F}{L} = \left[ \frac{2\eta}{h^3} \frac{dh}{dt} x^3 + \left[ \left( \frac{P_{\text{atm}} - P_{\text{ctr}}}{L} \right) - \frac{3\eta}{2h^3} \frac{dh}{dt} L \right] x^2 + P_{\text{ctr}} x - P_{\text{atm}} x \right]_{-L/2}^{L/2}$$

$$\begin{aligned} \frac{F}{L} = & \left[ \frac{\eta}{4h^3} \frac{dh}{dt} L^3 + \left[ \left( \frac{P_{\text{atm}} - P_{\text{ctr}}}{L} \right) - \frac{3\eta}{2h^3} \frac{dh}{dt} L \right] \left( \frac{L^2}{4} \right) + \frac{P_{\text{ctr}} L}{2} - \frac{P_{\text{atm}} L}{2} \right] \\ & - \left[ -\frac{\eta}{4h^3} \frac{dh}{dt} L^3 + \left[ \left( \frac{P_{\text{atm}} - P_{\text{ctr}}}{L} \right) - \frac{3\eta}{2h^3} \frac{dh}{dt} L \right] \left( \frac{L^2}{4} \right) - \frac{P_{\text{ctr}} L}{2} + \frac{P_{\text{atm}} L}{2} \right] \end{aligned}$$

$$\text{Force per unit area of plate} = \frac{F}{L^2} = \frac{\eta}{2h^3} \frac{dh}{dt} L^2 + (P_{\text{ctr}} - P_{\text{atm}})$$

## Problem 33

**A circular disc of radius $R$ is at a distance $h(t)$ from a solid surface. The disc is moving towards the solid surface at a constant velocity of $V$ through a body of incompressible Newtonian fluid at bulk fluid pressure $P_0$ and of density $\rho$ and viscosity $\eta$. The flow pattern of the fluid contained between the disc and the bottom surface may be assumed fully developed, where end effects are negligible. Gravity effects may also be ignored.**

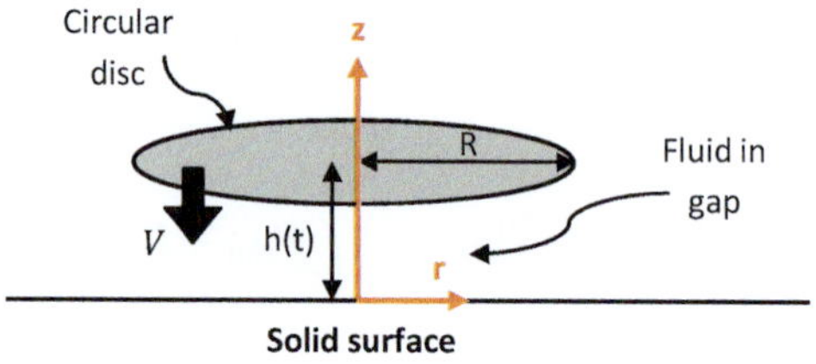


(a) **Comment on the assumptions necessary to model this fluid flow using the creeping flow approximation. Using the Navier–Stokes equation, find the radial velocity profile $v_r$ assuming creeping flow.**
(b) **Determine the mean value of the radial velocity component, $\langle v_r \rangle$.**
(c) **Derive another expression for $\langle v_r \rangle$ using the continuity equation.**
(d) **Express pressure in the separation space as a function of the $r$-coordinate.**
(e) **Find the force exerted by the fluid on the bottom surface.**
(f) **Evaluate the following two expressions and compare them to the pressure gradient in r. Comment on the results.**

$$\rho\langle v_r\rangle \frac{\partial\langle v_r\rangle}{\partial r} \text{ and } \eta\frac{1}{r}\left(\frac{\partial}{\partial r}\left(r\frac{\partial\langle v_r\rangle}{\partial r}\right)\right)$$

## Solution 33

(a) In order to assume creeping flow, we require the following conditions so that we may ascertain slow flow whereby inertial contributions may be neglected, and viscous effects dominate:

- The gap is narrow compared to the length scale of the disc, i.e., $h \ll R$. Consequently, the derivatives in the $z$ direction dominate while derivatives in the $r$ direction are small. Also, the pressure gradient in the $z$ direction is negligible as the flow consists of almost straight and parallel (to the $r$ axis) streamlines.
- Re for the flow is low, i.e., Re $\sim \frac{Vh\rho}{\eta} \ll 1$.

To find $v_r$, let us list the assumptions that help us simplify the Navier–Stokes equations

- Constant viscosity for a Newtonian fluid.
- The liquid is incompressible, $\rho$ is constant.
- The flow is axisymmetric; hence $\frac{\partial}{\partial\theta} = 0$.
- The fluid flow is at steady state; hence $\frac{\partial}{\partial t} = 0$.
- The fluid is not in rotational flow; hence $v_\theta = 0$. Also, $v_r$ is the dominant flow direction; hence we assume $v_z = 0$.
- Gravity effects are absent and inertial effects are ignored under the creeping flow (i.e., convective derivative portion of the Navier–Stokes equation is zero).

Applying the assumptions to the $r$-component equation, we have the following expression.

$$0 = -\frac{\partial P}{\partial r} + \eta\frac{\partial^2 v_r}{\partial z^2}$$

For the sake of discussion, let us do a simple order of magnitude analysis that will reaffirm our result above. Starting with a more general form of the Navier–Stokes equation, we may see how some of the terms are removed due to their relative small values.

$$\rho\left(\frac{\partial v_r}{\partial t}+v_r\frac{\partial v_r}{\partial r}+\frac{v_\theta}{r}\frac{\partial v_r}{\partial \theta}-\frac{{v_\theta}^2}{r}+v_z\frac{\partial v_r}{\partial z}\right)=\frac{\partial P}{\partial r}$$
$$+\eta\left[\frac{\partial}{\partial r}\left(\frac{1}{r}\left(\frac{\partial(rv_r)}{\partial r}\right)\right)+\frac{1}{r^2}\frac{\partial^2 v_r}{\partial \theta^2}-\frac{2}{r^2}\frac{\partial^2 v_\theta}{\partial \theta^2}+\frac{\partial^2 v_r}{\partial z^2}\right]$$
$$+\rho g_r$$

| Terms that were "cancelled off" | Reasons |
|---|---|
| $\frac{\partial v_r}{\partial t}$ | Steady state, therefore equal to zero. |
| $\frac{v_\theta}{r}\frac{\partial v_r}{\partial \theta}$<br>$\frac{{v_\theta}^2}{r}$ | $v_\theta = 0$ since fluid is not in rotational flow. |
| $\frac{1}{r^2}\frac{\partial^2 v_r}{\partial \theta^2}$<br>$\frac{2}{r^2}\frac{\partial^2 v_\theta}{\partial \theta^2}$ | Flow is axisymmetric; hence derivatives in $\theta$ direction are zero. |
| $\rho g_r$ | No gravitational component in $r$ direction. |

After the above removal of terms, we are left with the following equation.

$$\underbrace{\rho\left(v_r\frac{\partial v_r}{\partial r}+v_z\frac{\partial v_r}{\partial z}\right)}_{\text{Inertial terms}}=\underbrace{\frac{\partial P}{\partial r}}_{\text{Pressure term}}+\eta\left[\underbrace{\frac{\partial}{\partial r}\left(\frac{1}{r}\left(\frac{\partial(rv_r)}{\partial r}\right)\right)}_{\text{Viscous term in r}}+\underbrace{\frac{\partial^2 v_r}{\partial z^2}}_{\text{Viscous term in z}}\right]$$

Let us examine each term closely using order of magnitude (OOM) analysis, and see how more terms (viscous term in $r$, and inertial terms) can also be neglected.

*Viscous term in r:*

$$\frac{\partial}{\partial r}\left(\frac{1}{r}\left(\frac{\partial(rv_r)}{\partial r}\right)\right)=\frac{\partial}{\partial r}\left(\frac{1}{r}\left(r\frac{\partial v_r}{\partial r}\right)\right)+\frac{\partial}{\partial r}\left(\frac{1}{r}\left(v_r\frac{\partial r}{\partial r}\right)\right)$$
$$=\frac{\partial^2 v_r}{\partial r^2}+\frac{\partial}{\partial r}\left(\frac{v_r}{r}\right)=\frac{\partial^2 v_r}{\partial r^2}+\frac{1}{r}\frac{\partial v_r}{\partial r}-\frac{v_r}{r^2}\sim\frac{v_r}{r^2}$$

We know that $h \ll r$; therefore $\frac{v_r}{h^2} \gg \frac{v_r}{r^2}$. Each of the terms in the above are small; hence this term $\frac{\partial}{\partial r}\left(\frac{1}{r}\left(\frac{\partial(rv_r)}{\partial r}\right)\right)$ may be neglected.

*Inertial terms*:

We know from the continuity equation that $\pi r^2 V = 2\pi r h v_r$; therefore $v_r \sim \frac{rV}{h}$ and $\frac{\partial v_r}{\partial r} \sim \frac{V}{h}$

$$\rho v_r \frac{\partial v_r}{\partial r} \sim \left(\frac{rV}{h}\right)\left(\frac{V}{h}\right) = \frac{\rho r V^2}{h^2}$$

$$\rho v_z \frac{\partial v_r}{\partial z} \sim \rho V \frac{\partial}{\partial z}\left(\frac{rV}{h}\right) \sim \frac{\rho r V^2}{h^2}$$

$$\rho v_r \frac{\partial v_r}{\partial r} + \rho v_z \frac{\partial v_r}{\partial z} \sim \frac{2\rho r V^2}{h^2} \quad \text{(factor 2 can be dropped for OOM analysis)}$$

Let us compare the relative magnitude of these inertial terms compared to the viscous term $\eta \frac{\partial^2 v_r}{\partial z^2}$

$$\eta \frac{\partial^2 v_r}{\partial z^2} \sim \eta \frac{v_r}{h^2} \sim \eta \frac{rV}{h^3}$$

$$\frac{\text{Inertial terms}}{\text{Viscous terms}} \sim \frac{\frac{\rho r V^2}{h^2}}{\eta \frac{rV}{h^3}} = \frac{hV\rho}{\eta} = \text{Re} \ll 1$$

Therefore, when we assumed creeping flow, where Re is low, we have also assumed that inertial terms may be neglected as verified here. Therefore we arrive at the same simplified expression as shown in part a.

$$0 = -\frac{\partial P}{\partial r} + \eta \frac{\partial^2 v_r}{\partial z^2}$$

(b) In order to obtain the mean value of $v_r$, we need to first figure out the $v_r$ profile by integrating the differential equation in a, using suitable boundary conditions.

$$\frac{1}{\eta}\frac{\partial P}{\partial r} = \frac{\partial^2 v_r}{\partial z^2}$$

$$\frac{\partial v_r}{\partial z} = \frac{z}{\eta}\frac{\partial P}{\partial r} + c_1$$

$$v_r = \frac{z^2}{2\eta}\frac{\partial P}{\partial r} + c_1 z + c_2$$

At $z = 0$ and $z = h$, $v_r = 0$. Therefore, $c_2 = 0$ and $c_1 = -\frac{h}{2\eta}\frac{\partial P}{\partial r}$

$$v_r = \frac{z^2}{2\eta}\frac{\partial P}{\partial r} - \frac{h}{2\eta}\frac{\partial P}{\partial r} z = \frac{1}{2\eta}\left(-\frac{\partial P}{\partial r}\right) z(h - z)$$

$$\begin{aligned}\langle v_r\rangle &= \frac{1}{h}\int_0^h v_r dz = \frac{1}{h}\int_0^h \frac{1}{2\eta}\left(-\frac{\partial P}{\partial r}\right) z(h-z)dz \\ &= \frac{1}{2\eta h}\left(-\frac{\partial P}{\partial r}\right)\left[\frac{z^2 h}{2}-\frac{z^3}{3}\right]_0^h = \frac{h^2}{12\eta}\left(-\frac{\partial P}{\partial r}\right)\end{aligned}$$

(c) Using the continuity equation, we have

$$\pi r^2 V = 2\pi r h\langle v_r\rangle$$

$$\langle v_r\rangle = \frac{rV}{2h}$$

(d)

$$\langle v_r\rangle = \frac{h^2}{12\eta}\left(-\frac{\partial P}{\partial r}\right)$$

$$\frac{\partial P}{\partial r} = -\frac{12\eta\langle v_r\rangle}{h^2} = -\frac{6\eta r V}{h^3}$$

$$P = -\frac{3\eta r^2 V}{h^3} + c_1$$

At $r = R$, $P = P_0$; therefore $c_1 = P_0 + \frac{3\eta R^2 V}{h^3}$

$$P = -\frac{3\eta r^2 V}{h^3} + P_0 + \frac{3\eta R^2 V}{h^3} = P_0 + \frac{3\eta V}{h^3}\left(R^2 - r^2\right)$$

(e) The force exerted by the fluid on the bottom solid surface is the sum of pressure force over the contact surface area

$$\begin{aligned}\text{Force on plate} &= \int_0^R (P-P_0)2\pi r dr = \frac{6\eta\pi V}{h^3}\int_0^R \left(R^2 - r^2\right) r dr \\ &= \frac{6\eta\pi V}{h^3}\left[\frac{R^2 r^2}{2}-\frac{r^4}{4}\right]_0^R = \frac{3\eta\pi V R^4}{2h^3}\end{aligned}$$

(f) We found from earlier that a simple form of expression for mean radial velocity is as follows.

$$\langle v_r\rangle = \frac{rV}{2h}$$

Hence we can use this result to evaluate the two expressions given,

$$\rho\langle v_r\rangle\frac{\partial\langle v_r\rangle}{\partial r}=\rho\left(\frac{rV}{2h}\right)\left(\frac{V}{2h}\right)=\frac{\rho rV^2}{4h^2} \tag{1}$$

$$\eta\frac{1}{r}\left(\frac{\partial}{\partial r}\left(r\frac{\partial\langle v_r\rangle}{\partial r}\right)\right)=\eta\frac{1}{r}\left(\frac{\partial}{\partial r}\left(\frac{rV}{2h}\right)\right)=\frac{\eta V}{2hr} \tag{2}$$

We can now compare the two terms with the pressure gradient in $r$, i.e., $\frac{\partial P}{\partial r}$, whereby expression (1) is the inertial term in $r$, and expression (2) is the viscous term in $r$.

$$\frac{\partial P}{\partial r}=-\frac{6\eta rV}{h^3}\ \text{(from earlier result)}$$

$$\frac{\rho\langle v_r\rangle\frac{\partial\langle v_r\rangle}{\partial r}}{\frac{\partial P}{\partial r}}=\frac{\frac{\rho rV^2}{4h^2}}{-\frac{6\eta rV}{h^3}}=-\frac{\rho Vh}{24\eta}=-\frac{1}{24}\text{Re}$$

where Re $\ll 1$ under the creeping flow assumption. Hence the inertial term is much smaller than the pressure gradient term.

$$\frac{\eta\frac{1}{r}\left(\frac{\partial}{\partial r}\left(r\frac{\partial\langle v_r\rangle}{\partial r}\right)\right)}{\frac{\partial P}{\partial r}}=\frac{\frac{\eta V}{2hr}}{-\frac{6\eta rV}{h^3}}=-\frac{h^2}{12r^2}$$

where $\frac{h}{r}\ll 1$; hence the viscous term in $r$ is also much smaller than the pressure gradient term. This result is consistent with the analysis in part a, whereby the inertial term in $r$ and viscous term in $r$ were neglected when simplifying the Navier–Stokes equation. This was based on the creeping flow and narrow gap assumptions.

## Problem 34

**We have a porous bed that has permeability $k$ and pressure gradient $\nabla P$. The superficial velocity of a fluid in the porous bed is given as:**

$$v=-\frac{k}{\eta}\nabla P$$

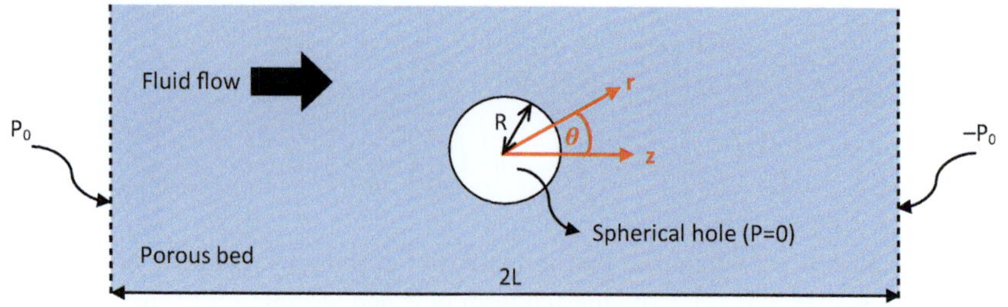


(a) **Show that pressure obeys the Laplace equation $\nabla^2 P = 0$. Then express the Laplace equation for pressure distribution as a function of $r$ and $\theta$ using spherical coordinates.**

(b) **A fluid flows through the porous bed with a hole in the midpoint of the length $2L$ of the bed. Assume that the hole does not attenuate the fluid flow and $R \ll L$. Use the trial solution for pressure profile $P = \left(mr + \frac{n}{r^2}\right)\cos\theta$ and verify if it is consistent with results in part a. Also determine the parameters $m$ and $n$ in terms of $\theta$, $L$ and $R$.**

(c) **Determine the volumetric flow rate of fluid through the hole.**

## Solution 34

### *Worked Solution*

(a) Before we begin solving the problem, let's recall some relevant concepts.

---

### *Background Concepts*

Recall that superficial velocity refers to the velocity assuming the phase occupies the entire cross section (i.e., in the absence of particles), and is equivalent to $\frac{Q}{A}$, where $Q$ is the volumetric flow rate of the fluid and $A$ is the cross-sectional area of the packed/porous bed.

For this problem, it is easier to work with spherical coordinates (and as advised in the problem) which are defined by the coordinates $r$, $\theta$ and Ø as shown below.

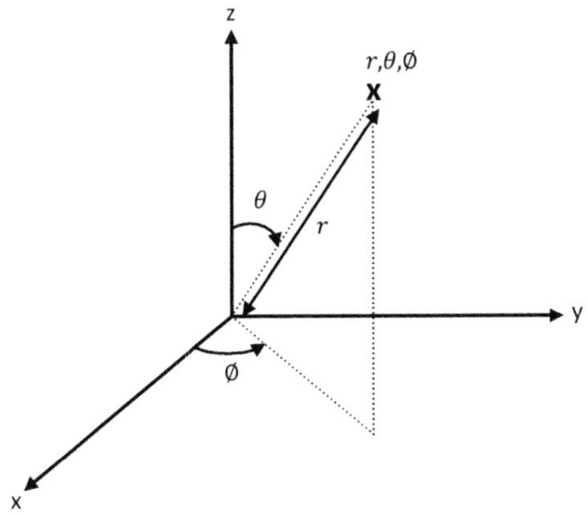


Some important expressions in spherical coordinates are as follows, which can be found from data booklets.

- The gradient operator $\nabla$ in spherical coordinates is given below, where *f* is an arbitrary scalar field.

$$\nabla \mathrm{f} = \frac{\partial}{\partial r} f_r + \frac{1}{r}\frac{\partial}{\partial \theta} f_\theta + \frac{1}{r \sin\theta}\frac{\partial}{\partial \emptyset} f_\emptyset$$

- The Laplacian operator $\nabla^2$ in spherical coordinates is given by

$$\nabla^2 \mathrm{f} = \frac{1}{r^2}\frac{\partial}{\partial r} r^2 \frac{\partial f}{\partial r} + \frac{1}{r^2 \sin\theta}\frac{\partial}{\partial \theta} \sin\theta \frac{\partial f}{\partial \theta} + \frac{1}{r^2 \sin^2\theta}\frac{\partial^2 f}{\partial \emptyset^2}$$

---

Let us now return to the problem, to show that pressure obeys Laplace's equation. Using the continuity equation, and assuming a Newtonian fluid with constant viscosity $\eta$ and constant density $\rho$ (i.e., incompressible), we have the following.

$$0 = \nabla \cdot \boldsymbol{v}$$

$$\text{Given that } \boldsymbol{v} = -\frac{k}{\eta}\nabla P$$

$$0 = \nabla \cdot \left(-\frac{k}{\eta}\nabla P\right) = \nabla^2 P, \text{as } \frac{k}{\eta} \text{ is a constant}$$

Now let's proceed to expressing the Laplace equation in spherical coordinates. We know that the flow is axisymmetric in the Ø direction; hence $\frac{\partial}{\partial \emptyset} = 0$.

$$\nabla^2 P = \frac{1}{r^2}\frac{\partial}{\partial r}r^2\frac{\partial P}{\partial r} + \frac{1}{r^2 \sin\theta}\frac{\partial}{\partial \theta}\sin\theta\frac{\partial P}{\partial \theta} = 0$$

$$\sin\theta\frac{\partial}{\partial r}r^2\frac{\partial P}{\partial r} + \frac{\partial}{\partial \theta}\sin\theta\frac{\partial P}{\partial \theta} = 0 \text{ (as shown)}$$

(b) Starting from the trial solution, let us work out the terms in the equation shown in part a.

$$\sin\theta\frac{\partial}{\partial r}r^2\frac{\partial P}{\partial r} + \frac{\partial}{\partial \theta}\sin\theta\frac{\partial P}{\partial \theta} = 0$$

Since $P = \left(mr + \frac{n}{r^2}\right)\cos\theta$, therefore

$$\frac{\partial P}{\partial r} = \left(m - \frac{2n}{r^3}\right)\cos\theta$$

$$\frac{\partial P}{\partial \theta} = -\left(mr + \frac{n}{r^2}\right)\sin\theta$$

Putting the expressions together

$$\sin\theta\frac{\partial}{\partial r}r^2\left(\left(m - \frac{2n}{r^3}\right)\cos\theta\right) + \frac{\partial}{\partial \theta}\sin\theta\left(-\left(mr + \frac{n}{r^2}\right)\sin\theta\right) = 0$$

$$\cos\theta\sin\theta\left(2mr + \frac{2n}{r^2}\right) - 2\cos\theta\sin\theta\left(mr + \frac{n}{r^2}\right) = 0$$

The left-hand side of this equation cancels itself out; hence the equation is valid which means the Laplace equation for pressure distribution holds for this trial solution.

To determine the parameters in the trial solution, we need to integrate the pressure distribution differential equation using appropriate boundary conditions. We know that at $z = -L$ and $z = L$, $P = P_0$ and $P = -P_0$ respectively. Also at $r = R$, $P = 0$.

For the boundary conditions in $z$ coordinates, we need to link the trial solution (absent of $z$) to $z$, using $z = r\cos\theta$.

$$P = \left(mr + \frac{n}{r^2}\right)\cos\theta = \left(m + \frac{n}{r^3}\right)z$$

$$P_0 = \left(m + \frac{n}{r^3}\right)(-L) \tag{1}$$

$$-P_0 = \left(m + \frac{n}{r^3}\right)(L) \tag{2}$$

Equations (1) and (2) are essentially the same. And they tell us that when $r$ is large, i.e., at $z = \pm L$, $P_0 = -mL \; or \; m = -\frac{P}{L}$. This result can be used when we apply the third boundary condition for $r$ as below,

$$0 = \left(mR + \frac{n}{R^2}\right)\cos\theta \rightarrow mR + \frac{n}{R^2} = 0 \; or \; \cos\theta = 0$$

$$n = -mR^3 = \frac{R^3 P}{L}$$

Therefore we have obtained $m = -\frac{P}{L}$ and $n = \frac{R^3 P}{L}$.

(c) We are given the velocity field as shown below, we can then find out the $r$ and $\theta$ components from the definition of the del operator in spherical coordinates.

$$\boldsymbol{v} = -\frac{k}{\eta}\nabla P$$

$$v_r = -\frac{k}{\eta}\frac{\partial P}{\partial r} \text{ and } v_\theta = -\frac{k}{\eta}\frac{1}{r}\frac{\partial P}{\partial \theta}$$

We also know from earlier that $P = \left(mr + \frac{n}{r^2}\right)\cos\theta$, which when we substitute the expressions for $m$ and $n$ found earlier gives us

$$P = \left(\frac{R^3}{r^2} - r\right)\left(\frac{P}{L}\right)\cos\theta$$

$$\begin{aligned} v_r\big|_{r=R} &= -\frac{k}{\eta}\frac{\partial P}{\partial r} = -\frac{k}{\eta}\left(\frac{P}{L}\right)\cos\theta\left(-\frac{2R^3}{r^3} - 1\right) \\ &= 3\frac{k}{\eta}\left(\frac{P}{L}\right)\cos\theta \end{aligned}$$

$$\begin{aligned} v_\theta\big|_{r=R} &= -\frac{k}{\eta}\frac{1}{r}\frac{\partial P}{\partial \theta} = \frac{k}{\eta}\frac{1}{R}\left(\frac{R^3}{R^2} - R\right)\left(\frac{P}{L}\right)\sin\theta \\ &= 0 \text{ (no rotational flow)} \end{aligned}$$

The volumetric flowrate through the hole, $\dot{Q}$ can be found by integrating velocity in the $r$ component across the relevant surface area. The area element in spherical coordinates at the surface of the hole is $R^2 \sin\theta d\theta d\varnothing$. Note that this definition of differential area assumes $\theta$ is the angle measured from $z$, the polar coordinate.

In general, the physical limits of the polar coordinate $\theta$ in spherical coordinates is $\pi$. In this problem, we integrate from $\theta = 0$ to $\frac{\pi}{2}$ as we only need to take one side of the direction of flow to determine flow rate through the hole.

Also note that the differential area from $\theta = 0$ to $\pi$ is $R^2 \sin \theta d\theta d\text{Ø}$ (i.e., northern hemisphere) while that from $\theta = \pi \ to \ 2\pi$ is $-R^2 \sin \theta d\theta d\text{Ø}$, due to symmetry about $\theta = 0$. For the Ø direction, we integrate from Ø $= 0$ to $2\pi$.

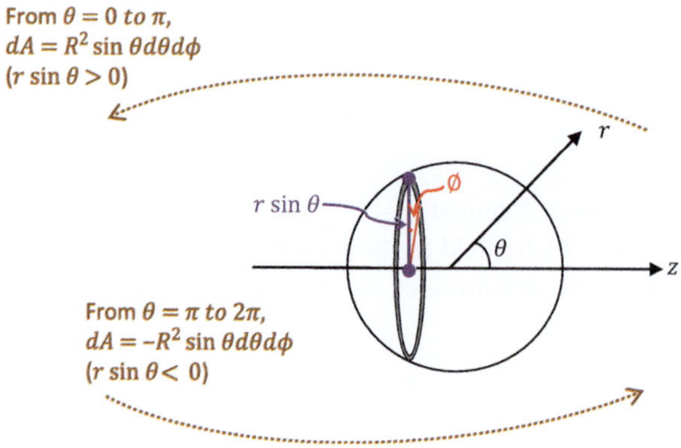


$$\dot{Q} = \int_{\text{Ø}=0}^{\text{Ø}=2\pi} \int_{\theta=0}^{\theta=\pi/2} \left( v_r|_{r=R} \right) R^2 \sin \theta d\theta d\text{Ø}$$

$$= \frac{3R^2 Pk}{L\eta} \int_0^{\pi/2} 2\pi \cos \theta \sin \theta d\theta = \frac{3\pi R^2 Pk}{L\eta} \int_0^{\pi/2} \sin 2\theta d\theta$$

$$= \frac{3\pi R^2 Pk}{L\eta} \left[ -\frac{1}{2} \cos 2\theta \right]_0^{\pi/2} = \frac{3\pi R^2 Pk}{L\eta}$$

## Problem 35

**Consider a fluid that flows in the space contained between two concentric spheres of radii $R_1$ and $R_2$ as shown below. The fluid is Newtonian, and the flow may be assumed fully developed and at a steady state, with a sufficiently low Re to use the creeping flow approximation.**

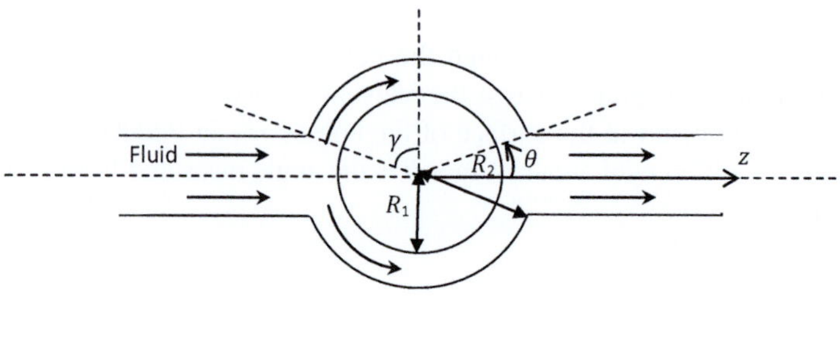

(a) **Show that $v_\theta = \frac{G(r)}{\sin\theta}$ where $G(r)$ is a function of r.**
(b) **Show that pressure is independent of the radial distance r.**
(c) **Find pressure gradient $\frac{\partial P}{\partial \theta}$ and show that the following is true:**

$$\frac{r}{3}\frac{\partial^3 G}{\partial r^3} + \frac{\partial^2 G}{\partial r^2} = 0$$

(d) **Find solutions of the form $G \sim r^\alpha$. Derive the general form of the solution for $G$, and using this general solution, determine $\frac{\partial P}{\partial \theta}$.**
(e) **Express $v_\theta$ in terms of the pressure difference between the fluid entry and exit points, given that:**

$$\int \frac{d\theta}{\sin\theta} = \frac{1}{2}\ln\left(\frac{1-\cos\theta}{1+\cos\theta}\right), \quad R_2 = 1.2R_1$$

## Solution 35

### *Worked Solution*

(a) Before we begin, let us state some key simplifying assumptions

- Constant viscosity for a Newtonian fluid.
- The liquid is incompressible, $\rho$ is constant.
- The flow is symmetric in the Ø direction; hence $\frac{\partial}{\partial Ø} = 0$.
- The fluid flow is at steady state; hence $\frac{\partial}{\partial t} = 0$.
- $v_\theta$ is the dominant flow direction; hence we assume $v_Ø = v_r = 0$. The flow is fully developed; hence $v_\theta \neq v_\theta(\theta)$
- Gravity effects are absent and inertial effects are ignored under the creeping flow (i.e., convective derivative portion of the Navier–Stokes equation is zero).

From the continuity equation in spherical coordinates, we can obtain the following

$$\frac{1}{r\sin\theta}\frac{\partial(v_\theta \sin\theta)}{\partial\theta} = 0$$

$$v_\theta \sin\theta = G(r)$$

$$v_\theta = \frac{G(r)}{\sin\theta}$$

(b) To find out the pressure gradient in the $r$ direction, let us look at the $r$-component Navier–Stokes equation

$$0 = -\frac{\partial P}{\partial r} + \eta\left[-\frac{2}{r^2}\frac{\partial v_\theta}{\partial \theta} - \frac{2}{r^2}v_\theta \cot\theta\right]$$

From our result in part a, we know the following expression

$$\frac{\partial v_\theta}{\partial \theta} = -G(r)\frac{\cos\theta}{\sin^2\theta}$$

Substituting into the pressure gradient equation, we show here that $P$ is not dependent on $r$.

$$\frac{\partial P}{\partial r} = \eta\left[\frac{2G(r)\cos\theta}{r^2\sin^2\theta} - \frac{2}{r^2}\frac{G(r)}{\sin\theta}\frac{\cos\theta}{\sin\theta}\right] = 0$$

(c) To find pressure gradient $\frac{\partial P}{\partial \theta}$, let us look at the $\theta$ component Navier–Stokes equation which will be simplified to the following after applying the earlier-stated assumptions

$$0 = -\frac{1}{r}\frac{\partial P}{\partial \theta} + \eta\left[\frac{1}{r^2}\frac{\partial}{\partial r}\left(r^2\frac{\partial v_\theta}{\partial r}\right) + \frac{1}{r^2\sin\theta}\frac{\partial}{\partial \theta}\left(\sin\theta\frac{\partial v_\theta}{\partial \theta}\right) - \frac{v_\theta}{r^2\sin^2\theta}\right]$$

$$\frac{1}{r}\frac{\partial P}{\partial \theta} = \eta\left[\frac{1}{r^2}\frac{\partial}{\partial r}\left(r^2\frac{G'(r)}{\sin\theta}\right) + \frac{1}{r^2\sin\theta}\frac{\partial}{\partial \theta}\left(-G(r)\frac{\cos\theta}{\sin\theta}\right) - \frac{G(r)}{r^2\sin^3\theta}\right]$$

$$\frac{1}{r}\frac{\partial P}{\partial \theta} = \eta\left[\frac{1}{r^2}\left(r^2\frac{G''(r)}{\sin\theta} + 2r\frac{G'(r)}{\sin\theta}\right) + \frac{-G(r)}{r^2\sin\theta}\left(\frac{-\sin^2\theta - \cos^2\theta}{\sin^2\theta}\right) - \frac{G(r)}{r^2\sin^3\theta}\right]$$

$$\frac{1}{r}\frac{\partial P}{\partial \theta} = \eta\left[\left(\frac{G''(r)}{\sin\theta} + \frac{2}{r}\frac{G'(r)}{\sin\theta}\right) + \frac{G(r)}{r^2\sin^3\theta} - \frac{G(r)}{r^2\sin^3\theta}\right]$$

$$\frac{\partial P}{\partial \theta} = \frac{\eta}{\sin\theta}[rG''(r) + 2G'(r)]$$

We know from earlier that $P$ is not a function of $r$; therefore

$$0 = \frac{\partial}{\partial r}\left(\frac{\partial P}{\partial \theta}\right) = \frac{\partial}{\partial r}\left[\frac{\eta}{\sin\theta}(rG''(r) + 2G'(r))\right]$$

$$rG'''(r) + G''(r) + 2G''(r) = rG'''(r) + 3G''(r) = 0$$

$$\frac{r}{3}\frac{\partial^3 G}{\partial r^3} + \frac{\partial^2 G}{\partial r^2} = 0$$

(d) Before we find the general form of the solution to this differential equation, let us explore trial solutions. One possible form of a trial solution is $G \sim r^\alpha$ as given.

$$\frac{\partial^3 G}{\partial r^3} \sim \alpha(\alpha-1)(\alpha-2)r^{\alpha-3}$$

$$\frac{\partial^2 G}{\partial r^2} \sim \alpha(\alpha-1)r^{\alpha-2}$$

$$\frac{r}{3}\alpha(\alpha-1)(\alpha-2)r^{\alpha-3} + \alpha(\alpha-1)r^{\alpha-2} = 0$$

$$\alpha(\alpha-1)\left(r^{\alpha-2}\right)[\alpha+1] = 0$$

$$\alpha = 0{,}1 \ or -1$$

$$G = c_1 + c_2 r + \frac{c_3}{r}$$

Hence, we can proceed to find $\frac{\partial P}{\partial \theta}$

$$\frac{\partial P}{\partial \theta} = \frac{\eta}{\sin\theta}[rG''(r) + 2G'(r)] = \frac{\eta}{\sin\theta}\left[r\left(\frac{2c_3}{r^3}\right) + 2\left(c_2 - \frac{c_3}{r^2}\right)\right] = \frac{2c_2\eta}{\sin\theta}$$

(e) To find pressure difference between the entry and exit points, let us translate this into boundary conditions.

Considering just the upper hemisphere (since pressure difference is symmetrically identical for the bottom hemisphere), at fluid enters at $\theta = \frac{\pi}{2} + \gamma$ and exits at $\theta = \frac{\pi}{2} - \gamma$

$$\frac{\partial P}{\partial \theta} = \frac{2c_2\eta}{\sin\theta}$$

$$\text{entry pressure} - \text{exit pressure} = \Delta P = \int_{\text{exit}}^{\text{entry}} \frac{2c_2\eta}{\sin\theta} d\theta = c_2\eta\left[\ln\left(\frac{1-\cos\theta}{1+\cos\theta}\right)\right]_{\frac{\pi}{2}-\gamma}^{\frac{\pi}{2}+\gamma}$$

$$= c_2\eta\left[\ln\left(\frac{1-\cos\left(\frac{\pi}{2}+\gamma\right)}{1+\cos\left(\frac{\pi}{2}+\gamma\right)}\right) - \ln\left(\frac{1-\cos\left(\frac{\pi}{2}-\gamma\right)}{1+\cos\left(\frac{\pi}{2}-\gamma\right)}\right)\right]$$

$$= c_2\eta\left[\ln\left(\frac{1+\sin\gamma}{1-\sin\gamma}\right) - \ln\left(\frac{1-\sin\gamma}{1+\sin\gamma}\right)\right] = 2c_2\eta\left[\ln\left(\frac{1+\sin\gamma}{1-\sin\gamma}\right)\right]$$

We found earlier that $v_\theta = \frac{G(r)}{\sin\theta} = \frac{1}{\sin\theta}\left(c_1 + c_2 r + \frac{c_3}{r}\right)$, so we can apply boundary conditions, when $r = R_2 = \frac{6}{5}R_1$, $v_\theta = 0$ and when $r = R_1$, $v_\theta = 0$.

$$0 = c_1 + c_2 R_1 + \frac{c_3}{R_1} \tag{1}$$

$$0 = c_1 + \frac{6}{5}c_2R_1 + \frac{5c_3}{6R_1} \tag{2}$$

Equation (2) subtract equation (1) gives

$$0 = \frac{1}{5}c_2R_1 - \frac{1}{6}\frac{c_3}{R_1} \rightarrow c_3 = \frac{6}{5}c_2{R_1}^2 \tag{3}$$

Substitute result (3) back into equation (1), we can find $c_1$

$$c_1 = -c_2R_1 - \frac{6}{5}c_2R_1$$

Finally we get $v_\theta$ in terms of

$$v_\theta = \frac{-c_2R_1 - \frac{6}{5}c_2R_1 + c_2r + \frac{\frac{6}{5}c_2{R_1}^2}{r}}{\sin\theta} = \frac{-\frac{11}{5}c_2R_1 + c_2r + \frac{\frac{6}{5}c_2{R_1}^2}{r}}{\sin\theta}$$

We know from earlier that

$$\Delta P = 2c_2\eta\left[\ln\left(\frac{1+\sin\gamma}{1-\sin\gamma}\right)\right] \rightarrow c_2 = \frac{\Delta P}{2\eta\left[\ln\left(\frac{1+\sin\gamma}{1-\sin\gamma}\right)\right]}$$

Therefore we can express $v_\theta$ in terms of the pressure difference $\Delta P$

$$v_\theta = \frac{\Delta P}{2\eta\left[\ln\left(\frac{1+\sin\gamma}{1-\sin\gamma}\right)\right]}\left[\frac{\frac{11}{5}R_1 + r + \frac{\frac{6}{5}{R_1}^2}{r}}{\sin\theta}\right]$$

## Problem 36

**Consider a solid sphere of radius *R* that is submerged stationary in an infinite body of fluid. Fluid flows past the sphere in a creeping flow and at a steady rate. At positions far upstream and downstream from the sphere, the velocity is constant at free stream velocity *U*, parallel to the *x* axis. Assume that the fluid is Newtonian and incompressible, and gravity effects may be ignored.**

**Given that the velocity profile of the fluid flow past the sphere in the *r* and *θ* directions are as follows:**

$$v_r = U\left(1 - \frac{3R}{2r} + \frac{R^3}{2r^3}\right)\cos\theta$$

$$v_\theta = -U\left(1 - \frac{3R}{4r} - \frac{R^3}{4r^3}\right)\sin\theta$$

(a) **Check that boundary conditions are satisfied by the velocity profile given above.**
(b) **Find the pressure profile for the fluid flow at the surface of the sphere. Assume that pressure at $\theta = 0$ is $P_c$.**
(c) **Find the form drag exerted by the fluid on the sphere.**

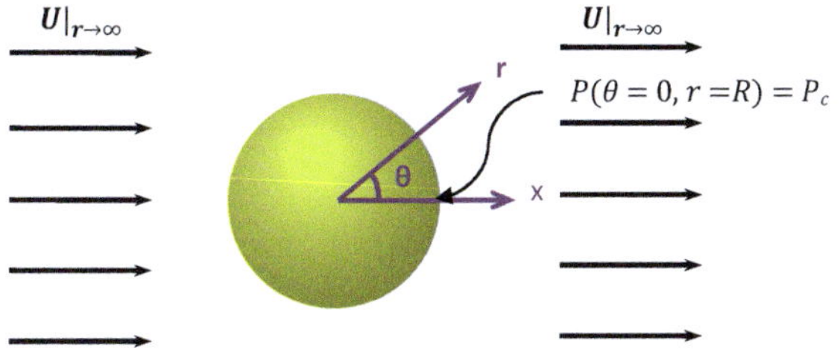


**Solution 36**

***Worked Solution***

(a) Before we begin, it is good practice to first list down assumptions for this flow as they may help to simplify our expressions (e.g., Navier–Stokes or continuity equations) later.

- Constant viscosity for a Newtonian fluid.
- The liquid is incompressible, $\rho$ is constant.
- In terms of coordinates, the flow is symmetric in the Ø direction; hence $\frac{\partial}{\partial Ø} = 0$.
- The fluid flow is at steady state; hence $\frac{\partial}{\partial t} = 0$.
- The dominant velocity components are in the $r$ and $\theta$ directions; hence $v_Ø = 0$.
- Gravity effects are absent and inertial effects are ignored under the creeping flow (i.e., convective derivative portion of the Navier–Stokes equation is zero). For creeping flow, $\mathrm{Re} = \frac{\rho UD}{\eta} \ll 1$.

The boundary conditions occur at the surface of the sphere, when $r = R$, $v_r = v_\theta = 0$. When $r = \infty$, $v_r = U\cos\theta$ and $v_\theta = -U\sin\theta$. These boundary conditions are satisfied by the velocity profiles given.

In the $r$ component,

$$v_r = U\left(1 - \frac{3R}{2r} + \frac{R^3}{2r^3}\right)\cos\theta$$

$$v_r = 0 = U\left(1 - \frac{3R}{2R} + \frac{R^3}{2R^3}\right)\cos\theta \text{ at } r = R$$

$$v_r = U\cos\theta = U(1 - 0 + 0)\cos\theta \text{ at } r = \infty$$

In the $\theta$ component,

$$v_\theta = -U\left(1 - \frac{3R}{4r} - \frac{R^3}{4r^3}\right)\sin\theta$$

$$v_\theta = 0 = -U\left(1 - \frac{3R}{4R} - \frac{R^3}{4R^3}\right)\sin\theta \text{ at } r = R$$

$$v_\theta = -U\sin\theta = -U(1 - 0 - 0)\sin\theta \text{ at } r = \infty$$

(b) Let us look at the $r$ and $\theta$ component Navier–Stokes equations. Using the velocity profiles provided, we can see that some of the terms would become zero and hence we obtain the following simplified equations.

In the $r$ component,

$$0 = -\frac{\partial P}{\partial r} + \eta\left[\frac{1}{r^2}\frac{\partial}{\partial r}\left(r^2\frac{\partial v_r}{\partial r}\right)\right]$$

$$\frac{\partial P}{\partial r} = \frac{\eta}{r^2}\frac{\partial}{\partial r}\left(r^2\frac{\partial v_r}{\partial r}\right)$$

Note that we can obtain $\frac{\partial v_r}{\partial r}$ from the velocity profile $v_r$ given.

$$\frac{\partial v_r}{\partial r} = U\cos\theta\left(\frac{3R}{2r^2} - \frac{3R^3}{2r^4}\right)$$

Therefore we can substitute the velocity gradient back into the pressure differential equation.

$$\frac{\partial P}{\partial r} = \frac{\eta}{r^2}\frac{\partial}{\partial r}\left(U\cos\theta\left(\frac{3R}{2} - \frac{3R^3}{2r^2}\right)\right)$$

$$= \left(\frac{\eta}{r^2}\right)\frac{3UR^3\cos\theta}{r^3} = \frac{3\eta UR^3\cos\theta}{r^5}$$

$$\left.\frac{\partial P}{\partial r}\right|_{r=R} = \frac{3\eta U \cos\theta}{R^2} \text{ (independent of } r\text{)}$$

$$P = P(\theta)$$

This tells us some information (but not all) about the pressure profile. It will be a function of $\theta$ only so we will need to look further into the $\theta$ component Navier–Stokes equation and apply boundary conditions in $\theta$.

$$0 = -\frac{1}{r}\frac{\partial P}{\partial \theta} + \eta\left[\frac{1}{r^2}\frac{\partial}{\partial r}\left(r^2\frac{\partial v_\theta}{\partial r}\right)\right]$$

$$\frac{\partial P}{\partial \theta} = \frac{\eta}{r}\frac{\partial}{\partial r}\left(r^2\frac{\partial v_\theta}{\partial r}\right)$$

Similarly, we can obtain $\frac{\partial v_\theta}{\partial r}$ from the velocity profile $v_\theta$ given.

$$\frac{\partial v_\theta}{\partial r} = -U\sin\theta\left(\frac{3R}{4r^2} + \frac{3R^3}{4r^4}\right)$$

Therefore we can substitute the velocity gradient back into the pressure differential equation.

$$\frac{\partial P}{\partial \theta} = \frac{\eta}{r}(-U\sin\theta)\frac{\partial}{\partial r}\left(\frac{3R}{4} + \frac{3R^3}{4r^2}\right)$$

$$= \frac{\eta}{r}(U\sin\theta)\frac{3R^3}{2r^3}$$

$$\left.\frac{\partial P}{\partial \theta}\right|_{r=R} = \frac{3\eta}{2R}(U\sin\theta)$$

$$P = \frac{3\eta}{2R}(-U\cos\theta) + c_1$$

When $\theta = 0$, $P = P_c$; therefore $c_1 = P_c + \frac{3\eta U}{2R}$ and the pressure profile is found as shown below.

$$P = \frac{3\eta U}{2R}(1 - \cos\theta) + P_c$$

(c) The form drag exerted on the sphere is caused by normal stresses (acting perpendicular to tangential plane); hence form drag comes from pressure term $P$. This is as opposed to shear stress (e.g., $\tau_{r\theta}$) that acts tangent to the surface resulting in skin/friction drag. Total drag force is the sum of form and skin drags.

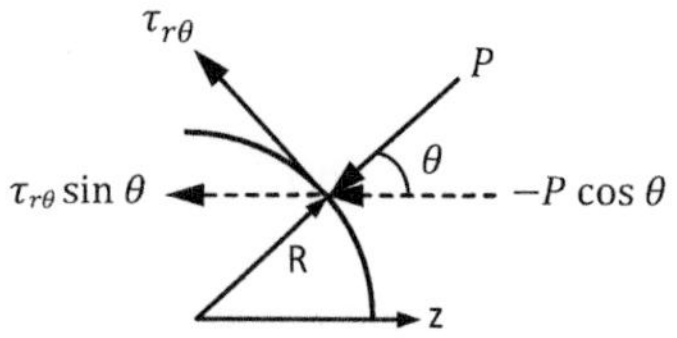


In spherical coordinates, a differential element will have area of $R^2 \sin\theta d\theta d\emptyset$, where $\emptyset$ is from 0 to $2\pi$, while $\theta$ is from 0 to $\pi$.

The form drag per unit area exerted by fluid on sphere is as follows. It is negative as it acts in the negative $z$ direction.

$$\frac{F_{\text{form}}}{A} = -P\cos\theta$$

$$F_{\text{form}}\big|_{r=R} = \int_{\emptyset=0}^{\emptyset=2\pi}\int_{\theta=0}^{\theta=\pi}(-P\cos\theta)\left(R^2\sin\theta d\theta d\emptyset\right)$$

Substituting the expression for $P$ and integrating with respect to $\emptyset$, we have the following. Note that $\sin 2\theta = 2\sin\theta\cos\theta$ by double-angle formula.

$$F_{\text{form}}\big|_{r=R} = -2\pi R^2\int_{\theta=0}^{\theta=\pi}\left(\frac{3\eta U}{2R}(1-\cos\theta)+P_c\right)\cos\theta\sin\theta d\theta$$

$$= -2\pi R^2\int_{\theta=0}^{\theta=\pi}\left(\frac{3\eta U}{2R}(1-\cos\theta)+P_c\right)\left(\frac{\sin 2\theta}{2}\right)d\theta$$

$$= -2\pi R^2\left[\int_{\theta=0}^{\theta=\pi}\left(\left(\frac{3\eta U}{2R}+P_c\right)\left(\frac{\sin 2\theta}{2}\right)\right)d\theta - \int_{\theta=0}^{\theta=\pi}\frac{3\eta U}{2R}\cos\theta\left(\frac{\sin 2\theta}{2}\right)d\theta\right]$$

$$= -2\pi R^2\left(\frac{3\eta U}{2R}\right)\left[-\int_{\theta=0}^{\theta=\pi}\cos^2\theta\sin\theta d\theta\right] = -3\pi R\eta U\left[\frac{\cos^3\theta}{3}\right]_0^{\pi}$$

$$= 2\pi R\eta U$$

Therefore the form drag exerted by the fluid on the sphere at the surface is $2\pi R\eta U$.

## Problem 37

**Consider a straight horizontal channel with a cross section shaped as an equilateral triangle with each side of length $\alpha$. There is fluid flow along the channel in the $z$ direction and the flow can be described as laminar, fully developed and steady. The fluid is Newtonian and incompressible, with viscosity**

**$\eta$. Gravity effects may be ignored. A trial form of the solution for velocity $v_z$ is $v_z = \lambda y\left(y - \sqrt{3}x\right)\left(y + \sqrt{3}x - \sqrt{3}\alpha\right)$.**

(a) **Sketch a diagram to illustrate this system, and comment on the driving force for this flow, and the other velocity components.**
(b) **Verify if the trial solution is acceptable, and determine $\lambda$.**
(c) **Determine the shear stress (per unit length of channel) along one side of the triangle. Comment on the values of mean shear stress and maximum shear stress along the chosen side.**

## Solution 37

*Worked Solution*

(a) The driving force for this flow is pressure, and it is used to overcome viscous effects arising from the fluid viscosity and related shear stress. The velocity components in the Cartesian coordinates are $v_x$, $v_y$ and $v_z$. It may be assumed that $v_x = v_y = 0$ and the dominant direction of flow is in the $z$ direction.

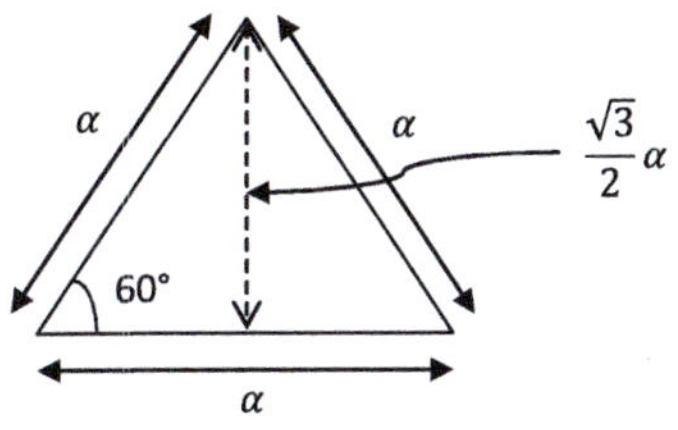


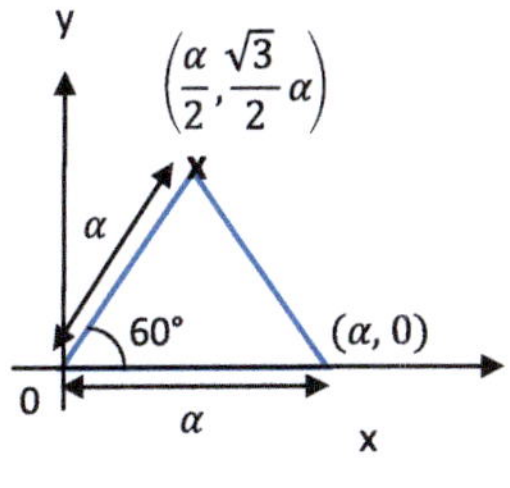


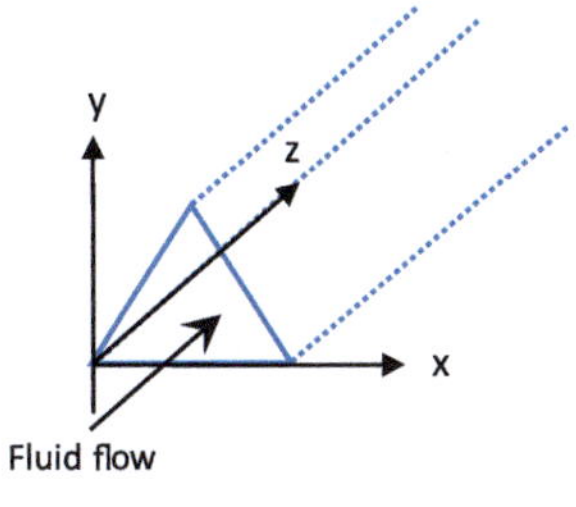


(b) Before we look at the trial solution, let us first state some assumptions about the flow which may help us to simplify equations later.

- Constant viscosity for a Newtonian fluid.
- The liquid is incompressible, $\rho$ is constant.
- The flow is at steady state ($\frac{\partial}{\partial t} = 0$).
- The dominant flow direction is in the $z$ direction; hence $v_x = v_y = 0$.

- The flow is fully developed where entry/end effects are assumed negligible; hence $v_z \neq v_z(z)$. This assumption is valid when the channel is long compared to the length scale of the cross section (e.g., $\alpha$ in this case).
- Gravity effects are absent as the channel is horizontal.

We are given a trial solution of the form shown below, and we can test if it is valid by imposing boundary conditions at each side of the triangle where velocity is zero.

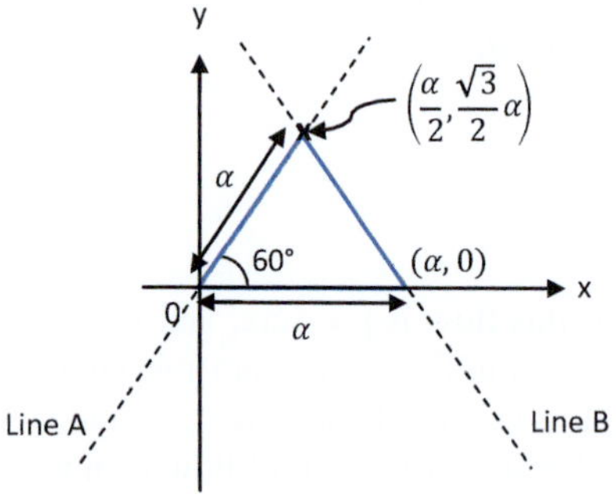


We can first figure out the straight line equations of Lines A and B marked. In general, we can find the equation of a straight line once we have the coordinates of two points on the line via $\frac{y-y_1}{x-x_1} = \frac{y_2-y_1}{x_2-x_1}$.

So the equation of Line A is

$$\frac{y-0}{x-0} = \frac{\frac{\sqrt{3}}{2}\alpha}{\frac{\alpha}{2}}$$

$$y = \sqrt{3}x$$

Similarly, the equation of Line B is

$$\frac{y-0}{x-\alpha} = \frac{\frac{\sqrt{3}}{2}\alpha - 0}{\frac{\alpha}{2} - \alpha} = -\sqrt{3}$$

$$y = -\sqrt{3}x + \sqrt{3}\alpha$$

Therefore, the boundary conditions are when $y = \sqrt{3}x$, $v_z = 0$, when $y = -\sqrt{3}x + \sqrt{3}\alpha$, $v_z = 0$ and when $y = 0$, $v_z = 0$. We observe that these conditions are fulfilled by the trial solution; hence it is valid.

$$v_z = \lambda y\left(y - \sqrt{3}x\right)\left(y + \sqrt{3}x - \sqrt{3}\alpha\right)$$

In order to find $\lambda$, let us consider the $z$ component Navier–Stokes equation

$$0 = -\frac{\partial P}{\partial z} + \eta\left[\frac{\partial^2 v_z}{\partial x^2} + \frac{\partial^2 v_z}{\partial y^2}\right]$$

$$\frac{\partial^2 v_z}{\partial x^2} = \frac{\partial}{\partial x}\left[\lambda y\sqrt{3}\left(y - \sqrt{3}x\right) + \left(-\lambda y\sqrt{3}\right)\left(y + \sqrt{3}x - \sqrt{3}\alpha\right)\right]$$

$$= \frac{\partial}{\partial x}(3\lambda y\alpha - 6\lambda yx) = -6\lambda y$$

$$\frac{\partial^2 v_z}{\partial y^2} = \frac{\partial}{\partial y}\left[\lambda y\left(y - \sqrt{3}x\right)(1) + \left(y + \sqrt{3}x - \sqrt{3}\alpha\right)\left(2\lambda y - \lambda\sqrt{3}x\right)\right]$$

$$= \frac{\partial}{\partial y}\left[\lambda y^2 - \lambda y\sqrt{3}x + 2\lambda y^2 - \lambda\sqrt{3}xy + 2\lambda y\sqrt{3}x - 3\lambda x^2 - 2\sqrt{3}\alpha\lambda y + 3\alpha\lambda x\right]$$

$$= \frac{\partial}{\partial y}\left[\lambda y^2 + 2\lambda y^2 - 3\lambda x^2 - 2\sqrt{3}\alpha\lambda y + 3\alpha\lambda x\right] = 2\lambda y + 4\lambda y - 2\sqrt{3}\alpha\lambda$$

$$= \lambda\left(6y - 2\sqrt{3}\alpha\right)$$

Therefore we substitute the second-order derivatives back into the Navier–Stokes equation to get

$$0 = -\frac{\partial P}{\partial z} + \eta\left[-6\lambda y + \lambda\left(6y - 2\sqrt{3}\alpha\right)\right]$$

$$-6\lambda y + \lambda\left(6y - 2\sqrt{3}\alpha\right) = \frac{1}{\eta}\frac{\partial P}{\partial z}$$

$$\lambda = -\frac{1}{2\sqrt{3}\alpha\eta}\frac{\partial P}{\partial z}$$

(c) Let us now find out the shear stress along one side (bottom face arbitrarily chosen here, whereby the plane will be the $y$ plane to which the $y$ axis is normal) of the triangular channel by relating shear stress $\tau$ to the velocity gradient.

$$\tau_{yz}\big|_{y=0} = \eta\frac{\partial v_z}{\partial y} = \eta\left(-3\lambda x^2 + 3\alpha\lambda x\right) = \frac{\sqrt{3}}{2\alpha}\frac{\partial P}{\partial z}x(x - \alpha)$$

Note that this is a parabolic shear stress profile which is in line with a pressure driven flow. You may observe that at the bottom face, when $x = 0$ and $x = \alpha$, shear stress is zero. Hence fouling materials will tend to accumulate at these two corners.

This parabola reaches a maximum point of shear stress at $\frac{\partial \tau_{yz}}{\partial x} = 0$ where $x = \frac{\alpha}{2}$.

$$\tau_{yz,\max}\Big|_{y=0} = \frac{\sqrt{3}}{2\alpha}\frac{\partial P}{\partial z}\frac{\alpha}{2}\left(-\frac{\alpha}{2}\right) = -\frac{\alpha\sqrt{3}}{8}\frac{\partial P}{\partial z}$$

We can also find the mean value of shear stress as follows.

$$\left\langle \tau_{yz}\Big|_{y=0}\right\rangle = \frac{1}{\alpha}\int_0^\alpha \frac{\sqrt{3}}{2\alpha}\frac{\partial P}{\partial z}x(x-\alpha)dx = \frac{\sqrt{3}}{2\alpha^2}\frac{\partial P}{\partial z}\left[\frac{x^3}{3} - \frac{\alpha x^2}{2}\right]_0^\alpha = -\frac{\alpha\sqrt{3}}{12}\frac{\partial P}{\partial z}$$

Therefore we can also observe that $\tau_{\max}\Big|_{y=0} = \frac{3}{2}\left\langle \tau_{yz}\Big|_{y=0}\right\rangle$ for this parabolic function.

## Problem 38

**Consider a Newtonian fluid of viscosity $\eta$ and constant density $\rho$ which is contained in a cylinder of length $L$ and radius $R$. Assume that $L \gg R$. The cylinder is horizontal such that gravity effects on fluid flow are negligible. At time $t < 0$, the cylinder and fluid rotate at a constant angular velocity of $\omega$. At time $t = 0$, the cylinder's rotation was stopped. You may assume fluid flow in the cylinder is slow where Re $\ll 1$.**

(a) **Starting with the original set of Navier–Stokes equations in cylindrical coordinates, derive the simplified equation that describes the fluid flow from $t = 0$.**

(b) **Using separation of variables between time and position, derive a differential equation in $r$. Show further that this equation can be expressed in the form below which is that of a Bessel function of order 1, where $R$ is a function of $x$, i.e., $R = R(x)$.**

$$R''x^2 + R'x + R(x^2 - 1) = 0$$

**Show that the velocity profile in $\theta$ can be expressed in the form as follows:**

$$v_\theta(r,t) = \sum_{n=1}^{\infty} J_1\left(\sqrt{\frac{\alpha_n}{\nu}}r\right)c_n e^{-\alpha_n t}$$

## Solution 38

*Worked Solution*

(a) Let us start with the $\theta$-component Navier–Stokes equation.

Fluid flow occurs predominantly in the $\theta$ direction; hence $v_r = v_z = 0$. Also since the cylinder is horizontal, there is no gravitational component contributing to flow; hence $g_\theta = 0$. Hence the terms highlighted are negligible and can be removed from consideration.

$$\rho\left(\frac{\partial v_\theta}{\partial t} + v_r\frac{\partial v_\theta}{\partial r} + \frac{v_\theta}{r}\frac{\partial v_\theta}{\partial \theta} + \frac{v_r v_\theta}{r} + v_z\frac{\partial v_\theta}{\partial z}\right)$$
$$= -\frac{1}{r}\frac{\partial P}{\partial \theta} + \eta\left[\frac{\partial}{\partial r}\left(\frac{1}{r}\frac{\partial (r v_\theta)}{\partial r}\right) + \frac{1}{r^2}\frac{\partial^2 v_\theta}{\partial \theta^2} + \frac{2}{r^2}\frac{\partial v_r}{\partial \theta} + \frac{\partial^2 v_\theta}{\partial z^2}\right] + \rho g_\theta$$

Also, due to axisymmetry about the $z$ axis, we know that $\frac{\partial}{\partial \theta} = 0$. The terms highlighted below can be removed.

$$\rho\left(\frac{\partial v_\theta}{\partial t} + \frac{v_\theta}{r}\frac{\partial v_\theta}{\partial \theta}\right) = -\frac{1}{r}\frac{\partial P}{\partial \theta} + \eta\left[\frac{\partial}{\partial r}\left(\frac{1}{r}\frac{\partial (r v_\theta)}{\partial r}\right) + \frac{1}{r^2}\frac{\partial^2 v_\theta}{\partial \theta^2} + \frac{\partial^2 v_\theta}{\partial z^2}\right]$$

Next we note that the cylinder is long; hence any entry or end effects in the $z$ axis are negligible and $\frac{\partial}{\partial z} = 0$. Hence the terms highlighted below may be further ignored.

$$\rho\left(\frac{\partial v_\theta}{\partial t}\right) = \eta\left[\frac{\partial}{\partial r}\left(\frac{1}{r}\frac{\partial (r v_\theta)}{\partial r}\right)\right]$$

We may now expand the differential on the right-hand side using product rule.

$$\rho\left(\frac{\partial v_\theta}{\partial t}\right) = \eta\left[\frac{\partial}{\partial r}\left(\frac{1}{r}\left(r\frac{\partial v_\theta}{\partial r} + v_\theta\right)\right)\right]$$
$$= \eta\left[\frac{\partial}{\partial r}\left(\frac{\partial v_\theta}{\partial r} + \frac{v_\theta}{r}\right)\right]$$
$$= \eta\left[\frac{\partial^2 v_\theta}{\partial r^2} + \frac{1}{r}\frac{\partial v_\theta}{\partial r} - \frac{v_\theta}{r^2}\right]$$

Let us similarly examine the $r$-component Navier–Stokes equation. For the same reasons as the $\theta$-component equation, we eliminate the gravity term, and assume $v_r = v_z = 0$, $\frac{\partial}{\partial z} = \frac{\partial}{\partial \theta} = 0$. Therefore the terms highlighted below may be removed.

$$\rho\left(\frac{\partial v_r}{\partial t}+v_r\frac{\partial v_r}{\partial r}+\frac{v_\theta}{r}\frac{\partial v_r}{\partial \theta}-\frac{{v_\theta}^2}{r}+v_z\frac{\partial v_r}{\partial z}\right)$$
$$=-\frac{\partial P}{\partial r}+\eta\left[\frac{\partial}{\partial r}\left(\frac{1}{r}\frac{\partial(rv_r)}{\partial r}\right)+\frac{1}{r^2}\frac{\partial^2 v_r}{\partial \theta^2}-\frac{2}{r^2}\frac{\partial v_\theta}{\partial \theta}+\frac{\partial^2 v_r}{\partial z^2}\right]+\rho g_r$$

This gives us an equation that tells us that the pressure gradient in $r$ direction balances the centripetal force caused by the rotation.

$$-\frac{\rho {v_\theta}^2}{r}=-\frac{\partial P}{\partial r}$$

Finally, we observe that the $z$-component Navier–Stokes equation is completely eliminated when the same assumptions are applied, and hence does not provide useful information for our understanding of the problem. For completeness, the $z$-component equation is shown here for reference.

$$\rho\left(\frac{\partial v_z}{\partial t}+v_r\frac{\partial v_z}{\partial r}+\frac{v_\theta}{r}\frac{\partial v_z}{\partial \theta}+v_z\frac{\partial v_z}{\partial z}\right)=-\frac{\partial P}{\partial z}+\eta\left[\frac{1}{r}\frac{\partial}{\partial r}\left(r\frac{\partial v_z}{\partial r}\right)+\frac{1}{r^2}\frac{\partial^2 v_z}{\partial \theta^2}+\frac{\partial^2 v_z}{\partial z^2}\right]$$
$$+\rho g_z$$

(b) Next we are required to solve the equation using separation of variables. We note that the key equation to be solved is that for the dominant flow direction, i.e., $v_\theta$.

$$\rho\left(\frac{\partial v_\theta}{\partial t}\right)=\eta\left[\frac{\partial^2 v_\theta}{\partial r^2}+\frac{1}{r}\frac{\partial v_\theta}{\partial r}-\frac{v_\theta}{r^2}\right]$$

Let the trial solution take the form as follows, where the variables time $t$ and radial position $r$ are separated into two independent functions $R$ and $T$. This can be done since the two functions are each a function of independent variables $r$ and $t$.

$$v_\theta=R(r)T(t)$$
$$RT'=\frac{\eta}{\rho}\left(R''T+\frac{R'}{r}T-\frac{RT}{r^2}\right)$$

The below has to equate to a constant (let it be $\lambda$) since that is the only case when a function in $t$ only is equivalent to a function in $r$ only.

$$\frac{T'}{T} = \frac{\eta}{\rho}\left(\frac{R''}{R} + \frac{R'}{Rr} - \frac{1}{r^2}\right) = \text{constant}$$

A type of function, e.g., $T$ that when differentiated, gives back itself multiplied by a constant is the exponential function. For example, if $T(t) = e^{2t}$, $T' = \frac{d}{dt}(e^{2t}) = 2e^{2t} = 2T$. Another example $T(t) = 3e^{-2t}$, $T' = \frac{d}{dt}(3e^{-2t}) = -2(3e^{-2t}) = -2T$. In fact, $T$ can take on the more general form of $T = Ae^{Bx}$, where $A$ and $B$ are constants. Therefore, we can guess that $T$ is an exponential function as follows, where $\alpha > 0$. We know that as time increases, the energy provided by the initial rotation is dissipated; hence velocity decays with time as $t \to \infty$, $v_\theta = RT = R(0) = 0$. That is why we have added a negative sign before $\alpha$.

$$T = c_1 e^{-\alpha t}$$

Going back to our equation and looking at only the $R$ part, we have

$$\nu\left(\frac{R''}{R} + \frac{R'}{Rr} - \frac{1}{r^2}\right) = -\alpha$$

$$\frac{\alpha}{\nu} = -\frac{R''}{R} - \frac{R'}{Rr} + \frac{1}{r^2}$$

Let $\lambda = \sqrt{\frac{\alpha}{\nu}}$, and multiply throughout by $Rr^2$ therefore the equation becomes

$$\lambda^2 R r^2 = -R'' r^2 - R' r + R$$

$$R'' r^2 + R' r + R\left[(\lambda r)^2 - 1\right] = 0$$

If we define a new variable $x = \lambda r$, $dx = \lambda dr$, $R'(r) = \frac{dR}{dr} = \lambda \frac{dR}{dx} = \lambda R'(x)$ and $R''(r) = \lambda^2 R''(x)$. Therefore assuming now we perform a variable change such that $R$ is now a function of $x$ instead of $r$, we get,

$$R'' x^2 + R' x + R\left(x^2 - 1\right) = 0$$

This expression is the form of the Bessel function of order 1, which has pre-established solutions of known values that we can look up in data booklets. Therefore, it is a useful function to use in solving problems especially those with cylindrical symmetry.

Before continuing, let us recall some basic concepts about Bessel functions in general. Bessel functions are solutions of second-order differential equations of the form below where $n$ is a constant which determines the order of the function. Typical values of $n$ for similar flow patterns as above are $n = 0$ and $n = 1$.

$$x^2\frac{d^2y}{dx^2}+x\frac{dy}{dx}+\left(x^2-n^2\right)y=0$$

In the case of $n = 0$ or $n = 1$, the Bessel function has two independent solutions which are commonly denoted as $J_n(x)$ and $Y_n(x)$, and the general solution to the Bessel equation can be written as a sum of the two, i.e., a linear combination of both, with the coefficients determined by boundary conditions of the specific problem.

General solution to Bessel equation: $y = AJ_n(x) + BY_n(x)$, where $A$ and $B$ are constants. The solutions can be expressed as power series, e.g., $J_n(x)=\left(\frac{1}{2}x\right)^n\sum_{m=0}^{\infty}\frac{\left(-\frac{1}{4}x^2\right)^m}{m!(n+m)!}$. However, note that the solution for $Y_n(x)$ is ill-defined at origin, i.e., when $x \to 0$, a singularity occurs and $B = 0$ if this is not acceptable. For example, in this problem, when $x = 0$, i.e., at the centerline of the cylinder, flow still exists and this singularity is not acceptable. Note that there are other kinds of Bessel functions, and $n$ can be any constant.

Now back to solving our earlier differential equation in the form of a Bessel function of order 1, where $x=\sqrt{\frac{\alpha}{\nu}}r$,

$$R''x^2+R'x+R\left(x^2-1\right)=0$$

$$R(x)=AJ_n(x)+BY_n(x)$$

$$R(r)=AJ_1\left(\sqrt{\frac{\alpha}{\nu}}r\right)+BY_1\left(\sqrt{\frac{\alpha}{\nu}}r\right)$$

We note that there should be a physical solution at $r = 0$; therefore $B = 0$ due to the singularity of $Y_1$ at this position. Therefore, $R(r)=AJ_1\left(\sqrt{\frac{\alpha}{\nu}}r\right)$.

To solve this, we need to impose boundary conditions, which are that at $r = R$, $v_\theta = 0$. For $J_1\left(\sqrt{\frac{\alpha}{\nu}}R\right)=0$, we can refer to the data booklet and find that this gives a series of solutions for $\sqrt{\frac{\alpha}{\nu}}R=x_n$, where $n = 1, 2, 3..$, i.e., $x_1 = 3.82$, $x_2 = 7.02$ etc.

Now we can combine the solution for both the position variable (i.e., $r$) and time variable ($t$):

$$v_\theta(r,t)=R(r)T(t)$$

$$=\sum_{n=1}^{\infty}J_1\left(\sqrt{\frac{\alpha_n}{\nu}}r\right)c_n\mathrm{e}^{-\alpha_n t}$$

Note that in our problem, since at $t = 0$ and $r = R$, $v_\theta = R\omega$. Therefore

$$v_\theta(R,0) = \sum_{n=1}^{\infty} J_1\left(\sqrt{\frac{\alpha_n}{\nu}}R\right)c_n(1) = R\omega$$

$$c_n = \frac{R\omega}{\sum\limits_{n=1}^{\infty} J_1\left(\sqrt{\frac{\alpha_n}{\nu}}R\right)}$$

## Problem 39

**Consider steady flow of a viscous fluid in a cylindrical pipe of radius $R$. The flow is laminar, and gravity effects may be ignored. The fluid has a constant density $\rho$ and viscosity $\eta$. This fluid flow may be modeled using cylindrical coordinates where $z$ refers to the axial direction.**

(a) **Find the axial velocity profile of the fluid.**

**Now, instead of steady-state flow, we introduce variations with time. At time $t < 0$, no pressure gradient is exerted on the fluid and it remains still. From $t = 0$ onwards, a constant pressure difference is applied to the fluid, causing a velocity (time-dependent) in the axial direction.**

(b) **Using appropriate boundary conditions for this scenario, find the differential equation for the time-dependent portion of axial velocity.**
(c) **Using the method of separation of variables, show how the differential equation in part b can be simplified to the form of a Bessel function of order zero.**

## Solution 39

***Worked Solution***

(a) The assumptions that apply for this steady flow are as follows:

- Constant viscosity for a Newtonian fluid.
- The liquid is incompressible, $\rho$ is constant.
- The flow is at steady state ($\frac{\partial}{\partial t} = 0$).
- The dominant flow direction is in the $z$ direction; hence $v_r = v_\theta = 0$.
- The flow is fully developed where entry/end effects are assumed negligible; hence $v_z \neq v_z(z)$.
- There is axisymmetry; therefore $\frac{\partial}{\partial \theta} = 0$
- Gravity effects are absent.

Let us now simplify the $z$-component Navier–Stokes equation

$$0 = -\frac{\partial P}{\partial z} + \eta\left[\frac{1}{r}\frac{\partial}{\partial r}\left(r\frac{\partial v_z}{\partial r}\right)\right]$$

$$\frac{r}{\eta}\frac{\partial P}{\partial z} = \frac{\partial}{\partial r}\left(r\frac{\partial v_z}{\partial r}\right)$$

$$\frac{r^2}{2\eta}\frac{\partial P}{\partial z} + c_1 = r\frac{\partial v_z}{\partial r}$$

$$\frac{\partial v_z}{\partial r} = \frac{r}{2\eta}\frac{\partial P}{\partial z} + \frac{c_1}{r}$$

$$v_z = \frac{r^2}{4\eta}\frac{\partial P}{\partial z} + c_1 \ln r + c_2$$

When $r = 0$, velocity should be defined; therefore $c_1 = 0$. When $r = R$, $v_z = 0$. $c_2 = -\frac{R^2}{4\eta}\frac{\partial P}{\partial z}$. Therefore the axial velocity profile is

$$v_z = \frac{1}{4\eta}\left(-\frac{\partial P}{\partial z}\right)(R^2 - r^2)$$

(b) In order to find the velocity profile that also varies with time, let us consider boundary conditions for time as well.

In terms of time, at $t = 0$, $v_z = 0$. As time $t \to \infty$, $v_z = 0$. In terms of position, at $r = R$, $v_z = 0$.

In order to make use of our result in part a, let us define the time-dependent velocity as a sum of the steady-state component from part a (denoted by $v_{z,\,ss}$) and the deviation introduced by time dependence (denoted by $v_z'$):

$$v_z = v_{z,\,ss} + v_z'$$

Let us return to the $z$-component Navier–Stokes equation, where we no longer ignore any terms that are time-dependent.

$$\rho\frac{\partial v_z}{\partial t} = -\frac{\partial P}{\partial z} + \eta\left[\frac{1}{r}\frac{\partial}{\partial r}\left(r\frac{\partial v_z}{\partial r}\right)\right]$$

$$\rho\frac{\partial v_{z,\,ss}}{\partial t} + \rho\frac{\partial v_z'}{\partial t} = -\frac{\partial P}{\partial z} + \eta\left[\frac{1}{r}\frac{\partial}{\partial r}\left(r\frac{\partial v_{z,\,ss}}{\partial r}\right)\right] + \eta\left[\frac{1}{r}\frac{\partial}{\partial r}\left(r\frac{\partial v_z'}{\partial r}\right)\right]$$

Since $v_{z,\,ss}$ describes velocity at steady state, hence this term $\rho\frac{\partial v_{z,\,ss}}{\partial t} = 0$. We observe further that the terms highlighted in yellow are in fact the same equation in part (a) for steady-state flow, which means they equate to zero.

$$\rho\frac{\partial v_z'}{\partial t}=-\frac{\partial P}{\partial z}+\eta\left[\frac{1}{r}\frac{\partial}{\partial r}\left(r\frac{\partial v_{z,\mathrm{ss}}}{\partial r}\right)\right]+\eta\left[\frac{1}{r}\frac{\partial}{\partial r}\left(r\frac{\partial v_z'}{\partial r}\right)\right]$$

$$\rho\frac{\partial v_z'}{\partial t}=\eta\left[\frac{1}{r}\frac{\partial}{\partial r}\left(r\frac{\partial v_z'}{\partial r}\right)\right]$$

$$\frac{\partial v_z'}{\partial t}=\frac{\eta}{\rho}\left[\frac{1}{r}\frac{\partial}{\partial r}\left(r\frac{\partial v_z'}{\partial r}\right)\right]$$

(c) Let us express the time-dependent velocity component as a product of two independent functions $R$ and $T$, each being a function of variables $r$ (radial position) and $t$ (time) which are also independent variables.

$$v_z'=R(r)T(t)$$

We can express the differential equation found in part b as follows, where $\alpha$ is a positive constant.

$$RT'=\frac{\nu}{r}\frac{\partial}{\partial r}(rR'T)=\frac{\nu}{r}T\frac{\partial}{\partial r}(rR')=\frac{\nu}{r}T(rR''+R')=\nu T\left(R''+\frac{R'}{r}\right)$$

$$\frac{T'}{T}=\nu\left(\frac{R''}{R}+\frac{R'}{Rr}\right)=\frac{\nu}{Rr}(R''r+R')=-\alpha$$

We recall our earlier boundary condition for time, as time increases, the energy causing the fluid flow is dissipated, hence velocity $v_z'$ should decay with time as $t\rightarrow\infty$. Therefore we included a negative sign before $\alpha$.

$$-\frac{\alpha}{\nu}rR=R''r+R'$$

$$R''r^2+R'r+\frac{\alpha}{\nu}r^2R=0$$

Let $x=\sqrt{\frac{\alpha}{\nu}}r$, therefore $dx=\sqrt{\frac{\alpha}{\nu}}dr$, $R'(r)=\frac{dR}{dr}=\sqrt{\frac{\alpha dR}{\nu dx}}=\sqrt{\frac{\alpha}{\nu}}R'(x)$, and $R''(r)=\left(\sqrt{\frac{\alpha}{\nu}}\right)^2R''(x)$. This change of variable from $r$ to $x$, whereby the function $R$ is now a function of $x$ instead of $r$, gives us the form of differential equation that can be solved using Bessel function of order zero.

$$R''x^2+R'x+R\left(x^2-0\right)=0$$

## Problem 40

**In fluid mechanics, we often come across the concept of a boundary layer. This is a region near to the no-slip boundary (e.g., pipe wall or solid planar surface) where wall effects are significant and viscous forces are dominant. This is as opposed to inertial forces which are more dominant as we go further from the wall.**

(a) **Show how the boundary layer may be described by the Navier–Stokes equation**

$$\rho\left(v_x\frac{\partial v_x}{\partial x}+v_z\frac{\partial v_x}{\partial z}\right)=-\frac{\partial P}{\partial x}+\eta\frac{\partial^2 v_x}{\partial z^2}$$

**for the system illustrated below. You may assume steady laminar flow of a Newtonian, incompressible fluid.**

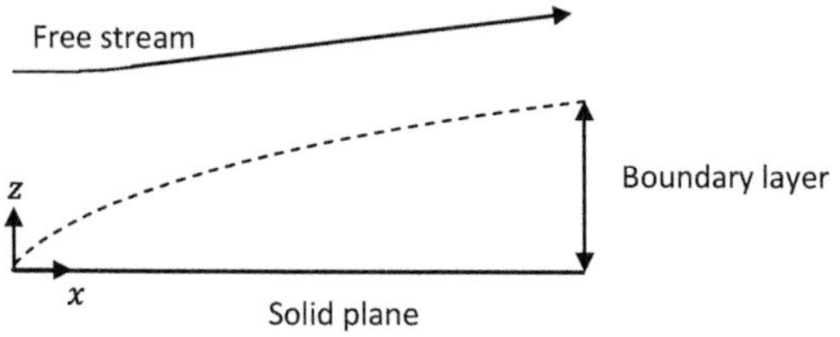


(b) **Assuming that we now have a free stream velocity approaching the solid plane at a constant value of $U$ that is parallel to the plane, and there is a small seepage of fluid into the solid planar surface at a velocity of $v = -v_s$ at $z = 0$. At positions further downstream, the velocity profile is fully developed in the $x$ direction. Find an expression for velocity $v_s$.**
(c) **Find an expression for $v_x$ in the boundary layer of this scenario of seepage, assuming that pressure remains constant in the $x$ direction.**
(d) **Evaluate the value of $z$ when $v_x$ is at at value of $0.7U$.**
(e) **Determine the stream function for this flow, and sketch the streamlines.**

## Solution 40

### *Worked Solution*

(a) The following assumptions apply for the boundary layer velocity profile

- Length scale of boundary layer is small (i.e., thickness as measured from $z = 0$). Viscous stresses dominate in this layer and hence velocity gradient is steep (velocity varies rapidly across the boundary layer as $z$ increases). Therefore velocities in both directions $x$ and $z$ should be considered.

- In the boundary layer region, the flow is not fully developed (entry effects are significant), i.e., the position $x$ matters for the velocity at that point, and hence $\frac{\partial v_x}{\partial x} \neq 0$.
- Constant viscosity for a Newtonian fluid.
- The liquid is incompressible, $\rho$ is constant.
- The flow is at steady state ($\frac{\partial}{\partial t} = 0$).
- Gravity effects are absent.
- Fluid flow is in the $x$ direction; hence the pressure driving force is dominant in this direction and $\frac{\partial P}{\partial z}$ assumed negligible.
- We assume 2D flow whereby $v_y = 0$.

Let us look at the Navier–Stokes equation in the $x$ and $z$ directions,

$$\rho\left(v_x\frac{\partial v_x}{\partial x} + v_z\frac{\partial v_x}{\partial z}\right) = -\frac{\partial P}{\partial x} + \eta\left(\frac{\partial^2 v_x}{\partial x^2} + \frac{\partial^2 v_x}{\partial z^2}\right) \tag{1}$$

$$\rho\left(v_x\frac{\partial v_z}{\partial x} + v_z\frac{\partial v_z}{\partial z}\right) = \eta\left(\frac{\partial^2 v_z}{\partial x^2} + \frac{\partial^2 v_z}{\partial z^2}\right) \tag{2}$$

We noted from one of the earlier assumptions that the velocity gradient in the $z$ direction is steep in the boundary layer, due to the significant shear stress. This is further exacerbated by the narrow boundary layer. Therefore we know that $\frac{\partial^2 v_x}{\partial z^2} \gg \frac{\partial^2 v_x}{\partial x^2}$. However note that we cannot cancel the term $v_x\frac{\partial v_x}{\partial x}$ because $\frac{\partial v_x}{\partial z} \gg \frac{\partial v_x}{\partial x}$ is small, because the coefficient $v_x$ in $v_x\frac{\partial v_x}{\partial x}$ is significant. Therefore equation (1) is shown to simplify to the following boundary layer expression as provided in the problem.

$$\rho\left(v_x\frac{\partial v_x}{\partial x} + v_z\frac{\partial v_x}{\partial z}\right) = -\frac{\partial P}{\partial x} + \eta\frac{\partial^2 v_x}{\partial z^2}$$

(b) Since seepage into the bottom surface occurs in the $-z$ direction, there is a negative sign before $v_s$ in the condition $v_z = -v_s$ at $z = 0$.

Using the mass continuity equation, we can relate $v_x$ to $v_z$

$$\frac{dv_x}{dx} + \frac{dv_z}{dz} = 0$$

At large values of $x$, we have fully developed flow, which means that $\frac{dv_x}{dx} = 0$. Substituting this into the continuity equation, we find that $\frac{dv_z}{dz} = 0$ which means that $v_z = v_z(x)$ only. We know further that in order for the boundary condition at $z = 0$ to hold for all values of $x$, $v_z = \text{constant} = -v_s$.

(c) To find an expression for $v_x$, we return to the boundary layer equation found in part a.

$$\rho\left(v_x\frac{\partial v_x}{\partial x}+v_z\frac{\partial v_x}{\partial z}\right)=-\frac{\partial P}{\partial x}+\eta\frac{\partial^2 v_x}{\partial z^2}$$

Knowing from earlier that $\frac{\partial v_x}{\partial x}=0$, and $v_z=-v_s$, and noting that the problem stated that $\frac{\partial P}{\partial x}=0$. Therefore we simplify the equation to the following where kinematic viscosity $\nu=\frac{\eta}{\rho}$.

$$\nu\frac{\partial^2 v_x}{\partial z^2}=-v_s\frac{\partial v_x}{\partial z}$$

$$\frac{\partial^2 v_x}{\partial z^2}=-\frac{v_s}{\nu}\frac{\partial v_x}{\partial z}$$

A suitable function whereby its derivatives give back itself multiplied by a constant is the exponential function. Therefore we adopt the trial solution

$$v_x=-c_1\mathrm{e}^{\alpha z}+c_2$$

$$\frac{\partial v_x}{\partial z}=-\alpha c_1\mathrm{e}^{\alpha z}$$

$$\frac{\partial^2 v_x}{\partial z^2}=-\alpha^2 c_1\mathrm{e}^{\alpha z}=\alpha\frac{\partial v_x}{\partial z}\rightarrow\alpha=-\frac{v_s}{\nu}$$

$$v_x=-c_1\mathrm{e}^{\left(-\frac{v_s}{\nu}\right)z}+c_2$$

We know the boundary condition for this boundary layer at $z\rightarrow\infty$, $v_x=U$. Therefore, $c_2=U$. Also at $z=0$, $v_x=0$ (no slip in the $x$ direction at solid surface). Therefore, $c_1=U$.

$$v_x=-U\mathrm{e}^{\left(-\frac{v_s}{\nu}\right)z}+U=U\left(1-\mathrm{e}^{\left(-\frac{v_s}{\nu}\right)z}\right)$$

(d) When $\frac{v_x}{U}=0.7$,

$$1-\mathrm{e}^{\left(-\frac{v_s}{\nu}\right)z}=0.7$$

$$\left(-\frac{v_s}{\nu}\right)z=\ln 0.3$$

$$\frac{v_s}{\nu}z=\ln 3.3$$

$$z=\frac{\nu}{v_s}\ln 3.3$$

(e) The stream function ψ(x, y) is characterized by two key equations:

$$v_x = \frac{\partial \psi}{\partial y}$$

$$v_y = -\frac{\partial \psi}{\partial x}$$

The above two equations are derived from a mass balance between two points at different streamlines.

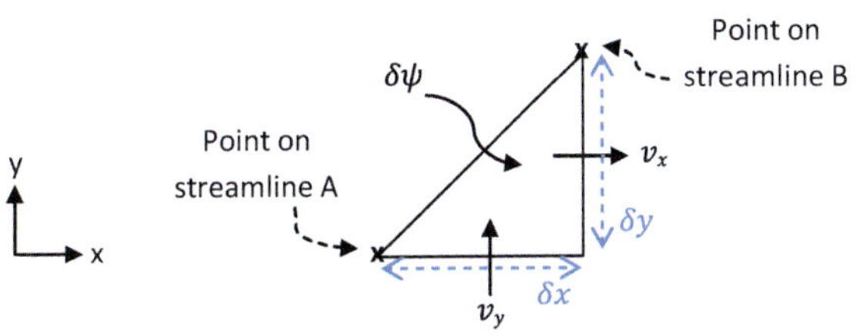


$$\text{In} - \text{Out} = 0$$

$$\delta\psi + v_y \delta x - v_x \delta y = 0$$

$$\delta\psi = -v_y \delta x + v_x \delta y = \frac{\partial \psi}{\partial x}\delta x + \frac{\partial \psi}{\partial y}\delta y$$

Therefore, $v_x = \frac{\partial \psi}{\partial y}, v_y = -\frac{\partial \psi}{\partial x}$ is shown.

---

***Background Concepts***

Let us recall some key aspects of the stream function before solving this problem.

- Lines of constant $\psi$ value in a steady flow are called streamlines. Fluid elements flow along the streamline and not across the streamline.
- The value of the stream function between two streamlines represents the region of the flow contained between these two streamlines. Volume flux between any two points of different stream function values can be visualized through a line connecting the two points. In the diagram below, we have a line connecting points $A$ and $B$, giving the volume flux $Q = \psi_B - \psi_A$, where $Q$ is the volumetric flow rate per unit depth. Note that $Q_1 = Q_2 = Q$ (constant).

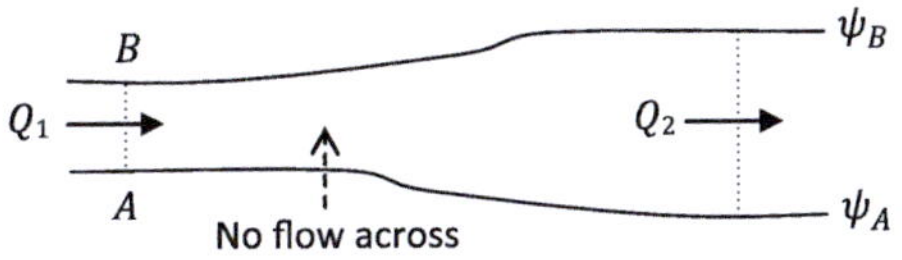


---

Let us now return to the problem. In order to determine the stream function, we can relate it to the velocity profile. Note that in our problem, the vertical coordinate is $z$ and not $y$.

Starting from the $z$-component velocity, we can integrate with respect to $x$ to obtain the stream function as shown below, where $f$ is a function of $z$.

$$v_z = -\frac{\partial \psi}{\partial x} = -v_s, \quad \psi = v_s x + f(z)$$

For the $x$-component velocity, we can similarly integrate with respect to $z$ to obtain the stream function below, where $g$ is a function of $x$.

$$v_x = \frac{\partial \psi}{\partial z} = U\left(1 - e^{\left(-\frac{v_s}{\nu}\right)z}\right), \quad \psi = U\left(z + \frac{\nu}{v_s} e^{\left(-\frac{v_s}{\nu}\right)z}\right) + g(x)$$

Combining both results, we have the stream function as follows.

$$\psi = v_s x + U\left(z + \frac{\nu}{v_s} e^{\left(-\frac{v_s}{\nu}\right)z}\right)$$

To sketch streamlines, we need an equation of $x$ vs $z$. We can sketch streamline by streamline, for a range of known values of stream function $\psi$. We substitute each $\psi$ value into the stream function equation, and for a series of specified $x$ values (i.e., plot range for $x$), we can evaluate corresponding $z$ values. We then repeat the process for different streamlines. Assuming $\psi_1$ is specified, we substitute known $x$ values to obtain corresponding values of $z$ from the equation as shown below.

$$x = \frac{\psi_1}{v_s} - \frac{U}{v_s}\left(z + \frac{\nu}{v_s} e^{\left(-\frac{v_s}{\nu}\right)z}\right)$$

$$\frac{dx}{dz} = -\frac{U}{v_s} + \frac{U}{v_s} e^{\left(-\frac{v_s}{\nu}\right)z}$$

From the above equation, we note that when $z = 0$, $\frac{dx}{dz} =$ constant. Therefore the plot becomes a vertical near the $x$-axis. Also, as $z$ increases, $\frac{dx}{dz}$ becomes more negative. Hence, close to the vertical $z$-axis, the gradient has the shape as shown below.

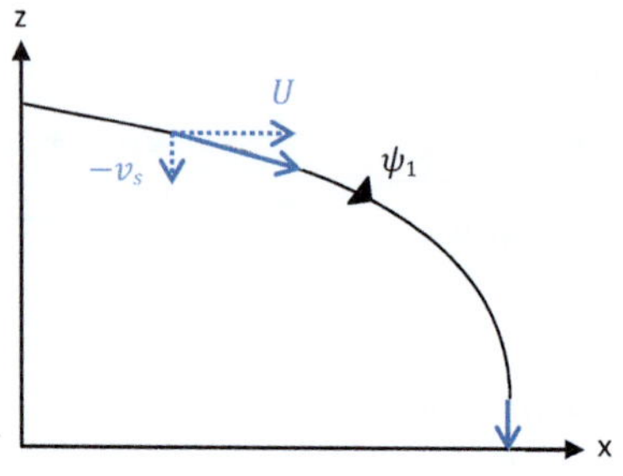

# Index





X. W. Ng, *Engineering Problems for Undergraduate Students*,
https://doi.org/10.1007/978-3-030-13856-1

CPI Antony Rowe
Eastbourne, UK
August 06, 2019